Communications
in Computer and Information Science   1237

Commenced Publication in 2007
Founding and Former Series Editors:
Simone Diniz Junqueira Barbosa, Phoebe Chen, Alfredo Cuzzocrea,
Xiaoyong Du, Orhun Kara, Ting Liu, Krishna M. Sivalingam,
Dominik Ślęzak, Takashi Washio, Xiaokang Yang, and Junsong Yuan

More information about this series at http://www.springer.com/series/7899

Marie-Jeanne Lesot · Susana Vieira ·
Marek Z. Reformat · João Paulo Carvalho ·
Anna Wilbik · Bernadette Bouchon-Meunier ·
Ronald R. Yager (Eds.)

Information Processing and Management of Uncertainty in Knowledge-Based Systems

18th International Conference, IPMU 2020
Lisbon, Portugal, June 15–19, 2020
Proceedings, Part I

 Springer

Editors
Marie-Jeanne Lesot
LIP6-Sorbonne University
Paris, France

Marek Z. Reformat
University of Alberta
Edmonton, AB, Canada

Anna Wilbik
Eindhoven University of Technology
Eindhoven, The Netherlands

Ronald R. Yager
Iona College
New Rochelle, NY, USA

Susana Vieira
IDMEC, IST, Universidade de Lisboa
Lisbon, Portugal

João Paulo Carvalho
INESC, IST, Universidade de Lisboa
Lisbon, Portugal

Bernadette Bouchon-Meunier
CNRS-Sorbonne University
Paris, France

ISSN 1865-0929   ISSN 1865-0937 (electronic)
Communications in Computer and Information Science
ISBN 978-3-030-50145-7   ISBN 978-3-030-50146-4 (eBook)
https://doi.org/10.1007/978-3-030-50146-4

This Springer imprint is published by the registered company Springer Nature Switzerland AG
The registered company address is: Gewerbestrasse 11, 6330 Cham, Switzerland

Preface

We are very pleased to present you with the proceedings of the 18th International Conference on Information Processing and Management of Uncertainty in Knowledge-Based Systems (IPMU 2020), held during June 15–19, 2020. The conference was scheduled to take place in Lisbon, Portugal, at the Instituto Superior Técnico, University of Lisbon, located in a vibrant renovated area 10 minutes from downtown. Unfortunately, due to the COVID-19 pandemic and international travel restrictions around the globe, the Organizing Committee made the decision to make IPMU 2020 a virtual conference taking place as scheduled.

The IPMU conference is organized every two years. Its aim is to bring together scientists working on methods for the management of uncertainty and aggregation of information in intelligent systems. Since 1986, the IPMU conference has been providing a forum for the exchange of ideas between theoreticians and practitioners working in these areas and related fields. In addition to many contributed scientific papers, the conference has attracted prominent plenary speakers, including the Nobel Prize winners Kenneth Arrow, Daniel Kahneman, and Ilya Prigogine.

A very important feature of the conference is the presentation of the *Kampé de Fériet Award* for outstanding contributions to the field of uncertainty and management of uncertainty. Past winners of this prestigious award are Lotfi A. Zadeh (1992), Ilya Prigogine (1994), Toshiro Terano (1996), Kenneth Arrow (1998), Richard Jeffrey (2000), Arthur Dempster (2002), Janos Aczel (2004), Daniel Kahneman (2006), Enric Trillas (2008), James Bezdek (2010), Michio Sugeno (2012), Vladimir N. Vapnik (2014), Joseph Y. Halpern (2016), and Glenn Shafer (2018). This year, the recipient of the *Kampé de Fériet Award* is Barbara Tversky. Congratulations!

The IPMU 2020 conference offers a versatile and comprehensive scientific program. There were four invited talks given by distinguished researchers: Barbara Tversky (Stanford University and Columbia University, USA), Luísa Coheur (Universidade de Lisboa, Instituto Superior Técnico, Portugal), Jim Keller (University of Missouri, USA), and Björn Schuller (Imperial College London, UK). A special tribute was organized to celebrate the life and achievements of Enrique Ruspini who passed away last year. He was one of the fuzzy-logic pioneers and researchers who contributed enormously to the fuzzy sets and systems body of knowledge. Two invited papers are dedicated to his memory. We would like to thank Rudolf Seising, Francesc Esteva, Lluís Godo, Ricardo Oscar Rodriguez, and Thomas Vetterlein for their involvement and contributions.

The IPMU 2020 program consisted of 22 special sessions and 173 papers authored by researchers from 34 different countries. All 213 submitted papers underwent the thorough review process and were judged by at least three reviewers. Many of them were reviewed by more – even up to five – referees. Furthermore, all papers were examined by the program chairs. The review process respected the usual conflict-of-interest standards, so that all papers received multiple independent evaluations.

Organizing a conference is not possible without the assistance, dedication, and support of many people and institutions.

We are particularly thankful to the organizers of special sessions. Such sessions, dedicated to variety of topics and organized by experts, have always been a characteristic feature of IPMU conferences. We would like to pass our special thanks to Uzay Kaymak, who helped evaluate many special session proposals.

We would like to acknowledge all members of the IPMU 2020 Program Committee, as well as multiple reviewers who played an essential role in the reviewing process, ensuring a high-quality conference. Thank you very much for all your work and efforts.

We gratefully acknowledge the technical co-sponsorship of the IEEE Computational Intelligence Society and the European Society for Fuzzy Logic and Technology (EUSFLAT).

A huge thanks and appreciation to the personnel of Lisbon's Tourism Office 'Turismo de Lisboa' (www.visitlisboa.com) for their eagerness to help, as well as their enthusiastic support.

Our very special and greatest gratitude goes to the authors who have submitted results of their work and presented them at the conference. Without you this conference would not take place. Thank you!

We miss in-person meetings and discussions, yet we are privileged that despite these difficult and unusual times all of us had a chance to be involved in organizing the virtual IPMU conference. We hope that these proceedings provide the readers with multiple ideas leading to numerous research activities, significant publications, and intriguing presentations at future IPMU conferences.

April 2020

Marie-Jeanne Lesot
Marek Z. Reformat
Susana Vieira
Bernadette Bouchon-Meunier
João Paulo Carvalho
Anna Wilbik
Ronald R. Yager

Organization

General Chair

João Paulo Carvalho   INESC-ID, Instituto Superior Técnico,
         Universidade de Lisboa, Portugal

Program Chairs

Marie-Jeanne Lesot   LIP6, Sorbonne Université, France
Marek Z. Reformat   University of Alberta, Canada
Susana Vieira   IDMEC, Instituto Superior Técnico,
         Universidade de Lisboa, Portugal

Executive Directors

Bernadette   LIP6, CNRS, France
   Bouchon-Meunier
Ronald R. Yager   Iona College, USA

Special Session Chair

Uzay Kaymak   Technische Universiteit Eindhoven, The Netherlands

Publication Chair

Anna Wilbik   Technische Universiteit Eindhoven, The Netherlands

Sponsor and Publicity Chair

João M. C. Sousa   IDMEC, Instituto Superior Técnico,
         Universidade de Lisboa, Portugal

Web Chair

Fernando Batista   INESC-ID, Instituto Superior Técnico,
         Universidade de Lisboa, Portugal

Keeley Crockett	Manchester Metropolitan University, UK
Giuseppe D'Aniello	University of Salerno, Italy
Bernard De Baets	Ghent University, Belgium
Martine De Cock	University of Washington, USA
Guy De Tré	Ghent University, Belgium
Sébastien Destercke	CNRS, UMR Heudiasyc, France
Antonio Di Nola	University of Salerno, Italy
Scott Dick	University of Alberta, Canada
Didier Dubois	IRIT, RPDMP, France
Fabrizio Durante	Free University of Bozen-Bolzano, Italy
Krzysztof Dyczkowski	Adam Mickiewicz University, Poland
Zied Elouedi	Institut Supérieur de Gestion de Tunis, Tunisia
Francesc Esteva	IIIA-CSIC, Spain
Dimitar Filev	Ford Motor Company, USA
Matteo Gaeta	University of Salerno, Italy
Sylvie Galichet	LISTIC, Université de Savoie, France
Jonathan M. Garibaldi	University of Nottingham, UK
Lluis Godo	IIIA-CSIC, Spain
Fernando Gomide	University of Campinas, Brazil
Gil González-Rodríguez	University of Oviedo, Spain
Przemysław Grzegorzewski	Systems Research Institute, Polish Academy of Sciences, Poland
Lawrence Hall	University of South Florida, USA
Istvan Harmati	Széchenyi István Egyetem, Hungary
Timothy Havens	Michigan Technological University, USA
Francisco Herrera	University of Granada, Spain
Enrique Herrera-Viedma	University of Granada, Spain
Ludmila Himmelspach	Heirich Heine Universität Düsseldorf, Germany
Eyke Hüllemeier	Paderborn University, Germany
Michal Holčapek	University of Ostrava, Czech Republic
Janusz Kacprzyk	Systems Research Institute, Polish Academy of Sciences, Poland
Uzay Kaymak	Eindhoven University of Technology, The Netherlands
Jim Keller	University of Missouri, USA
Frank Klawonn	Ostfalia University of Applied Sciences, Germany
László T. Kóczy	Budapest University of Technology and Economics, Hungary
John Kornak	University of California, San Francisco, USA
Vladik Kreinovich	University of Texas at El Paso, USA
Ondrej Krídlo	University of P. J. Safarik in Kosice, Slovakia
Rudolf Kruse	University of Magdeburg, Germany
Christophe Labreuche	Thales R&T, France
Jérôme Lang	CNRS, LAMSADE, Université Paris-Dauphine, France
Anne Laurent	LIRMM, UM, France
Chang-Shing Lee	National University of Tainan, Taiwan

Henrik Legind Larsen	Legind Technologies, Denmark
Marie-Jeanne Lesot	LIP6, Sorbonne Université, France
Weldon Lodwick	University of Colorado, USA
Edwin Lughofer	Johannes Kepler University Linz, Austria
Luis Magdalena	Universidad Politécnica de Madrid, Spain
Christophe Marsala	LIP6, Sorbonne Université, France
Trevor Martin	University of Bristol, UK
Sebastià Massanet	University of the Balearic Islands, Spain
Marie-Hélène Masson	Université de Picardie Jules Verne (Heudiasyc), France
Jesús Medina	University of Cádiz, Spain
Patricia Melin	Tijuana Institute of Technology, Mexico
Jerry Mendel	University of Southern California, USA
Radko Mesiar	STU, Slovakia
Enrique Miranda	University of Oviedo, Spain
Javier Montero	Universidad Complutense de Madrid, Spain
Susana Montes	University of Oviedo, Spain
Jacky Montmain	École des Mines d'Alès, France
Juan Moreno Garcia	Universidad de Castilla-La Mancha, Spain
Petra Murinová	University of Ostrava IT4Innovations, Czech Republic
Yusuke Nojima	Osaka Prefecture University, Japan
Vilém Novák	University of Ostrava, Czech Republic
Hannu Nurmi	University of Turku, Finland
Manuel Ojeda-Aciego	University of Malaga, Spain
Nikhil Pal	ISI, India
Gabriella Pasi	University of Milano-Bicocca, Italy
David Pelta	University of Granada, Spain
Irina Perfilieva	University of Ostrava, Czech Republic
Fred Petry	Naval Research Lab, USA
Davide Petturiti	University of Perugia, Italy
Vincenzo Piuri	University of Milan, Italy
Olivier Pivert	IRISA, ENSSAT, France
Henri Prade	IRIT, CNRS, France
Raúl Pérez-Fernández	Universidad de Oviedo, Spain
Anca Ralescu	University of Cincinnati, USA
Dan Ralescu	University of Cincinnati, USA
Marek Z. Reformat	University of Alberta, Canada
Adrien Revault d'Allonnes	LIASD, France
Agnès Rico	LIRIS, Université Claude Bernard Lyon 1, France
M. Dolores Ruiz	University of Cádiz, Spain
Thomas A. Runkler	Siemens Corporate Technology, Germany
Mika Sato Illic	University of Tsukuba, Japan
Daniel Sanchez	University of Granada, Spain
Glen Shafer	Rutgers University, USA
Grégory Smits	IRISA, University of Rennes 1, France
João Sousa	TU Lisbon, IST, Portugal

Martin Štěpnička   IRAFM, University of Ostrava, Czech Republic
Umberto Straccia   ISTI-CNR, Italy
Olivier Strauss   LIRMM, France
Michio Sugeno   Tokyo Institute of Technology, Japan
Eulalia Szmidt   Systems Research Institute, Polish Academy
                              of Sciences, Poland
Marco Tabacchi   Università degli Studi di Palermo, Italy
Vicenc Torra   Maynooth University, Ireland
Linda C. van der Gaag   Utrecht University, The Netherlands
Barbara Vantaggi   Sapienza University of Rome, Italy
José Luis Verdegay   University of Granada, Spain
Thomas Vetterlein   Johannes Kepler University Linz, Austria
Susana Vieira   Universidade de Lisboa, Portugal
Christian Wagner   University of Nottingham, UK
Anna Wilbik   Eindhoven University of Technology, The Netherlands
Sławomir Zadrożny   Systems Research Institute, Polish Academy
                              of Sciences, Poland

Additional Members of the Reviewing Committee

Raoua Abdelkhalek   Yurilev Chalco-Cano
Julien Alexandre Dit Sandretto   Manuel Chica
Zahra Alijani   Panagiotis Chountas
Alessandro Antonucci   Davide Ciucci
Jean Baratgin   Frank Coolen
Laécio C. Barros   Maria Eugenia Cornejo Piñero
Leliane N. Barros   Cassio P. de Campos
Libor Behounek   Gert De Cooman
María José Benítez Caballero   Laura De Miguel
Kyle Bittner   Jean Dezert
Jan Boronski   J. Angel Diaz-Garcia
Reda Boukezzoula   Graçaliz Dimuro
Ross Boylan   Paweł Drygaś
Andrey Bronevich   Hassane Essafi
Petr Bujok   Javier Fernandez
Michal Burda   Carlos Fernandez-Basso
Rafael Cabañas de Paz   Juan Carlos Figueroa-García
Inma P. Cabrera   Marcelo Finger
Tomasa Calvo   Tommaso Flaminio
José Renato Campos   Robert Fullér
Andrea Capotorti   Marek Gagolewski
Diego Castaño   Angel Garcia Contreras
Anna Cena   Michel Grabisch
Mihir Chakraborty   Karel Gutierrez

Allel Hadjali
Olgierd Hryniewicz
Miroslav Hudec
Ignacio Huitzil
Seong Jae Hwang
Atsushi Inoue
Vladimir Janis
Balasubramaniam Jayaram
Richard Jensen
Luis Jimenez Linares
Katarzyna Kaczmarek
Martin Kalina
Hiroharu Kawanaka
Alireza Khastan
Martins Kokainis
Ryszard Kowalczyk
Maciej Krawczak
Jiri Kupka
Serafina Lapenta
Ulcilea Leal
Antonio Ledda
Eric Lefevre
Nguyen Linh
Nicolas Madrid
Arnaud Martin
Denis Maua
Gilles Mauris
Belen Melian
María Paula Menchón
David Mercier
Arnau Mir
Soheyla Mirshahi
Marina Mizukoshi
Jiří Močkoř
Miguel Molina-Solana
Ignacio Montes
Serafin Moral
Tommaso Moraschini
Andreia Mordido
Juan Antonio Morente-Molinera
Fred Mubang
Vu-Linh Nguyen
Radoslaw Niewiadomski

Carles Noguera
Pavels Orlovs
Daniel Ortiz-Arroyo
Jan W. Owsinski
Antonio Palacio
Manuel J. Parra Royón
Jan Paseka
Viktor Pavliska
Renato Pelessoni
Barbara Pękala
Benjamin Quost
Emmanuel Ramasso
Eloisa Ramírez Poussa
Luca Reggio
Juan Vicente Riera
Maria Rifqi
Luis Rodriguez-Benitez
Guillaume Romain
Maciej Romaniuk
Francisco P. Romero
Clemente Rubio-Manzano
Aleksandra Rutkowska
Juan Jesus Salamanca Jurado
Teddy Seidenfeld
Mikel Sesma-Sara
Babak Shiri
Amit Shukla
Anand Pratap Singh
Damjan Skulj
Sotir Sotirov
Michal Stronkowski
Andrea Stupnánová
Matthias Troffaes
Dana Tudorascu
Leobardo Valera
Arthur Van Camp
Paolo Vicig
Amanda Vidal Wandelmer
Joaquim Viegas
Jin Hee Yoon
Karl Young
Hua-Peng Zhang

Special Session Organizers

Javier Andreu	University of Essex, UK
Michał Baczyński	University of Silesia in Katowice, Poland
Isabelle Bloch	Télécom ParisTech, France
Bernadette Bouchon-Meunier	LIP6, CNRS, France
Reda Boukezzoula	Université de Savoie Mont-Blanc, France
Humberto Bustince	Public University of Navarra, Spain
Tomasa Calvo	University of Alcalá, Spain
Martine Ceberio	University of Texas at El Paso, USA
Yurilev Chalco-Cano	University of Tarapacá at Arica, Chile
Giulianella Coletti	Università di Perugia, Italy
Didier Coquin	Université de Savoie Mont-Blanc, France
M. Eugenia Cornejo	University of Cádiz, Spain
Bernard De Baets	Ghent University, Belgium
Guy De Tré	Ghent University, Belgium
Graçaliz Dimuro	Universidade Federal do Rio Grande, Brazil
Didier Dubois	IRIT, Université Paul Sabatier, France
Hassane Essafi	CEA, France
Carlos J. Fernández-Basso	University of Granada, Spain
Javier Fernández	Public University of Navarra, Spain
Tommaso Flaminio	Spanish National Research Council, Spain
Lluis Godo	Spanish National Research Council, Spain
Przemyslaw Grzegorzewski	Warsaw University of Technology, Poland
Rajarshi Guhaniyogi	University of California, Santa Cruz, USA
Karel Gutiérrez Batista	University of Granada, Spain
István Á. Harmati	Széchenyi István University, Hungary
Michal Holčapek	University of Ostrava, Czech Republic
Atsushi Inoue	Eastern Washington University, USA
Balasubramaniam Jayaram	Indian Institute of Technology Hyderabad, India
Janusz Kacprzyk	Systems Research Institute, Polish Academy of Sciences, Poland
Hiroharu Kawanaka	Mie University, Japan
László T. Kóczy	Budapest University of Technology and Economics, Hungary
John Kornak	University of California, San Francisco, USA
Vladik Kreinovich	University of Texas at El Paso, USA
Henrik Legind Larsen	Legind Technologies, Denmark
Weldon Lodwick	Federal University of São Paulo, Brazil
Maria Jose Martín-Bautista	University of Granada, Spain
Sebastia Massanet	University of the Balearic Islands, Spain
Jesús Medina	University of Cádiz, Spain
Belén Melián-Batista	University of La Laguna, Spain
Radko Mesiar	Slovak University of Technology, Slovakia
Enrique Miranda	University of Oviedo, Spain

Ignacio Montes	University of Oviedo, Spain
Juan Moreno-Garcia	University of Castilla-La Mancha, Spain
Petra Murinová	University of Ostrava, Czech Republic
Vílem Novák	University of Ostrava, Czech Republic
David A. Pelta	University of Granada, Spain
Raúl Pérez-Fernández	University of Oviedo, Spain
Irina Perfilieva	University of Ostrava, Czech Republic
Henri Prade	IRIT, Université Paul Sabatier, France
Anca Ralescu	University of Cincinnati, USA
Eloísa Ramírez-Poussa	University of Cádiz, Spain
Luis Rodriguez-Benitez	University of Castilla-La Mancha, Spain
Antonio Rufian-Lizana	University of Sevilla, Spain
M. Dolores Ruiz	University of Granada, Spain
Andrea Stupnanova	Slovak University of Technology, Slovakia
Amanda Vidal	Czech Academy of Sciences, Czech Republic
Aaron Wolfe Scheffler	University of California, San Francisco, USA
Adnan Yazici	Nazarbayev University, Kazakhstan
Sławomir Zadrożny	Systems Research Institute Polish Academy of Sciences, Poland

List of Special Sessions

Fuzzy Interval Analysis

Antonio Rufian-Lizana	University of Sevilla, Spain
Weldon Lodwick	Federal University of São Paulo, Brazil
Yurilev Chalco-Cano	University of Tarapacá at Arica, Chile

Theoretical and Applied Aspects of Imprecise Probabilities

Enrique Miranda	University of Oviedo, Spain
Ignacio Montes	University of Oviedo, Spain

Similarities in Artificial Intelligence

Bernadette Bouchon-Meunier	LIP6, CNRS, France
Giulianella Coletti	Università di Perugia, Italy

Belief Function Theory and Its Applications

Didier Coquin	Université de Savoie Mont-Blanc, France
Reda Boukezzoula	Université de Savoie Mont-Blanc, France

Aggregation: Theory and Practice

Tomasa Calvo	University of Alcalá, Spain
Radko Mesiar	Slovak University of Technology, Slovakia
Andrea Stupnánová	Slovak University of Technology, Slovakia

Aggregation: Pre-aggregation Functions and Other Generalizations

Humberto Bustince	Public University of Navarra, Spain
Graçaliz Dimuro	Universidade Federal do Rio Grande, Brazil
Javier Fernández	Public University of Navarra, Spain

Aggregation: Aggregation of Different Data Structures

Bernard De Baets	Ghent University, Belgium
Raúl Pérez-Fernández	University of Oviedo, Spain

Fuzzy Methods in Data Mining and Knowledge Discovery

M. Dolores Ruiz	University of Granada, Spain
Karel Gutiérrez Batista	University of Granada, Spain
Carlos J. Fernández-Basso	University of Granada, Spain

Computational Intelligence for Logistics and Transportation Problems

David A. Pelta	University of Granada, Spain
Belén Melián-Batista	University of La Laguna, Spain

Fuzzy Implication Functions

Michał Baczyński	University of Silesia in Katowice, Poland
Balasubramaniam Jayaram	Indian Institute of Technology Hyderabad, India
Sebastià Massanet	University of the Balearic Islands, Spain

Soft Methods in Statistics and Data Analysis

Przemysław Grzegorzewski	Warsaw University of Technology, Poland

Image Understanding and Explainable AI

Isabelle Bloch	Télécom ParisTech, France
Atsushi Inoue	Eastern Washington University, USA
Hiroharu Kawanaka	Mie University, Japan
Anca Ralescu	University of Cincinnati, USA

Fuzzy and Generalized Quantifier Theory

Vilém Novák	University of Ostrava, Czech Republic
Petra Murinová	University of Ostrava, Czech Republic

Mathematical Methods Towards Dealing with Uncertainty in Applied Sciences

Irina Perfilieva	University of Ostrava, Czech Republic
Michal Holčapek	University of Ostrava, Czech Republic

Statistical Image Processing and Analysis, with Applications in Neuroimaging

John Kornak	University of California, San Francisco, USA
Rajarshi Guhaniyogi	University of California, Santa Cruz, USA
Aaron Wolfe Scheffler	University of California, San Francisco, USA

Interval Uncertainty

Martine Ceberio	University of Texas at El Paso, USA
Vladik Kreinovich	University of Texas at El Paso, USA

Discrete Models and Computational Intelligence

László T. Kóczy	Budapest University of Technology and Economics, Hungary
István Á. Harmati	Széchenyi István University, Hungary

Current Techniques to Model, Process and Describe Time Series

Juan Moreno-Garcia	University of Castilla-La Mancha, Spain
Luis Rodriguez-Benitez	University of Castilla-La Mancha, Spain

Mathematical Fuzzy Logic and Graded Reasoning Models

Tommaso Flaminio	Spanish National Research Council, Spain
Lluís Godo	Spanish National Research Council, Spain
Vílem Novák	University of Ostrava, Czech Republic
Amanda Vidal	Czech Academy of Sciences, Czech Republic

Formal Concept Analysis, Rough Sets, General Operators and Related Topics

M. Eugenia Cornejo	University of Cádiz, Spain
Didier Dubois	IRIT, Université Paul Sabatier, France
Jesús Medina	University of Cádiz, Spain
Henri Prade	IRIT, Université Paul Sabatier, France
Eloísa Ramírez-Poussa	University of Cádiz, Spain

Computational Intelligence Methods in Information Modelling, Representation and Processing

Guy De Tré	Ghent University, Belgium
Janusz Kacprzyk	Systems Research Institute, Polish Academy of Sciences, Poland
Adnan Yazici	Nazarbayev University, Kazakhstan
Sławomir Zadrożny	Systems Research Institute Polish Academy of Sciences, Poland

Abstracts of Invited Talks

How Action Shapes Thought

Barbara Tversky

Columbia Teachers College and Stanford University
btversky@stanford.edu

When you ask someone a question they can't answer, the response is often a shrug of the shoulders, arms outstretched, elbows bent, palms up. Translated into words, that shrug means "dunno" or "who knows?" An expression of uncertainty. It's instantly understood that way as well. No need for translation to words, the meaning of the gesture is clear. Now consider another gesture, one made by a preschooler known to shrug her shoulders on other occasions, asking about her day. The answer: not a shrug, but a hand outspread horizontally, teeter-tottering between thumb and baby finger. Or, on another occasion, one thumb up, one thumb down. The shrug seems to say, there's an answer, but I don't know it. The information is in the air, but I haven't caught it. The teeter-tottering hand and up and down thumbs seem to express a different kind of uncertainty, I have the information but it's not decisive, it goes both ways, It goes up and down, back and forth; it's balanced. Now I step out of my usual role as a cognitive psychologist and adopt the role of a linguist, where anecdotes are the stuff of thought and analysis. This preschooler distinguishes two fundamental kinds of uncertainty, one where the information might (or might not) be out there but I don't have it and the other where I have the relevant information but I can't decide one way or another, the information tilts both ways, Not only does this preschooler know the distinction between the two types of uncertainty, she can express them.

To express either kind of uncertainty –and many other thoughts– she doesn't use words, she uses gestures. Gestures come faster than words, are more direct than words, and more precise than words. Let's start with the simplest of gestures, pointing. Babies point long, in baby-time, before they speak. Points direct the eyes to pin-point spots in the world; "there" can't do that unless accompanied by a string of spatial descriptors that are likely to be vague or wrong or both. From where to how, contrast showing how to open a jar or insert a drawer to explaining how to open a jar or insert a drawer. Gestures truncate and abstract actions in the world to convey actions on things. They also use abstractions of actions to convey actions on thought, raising arguments for and against and placing them on sides of the body, an imaginary whiteboard, then pointing to indicate each side in turn. You have undoubtedly seen speakers do this, you have likely done it yourself; those two sides in space, on your right and on your left, help you keep track of the pros and cons whether you are speaker or listener. Gestures help both speakers and listeners to think and to talk. When asked to sit on their hands, speakers flounder finding words. When people are asked to study and remember descriptions of spatial layouts or actions of mechanical systems, most spontaneously gesture. Their gestures make models of the space or of the actions. When asked to sit on their hands while studying, people remember less and realize fewer of the inferences needed for deep understanding. Thus gestures, abstractions of actions on objects used to represent actions on thought, enable thought and embody thought both for thinkers and for their audiences.

Gestures can be regarded, justly, as diagrams in the air. Gestures are fleeting; transforming them to a page keeps them, and allows scrutinizing them, drawing inferences from them, revising them, by individuals or by groups. Like gestures, graphics use marks in space and place in space to convey meanings more directly than words. Points stand for places or ideas; lines connect them, showing relationships; arrows show asymmetric relations; boxes contain a related set of ideas and separate those from others. Ideas that are close in space are close on any dimension; ideas high in space are high on any dimension, ideas that are central are just that, central. Concepts and relations that are created and understood immediately, in contrast to words, whose meanings are mediated.

Our unnamed preschooler spontaneously expressed two basic senses of uncertainty in her gestures, uncertainty due to absence of information and uncertainty due to indecisive information. Conveying these forms of uncertainty, and perhaps others, for different content in diagrams is still finding its way. Error bars and fuzzy lines are some of the ways diagrams express imprecise quantitative information. Expressing absent or imprecise or undecisive information for qualitative information has been challenging.

Language, too, carries these spatial meanings. We've grown closer, or farther apart. The central argument is… Someone's on the top of the heap or fallen into a depression. That space is wide open, To mix spatial metaphors: navigating the crisis will be a delicate balance.

Spatial thinking is the foundation of all thought. Not the entire edifice but the foundation. All creatures must move in space and interact with things in space to survive. Even plants must move in response to wind, rain, and sun. The evidence comes from many places, from gesture, from language, from diagrams and sketches. It also comes from neuroscience: the same places in hippocampus that represent places are used to represent people, events, and ideas. The same places in entorhinal cortex that map spatial relations also map temporal, social, and conceptual relations, In humans, for the most part, in real space, feet do the navigation and hands do the interaction with things. In conceptual spaces, it's fingers and hands that navigate in the air or on the screen just as it's fingers and hands that interact with points in conceptual spaces in the air or on the screen.

Thus, actions in real space on objects in real space get truncated and abstracted to form gestures that express actions on ideas in spaces in the air. The same truncated abstracted actions create actions on ideas on the space of the page. This cycle of actions in space that are transformed to gestures that create abstractions in the air or to marks that create abstractions on the page can be unified in the concept, *spraction*, a contraction for the never-ending cycle of space, action, and abstraction.

Reference

Tversky, B.: Mind in Motion: How Action Shapes Thought. Basic, NY (2019)

Biography

Barbara Tversky studied cognitive psychology at the University of Michigan. She held positions first at the Hebrew University in Jerusalem and then at Stanford, from 1978–2005 when she took early retirement. She is an active Emerita Professor of Psychology at Stanford and Professor of Psychology at Columbia Teachers College. She is a fellow of the Association for Psychological Science, the Cognitive Science Society, the Society for Experimental Psychology, the Russell Sage Foundation, and the American Academy of Arts and Science. She has been on the Governing Boards of the Psychonomic Society, the Cognitive Science Society and the International Union of Psychological Science. She is Past-President of the Association for Psychological Science. She has served on the editorial boards of many journals and the organizing committees of dozens of international interdisciplinary meetings.

Her research has spanned memory, categorization, language, spatial cognition, event perception and cognition, diagrammatic reasoning, sketching, creativity, design, and gesture. The overall goals have been to uncover how people think about the spaces they inhabit and the actions they perform and see and then how people use the world, including their own actions and creations, to remember, to think, to create, to communicate. A recent book, Mind in Motion: How Action Shapes Thought, Basic Books, overview that work. She has collaborated widely, with linguists, philosophers, neuroscientists, computer scientists, chemists, biologists, architects, designers, and artists.

Making Sense Out of Activity Sensing in Eldercare

Jim Keller

Electrical Engineering and Computer Science Department,
University of Missouri
KellerJ@missouri.edu

With the increase in the population of older adults around the world, a significant amount of work has been done on in-home sensor technology to aid the elderly age independently. However, due to the large amounts of data generated by the sensors, it takes a lot of effort and time for the clinicians to makes sense of this data. In this talk, I will survey two connected approaches to provide explanations of these complex sensor patterns as they relate to senior health. Abnormal sensor patterns produced by certain resident behaviors could be linked to early signs of illness. In seven eldercare facilities around Columbia, MO operated by Americare, we have deployed an intelligent elderly monitoring system with summarization and symptom suggesting capabilities for 3 years.

The first procedure starts by identifying important attributes in the sensor data that are relevant to the health of the elderly. We then develop algorithms to extract these important health related features from the sensor parameters and summarize them in natural language, with methods grounded in fuzzy set theory. We focus on making the natural language summaries to be informative, accurate and concise, and have conducted numerous surveys of experts to validate our choices. While our initial focus is on producing summaries that are informative to healthcare personnel, a recent grant centers on providing feedback to the elders and their families. The Amazon Echo Show is used as the communication device to provide simplified graphics and linguistic health messages.

The second approach is a framework for detecting health patterns utilizing sensor sequence similarity and natural language processing (NLP). A context preserving representation of daily activities is used to measure the similarity between the sensor sequences of different days. Medical concepts are extracted from nursing notes that allows us to impute potential reasons for health alerts based on the activity similarity. Joining these two approaches provide a powerful XAI description of early illness recognition for elders.

Biography

James M. Keller received the Ph.D. in Mathematics in 1978. He is now the Curators' Distinguished Professor Emeritus in the Electrical Engineering and Computer Science Department at the University of Missouri. Jim is an Honorary Professor at the University of Nottingham. His research interests center on computational intelligence: fuzzy set theory and fuzzy logic, neural networks, and evolutionary computation with a focus on problems in computer vision, pattern recognition, and information fusion including bioinformatics, spatial reasoning in robotics, geospatial intelligence, sensor and information analysis in technology for eldercare, and landmine detection.

His industrial and government funding sources include the Electronics and Space Corporation, Union Electric, Geo-Centers, National Science Foundation, the Administration on Aging, The National Institutes of Health, NASA/JSC, the Air Force Office of Scientific Research, the Army Research Office, the Office of Naval Research, the National Geospatial Intelligence Agency, the U.S. Army Engineer Research and Development Center, the Leonard Wood Institute, and the Army Night Vision and Electronic Sensors Directorate. Professor Keller has coauthored over 500 technical publications.

Jim is a Life Fellow of the Institute of Electrical and Electronics Engineers (IEEE), a Fellow of the International Fuzzy Systems Association (IFSA), and a past President of the North American Fuzzy Information Processing Society (NAFIPS). He received the 2007 Fuzzy Systems Pioneer Award and the 2010 Meritorious Service Award from the IEEE Computational Intelligence Society (CIS). He has been a distinguished lecturer for the IEEE CIS and the ACM. Jim finished a full six year term as Editor-in-Chief of the IEEE Transactions on Fuzzy Systems, followed by being the Vice President for Publications of the IEEE Computational Intelligence Society from 2005–2008, then as an elected CIS Adcom member, and is in another term as VP Pubs (2017–2020). He was the IEEE TAB Transactions Chair as a member of the IEEE Periodicals Committee, and is a member of the IEEE Publication Review and Advisory Committee from 2010 to 2017. Among many conference duties over the years, Jim was the general chair of the 1991 NAFIPS Workshop, the 2003 IEEE International Conference on Fuzzy Systems, and co-general chair of the 2019 IEEE International Conference on Fuzzy Systems.

From Eliza to Siri and Beyond

Luísa Coheur[1,2]

[1] INESC-ID Lisboa
[2] Instituto Superior Técnico/Universidade de Lisboa
luisa.coheur@inesc-id.pt

Since Eliza, the first chatbot ever, developed in the 60s, researchers try to make machines understand (or mimic the understanding) of Natural Language input. Some conversational agents target small talk, while others are more task-oriented. However, from the earliest rule-based systems to the recent data-driven approaches, although many paths were explored with more or less success, we are not there yet. Rule-based systems require much manual work; data-driven systems require a lot of data. Domain adaptation is (again) a current hot-topic. The possibility to add emotions to the conversational agents' responses, or to make their answers capture their "persona", are some popular research topics. This paper explains why the task of Natural Language Understanding is so complicated, detailing the linguistic phenomena that lead to the main challenges. Then, the long walk in this field is surveyed, from the earlier systems to the current trends.

Biography

Luísa Coheur graduated in Applied Mathematics and Computation and has an M.Sc. degree in Electrical and Computer Engineering, both from Instituto Superior Técnico (IST). In 2004, she concluded her Dual degree Ph.D in Computer Science and Engineering (IST), and Linguistique, Logique et Informatique (Université Blaise-Pascal). She is a researcher at INESC-ID since 2001, and a lecturer at IST since March 2006. Luísa Coheur has been working in the Natural Language Processing field since her Master's thesis. Her main research interest is Natural Language Understanding, being Question/Answering, Dialogue Systems and Machine Translation her key application scenarios. She strongly believes that science should be in service to the public good, and she is currently building a prototype that translates European Portuguese into LGP (Língua Gestual Portuguesa), using an avatar. She participated in several national and international projects; she supervised and/or co-supervised 55 masters' and 6 Ph.D students. Luísa Coheur is also a part-time writer. She has 3 published books and two short stories, which won literature prizes.

Average Jane, Where Art Thou? – Recent Avenues in Efficient Machine Learning Under Subjectivity Uncertainty

Björn W. Schuller[1,2,3] (iD)

[1] GLAM, Imperial College London, SW7 2AZ London, UK
[2] Chair of EIHW, University of Augsburg, 86159 Augsburg, Germany
[3] audEERING, 82205 Gilching, Germany
schuller@ieee.org

In machine learning tasks an actual 'ground truth' may not be available. Then, machines often have to rely on human labelling of data. This becomes challenging the more subjective the learning task is, as human agreement can be low. To cope with the resulting high uncertainty, one could train individual models reflecting a single human's opinion. However, this is not viable, if one aims at mirroring the general opinion of a hypothetical 'completely average person' – the 'average Jane'. Here, I summarise approaches to optimally learn efficiently in such a case. First, different strategies of reaching a single learning target from several labellers will be discussed. This includes varying labeller trustability and the case of time-continuous labels with potential dynamics. As human labelling is a labour-intensive endeavour, active and cooperative learning strategies can help reduce the number of labels needed. Next, sample informativeness can be exploited in teacher-based algorithms to additionally weigh data by certainty. In addition, multi-target learning of different labeller tracks in parallel and/or of the uncertainty can help improve the model robustness and provide an additional uncertainty measure. Cross-modal strategies to reduce uncertainty offer another view. From these and further recent strategies, I distil a number of future avenues to handle subjective uncertainty in machine learning. These comprise bigger, yet weakly labelled data processing basing amongst other on reinforcement learning, lifelong learning, and self-learning. Illustrative examples stem from the fields of Affective Computing and Digital Health – both notoriously marked by subjectivity uncertainty.

Biography

Björn W. Schuller received his diploma, doctoral degree, habilitation, and Adjunct Teaching Professor in Machine Intelligence and Signal Processing all in EE/IT from TUM in Munich/Germany. He is Full Professor of Artificial Intelligence and the Head of GLAM at Imperial College London/UK, Full Professor and Chair of Embedded Intelligence for Health Care and Wellbeing at the University of Augsburg/Germany, co-founding CEO and current CSO of audEERING – an Audio Intelligence company based near Munich and in Berlin/Germany, and permanent Visiting Professor at HIT/China amongst other Professorships and Affiliations. Previous stays include Full Professor at the University of Passau/Germany, and Researcher at Joanneum Research in Graz/Austria, and the CNRS-LIMSI in Orsay/France.

He is a Fellow of the IEEE, Fellow of the ISCA, Golden Core Awardee of the IEEE Computer Society, President-Emeritus of the AAAC, and Senior Member of the ACM. He (co-)authored 900+ publications (h-index = 79), is Field Chief Editor of Frontiers in Digital Health and was Editor in Chief of the IEEE Transactions on Affective Computing, General Chair of ACII 2019, ACII Asia 2018, and ACM ICMI 2014, and a Program Chair of Interspeech 2019, ACM ICMI 2019/2013, ACII 2015/2011, and IEEE SocialCom 2012 amongst manifold further commitments and service to the community. His 40+ awards include having been honoured as one of 40 extraordinary scientists under the age of 40 by the WEF in 2015. He served as Coordinator/PI in 15+ European Projects, is an ERC Starting Grantee, and consultant of companies such as Barclays, GN, Huawei, or Samsung.

Contents – Part I

Decision Making, Preferences and Votes

Optimization and Uncertainty

Games

Real World Applications

Knowledge Processing and Creation

XAI

Image Processing

Temporal Data Processing

Text Analysis and Processing

Contents - Part II

Similarities in Artificial Intelligence

Belief Function Theory and Its Applications

**Aggregation: Pre-aggregation Functions and Other
Generalizations of Monotonicity**

Aggregation: Aggregation of Different Data Structures

Fuzzy Methods in Data Mining and Knowledge Discovery

**Computational Intelligence for Logistics
and Transportation Problems**

Fuzzy Implication Functions

Contents - Part III

Soft Methods in Statistics and Data Analysis

Image Understanding and Explainable AI

Fuzzy and Generalized Quantifier Theory

Mathematical Methods Towards Dealing with Uncertainty in Applied Sciences

**Statistical Image Processing and Analysis, with Applications
in Neuroimaging**

Interval Uncertainty

Discrete Models and Computational Intelligence

Homage to Enrique Ruspini

On Ruspini's Models of Similarity-Based Approximate Reasoning

Francesc Esteva[1], Lluís Godo[1(✉)], Ricardo Oscar Rodriguez[2],
and Thomas Vetterlein[3]

[1] Artificial Intelligence Research Institute (IIIA-CSIC),
Campus de la UAB, 08193 Bellaterra, Barcelona, Spain
{esteva,godo}@iiia.csic.es
[2] Universidad de Buenos Aires, FCEN, Dept. de Ciencias de la Computación,
UBA-CONICET, Instituto de Ciencias de la Computación, Buenos Aires, Argentina
ricardo@dc.uba.ar
[3] Department of Knowledge-Based Mathematical Systems,
Johannes Kepler University, Linz, Austria
Thomas.Vetterlein@jku.at

Abstract. In his 1991 seminal paper, Enrique H. Ruspini proposed a similarity-based semantics for fuzzy sets and approximate reasoning which has been extensively used by many other authors in various contexts. This brief note, which is our humble contribution to honor Ruspini's great legacy, describes some of the main developments in the field of logic that essentially rely on his ideas.

Keywords: Fuzzy similarity · Approximate reasoning · Graded entailments · Modal logic

1 Introduction

Similarity is a relevant notion in the context of at least three cognitive tasks: classification, case-based reasoning, and interpolation [1]. For classification tasks, objects are put together in the same class when they are indistinguishable with respect to some suitable criteria. Furthermore, case-based reasoning exploits the similarity between already solved problems and a new given problem to be solved in order to build up a solution to it. Finally, interpolation mechanisms estimate the value of a partially unknown function at a given point of a space by exploiting the proximity or closeness of this point to other points for which the value of the function is known.

It was Ruspini in [13] (cf. also [14]) who started the task of formalising approximate reasoning underlying these and other cognitive tasks in a logical setting. He elaborated on the notion of fuzzy similarity, as suggested by Zadeh's theory of approximate reasoning [19]. According to the approach originally proposed by Ruspini to model fuzzy similarity-based reasoning, the set W of interpretations or possible worlds is, in a first step, equipped with a map

© Springer Nature Switzerland AG 2020
M.-J. Lesot et al. (Eds.): IPMU 2020, CCIS 1237, pp. 3–13, 2020.
https://doi.org/10.1007/978-3-030-50146-4_1

$S : W \times W \mapsto [0,1]$ supposed to fulfil the basic properties of fuzzy or graded similarity relation:

Reflexivity: $S(u,u) = 1$ for all $u \in W$
Separability: $S(u,v) = 1$ iff $u = v$, for all $u,v \in W$
Symmetry: $S(u,v) = S(v,u)$ for all $u,v \in W$
$\otimes$-**Transivity:** $S(u,v) \otimes S(v,w) \leq S(u,w)$ for all $u,v,w \in W$

where $\otimes$ is a t-norm.

Reflexive and symmetric fuzzy relations are often called *closeness* relations, while those further satisfying $\otimes$-transitivity are usually called $\otimes$-*similarity rela-tions*, first introduced by Trillas in [15] under the name of T-indistinguishability relations. Sometimes, the name similarity relation is actually also used to denote $\otimes$-similarity relations where $\otimes = \min$. These min-similarity relations have the remarkable property that their level cuts $S_\alpha = \{(u,v) \in W \times W \mid S(u,v) \geq \alpha\}$, for any $\alpha \in [0,1]$, are equivalence relations. See Recasens' monograph [11] for any question related to fuzzy similarity relations.

The notion of similarity can be regarded as a dual to the notion of a gener-alised (bounded) metric, in the sense that if S measures resemblance between possible worlds, $\delta = 1 - S$ measures how distant they are. Then the $\otimes$-transitivity property corresponds to a generalised triangular inequality property for δ. In the particular case of $\otimes$ being Łukasiewicz t-norm, δ is a bounded metric, while δ becomes an ultrametric when $\otimes = \min$.

Given the set of possible worlds or interpretations together with a fuzzy similarity relation, Ruspini built up, in a second step, a basic framework to define possibilistic structures and concepts by quantifying proximity, closeness, or resemblance between pairs of (classical) logical statements. Since in classical logic we may identify propositions with sets of worlds, this problem reduces to the question how to extend a similarity between worlds to a measure of similarity between sets of worlds. As is well-known in the case of metric spaces, a metric between points does not univocally extend to a meaningful metric between sets of points. Ruspini defined in [13] two measures,

$$I_S(p \mid q) = \inf_{w \models q} \sup_{w' \models p} S(w,w') \qquad \text{and} \qquad C_S(p \mid q) = \sup_{w \models q} \sup_{w' \models p} S(w,w'),$$

called *implication* and *consistency*, which are the lower and upper bounds, respec-tively, of the resemblance or proximity degree between p and q, from the per-spective of q. Actually, if one defines the fuzzy set approx(p) of worlds *close* to those of p by the membership function

$$\mu_{\text{approx}(p)}(w) = \sup\{S(w,w') \mid w' \models p\},$$

then we can write $I_S(p \mid q) = \inf_{w \models q} \mu_{\text{approx}(p)}(w)$ and it becomes clear that $I_S(p \mid q)$ is a measure of inclusion of the (crisp) set of q-worlds into the (fuzzy) set approx(p) of worlds close to p. Similarly, we can write $C_S(p \mid q) = \sup_{w \models q} \mu_{\text{approx}(p)}(w)$ and thus $C_S(p \mid q)$ is a measure of intersection between the set of q-worlds with the set of worlds close to p. Observe that, when the propositional language only contains finitely many propositional symbols and q is equivalent to a maximal consistent set of propositions, both measures coincide because there is a unique world w such that $w \models q$. In such a case, $I_S(p \mid q) = C_S(p|q) = \mu_{\mathrm{approx}(p)}(w)$.[1]

With the implication measures I_S, Ruspini's aim was to capture approximate inference patterns related to the so-called generalised modus ponens. The value of $I_S(p \mid q)$ provides the measure to what extent p is close to be true given q for granted. In particular, when the similarity relation S is separating and the set of worlds is finite then, $I_S(p \mid q) = 1$ iff $q \models p$. Moreover, if S is $\otimes$-transitive, for a t-norm $\otimes$, then I_S is $\otimes$-transitive as well [13], i.e. the inequality

$$I_S(r \mid p) \otimes I_S(p \mid q) \leq I_S(r \mid q)$$

holds for any propositions p, q, and r. This property allows to formulate a kind of generalized resolution rule:

from: $I_S(r \mid p) \geq \alpha$ and $I_S(p \mid q) \geq \beta$
infer: $I_S(r \mid q) \geq \alpha \otimes \beta$.

On the other hand, the value of $C_S(p \mid q)$ provides the measure to what extent p can be considered compatible with the available knowledge represented by q. In particular, in the finite case and with S satisfying the separation property, $C_S(p \mid q) = 1$ iff $q \not\models \neg p$.

Implication and consistency measures have quite different properties, apart from the fact that both I_S and C_S are reflexive, i.e., $I_S(p \mid p) = C_S(p \mid p) = 1$, and non-decreasing in the first variable: i.e., if $p \models r$, then $I_S(p \mid q) \leq I_S(r \mid q)$ and $C_S(p \mid q) \leq C_S(r \mid q)$. But w.r.t. to the second variable, I_S is non-increasing while C_S keeps being non-decreasing. Moroever, unlike I_S, C_S is a symmetric measure, i.e. $C_S(p \mid q) = C_S(q \mid p)$, and it is not $\otimes$-transitive in general. On the other hand, it is easy to show that, for a fixed proposition r, the measure $C_S(\cdot \mid r)$ is in fact a possibility measure [2] since the following identities hold true:

(C1) $C_S(\top \mid r) = 1$
(C2) $C_S(\bot \mid r) = 0$
(C3) $C_S(p \vee q \mid r) = \max(C_S(p \mid r), C_S(q \mid r))$.

The counterpart of the last property for implication measures is the following one:

(I3) $I_S(p \mid q \vee r) = \min(I_S(p \mid q), I_S(p \mid r))$,

that is related to the so-called *Left-Or* property of consequence relations. We will return to this consideration in Sect. 2.

Note that conditional versions of the I_S and C_S measures were already considered by Ruspini in [13], and then further elaborated in [6] and [3] in order

---

[1] By an abuse of notation, in this case we will also write $I_S(p \mid w)$ or $C_S(p \mid w)$.

to cast different forms of the generalized modus ponens inference pattern under the frame of similarity-based reasoning.

All these seminal ideas of Ruspini have been very fruitful in the foundations of approximate reasoning. In particular, one can find in the literature a number of approaches addressing the formalisation of similarity-based reasoning from a logical perspective. Due to space restrictions, in the rest of this short paper we restrict ourselves to review two main lines of developments in this area, namely,

– Graded similarity-based entailments, and
– Formalisations as conditional logics and as modal logics.

2   Graded Similarity-Based Entailments

Let W be the set of classical interpretations (or worlds) of a propositional language. The rules of classical logic allows us to unambiguously decide whether a given proposition p is true or false in each of the worlds. We write $w \models p$ to denote that p is true at $w \in W$ (or that w satisfies p, or that w is a model of p), and $w \not\models p$ to denote that p is false at w. In other words, each world partitions the set of proposition into two classes: those that are true and those that are false.

Assume now we have a $\otimes$-similarity relation S on the set W. This allows us to be more fine-grained when classifying propositions, since even two propositions p and q can be both false at a given world w, it may be the case that w is closer to the set of models of p than to those of q. In more precise terms, even if $w \not\models p$ and $w \not\models q$, it can be the case that

$$\mu_{\mathrm{approx}(p)}(w) > \mu_{\mathrm{approx}(q)}(w).$$

In such a case one can say that, in the world w, p is *closer to be true* than q, or that p is more *truthlike* than q, in the sense of [10].

In the rest of this section, we will overview three different ways of how this idea of having worlds more or less close to others can be used in a logical setting to introduce different kinds of graded similarity-based entailments [1,7].

2.1   Approximate Entailment

Given a $\otimes$-similarity relation S on the set W of classical interpretations of a propositional language, one starts by defining for each $\alpha \in [0,1]$ a (graded) approximate satisfaction relation $\models_S^\alpha$, by stipulating for each $w \in W$ and proposition p:

$$w \models_S^\alpha p \text{ iff there exists a model } w' \text{ of } p \text{ which is } \alpha\text{-similar to } w,$$
$$\text{i.e. such that } w' \models p \text{ and } S(w, w') \geq \alpha,$$
$$\text{i.e. } \mu_{\mathrm{approx}(p)}(w) \geq \alpha.$$

If $w \models_S^\alpha p$ we say that w is an *approximate model* (at level α) of p. The approximate satisfaction relation can be extended over to an approximate entailment relation in the usual way: a proposition p entails a proposition q at degree α, written $p \models_S^\alpha q$, if each model of p is an approximate model of q at level α, that is,

$$p \models_S^\alpha q \text{ iff } w \models_S^\alpha q \text{ for all model } w \text{ of } p, \text{i.e.}$$
$$\text{iff } I_S(q \mid p) \geq \alpha$$

Then $p \models_S^\alpha q$ stands for "q *approximately follows from* p" and α is a level of strength. Under this perspective p, together with the similarity relation $S :$ $W \times W \rightarrow [0, 1]$ on the set of interpretations, represents an epistemic state accounting for the factual information about the world. Then, we can know, not only what are the consequences we can infer from p using classical reasoning, but also those propositions which are approximate consequences of p, in the sense that they are close to some other proposition which is indeed a classical consequence of p.

In the case the propositional language is finitely generated, the following properties characterise these graded entailment relations $\models_S^\alpha$, see [1]:

(1) **Nestedness:**   if $p \models^\alpha q$ and $\beta \leq \alpha$ then $p \models^\beta q$;
(2) **⊗-Transitivity:**   if $p \models^\alpha r$ and $r \models^\beta q$ then $p \models^{\alpha \otimes \beta} q$;
(3) **Reflexivity:** $p \models^1 p$;
(4) **Rightweakening:**   if $p \models^\alpha q$ and $q \models r$ then $p \models^\alpha r$;
(5) **Leftstrengthening:** if $p \models r$ and $r \models^\alpha q$ then $p \models^\alpha q$;
(6) **Left-Or:** $p \vee r \models^\alpha q$ iff $p \models^\alpha q$ and $r \models^\alpha q$;
(7) **Right-Or:**   if r has a single model,
                            $r \models^\alpha p \vee q$ iff $r \models^\alpha p$ or $r \models^\alpha q$.

The ⊗-transitivity property is weaker than usual and the graceful degradation of the strength of entailment it expresses, when $\otimes \neq \min$, is rather natural. The fourth and fifth properties are consequences of the transitivity property (since $q \models r$ entails $q \models^1 r$) and express a form of monotonicity. It must be noticed that $\models^\alpha$ does not satisfy the *Right-And* property, i.e. from $p \models^\alpha q$ and $p \models^\alpha r$ it does not follow in general that $p \models^\alpha q \wedge r$. Hence the set of α-approximate consequences of p in the sense of $\models^\alpha$, for $\alpha < 1$, will not be deductively closed in general. The Left-Or shows how disjunctive information is handled, while the Right-Or reflects the decomposability of the approximate satisfaction relation with respect to the $\vee$ connective only in the case the premise has a single model.

In the case where some (imprecise) *background knowledge* about the world is known and described under the form of some proposition K (i.e. the actual world is in the set of worlds satisfying K), then an approximate entailment relative to K can be straightforwardly defined as

$$p \models_{S,K}^\alpha q \text{ iff } p \wedge K \models_S^\alpha q \text{ iff } I_S(q \mid p \wedge K) \geq \alpha$$

See [1] for more details and properties of this derived notion of relative entailment.

2.2   Proximity Entailment

The above approximate satisfaction relation $w \models_S^\alpha p$ can be also extended over another entailment relation $\models_S$ among propositions as follows: $p \models_S^\alpha q$ holds whenever each approximate model of p at a given level β is also an approximate model of q but at a possibly lower level $\alpha \otimes \beta$. Formally:

$$p \models_S^\alpha q \text{ holds iff, for each } w, w \models_S^\beta p \text{ implies } w \models_S^{\alpha \otimes \beta} q$$

Now, $p \models_S^\alpha q$ means "approximately-p entails approximately-q" and α is a level of strength, or in other words, when worlds in the vicinity of p-worlds are also in the vicinity (but possibly a bit farther) of q-worlds. This notion of entailment, called *proximity entailment* in [1], also admits a characterization in terms of another similarity-based measure

$$J_S(q \mid q) = \inf_w \{ I_S(p \mid w) \Rightarrow I_S(q \mid w) \},$$

where $\Rightarrow$ is the residuum of the (left-continuous) t-norm $\otimes$ and $I_S(p \mid w) = \sup_{w' \models p} S(w, w')$. Indeed, one can easily check that $p \models_S^\alpha q$ holds iff $J_S(q \mid p) \geq \alpha$. This notion of approximate entailment relation can be easily made relative to a context or background knowldge, described by a (finite) set of propositions K, by defining

$$p \models_{S,K}^\alpha q \text{ iff, } \text{ for each } w \text{ model of } K, w \models_S^\beta p \text{ implies } w \models_S^{\alpha \otimes \beta} q.$$

One can analogously characterize this entailment by a generalized measure $J_{S,K}$, namely it holds that

$$p \models_{K,S}^\alpha q \text{ iff } J_{K,S}(q \mid p) = \inf_{w:w \models K} \{ I_S(p \mid w) \Rightarrow I_S(q \mid w) \} \geq \alpha.$$

The entailment $\models_{S,K}^\alpha$ satisfies similar properties to those satisfied by $\models_{S,K}^\alpha$. Characterizations of both similarity-based graded entailments in terms of these properties are given in [1]. It is also shown there that $\models_S^\alpha$ and $\models_S^\alpha$ actually coincide, i.e. when there is no background knowledge K, or equivalently when K is a tautology. However, when K is not a tautology, $\models_{S,K}^\alpha$ is generally stronger than $\models_{S,K}^\alpha$.

2.3   Strong Entailment

Finally, the notion of graded satisfiability $w \models_S^\alpha p$, can be also used for supporting a strong entailment relation with the following intended meaning: a proposition p strongly entails a proposition q at degree α, written $p \approx_S^\alpha q$, if each approximate model of p at level α is a model of q that is,

$$p \approx_S^\alpha q \text{ iff, } \text{ for all } w, w \models_S^\alpha p \text{ implies } w \models q.$$

This stronger form of entailment is a sort of dual of the approximate entailment, as it denotes a notion of entailment that is robust to small (up to level α) deformations of the antecedent, while still entailing the consequent. In a similar way the approximate entailment was linked to the implication measure I_S, this strong graded entailment is related to the consistency measure C_S, in the following way:

$$p \mathrel{\mid\approx}_S^\alpha q \quad \text{iff} \quad C_S(\neg q | p) < \alpha,$$

by assuming the language is finitely generated and $\alpha > 0$. Moreover, a characterization of this strong entailment in terms of some nice properties is given in [7].

3 Logical Formalisations

3.1 Conditional-Like Logics of Graded Approximate and Strong Entailments

In a series of papers [7,16–18], the authors have been concerned with logics to reason about graded entailments. Graded approximate and strong entailments are taken as primitive objects of a propositional language. Let us briefly describe here the main features of the Logic of Approximate Entailment (LAE) from [7].

The basic building block of LAE are graded implications of the form

$$\phi >_\alpha \psi,$$

where ϕ, ψ are classical propositional formulas and α belongs to a suitable scale V of similarity values. The set of similarity values is endowed with a monoidal operation $\otimes$, which in case of the real unit interval is a t-norm. Furthermore, the language of LAE is built up from graded implications and constants $\bot$, $\top$ by means of the classical binary operators $\wedge$ and $\vee$ and the unary operator $\neg$.

The semantics is the expected one: models are pairs $\langle M, e \rangle$, where $M = (W, S)$ is a similarity space, e is an evaluation that maps propositional formulas into subsets of W, interpreting $\wedge, \vee, \neg$ by set intersection, union and complementation, respectively. Given a similarity space M, the satisfaction of a formula by an evaluation e is inductively defined as follows. For graded implications, one defines:

$$\langle M, e \rangle \models \varphi >_\alpha \psi \quad \text{if} \quad e(\varphi) \subseteq U_\alpha(e(\psi)),$$

where, for each $A \subseteq W$, $U_\alpha(A) = \{ w \in W \mid S(w, w') \geq \alpha, \text{ for some } w' \in A \}$ is the α-neighbourhood of A. Moreover, if Φ is a Boolean combination of graded implications, $\langle M, e \rangle \models \Phi$ is defined in accordance with the rules of classical propositional logic (CPL).

This gives rise to the following notion of logical consequence: for each subset of LAE-formulas $\mathcal{T} \cup \{\Phi\}$,

$$\mathcal{T} \models_{LAE} \Phi \quad \text{iff} \quad \begin{array}{l} \text{for any similarity space } M = (W, S) \text{ and any evaluation } e, \\ \text{if } \langle M, e \rangle \text{ satisfies all formulas of } \mathcal{T}, \text{ then it also satisfies } \phi. \end{array}$$

In the finitary case, i.e., when the propositional formulas are built up from a finite set of propositional variables, the logic LAE defined in [7] is the system consisting of the following axioms and rule:

(A1) $\phi >_1 \psi$, where ϕ, ψ are such that $\phi \to \psi$ is a tautology of CPL
(A2) $(\phi >_\alpha \psi) \to (\phi >_\beta \psi)$, where $\alpha \geq \beta$
(A3) $\neg(\psi >_1 \bot) \to (\phi >_0 \psi)$
(A4) $(\phi >_\alpha \bot) \to (\phi >_1 \bot)$
(A5) $(\phi >_\alpha \chi) \wedge (\psi >_\alpha \chi) \to (\phi \vee \psi >_\alpha \chi)$
(A6) $(\phi >_1 \psi) \to (\phi \wedge \neg \psi >_1 \bot)$
(A7) $(\phi >_\alpha \psi) \wedge (\psi >_\beta \chi) \to (\phi >_{\alpha \otimes \beta} \chi)$
(A8) $\neg(\delta >_1 \bot) \to ((\delta >_\alpha \epsilon) \to (\epsilon >_\alpha \delta))$, where δ, ϵ are m.e.c.'s
(A9) $(\epsilon >_\alpha \phi \vee \psi) \to (\epsilon >_\alpha \phi) \vee (\epsilon >_\alpha \psi)$, where ϵ is a m.e.c.
(A10) LAE-formulas obtained by uniform replacements of variables in CPL-tautologies by LAE graded conditionals
(MP) Modus Ponens

Here, m.e.c. means maximal elementary conjunction, i.e., a conjunction where every propositional variable appears, either in positive or negative form. It turns out that, as proved in [7], this axiomatic system provides a sound and complete axiomatisation of the semantic $\models_{LAE}$.

In [16,17], we have proposed a simplified proof system for a variant of LAE. Namely, we have focused on the case of $\otimes$-similarity relations, where $\otimes$ is the product t-norm. The concept of a m.e.c., which occurs in axioms (A8) and (A9) and plays an essential role in the above approach, could be dropped. The notion of an α-neighbourhood of a set A is in this context to be slightly adapted: $U_\alpha(A) = \{w \in W \mid S(w, A) \geq \alpha\}$, where $S(w, A) = \sup_{a \in A} S(w, a)$. Consider the axioms and rule (A1), (A2), (A4), (A5), (A7), and (MP), as well as

(A11) $(\phi >_1 \psi) \to (\phi \wedge \chi >_1 \psi \wedge \chi)$

A proposition Φ is valid in the logic of approximate entailment based on the product t-norm if and only if Φ is provable by means of the indicated axioms and rule.

The proof of this completeness theorem is involved and consists of two parts. In [16], we have shown a similar statement but without the assumption that the similarity relation is symmetric, and we have represented proofs by weighted directed forests. In [17], we have established that spaces based on a possibly non-symmetric similarity relation can, in a certain sense, be embedded into a space based on a similarity relation in the usual sense. Both results combined lead to the completeness theorem mentioned.

The logic LAE has been further developed in a different direction in [18] to account for additional nice features that the approximate entailment has when assuming the language talks about properties on (products of) linearly ordered domains.

Finally, it is worth mentioning that similar syntactical characterisations for strong and proximity entailments can be envisaged. Indeed, in [7] a logic of graded strong entailment, called LSE, is introduced by considering similar graded conditionals $\varphi \succ_\alpha \psi$ with the following semantics:

$$\langle M, e \rangle \models \varphi \succ_\alpha \psi \text{ if } U_\alpha(e(\varphi)) \subseteq e(\psi),$$

As for the proximity entailment, in [8] a corresponding logic with graded conditionals $\varphi \gg_\alpha \psi$ is also introduced with a somewhat more involved semantics:

$$\langle M, e \rangle \models \varphi \gg_\alpha \psi \text{ if } \forall \beta : U_\beta(e(\varphi)) \subseteq U_{\alpha \otimes \beta}(e(\psi)).$$

3.2 Modal Logic Connections

In his original work, Ruspini mentions the use of modal concepts to explain his similarity-based possibilistic structures but he never studied in detail the underlying modal logics. In fact, this was done in Rodriguez's PhD thesis [12] following his suggestion, and also reported in [4,5,8,9]. In this section we want to summarise the main results which appear there. According to Ruspini's intuition, it makes sense to consider a modal approach to similarity-based reasoning based on Kripke structures of the form

$$M = (W, S, e),$$

where W is a set of possible worlds, $S : W \times W \to [0,1]$ a similarity relation between worlds, and e a classical two-valued truth assignment of propositional variables in each world $e : W \times \text{Var} \to \{0,1\}$. Then, for each $\alpha \in [0,1]$ one can consider the α-cut of S, $S_\alpha = \{(w,w') \in W \times W \mid S(w,w') \geq \alpha\}$, as a classical accessibility relation on $W \times W$, which gives meaning to a pair of dual possibility and necessity modal operators $\Diamond_\alpha$ and $\Box_\alpha$:

$$(M, w) \models \Diamond_\alpha \varphi \text{ if there is } w' \in W \text{ s.t. } (w, w') \in S_\alpha \text{ and } (M, w') \models \varphi.$$

This defines, in fact, a multi-modal logical framework (with as many modalities as level cuts of the similarity relations). Such a multimodal logic setting is systematically developed in [5].

Note that, if W is the set of classical interpretations of a propositional language $\mathcal{L}$, then the above notion of modal satisfiability for the possibility operators $\Diamond_\alpha$ captures precisely the notion of approximate satisfiability considered in Sect. 2, in the sense that, for any non-modal proposition p, $(M, w) \models \Diamond_\alpha p$ holds iff $w \models_S^\alpha p$ holds. Moreover, as already intuitively pointed out by Ruspini in [13], the *approximate entailment* $p \models_S^\alpha q$ can also be captured by the formula

$$p \to \Diamond_\alpha q,$$

in the sense that $p \models_S^\alpha q$ holds iff $p \to \Diamond_\alpha q$ is valid in $M = (W, S, e)$. Analogously, the strong entailment $p \approx_S^\alpha q$ can be captured by the formula

$$\Diamond_\alpha p \to q.$$

As for the *proximity entailments* $\models^{\alpha}_{S}$, recall that $p \models^{\alpha}_{S} q$ holds iff for all $w \in W$ and for all β, $w \models^{\beta}_{S} p$ implies $w \models^{\alpha \otimes \beta}_{S} q$. Therefore, it cannot be represented in the multi-modal framework unless the similarity relations are forced to have a fixed, predefined set G of finitely-many different levels, say $\{0,1\} \subseteq G \subset [0,1]$. In that case, the validity of the formula

$$\bigwedge_{\beta \in G} \Diamond_{\alpha} p \rightarrow \Diamond_{\alpha \otimes \beta} q$$

in the model (W, S, e) is equivalent to the entailment $p \models^{\alpha}_{S} q$. Obviously, when G is not finite, for instance when $G = [0,1]$, this representation is not suitable any longer. However, the underlying modal logic can still be formalised by introducing further modal operators accounting for the open cuts of the similarty relation in the models, that is considering the operators $\Diamond^{c}_{\alpha}$ and $\Diamond^{o}_{\alpha}$ for each rational $\alpha \in G \cap \mathbb{Q}$ with the following semantics:

$$(M, w) \models \Diamond^{c}_{\alpha}\varphi \text{ if } I_{S}(\varphi \mid w) \geq \alpha,$$
$$(M, w) \models \Diamond^{o}_{\alpha}\varphi \text{ if } I_{S}(\varphi \mid w) > \alpha.$$

Obviously, when G is finite, $\Diamond^{c}_{\alpha}$ and $\Diamond^{o}_{\alpha}$ are interdefinable. In any case, different multimodal systems can be axiomatized as it is shown in [5,8].

4   Conclusions and Dedication

This paper contains a brief summary of some developments in the research field of similarity-based approximate reasoning models and their logical formalisations, where Ruspini's inspiring ideas have been very fruitful and decisive. It is our humble homage to Enrique, an excellent researcher and even better person. The authors are very grateful to him to have had the chance to enjoy his friendship and shared with him many interesting scientific discussions.

Acknowledgments. Esteva and Godo acknowledge partial support by the Spanish MINECO/FEDER project RASO (TIN2015- 71799-C2-1-P). Rodriguez acknowledges support by the CONICET research project PIP 112-20150100412CO *Desarrollo de Herramientas Algebraicas y Topológicas para el Estudio de Lógicas de la Incertidumbre y la Vaguedad* (DHATELIV). Additionally, he has also been funded by the projects UBA-CyT 20020150100002BA and PICT/O $N°$ 2016-0215.

References

1. Dubois, D., Esteva, F., Garcia, P., Godo, L., Prade, H.: A logical approach to interpolation based on similarity relations. Int. J. Approx. Reason. **17**, 1–36 (1997)
2. Dubois, D., Lang, J., Prade, H.: Possibilistic logic. In: Gabbay, D.M., Hogger, C.J., Robinson, J.A. (eds.) Handbook of Logic in Artificial Intelligence and Logic Programming. Nonmonotonic Reasoning and Uncertain Reasoning, vol. 3, pp. 439–513. Oxford University Press (1994)

3. Esteva, F., Garcia, P., Godo, L.: On conditioning in similarity logic. In: Bouchon-Meunier, B., Yager, R., Zadeh, L.A. (eds.) Fuzzy Logic and Soft Computing, Advances in Fuzzy Systems-Applications and Theory, vol. 4, pp. 300–309. World Scientific (1995)
4. Esteva, F., Garcia, P., Godo, L.: About similarity-based logical systems. In: Dubois, D., Prade, H., Klement, E.P. (eds.) Fuzzy Sets, Logics and Reasoning About Knowledge. APLS, vol. 15, pp. 269–288. Springer, Dordrecht (1999). https://doi.org/10.1007/978-94-017-1652-9_18
5. Esteva, F., Garcia, P., Godo, L., Rodriguez, R.O.: A modal account of similarity-based reasoning. Int. J. Approx. Reason. 16(3–4), 235–260 (1997)
6. Esteva, F., Garcia, P., Godo, L., Ruspini, E., Valverde, L.: On similarity logic and the generalized modus ponens. In: Proceedings of the IEEE International Conference on Fuzzy Systems (FUZZ-IEEE 1994), Orlando, USA, pp. 1423–1427. IEEE Press (1994)
7. Esteva, F., Godo, L., Rodriguez, R.O., Vetterlein, T.: Logics for approximate and strong entailment. Fuzzy Sets Syst. 197, 59–70 (2012)
8. Godo, L., Rodriguez, R.O.: Logical approaches to fuzzy similarity-based reasoning: an overview. In: Della Riccia, G., et al. (eds.) Preferences and Similarities. CISM, vol. 504, pp. 75–128. Springer, Berlin (2008). https://doi.org/10.1007/978-3-211-85432-7_4
9. Godo, L., Rodríguez, R.O.: A fuzzy modal logic for similarity reasoning. In: Chen, G., Ying, M., Cai, K.Y. (eds.) Fuzzy Logic and Soft Computing. ASIS, vol. 6, pp. 33–48. Springer, Boston (1999). https://doi.org/10.1007/978-1-4615-5261-1_3
10. Niiniluoto, I.: Truthlikeness. Synthese Library, vol. 185, 1st edn. D. Reidel Publishing Company (1987)
11. Recasens, J.: Indistinguishability Operators: Modelling Fuzzy Equalities and Fuzzy Equivalence Relations. Studies in Fuzziness and Soft Computing, vol. 260. Springer, Heidelberg (2011). https://doi.org/10.1007/978-3-642-16222-0
12. Rodriguez, R.O.: Aspectos formales en el Razonamiento basado en Relaciones de Similitud Borrosas. Ph.D. dissertation, Technical University of Catalonia (UPC) (2002). (in Spanish)
13. Ruspini, E.H.: On the semantics of fuzzy logic. Int. J. Approx. Reason. 5, 45–88 (1991)
14. Ruspini, E.H.: A logic-based view of similarities and preferences. In: Della Riccia, G., et al. (eds.) Preferences and Similarities. CISM, vol. 504, pp. 23–46. Springer, Berlin (2008). https://doi.org/10.1007/978-3-211-85432-7_2
15. Trillas, E.: Assaig sobre les relacions d'indistingibilitat. Actes del Congrés Català de Lògica, pp. 51–59 (1982). (in Catalan)
16. Vetterlein, T.: Logic of approximate entailment in quasimetric spaces. Int. J. Approx. Reason. 64, 39–53 (2015)
17. Vetterlein, T.: Logic of approximate entailment in quasimetric and in metric spaces. Soft Comput. 21(17), 4953–4961 (2017)
18. Vetterlein, T., Esteva, F., Godo, L.: Logics for approximate entailment in ordered universes of discourse. Int. J. Approx. Reason. 71, 50–63 (2016)
19. Zadeh, L.A.: A theory of approximate reasoning. In: Hayes, J.E., Michie, D., Mikulich, L.I. (eds.) Machine Intelligence, vol. 9, pp. 149–194. Elsevier, New York (1979)

Fuzzy Memories of Enrique Hector Ruspini (1942–2019)

Rudolf Seising[(✉)] [iD]

Research Institute for the History of Science and Technology,
Deutsches Museum, Museumsinsel 1, 80538 Munich, Germany
r.seising@deutsches-museum.de

Abstract. This paper is a personal obituary of Enrique Hector Ruspini, an extraordinary scientist, tireless researcher, a smart science organizer and a great man. It can only mention a few of his scientific and private interests, achievements and achievements.

Keywords: Enrique Ruspini · Lotfi Zadeh · Jim Bezdek · Fuzzy sets · Fuzzy clustering

1 Introduction

Enrique Hector Ruspini was an extraordinary scientist, a tireless researcher, an immensely interesting person, but above all, he was a good friend! I knew him for about 20 years, where we met very often and often ate and drank well. He loved good food and fine wine. When I was in California, I used to visit him in Palo Alto. The first time we met for lunch and an interview on 30 July 2002 in Palo Alto. He was then employed at the Stanford research Institute an I was with the University of Vienna in Austria. 7 years later, when I was an adjoint researcher in the *European Centre for Soft Computing* (ECSC) in Mieres, Asturias (Spain) he became a principal researcher at that center and we saw us almost daily. After he left the ECSC in 2013, we met at conferences and I visited him and his wife Susana in California when I was there in Berkeley with Lotfi Zadeh to work on the history of fuzzy set theory. Enrique introduced me to the *Computational Intelligence Society (CIS)* of *IEEE* and about 10 years ago he, Jim Bezdek and Jim Keller accepted me as a member of the *CIS History Committee*. Enrique suggested that I continue the series of video interviews with pioneers and other researchers of CI, which he, the two Jims and others had started earlier.

In 2017, I planned two trips to California and to Lotfi and Enrique. Enrique was invited to attend the dedication ceremony of the robot Shakey to the *Computer History Museum* in Mountain View at February 16. He offered to let me join him. It was a memorable event and we had the opportunity to view the entire exhibition (see Figs. 1 and 8).

© Springer Nature Switzerland AG 2020
M.-J. Lesot et al. (Eds.): IPMU 2020, CCIS 1237, pp. 14–26, 2020.
https://doi.org/10.1007/978-3-030-50146-4_2

The next day I asked Enrique to accompany me to meet Lotfi and he agreed. Lotfi was already very weak then but we could talk for about an hour in threes over tea and cookies. It was the last time the two of them saw each other and I photographed them while they were talking (see Fig. 2).

Fig. 1. left: Poster of the dedication ceremony of robot Shakey to the *Computer History Museum* **right:** Enrique at the *Computer History Museum* at February 16, 2017. (Photo: Rudolf Seising)

During my second visit in this year in Berkeley Lotfi passed away at 6 September. Some days later, at September 15, I drove to Palo Alto to interview Enrique at his home. The weather was nice, and we sat outside in his garden (see Fig. 3). The video of this interview is about an hour long but unfortunately, there was a lot of noise because of some birds in the garden and some planes in the air. Because of the strong noise, we have not yet included the video of the interview in the official collection of ieee.tv (https://ieeetv.ieee.org/channels/cis-oral-history?tab=allvideos), but it is available on the net: [1].

Fig. 2. Enrique and the author during the interview, September 15, 2017. (Photo: Rudolf Seising)

In 2018, Enrique was more and more suffering from a creeping illness. In this year, he should receive the *Frank Rosenblatt Award* "for fundamental contributions to the understanding of fuzzy logic concepts and their applications". However, he could not travel to Rio to receive this award. Fortunately, in June 2019, he could attend the *2019 IEEE International Conference on Fuzzy Systems* in New Orleans, LA., and he could take the award (Fig. 4). Some weeks later, at October 15 in 2019 Enrique passed away.

Fig. 3. Enrique's and Lotfi's meeting in February 2017 in Lotfi's house. (Photo: Rudolf Seising)

2 Facts and Data

Enrique Hector Ruspini was born at December 20, 1942 in Buenos Aires, Argentina. He received the Licenciado en Ciencias Matemáticas (bachelor's degree in Mathematics) from the *University of Buenos Aires*, Argentina, in 1965 and his doctoral degree in System Science from the *University of California* at Los Angeles in 1977 [2]. Dr. Ruspini had held positions at the *University of Buenos Aires*, the *University of Southern California*, UCLA's *Brain Research Institute*, Hewlett-Packard Laboratories, the *SRI International Artificial Intelligence Center*, and the *European Center for Soft Computing* in Asturias (Spain). He was also a Distinguished Lecturer of the *IEEE Computational Intelligence Society*.

Dr. Ruspini, who was the recipient of the 2009 Fuzzy Systems Pioneer Award of the *IEEE Computational Intelligence Society*, received in 2004 the Meritorious Service Award of the *IEEE Neural Networks Society* for leading the transition of the *Neural Networks Council* into Society status. He was one of the founding members of the *North American Fuzzy Information Processing Society* (NAFIPS) and the recipient of that society's King-Sun Fu Award. He was an IFSA First Fellow, an IEEE Life Fellow and a former member of the IEEE Board of Directors and past President of the *IEEE Neural Networks Council (now IEEE Computational Intelligence Society)*. In 2018, he was laureate of the IEEE Frank Rosenblatt Technical Field Award of the IEEE Computational Intelligence Society. He was a First Fellow of the International Fuzzy Systems Association, a Fulbright Scholar, a European Union Marie Curie Fellow, and a SRI International Fellow. Dr. Ruspini was the Editor in Chief (together with Piero P. Bonissone and Witold Pedrycz) of the *Handbook of Fuzzy Computation*. He was a member of the Advisory and Editorial Boards of numerous professional journals, e.g. *IEEE Transactions on Fuzzy Systems*, *International Journal of Fuzzy Systems*, *International Journal of Uncertainty, Fuzziness, and Knowledge-Based Systems*, *Fuzzy Sets and Systems*, *Mathware and Soft Computing*, and the *Journal of Advanced Computational Intelligence and Intelligent Informatics*. He published more than 100 research papers.

3 The 1960s

Enrique studied physics in Argentina when he encountered one of the first computers in his home country. He switched to mathematics and to the *Instituto de Cálculo* where he was concerned with the numerical solution of differential equations. He was interested in novel applications of computer science, e.g. in the field of biomedical engineering and in modeling and simulation in mathematical economics and he had a position paid by a group associated with the University of Buenos Aires that worked in a children's hospital. There was a group of neuroscientists doing research on neural signals and another group of colleagues was concerned with numerical taxonomy and classification of biological species. In 1964 he started planning to go abroad to get a Ph D. and then coming back to Argentina to get a professorship as it was usual. He wrote to many people and to Richard Ernest Bellman who as then professor of mathematics, electrical engineering and medicine. In 1965 or 1966, Enrique got an offer from Bellman to work at the *Space Biology Laboratory* in the *Brain Research Institute* at the *University of South California* (USC) that he accepted and in the academic year of 1966 he started a new job at USC in Los Angeles. Already in the late year of 1965 he read the seminal paper "Fuzzy Sets" by Lotfi Zadeh and at the same time, he was asked to use anyone of the new computer classification techniques. So, clustering came into his life at the same time as fuzzy sets and this was the beginning of fuzzy clustering. Very early, after he came to the US, he could publish his article "A New Approach to Clustering" [3].

Richard Bellman and his associate Robert Kalaba were coauthors of Zadeh's first memo on fuzzy sets[1]. Enrique remembered Bellman as a most interesting person and obviously one of the brightest applied mathematicians that ever lived. (8:25). There appeared many papers on control theory, differential systems and other areas. By Bellman, Kalaba and Ruspini in the 1960s. Enrique worked together mainly with Bob Kalaba and once in the late 1960s, when they were having lunch in the faculty club of USC Lotfi Zadeh came in to say Hello. Enrique had sent him a draft of his paper on fuzzy clustering and that afternoon they had some discussion on fuzzy clustering. This was the starting point of an association that lasted about half a century until Lotfi's death in 2017. In my interview, Enrique called Zadeh a great mentor ([1], 14:51).

4 The 1970s and 1980s

Many people received Enrique's paper on Fuzzy Clustering [2] and by people that were reluctant. In my interview, he said that this gave him an impulse to fight the unbelievers. He kept working on applications of fuzzy sets and at the same time on problems associated with medicine, and he also could combine both research areas e.g. in the paper on "A test of sleep staging systems in the unrestrained chimpanzee" in 1972 [5].

Jim Bezdek, who finished his Ph D thesis "Fuzzy Mathematics in Pattern Classification" in 1973 at Cornell University, Ithaca, New York, had read Enrique's paper on fuzzy clustering and he arranged an invitation to Enrique to give a talk at the *Seventh Annual Meetings of The Classification Society*, North American Branch. This meetings were held at the University of Rochester, May 23–25, 1976 an they both gave talks in the topical session "(4) Fuzzy clustering algorithms" [6]: Enrique lectured on "Fuzzy clustering as an optimum mapping between metric spaces" and Jim on" Feature selection for binary data with fuzzy ISODATA". Because many of the researchers in that society liked the simplicity and the power that the fuzzy approach brought to classification problems. Enrique and Jim were close friends from that time on. Jim interviewed Enrique during the *IEEE Congress on Evolutionary Computing* in Cancun, Mexico at June 20, 2011. This video of this interview is available as a video of the *CIS Oral History project* [7]. In 2015 I interviewed the two together via e-mail and the text of this interview is available online [8, 9].

At the *International Congress on Applied Systems Research and Cybernetics* in Acapulco, Mexico, in 1980, Ron R. Yager had organized a number of sessions on fuzzy sets and fuzzy systems, possibility theory and special topics in systems research

---

[1] For details on the history of fuzzy sets, see the author's book [4].

[10]. Here, Enrique and Ebrahim Mamdani met for the first time and he learned about his combination of rule-based systems and fuzzy control. On the flight returning to the USA from Acapulco Enrique Madan Gupta and Jim Bezdek were still fascinated by the impression of having their own forum for the exchange of ideas in the field of fuzzy sets. It was tempting to plan such a forum. Therefore, Jim and Enrique became founding members of the *North American Fuzzy Information Processing Society* (*NAFIPS*). Jim Bezdek organized the first *NAFIP*-meeting (in that time without the "s" at the end) in Logan, Utah[2].

Fig. 4. Jim Bezdek interviews Enrique during the IEEE Congress on Evolutionary Computing in Cancun, Mexico June 20, 2011. (Photo from the video interview on the IEEE CIS history website, https://history.ieee-cis.sightworks.net/)

From Los Angeles Enrique moved to the *Hewlett Packard Laboratories* in Palo Alto in 1982 where he was concerned with databased systems and data analysis. However, in that time HP reorganized the company and Enrique cold not do as much research as he wanted to and therefore, he left HP to years later to join the *SRI International's Artificial Intelligence Center*, in Menlo Park, California, as a principal scientist. Sometime after, he had met John Lowrance, a principal scientist at SRI in Lotfi's seminar at UC Berkeley. Lowrance was very interested in evidential reasoning, especially in the theory of Dempster and Shafer, but also in other related ideas. Because of Enrique's skills in these epistemological issues, he hired him for SRI. With Alessandro Saffioti, and others Enrique developed a fuzzy controller, which they implemented in the autonomous mobile robot named FLAKEY, the successor to SHAKEY (see Fig. 5) [12, 13].

---

[2] See the whole story and a picture of some of the 41 participants to NAFIP-1- meeting, and the poster of this conference with many of the participants' signatures in [11].

Enrique "confessed" in the interview that in the beginning he "didn't pay much that attention to the semantic questions" of fuzzy sets [...] but later, it became necessary. He was approached by people who wanted to better understand this theory and consequently he started research on "conceptual relations that synthesizes utilitarian and logical concepts" [14], on "the basic conceptual differences between probabilistic and possibilistic approaches" [15] and "On the Semantics of Fuzzy Logic" at *SRI* in the 1990s [16] ([1], 24:28-25:10). Many years later, he summarized these investigations in the contribution "On the Meaning of Fuzziness" for the two-volume collection "On Fuzziness. A Homage to Lotfi A. Zadeh" [17] Enrique worked for *SRI International's AI Center* for 25 years until he retired in 2009, and then he started a new phase at the ECSC in Spain.

Fig. 5. Flakey the robot, developed around 1985 at SRI International

When and how did Enrique arrive at the artificial neural networks (ANN)? In my interview, he said that in a sense he "kind of never had" (41:48). I insisted that he was the president of the *IEEE Neural Networks Council*, which is now *IEEE Computational Intelligence Society* and he answered: "Yeah, that's right but only because they incorporated fuzzy logic and evolutionary computation" ([1], 41:56). Enrique had never worked on artificial neural networks but he was a member of *IEEE*. At the end of the 1980s when there was the renaissance of ANN the *IEEE* formed the *Neural Networks Committee* and they organized a number of very successful conferences. Therefore, on November 17, 1989, *IEEE* decided to let the committee become the *Neural Networks Council* (NNC).

5 From the 1990s Until Today

In 1991, Jim Bezdek had called Enrique because the *IEEE* had signaled their interest in Fuzzy Logic to Lotfi Zadeh. They wanted to give researchers in Fuzzy Logic a major platform because there were a large number of hybrid methods combining neural networks and fuzzy logic. Lotfi had recommended contacting Jim, Piero and Enrique and they had achieved the establishment of the new series of *IEEE International Conferences on Fuzzy Systems* (*FUZZ-IEEE*) that continues until today. The first of these conferences took place in San Diego and the *IEEE NNC* sponsored this event with Lotfi as honorary chair, Jim as conference chair and Enrique as tutorial chair. The *Second IEEE International Conference on Fuzzy Systems* should be in San Francisco, in spring 1993; and because of organizational problems, it was held in conjunction with the 1993 International Conference on Neural Networks. Piero was program chair and Enrique was general chair of the conference. In my interview, he suspected that this was probably why his name was associated to neural networks ([1], 48:46). In 2001 he became president of the NNC and in this year he led the Council (NNC) into the Society (NNC). Two years later, in November 2003, it changed its name to its current one, the IEEE *Computational Intelligence Society* (CIS) and I asked in my interview who chose this name. Enrique answered: "we needed a name that was descriptive for the new society or the new council" – and here he emphasized that contrary to the opinion of many, it was not an easy way from this Neural Networks Council (NNC) to the Neural Networks Society (NNS) they formed on November 21, 2001 – "it took a long time" ([1], 49:52). He continued, "we needed a name that would differentiate us from just pure applied AI." (50:14). They were looking for ideas and then it was Jim Bezdek who suggested "computational intelligence" (CI).

In my interview Enrique said that "curacy enough until this day it is very difficult to explain to people what computational intelligence is without listing each one of the components" (fuzzy sets, artificial neural networks, evolutionary algorithms) (51:41) and he mentioned the political problems to be able to cooperate that into a description of our scope. He also mentioned that researcher in classical AI "have the same problem in describing what artificial intelligence is" ([1], 52:15).

Another term is soft computing (SC), introduced by Lotfi Zadeh in the 1990s. It was part of the name of both the *BISC* (*Berkeley Initiative in Soft Computing*) at the University of California and the *ECSC* (*European Centre for Soft Computing*) in Spain. The latter was the last academic institution Enrique worked for, and it turned out that way: In 2007 during the *FUZZ-IEEE 2007 conference* at the *Imperial College* in London Enric Trillas and Enrique had lunch together. After his retirement from academia in Spain Enric was an Emeritus Researcher at the *ECSC* and now, they discussed the eventualities of Enrique becoming a Principal Investigator at this place. Two years later, they realized this plan. In his contribution to the commemorative publication for Enric Trillas on his 75th birthday, Enrique tried to remember when the two might have first met:

"I do not believe that I met Enric before 1985 although it might have been a bit earlier. It was certainly after 1980 because I had not yet travel to Europe and I do not remember him visiting California before then. I recall, however, delivering a lecture at the School of Architecture of the Polytechnic University of Barcelona in the early 90's where he was present and where I took a picture of the audience with my brand new Cannon Photura: one of the first cameras featuring a fuzzy logic autofocus system. Our acquaintance dates, perhaps, to the time when I visited Spain to participate in the First IFSA Congress in Mallorca. I have tried to do a bit of detailed research about our initial meeting to give a more precise account of its circumstances but could not find any additional information to remove remaining ambiguities. The reader may certainly wonder why this quest for accuracy about either the author and its subject—both notorious examples of fuzzy scientists—matters at all. My obsession with this personal milestone, stems, however, from the realization that I was well along in my career before I met this remarkable man with whom I have had so many fruitful and continued interactions since our first encounter.

The photographic evidence shows that, by 1996, we had developed a friendship that allowed us to engage on merry pranks during lighter moments of serious scientific meetings, as seen in Fig. 6, when we posed as the never finished statues of a magnificent building originally constructed to house a technical labor university." [18].

Supported by the *Foundation for the Advancement of Soft Computing*, which was a private non-profit foundation, the ECSC was launched by the beginning of the year 2006. The most important goal of the center was the basic and applied research in soft computing and the technology transfer in industrial applications of intelligent systems design for the resolution of real-world problems. For almost ten years, the Center was a meeting point for experts in CI and SC all over the world.

Fig. 6. "Lost" statues at the Universidad Laboral de Gijon, Asturias, Spain, 1996: Enric Trillas, Enrique Ruspini, and José Luis Verdegay, then Professor at the *Department of Computer Science and Artificial Intelligence (DECSAI)*, of the University of Granada, Spain.

In the ECSC Enrique headed the research unit on Collaborative intelligent systems, he established a connection with chemists who were interested in fuzzy logic, he was working in distributed AI. At the Open Workshop "Fuzziness and Medicine" as part of the *I. International Symposium on Fuzziness, Philosophy and Medicine* I organized at the ECSC on 23–25 March 2011, Enrique gave the keynote address "First Steps on Fuzzy Sets in Medicine." In this talk he combined his early research in biomedicine and brain research with his later work on the semantics of fuzzy logic. (see Fig. 7). Two and a half years later, Enrique left the center in November 2013 and due to the economic crisis, the ECSC had to be closed in 2016.

Fig. 7. Enrique during his talk "First Steps on Fuzzy Sets in Medicine" for the Open Workshop "Fuzziness and Medicine" at March 24 2013, *ECSC*, Asturias, Spain. (Photo: Rudolf Seising)

When I asked about the future of fuzzy logic and soft computing, also in light of the recently deceased Lotfi Zadeh, Enrique answered very optimistically in the interview: "We are going to continue!" Referring to the many technical applications with embedded CI technology, he pointed to the video camera pointed at him, he pointed out "anywhere we can see it, in those cameras that are filming us; in cars in appliances we have fuzzy logic. So, it had become part of the family. People who were skeptic about fuzzy logic, professors who were sceptic, now their students are applying fuzzy logic to of fuzzy logic, professors who were sceptic they have students who are applying fuzzy logic to problems. So, it has become imbedded and there is still a huge number of problems; for example in language understanding that would benefit from that. The important thing for people in academia by now is to keep on using fuzzy logic in their tool kit. ([1], 59:55-1:00:46) As a kind of tribute to Lotfi Zadeh and Richard Bellman he finished our interview. Regarding Lotfi Zadeh he said: "He was never somebody who said, well, here is fuzzy logic and that's it what you should use [...] No, he always said, here you have all these tools, probabilities, calculus of evidence, fuzzy logic, all sorts of methods. And that was incidentally, what I remember about Richard Bellman: He said: It is always better to have lots of methods and to try different methods and to see what kind of solutions you get. That gives you so much. So, fuzzy logic should remain in the tool kit and I see that it is entrenching there" ([1], 1:01:26).

Enrique was passionately committed to the various institutions of his scientific discipline, as treasurer, president, vice president and conference organizer. He was also a member of the CIS History Committee from the very beginning, because it was important to him that the historical development of computational intelligence not be forgotten and that young researchers learn about the history of their topics.

Fig. 8. Enrique Trillas at the entrance to the *Computer History Museum* in Mountain View, CA at February 16, 2017. (Photo: Rudolf Seising)

References

1. Seising, R.: Interview with Enrique Ruspini in Palo Alto. https://ieeetv.ieee.org/ns/ieeetvdl/CIS/Ruspini_converted.mp4
2. Ruspini, E.H.: A theory of mathematical classification, Dissertation: Thesis (Ph.D.)–University of California, Los Angeles—Engineering (1977)

3. Ruspini, E.H.: A new approach to clustering. Inf. Control **15**, 22–32 (1969)
4. Seising, R.: The Fuzzification of Systems. The Genesis of Fuzzy Set Theory and Its Initial Applications – The Developments up to the 1970s (STUDFUZZ 216) Springer, Berlin (2007). https://doi.org/10.1007/978-3-540-71795-9
5. Larsen, L.E., Ruspini, E.H., McNew, J.J., Walter, D.O.: A test of sleep staging systems in the unrestrained chimpanzee. Brain Res. **40**, 319–343 (1972)
6. International Classification, vol. **3**(2), pp. 98 (1976). Reports and Communications
7. Bezdek, J.: Interview with Enrique Ruspini, CIS Oral History Project Video (2011). https://ieeetv.ieee.org/history/cis-history-enrique-ruspini-2011?rf=series|5&
8. Seising, R.: On the History of Fuzzy Clustering: An Interview with Jim Bezdek and Enrique Ruspini, Archives for the Philosophy and History of Soft Computing, vol. 2, pp. 1–14 (2014). https://www.unipapress.it/it/book/aphsc-|-2-2014_168/
9. Seising, R.: On the history of fuzzy clustering: an interview with Jim Bezdek and Enrique Ruspini. IEEE Syst. Man Cybern. Mag. **1**(1), 20–48 (2015)
10. Lasker, G.E. (ed.): Applied Systems and Cybernetics: Proceedings of the International Congress on Applied System Research and Cybernetics. Pergamon Press, Oxford (1981)
11. Bezdek, J., Ruspini, E.H.: The story of the NAFIP-1 poster. Arch. Philos. Hist. Soft Comput. **1**(1), 1–5 (2018)
12. Ruspini, E.H.: Fuzzy logic in the Flakey robot. In: Proceedings of the International Conference on Fuzzy Logic and Neural Networks, Iizuka (Japan) Fukuoka, pp. 767–770 (1990)
13. Saffiotti, A., Ruspini, E., Konolige, K.G.: A fuzzy controller For Flakey, an autonomous mobile robot. In: Reusch, B. (Hrsg.) Fuzzy Logic: Theorie und Praxis, 3. Dortmunder Fuzzy-Tage Dortmund, 7–9, pp. 3–12. Springer, Heidelberg (1993). https://doi.org/10.1007/978-3-642-78694-5_1
14. Ruspini, E.H.: Truth as utility: a conceptual synthesis. In: Proceedings of the Conference on Uncertainty in Artificial Intelligence, Los Angeles, CA, pp. 316–322 (1991)
15. Ruspini, E.H.: Approximate Reasoning: Past, Present, Future. Technical Note No. 4D2, Artificial In-telligence Center, SRI International, Menlo Park, California, 19\10 (1991)
16. Ruspini, E.H.: On the semantics of fuzzy logic. Int. J. Approximate Reasoning **5**, 45–88 (1991)
17. Seising, R., Trillas, E., Moraga, C., Termini, S. (eds.): On Fuzziness. A Homage to Lotfi A. Zadeh volume 2, chapter 87, (STUDFUZZ 299), pp. 598–609. Springer, Heidelberg (2013). https://doi.org/10.1007/978-3-642-35644-5
18. Ruspini, E.H.: ENRIC by Enrique. In: Seising, R. (ed.) Accuracy and Fuzziness. A Life in Science and Politics. SFSC, vol. 323, pp. 313–317. Springer, Cham (2015). https://doi.org/10.1007/978-3-319-18606-1_30

Invited Talks

From Eliza to Siri and Beyond

Luísa Coheur[1,2]([✉])([ID])

[1] INESC-ID Lisboa, Lisbon, Portugal
`luisa.coheur@inesc-id.pt`
[2] Instituto Superior Técnico/Universidade de Lisboa, Lisbon, Portugal

Abstract. Since Eliza, the first chatbot ever, developed in the 60s, researchers try to make machines understand (or mimic the understanding) of Natural Language input. Some conversational agents target small talk, while others are more task-oriented. However, from the earliest rule-based systems to the recent data-driven approaches, although many paths were explored with more or less success, we are not there yet. Rule-based systems require much manual work; data-driven systems require a lot of data. Domain adaptation is (again) a current hot-topic. The possibility to add emotions to the conversational agents' responses, or to make their answers capture their "persona", are some popular research topics. This paper explains why the task of Natural Language Understanding is so complicated, detailing the linguistic phenomena that lead to the main challenges. Then, the long walk in this field is surveyed, from the earlier systems to the current trends.

Keywords: Natural language processing · Natural Language Understanding · Chatbots

1 Introduction

Since the first chatbots from the 60s (such as ELIZA [42]) to the current virtual assistants (such as SIRI[1]), many things have changed. However, and despite the incredible achievements done in Natural Language Processing (NLP), we are far from creating a machine capable of **understanding natural language**, a long-standing goal of **Artificial Intelligence (AI)** and probably the ultimate goal of NLP.

Understanding natural language is an extremely complex task. Researchers in NLP have been struggling with it since the early days. Even the concept of "understanding" is not consensual. Some authors consider that whatever the implemented methods are, if a system is capable of providing a correct answer to some natural language input, then we can say that it was able to correctly interpret that input. Other authors assume that the mapping of the input sentence into some **semantic representation**, which captures its meaning, is necessary.

---

[1] https://www.apple.com/siri/.

© Springer Nature Switzerland AG 2020
M.-J. Lesot et al. (Eds.): IPMU 2020, CCIS 1237, pp. 29–41, 2020.
https://doi.org/10.1007/978-3-030-50146-4_3

The process of mapping the natural language input into the semantic representation is called **semantic parsing**.

Deep Learning brought a new way of doing things and also an extra verve to the field. However, independently of the approach followed, and even if some interesting and accurate dialogues with machines can be achieved in some limited domains, we can quickly find out inconsistencies in conversational agents responses. In fact, no current system genuinely understands language.

This paper is organized as follows: Sect. 2 surveys some of the main challenges when dealing with natural language understanding, and Sect. 3 give some historical perspective of the main advances in the area. Then, Sect. 4 discuss current trends and Sect. 5 concludes and points to some future work.

2 Challenges

Most researchers do not realize that language is this complex until embracing a NLP task involving understanding natural language. Indeed, we manage to communicate with some success. Consequently, we are not aware that the sequences of words that we produce and interpret within our dialogues are extremely variable (despite obeying to certain syntax rules), often ambiguous (several interpretations are often possible) and that, sometimes, sophisticated reasoning (sometimes considering features that go beyond natural language) needs to be applied so that a fully understanding can be achieved.

In fact, a factor that makes the computational processing of our language terribly complex is **language variability**, that is, our ability to say the same thing in so many different ways (e.g., *yes, right, Ok, Okay, Okie dokie, looks good, absolutely, of course* are just some ways of expressing agreement). If a bot operates in a strictly closed domain it is possible to gather the semantics of the most common questions that will be posed to it[2]. Still, the main problem is not the semantics of the most common questions, but their form, which can vary a lot. For instance, in some closed domains we can build a list of FAQs representing what will be asked to a virtual agent. However, we will hardly have a list of all the paraphrases (sentences with the same meaning) of those questions. As an example, consider the following sentences (from [12]):

1. *Symptoms of influenza include fever and nasal congestion;*
2. *Fever and nasal congestion are symptoms of influenza;*
3. *A stuffy nose and elevated temperature are signs you may have the flu.*

Sentences (1) and (2) can be easily identified as paraphrases, by simply considering the lexical units in common of both sentences. However, sentence (3) will only be identified as a paraphrase of (1) and (2) if we know that *fever* and *elevated temperature* are similar or equal concepts and the same between *nasal congestion* and *stuffy nose*. WordNet [20] and current word embeddings do, indeed, solve some of these problems, but not all.

---

[2] Nevertheless, we can hardly predict the flow of the dialogue, unless the machine takes the initiative and do not leave much room for innovation to the human.

Another factor that makes language so complicated is the fact that it is inherently **ambiguous**. Sentences with different meanings emerge from **ambiguity at the lexical level**. For instance, some meanings of the word *light* are[3]:

1. *comparatively little physical weight or density*
2. *visual effect of illumination on objects or scenes as created in pictures.*

This leads to ambiguous sentences like *I will take the light suit.*

There is a well known NLP task called **Word Sense Disambiguation** that targets the specific problem of lexical ambiguity. Nonetheless, lexical entities are not the only source of ambiguity. A good example of **syntactic ambiguity** is the classical sentence *I saw a man in the hill with a telescope.* Who had the telescope? Who was on the hill? Many interpretations are possible. Another good example of ambiguity is the sentence *João and Maria got married.* We do not even notice that this sentence is ambiguous. However, did they marry each other or with other people?

In most cases we can find the correct interpretation of a sentence by considering the **context** in which it is uttered. Many works (including recent ones, such as the work described in [37]) propose different ways of dealing with context, which is still a popular research topic. Yet, to complicate things further, although context can help to dismantle some productions, it can also lead to more interpretations of a sentence. For instance, *I found it hilarious* can be detected by a *Sentiment Analyser* as a positive comment, which is correct if we are evaluating a comic film, but probably not if we refer to a horror movie or a drama. Another example: *I'll be waiting for you at 4.p.m. outside school* can be a typical line of a dialogue between mother and son, but can also be a bullying threat. Context is everything.

A further difficulty is that it is impossible to predict all the different sentences that will be posed to a virtual agente, unless, as previously mentioned, it operates in a really strictly closed domain. Therefore, a bottleneck of conversational agents are Out-of-Domain (OOD) requests. For instance, as reported in [2], Edgar Smith [11], a virtual butler capable of answering questions about Monserrate Palace in Sintra, was reasonably effective when the user was asking **In-Domain questions**. However, people kept asking it **OOD questions**, such as *Do you like Cristiano Ronaldo?*, *Are you married?*, *Who is your favourite actress?*. Although it might be argued that, in light of their assistive nature, such systems should be focused in their domain-specific functions, the fact is that people become more engaged with these applications if OOD requests are addressed [5, 24].

Several other linguistica phenomena make dialogues even more difficult to follow by a machine. For instance, for interpreting the sentences *Rebelo de Sousa and Costa went to Spain. The president went to Madrid and the prime-minister to Barcelona.*, we need to know that Marcelo is the president and that Costa is the prime-minister. The NLP task that deals with this phenomenon is **Coreference Resolution**. Also, **elliptical constructions** – omission of one or more

---

[3] According with https://muse.dillfrog.com/meaning/word/light.

words in a sentence that can be inferred – make things extremely difficult to machines (e.g. *Is Gulbenkian's museum open on Sundays? And on Saturdays?*). In addition, we use **idioms** (e.g., *it's raining cats and dogs*), **colocations** and many multi-word expressions whose meaning is not necessary related with the meaning of its parts. For instance, colocations are sequences up to three words that we learn to use since ever, but that we do not really know why we say it that way (e.g., in Portuguese we say *perdi o avião* (literally *I lost the plane*), to say *I missed my flight*). Why the verb *perdi?* (by the same token, why *miss?*). If for a foreign language learner the production (and understanding) of these expressions is a real challenge, it certainly is too for a computer. Moreover, we cannot ignore **humor** and **sarcasm**, which further complicate the machine tasks of understanding language (e.g., *Those who believe in telekinetics, raise my hand. – Kurt Vonnegut*).

Furthermore, the simple fact that we might be using speech when interacting with the machine (not to mention sign language) adds an extra layer of problems. For instance, a noisy environment or a strong accent can be enough to unable our understanding of what is being said (the same for a virtual agent). Also, the way we say something influences the interpretation of a sentence. For instance, *Good morning* can be said, in an unpleasant tone, to those who arrive late at class, meaning *Finally!*.

To conclude, the way we express ourselves has to do with our mastery of language, which, in turn, is influenced by the time in which we live, by our age, social condition, occupation, region where we were born and/or inhabit, emotional state, among others. All those features make each one of us a special case (and computers prefer archetypal patterns). A truly robust application should be able to handle all of our productions. Therefore, it must be equipped with some reasoning ability, which must also take into account all non-verbal elements involved in a conversation (e.g., the interlocutor's facial expression, the scenario, etc.). And the truth is that we are still far from being able to integrate all these variables in the (natural language) understanding process.

3   Historical Perspective

As previously said, ELIZA is considered to be the first chatbot ever, developed in the 1960s, with the aim of simulating a psychotherapist. Although it was able to establish a conversation, simulating it was a human being, its virtual model was based in rephrasing the user input, whenever it matched a set of hand-crafted rules and also in providing content-free remarks (such as *Please go on.*) in the absence of a matching. For instance, ELIZA could have a rule like the following one, in which * is the wildcard and matches every sequence of words; the (2) means that the sequence of words captured by the second wildcard would be returned in ELIZA's answer:

Rule: * *you are* */ What makes you think I am (2)?*

If the user input was, for instance, *I think you are bright*, it would answer *What makes you think I am bright?*. At that time, many people believed they were talking with another human (the "ELIZA effect"). Having no intention of modelling the human cognitive process and despite its simplicity, ELIZA showed how a software program can cause a huge impact by the mere illusion of understanding. Nowadays, ELIZA is still one of the most widely known applications in AI. ELIZA and subsequent bots (such as the paranoid mental patient PARRY [9] or JABBERWACKY [8]), definitely provided the seeds to many different directions to explore. For instance, the idea of having a virtual agent with a "persona" that explains the flow (and the flaws) of the conversation continues to be widely used. Another idea that is due to the first chatbots and that is still being explored today is that of "learning by talking". For example, if a user asks *What do you think of Imagine Dragons?* and the bot does not know how to answer, it will record the question. The next time it interacts with a human, that same question will be asked by the bot and (hopefully) a possible answer will be gathered. Of course, this approach can go wrong and a very (recent) popular case was that of a chatbot that was willing to learn by interacting with humans. Within hours of being in use, the chatbot was racist, nazi and vulgar, and had to be turned off. A survey on the early chatbots and their contributions can be found in [25].

Another important line of research emerged in the beginning of the 70s, due to three seminal papers of Richard Montague: *English as a Formal Language* [21], *The Proper Treatment of Quantification in Ordinary English* [22], and *Universal Grammar* [23]. Montague proposed a formal framework to map syntactic structures into semantic representations, providing a systematic and **compositional** way of doing it. Although Montague's work was limited to a very small subset of the English language and difficult to extend, the idea of taking advantage of syntax to build semantic forms in a compositional way was used in the many **Natural Language Interfaces to Databases (NLIDB)** that popped-up by that time (mostly during the 80s), and that are still being developed nowadays, although with different techniques ([3] surveys classical NLIDBs and [1] more recent ones).

Then, **Question/Answering (QA) (and Dialogue)** systems started to come out. Contrary to NLIDB, QA systems knowledge sources are not (necessarily) ground in databases; they are quite similar otherwise. On the subject of Dialogue systems, they are designed to engage in a dialogue with the user to get the information they need to complete a task (for instance), and, thus, there is more than just a question and an answer involved. The strong development of the former in the late 90s and beginning of the XXI century was partially due evaluation *fora*, such as TREC[4] (since 99) and CLEF[5] (since 2003), which provided tasks entirely dedicated to QA, such as QA@CLEF[6]. This allowed researchers to straightforwardly compare their systems, as everybody was evaluated with the same test sets. A side effect of these competitions was the release of data that become usually available to the whole community. This certainly also consolidated the rise of Machine Learning techniques against traditional rule-based

---

[4] https://trec.nist.gov.
[5] http://www.clef-initiative.eu/home.
[6] http://www.clef-initiative.eu/track/qaclef.

approaches. Nowadays, an important task in NLP is **Machine Reading Comprehension** that targets to understand unstructured text in order to answer questions about it, but new related challenges are still emerging, such as the Conversational Question Answering Challenge (CoQA) [30].

In what concerns Dialogue Systems, several domains were explored from the early days. Particularly prolific were the conversational agents targeting the concept of Edutainment, that is, education through entertainment. Following this strategy, several bots have animated museums all over the world: the 3D animated Hans Christian Andersen [4] established multimodal conversations about the writer's life and tales, Max [27] was employed as guide in the Heinz Nixdorf Museums Forum, Sergeant Blackwell [31], installed in the Cooper-Hewitt National Design Museum in New York, was used by the U.S. Army Recruiting Command as a hi-tech attraction and information source, and the previously mentioned **Edgar Smith** (Fig. 1).

Fig. 1. Edgar Smith in Monserrate.

In some of these systems the agent's knowledge base was constituted of pairs of sentences (S1, S2), where S2 (the answer) is a response to S1 (the trigger). Their "understanding" process was based on a **retrieval approach**: if the user says something that is "close" to some trigger that the agent has in its knowledge base (that is, if the user input matches or rephrases a trigger), then it will return the correspondent answer. Others based their approach in information extraction techniques, capable of detecting the user intentions and extract relevant entities from the dialogues in order to capture the "meaning" of the sentence. A typical general architecture of the latter systems combined natural language understanding and generation modules (sometimes template-based). A dialogue manager was present in most approaches. Nowadays, **end-to-end**

data-driven systems – that is systems trained directly from virtual data to optimize an objective function [17] –, have replaced or at least try to replace these architectures.

Then, things started to happen fast: Watson wins Jeopardy! in 2011 (a big victory to NLP), Apple releases Siri, also in 2011, Google Now appears in 2012, and the world faces a new level of conversational agents: the virtual assistants built by these colossal companies[7]. In the meantime, Deep Learning starts to win in all fronts, and Sequence to Sequence (seq2seq) models start to be successfully applied to Machine Translation [36]. Considering that these models take as input a sequence of words in one language and output a sequence of words in another language, the first generative dialogue systems based on these models did not take long to appear, and the already mentioned **end-to-end (dialogue) systems** came to light (v.g. [18,35,37,39,44,45,47]). These systems differ from retrieval-based systems as in the latter pre-defined responses are given to the user; in the generative-based approaches responses are generated in run-time. The majority of these systems are trained to engage in general conversations (chit-chat agents) and, therefore, make use of movie subtitles (or Reddit[8] data).

Besides seq2seq architectures, two concepts are responsible for many of the latest achievements: **Neural Word Embeddings** and **Pre-trained Language Models**. Machine learning algorithms cannot usually directly deal with plain text. Therefore, the idea of converting words into vectors has been explored for a long time. Word Embeddings are functions that map words (or characters, paragraphs or even documents) into vectors (and neural networks can learn these mappings). More recently deep learning led to the creation of several embedding types, from context-free to contextual models, from character- to word-based, etc. Considering context-free embeddings, each word form is associated with a single embedding, and, thus, the word *light* will have a single embedding associated. In contextual models, the embedding captures a specific meaning of the word, and, therefore, *light* will have several vectors associated, according with the context in which it occurs. Examples of context-free embeddings are the ones created with Word2Vec models [19]; examples of the latter are ELMo [26] and BERT [10]. A simple way to use these models in dialogue systems is, for instance, in a retrieval-base approach, calculate the embedding of the agent's knowledge base sentences and the embedding associated with the user given sentence. Then, find the cosine similarity between the embedding of the latter and the ones from the knowledge base, and return the one with the highest value. In what concerns language models, these can be seen as models that are trained to predict the next word (or character) in a sequence. For instance, by counting n-grams (sequences of n tokens) we can build a language model. These have many application scenarios. For instance, in a translation setup, decide which, from possible translations of a source sentence, is more probable, considering the target language (and the n-grams observed in that target language). An example of a language model is

---

[7] Microsoft's Cortana and Alexa were born in 2014.
[8] https://www.reddit.com.

GPT-2 [29], trained in 40GB of text data, and developed by OPenAI[9]. It should be said that behind some of the most successful models (as BERT and GPT-2), there is a Transformer-base model. Transformers [38] were introduced in 2017 and are enjoying great success in NLP.

Several problems are still under research. The next section presents some of the current trends.

4   Current Trends

Several challenges lie ahead of the current end-to-end conversational agents. Just to name a few, some systems are unable to track the topic of the conversation, or are prone to generate trivial (and universal) responses such as "ok" or "I don't know" (the "universal answer" drawback). Some authors propose neural models that enable dialogue state tracking (v.g. [48]), or new methods that inject diversity in the generated responses (v.g. [33]).

Current systems try to take into account important features of the user request, as for instance, their sentiment [6, 14, 34]. This is particularly important for support bots, as, for instance, a very unhappy customer (negative polarity) should not receive an answer starting with *Hello, we are so happy to hear from you!!!*. Researchers also explore how to add declarative knowledge to neural network architectures. As an example, in [16], neural networks are augmented with First-Order Logic. Domain adaptation is also, as previously said, a hot-topic (v.g [28]). Some current research follows in the transfer-learning paradigm: pre-trained Language Models are fine-tuned with specific data. For instance, in [7], GPT-2 is used as a pre-trained model and fine-tuned in task-oriented dialogue systems; in [43] a model is pre-trained on the BookCorpus dataset [49], and, then, tuned on the PERSONA-CHAT corpus [46]. The latter corpus was created to allow the building of chit-chat conversational agents, with a configurable, but persistent persona [46]. This idea of creating a bot with a consistent persona is also the topic of research of many current works [13, 15].

Many more proposals, not necessarily following in the end-to-end paradigm, are also worth to be mentioned. For instance, in [41] the authors propose the building of a semantic parser overnight, in which crowdsourcing is used to paraphrase canonical utterances (automatically built), into natural language questions; in [40] the computer learns, from scratch, the language used by people playing a game in the blocks world. Users can use whatever language they want, and they can even invent one.

5   Main Conclusions and Future Work

Since its early days, the NLP community has embraced the task of building conversational agents. However, and despite all the recent achievements (mainly due to Deep Learning), a short conversation with these systems quickly exposes their

---

[9] https://openai.com.

weaknesses [32], including the lack of a consistent personality. The community is pretty aware of these limitations, and recent work is focusing on boosting the conversational agents' capabilities. Pre-trained models will certainly continue to be explored, as well as ways to enrich the model training with different types of knowledge.

The adaptation of a bot to a specific user is also something to explore, as the given answer will probably differ regardless of whether we interact with the bot once (for instance, when buying tickets for visiting a specific monument) or regularly (for instance, to reserve a hotel in a particular platform). In the latter case we want our assistant to remember some information about the user.

Acknowledgements. I would like to express my gratitude to Vânia Mendonça, who gave me very detailed comments about this document. However, the responsibility for any imprecision lies with me.

References

1. Affolter, K., Stockinger, K., Bernstein, A.: A comparative survey of recentnatural language interfaces for databases. VLDB J. **28**(5), 793–819 (2019). https://doi.org/10.1007/s00778-019-00567-8
2. Ameixa, D., Coheur, L., Fialho, P., Quaresma, P.: Luke, I am your father: dealing with out-of-domain requests by using movies subtitles. In: Bickmore, T., Marsella, S., Sidner, C. (eds.) IVA 2014. LNCS (LNAI), vol. 8637, pp. 13–21. Springer, Cham (2014). https://doi.org/10.1007/978-3-319-09767-1_2
3. Androutsopoulos, I., Ritchie, G., Thanisch, P.: Natural language interfaces to databases - an introduction. Nat. Lang. Eng. **1**(1), 29–81 (1995). https://doi.org/10.1017/S135132490000005X
4. Bernsen, N.O., Dybkjær, L.: Meet Hans Christian Andersen. In: Proceedings of Sixth SIGdial Workshop on Discourse and Dialogue, pp. 237–241 (2005)
5. Bickmore, T., Cassell, J.: How about this weather? Social dialogue with embodied conversational agents. In: Socially Intelligent Agents: The Human in the Loop, pp. 4–8. AAAI Press, Menlo Park (2000)
6. Bothe, C., Magg, S., Weber, C., Wermter, S.: Dialogue-based neural learning to estimate the sentiment of a next upcoming utterance. In: Lintas, A., Rovetta, S., Verschure, P.F.M.J., Villa, A.E.P. (eds.) ICANN 2017. LNCS, vol. 10614, pp. 477–485. Springer, Cham (2017). https://doi.org/10.1007/978-3-319-68612-7_54
7. Budzianowski, P., Vulic, I.: Hello, it's GPT-2 - how can i help you? Towards the use of pretrained language models for task-oriented dialogue systems (2019). https://doi.org/10.17863/CAM.47097
8. Carpenter, R., Freeman, J.: Computing Machinery and the Individual: The Personal Turing Test (2005). http://www.jabberwacky.com
9. Colby, K.M.: Ten criticisms of PARRY. SIGART Newslett. 5–9 (1974)
10. Devlin, J., Chang, M.W., Lee, K., Toutanova, K.: BERT: pre-training of deep bidirectional transformers for language understanding. In: Proceedings of the 2019 Conference of the North American Chapter of the Association for Computational Linguistics: Human Language Technologies, Minneapolis, Minnesota, Volume 1 (Long and Short Papers), pp. 4171–4186. Association for Computational Linguistics, June 2019. https://doi.org/10.18653/v1/N19-1423. https://www.aclweb.org/anthology/N19-1423

11. Fialho, P., et al.: Meet EDGAR, a tutoring agent at MONSERRATE. In: Proceedings of the 51st Annual Meeting of the Association for Computational Linguistics: System Demonstrations, Sofia, Bulgaria, pp. 61–66. Association for Computational Linguistics, August 2013. https://www.aclweb.org/anthology/P13-4011

12. Fialho, P., Coheur, L., Quaresma, P.: From lexical to semantic features in paraphrase identification. In: Rodrigues, R., Janousek, J., Ferreira, L., Coheur, L., Batista, F., Oliveira, H.G. (eds.) 8th Symposium on Languages, Applications and Technologies (SLATE 2019). OpenAccess Series in Informatics (OASIcs), Dagstuhl, Germany, vol. 74, pp. 9:1–9:11. Schloss Dagstuhl-Leibniz-Zentrum fuer Informatik (2019). https://doi.org/10.4230/OASIcs.SLATE.2019.9. http://drops.dagstuhl.de/opus/volltexte/2019/10876

13. Herzig, J., Shmueli-Scheuer, M., Sandbank, T., Konopnicki, D.: Neural response generation for customer service based on personality traits. In: Proceedings of the 10th International Conference on Natural Language Generation, Santiago de Compostela, Spain, pp. 252–256. Association for Computational Linguistics, September 2017. https://doi.org/10.18653/v1/W17-3541. https://www.aclweb.org/anthology/W17-3541

14. Kong, X., Li, B., Neubig, G., Hovy, E., Yang, Y.: An adversarial approach to high-quality, sentiment-controlled neural dialogue generation. In: AAAI 2019 Workshop on Reasoning and Learning for Human-Machine Dialogues (DEEP-DIAL 2019), Honolulu, Hawaii, January 2019. https://arxiv.org/abs/1901.07129

15. Li, J., Galley, M., Brockett, C., Spithourakis, G., Gao, J., Dolan, B.: A persona-based neural conversation model. In: Proceedings of the 54th Annual Meeting of the Association for Computational Linguistics, Berlin, Germany, (Volume 1: Long Papers), pp. 994–1003. Association for Computational Linguistics, August 2016. https://doi.org/10.18653/v1/P16-1094. https://www.aclweb.org/anthology/P16-1094

16. Li, T., Srikumar, V.: Augmenting neural networks with first-order logic. In: Proceedings of the 57th Annual Meeting of the Association for Computational Linguistics, Florence, Italy, pp. 292–302. Association for Computational Linguistics, July 2019. https://doi.org/10.18653/v1/P19-1028. https://www.aclweb.org/anthology/P19-1028

17. Lowe, R.T., Pow, N., Serban, I.V., Charlin, L., Liu, C., Pineau, J.: Training end-to-end dialogue systems with the ubuntu dialogue corpus. D&D 8(1), 31–65 (2017). http://dad.uni-bielefeld.de/index.php/dad/article/view/3698

18. Lu, Y., Keung, P., Zhang, S., Sun, J., Bhardwaj, V.: A practical approach to dialogue response generation in closed domains. CoRR abs/1703.09439 (2017)

19. Mikolov, T., Chen, K., Corrado, G., Dean, J.: Efficient estimation of word representations in vector space. In: Bengio, Y., LeCun, Y. (eds.) 1st International Conference on Learning Representations, ICLR 2013, Scottsdale, Arizona, USA, 2–4 May 2013, Workshop Track Proceedings (2013). http://arxiv.org/abs/1301.3781

20. Miller, G.A.: WordNet: a lexical database for English. Commun. ACM 38, 39–41 (1995)

21. Montague, R.: English as a formal language. In: Thomason, R. (ed.) Formal Philosophy. Selected papers of Richard Montague, pp. 188–221. Yale University Press, New Haven (1974)

22. Montague, R.: The proper treatment of quantification in ordinary English. In: Thomason, R. (ed.) Formal Philosophy. Selected papers of Richard Montague, pp. 247–270. Yale University Press, New Haven (1974)

23. Montague, R.: Universal grammar. In: Thomason, R. (ed.) Formal Philosophy. Selected papers of Richard Montague, pp. 223–246. Yale University Press, New Haven (1974)
24. Patel, R., Leuski, A., Traum, D.: Dealing with out of domain questions in virtual characters. In: Gratch, J., Young, M., Aylett, R., Ballin, D., Olivier, P. (eds.) IVA 2006. LNCS (LNAI), vol. 4133, pp. 121–131. Springer, Heidelberg (2006). https://doi.org/10.1007/11821830_10
25. Pereira, M.J., Coheur, L., Fialho, P., Ribeiro, R.: Chatbots' greetings to human-computer communication. CoRR abs/1609.06479 (2016). http://arxiv.org/abs/1609.06479
26. Peters, M., et al.: Deep contextualized word representations. In: Proceedings of the 2018 Conference of the North American Chapter of the Association for Computational Linguistics: Human Language Technologies, New Orleans, Louisiana, Volume 1 (Long Papers), pp. 2227–2237. Association for Computational Linguistics, June 2018. https://doi.org/10.18653/v1/N18-1202. https://www.aclweb.org/anthology/N18-1202
27. Pfeiffer, T., Liguda, C., Wachsmuth, I., Stein, S.: Living with a virtual agent: seven years with an embodied conversational agent at the heinz nixdorf museumsforum. In: Proceedings of the International Conference Re-thinking Technology in Museums 2011 - Emerging Experiences, pp. 121–131. Thinkk Creative & The University of Limerick (2011)
28. Qian, K., Yu, Z.: Domain adaptive dialog generation via meta learning. In: Proceedings of the 57th Annual Meeting of the Association for Computational Linguistics, Florence, Italy, pp. 2639–2649. Association for Computational Linguistics, July 2019. https://doi.org/10.18653/v1/P19-1253. https://www.aclweb.org/anthology/P19-1253
29. Radford, A., Wu, J., Child, R., Luan, D., Amodei, D., Sutskever, I.: Language models are unsupervised multitask learners (2019)
30. Reddy, S., Chen, D., Manning, C.D.: CoQA: a conversational question answering challenge. Trans. Assoc. Comput. Linguist. 7, 249–266 (2019). https://www.aclweb.org/anthology/Q19-1016
31. Robinson, S., Traum, D., Ittycheriah, M., Henderer, J.: What would you ask a conversational agent? Observations of human-agent dialogues in a museum setting. In: International Conference on Language Resources and Evaluation (LREC), Marrakech, Morocco (2008)
32. Serban, I.V., Lowe, R., Henderson, P., Charlin, L., Pineau, J.: A survey of available corpora for building data-driven dialogue systems: the journal version. Dialogue Discourse 9(1) (2018)
33. Shao, Y., Gouws, S., Britz, D., Goldie, A., Strope, B., Kurzweil, R.: Generating high-quality and informative conversation responses with sequence-to-sequence models. In: Proceedings of the 2017 Conference on Empirical Methods in Natural Language Processing (EMNLP), Copenhagen, Denmark, pp. 2210–2219 (2017)
34. Shi, W., Yu, Z.: Sentiment adaptive end-to-end dialog systems. In: Proceedings of the 56th Annual Meeting of the Association for Computational Linguistics, Melbourne, Australia, (Volume 1: Long Papers), pp. 1509–1519. Association for Computational Linguistics, July 2018. https://doi.org/10.18653/v1/P18-1140. https://www.aclweb.org/anthology/P18-1140
35. Song, Y., Yan, R., Feng, Y., Zhang, Y., Zhao, D., Zhang, M.: Towards a neural conversation model with diversity net using determinantal point processes. In: Proceedings of the Thirty-Second AAAI Conference on Artificial Intelligence, New Orleans, Louisiana, USA, 2–7 February 2018 (2018)

36. Sutskever, I., Vinyals, O., Le, Q.V.: Sequence to sequence learning with neural networks. In: Ghahramani, Z., Welling, M., Cortes, C., Lawrence, N.D., Weinberger, K.Q. (eds.) Advances in Neural Information Processing Systems 27, pp. 3104–3112. Curran Associates, Inc. (2014). http://papers.nips.cc/paper/5346-sequence-to-sequence-learning-with-neural-networks.pdf
37. Tian, Z., Yan, R., Mou, L., Song, Y., Feng, Y., Zhao, D.: How to make context more useful? An empirical study on context-aware neural conversational models. In: Proceedings of the 55th Annual Meeting of the Association for Computational Linguistics, Vancouver, Canada, (Volume 2: Short Papers), pp. 231–236. Association for Computational Linguistics, July 2017
38. Vaswani, A., et al.: Attention is all you need. In: Proceedings of the 31st International Conference on Neural Information Processing Systems, NIPS 2017, pp. 6000–6010. Curran Associates Inc., Red Hook (2017)
39. Vinyals, O., Le, Q.V.: A neural conversational model. CoRR abs/1506.05869 (2015)
40. Wang, S.I., Liang, P., Manning, C.D.: Learning language games through interaction. In: Proceedings of the 54th Annual Meeting of the Association for Computational Linguistics, Berlin, Germany, (Volume 1: Long Papers), pp. 2368–2378. Association for Computational Linguistics, August 2016. https://doi.org/10.18653/v1/P16-1224. https://www.aclweb.org/anthology/P16-1224
41. Wang, Y., Berant, J., Liang, P.: Building a semantic parser overnight. In: Proceedings of the 53rd Annual Meeting of the Association for Computational Linguistics and the 7th International Joint Conference on Natural Language Processing, Beijing, China, (Volume 1: Long Papers), pp. 1332–1342. Association for Computational Linguistics, July 2015. https://doi.org/10.3115/v1/P15-1129. https://www.aclweb.org/anthology/P15-1129
42. Weizenbaum, J.: ELIZA - a computer program for the study of natural language communication between man and machine. Commun. ACM 9, 36–45 (1966)
43. Wolf, T., Sanh, V., Chaumond, J., Delangue, C.: TransferTransfo: a transfer learning approach for neural network based conversational agents. CoRR abs/1901.08149 (2019). http://arxiv.org/abs/1901.08149
44. Yao, L., Zhang, Y., Feng, Y., Zhao, D., Yan, R.: Towards implicit content-introducing for generative short-text conversation systems. In: Empirical Methods in Natural Language Processing (EMNLP), pp. 2190–2199. Association for Computational Linguistics, Copenhagen (2017)
45. Yin, J., Jiang, X., Lu, Z., Shang, L., Li, H., Li, X.: Neural generative question answering. In: International Joint Conference on Artificial Intelligence IJCAI, pp. 2972–2978. IJCAI/AAAI Press, New York (2016)
46. Zhang, S., Dinan, E., Urbanek, J., Szlam, A., Kiela, D., Weston, J.: Personalizing dialogue agents: I have a dog, do you have pets too? In: Proceedings of the 56th Annual Meeting of the Association for Computational Linguistics, Melbourne, Australia, (Volume 1: Long Papers), pp. 2204–2213. Association for Computational Linguistics, July 2018. https://doi.org/10.18653/v1/P18-1205. https://www.aclweb.org/anthology/P18-1205
47. Zhao, T., Eskenazi, M.: Towards end-to-end learning for dialog state tracking and management using deep reinforcement learning. In: Proceedings of the 17th Annual Meeting of the Special Interest Group on Discourse and Dialogue, Los Angeles, pp. 1–10. Association for Computational Linguistics, September 2016. https://doi.org/10.18653/v1/W16-3601. https://www.aclweb.org/anthology/W16-3601

48. Zhao, T., Zhao, R., Eskenazi, M.: Learning discourse-level diversity for neural dialog models using conditional variational autoencoders. In: Proceedings of the 55th Annual Meeting of the Association for Computational Linguistics, Vancouver, Canada, (Volume 1: Long Papers), pp. 654–664. Association for Computational Linguistics, July 2017. https://doi.org/10.18653/v1/P17-1061

49. Zhu, Y., et al.: Aligning books and movies: towards story-like visual explanations by watching movies and reading books. In: Proceedings of the 2015 IEEE International Conference on Computer Vision (ICCV), ICCV 2015, pp. 19–27. IEEE Computer Society, USA (2015). https://doi.org/10.1109/ICCV.2015.11. https://doi.org/10.1109/ICCV.2015.11

Average Jane, Where Art Thou? – Recent Avenues in Efficient Machine Learning Under Subjectivity Uncertainty

Georgios Rizos[1] and Björn W. Schuller[1,2,3(✉)]

[1] GLAM, Imperial College London, London SW7 2AZ, UK
{rizos,schuller}@ieee.org
[2] Chair of EIHW, University of Augsburg, 86159 Augsburg, Germany
[3] audEERING, 82205 Gilching, Germany

Abstract. In machine learning tasks an actual 'ground truth' may not be available. Then, machines often have to rely on human labelling of data. This becomes challenging the more subjective the learning task is, as human agreement can be low. To cope with the resulting high uncertainty, one could train individual models reflecting a single human's opinion. However, this is not viable, if one aims at mirroring the general opinion of a hypothetical 'completely average person' – the 'average Jane'. Here, I summarise approaches to optimally learn efficiently in such a case. First, different strategies of reaching a single learning target from several labellers will be discussed. This includes varying labeller trustability and the case of time-continuous labels with potential dynamics. As human labelling is a labour-intensive endeavour, active and cooperative learning strategies can help reduce the number of labels needed. Next, sample informativeness can be exploited in teacher-based algorithms to additionally weigh data by certainty. In addition, multi-target learning of different labeller tracks in parallel and/or of the uncertainty can help improve the model robustness and provide an additional uncertainty measure. Cross-modal strategies to reduce uncertainty offer another view. From these and further recent strategies, I distil a number of future avenues to handle subjective uncertainty in machine learning. These comprise bigger, yet weakly labelled data processing basing amongst other on reinforcement learning, lifelong learning, and self-learning. Illustrative examples stem from the fields of Affective Computing and Digital Health – both notoriously marked by subjectivity uncertainty.

Keywords: Machine learning · Uncertainty · Subjectivity · Active learning · Cooperative learning

1 Subjectivity and AI

In many machine learning applications of interest, the ground truth reflects some inherently human-centric capacity, like affect [1], or corresponds to an

M.-J. Lesot et al. (Eds.): IPMU 2020, CCIS 1237, pp. 42–55, 2020.
https://doi.org/10.1007/978-3-030-50146-4_4

expert assessment, as in the case of health informatics [2]. In such cases, ground truth cannot be collected automatically via crawling or sensors, and **a human annotation step must be injected in the machine learning pipeline**.

Considering, for example, the differing reports made by experienced healthcare professionals when assessing a medical image [3], it is not hard to imagine that certain annotation tasks can exhibit great subjectivity. Given a photo of a man wearing a slight frown, is he sad, angry, or is this maybe his natural expression? What about a photo of a woman [4]? Is the gender of the person making the assessment [5] relevant? What about their age or cultural background [6]? In the online social media setting, is the mention of curse words (or even slurs) in a tweet considered offensive if it is used between friends [7]? Or, in the age of alternative facts, how can one be certain that an online post does not contain fake news [8]? Perhaps the task under consideration is **inherently ambiguous**, as in asking whether a scene depicted in a photo is warm or cold [9], or **simply difficult even for experts**, as in the identification of volcanoes on a photo of a $75 \cdot 75\,\text{km}^2$ surface patch of Venus [10]. Such ambiguities become exacerbated when the level of expertise and trustworthiness of the annotators enter consideration [11].

In subjective perception studies like those in the above non-comprehensive list, we observe that if we were to ask multiple humans to annotate a single sample, we would often receive disagreeing evaluations. In fact, we typically would *require* multiple raters in an attempt to eliminate the possible bias of one single annotator. We are now faced with a different problem however: **our ground truth, instead of providing clear answers, introduces uncertainty, and disagreement as well; which opinion is correct, if any?** Here we present an overview of approaches utilised to address the issue of ground-truth uncertainty due to subjectivity, a discussion of current state-of-the-art methods, persistent challenges, and an outline of promising future directions.

2 The Dimensions of AI for Subjective Data

The presence of multiple human evaluations per sample forms a constellation of challenges and opportunities, and certain approaches to address some of the former may fail to exploit the latter. Each approach is delineated by the decisions made to accommodate a series of problem dimensions. We discuss the principal dimension that allows for a more high-level clustering in the following subsection.

2.1 Adapting the Labels vs. Adapting the Algorithms

We fundamentally must make a decision on whether we desire to work on the traditional setting, in which samples assume **hard labels** after a fusion of the original, distinct opinions, or we want to introduce the additional modelling complexity to accommodate the particularities of handling subjective data.

We treat the two extreme philosophies as **treating subjectivity as a problem** (see Sect. 3), or **embracing subjectivity and potentially leveraging the opportunities it offers** for better learning (see Sect. 4), respectively.

2.2 The Many Considerations of Working with Subjective Data

There are many permutations in the placement of relevant approaches with respect to the below dimensions, and in certain cases a middle-way is proposed. **It is important to keep in mind what assumptions are implicitly being made by the methods summarised in this overview study.**

Subjectivity as a Source of Information: Whereas rater disagreement does indeed inject noise in the ground-truth that requires extra steps to accommodate, it is also true that it can be quantified and also be used as an additional source of information or algorithm feature if quantified [12–14].

Bad Data or *Interesting* Data? Taking an example from affective computing, certain speech utterances may be characterised as being **prototypical**, i.e., they are clear examples of one emotion, or can be **ambiguous**, [15,16], i.e., at the fuzzy border of more than one emotions. If we subscribe to the first extreme, then we assume that a hard label is the true label of a sample, regardless of whether we use one, and that disagreeing labels constitute observation noise due to rater bias, that needs to be removed by means of denoising. Very often, **data that exhibit high rater disagreement are assumed to be less informative**, due to the inability of raters to come to a consensus. On the other hand, **disagreeing labels may actually capture a separate mode of a true soft label distribution**.

Feasibility of Modelling Raters: Further to the above, in certain cases we have knowledge of the set of labels produced by each (anonymised) rater, and we can use this information to estimate the trustworthiness of each either by proxy of inter-rater reliability, or by using an ensemble of models, where each corresponds to a different rater. Inversely, usually in the case of crowdsourcing, we either do not know which rater provided which label, or we simply have very little overlap between raters to compare their performances. In such cases, we can only attempt to model rater types [17,18].

Predicting With Uncertainty: Depending on the application, it may be desirable to provide a measure of subjectivity uncertainty (or, inversely, confidence) alongside the prediction of each test sample. Tasks that require risk-aware AI, may be related to healthcare, self-driving car technology, or simply cases where catastrophic performance translates to great financial costs. Being able to predict subjectivity uncertainty may also be the primary task of interest [3].

Interactive Learning: One approach to reducing the rater disagreement for a label, is to repeatedly label it, leading to the need for a thorough study of how the disagreement is treated and utilised in such a process. Furthermore, even in regular **active learning**, rater disagreement may be a significant information modality.

2.3   Related Problems in AI

We treat the subjectivity issue as being distinct from other cases that imply more than one label per sample, such as the presence of *hierarchical labels*, or applications like *acoustic event detection*, in which many events are present in one audio recording (possibly multiple times). The case of *multi-label classification* is related to a degree, if we assume that a data sample can have a soft label, i.e., a distribution over categorical labels. However, even in the soft label approach, we focus more on capturing the human subjectivity in labelling each sample.

3   Addressing Subjectivity

The issue we discuss here is **truth inference**, i.e., the extraction of a single hard label per sample, by treating the opinions of various raters as noisy observations of a true value.

3.1   Simple Fusion of Labels

Under the assumption that disagreeing, minority voices in annotation should be considered as mistakes, or random observation noise, we are naturally interested in getting a single, hard label per sample by performing a denoising fusion of the original labels, where each is given the same weight. For categorical labels, this amounts to majority voting, and for continuous ones to mean, or median averaging. A major motivation for adopting such an outlook, would be that the reduction to hard labels allows for the application of established AI literature on the application of interest.

This approach was used to leverage the utility of crowdsourcing, in order to cheaply annotate datasets, including many that became milestones in Deep Learning (DL) research, such as ImageNet [19], MS COCO [20], the MIT Places database [21], the SUN attribute dataset [9], and is still very popular [7,22].

3.2   Weighted Fusion of Labels

The output quality of annotators can vary widely due to differing levels of expertise [23] or personality characteristics, with some of them 'spamming' random answers [17]. As such, whereas the usage of crowdsourcing allows for fast and cheap solution to the annotation of large datasets, the observation noise that may be injected by the process introduces a requirement for increased quality control. The aim here is to estimate and assign trustworthiness values to each rater based on their performance with respect to the annotation task and/or other raters, such that their opinions are weighed differently.

The Evaluator Weighted Estimator (EWE) approach [24] is based on the calculation of rater-specific inter-reliability scores for weighing each rater's scores. EWE has successfully been used for improving the quality of affect recognition [25]. This approach is only meaningful if there is significant overlap among raters with respect to samples, such that the aforementioned scores can be calculated with confidence. This is very often the case in data annotation in the laboratory setting, but not always; especially in the crowdsourcing case. In that case, the rater-specific scores cannot be based on inter-rater reliability, and instead can be calculated based on the performance consistency of each rater on the task, according to the Weighted Trustability Evaluator (WTE) [26] method.

The Case of Time-Continuous Labels. It is of great use to model affect in a continuous-valued, and continuous-time manner. When raters provide **sequences of continuous-valued emotion dimensions**, there is one more challenge to consider: **rater reaction lag**, which is often specific to the person. In these cases, the weighted fusion must be performed in a manner that accounts for lag-based discrepancies between raters as well.

In a study performed in [27], the authors proposed to weigh each rater based on an inter-rater correlation based score, as well as manually experimented with different lags per segment. In [28], a Canonical Correlation Analysis (CCA) approach is proposed in conjunction with time-warping on the latent space to accommodate lag discrepancies. Time warping with additional rank based annotations that reduce the subjectivity of continuous values was proposed in [29]. This issue has also been approached via Expectation-Maximisation (EM) [30], by assuming that the sequences provided by each rater are perturbed versions of a common ground truth. By assuming knowledge of which sequences were provided by which raters, reaction lag can be modelled specifically in a rater-specific manner [31]. Outside continuous affect recognition, a Bayesian dynamical model that models noisy crowdsourced time-series labels of financial data was proposed in [32].

4 Embracing Subjectivity

The main question now becomes whether it is preferable to model for a hypothetical "Average Jane", or whether we can approach the problem by explicitly taking into account the presence of very unique voices. This might mean that we assume that samples are inherently non-prototypical and ambiguous, or that we want to model individual raters to capture unique groups of thought, or even that rater disagreement is one more attribute of the data; to be learnt and predicted.

4.1 Incorporating Subjectivity in the Algorithm

As an intermediate step before fully embracing subjectivity as a source of opportunities instead of simply viewing it as a challenge to be solved before proceeding to the modelling stage, we now discuss the case according to which we indeed extract a single hard label per sample, as well as a measure of rater disagreement that is to be used as an additional input, target, or algorithm feature to improve learning.

Learning with priviledged information, or **master-class learning** [33] is a machine learning paradigm according to which each sample has additional information that facilitates learning during the training phase, but is not required for making predictions during testing. We consider the rater disagreement to be such priviledged information. In the study performed in [34], the sample annotation agreement (prototypicality in context), was used to weigh positively the loss of the corresponding samples, and in fact, samples that exhibited high disagreement were discarded. Similarly, less emphasis is placed in samples with less annotation confidence in the Gaussian Process (GP) based method proposed in [12].

The above methods assume that low rater agreement implies that the sample is inherently useless, and will hinder the training process. However, whereas the high disagreement implies a very subjective sample, deep learning is known to be robust to massive random observation noise [35], and perhaps simply downweighing, or removing such samples naively is not the best approach. To this end, a more appropriate solution might be **to learn what high rater disagreement means in the context of a particular dataset.** A multi-task framework was adopted by [13,36], in which the first task is the prediction of the fused hard emotion label, and the second is the prediction of the inter-rater disagreement of a sample. By adopting such a framework, the predictive performance of the first task is improved. Further to the subject of deciding which samples are more informative in the context of a dataset and task, the authors of [14] utilised, among others, annotator disagreement as an input to a teacher model that makes decisions on which samples are more informative towards training a base predictor.

Assuming that subjectivity is inherent in the application of interest, we might be interested in **being able to predict the inter-rater disagreement of a test sample.** This has been shown as a possibility [37], and has been used in the context of active learning [38] for emotion recognition. Such an approach was adopted in [3] in the digital health domain, for the identification of samples that would most benefit from a medical second opinion.

4.2   Assuming Soft Labels

Instead of fusing the differing labels into a single, hard label, one can calculate the empirical distribution thereof, define a soft label distribution per sample, and train a model on that. The soft label encodes the ambiguousness in ground truth for each sample, and allows for the model to learn label correlations as well, something that has been cited to be a significant contribution to the good performance of model distillation techniques [39]. This is an approach that has been adopted extensively, as in the studies performed in [16,40–42], and as part of [43]. In certain cases, samples for which a hard label cannot be extracted with majority voting due to a lack of a consensus are discarded, however the authors of [16] utilise these ambiguous samples in a soft label framework and observe a competitive performance using only the ambiguous data, and a clear improvement if all data are used.

It is also worth exploring the usefulness of inter-label correlations. The presence of multiple labels per sample allows for richer information available in the estimation of their positions in a low-dimensional manifold [44–46].

4.3   Explicitly Modelling a Rater

When we know the correspondence between labels and raters, we are given the opportunity to model the evaluation process of each particular rater, by utilising an ensemble of models. One of the motivations behind this approach is that in the case where one of the raters is an expert and the others are novices or spammers, the voice of the latter will outweigh that of the former regardless of whether we are using hard labels or soft.

In the methods proposed in [8,11,43,47–49] the authors estimate the model parameters, and a measure of rater trustworthiness in a joint manner. In cases where the number of raters is prohibitively high, there have been attempts to model schools of thought [18]. Using a separate model for explicitly modelling a rater has also been utilised for machine translation [50], and emotion recognition [51]. In the latter study, the model has a common base but multiple heads, each aimed at modelling a hard label as output by a different rater, and then a fusion is applied such that a soft label is predicted. A similar approach was used more recently, in a study about machine vision on biomedical images [52], in which the authors additionally propose that learning a sample-specific weighted averaging of raters' inputs, motivated by the fact that certain raters may be better at annotating certain types of data.

5   Active Learning Under Subjectivity

In the presence of rater disagreement, one possible avenue to improve the quality of the data would be to apply *repeated labelling*, i.e., request additional raters to label the data points, with the purpose of reducing the impact of each individual voice. In [41] it was shown that the simple strategy of relabelling all samples resulted to much better data quality, leading to better test performance; however, it was shown that by focusing the relabelling process on a small set of samples was a much better choice in the interest of budget constraints. The above technique is important to consider in cases where acquiring entirely new examples is more expensive than simply acquiring additional labels for existing ones.

Another related concept is that of **self-healing** [41,53]; in [53], repeated labelling is used to request additional samples to improve the label signal of a selected subset of the already labelled data. Self-healing is the adaptation of the labels of the rest according to the newly updated predictions of the classifier.

Even in the case where we need to label new samples through active learning, the rater agreement is important information to have; in [38], the authors have trained rater uncertainty models, and utilised the output value on unlabelled samples as a proxy of *data informativeness* for selecting samples to be annotated. Alternatively, in [54], the authors utilised the predictive uncertainty output of

Support Vector Machines (SVMs) to select both new examples and the number of raters required to achieve a desired level of agreement.

Most importantly, exploiting subjectivity and rater disagreement is an opportunity for the improvement of active learning. By using a GP that models each rater explicitly in [55], and by jointly modelling each rater's trustworthiness [56] better representation of uncertainty is achieved for improved active learning.

In certain cases, during the data annotation process, an **unsure option** was also provided to the raters for assigning to the most ambiguous examples, something that lightens their workload, as well. In [57], an active learning framework can learn on samples with unsure labels, albeit it does not make predictions that a sample is unsure.

Explicitly utilising explicit rater models that also estimate rater trustworthiness has been used successfully to improve active learning from crowds in [58].

In order to model complex, high dimensionality data, such as text, audio, images, and graphs, deep learning has been very successful; Bayesian deep learning [59] has allowed for the principled estimation of model parameters by defining a weight prior, as well as incorporating knowledge from the evidence in these domains. In [60], the authors use Bayesian deep active learning in the context of multiple annotators on an Amazon Alexa dateset.

6   Learning with Subjectivity as the Norm?

Despite the widespread proof of more sophisticated methods being better than unweighted label fusion, majority voting in order to extract a hard label is still being widely applied: for example, in the subjective application of discriminating between hate speech and simply offensive language on online social media [7] and even more recent work on the subject [61]. The output of the crowdsourcing workers in the context of [7] indicate that very often there is a confusion in the human perception between hate-speech and offensive language, indicative of an ambiguous task, or at the very least, multiple ambiguous samples.

We believe that there is great value in adopting subjectivity-aware AI methods as the norm, and discuss several possible frontiers, as well as related fields with which we believe a collusion would be profitable.

6.1   Subjectivity-Aware AI Pipeline

In order to fully embrace subjectivity in the AI pipeline, there is great value in adapting each stage such that it can accommodate it. Given our knowledge that rater performance can be variable [17] and task- and data- specific [62], expecting them to provide hard labels for samples they are unsure about, inevitably leads to lower data quality. The *unsure option* [57,63] for raters has been shown to be one possible addition to the annotation process that may provide the downstream stages with valuable **ambiguity ground-truth**.

Furthermore, by considering financial budgeting behind crowdsourcing, it might be possible to use a mixture of experts and novices. One more way to generate ground-truth ambiguity information is by using improved **interactive learning techniques** that allow for experts to evaluate a limited amount novice labels [64–66].

Recently, computational ways of modelling a rater's attention during the annotation process, for counteracting the **rater drift** problem have been applied to model annotation quality [62].

6.2 Uncertainty-Aware Deep Learning

There has been recent interest in quantifying predictive uncertainty using deep learning [59,67], not necessarily in the multiple annotator case. The authors of [59] propose a method that decomposes uncertainty of a test sample into two different factors, i.e., epistemic uncertainty that is due to lack of observed data at that area in data space, and aleatory uncertainty, that is representative of observation noise in labelling. They show how explicitly modelling for such uncertainty factors improves learning in both classification and regression computer vision tasks such as segmentation and depth prediction, and discuss the **explanatory capacities** of such an approach. The importance of, and an approach for deep uncertainty decomposition with explainability tangents have also been discussed in the digital health domain [68].

We believe that such methods can naturally be applied in the multiple annotator setting, and the decomposition of uncertainty can provide valuable insight towards understanding the degree to which a sample is mislabelled by certain raters, or whether it is inherently ambiguous, something that should have profound impact in the repeated labelling, and active learning from crowds domains. In fact, the authors of an earlier study in repeated labelling have made initial explorations towards using different definitions of uncertainty [41]. Keeping more recent developments in mind [59,68], an interesting question is: what is the relation between annotation subjectivity uncertainty and predictive aleatory uncertainty?

6.3 The Information Value of Data

It is important to keep in mind the assumptions made by adopting any of the aforementioned approaches. In certain approaches that use rater disagreement as priviledged information (see Subsect. 4.1), high disagreement is treated as an indicator for low sample quality, motivating the discarding of such data.

The relation between uncertainty and data informativeness is a decision that should be made in a **dataset- and task-dependent** manner [14,69,70], given that in certain cases, training is focused on hard samples in order to improve training [71], in others easy samples are utilised in the beginning of curriculum learning [72], and finally, in other cases, the middle-way is adopted [38].

The **quantification of the information value of data** is very impactful towards active learning [70], should be performed with labelling subjectivity in mind [14], and should be performed in a dataset-specific mannner. For example, a framework for achieving such a joint quantification of sample value during active learning in an online manner through reinforcement learning has been proposed in [69].

6.4   Fairness in AI

It is important to develop AI frameworks that do not reinforce or reflect biases present in data. By modelling individual raters, or schools of thought (see Subsect. 4.3), greater capacity for capturing certain dimensions of bias is provided. We believe that a more thorough exploration of bias-aware methods [73] on subjective tasks is an avenue that will be explored to a great degree in the future.

7   Conclusions

Even though subjectivity is a well-known quality in certain applications and data and many approaches have been developed to address it, we feel that a paradigm shift towards treating it like an opportunity for improved modelling should be undertaken. We have summarised various groups of work that accommodate for the presence of multiple raters, based on their underlying philosophies, and have built upon them to incite discussion towards possible future opportunities.

References

1. Schuller, B.W.: Speech emotion recognition: two decades in a nutshell, benchmarks, and ongoing trends. Commun. ACM **61**(5), 90–99 (2018)
2. Esteva, A., et al.: A guide to deep learning in healthcare. Nat. Med. **25**(1), 24–29 (2019)
3. Raghu, M., et al.: Direct uncertainty prediction for medical second opinions. In: Proceedings of the International Conference on Machine Learning, pp. 5281–5290 (2019)
4. Deutsch, F.M., LeBaron, D., Fryer, M.M.: What is in a smile? Psychol. Women Q. **11**(3), 341–352 (1987)
5. Fischer, A.H., Kret, M.E., Broekens, J.: Gender differences in emotion perception and self-reported emotional intelligence: a test of the emotion sensitivity hypothesis. PloS One **13**(1) (2018)
6. McCluskey, K.W., Albas, D.C.: Perception of the emotional content of speech by Canadian and Mexican children, adolescents, and adults. Int. J. Psychol. **16**(1–4), 119–132 (1981)
7. Davidson, T., Warmsley, D., Macy, M., Weber, I.: Automated hate speech detection and the problem of offensive language. In: Proceedings of the International AAAI Conference on Web and Social Media (2017)
8. Tschiatschek, S., Singla, A., Gomez Rodriguez, M., Merchant, A., Krause, A.: Fake news detection in social networks via crowd signals. In: Companion Proceedings of the the Web Conference, pp. 517–524 (2018)
9. Patterson, G., Xu, C., Su, H., Hays, J.: The sun attribute database: beyond categories for deeper scene understanding. Int. J. Comput. Vis. **108**(1–2), 59–81 (2014)

10. Smyth, P., Fayyad, U.M., Burl, M.C., Perona, P., Baldi, P.: Inferring ground truth from subjective labelling of venus images. In: Proceedings of Advances in Neural Information Processing Systems, pp. 1085–1092 (1995)
11. Raykar, V.C., et al.: Learning from crowds. J. Mach. Learn. Res. **11**(Apr), 1297–1322 (2010)
12. Sharmanska, V., Hernández-Lobato, D., Miguel Hernandez-Lobato, J., Quadrianto, N.: Ambiguity helps: classification with disagreements in crowdsourced annotations. In: Proceedings of the IEEE Conference on Computer Vision and Pattern Recognition, pp. 2194–2202 (2016)
13. Han, J., Zhang, Z., Schmitt, M., Pantic, M., Schuller, B.: From hard to soft: towards more human-like emotion recognition by modelling the perception uncertainty. In: Proceedings of the ACM International Conference on Multimedia, pp. 890–897. ACM (2017)
14. Rizos, G., Schuller, B.: Modelling sample informativeness for deep affective computing. In: Proceedings of the IEEE International Conference on Acoustics, Speech and Signal Processing, pp. 3482–3486. IEEE (2019)
15. Cowen, A.S., Keltner, D.: Self-report captures 27 distinct categories of emotion bridged by continuous gradients. Proc. Natl. Acad. Sci. **114**(38), E7900–E7909 (2017)
16. Ando, A., Kobashikawa, S., Kamiyama, H., Masumura, R., Ijima, Y., Aono, Y.: Soft-target training with ambiguous emotional utterances for DNN-based speech emotion classification. In: Proceedings of the International Conference on Acoustics, Speech and Signal Processing, pp. 4964–4968. IEEE (2018)
17. Kazai, G., Kamps, J., Milic Frayling, N.: Worker types and personality traits in crowdsourcing relevance labels. In: Proceedings of the ACM International Conference on Information and Knowledge Management, pp. 1941–1944 (2011)
18. Tian, Y., Zhu, J.: Learning from crowds in the presence of schools of thought. In: Proceedings of the ACM International Conference on Knowledge Discovery and Data Mining, pp. 226–234 (2012)
19. Russakovsky, O., et al.: ImageNet large scale visual recognition challenge. Int. J. Comput. Vis. **115**(3), 211–252 (2015)
20. Lin, T.-Y., et al.: Microsoft COCO: common objects in context. In: Fleet, D., Pajdla, T., Schiele, B., Tuytelaars, T. (eds.) ECCV 2014. LNCS, vol. 8693, pp. 740–755. Springer, Cham (2014). https://doi.org/10.1007/978-3-319-10602-1_48
21. Zhou, B., Lapedriza, A., Xiao, J., Torralba, A., Oliva, A.: Learning deep features for scene recognition using places database. In: Proceedings of Advances in Neural Information Processing Systems, pp. 487–495 (2014)
22. Li, Y., Tao, J., Schuller, B., Shan, S., Jiang, D., Jia, J.: MEC 2016: the multimodal emotion recognition challenge of CCPR 2016. In: Tan, T., Li, X., Chen, X., Zhou, J., Yang, J., Cheng, H. (eds.) CCPR 2016. CCIS, vol. 663, pp. 667–678. Springer, Singapore (2016). https://doi.org/10.1007/978-981-10-3005-5_55
23. Zhang, C., Chaudhuri, K.: Active learning from weak and strong labelers. In: Proceedings of Advances in Neural Information Processing Systems, pp. 703–711 (2015)
24. Grimm, M., Kroschel, K.: Evaluation of natural emotions using self assessment manikins. In: Proceedings of the IEEE Workshop on Automatic Speech Recognition and Understanding, pp. 381–385. IEEE (2005)
25. Schuller, B., Hantke, S., Weninger, F., Han, W., Zhang, Z., Narayanan, S.: Automatic recognition of emotion evoked by general sound events. In: Proceedings of the IEEE International Conference on Acoustics, Speech and Signal Processing, pp. 341–344. IEEE (2012)

26. Hantke, S., Marchi, E., Schuller, B.: Introducing the weighted trustability evaluator for crowdsourcing exemplified by speaker likability classification. In: Proceedings of the International Conference on Language Resources and Evaluation, pp. 2156–2161 (2016)
27. Nicolaou, M.A., Gunes, H., Pantic, M.: Continuous prediction of spontaneous affect from multiple cues and modalities in valence-arousal space. IEEE Trans. Affect. Comput. 2(2), 92–105 (2011)
28. Nicolaou, M.A., Pavlovic, V., Pantic, M.: Dynamic probabilistic CCA for analysis of affective behavior and fusion of continuous annotations. IEEE Trans. Pattern Anal. Mach. Intell. 36(7), 1299–1311 (2014)
29. Booth, B.M., Mundnich, K., Narayanan, S.S.: A novel method for human bias correction of continuous-time annotations. In: Proceedings of the IEEE International Conference on Acoustics, Speech and Signal Processing, pp. 3091–3095. IEEE (2018)
30. Gupta, R., Audhkhasi, K., Jacokes, Z., Rozga, A., Narayanan, S.S.: Modeling multiple time series annotations as noisy distortions of the ground truth: an expectation-maximization approach. IEEE Trans. Affect. Comput. 9(1), 76–89 (2016)
31. Mariooryad, S., Busso, C.: Correcting time-continuous emotional labels by modeling the reaction lag of evaluators. IEEE Trans. Affect. Comput. 6(2), 97–108 (2014)
32. Bakhtiari, B., Yazdi, H.S.: Bayesian filter based on the wisdom of crowds. Neurocomputing 283, 181–195 (2018)
33. Vapnik, V., Izmailov, R.: Learning using privileged information: similarity control and knowledge transfer. J. Mach. Learn. Res. 16(2023–2049), 2 (2015)
34. Kim, Y., Provost, E.M.: Leveraging inter-rater agreement for audio-visual emotion recognition. In: Proceedings of the International Conference on Affective Computing and Intelligent Interaction, pp. 553–559. IEEE (2015)
35. Veit, A., Alldrin, N., Chechik, G., Krasin, I., Gupta, A., Belongie, S.: Learning from noisy large-scale datasets with minimal supervision. In: Proceedings of the IEEE Conference on Computer Vision and Pattern Recognition, pp. 839–847 (2017)
36. Eyben, F., Wöllmer, M., Schuller, B.: A multitask approach to continuous five-dimensional affect sensing in natural speech. ACM Trans. Interact. Intell. Syst. 2(1), 1–29 (2012)
37. Steidl, S., Batliner, A., Schuller, B., Seppi, D.: The hinterland of emotions: facing the open-microphone challenge. In: Proceedings of the International Conference on Affective Computing and Intelligent Interaction and Workshops, pp. 1–8. IEEE (2009)
38. Zhang, Z., Deng, J., Marchi, E., Schuller, B.: Active learning by label uncertainty for acoustic emotion recognition. In: Proceedings of the Annual Conference of the International Speech Communication Association (2013)
39. Hinton, G., Vinyals, O., Dean, J.: Distilling the knowledge in a neural network. arXiv preprint arXiv:1503.02531 (2015)
40. Jin, R., Ghahramani, Z.: Learning with multiple labels. In: Proceedings of Advances in Neural Information Processing Systems, pp. 921–928 (2003)
41. Ipeirotis, P.G., Provost, F., Sheng, V.S., Wang, J.: Repeated labeling using multiple noisy labelers. Data Min. Knowl. Disc. 28(2), 402–441 (2014)
42. Kim, Y., Kim, J.: Human-like emotion recognition: multi-label learning from noisy labeled audio-visual expressive speech. In: Proceedings of the IEEE International Conference on Acoustics, Speech and Signal Processing, pp. 5104–5108. IEEE (2018)

43. Chou, H.-C., Lee, C.-C.: Every rating matters: joint learning of subjective labels and individual annotators for speech emotion classification. In: Proceedings of the IEEE International Conference on Acoustics, Speech and Signal Processing, pp. 5886–5890. IEEE (2019)
44. Zhang, H., Jiang, L., Xu, W.: Multiple noisy label distribution propagation for crowdsourcing. In: Proceedings of the International Joint Conference on Artificial Intelligence, pp. 1473–1479. AAAI Press (2019)
45. Zhang, J., Sheng, V.S., Wu, J.: Crowdsourced label aggregation using bilayer collaborative clustering. IEEE Trans. Neural Netw. Learn. Syst. **30**(10), 3172–3185 (2019)
46. Liu, Y., Zhang, W., Yu, Y., et al.: Truth inference with a deep clustering-based aggregation model. IEEE Access **8**, 16 662–16 675 (2020)
47. Yan, Y., et al.: Modeling annotator expertise: learning when everybody knows a bit of something. In: Proceedings of the International Conference on Artificial Intelligence and Statistics, pp. 932–939 (2010)
48. Rodrigues, F., Pereira, F.C.: Deep learning from crowds. In: Proceedings of the AAAI Conference on Artificial Intelligence (2018)
49. Morales-Álvarez, P., Ruiz, P., Santos-Rodríguez, R., Molina, R., Katsaggelos, A.K.: Scalable and efficient learning from crowds with gaussian processes. Inf. Fusion **52**, 110–127 (2019)
50. Cohn, T., Specia, L.: Modelling annotator bias with multi-task Gaussian processes: an application to machine translation quality estimation. In: Proceedings of the Annual Meeting of the Association for Computational Linguistics, pp. 32–42 (2013)
51. Fayek, H.M., Lech, M., Cavedon, L.: Modeling subjectiveness in emotion recognition with deep neural networks: ensembles vs soft labels. In: Proceedings of the International Joint Conference on Neural Networks, pp. 566–570. IEEE (2016)
52. Guan, M.Y., Gulshan, V., Dai, A.M., Hinton, G.E.: Who said what: modeling individual labelers improves classification. In: Proceedings of the AAAI Conference on Artificial Intelligence (2018)
53. Shu, Z., Sheng, V.S., Li, J.: Learning from crowds with active learning and self-healing. Neural Comput. Appl. **30**(9), 2883–2894 (2018)
54. Zhang, Y., Coutinho, E., Zhang, Z., Quan, C., Schuller, B.: Dynamic active learning based on agreement and applied to emotion recognition in spoken interactions. In: Proceedings of the ACM International Conference on Multimodal Interaction, pp. 275–278 (2015)
55. Rodrigues, F., Pereira, F., Ribeiro, B.: Gaussian process classification and active learning with multiple annotators. In: Proceedings of the International Conference on Machine Learning, pp. 433–441 (2014)
56. Long, C., Hua, G.: Multi-class multi-annotator active learning with robust Gaussian process for visual recognition. In: Proceedings of the IEEE International Conference on Computer Vision, pp. 2839–2847 (2015)
57. Zhong, J., Tang, K., Zhou, Z.-H.: Active learning from crowds with unsure option. In: Proceedings of the International Joint Conference on Artificial Intelligence (2015)
58. Calma, A., Sick, B.: Simulation of annotators for active learning: uncertain oracles. In: Proceedings of the ECML PKDD Interactive Adaptive Learning Workshop, p. 49 (2017)
59. Kendall, A., Gal, Y.: What uncertainties do we need in Bayesian deep learning for computer vision? In: Proceedings of Advances in Neural Information Processing Systems, pp. 5574–5584 (2017)

60. Yang, J., Drake, T., Damianou, A., Maarek, Y.: Leveraging crowdsourcing data for deep active learning an application: learning intents in alexa. In: Proceedings of the World Wide Web Conference, pp. 23–32 (2018)
61. Rizos, G., Hemker, K., Schuller, B.: Augment to prevent: short-text data augmentation in deep learning for hate-speech classification. In: Proceedings of the ACM International Conference on Information and Knowledge Management, pp. 991–1000 (2019)
62. Tu, J., Yu, G., Wang, J., Domeniconi, C., Zhang, X.: Attention-aware answers of the crowd. In: Proceedings of the 2020 SIAM International Conference on Data Mining, pp. 451–459. SIAM (2020)
63. Takeoka, K., Dong, Y., Oyamada, M.: Learning with unsure responses. In: Proceedings of the AAAI Conference on Artificial Intelligence. AAAI (2020)
64. Hu, Q., He, Q., Huang, H., Chiew, K., Liu, Z.: Learning from crowds under experts' supervision. In: Tseng, V.S., Ho, T.B., Zhou, Z.-H., Chen, A.L.P., Kao, H.-Y. (eds.) PAKDD 2014. LNCS (LNAI), vol. 8443, pp. 200–211. Springer, Cham (2014). https://doi.org/10.1007/978-3-319-06608-0_17
65. Liu, M., Jiang, L., Liu, J., Wang, X., Zhu, J., Liu, S.: Improving learning-from-crowds through expert validation. In: Proceedings of the International Joint Conferences on Artificial Intelligence, pp. 2329–2336 (2017)
66. Liu, S., Chen, C., Lu, Y., Ouyang, F., Wang, B.: An interactive method to improve crowdsourced annotations. IEEE Trans. Vis. Comput. Graph. **25**(1), 235–245 (2018)
67. Rodrigues, F., Pereira, F.C.: Beyond expectation: deep joint mean and quantile regression for spatiotemporal problems. IEEE Trans. Neural Netw. Learn. Syst. (2020)
68. Kwon, Y., Won, J.-H., Kim, B.J., Paik, M.C.: Uncertainty quantification using bayesian neural networks in classification: application to biomedical image segmentation. Comput. Stat. Data Anal. **142**, 106816 (2020)
69. Haußmann, M., Hamprecht, F., Kandemir, M.: Deep active learning with adaptive acquisition. In: Proceedings of the International Joint Conference on Artificial Intelligence, pp. 2470–2476. AAAI Press (2019)
70. Ghorbani, A., Zou, J.: Data shapley: equitable valuation of data for machine learning. In: Proceedings of the International Conference on Machine Learning, pp. 2242–2251 (2019)
71. Shrivastava, A., Gupta, A., Girshick, R.: Training region-based object detectors with online hard example mining. In: Proceedings of the IEEE Conference on Computer Vision and Pattern Recognition, pp. 761–769 (2016)
72. Pentina, A., Sharmanska, V., Lampert, C.H.: Curriculum learning of multiple tasks. In: Proceedings of the IEEE Conference on Computer Vision and Pattern Recognition, pp. 5492–5500 (2015)
73. Kim, B., Kim, H., Kim, K., Kim, S., Kim, J.: Learning not to learn: training deep neural networks with biased data. In: Proceedings of the IEEE Conference on Computer Vision and Pattern Recognition, pp. 9012–9020 (2019)

Foundations and Mathematics

Why Spiking Neural Networks
Are Efficient: A Theorem

Michael Beer[1], Julio Urenda[2], Olga Kosheleva[2], and Vladik Kreinovich[2(✉)]

[1] Leibniz University Hannover, 30167 Hannover, Germany
beer@irz.uni-hannover.de
[2] University of Texas at El Paso, El Paso, TX 79968, USA
{jcurenda,olgak,vladik}@utep.edu

Abstract. Current artificial neural networks are very successful in many machine learning applications, but in some cases they still lag behind human abilities. To improve their performance, a natural idea is to simulate features of biological neurons which are not yet implemented in machine learning. One of such features is the fact that in biological neural networks, signals are represented by a train of spikes. Researchers have tried adding this spikiness to machine learning and indeed got very good results, especially when processing time series (and, more generally, spatio-temporal data). In this paper, we provide a possible theoretical explanation for this empirical success.

Keywords: Spiking neural networks · Shift-invariance · Scale-invariance

1 Formulation of the Problem

Why Spiking Neural Networks: A Historical Reason. At this moment, artificial neural networks are the most successful – and the most promising – direction in Artificial Intelligence; see, e.g., [3].

Artificial neural networks are largely patterned after the way the actual biological neural networks work; see, e.g., [2,3,6]. This patterning makes perfect sense: after all, our brains are the result of billions of years of improving evolution, so it is reasonable to conclude that many features of biological neural networks are close to optimal – not very efficient features would have been filtered out in this long evolutionary process.

However, there is an important difference between the current artificial neural networks and the biological neural networks:

– when some processing of the artificial neural networks is implemented in hardware – by using electronic or optical transformation – each numerical value is represented by the intensity (amplitude) of the corresponding signal;

This work was supported in part by the US National Science Foundation grants 1623190 (A Model of Change for Preparing a New Generation for Professional Practice in Computer Science) and HRD-1242122 (Cyber-ShARE Center of Excellence).

M.-J. Lesot et al. (Eds.): IPMU 2020, CCIS 1237, pp. 59–69, 2020.
https://doi.org/10.1007/978-3-030-50146-4_5

– in contrast, in the biological neural networks, each value – e.g., the intensity of the sound or of the light – is represented by a series of instantaneous spikes, so that the original value is proportional to the frequency of these spikes.

Since simulating many other features of biological neural networks has led to many successes, a natural idea is to also try to emulate the spiking character of the biological neural networks.

Spiking Neural Networks Are Indeed Efficient. Interestingly, adding spiking to artificial neural networks has indeed led to many successful applications, especially in processing temporal (and even spatio-temporal) signals; see, e.g., [4] and references therein.

But Why? A biological explanation of the success of spiking neural networks – based on the above evolution arguments – makes perfect sense, but it would be nice to supplement it with a clear mathematical explanation – especially since, in spite of all the billions years of evolution, we humans are not perfect as biological beings, we need medicines, surgeries, and other artificial techniques to survive, and our brains often make mistakes.

What We Do in This Paper. In this paper, we consider the question of signal representation from the mathematical viewpoint, and we show that the spiking representation is indeed optimal in some reasonable sense.

Comment. Some of the arguments that we present in this paper are reasonably well-known in the signal processing community. However, since neural networks – in particular, spiking neural networks – are used in other applications as well, we included these arguments in this paper anyway, so that our paper will be more easily accessible (and more convincing) to anyone interested in neural networks and their applications.

2   Analysis of the Problem and the First Result

Looking for Basic Functions. In general, to represent a signal $x(t)$ means to approximate it as a linear combination of some basic functions. For example, it is reasonable to represent a periodic signal as a linear combination of sines and cosines. In more general cases – e.g., when analyzing weather – it makes sense to represent the observed values as a linear combination of functions t, t^2, etc., representing the trend and sines and cosines that describe the periodic part of the signal. To get a more accurate presentation, we need to take into account that the amplitudes of the periodic components can also change with time, so we end up with terms of the type $t \cdot \sin(\omega \cdot t)$.

If we analyze how radioactivity of a sample changes with time, a reasonable idea is to describe the measured values $x(t)$ as a linear combination of exponentially decaying functions $\exp(-k \cdot t)$ representing the decay of different isotopes, etc.

So, in precise terms, selecting a representation means selecting an appropriate family of basic functions. In general, we may have several parameters $c_1, \ldots, c_n$

characterizing functions from each family. Sometimes, there is only one parameter, as in sines and cosines. In other cases, we can have several parameters – e.g., in control applications, it makes sense to consider decaying periodic signals of the type $\exp(-k \cdot t) \cdot \sin(\omega \cdot t)$, with two parameters k and ω. In general, elements $b(t)$ of each such family can be described by a formula $b(t) = B(c_1, \ldots, c_n, t)$ corresponding to different tuples $c = (c_1, \ldots, c_n)$.

Dependence on Parameters Must Be Continuous in Some Reasonable Sense. We want the dependence $B(c_1, \ldots, c_n, t)$ to be computable, and it is known that all computable functions are, in some reasonable sense, continuous; see, e.g., [7].

Indeed, in real life, we can only determine the values of all physical quantities c_i with some accuracy: measurements are always not 100% accurate, and computations always involve some rounding. For any given accuracy, we can provide the value with this accuracy – but it will practically never be the exact value. Thus, the approximate values of c_i are the only thing that our computing algorithm can use when computing the value $B(c_1, \ldots, c_n, t)$. This algorithm can ask for more and more accurate values of c_i, but at some point it must produce the result. At this point, we only known approximate values of c_i, i.e., we only know the interval of possible values of c_i. And for all the values of c_i from this interval, the result of the algorithm provides, with the given accuracy, the approximation to the desired value $B(c_1, \ldots, c_n, t)$. This is exactly what continuity is about!

One has to be careful here, since the real-life processes may actually be, for all practical purposes, discontinuous. Sudden collapses, explosions, fractures do happen.

For example, we want to make sure that a step-function which is equal to 0 for $t < 0$ and to 1 for $t \geq 0$ is close to an "almost" step function which is equal to 0 for $t < 0$, to 1 for $t \geq \varepsilon$ (for some small ε) and to t/ε for $t \in (0, \varepsilon)$.

In such situations, we cannot exactly describe the value at moment t – since the moment t is also measured approximately, but what we can describe is its values at a moment close to t. In other words, we can say that the two functions $a_1(t)$ and $a_2(t)$ are ε-close if:

- for every moment t_1, there exists moments t_{21} and t_{22} which are ε-close to t_1 (i.e., for which $|t_{2i} - t_1| \leq \varepsilon$) and for which $a_1(t_1)$ is ε-close to a convex combination of values $a_2(t_{2i})$, and
- for every moment t_2, there exists moments t_{11} and t_{12} which are ε-close to t_2 and for which $a_2(t_2)$ is ε-close to a convex combination of values $a_1(t_{1i})$.

Additional Requirement. Since we consider linear combinations of basic functions, it does not make sense to have two basic functions that differ only by a constant: if $b_2(t) = C \cdot b_1(t)$, then there is no need to consider the function $b_2(t)$ at all; in each linear combination we can replace $b_2(t)$ with $C \cdot b_1(t)$.

We Would Like to Have the Simplest Possible Family of Basic Functions. How many parameters c_i do we need? The fewer parameters, the easier it is to adjust the values of these parameters, and the smaller the probability of *overfitting* – a known problem of machine learning in particular and of data analysis in general, when we fit the formula to the observed data and its random fluctuations too well and this make it much less useful in other cases where random fluctuations will be different.

We cannot have a family with no parameters at all – that would mean, in effect, that we have only one basic function $b(t)$ and we approximate every signal by an expression $C \cdot b(t)$ obtained by its scaling. This will be a very lousy approximation to real-life processes – since these processes are all different, they do not resemble each other at all.

So, we need at least one parameter. Since we are looking for the simplest possible family, we should therefore consider families depending on a single parameter c_1, i.e., families consisting of functions $b(t) = B(c_1, t)$ corresponding to different values of the parameter c_1.

Most Observed Processes Are Limited in Time. From our viewpoint, we may view astronomical processes as going on forever – although, in reality, even they are limited by billions of years. However, in general, the vast majority of processes that we observe and that we want to predict are limited in time: a thunderstorm stops, a hurricane ends, after-shocks of an earthquake stop, etc.

From this viewpoint, to get a reasonable description of such processes, it is desirable to have basic functions which are also limited in time, i.e., which are equal to 0 outside some finite time interval. This need for finite duration is one of the main reasons in many practical problems, a decomposition into wavelets performs much better that a more traditional Fourier expansion into linear combinations of sines and cosines; see, e.g., [1] and references therein.

Shift- and Scale-Invariance. Processes can start at any moment of time. Suppose that we have a process starting at moment 0 which is described by a function $x(t)$. What if we start the same process t_0 moments earlier? At each moment t, the new process has been happening for the time period $t + t_0$. Thus, at the moment t, the new process is at the same stage as the original process will be at the future moment $t + t_0$. So, the value $x'(t)$ of a quantity characterizing the new process is equal to the value $x(t + t_0)$ of the original process at the future moment of time $t + t_0$.

There is no special starting point, so it is reasonable to require that the class of basic function not change if we simply change the starting point. In other words, we require that for every t_0, the shifted family $\{B(c_1, t + t_0)\}_{c_1}$ coincides with the original family $\{B(c_1, t)\}_{c_1}$.

Similarly, processes can have different speed. Some processes are slow, some are faster. If a process starting at 0 is described by a function $x(t)$, then a λ times faster process is characterized by the function $x'(t) = x(\lambda \cdot t)$. There is no special speed, so it is reasonable to require that the class of basic function not change if we simply change the process's speed. In other words, we require that for every $\lambda > 0$, the "scaled" family $\{B(c_1, \lambda \cdot t)\}_{c_1}$ coincides with the original family $\{B(c_1, t)\}_{c_1}$.

Now, we are ready for the formal definitions.

Definition 1. *We say that a function $b(t)$ is* limited in time *if it equal to 0 outside some interval.*

Definition 2. *We say that a function $b(t)$ is a* spike *if it is different from 0 only for a single value t. This non-zero value is called the* height *of the spike.*

Definition 3. *Let $\varepsilon > 0$ be a real number. We say that the numbers a_1 and a_2 are ε-close if $|a_1 - a_2| \leq \varepsilon$.*

Definition 4. *We say that the functions $a_1(t)$ and $a_2(t)$ are ε-close if:*

- *for every moment t_1, there exists moments t_{21} and t_{22} which are ε-close to t_1 (i.e., for which $|t_{2i} - t_1| \leq \varepsilon$) and for which $a_1(t_1)$ is ε-close to a convex combination of values $a_2(t_{2i})$, and*
- *for every moment t_2, there exists moments t_{11} and t_{12} which are ε-close to t_2 and for which $a_2(t_2)$ is ε-close to a convex combination of values $a_1(t_{1i})$.*

Comment. One can check that this definition is equivalent to the inequality $d_H(A_1, A_2) \leq \varepsilon$ bounding the Hausdorff distance $d_H(A_1, A_2)$ between the two sets A_i each of which is obtained from the closure C_i of the graphs of the corresponding function $a_i(t)$ by adding the whole vertical interval $t \times [a, b]$ for every two points (t, a) and (t, b) with the same first coordinate from the closure C_i.

Definition 5. *We say that a mapping $B(c_1, t)$ that assigns, to each real number c_1, a function $b(t) = B(c_1, t)$ is* continuous *if, for every value c_1 and for every $\varepsilon > 0$, there exists a real number $\delta > 0$ such that, if c_1' is δ-close to c_1, then the function $b(t) = B(c_1, t)$ is ε-close to the function $b'(t) = B(c_1', t)$.*

Definition 6. *By a* family of basic functions, *we mean a continuous mapping for which:*

- *for each c_1, the function $b(t) = B(c_1, t)$ is limited in time, and*
- *if c_1 and c_1' are two different numbers, then the functions $b(t) = B(c_1, t)$ and $b'(t) = B(c_1', t)$ cannot be obtained from each other by multiplication by a constant.*

Definition 7. *We say that a family of basic functions $B(c_1, t)$ is* shift-invariant *if for each t_0, the following two classes of functions of one variable coincide:*

$$\{B(c_1, t)\}_{c_1} = \{B(c_1, t + t_0)\}_{c_1}.$$

Definition 8. *We say that a family of basic functions $B(c_1, t)$ is* scale-invariant *if for each $\lambda > 0$, the following two classes of functions of one variable coincide:*

$$\{B(c_1, t)\}_{c_1} = \{B(c_1, \lambda \cdot t)\}_{c_1}.$$

Proposition 1. *If a family of basic functions $B(c_1, t)$ is shift- and scale-invariant, then for every c_1, the corresponding function $b(t) = B(c_1, t)$ is a spike, and all these spikes have the same height.*

Discussion. This result provides a possible explanation for the efficiency of spikes: namely, a family of spikes is the only one which satisfies the reasonable conditions of shift- and scale-invariance, i.e., the only family that does not change if we change the starting point of the process and/or change the process's speed.

It should be emphasized that the only thing we explain by this result is the use of spikes. Of course, just like for all other computational techniques, there are many other factors contributing to the empirical success of spiking neural networks – e.g., in this case, efficient algorithms for processing the spikes.

Proof. Let us assume that the family of basic functions $B(c_1, t)$ is shift- and scale-invariant. Let us prove that all the functions $b(t) = B(c_1, t)$ are spikes.

$1°$. First, we prove that none of the functions $B(c_1, t)$ is identically 0.

Indeed, the zero function can be contained from any other function by multiplying that other function by 0 – and this would violate the second part of Definition 6 (of a family of basic functions).

$2°$. Let us prove that each function from the given family is a spike.

Indeed, each of the functions $b(t) = B(c_1, t)$ is not identically zero, i.e., it attains non-zero values for some t. By the Definition 6 of a family of basic functions, each of these functions is limited in time, i.e., the values t for which the function $b(t)$ is non-zero are bounded by some interval. Thus, the values $t_- \stackrel{\text{def}}{=} \inf\{t : b(t) \neq 0\}$ and $t_+ \stackrel{\text{def}}{=} \sup\{t : b(t) \neq 0\}$ are finite, with $t_- \leq t_+$.

Let us prove that we cannot have $t_- < t_+$. Indeed, in this case, the interval $[t_-, t_+]$ is non-degenerate. Thus, by an appropriate combination of shift and scaling, we will be able to get this interval from any other non-degenerate interval $[a, b]$, with $a < b$: indeed, it is sufficient to take the transformation $t \to \lambda \cdot t + t_0$, where $\lambda = \dfrac{t_+ - t_-}{b - a}$ and $t_0 = \lambda \cdot a - t_-$. For each of these transformations, due to shift- and scale-invariance of the family, the correspondingly re-scaled function $b'(t) = b(\lambda \cdot t + t_0)$ also belongs to the family $B(c_1, t)$, and for this function, the corresponding values t'_- and t'_+ will coincide with a and b. All these functions are different – so, we will have a 2-dimensional family of functions (i.e., a family depending on 2 parameters), which contradicts to our assumption that the family $B(c_1, t)$ is one-dimensional.

The fact that we cannot have $t_- < t_+$ means that we should have $t_- = t_+$, i.e., that every function $b(t)$ from our family is indeed a spike.

$3°$. To complete the proof, we need to prove that all the spikes that form the family $B(c_1, t)$ have the same height.

Let us describe this property in precise terms. Let $b_1(t)$ and $b_2(t)$ be any two functions from the family. According to Part 2 of this proof, both functions are spikes, so:

- the value $b_1(t)$ is only different from 0 for some value t_1; let us denote the corresponding height $b_1(t_1)$ by h_1;
- similarly, the value $b_2(t)$ is only different from 0 for some value t_2; let us denote the corresponding height $b_2(t_2)$ by h_2.

We want to prove that $h_1 = h_2$.

Indeed, since the function $b_1(t)$ belongs to the family, and the family is shift-invariant, then for $t_0 \stackrel{\text{def}}{=} t_1 - t_2$, the shifted function $b_1'(t) \stackrel{\text{def}}{=} b_1(t + t_0)$ also belongs to this family. The shifted function is non-zero when $t + t_0 = t_1$, i.e., when $t = t_1 - t_0 = t_2$, and it has the same height h_1.

If $h_1 \neq h_2$, this would contradict to the second part of Definition 6 (of the family of basic functions) – because then we would have two functions $b_1'(t)$ and $b_2(t)$ in this family, which can be obtained from each other by multiplying by a constant. Thus, the heights must be the same.

The proposition is proven.

3 Main Result: Spikes Are, in Some Reasonable Sense, Optimal

It Is Desirable to Check Whether Spiked Neurons Are Optimal. In the previous section, we showed that spikes naturally appear if we require reasonable properties like shift- and scale-invariance. This provides some justification for the spiked neural networks.

However, the ultimate goal of neural networks is to solve practical problems. From this viewpoint, we need to take into account that a practitioner is not interested in invariance or other mathematical properties, a practitioner wants to optimize some objective function. So, from the practitioner's viewpoint, the main question is: are spiked neurons optimal?

Different Practitioners Have Different Optimality Criteria. The problem is that, in general, different practitioners may have different optimality criteria. In principle, we can pick one such criterion (or two or three) and analyze which families of basic functions are optimal with respect to these particular criterion – but this will not be very convincing to a practitioner who has a different optimality criterion.

An ideal explanation should work for *all* reasonable optimality criteria. This is what we aim at in this section. To achieve this goal, let us analyze what we mean by an optimality criterion, and which optimality criteria can be considered reasonable. In this analysis, we will follow a general analysis of computing-related optimization problems performed in [5].

What Is an Optimality Criterion: Analysis. At first glance, the answer to this question may sound straightforward: we have an objective function $J(a)$ that assigns, to each alternative a, a numerical value $J(a)$, and we want to select an alternative for which the value of this function is the largest possible (or, if we are interested in minimizing losses, the smallest possible).

This formulation indeed describes many optimality criteria, but not all of them. Indeed, assume, for example, that we are looking for the best method a for approximating functions from a given class. A natural criterion may be to minimize the mean squared approximation error $J(a)$ of the method a. If there is only one method with the smallest possible mean squared error, then this method is selected. But what if there are several different methods with the same mean squared error – and this is often the case. In this case, we can use this non-uniqueness to optimize something else: e.g., select, out of several methods with the same mean squared error, the method for which the average computation time $T(a)$ is the smallest.

In this situation, the optimality criterion cannot be described by single objective function, it takes a more complex form. Namely, we say that a method a' is better than a method a if:

- either $J(a) < J(a')$,
- or $J(a) = J(a')$ and $T(a) < T(a')$.

This additional criterion may still leave us with several equally good methods. We can use this non-uniqueness to optimize yet another criterion: e.g., worst-case computation time, etc.

The only thing which is needed to describe an optimality criterion is that this criterion must enable us to compare the quality of different alternatives. In mathematical terms, this criterion must enable us to decide which alternatives are better (or of the same quality); let us denote this by $a \leq a'$. Clearly, if a' is better than a (i.e., $a \leq a'$) and a'' is better than a' ($a' \leq a''$), then a'' is better than a ($a \leq a''$), so the relation $\leq$ must be transitive. Such relations are known as *pre-orders*.

Comment. Not all such relations are orders: that would require an additional property that if $a \leq b$ and $b \leq a$, then $a = b$, and, as we have mentioned earlier, this is not necessarily true.

An Optimality Criterion Must Be Final. In terms of the relation $\leq$, optimal means better than (or of the same quality as) all other alternatives: $a \leq a_{\mathrm{opt}}$ for all a.

As we have mentioned earlier, if we have several different optimal alternatives, then we can use this non-uniqueness to optimize something else – i.e., in effect, to modify the corresponding optimality criterion. Thus, when the optimality criterion allows several different optimal alternatives, this means that this criterion is *not* final, it has to be modified. For a *final* criterion, we should have only one optimal alternative.

An Optimality Criterion Must Be Invariant. In real life, we deal with real-life processes $x(t)$, in which values of different quantities change with time t. The corresponding numerical values of time t depend on the starting point that we use for measuring time and on the measuring unit: e.g., 1 h is equivalent to 60 min; numerical values are different, but from the physical viewpoint, this is the same time interval.

We are interested in a universal technique for processing data. It is therefore reasonable to require that the relative quality of different techniques should not change if we simply change the starting point for measuring time or a measuring unit.

Let us describe all this in precise terms.

Definition 9. *Let a set A be given; its elements will be called* alternatives.

- *By an* optimality criterion $\leq$ *on the set A, we mean a transitive relation (i.e., a pre-order) on this set.*
- *An element a_{opt} is called* optimal *with respect to the criterion $\leq$ is for all $a \in A$, we have $a \leq a_{\text{opt}}$.*
- *An optimality criterion is called* final *if, with respect to this criterion, there exists exactly one optimal alternative.*

Definition 10. *For each family of basic functions $B(c_1, t)$ and for each value t_0, by its shift $T_{t_0}(B)$, we mean a family that assigns, to each number c_1, a function $B(c_1, t + t_0)$.*

Definition 11. *We say that an optimality criterion on the class of all families of basic functions is* shift-invariant *if for every two families B and B' and for each t_0, $B \leq B'$ implies that $T_{t_0}(B) \leq T_{t_0}(B')$.*

Definition 12. *For each family of basic functions $B(c_1, t)$ and for each value $\lambda > 0$, by its scaling $S_\lambda(B)$, we mean a family that assigns, to each number c_1, a function $B(c_1, \lambda \cdot t)$.*

Definition 13. *We say that an optimality criterion on the class of all families of basic functions is* scale-invariant *if for every two families B and B' and for each $\lambda > 0$, $B \leq B'$ implies that $S_\lambda(B) \leq S_\lambda(B')$.*

Now, we are ready to formulate our main result.

Proposition 2. *For every final shift- and scale-invariant optimality criterion on the class of all families of basic functions, all elements of the optimal family are spikes of the same height.*

Discussion. We started this paper with mentioning that in many practical problems, techniques based on representing signals as a linear combination of spikes are known to be very efficient. In different applications, efficiency mean different things:

- In signal processing, efficiency may mean that the corresponding computations finish earlier than when we use other techniques – e.g., Fourier transform techniques, when we represent a signal as a linear combination of sinusoids, or wavelet techniques, when we represent a signal as a linear combination of wavelets.
- In machine learning, efficiency may mean the same – faster computations, but it may also mean that the trained neural network leads to more accurate predictions than networks based on different representations of signals.

In other words, in different situations, we may have different optimality criteria that describe which methods we consider to be better and which methods we consider to be worse. Our result – the above Proposition 2 – shows that no matter what optimality criterion we use, as long as this criterion satisfies natural properties of scale- and shift-invariance (i.e., independence on the choice of the relative speed and/or starting point), the only family which is optimal in the sense of this criterion is the family of spikes.

Thus, this result provides a possible explanation of why spiking neural networks have been efficient in several different situations – situations in each of which, in general, efficiency was understood somewhat differently.

Proof. Let us prove that the optimal family B_{opt} is itself shift- and scale-invariant; then this result will follow from Proposition 1.

Indeed, let us consider any transformation T – be it shift or scaling. By definition of optimality, B_{opt} is better than (or is of the same quality) as any other family B: $B \leq B_{\text{opt}}$. In particular, for every B, this is true for the family $T^{-1}(B)$, i.e., $T^{-1}(B) \leq B_{\text{opt}}$, where, as usual, T^{-1} denotes the inverse transformation.

Due to invariance of the optimality criterion, $T^{-1}(B) \leq B_{\text{opt}}$ implies that $T(T^{-1}(B)) \leq T(B_{\text{opt}})$, i.e., that $B \leq T(B_{\text{opt}})$. This is true for each family B, thus the family $T(B_{\text{opt}})$ is optimal. However, we assumed that our optimality criterion is final, which means that there is only one optimal family. Thus, we have $T(B_{\text{opt}}) = B_{\text{opt}}$, i.e., the optimal family B_{opt} is indeed invariant with respect to any of the shifts and scalings. Now, by applying Proposition 1, we conclude the proof of this proposition.

4   Conclusions

A usual way to process signals is to approximate each signal by a linear combinations of basic functions – e.g., sinusoids or wavelets. In the last decades, a new approximation turned out to be very efficient in many practical applications – an approximation of a signal by a linear combination of spikes. In this paper, we provide a possible theoretical explanation for this empirical success – to be more precise, we provide two related theoretical explanations.

Our first explanation is that spikes are the only family of basic functions which does not change if we change the relative speed and/or change the starting point of the analyzed processes – and is, thus, the only family which is equally applicable to all possible speeds and all possible starting points. In mathematical terms, spikes are the only family which is scale- and shift-invariant.

Our second explanation is that for every reasonable optimality criterion on the class of all possible families of basic functions, the optimal family is the family of spikes – provided that the optimality criterion itself satisfies the natural properties of scale- and shift-invariance.

Acknowledgments. The authors are greatly thankful to Nikola Kasabov and to all the participants of the 2019 IEEE Series of Symposia on Computational Intelligence (Xiamen, China, December 4–6, 2019) for valuable discussions, and to the anonymous referees for helpful suggestions.

References

1. Addison, P.S.: The Illustrated Wavelet Transform Handbook: Introductory Theory and Applications in Science, Engineering, Medicine and Finance. CRC Press, Boca Raton (2016)
2. Bishop, C.M.: Pattern Recognition and Machine Learning. Springer, New York (2006)
3. Goodfellow, I., Bengio, Y., Courville, A.: Deep Leaning. MIT Press, Cambridge (2016)
4. Kasabov, N.K. (ed.): Time-Space, Spiking Neural Networks and Brain-Inspired Artificial Intelligence. Springer, Cham (2019). https://doi.org/10.1007/978-3-662-57715-8
5. Nguyen, H.T., Kreinovich, V.: Applications of Continuous Mathematics to Computer Science. Kluwer, Dordrecht (1997)
6. Reed, S.K.: Cognition: Theories and Application. Wadsworth Cengage Learning, Belmont (2010)
7. Weihrauch, K.: Computable Analysis: An Introduction. Springer, Heidelberg (2000). https://doi.org/10.1007/978-3-642-56999-9

Which Distributions (or Families of Distributions) Best Represent Interval Uncertainty: Case of Permutation-Invariant Criteria

Michael Beer[1], Julio Urenda[2], Olga Kosheleva[2]⬤, and Vladik Kreinovich[2](✉)⬤

[1] Leibniz University Hannover, 30167 Hannover, Germany
beer@irz.uni-hannover.de
[2] University of Texas at El Paso, El Paso, TX 79968, USA
{jcurenda,olgak,vladik}@utep.edu

Abstract. In many practical situations, we only know the interval containing the quantity of interest, we have no information about the probabilities of different values within this interval. In contrast to the cases when we know the distributions and can thus use Monte-Carlo simulations, processing such interval uncertainty is difficult – crudely speaking, because we need to try all possible distributions on this interval. Sometimes, the problem can be simplified: namely, for estimating the range of values of some characteristics of the distribution, it is possible to select a single distribution (or a small family of distributions) whose analysis provides a good understanding of the situation. The most known case is when we are estimating the largest possible value of Shannon's entropy: in this case, it is sufficient to consider the uniform distribution on the interval. Interesting, estimating other characteristics leads to the selection of the same uniform distribution: e.g., estimating the largest possible values of generalized entropy or of some sensitivity-related characteristics. In this paper, we provide a general explanation of why uniform distribution appears in different situations – namely, it appears every time we have a permutation-invariant optimization problem with the unique optimum. We also discuss what happens if we have an optimization problem that attains its optimum at several different distributions – this happens, e.g., when we are estimating the smallest possible value of Shannon's entropy (or of its generalizations).

Keywords: Interval uncertainty · Maximum Entropy approach · Uniform distribution · Sensitivity analysis

This work was supported in part by the US National Science Foundation grants 1623190 (A Model of Change for Preparing a New Generation for Professional Practice in Computer Science) and HRD-1242122 (Cyber-ShARE Center of Excellence).

© Springer Nature Switzerland AG 2020
M.-J. Lesot et al. (Eds.): IPMU 2020, CCIS 1237, pp. 70–79, 2020.
https://doi.org/10.1007/978-3-030-50146-4_6

1 Formulation of the Problem

Interval Uncertainty is Ubiquitous. When an engineer designs an object, the original design comes with exact numerical values of the corresponding quantities, be it the height of ceiling in civil engineering or the resistance of a certain resistor in electrical engineering. Of course, in practice, it is not realistic to maintain the exact values of all these quantities, we can only maintain them with some tolerance. As a result, the engineers not only produce the desired ("nominal") value x of the corresponding quantity, they also provide positive and negative tolerances $\varepsilon_+ > 0$ and $\varepsilon_- > 0$ with which we need to maintain the value of this quantity. The actual value must be in the interval $\mathbf{x} = [\underline{x}, \overline{x}]$, where $\underline{x} \stackrel{\text{def}}{=} x - \varepsilon_-$ and $\overline{x} \stackrel{\text{def}}{=} x + \varepsilon_+$.

All the manufacturers need to do is to follow these interval recommendations. There is no special restriction on probabilities of different values within these intervals – these probabilities depends on the manufacturer, and even for the same manufacturer, they may change every time the manufacturer makes some adjustments to the manufacturing process.

Data Processing Under Interval Uncertainty is Often Difficult. Because of the ubiquity of interval uncertainty, many researchers have considered different data processing problems under this uncertainty; this research area is known as *interval computations*; see, e.g., [5,10,11,14].

The problem is that the corresponding computational problems are often very complex, much more complex than solving similar problems under *probabilistic* uncertainty – when we know the probabilities of different values within the corresponding intervals. For example, while for the probabilistic uncertainty, we can, in principle, always use Monte-Carlo simulations to understand how the input uncertainty affects the result of data processing, a similar problem for interval uncertainty is NP-hard already for the simplest nonlinear case when the whole data processing means computing the value of a quadratic function – actually, it is even NP-hard if we want to find the range of possible values of variance in a situation when inputs are only known with interval uncertainty [8,13].

This complexity is easy to understand: interval uncertainty means that we may have different probability distributions on the given interval. So, to get guaranteed estimates, we need, in effect, to consider all possible distributions – which leads to very time-consuming computations. For some problems, this time can be sped up, but in general, the problems remain difficult.

It is Desirable to Have a Reasonably Small Family of Distributions Representing Interval Uncertainty. In the ideal world, we should always take into account interval uncertainty – i.e., take into account that, in principle, all mathematically possible probability distributions on the given interval are actually possible.

However, as we have just mentioned, many of the corresponding interval computation problems are NP-hard. In practical terms, this means that the corresponding computations will take forever.

Since in such situations, it is not possible to exactly take interval uncertainty into account – i.e., we cannot consider *all* possible distributions on the interval – a natural idea is to consider *some* typical distributions. This can be a finite-dimensional family of distributions, this can be even a finite set of distributions – or even a single distribution. For example, in measurements, practitioners often use uniform distributions on the corresponding interval; this selection is even incorporated in some international standards for processing measurement results; see, e.g., [14].

Of course, we need to be very careful which family we choose: by limiting the class of possible distributions, we introduce an artificial "knowledge", and thus, modify the data processing results. So, we should select the family depending on what characteristic we want to estimate – and beware that a family that works perfectly well for one characteristic may produce a completely misleading result when applied to some other desired characteristic. Examples of such misleading results are well known – and we will present some such results later.

Continuous vs. Discrete Distributions: Idealized Mathematical Description vs. Practical Description. Usually, in statistics and in measurement theory, when we say that the actual value x belongs to the interval $[a, b]$, we assume that x can take any real value between a and b. However, in practice, even with the best possible measuring instruments, we can only measure the value of the physical quantity x with some uncertainty h. Thus, from the practical viewpoint, it does not make any sense to distinguish between, e.g., the values a and $a + h$ – even with the best measuring instruments, we will not be able to detect this difference.

From the practical viewpoint, it makes sense to divide the interval $[a, b]$ into small subintervals $[a, a + h], [a + h, a + 2h], \ldots$ within each of which the values of x are practically indistinguishable.

Correspondingly, to describe the probabilities of different values x, it is sufficient to find the probabilities $p_1, p_2, \ldots, p_n$ that the actual value x is in one of these small subintervals:

– the probability p_1 that x is in the first small subinterval $[a, a + h]$;
– the probability p_2 that x is in the first small subinterval $[a + h, a + 2h]$; etc.

These probabilities should, of course, add up to 1: $\sum_{i=1}^{n} p_i = 1$.

In the ideal case, when we get more and more accurate measuring instruments – i.e., when $h \to 0$ – the corresponding discrete probability distributions will tend to the corresponding continuous distribution. So, from this viewpoint:

– selecting a probability distribution means selecting a tuple of values $p = (p_1, \ldots, p_n)$, and
– selecting a family of probability distributions means selecting a family of such tuples.

First Example of Selecting a Family of Distributions: Estimating Maximum Entropy. Whenever we have uncertainty, a natural idea is to provide a numerical estimate for this uncertainty. It is known that one of the natural measures of uncertainty is Shannon's entropy $-\sum_{i=1}^{n} p_i \cdot \log_2(p_i)$; see, e.g., [6,13]. When we know the probability distribution, i.e., when we know all the values p_i, then the above formula enables us to uniquely determine the corresponding entropy.

However, in the case of interval uncertainty, we can have several different tuples, and, in general, for different tuples, entropy is different. As a measure of uncertainty of the situation, it is reasonable to take the largest possible value. Indeed, Shannon's entropy can be defined as the average number of binary ("yes"-"no") questions that are needed to uniquely determine the situation: the larger this number, the larger the initial uncertainty. Thus, it is natural to take the largest number of such questions as a characteristic of interval uncertainty.

For this characteristic, we want to select a distribution – or, if needed, a family of distributions – whose entropy is equal to the largest possible entropy of all possible probability distributions on the interval. Selecting such a "most uncertain" distribution is known as the *Maximum Entropy approach*; this approach has been successfully used in many practical applications; see, e.g., [6].

It is well known that out of all possible tuples with $\sum_{i=1}^{n} p_i = 1$, the entropy is the largest possible when all the probabilities are equal to each other, i.e., when

$$p_1 = \ldots = p_n = 1/n.$$

In the limit $h \to 0$, such distributions tend to the uniform distribution on the interval $[a, b]$. This is one of the reasons why, as we have mentioned, uniform distributions are recommended in some measurement standards.

Modification of This Example. In addition to Shannon's entropy, there are other measures of uncertainty – which are usually called *generalized entropy*. For example, in many applications, practitioners use the quantity $-\sum_{i=1}^{n} p_i^\alpha$ for some $\alpha \in (0,1)$. It is known that when $\alpha \to 0$, this quantity, in some reasonable sense, tends to Shannon's entropy – to be more precise, the tuple at which the generalized entropy attains its maximum under different condition tends to the tuple at which Shannon's entropy attains its maximum.

The maximum of this characteristic is also attained when all the probabilities p_i are equal to each other.

Other Examples. The authors of [4] analyzed how to estimate sensitivity of Bayesian networks under interval uncertainty. It also turned out that if, for the purpose of this estimation, we limit ourselves to a single distribution, then the most adequate result also appears if we select a uniform distribution, i.e., in effect, the values $p_1 = \ldots = p_n$; see [4] for technical details.

Idea. The fact that the same uniform distribution appears in many different situations, under different optimality criteria, make us think that there must be a general reason for this distribution. In this paper, we indeed show that there is such a reason.

Beyond the Uniform Distribution. For other characteristics, other possible distributions provide a better estimate. For example, if instead of estimating the *largest* possible value of the entropy, we want to estimate the *smallest* possible value of the entropy, then the corresponding optimal value 0 is attained for several different distributions. Specifically, there are n such distributions corresponding to different values $i_0 = 1, \ldots, n$. In each of these distributions, we have $p_{i_0} = 1$ and $p_i = 0$ for all $i \neq i_0$.

In the continuous case $h \to 0$, these probability distributions correspond to point-wise probability distributions in which a certain value x_0 appears with probability 1.

Similar distributions appear for several other optimality criteria: e.g., when we minimize generalized entropy instead of minimizing Shannon's entropy. A natural question is: how can we explain that these distributions appear as solutions to different optimization problems? Similar to the uniform case, there should also be a general explanation – and a simple general explanation will indeed be provided in this paper.

2 Analysis of the Problem

What Do Entropy, Generalized Entropy, etc. Have in Common? We would like to come up with a general result that generalizes both the maximum entropy, the maximum generalized entropy, and other cases. To come up with such a generalization, it is reasonable to analyze what these results have in common.

Let Us Use Symmetries. In general, our knowledge is based on *symmetries*, i.e., on the fact that some situations are similar to each other. Indeed, if all the world's situations were completely different, we would not be able to make any predictions. Luckily, real-life situations have many features in common, so we can use the experience of previous situations to predict future ones.

The idea of using symmetries is well-known to many readers. However, since not everyone is very familiar with this idea, we added a brief explanation in this subsection. Readers who are well familiar with the idea of symmetry are welcome to skip the rest of this subsection, and go straight to the subsection about permutations.

So here is our brief explanation. For example, when a person drops a pen, it starts falling down to Earth with the acceleration of $9.81 \, \text{m/s}^2$. If this person moves to a different location and repeats the same experiment, he or she will get the exact same result. This means that the corresponding physics is invariant with respect to shifts in space.

Similarly, if the person repeats this experiment in a year, the result will be the same. This means that the corresponding physics is invariant with respect to shifts in time.

Alternatively, if the person turns around a little bit, the result will still be the same. This means that the underlying physics is also invariant with respect to rotations, etc.

This is a very simple example, but such symmetries are invariances are actively used in modern physics (see, e.g., [1,15]) – and moreover, many previously proposed fundamental physical theories such as:

- Maxwell's equations that describe electrodynamics,
- Schroedinger's equations that describe quantum phenomena,
- Einstein's General Relativity equation that describe gravity,

can be derived from the corresponding invariance assumptions; see, e.g., [2,3,7,9].

Symmetries also help to explain many empirical phenomena in computing; see, e.g., [12]. From this viewpoint, a natural way to look for what the two examples have in common is to look for invariances that they have in common.

Permutations – Natural Symmetries in the Entropy Example. We have n probabilities $p_1, \ldots, p_n$. What can we do with them that would preserve the entropy? In principle, we can transform the values into something else, but the easiest possible transformations is when we do not change the values themselves, just swap them.

Bingo! Under such swap, the value of the entropy does not change. In precise terms, both the objective function $S = -\sum\limits_{i=1}^{n} p_i \cdot \ln(p_i)$ and the constraint $\sum\limits_{i=1}^{n} p_i = 1$ do not change is we perform any permutation

$$\pi : \{1, \ldots, n\} \to \{1, \ldots, n\},$$

i.e., replace the values $p_1, \ldots, p_n$ with the permuted values $p_{\pi(1)}, \ldots, p_{\pi(n)}$.

Interestingly, the above-described generalized entropy is also permutation-invariant. Thus, we are ready to present our general results.

3   Our Results

Definition 1

- *We say that a function $f(p_1, \ldots, p_n)$ is* permutation-invariant *if for every permutation $\pi : \{1, \ldots, n\} \to \{1, \ldots, n\}$, we have*

$$f(p_1, \ldots, p_n) = f(p_{\pi(1)}, \ldots, p_{\pi(n)}).$$

- *By a* permutation-invariant optimization problem, *we mean a problem of optimizing a permutation-invariant function $f(p_1, \ldots, p_n)$ under constraints of the type $g_i(p_1, \ldots, p_n) = a_i$ or $h_j(p_1, \ldots, p_n) \geq b_j$ for permutation-invariant functions g_i and h_j.*

Comment. In other words, we consider the following problem:

- *given* permutation-invariant functions $f(p_1, \ldots, p_n)$, $g_1(p_1, \ldots, p_n)$, $g_2(p_1, \ldots, p_2)$, $\ldots$, $h_1(p_1, \ldots, p_n)$, $h_2(p_1, \ldots, p_2)$, $\ldots$, and values $a_1, a_2, \ldots, b_1, b_2, \ldots$;

– *find:* among all tuples $p = (p_1, \ldots, p_n)$ that satisfy the conditions $\sum_{i=1}^{n} p_i = 1$,

$$g_1(p_1, \ldots, p_n) = a_1, \quad g_2(p_1, \ldots, p_n) = a_2, \quad \ldots,$$

and

$$h_1(p_1, \ldots, p_n) \geq b_1, \quad h_2(p_1, \ldots, p_n) \geq b_2, \quad \ldots$$

find the tuple with the largest (or smallest) possible value of the objective function $f(p_1, \ldots, p_n)$.

Proposition 1. *If a permutation-invariant optimization problem has only one solution, then for this solution, we have $p_1 = \ldots = p_n$.*

Discussion. This explains why we get the uniform distribution in several cases: in the maximum entropy case, in the maximum generalized entropy case, etc.

Proof. We will prove this result by contradiction. Suppose that the values p_i are not all equal. This means that there exist i and j for which $p_i \neq p_j$. Let us swap p_i and p_j, and denote the corresponding values by p_i', i.e.:

– we have $p_i' = p_j$,
– we have $p_j' = p_i$, and
– we have $p_k' = p_k$ for all other k.

Since the values p_i satisfy all the constraints, and all the constraints are permutation-invariant, the new values p_i' also satisfy all the constraints. Since the objective function is permutation-invariant, we have $f(p_1, \ldots, p_n) = f(p_1', \ldots, p_n')$. Since the values $(p_1, \ldots, p_n)$ were optimal, the values $(p_1', \ldots, p_n') \neq (p_1, \ldots, p_n)$ are thus also optimal – which contradicts to the assumption that the original problem has only one solution.

This contradiction proves for the optimal tuple $(p_1, \ldots, p_n)$ that all the values p_i are indeed equal to each other. The proposition is proven.

Discussion. What is the optimal solution is not unique? We can have a case when we have a small finite number of solutions.

We can also have a case when we have a 1-parametric family of solutions – i.e., a family depending on one parameter. In our discretized formulation, each parameter has n values, so this means that we have n possible solutions. Similarly, a 2-parametric family means that we have n^2 possible solutions, etc.

Here are precise definitions and related results.

Definition 2

– *We say that a permutation-invariant optimization problem with n unknowns $p_1, \ldots, p_n$ has a small finite number of solutions if it has fewer than n solutions.*
– *We say that a permutation-invariant optimization problem with n unknowns $p_1, \ldots, p_n$ has a d-parametric family of solutions if it has no more than n^d solutions.*

Proposition 2. *If a permutation-invariant optimization problem has a small finite number of solutions, then it has only one solution.*

Discussion. Due to Proposition 1, in this case, the only solution is the uniform distribution $p_1 = \ldots = p_n$.

Proof. Since $\sum p_i = 1$, there is only one possible solution for which $p_1 = \ldots = p_n$: the solution for which all the values p_i are equal to $1/n$.

Thus, if the problem has more than one solution, some values p_i are different from others – in particular, some values are different from p_1. Let S denote the set of all the indices j for which $p_j = p_1$, and let m denote the number of elements in this set. Since some values p_i are different from p_1, we have $1 \le m \le n-1$.

Due to permutation-invariance, each permutation of this solution is also a solution. For each m-size subset of the set of n-element set of indices $\{1, \ldots, n\}$, we can have a permutation that transforms S into this set and thus, produces a new solution to the original problem. There are $\binom{n}{m}$ such subsets. For all m from 1 to $n-1$, the smallest value of the binomial coefficient $\binom{n}{m}$ is attained when $m = 1$ or $m = n-1$, and this smallest value is equal to n. Thus, if there is more than one solution, we have at least n different solutions – and since we assumed that we have fewer than n solutions, this means that we have only one. The proposition is proven.

Proposition 3. *If a permutation-invariant optimization problem has a 1-parametric family of solutions, then this family of solutions is characterized by a real number $c \le 1/(n-1)$, for which all these solutions have the following form: $p_i = c$ for all i but one and $p_{i_0} = 1 - (n-1) \cdot c$ for the remaining value i_0.*

Discussion. In particular, for $c = 0$, we get the above-mentioned 1-parametric family of distributions for which Shannon's entropy (or generalized entropy) attain the smallest possible value.

Proof. As we have shown in the proof of Proposition 2, if in one of the solutions, for some value p_i we have m different indices j with this value, then we will have at least $\binom{n}{m}$ different solutions. For all m from 2 to $n-2$, this number is at least as large as $\binom{n}{2} = \dfrac{n \cdot (n-1)}{2}$ and is, thus, larger than n.

Since overall, we only have n solutions, this means that it is not possible to have $2 \le m \le n-2$. So, the only possible values of m are 1 and $n-1$.

If there was no group with $n-1$ values, this would means that all the groups must have $m = 1$, i.e., consist of only one value. In other words, in this case, all n values p_i would be different. In this case, each of $n!$ permutations would lead to a different solution – so we would have $n! > n$ solutions to the original problem – but we assumed that overall, there are only n solutions. Thus, this case is also impossible.

So, we do have a group of $n-1$ values with the same p_i. Then we get exactly one of the solutions described in the formulation of the proposal, plus solutions obtained from it by permutations – which is exactly the described family.

The proposition is proven.

4   Conclusions

Traditionally, in engineering, uncertainty is described by a probability distribution. In practice, we rarely know the exact distribution. In many practical situations, the only information we know about a quantity is the interval of possible values of this quantity – and we have no information about the probability of different values within this interval. Under such interval uncertainty, we cannot exclude any mathematically possible probability distribution. Thus, to estimate the range of possible values of the desired uncertainty characteristic, we must, in effect, consider all possible distributions. Not surprisingly, for many characteristics, the corresponding computational problem becomes NP-hard.

For some characteristics, we can provide a reasonable estimate for their desired range if instead of all possible distributions, we consider only distributions from some finite-dimensional family. For example, to estimate the largest possible value of Shannon's entropy (or of its generalizations), it is sufficient to consider only the uniform distribution. Similarly, to estimate the smallest possible value of Shannon's entropy or of its generalizations, it is sufficient to consider point-wise distributions, in which a single value from the interval appears with probability 1. The fact that different optimality criteria lead to the same distribution – or to the same family of distributions – made us think that there should be a general reason for the appearance of these families. In this paper, we show that indeed, the appearance of these distributions and these families can be explained by the fact that all the corresponding optimization problems are permutation-invariant.

Thus, in the future, if a reader encounters a permutation-invariant optimization problem for which it is known that there is a unique solution – or that there is only a 1-parametric family of solutions – then there is no need to actually solve the corresponding problem (which may be complex to directly solve). In such situations, it is possible to simply use our general symmetry-based results for finding the corresponding solution – and thus, for finding a distribution (or a family of distributions) that, for the corresponding characteristic, best represent interval uncertainty.

Acknowledgments. The authors are greatly thankful to all the participants of the 2019 IEEE Symposium on Computational Intelligence in Engineering Solutions CIES'2019 (Xiamen, China, December 6–9, 2019) for useful thought-provoking discussions and to the anonymous referees for valuable suggestions.

References

1. Feynman, R., Leighton, R., Sands, M.: The Feynman Lectures on Physics. Addison Wesley, Boston (2005)
2. Finkelstein, A.M., Kreinovich, V.: Derivation of Einstein's, Brans-Dicke and other equations from group considerations. In: Choque-Bruhat, Y., Karade, T.M. (eds.) On Relativity Theory. Proceedings of the Sir Arthur Eddington Centenary Symposium, Nagpur, India, vol. 2, pp. 138–146. World Scientific, Singapore (1985)
3. Finkel'shteĭn, A.M., Kreĭnovich, V.Y., Zapatrin, R.R.: Fundamental physical equations uniquely determined by their symmetry groups. In: Borisovich, Y.G., Gliklikh, Y.E., Vershik, A.M. (eds.) Global Analysis — Studies and Applications II. LNM, vol. 1214, pp. 159–170. Springer, Heidelberg (1986). https://doi.org/10.1007/BFb0075964
4. He, L., Beer, M., Broggi, M., Wei, P., Gomes, A.T.: Sensitivity analysis of prior beliefs in advanced Bayesian networks. In: Proceedings of the 2019 IEEE Symposium Series on Computational Intelligence SSCI 2019, Xiamen, China, 6–9 December 2019, pp. 775–782 (2019)
5. Jaulin, L., Kiefer, M., Didrit, O., Walter, E.: Applied Interval Analysis, with Examples in Parameter and State Estimation, Robust Control, and Robotics. Springer, London (2001). https://doi.org/10.1007/978-1-4471-0249-6
6. Jaynes, E.T., Bretthorst, G.L.: Probability Theory: The Logic of Science. Cambridge University Press, Cambridge (2003)
7. Kreinovich, V.: Derivation of the Schroedinger equations from scale invariance. Theor. Math. Phys. 8(3), 282–285 (1976)
8. Kreinovich, V., Lakeyev, A., Rohn, J., Kahl, P.: Computational Complexity and Feasibility of Data Processing and Interval Computations. Kluwer, Dordrecht (1998)
9. Kreinovich, V., Liu, G.: We live in the best of possible worlds: Leibniz's insight helps to derive equations of modern physics. In: Pisano, R., Fichant, M., Bussotti, P., Oliveira, A.R.E. (eds.) The Dialogue between Sciences, Philosophy and Engineering. New Historical and Epistemological Insights, Homage to Gottfried W. Leibnitz 1646–1716, pp. 207–226. College Publications, London (2017)
10. Mayer, G.: Interval Analysis and Automatic Result Verification. de Gruyter, Berlin (2017)
11. Moore, R.E., Kearfott, R.B., Cloud, M.J.: Introduction to Interval Analysis. SIAM, Philadelphia (2009)
12. Nguyen, H.T., Kreinovich, V.: Applications of Continuous Mathematics to Computer Science. Kluwer, Dordrecht (1997)
13. Nguyen, H.T., Kreinovich, V., Wu, B., Xiang, G.: Computing Statistics Under Interval and Fuzzy Uncertainty. Springer, Heidelberg (2012). https://doi.org/10.1007/978-3-642-24905-1
14. Rabinovich, S.G.: Measurement Errors and Uncertainties: Theory and Practice. Springer, New York (2005). https://doi.org/10.1007/0-387-29143-1
15. Thorne, K.S., Blandford, R.D.: Modern Classical Physics: Optics, Fluids, Plasmas, Elasticity, Relativity, and Statistical Physics. Princeton University Press, Princeton (2017)

A $L1$ Minimization Optimal Corrective Explanation Procedure for Probabilistic Databases

Marco Baioletti⦿ and Andrea Capotorti$^{(\boxtimes)}$⦿

Dipartimento di Matematica ed Informatica, Università degli Studi di Perugia,
Perugia, Italy
{marco.baioletti,andrea.capotorti}@unipg.it
http://www.dmi.unipg.it

Abstract. We propose to use a, recently introduced, efficient $L1$ distance minimization through mixed-integer linear programming for minimizing the number of valuations to be modified inside an incoherent probabilistic assessment. This is in line with one basic principle of optimal corrective explanation for decision makers.

A shrewd use of constraints and of slack variables permit to steer the correction of incoherent assessments towards aimed directions, like e.g. the minimal number of changes. Such corrective explanations can be searched alone, as minimal changes, or jointly with the property of being also inside the $L1$ distance minimizers (in a bi-optimal point of view).

The detection of such bi-optimal solutions can be performed efficiently by profiting from the geometric characterization of the whole set of $L1$ minimizers and from the properties of $L1$ topology.

Keywords: Incoherence corrections · $L1$ constrained minimization · Mixed Integer Programming · Optimal corrective explanation · Probabilistic databases

1 Introduction

Uncertain data are nowadays becoming increasingly important in probabilistic databases [21] and they can emerge from traditional sources (e.g., by data integration like in [22]) or from the so called "next generation" sources (e.g., by information extraction like in [18]). Such kind of data bring with themselves a crucial characteristic: they must be consistent with sound uncertainty measures to be used properly, and this is not always assured, especially whenever they come from different sources of information.

As well outlined in [6], the way data fusion problem is tackled depends on the way information is represented. Since one of the most familiar and adopted

---

Supported by project "Algebraic statistics in a coherent setting for Bayesian networks" - DMI Unipg - Ricerca di Base 2018.

M.-J. Lesot et al. (Eds.): IPMU 2020, CCIS 1237, pp. 80–92, 2020.
https://doi.org/10.1007/978-3-030-50146-4_7

measure of uncertainty is probability, consequently we focus here on the even more mandatory task of correcting inconsistent probabilistic databases (see, e.g., [20]).

The choice of correcting probability values reflects the willingness to maintain the probabilistic nature of the different sources of information. This because the agent who performs the fusion would preserve the expressive power of probability framework. Of course, a change of the uncertainty management paradigm could be possible by adopting more general degrees of belief to deal with ill-posed sentences, like e.g. Belief functions, Fuzzy Logic or possibility measures (there is a vast literature on this, see among the others [1,5,7,8,10,16]) but this would be a strong intervention on the information representation, with a possible loss in expressiveness (as defined in [16]), especially if the fusion process is performed by a "third party" with respect the original sources. Hence in the present contribution we describe a way to proceed when the fusion process is intended to follow probability rules.

Recently (see e.g. [3,4]), we have proposed an efficient method for correcting incoherent (i.e. inconsistent) probability assessments. This method is based on *L*1 distance minimization and Mixed Integer Programming (MIP) procedures can be designed to implement it, in line with what has been done in [13,14] where such technique was introduced for the check of coherence problem.

The need of incoherence correction can be originated by different needs and can have different goals, hence we have designed specific MIP procedures to:

- correct straight unconditional assessments [4, §3];
- merge inconsistent databases [4, §4];
- revise the belief in a dynamical setting [3, §5];
- solve the so called plain statistical matching problem [4, §6];
- solve its generalization of the statistical matching problem with missclassification errors [9, §5];
- minimize the number of valuations to be modified, in line with one basic principle of optimal corrective explanation for decision makers (this contribution).

We are going to illustrate the last item of the previous list as it will appear in Sect. 4. Before doing it, we are going to recall in Sect. 2 the basic notions of partial probability assessments with the associated property of being coherent and in Sect. 3 how MIP is used to solve incoherence through *L*1 distance minimization and to find the whole set of optimal solutions.

In Subsect. 4.1 the problem of finding all corrective explanations that are also at minimal *L*1 distance from the initial assessment is posed and solved.

Two prototypical examples will illustrate our procedure.

2   Probability Assessments

In this section we recall some notions about probability assessments and coherence.

Definition 1. *A probability assessment is a tuple* $\pi = (V, U, p, \mathfrak{C})$, *where:*

- $V = \{X_1, \ldots, X_k\}$ *is a finite set of propositional variables, representing any potential event of interest;*
- U *is a subset of V of cardinality n that contains the effective events taken into consideration;*
- $p : U \to [0,1]^n$ *is a vector which assigns a "potential" probability value $p(X_i)$, $i = 1, \ldots, n$, to each variable in U;*
- $\mathfrak{C}$ *is a finite set of logical constraints which are present among all the variables in V and that characterize the descriptions of the events of interest;*

A probability assessment represents a state of knowledge about the probability for events in U to be true. Such probability values can be evaluated (or "assessed", as more properly said into specific literature) on the base of observed data or of expert evaluations. Events are explicitly interconnected through the logical constraints $\mathfrak{C}$, which can be written involving all the potential events V, and not only those of U, to permit to extend an initial assessment to a larger domain without redefining the whole framework. The constraints $\mathfrak{C}$ are crucial to represent the domain of a database, especially whenever it is built by merging different sources of information, since they permit to represent any kind of compound event, i.e. any macro situation, as, for example, that an event is the conjunction of other two events, or the implications or incompatibilities among some elements of V.

As usually done in Boolean logic, $\mathfrak{C}$ can be expressed in conjunctive normal form (CNF), namely, $\mathfrak{C} = \{c_1, \ldots, c_m\}$ where each element c_i of $\mathfrak{C}$ is a disjunctive logical clause of the form:

$$c_i = \left(\bigvee_{h \in H_i} X_h \right) \vee \left(\bigvee_{l \in L_i} \neg X_l \right) \tag{1}$$

with X_h and X_l in V, for some subsets of indexes $H_i, L_i \subseteq \{1, \ldots, k\}$.

This form results particularly helpful in the implementation part of the correction procedure, as we will described in the next section.

Since a probabilistic assessment π is partial, i.e. not defined on a fully structured domain like a Boolean algebra, it may or not be coherent, that means consistent with a probability distribution. In literature there are different, but all equivalent, ways of defining coherence. They are based on semantic, syntactical or operational point of views (see, e.g., [11,12,15,19]), anyway a formulation in any approach can be easily translated into a formulation in a different one (see in particular the characterization results in [11,12]). For our proposal it is more fruitful to adopt the operational point of view already used in [2]. This will permit to face the coherence problem directly with mathematical programming tools.

Let us introduce now the notions of truth assignment and atom.

Definition 2. *A **truth assignment** on V is a function $\alpha : V \to \{0, 1\}$. We denote by 2^V the set of all truth assignments. We denote with $\alpha \models \phi$ the fact* *that the assignment* $\alpha \in 2^V$ *satisfies a Boolean expression* ϕ *(which means that replacing each variable* x *appearing in* ϕ *with the corresponding truth value* $\alpha(x)$, *the expression* ϕ *evaluates to* 1).

Definition 3. *A truth assignment* $\alpha \in 2^V$ *is called* **atom** *for a probability assessment* $\pi = (V, U, p, \mathfrak{C})$ *if* α *satisfies all the logical constraints* $c_i \in \mathfrak{C}$.

The coherence of a probability assessment is then defined as follows

Definition 4. *A probability assessment* $\pi = (V, U, p, \mathfrak{C})$ *is* **coherent** *if there exists a probability distribution* $\mu : 2^V \rightarrow [0, 1]$ *which satisfies the following properties*

1. *for each* $\alpha \in 2^V$, *if there exists a constraint* $c_i \in \mathfrak{C}$ *such that* $\alpha \not\models c_i$, *then* $\mu(\alpha) = 0$;
2. $\sum_{\alpha \in 2^V} \mu(\alpha) = 1$;
3. *for each* $X_j \in U$, $\sum_{\alpha \in 2^V, \alpha \models X_j} \mu(\alpha) = p(X_j)$.

Checking the coherence of a probability assessment is a computational hard problem for which there exist many algorithmic approaches. In this paper we are mainly interested in the approach firstly described in [13], where the problem is solved by means of a Mixed Integer Programming (MIP) approach.

3 Correcting Probability Assessments

Whenever a probability assessment $\pi = (V, U, p, \mathfrak{C})$ results to not be coherent, it is possible to "modify" it in order to obtain a coherent probability assessment π' which is "as close as possible" to π.

Our approach consists in revising only the probability values p, because we consider the logical constraints $\mathfrak{C}$ as fixed. Hence the modified coherent assessment is $\pi' = (V, U, p', \mathfrak{C})$, with p' the corrected probability values.

Closeness between p and p' is measured through a specific metric, and in our approach we are using the $L1$ distance between p and p', which is defined as

$$d_1(p, p') = \sum_{i=1}^{n} |p(X_i) - p'(X_i)|. \tag{2}$$

The use of this distance has two important properties. First of all, minimization of the displacements $|p(X_i) - p'(X_i)|$ respects the basic *principle of minimal change* in a numerical uncertainty setting. Secondly, there is a clear *computational advantage* with respect to other distances, like e.g. $L2$ or Kullback-Leibler divergence, because minimization of $L1$ distance can be solved by MIP programming that has nowadays consolidated efficient algorithms implemented in several solvers while others distances would need less efficient and less robust algorithms for nonlinear (quadratic, logarithmic, etc.) problems.

Given a probability assessment $\pi = (V, U, p, \mathfrak{C})$, we denote by $\mathcal{C}(\pi)$ the sets of all the $L1$-corrections of π.

The procedure to find the corrections of $\pi = (V, U, p, \mathfrak{C})$ is described in [4]. Here, for sake of completeness, we recall the most important points. The basic idea is that in order to find the corrections of a probability assessment π it is possible to solve a mixed integer program, which is denoted by $\mathcal{P}1$. The method of obtaining $\mathcal{P}1$ from π exploits the same idea proposed in [13].

An important property is that if a probability assessment is coherent, *there exists a sparse probability distribution μ* so that p' can be written as a *convex combination of at most $n+1$ atoms* $\alpha^{(1)}, \ldots, \alpha^{(n+1)}$. Hence it is possible to build a MIP problem $\mathcal{P}1$, whose size is polynomial with respect to the size of π.

The *non-negative* variables for $\mathcal{P}1$ are summarized in Table 1.

Table 1. List of variables in the MIP program $\mathcal{P}1$ used to correct an incoherent probabilistic assessments.

Name	Indexes	Size	Type
a_{ij}	$i = 1, \ldots, n$ $j = 1, \ldots, n+1$	$n(n+1)$	Binary
b_{ij}	$i = 1, \ldots, n$ $j = 1, \ldots, n+1$	$n(n+1)$	Real
q_j	$j = 1, \ldots, n+1$	$n+1$	Real
r_i	$i = 1, \ldots, n$	n	Real
s_i	$i = 1, \ldots, n$	n	Real

The linear constraints of $\mathcal{P}1$ are

– for $i = 1, \ldots, m$ and $j = 1, \ldots, n+1$

$$\sum_{h \in H_i} a_{h,j} + \sum_{l \in L_i} (1 - a_{l,j}) \geq 1 \tag{3}$$

– for each $i = 1, \ldots, n$,

$$\sum_{j=1}^{n+1} b_{ij} = p(X_i) + (r_i - s_i) \tag{4}$$

– for $i = 1, \ldots, n$ and $j = 1, \ldots, n+1$,

$$0 \leq b_{ij} \leq a_{ij}, \quad a_{ij} - 1 + q_j \leq b_{ij} \leq q_j \tag{5}$$

$$\sum_{i=1}^{n+1} q_j = 1 \tag{6}$$

- for $i = 1, \ldots, n$,

$$r_i \leq 1, \quad s_i \leq 1 \tag{7}$$

where, we recall, n is the cardinality of p, m the cardinality of $\mathfrak{C}$, while H_i and L_i are the indexes subsets in the disjunctive normal form (1) of each $c_i \in \mathfrak{C}$.

The variables a_{ij} are binary, i.e. constrained in $\{0, 1\}$. Each value a_{ij} corresponds to the atom component $\alpha^{(j)}(X_i)$, for $i = 1, \ldots, n$ and $j = 1, \ldots, n+1$. Indeed, the constraint (3)

$$\sum_{h \in H_i} a_{hj} + \sum_{l \in L_i} (1 - a_{lj}) \geq 1 \quad i = 1, \ldots, m$$

$$j = 1, \ldots, n+1$$

forces each atom $\alpha^{(j)}(a_{1j}, \ldots, a_{nj})$ in the solution to satisfy all the clauses $c_i \in \mathfrak{C}$.

The values $q_1, \ldots, q_{n+1}$ represent the coefficient of the convex combination which generates p', which also correspond to the probabilities $\mu(\alpha^{(1)}), \ldots, \mu(\alpha^{(n+1)})$.

The constraint (3)

$$0 \leq b_{ij} \leq a_{ij}, \quad a_{ij} - 1 + q_j \leq b_{ij} \leq q_j \quad i = 1, \ldots, n$$

$$j = 1, \ldots, n+1$$

is equivalent to the non linear constraint

$$b_{ij} = a_{ij} \cdot q_j \text{ for } i = 1, \ldots, n \text{ and } j = 1, \ldots, n+1.$$

Since $a_{ij} = 1$ if and only if $\alpha^{(j)}$ satisfies $\mathbf{X_i}$, the sum

$$\sum_{j=1}^{n+1} b_{ij}$$

is also equal to $\mathbf{p'(X_i)}$.

The variables r_i, s_i are slack variables, which represent the positive and the negative difference between $p(X_i)$ and $p'(X_i)$, that permit to translate the inequalities $p'(X_i) \leq p(X_i)$ $p'(X_i) \geq p(X_i)$ into equations (4):

$$\sum_{j=1}^{n+1} b_{ij} = p(X_i) + (r_i - s_i) \quad i = 1, \ldots, n$$

Finally, the program $\mathcal{P}1$ is composed by the constraints (1–5) and its objective function to be minimized is

$$\sum_{i=1}^{n} (r_i + s_i). \tag{8}$$

It is possible to prove that any solution of the linear program $\mathcal{P}1$ corresponds to a $L1$-correction p' of p [4].

In many situations $\mathcal{C}(\pi)$ has more than one element and the MIP problem is able to find just one solution, which could not be a good representative of all the elements of $\mathcal{C}(\pi)$, as happens when it is an extreme value.

Hence program $\mathcal{P}1$ must be associated to an other MIP program $\mathcal{P}2$ to generate the most "*baricentric*" point $\bar{p}$ of $\mathcal{C}(\pi)$, i.e. a solution that spreads as much uniformly as possible the numerical amount of the distance (2) amongst the n values in p (for details and formal proofs refer again to [4]).

In $\mathcal{P}2$ all the constraints and the variables of $\mathcal{P}1$ are reported and it contains a new real variable z, which is subject to the constraints $r_i + s_i \leq z$, for $i = 1, \ldots, n$ (hence $z \geq \max_{i=1,\ldots,n} (r_i + s_i)$), and the new additional constraint $\sum_{i=1}^{n}(r_i + s_i) = \delta$, where δ is the optimal value of the objective function in $\mathcal{P}1$.

In this way, the $\mathcal{P}2$ objective function to be minimized is simply z.

Using $\bar{p}$, it is possible to find the face F_1 of the polytope of coherent assessments $\mathcal{Q}$ where $\mathcal{C}(\pi)$ lies.

The face F_1 is itself a convex set with at most $n+1$ atoms as extremal points, which can be found as a part of the solutions of $\mathcal{P}2$ (i.e., the optimal values of a_{ij}).

By looking at the signs of $\bar{p}(X_i) - p(X_i)$, for $i = 1, \ldots, n$, it is also possible to determine the face F_2 of the $L1$ ball $\mathcal{B}_\pi(\delta)$ which contains $\mathcal{C}(\pi)$. Indeed, F_2 is a convex set with at most n extremal points of the form

$$p + sign(\bar{p}(X_j) - p(X_j)) \cdot \delta \cdot e_j. \tag{9}$$

The *whole set of corrections* will result as:

$$\mathcal{C}(\pi) = F_1 \cap F_2. \tag{10}$$

The computation of (10) can be done by using a procedure *FaceEnum*, which, given a polyhedron, finds its H representation formed by the half-spaces that contain its faces, and a procedure *VertexEnum*, which finds the V representation constituted by its extreme points or vertices. Examples of *FaceEnum* and *VertexEnum* are described, e.g., in [17].

The overall procedure is described in the following pseudo-code.

procedure Correct
Input: assessment $(V, U, p, \mathfrak{C})$
Output: extremal points W and minimum distance δ
begin
        prepare MIP program $\mathcal{P}1$
        solve $\mathcal{P}1$
        extract the optimal value δ
        if $\delta = 0$ **then**
                return $(\{p\}, 0)$
        else

```
            prepare MIP program P2
            solve P2
            extract the values aij, ri, si
            E1 :=columns of matrix aij
            compute p̄ from ri, si
            compute E2 with Formula (9)
            H1 := FaceEnum(E1)
            H2 := FaceEnum(E2)
            Q := VertexEnum(H1 ∪ H2)
            return (Q, δ)
        endif
end
```

4 Optimal Corrective Explanation

The approach to correcting probability assessments described in the previous section takes into account only the sum of the corrections made on the probability values.

If we consider that enforcing the coherence of a probability assessment can be seen as a process of constraint satisfaction, a different perspective can be used. Indeed, as in an interactive tasks of constraint programming, it could be helpful to detect a minimal number of events whose probability value should be modified. In this way, the correction needed to obtain a coherent assessment has a better explanation, because the modifications are concentrated to a small number of events, instead being spread on more events. In any case, also the sum of the corrections must be minimal.

Concretely, the objective function can take into account of both: the number of the affected probability values and the sum of all the modifications. A procedure to compute this result is based on the previously described procedure Correct, with some simple modifications.

First of all, the program $\mathcal{P}1$ is extended by adding specific variables and constraints that permit to "count" the modified probability values.

The new integer variables are I_i, for $i = 1, \ldots, n$, while the additional constraints are

$$r_i + s_i \leq I_i \leq 1, \quad i = 1, \ldots, n; \tag{11}$$

If we just search for a solution with a minimal number of changes, the new objective function is simply

$$\sum_{i=1}^{n} I_i. \tag{12}$$

Let us show its application with a simple numerical example:

Example 1. Let $U = V = \{X_1, X_2, X_3\}$

$\mathfrak{C} = (X_2 \vee X_3) \wedge (X_2 \vee \neg X_1) \wedge (X_3 \vee \neg X_1)$

$p = (.9, .8, .9)$
hence with $n = 3$, $m = 3$ and $H_1 = \{2, 3\}$, $L_1 = \emptyset$, $H_2 = \{2\}$, $L_2 = \{1\}$, $H_3 = \{3\}$ $L_3 = \{1\}$.

By plugging these values into constraints (3) - (7),(11) and by searching the optimal minimal value of the objective function (12) we obtain an optimal corrective explanation of $I_2 = 1$ with $r_2 = .2$. Hence only p_2 must be modified into $p'_2 = 1.0$.

Note that this correction actually minimize also the L1 distance objective function (8).

If, on the contrary, we start from the incoherent assessment $p = (.99, .8, .9)$, we obtain as optimal corrective explanation $I_2 = 1$ and $I_3 = 1$, with $r_2 = .2$ and $r_3 = .09$, so that $p'_2 = 1.0$ and $p'_3 = .99$. Note that also this correction minimize the L1 distance objective function.

4.1   Optimal Corrective Explanation at Minimal $L1$ Distance

From the previous example one could be tempted to induce that each optimal corrective explanation is also a L1 distance minimizer, but obviously it is not always the case since the polytope of coherent assessments Q could have inner points in the same direction $p \pm \delta \cdot (I_1, \ldots, I_n)$.

To be sure to find on optimal corrective explanation that has also the minimal L1 distance, it is enough to change only the objective function to

$$\sum_{i=1}^{n} r_i + s_i + I_i. \tag{13}$$

Is is immediate to prove the following

Proposition 1. Any assessment p' induced by the solution of the MIP program with objective function (13) has a minimal number of corrections and a minimal L1 distance w.r.t. the initial assessment p.

In fact, if there exists another correction p'' with the same number of changed components but with a smaller L1 distance, or with the same distance but with a smaller number of changed components, the objective function (13) would inevitably result smaller since the constraints that make the I_i equal to 1 whenever any one of the correction components r_i or s_i is different from zero.

Let us see an other toy example that shows what we can obtain.

Example 2. $U = V = \{X_1, X_2, X_3\}$
$\mathfrak{C} = (X_2 \vee X_3 \vee \neg X_1) \wedge (X_3 \vee \neg X_1 \vee \neg X_2) \wedge (\neg X_1 \vee \neg X_2 \vee \neg X_3) \wedge (X_1 \vee X_3 \vee \neg X_2)$
$p = (1.0, .3, .2)$
that could lead to an optimal corrective explanation of $I_1 = 1$ and $I_2 = 1$ with

$s_1 = 1.0$ and $s_2 = .1$ hence p_1 must be modified into $p'_1 = 0.0$ and p_2 must be modified into $p'_2 = 0.2$. Also this correction minimizes the $L1$ distance objective function (2) with

$$d_1(p, p') = 1.1 \qquad (14)$$

If we want to collect all such corrections that simultaneously minimize the number of changes and the $L1$ distance, let us denote such optimal corrections by $\mathcal{O}(\pi)$, we can resort to the reasoning illustrated in the previous section to find the whole set of minimal distance corrections $\mathcal{C}(\pi)$.

Since $\mathcal{C}(\pi)$ has been obtained in (10) as the intersection of the two faces F_1 and F_2, if it contains some extremal points of the face F_2, that we recall is a face of the $L1$ ball $\mathcal{B}_\pi(\delta)$, then these are the searched solutions. In fact, since any other correction p'' is a convex combination of such extremal points can have only a greater number of changes w.r.t. the extremes. In this case $\mathcal{O}(\pi)$ is composed of isolated points and has a cardinality of at most n.

If, on the contrary, $\mathcal{C}(\pi)$ does not contain any extremal point of F_2, the extremal points of the convex set $F_1 \cap F_2$ surely belong to the searched solutions $\mathcal{O}(\pi)$ as long as any other convex combination among those that change the same values. No other point in $\mathcal{C}(\pi)$ can belong to $\mathcal{O}(\pi)$, since it will result as convex combination of extremal points that changes different components and hence it will change the union of those. In such a case $\mathcal{O}(\pi)$ will result as some isolated vertex and some edges of $F_1 \cap F_2$.

Let us show the two different situations by referring to the previous introduced toy examples.

In Example 1 we have (see Fig. 1) that $\mathcal{C}(\pi) = F_1 \cap F_2$ it has four extremal points $p' = (.9, 1.0, .9)$, $p'' = (.7, .8, .9)$, $p''' = (.8, .8, 1.0)$, $p^{iv} = (.9, .9, 1.0)$, and contains two extremal points of F_2 so that $\mathcal{O}(\pi)$ is the isolated vertexes p' and p''.

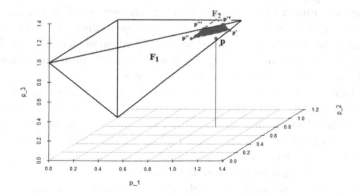

Fig. 1. Whole sets of corrections for Ex.1: $\mathcal{C}(\pi)$ is the colored area of $F_1 \cap F_2$ with extremal points $\{p', p'', p''', p^{iv}\}$; while $\mathcal{O}(\pi)$ is the two separated points $\{p', p''\}$ (Color figure online)

In Example 2 we have (see Fig. 2) that $C(\pi) = F_1 \cap F_2$ does not contain any extremal point of F_2 and it has five extremal points $p' = (0.0, .2, .2)$, $p'' = (.2, 0.0, .2)$, $p''' = (1.0, 0.0, 1.0)$, $p^{iv} = (.7, .3, 1.0)$, $p^v = (0.0, .3, .3)$; while $\mathcal{O}(\pi)$ is the isolated vertex p''' and the two edges $\overline{p'p''}$, and $\overline{p^{iv}p^v}$.

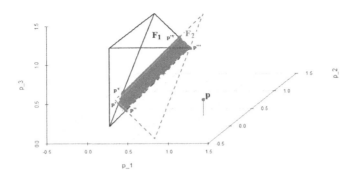

Fig. 2. Whole sets of corrections for Ex.2: $C(\pi)$ is the colored area of $F_1 \cap F_2$ with extremal points $\{p', p'', p''', p^{iv}, p^v\}$; while $\mathcal{O}(\pi)$ is p''' and the two solid edges $\overline{p'p''}$, and $\overline{p^{iv}p^v}$ (Color figure online)

5   Conclusion

With this contribution we have shown how Mixed Integer Programming can be profitable used to solve probabilistic databases conflicts.

In particular, using additional variables and constraints, it is possible to render linear most practical problems. Specifically, we focused on the problem of minimizing the number of corrections in an incoherent set of probability assessments, obtaining a so called *optimal corrective explanation*. A smart use of the involved slack variables and an immediate modification of the objective function has easily permitted to obtain a double goal: minimal number of changes at a minimal $L1$ distance, in a bi-objective prospective.

By profiting from geometric properties of the whole set of minimal $L1$ distance solutions, jointly with the topology of the $L1$ metric, it was possible to characterize the whole set of such bi-optimal solutions so that we easily proposed an efficient procedure to compute them. The knowledge of all the bi-optimal corrective explanations permits the decision maker to select the most appropriate adjustment of an inconsistent probabilistic database.

In next future we are going to implement the proposed method in a unique procedure in order to deal and to solve real practical problems of reasonable size.

References

1. Bacchus, F., Grove, A.J., Halpern, J.Y., Koller, D.: From statistical knowledge bases to degrees of belief. Artif. Intell. **87**(1–2), 75–143 (1996). https://doi.org/10.1016/S0004-3702(96)00003-3
2. Baioletti, M., Capotorti, A., Tulipani, S., Vantaggi, B.: Elimination of Boolean variables for probabilistic coherence. Soft Comput. **4**(2), 81–88 (2000). https://doi.org/10.1007/s005000000040
3. Baioletti, M., Capotorti, A.: Efficient L1-based probability assessments correction: algorithms and applications to belief merging and revision. In: Proceedings of the 9th International Symposium on Imprecise Probability: Theories and Applications (ISIPTA 2015), Pescara (IT), pp. 37–46. ARACNE (2015)
4. Baioletti, M., Capotorti, A.: A $L1$ based probabilistic merging algorithm and its application to statistical matching. Appl. Intell. **49**(1), 112–124 (2018). https://doi.org/10.1007/s10489-018-1233-z
5. Benferhat, S., Dubois, D., Prade, H.: How to infer from inconsistent beliefs without revising? In: Proceedings of the 14th International Joint Conference on Artificial Intelligence (IJCAI 1995), Montréal, Canada, 20–25 August, pp. 1449–1455 (1995)
6. Benferhat, S., Dubois, D., Prade, H.: From semantic to syntactic approaches to information combination in possibilistic logic. In: Bouchon-Meunier, B. (eds) Aggregation and Fusion of Imperfect Information, pp. 141–161. Physica-Verlag HD (1998). https://doi.org/10.1007/978-3-7908-1889-5_9
7. Benferhat, S., Sossai, C.: Reasoning with multiple-source information in a possibilistic logic framework. Inf. Fusion **7**, 80–96 (2006). https://doi.org/10.1016/j.inffus.2005.01.006
8. Bosc, P., Pivert, O.: Querying possibilistic databases: three interpretations. In: Yager, R., Abbasov, A., Reformat, M., Shahbazova, S. (eds.) Soft Computing: State of the Art Theory and Novel Applications. STUDFUZZ, vol. 291, pp. 161–176. Springer, Heidelberg (2013). https://doi.org/10.1007/978-3-642-34922-5_12
9. Buonunori, G., Capotorti, A.: Behavior of L_1-based probabilistic correction applied to statistical matching with misclassification information. In: Kratochvíl, V., Vejnarová, J. (eds.) Proceedings of the 11th Workshop on Uncertainty Processing (WUPES 2018), Trebon, Czech Republic, pp. 25–36. MatfyzPress, Czech Republic (2018)
10. Castro, J.L., Herrera, F., Verdegay, J.L.: Knowledge-based systems and fuzzy boolean programming. Intell. Syst. **9**(2), 211–225 (1994). https://doi.org/10.1002/int.4550090203
11. Coletti, G.: Coherent numerical and Ordinal probabilistic assessments. IEEE Trans. Syst. Man Cybern. **24**, 1747–1754 (1994). https://doi.org/10.1109/21.328932
12. Coletti, G., Scozzafava, R.: Probabilistic Logic in a Coherent Setting. Kluwer, Series "Trends in Logic", Dordrecht (2002). https://doi.org/10.1007/978-94-010-0474-9
13. Cozman, F.G., di Ianni, L.F.: Probabilistic satisfiability and coherence checking through integer programming. In: van der Gaag, L.C. (ed.) ECSQARU 2013. LNCS (LNAI), vol. 7958, pp. 145–156. Springer, Heidelberg (2013). https://doi.org/10.1007/978-3-642-39091-3_13
14. Cozman, F.G., di Ianni, L.F.: Probabilistic satisfiability and coherence checking through integer programming. Int. J. Approximate Reason. **58**, 57–70 (2015). https://doi.org/10.1016/j.ijar.2014.09.002

15. de Finetti, B.: Teoria della Probabilità. Torino Einaudi, 1970. Wiley, London (1974). https://doi.org/10.1002/9781119286387. (English translation Theory of probability)
16. Dubois, D., Liu, W., Ma, J., Prade, H.: The basic principles of uncertain information fusion. An organised review of merging rules in different representation frameworks. Inf. Fusion Part A **32**, 12–39 (2016). https://doi.org/10.1016/j.inffus.2016.02.006
17. Fukuda, K.: Lecture: Polyhedral Computation, Spring (2011). Institute for Operations Research and Institute of Theoretical Computer Science, ETH Zurich. https://inf.ethz.ch/personal/fukudak/lect/pclect/notes2015/PolyComp2015.pdf
18. Gupta, R., Sarawagi, S.: Creating probabilistic databases from information extraction models. In: Proceedings of the 32nd International Conference on Very Large Data Bases, pp. 965–976 (2006)
19. Lad, F.: Operational Subjective Statistical Methods: A Mathematical, Philosophical, and Historical Introduction. Wiley, New York (1996)
20. Lian, A.X., Chen, A.L., Song, A.S.: Consistent query answers in inconsistent probabilistic databases. In: Proceedings of the 2010 ACM SIGMOD International Conference on Management of Data, Indianapolis, Indiana, pp. 303–314 (2010). https://doi.org/10.1145/1807167.1807202
21. Suciu, D., Olteanu, D., Ré, C., Koch, C.: Probabilistic Databases, Synthesis Lectures On Data Management. Morgan & Claypool Publishers (2011). https://doi.org/10.2200/S00362ED1V01Y201105DTM016
22. Dong, X.L., Halevy, A., Yu, C.: Data integration with uncertainty. Very Large Data Bases J. **18**, 469–500 (2009). https://doi.org/10.1007/s00778-008-0119-9

Sufficient Solvability Conditions for Systems of Partial Fuzzy Relational Equations

Nhung Cao[ID] and Martin Štěpnička[(✉)][ID]

CE IT4Innovations – Institute for Research and Applications of Fuzzy Modeling,
University of Ostrava, 30. dubna 22, 701 03 Ostrava, Czech Republic
{nhung.cao,martin.stepnicka}@osu.cz

Abstract. This paper focuses on searching sufficient conditions for the solvability of systems of partial fuzzy relational equations. In the case of solvable systems, we provide solutions of the systems. Two standard systems of fuzzy relational equations – namely the systems built on the basic composition and on the Bandler-Kohout subproduct – are considered under the assumption of partiality. Such an extension requires to employ partial algebras of operations for dealing with undefined values. In this investigation, we consider seven most-known algebras of undefined values in partial fuzzy set theory such as the Bochvar, Bochvar external, Sobociński, McCarthy, Nelson, Kleene, and the Łukasiewicz algebra. Conditions that are sufficient for the solvability of the systems are provided. The crucial role will be played by the so-called boundary condition.

Keywords: Fuzzy relational equations · Partial fuzzy logics · Partial fuzzy set theory · Undefined values · Boundary condition

1 Introduction

Systems of fuzzy relational equations were initially studied by Sanchez in the 1970s [17] and later on, many authors have focused on this topic and it becomes an important topic in fuzzy mathematics especially in fuzzy control. The most concerned problem attracting a large number of researchers regards the solvability criterions or at least conditions sufficient for the solvability of the systems. The applications of the topic are various including in the dynamic fuzzy system [14], solving nonlinear optimization problems and covering problem [13,15], and many others. It is worth mentioning that the topic is still a point of the interest in the recent research [5,10,12].

Recently, investigations of the systems of fuzzy relational equations allowing the appearance of undefined values in the involved fuzzy sets were initiated

---

The authors announce the support of Czech Science Foundation through the grant 20-07851S.

© Springer Nature Switzerland AG 2020
M.-J. Lesot et al. (Eds.): IPMU 2020, CCIS 1237, pp. 93–106, 2020.
https://doi.org/10.1007/978-3-030-50146-4_8

[4,7]. Partial fuzzy logic which is considered as a generalization of the three-valued logic, and the related partial fuzzy set theory has been established [1–3,11]. Several well-known algebras were already generalized in partial fuzzy set theory such as the Bochvar algebra, the Sobociński algebra, the Kleene algebra, or Nelson algebra [2]. Lets us note that it seems there is no absolutely accepted general agreement on what types of undefined values are the particular algebras mostly appropriate for but they turned out to be useful in various areas and applications [9].

Recently, further algebras for partial fuzzy logics motivated by dealing with missing values were designed [6,18]. In [4], the initial investigation on the solvability of the systems of partial fuzzy relational equations was provided. The study was restricted on the equations with fully defined (non-partial) consequents. In [7], the problem was extended by considering the partially defined consequents, however, only the Dragonfly algebra [18] was considered and only one of the systems of equations was investigated. The article [7] provided readers with the particular shape of the solution however, under the assumption of the solvability. However, the solvability was not ensured, no criterion was provided. This article aims at paying this debt and focuses on the determination of the sufficient conditions for the solvability of both standard systems of partial fuzzy relational equations. Various kinds of algebras dealing with undefined values are considered, in particular the Bochvar, Sobociński, Kleene, McCarthy, Bochvar external, Nelson, and Łukasiewicz algebras.

2 Preliminaries

2.1 Various Kinds of Algebras of Undefined Values

In this subsection, we briefly recall the definitions of several algebras of undefined values we apply in this work. Let us consider a complete residuated lattice $\mathcal{L} = \langle [0,1], \wedge, \vee, \otimes, \rightarrow 0, 1 \rangle$ as the structure for the whole article and thus, all the used operations will be stemming from it. Let $\star$ denotes the undefined values regardless its particular semantic sub-type of the undefinedness [9]. Then the operations dealing with undefined values are defined on the support $L^\star = [0,1] \cup \{\star\}$, for more details we refer to [2]. Note that the operations on $L^\star$ applying to $a, b \in [0,1]$ are identical with the operations from the lattice $\mathcal{L}$. The following brief explanation of the role of $\star$ in particular algebras is based on Tables 1, 2 and 3.

The value $\star$ in the Bochvar (abbr. B when denoting the operations) algebra works as an annihilator and so, no matter which values $a \in L^\star$ is combined with it, the result is always $\star$. In the Sobociński (abbr. S) algebra, $\star$ acts like a neutral element for the conjunction and the disjunction as well. It means that the conjunctive/disjunctive combination of any value $a \in L^\star$ with $\star$ results in a. In the Kleene algebra (abbr. K), the operations combining $\star$ and 0 or 1 comply the ordering $0 \leq \star \leq 1$, otherwise they coincide with the Bochvar algebra operations when $\star$ is combined with $a \notin \{0,1\}$. The Łukasiewicz algebra (abbr. L) and the Nelson (abbr. N) algebra are identical with the Kleene algebra regarding their conjunctions and disjunctions however, the difference lies in the implication operations. In particular, in the Łukasiewicz case $\star \to_L \star = 1$ holds, and in the Nelson case the equalities $\star \to_N 0 = 1$ and $\star \to_N \star = 1$ hold, while in both cases, the Kleene implication results into $\star$ again. The McCarthy (abbr. Mc) algebra interestingly combines the Kleene and the Bochvar behavior with the distinction between the cases whether $\star$ appears in the first argument or in the second argument of the operation.

Let us recall two useful external ones [9]: $\downarrow$ is given by $\downarrow\alpha = 0$ if $\alpha = \star$ and $\downarrow\alpha = \alpha$ otherwise; and $\uparrow$ is given by $\uparrow\alpha = 1$ if $\alpha = \star$ and $\uparrow\alpha = \alpha$ otherwise. The external operations play a significant role in the so-called Bochvar external algebra (abbr. Be) as it applies operation $\downarrow$ to $\star$ and lowers it to 0 in any combinations with $a \in L^\star$.

Table 1. Conjunctive operations of distinct algebras $(\alpha, \beta \in (0,1])$.

		Bochvar	Bochvar external	Sobociński	Kleene	McCarthy	Nelson	Łukasiewicz
α	⋆	⋆	0	α	⋆	⋆	⋆	⋆
⋆	β	⋆	0	β	⋆	⋆	⋆	⋆
⋆	⋆	⋆	0	⋆	⋆	⋆	⋆	⋆
⋆	0	⋆	0	0	0	⋆	0	0
0	⋆	⋆	0	0	0	0	0	0

Table 2. Disjunctive operations of distinct algebras $(\alpha, \beta \in [0,1))$.

		Bochvar	Bochvar external	Sobociński	Kleene	McCarthy	Nelson	Łukasiewicz
α	⋆	⋆	α	α	⋆	⋆	⋆	⋆
⋆	β	⋆	β	β	⋆	⋆	⋆	⋆
⋆	⋆	⋆	0	⋆	⋆	⋆	⋆	⋆
⋆	1	⋆	1	1	1	⋆	1	1
1	⋆	⋆	1	1	1	1	1	1

Table 3. Implicative operations of distinct algebras $(\alpha \in (0,1],\ \beta \in (0,1))$.

		Bochvar	Bochvar external	Sobociński	Kleene	McCarthy	Nelson	Łukasiewicz
α	⋆	⋆	¬α	¬α	⋆	⋆	⋆	⋆
⋆	β	⋆	1	β	⋆	⋆	⋆	⋆
⋆	⋆	⋆	1	⋆	⋆	⋆	1	1
⋆	1	⋆	1	1	1	⋆	1	1
0	⋆	⋆	1	1	1	1	1	1
⋆	0	⋆	1	0	⋆	⋆	1	⋆

2.2   Systems of Fuzzy Relational Equations

Let us denote the set of all fuzzy sets on a universe U by $\mathcal{F}(U)$. Then two standard systems of fuzzy relational equations are provided in the forms:

$$A_i \circ R = B_i, \quad i = 1, 2, \ldots, m \tag{1}$$
$$A_i \lhd R = B_i, \quad i = 1, 2, \ldots, m \tag{2}$$

where $A_i \in \mathcal{F}(X)$, $B_i \in \mathcal{F}(Y), i = 1, \ldots, m$ for some universes X, Y. The *direct product* $\circ$ and the *Bandler-Kohout subproduct (BK-subproduct)* $\lhd$ in systems (1) and (2) are expanded as follows:

$$(A_i \circ R)(y) = \bigvee_{x \in X} \left(A_i(x) \otimes R(x, y) \right), \ (A_i \lhd R)(y) = \bigwedge_{x \in X} \left(A_i(x) \to R(x, y) \right).$$

In [8], the authors defined so-called *boundary condition* and shown, that it is a sufficient condition for the solvability of the direct product systems (1). In [16], using the so-called skeleton matrix, it was shown that it serves as the sufficient condition also for the solvability of (2) and in [19], an alternative proof not requiring the skeleton matrix was presented.

Definition 1. Let $A_i \in \mathcal{F}(X)$ for $i \in \{1, \ldots, m\}$ be normal. We say, that A_i meet the *boundary* condition if for each i there exists an $x_i \in X$ such that $A_i(x_i) = 1$ and $A_j(x_i) = 0$ for any $j \neq i$.

Theorem 1 [8,16]. *Let A_i fulfill the boundary condition. Then systems (1)–(2) are solvable and the following models are solutions of the systems, respectively:*

$$\hat{R}(x, y) = \bigwedge_{i=1}^{m} (A_i(x) \to B_i(y)), \ \check{R}(x, y) = \bigvee_{i=1}^{m} (A_i(x) \otimes B_i(y)).$$

3   Sufficient Conditions Under Partiality

As we have recalled above, the standard systems of fuzzy relational equations are solvable if the antecedents fulfil the *boundary condition* [8]. Of course, the question whether the solvability of partial fuzzy relational equations can be ensured by the same or similar condition appears seems natural. As we will demonstrate the answer is often positive. Moreover, we investigate some specific cases of solvable systems even if the boundary condition is not preserved.

Let $\mathcal{F}^{\star}(U)$ stands for the set of all partially defined fuzzy sets (partial fuzzy sets) on a universe U, i.e., let

$$\mathcal{F}^{\star}(U) = \{A \mid A : U \to L^{\star}\}.$$

The following denotations will be used in the article assuming that the right-hand side expressions hold for all $u \in U$:

$$A = \emptyset \quad \text{if} \quad A(u) = 0,$$
$$A = \emptyset^{\star} \quad \text{if} \quad A(u) = \star,$$
$$A = 1 \quad \text{if} \quad A(u) = 1.$$

Moreover, let us introduce the following denotations for particular parts of the universe U with respect to a given partial fuzzy set $A \in \mathcal{F}^*(U)$:

$$\text{Def}(A) = \{u \mid A(u) \neq \star\}, \qquad A_0 = \{u \in U \mid A_i(u) = 0\},$$
$$A_\star = \{u \in U \mid A_i(u) = \star\}, \qquad A_P = \{u \in U \mid A_i(u) \notin \{0, \star\}\}.$$

3.1 Bochvar Algebra and McCarthy Algebra

Let us first consider the use of the Bochvar operations in the systems:

$$A_i \circ_B R = B_i, \quad i = 1, \dots, m, \tag{3}$$
$$A_i \triangleleft_B R = B_i \quad i = 1, \dots, m. \tag{4}$$

We recall that in the Bochvar algebra the $\star$ behaves like an annihilator i.e., when it combines with any other values the result is always $\star$. Thus, when there is an $x \in X$ such that $A_i(x) = \star$ the inferred output B_i is a fuzzy set to which all the elements have an undefined membership degree, i.e., $B_i = \emptyset^\star$. It immediately leads to the following theorems with necessary conditions demonstrating that the solvability of both systems falls into trivial cases as long as the partial fuzzy sets appear on the inputs.

Theorem 2. *The necessary condition for the solvability of system (3) is that $B_j = \emptyset^\star$ for all such indexes $j \in \{1, \dots, m\}$ for which the corresponding antecedents $A_j \in \mathcal{F}^*(X) \setminus \mathcal{F}(X)$.*

Sketch of the proof: As there exists $x \in X$ such that $A_i(x) = \star$, one can check that the following holds for any $R \in \mathcal{F}^*(X \times Y)$:

$$(A_i \circ_B R)(y) = \star \vee_B \bigvee_{\substack{B \\ x \notin A_{i\star}}} (A_i(x) \otimes_B R(x, y)) = \star$$

which leads to that B_i has to be equal to $\emptyset^\star$. $\qquad\square$

Corollary 1. *If $B_i = \emptyset^\star$ for all $i \in \{1, \dots, m\}$ then system (3) is solvable.*

Sketch of the proof: Based on a simple demonstration that $R_B^\star \in \mathcal{F}^*(X \times Y)$ given by $R_B^\star(x, y) = \star$ is a solution. $\qquad\square$

Theorem 3. *The necessary condition for the solvability of system (4) is that $B_j = \emptyset^\star$ for all such indexes $j \in \{1, \dots, m\}$ for which the corresponding antecedents $A_j \in \mathcal{F}^*(X) \setminus \mathcal{F}(X)$.*

Sketch of the proof: The proof is similar to the proof of Theorem 2. $\qquad\square$

Corollary 2. *If $B_i = \emptyset^\star$ for all $i \in \{1, \dots, m\}$ then system (4) is solvable.*

Sketch of the proof: Based on a simple demonstration that $R_{\mathrm{B}}^{\star} \in \mathcal{F}^{\star}(X \times Y)$ given by $R_{\mathrm{B}}^{\star}(x, y) = \star$ is a solution. □

Theorems 2 and 3 are direct consequences of the "annihilating effect" of $\star$ in the Bochvar algebra. Whenever the input is undefined, the consequents have to be even fully undefined.

Now, we focus on the systems applying the McCarthy algebra:

$$A_i \circ_{\mathrm{Mc}} R = B_i, \; i = 1, \ldots, m, \tag{5}$$
$$A_i \lhd_{\mathrm{Mc}} R = B_i, \; i = 1, \ldots, m. \tag{6}$$

As the McCarthy operations provide the same result as the Bochvar operations whenever $\star$ appears in their first argument, we naturally come to results about solvability of (5)–(6) that are the analogous to the results about solvability of (3)–(4).

Theorem 4. *The necessary condition for the solvability of system (5) is that* $B_j = \emptyset^{\star}$ *for all such indexes* $j \in \{1, \ldots, m\}$ *for which the corresponding antecedents* $A_j \in \mathcal{F}^{\star}(X) \setminus \mathcal{F}(X)$.

Sketch of the proof: Analogous to Theorem 2. □

Theorem 5. *The necessary condition for the solvability of system (6) is that* $B_j = \emptyset^{\star}$ *for all such indexes* $j \in \{1, \ldots, m\}$ *for which the corresponding antecedents* $A_j \in \mathcal{F}^{\star}(X) \setminus \mathcal{F}(X)$.

Sketch of the proof: Analogous to Theorem 3. □

Although the necessary conditions formulated in Theorems 4 and 5 are identical for McCarthy and Bochvar algebra, the sufficient condition for the McCarthy algebra has to also take into account the differences in the operations of these two otherwise very similar algebras.

Corollary 3. *If* $B_i = \emptyset^{\star}$ *and* $A_i \neq \emptyset$ *for all* $i \in \{1, \ldots, m\}$ *then system (5) is solvable.*

Sketch of the proof: As in the case of Corollary 1 the proof is based on a simple demonstration that $R_{\mathrm{Mc}}^{\star} \in \mathcal{F}^{\star}(X \times Y)$ given by $R_{\mathrm{Mc}}^{\star}(x, y) = \star$ is a solution however, the case of the empty input that would lead to the empty output has to be eliminated from the consideration. □

Corollary 4. *If* $B_i = \emptyset^{\star}$ *and* $A_i \neq \emptyset$ *for all* $i \in \{1, \ldots, m\}$ *then system (6) is solvable.*

Sketch of the proof: As in the case of Corollary 2 the proof is based on a simple demonstration that $R_{\mathrm{Mc}}^{\star} \in \mathcal{F}^{\star}(X \times Y)$ given by $R_{\mathrm{Mc}}^{\star}(x, y) = \star$ is a solution however, the case of the empty input that would lead to the output constantly equal to 1, has to be eliminated from the consideration. □

3.2   Bochvar External Algebra and Sobociński Algebra

In this section, we present the investigation of the solvability of systems of partial fuzzy relational equations in the case of the Bochvar external algebra and in the case of Sobociński algebra. Let us start with the Bochvar external operations employed in the systems:

$$A_i \circ_{\mathrm{Be}} R = B_i, \quad i = 1 \ldots, m, \tag{7}$$

$$A_i \triangleleft_{\mathrm{Be}} R = B_i, \quad i = 1 \ldots, m. \tag{8}$$

Theorem 6. *Let A_i meet the boundary condition. Then*

$$(A_i \circ_{\mathrm{Be}} \hat{R}_{\mathrm{Be}})(y) = B_i(y), \quad \text{for } y \in \mathrm{Def}(B_i),$$

where

$$\hat{R}_{\mathrm{Be}}(x, y) = \bigwedge_{i=1}^{m}{}_{\mathrm{Be}} (A_i(x) \to_{\mathrm{Be}} B_i(y)).$$

Sketch of the proof: Based on the definition of the Bochvar external operations, $A_i(x) \to_{\mathrm{Be}} B_i(y) \neq \star$, no matter the choice of x, y, and hence:

$$\left(A_i \circ_{\mathrm{Be}} \hat{R}_{\mathrm{Be}}\right)(y) \leq \bigvee_{x \in X}{}_{\mathrm{Be}} (A_i(x) \otimes_{\mathrm{Be}} ((A_i(x)) \to_{\mathrm{Be}} B_i(y))).$$

We may split the right-hand side expression running over X into two expressions, one running over $A_{i0} \cup A_{i\star}$, the other one running over A_{iP} and show, that each of them is smaller or equal to B_i:

$$\bigvee_{x \in A_{i0} \cup A_{i\star}}{}_{\mathrm{Be}} (A_i(x) \otimes_{\mathrm{Be}} (A_i(x) \to_{\mathrm{Be}} B_i(y))) = 0 \leq B_i(y)$$

$$\bigvee_{x \in A_{iP}}{}_{\mathrm{Be}} (A_i(x) \otimes_{\mathrm{Be}} (A_i(x) \to_{\mathrm{Be}} B_i(y))) \leq B_i(y)$$

which implies $(A_i \circ_{\mathrm{Be}} \hat{R}_{\mathrm{Be}})(y) \leq B_i(y)$.

Now, we prove the opposite inequality. Based on the assumption of the boundary condition, let us pick x_i such that $A_i'(x_i) = 1$ and $A_j(x_i) = 0, j \neq i$. Then we may check

$$(A_i \circ_{\mathrm{Be}} \hat{R}_{\mathrm{Be}})(y) \geq A_i(x_i) \otimes_{\mathrm{Be}} \hat{R}_{\mathrm{Be}}(x_i, y) = B_i(y)$$

which completes the sketch of the proof. $\square$

If we assume that the output fuzzy sets B_i are fully defined we obtain the following corollary.

Corollary 5. *Let A_i meet the boundary condition and let $B_i \in \mathcal{F}(Y)$. Then system (7) is solvable and $\hat{R}_{\mathrm{Be}}$ is its solution.*

The following theorem and corollary provides us with similar results for the BK-subproduct system of partial fuzzy relational equations (8).

Theorem 7. *Let A_i meet the boundary condition. Then*

$$(A_i \lhd_{\mathrm{Be}} \check{R}_{\mathrm{Be}})(y) = B_i(y), \quad for\ y \in \mathrm{Def}(B_i)$$

where

$$\check{R}_{\mathrm{Be}}(x, y) = \bigvee_{i=1}^{m}{}_{\mathrm{Be}} \left(A_i(x) \otimes_{\mathrm{Be}} B_i(y) \right).$$

Sketch of the proof: Due to the external operations, $A_i(x) \otimes_{\mathrm{Be}} B_i(y) \neq \star$ holds independently on the choice of x and y. Jointly with the property $c \rightarrow_{\mathrm{Be}} a \leq c \rightarrow_{\mathrm{Be}} b$ that holds for $a \leq b$ it leads to the inequality

$$\left(A_i \lhd_{\mathrm{Be}} \check{R}_{\mathrm{Be}} \right)(y) \geq \bigwedge_{x \in X}{}_{\mathrm{Be}} \left(A_i(x) \rightarrow_{\mathrm{Be}} \left(A_i(x) \otimes_{\mathrm{Be}} B_i(y) \right) \right).$$

For $y \in \mathrm{Def}(B_i)$ we get the following inequalities

$$\bigwedge_{x \in A_{i0} \cup A_{i\star}}{}_{\mathrm{Be}} \left(A_i(x) \rightarrow_{\mathrm{Be}} \left(A_i(x) \otimes_{\mathrm{Be}} B_i(y) \right) \right) = 1 \geq B_i(y),$$

$$\bigwedge_{x \in A_{iP}}{}_{\mathrm{Be}} \left(A_i(x) \rightarrow_{\mathrm{Be}} \left(A_i(x) \otimes_{\mathrm{Be}} B_i(y) \right) \right) \geq B_i(y)$$

that jointly prove that $(A_i \lhd_{\mathrm{Be}} \check{R}_{\mathrm{Be}})(y) \geq B_i(y)$. In order to prove the opposite inequality, we again pick up the point x_i in order to use the boundary condition. □

If the consequents B_i in system (8) are fully defined we obtain the following corollary.

Corollary 6. *Let A_i meet the boundary condition and let $B_i \in \mathcal{F}(Y)$. Then system (8) is solvable and $\check{R}_{\mathrm{Be}}$ is its solution.*

Now let us focus on the following systems applying the Sobociński operations:

$$A_i \circ_{\mathrm{S}} R = B_i, \quad i = 1 \ldots, m, \tag{9}$$

$$A_i \lhd_{\mathrm{S}} R = B_i, \quad i = 1 \ldots, m. \tag{10}$$

Theorem 8. *Let A_i meet the boundary condition and let $B_i \in \mathcal{F}(Y)$. Then system (9) is solvable and the following fuzzy relation*

$$\hat{R}_{\mathrm{S}}(x, y) = \bigwedge_{i=1}^{m}{}_{\mathrm{S}} \left(A_i(x) \rightarrow_{\mathrm{S}} B_i(y) \right)$$

is its solution.

Sketch of the proof: The proof uses an analogous technique as the proof of Theorem 6.   □

Theorem 9. *Let A_i meet the boundary condition and let $B_i \in \mathcal{F}(Y)$. Then system (10) is solvable and the following fuzzy relation*

$$\check{R}_S(x, y) = \bigvee_{i=1}^{m} {}_S (A_i(x) \otimes_S B_i(y))$$

is its solution.

Sketch of the proof: The proof uses an analogous technique as the proof of Theorem 7.   □

Remark 1. *Let us mention that the fuzzy relations introduced in the theorems above as the solutions to the systems of partial fuzzy relational equations are not the only solutions. They are indeed the most expected solutions as their construction mimics the shape of the preferable solutions of fully defined fuzzy relational systems, but, for instance, fuzzy relation*

$$\check{R}'_S(x, y) = \bigvee_{i=1}^{m} {}_S (\uparrow A_i(x) \otimes_S B_i(y))$$

has been shown to be a solution of the system (10) under the assumption of its solvability [4]. And the solvability can ensured by the boundary condition, see Theorem 9.

3.3   Kleene Algebra, Łukasiewicz Algebra and Nelson Algebra

Let us start with the focus on the systems employing the Kleene operations:

$$A_i \circ_K R = B_i, i = 1 \ldots, m, \tag{11}$$
$$A_i \lhd_K R = B_i, i = 1 \ldots, m. \tag{12}$$

Theorem 10. *Let for all $j \in \{1, \ldots, m\}$ one of the following conditions holds*

(a) A_j is a normal fuzzy set and $B_j = 1$,
(b) $A_j \neq \emptyset$ and $B_j = \emptyset^*$.

Then system (11) is solvable and moreover, the following partial fuzzy relation

$$\hat{R}_K(x, y) = \bigwedge_{i=1}^{m} {}_K (A_i(x) \rightarrow_K B_i(y))$$

is one of the solutions.

Sketch of the proof: By proving that $\hat{R}_K$ is a solution we prove also the solvability of the system. Let us take arbitrary j and assume that condition (a) holds. Then, for $x' \in X$ such that $A_j(x') = 1$ we can prove that $A_j(x') \otimes_K \hat{R}_K(x', y) = 1$ and hence, the following holds

$$(A_j \circ_K \hat{R}_K)(y) = \bigvee_K\limits_{x \neq x'} \left(A_j(x) \otimes_K \hat{R}_K(x, y) \right) \vee_K 1 = 1.$$

Now, let us assume that (b) holds for the given j. Then independently on the choice of x and y, $\hat{R}_K(x, y) \in \{\star, 1\}$, and based on the following facts

$$\bigvee_K\limits_{x \in A_{j0}} \left(A_j(x) \otimes_K \hat{R}_K(x, y) \right) = 0, \quad \bigvee_K\limits_{x \in A_{j\star} \cup A_{iP}} \left(A_j(x) \otimes_K \hat{R}_K(x, y) \right) = \star$$

we may derive $(A_j \circ_K \hat{R}_K)(y) = \star$.

In both cases (a) and (b), the result of $(A_j \circ_K \hat{R}_K)$ was equal to the consequent B_j and the proof was made for arbitrarily chosen index j.   □

Theorem 11. *Let for all $j \in \{1, \ldots, m\}$ one of the following conditions holds*

(a) A_j is a normal fuzzy set and $B_j = \emptyset$,
(b) $A_j \neq \emptyset$ and $B_j = \emptyset^\star$.

Then system (12) is solvable and moreover, the following partial fuzzy relation

$$\check{R}_K(x, y) = \bigvee_K\limits_{i=1}^{m} (A_i(x) \otimes_K B_i(y))$$

is one of the solutions.

Sketch of the proof: Let us take an arbitrary j and assume that (a) holds. Then, for $x' \in X$ such that $A_j(x') = 1$ we can prove that $A_i(x') \to_K \check{R}_K(x, y) = 0$ and hence, the following holds

$$(A_i \vartriangleleft_K \check{R}_K)(y) = \bigwedge_K\limits_{x \in X \setminus \{x'\}} \left(A_i(x) \to_K \check{R}_K(x, y) \right) \wedge_K 0 = 0.$$

Now, let us assume that (b) holds for the given j. Then $\check{R}_K(x, y) \in \{\star, 1\}$ independently on the choice of x and y, and based on the following facts

$$\bigwedge_K\limits_{x \in A_{i0}} \left(A_i(x) \to_K \check{R}_K(x, y) \right) = 1, \quad \bigwedge_K\limits_{x \in A_{i\star} \cup A_{iP}} \left(A_i(x) \to_K \check{R}_K(x, y) \right) = \star$$

we may derive $(A_j \vartriangleleft_K \check{R}_K)(y) = \star$.

In both cases (a) and (b), the result of $(A_j \circ_K \hat{R}_K)$ was equal to the consequent B_j and the proof was made for arbitrarily chosen index j.   □

Theorem 12. *Let for all $j \in \{1, \ldots, m\}$ the following condition holds*

(c) $B_j = 1$.

Then system (12) is solvable and moreover, the following fuzzy relation

$$\check{R}'_K(x, y) = \bigvee_{i=1}^{m} {}_K (\uparrow A_i(x) \otimes_K B_i(y))$$

is one of the solutions.

Sketch of the proof: The proof is based on the following three equalities

$$\bigwedge_{x \in A_{i0}} {}_K \left(A_i(x) \to_K \check{R}'_K(x, y) \right) = 1$$

$$\bigwedge_{x \in A_{i*}} {}_K \left(A_i(x) \to_K \check{R}'_K(x, y) \right) = 1$$

$$\bigwedge_{x \in A_{iP}} {}_K \left(A_i(x) \to_K \bigvee_{i=1}^{m} {}_K (\uparrow A_i(x) \otimes_K B_i(y)) \right) \geq \bigwedge_{x \in A_{iP}} {}_K (A_i(x) \to_K A_i(x)) = 1.$$

$\square$

The use of the Lukasiewicz operations and the Nelson operations give the same results and very similar to the use of the Kleene operations. Therefore, we will study the system jointly for both algebras of operations, in particular, we will consider

$$A_i \circ_\gamma R = B_i, i = 1 \ldots, m, \tag{13}$$

$$A_i \triangleleft_\gamma R = B_i, i = 1 \ldots, m. \tag{14}$$

where $\gamma \in \{L, N\}$ will stand for the the Lukasiewicz and Nelson algebra, respectively. Therefore, the following results will hold for both algebras.

Theorem 13. *Let for all $j \in \{1, \ldots, m\}$ one of the following conditions holds*

(a) A_i *is a normal fuzzy set and* $B_i = 1$,
(b) $A_i \neq \emptyset$ *and* $B_i = \emptyset^\star$.

Then system (13) is solvable and moreover, the following partial fuzzy relation

$$\hat{R}_\gamma(x, y) = \bigwedge_{i=1}^{m} {}_\gamma (A_i(x) \to_\gamma B_i(y))$$

is one of the solutions.

Sketch of the proof: The proof uses an analogous technique as the proof of Theorem 10. $\square$

Theorem 14. *Let for all $j \in \{1, \ldots, m\}$ one of the following conditions holds*

(a) A_j is a normal fuzzy set and $B_i = \emptyset$,
(b) there exists $x \in X$ such that $A_j(x) \notin \{0, \star\}$ and $B_i = \emptyset^\star$,
(c) $B_i = 1$.

Then system (14) is solvable and the following partial fuzzy relation

$$\check{R}_\gamma(x, y) = \bigvee_{i=1}^{m}{}_\gamma \, (A_i(x) \otimes_\gamma B_i(y))$$

is one of the solutions.

Sketch of the proof: Under the assumption that (a) holds, the proof uses the same technique as the proof of Theorem 11.

When proving the theorem under the assumption of the preservation of (b), we stem from the fact that $A_j(x) \to_\gamma \check{R}_\gamma(x, y) = 1$ when $A_j(x) = \star$, and from the fact that $A_j(x) \to_\gamma \check{R}_\gamma(x, y) = \star$ when $A_j(x) \notin \{0, \star\}$, and hence, we come to the conclusion that $(A_j \vartriangleleft_\gamma \check{R}_\gamma)(y) = \star$ for any $y \in Y$.

Let us consider case (c). Using the fact that $A_j(x) \to_\gamma \check{R}_\gamma(x, y) = 1$ in case of $A_j(x) = \star$ and also for $A_j(x) \neq \star$ we come to the same conclusion, we prove that $(A_i \vartriangleleft_\gamma \check{R}_\gamma)(y) = 1$ for arbitrary $y \in Y$. $\qquad \square$

All the results presented above can be summarized in Table 4.

Table 4. Sufficient solvability conditions for systems of partial fuzzy relational equations: $A_i \circ_\tau R = B_i$, $A_i \vartriangleleft_\tau R = B_i$ where $\tau \in \{\text{B}, \text{Mc}, \text{Be}, \text{S}, \text{K}, \text{L}, \text{N}\}$.

Distinct algebras	$A_i \circ_\tau R = B_i$		$A_i \vartriangleleft_\tau R = B_i$	
	Sufficient conditions	Solutions	Sufficient conditions	Solutions
Bochvar	$B_i = \emptyset^\star$	$R_\tau^\star$	$B_i = \emptyset^\star$	$R_\tau^\star$
MaCarthy	$A_i \neq \emptyset, B_i = \emptyset^\star$	$R_\tau^\star$	$A_i \neq \emptyset, B_i = \emptyset^\star$	$R_\tau^\star$
Bochvar external and Sobociński	A_i – boundary and $B_i \in \mathcal{F}(Y)$	$\hat{R}_\tau$	A_i – boundary and $B_i \in \mathcal{F}(Y)$	$\check{R}_\tau$
Kleene	A_i – normal, $B_i = 1$	$\hat{R}_\tau$	A_i – normal, $B_i = \emptyset$	$\check{R}_\tau$
	$A_i \neq \emptyset, B_i = \emptyset^\star$		$A_i \neq \emptyset, B_i = \emptyset^\star$	
			$B_i = 1$	$\check{R}_\tau'$
Łukasiewicz and Nelson	A_i – normal, $B_i = 1$	$\hat{R}_\tau$	A_i – normal, $B_i = \emptyset$	$\check{R}_\tau$
	$A_i \neq \emptyset, B_i = \emptyset^\star$		$\exists x : A_i(x) \notin \{0, \star\}$ and $B_i = \emptyset^\star$	
			$B_i = 1$	

4 Conclusion and Future Work

We have attempted to find, formulate and prove sufficient conditions for the solvability of systems of partial fuzzy relational equations. Distinct well-known algebras dealing with undefined values have been considered, namely Bochvar, Bochvar external, Sobociński, Kleene, Nelson, Łukasiewicz, and McCarthy algebras. Let us recall that the choice of many algebras to apply to such a study was not random but to cover various types of undefined values and consequently, various areas of applications. We have obtained distinct sufficient conditions for distinct algebras. Some of the conditions seem to be rather flexible, e.g., for the case of Bochvar external and Sobociński, it was sufficient to consider the boundary condition met by the antecedent fuzzy sets. On the other hand, most cases showed that the solvability can be guaranteed under very restrictive conditions. Although apart from the Bochvar case, the conditions are not necessary but only sufficient, from the construction of the proofs and from the investigation of the behavior of the particular operations it is clear, that in such algebras, very mild conditions cannot be determined.

For the future work, we intend to complete the study by adding also necessary conditions and by considering also the Dragonfly and Lower estimation algebras that seem to be more promising for obtaining mild solvability conditions similarly to the case of Sobociński or Bochvar external algebra. Furthermore, there exist problems derived from the solvability modeling more practically oriented research that are not expected to be so demanding on the conditions such as the solvability itself. By this, we mean, for instance, modeling the partial inputs incorporated into the fully defined systems of fuzzy relational equations. Indeed, this models very natural situations when the knowledge (antecedents and consequents) is fully defined but the input is partly damaged by, e.g., containing missing values etc.

Finally, we plan to study the compatibility (or the sensitivity) of the used computational machinery with undefined values with respect to ranging values that can possibly replace the $\star$. This investigation should show us which algebras are the most robust ones when we know in advance that $\star$ belongs to a certain range or $\star$ is described using natural language such as "low values", or "big values", etc.

References

1. Běhounek, L., Daňková, M.: Towards fuzzy partial set theory. In: Carvalho, J.P., Lesot, M.-J., Kaymak, U., Vieira, S., Bouchon-Meunier, B., Yager, R.R. (eds.) IPMU 2016. CCIS, vol. 611, pp. 482–494. Springer, Cham (2016). https://doi.org/10.1007/978-3-319-40581-0_39
2. Běhounek, L., Novák, V.: Towards fuzzy partial logic. In: 2015 IEEE International Symposium on Multiple-Valued Logic (ISMVL), pp. 139–144. IEEE (2015)
3. Běhounek, L., Dvořák, A.: Fuzzy relational modalities admitting truth-valueless propositions. Fuzzy Sets Syst. **388**, 38–55 (2020)

4. Cao, N.: Solvability of fuzzy relational equations employing undefined values. In: Proceedings of EUSFLAT 2019. Atlantis Studies in Uncertainty Modelling (ASUM), pp. 227–234. Atlantis Press (2019)

5. Cao, N., Štěpnička, M.: Fuzzy relation equations with fuzzy quantifiers. In: Kacprzyk, J., Szmidt, E., Zadrożny, S., Atanassov, K.T., Krawczak, M. (eds.) IWIFSGN/EUSFLAT 2017. AISC, vol. 641, pp. 354–367. Springer, Cham (2018). https://doi.org/10.1007/978-3-319-66830-7_32

6. Cao, N., Štěpnička, M.: Compositions of partial fuzzy relations employing the lower estimation approach. In: The 10th International Conference on Knowledge and Systems Engineering (KSE 2018), HCM, Vietnam, pp. 146–151. IEEE (2018)

7. Cao, N., Štěpnička, M.: Fuzzy relational equations employing Dragonfly operations. In: 2019 11th International Conference on Knowledge and Systems Engineering (KSE), pp. 127–132. IEEE (2019)

8. Chung, F., Lee, T.: A new look at solving a system of fuzzy relational equations. Fuzzy Sets Syst. **88**(3), 343–353 (1997)

9. Ciucci, D., Dubois, D.: A map of dependencies among three-valued logics. Inf. Sci. **250**, 162–177 (2013)

10. Cornejo, M.E., Lobo, D., Medina, J.: On the solvability of bipolar max-product fuzzy relation equations with the product negation. J. Comput. Appl. Math. **354**, 520–532 (2018)

11. Daňková, M.: Fuzzy relations and fuzzy functions in partial fuzzy set theory. In: Kacprzyk, J., Szmidt, E., Zadrożny, S., Atanassov, K.T., Krawczak, M. (eds.) IWIFSGN/EUSFLAT -2017. AISC, vol. 641, pp. 563–573. Springer, Cham (2018). https://doi.org/10.1007/978-3-319-66830-7_50

12. Díaz-Moreno, J.C., Medina, J., Turunen, E.: Minimal solutions of general fuzzy relation equations on linear carriers. An algebraic characterization. Fuzzy Sets Syst. **311**, 112–123 (2017)

13. Ghodousian, A., Babalhavaeji, A.: An efficient genetic algorithm for solving non-linear optimization problems defined with fuzzy relational equations and Max-Lukasiewicz composition. Appl. Soft Comput. **69**, 475–492 (2018)

14. Kurano, M., Yasuda, M., Nakagami, J., Yoshida, Y.: A fuzzy relational equation in dynamic fuzzy systems. Fuzzy Sets Syst. **101**(3), 439–443 (1999)

15. Lin, J.L., Wu, Y.K., Guu, S.M.: On fuzzy relational equations and the covering problem. Inf. Sci. **181**(14), 2951–2963 (2011)

16. Perfilieva, I.: Finitary solvability conditions for systems of fuzzy relation equations. Inf. Sci. **234**, 29–43 (2013). https://doi.org/10.1016/j.bbr.2011.03.031

17. Sanchez, E.: Resolution of composite fuzzy relation equations. Inf. Control **30**, 38–48 (1976)

18. Štěpnička, M., Cao, N., Běhounek, L., Burda, M., Dolný, A.: Missing values and Dragonfly operations in fuzzy relational compositions. Int. J. Approx. Reason. **113**, 149–170 (2019)

19. Štěpnička, M., Jayaram, B.: Interpolativity of at-least and at-most models of monotone fuzzy rule bases with multiple antecedent variables. Fuzzy Sets Syst. **297**, 26–45 (2016)

Polar Representation of Bipolar Information: A Case Study to Compare Intuitionistic Entropies

Christophe Marsala[(⊠)] and Bernadette Bouchon-Meunier

Sorbonne Université, CNRS, LIP6, 75005 Paris, France
{Christophe.Marsala,Bernadette.Bouchon-Meunier}@lip6.fr

Abstract. In this paper, a new approach to compare measures of entropy in the setting of the intuitionistic fuzzy sets introduced by Atanassov. A polar representation is introduced to represent such bipolar information and it is used to study the three main intuitionistic fuzzy sets entropies of the literature. A theoretical comparison and some experimental results highlight the interest of such a representation to gain knowledge on these entropies.

Keywords: Entropy · Intuitionistic fuzzy set · Bipolar information

1 Introduction

Measuring information is a very crucial task in Artificial intelligence. First of all, one main challenge is to define what is information, as it is done by Lotfi Zadeh [15] who considers different approaches to define information: the probabilistic approach, the possibilistic one, and their combination. In the literature, we can also cite the seminal work by J. Kampé de Fériet who introduced a new way to consider information and its aggregation [10,11].

In previous work, we have focused on the monotonicity of entropy measures and highlighted the fact that there exist several forms of monotonicity [4,5]. But highlighting that two measures share the same monotonicity property is often not sufficient in an application framework: to choose between two measures, their differences in behaviour are usually more informative.

In this paper, we do not focus on defining information but we discuss on the comparison of measures of information in the particular case of Intuitionistic Fuzzy Sets introduced by Atanassov (AIFS) [2] and related measures of entropy (simply called hereafter AIFS entropies) that have been introduced to measure intuitionistic fuzzy set-based information.

In this case, we highlight the fact that trying to interpret such a measure according to variations of the AIFS could not be clearly understandable. Instead, we propose to introduce a polar representation of AIFS in order to help the understanding of the behaviour of AIFS entropies. As a consequence, in a more general context, we argue that introducing a polar representation for bipolar

© Springer Nature Switzerland AG 2020
M.-J. Lesot et al. (Eds.): IPMU 2020, CCIS 1237, pp. 107–116, 2020.
https://doi.org/10.1007/978-3-030-50146-4_9

information could be a powerful way to improve the understandability of the behaviour for related measures.

This paper is organized as follows: in Sect. 2, we recall the basis of intuitionistic fuzzy sets and some known measures of entropy in this setting. In Sect. 3, we propose an approach to compare measures of entropy of intuitionistic fuzzy sets that is based on a polar representation of intuitionistic membership degrees. In Sect. 4, some experiments are presented that highlight the analytical conclusions drawn in the previous section. The last section concludes the paper and presents some future works.

2   Intuitionistic Fuzzy Sets and Entropies

First of all, in this section, some basic concepts related to intuitionistic fuzzy sets are presented. Afterwards, existing AIFS entropies are recalled.

2.1   Basic Notions

Let $U = \{u_1, \ldots, u_n\}$ be a universe, an *intuitionistic fuzzy set introduced by Atanassov* (AIFS) A of U is defined [2] as:

$$A = \{(u, \mu_A(u), \nu_A(u)) | u \in U\}$$

with $\mu_A : U \to [0,1]$ and $\nu_A : U \to [0,1]$ such that $0 \le \mu_A(u) + \nu_A(u) \le 1$, $\forall u \in U$. Here, $\mu_A(u)$ and $\nu_A(u)$ represent respectively the membership degree and the non-membership degree of u in A.

Given an intuitionistic fuzzy set A of U, the *intuitionistic index of u to A* is defined for all $u \in U$ as: $\pi_A(u) = 1 - (\mu_A(u) + \nu_A(u))$. This index represents the margin of hesitancy lying on the membership of u in A or the lack of knowledge on A. In [6], an AIFS A such that $\mu_A(u) = \nu_A(u) = 0, \forall u \in U$ is called *completely intuitionistic*.

2.2   Entropies of Intuitionistic Fuzzy Sets

Existing Entropies. In the literature, there exist several definitions of the entropy of an intuitionistic fuzzy set and several works proposed different ways to define such entropy, for instance from divergence measures [12]. In this paper, in order to illustrate the polar representation, we focus on three classical AIFS entropies.

In [13], the entropy of the AIFS A is defined as:

$$E_1(A) = 1 - \frac{1}{2n} \sum_{i=1}^{n} |\mu_A(u_i) - \nu_A(u_i)|,$$

where n is the cardinality of the considered universe.

Other definitions are introduced in [6] based on extensions of the Hamming distance and the Euclidean distance to intuitionistic fuzzy sets. For instance, the following entropy is proposed:

$$E_2(A) = \sum_{i=1}^{n} \pi_A(u_i) = n - \sum_{i=1}^{n} (\mu_A(u_i) + \nu_A(u_i)).$$

In [9], another entropy is introduced:

$$E_3(A) = \frac{1}{2n} \sum_{i=1}^{n} \left(1 - |\mu_A(u_i) - \nu_A(u_i)|\right)(1 + \pi_A(u_i)).$$

Definitions of Monotonicity. All AIFS entropies share a property of monotonicity, but authors don't agree about a unique definition of monotonicity.

Usually, monotonicity is defined according to the definition of a partial order $\leq$ on AIFS. Main definitions of monotonicity for entropies that have been proposed are based on different definition of the partial order In the following, we show the definitions of the partial order *less fuzzy* proposed by [6,9,13].

Let $E(A)$ be the entropy of the AIFS A. The following partial orders $(M1)$ or $(M2)$ can be used:

(M1) $E(A) \leq E(B)$, if A is *less fuzzy* than B.
    i.e. $\mu_A(u) \leq \mu_B(u)$ and $\nu_A(u) \geq \nu_B(u)$ when $\mu_B(u) \leq \nu_B(u)$, $\forall u \in U$, or $\mu_A(u) \geq \mu_B(u)$ and $\nu_A(u) \leq \nu_B(u)$ when $\mu_B(u) \geq \nu_B(u)$, $\forall u \in U$.

(M2) $E(A) \leq E(B)$ if $A \leq B$
    i.e. $\mu_A(u) \leq \mu_B(u)$ and $\nu_A(u) \leq \nu_B(u)$, $\forall u \in U$.

Each definition of the monotonicity produces the definition of a particular form of E:

- it is easy to show that E_1 satisfies *(M1)*;
- E_2 has been introduced by [6] to satisfy *(M2)*;
- E_3 has been defined by [9] from *(M1)*.

Indeed, these three AIFS entropies are different by definition as they are based on different definitions of monotonicity. However, if we want to choose the best entropy to use for a given application, it may not be so clear. A comparative study as those presented in Fig. 2 do not bring out much information about the way they are different. In the following section, we introduce a new approach to better highlight differences in the behaviour of these entropies.

3 Comparing AIFS Entropies

Usually, the study of measures, either entropies or other kinds of measures, is done by means of a given set of properties. In the previous section, we focused on the property of monotonicity showing that several definitions exist and could be used. Thus, if any measure could be built according to a given definition of monotonicity, at the end, this could not be very informative to understand clearly their differences in behaviour.

A first approach focuses on the study of the variations of an entropy according to the variations of each of its AIFS components μ and ν. However, this can only highlight the "horizontal" variations (when μ varies) or the "vertical" variations (when ν varies) and fails to enable a good understandability when *both* quantities vary.

In this paper, we introduce a polar representation of AIFS in order to be able to understand more clearly the dual influence of this bipolar information. Indeed, we show that the comparison of measures can be made easier with such a representation. We focus on AIFS, but we believe that such a study can also be useful for other bipolar information measures.

3.1   Polar Representation of an AIFS

In this part, we introduce a polar representation of an AIFS and we represent an AIFS as a complex number. In [3], this kind of representation is a way to show a geometric representation of an AIFS. In a more analytic way, such a representation could also be used to represent basic operations (intersection, union,...) on AIFS [1, 14], or for instance on the Pythagorean fuzzy sets [7].

Indeed, as each $u \in U$ is associated with two values $\mu_A(u)$ and $\nu_A(u)$, the membership of u to A can thus be represented as a point in a 2-dimensional space. In this sense, $\mu_A(u)$ and $\nu_A(u)$ represent the Cartesian coordinates of this point. We can then think of a complex number representation as we did in [5] or, equivalently, a representation of such a point by means of polar coordinates. In the following, we show that such a representation makes easier specific studies of these measures.

The AIFS A is defined for $u \in U$ as $\mu_A(u)$ and $\nu_A(u)$, that can be represented as the complex number $z_A(u) = \mu_A(u) + i\nu_A(u)$ (see Fig. 1). Thus, for this u, an AIFS is a point under (or on) the straight line $y = 1 - x$. When it belongs to the line $y = 1 - x$, it corresponds to the special case of a fuzzy set.

Another special case corresponds to the straight line $y = x$ that corresponds to AIFS such that $\mu_A(u) = \nu_A(u)$: AIFS above this line are such that $\mu_A(u) \leq \nu_A(u)$ and those under this line are such that $\mu_A(u) \geq \nu_A(u)$.

Hereafter, using classical notation from complex numbers, given $z_A(u)$, we note $\theta_A(u) = \arg(z_A(u))$ and $r_A(u) = |z_A(u)| = \sqrt{\mu_A(u)^2 + \nu_A(u)^2}$ (see Fig. 1).

The values $r_A(u)$ and $\theta_A(u)$ provide the polar representation of the AIFS $(u, \mu_A(u), \nu_A(u))$ for all $u \in U$. Following classical complex number theory, we have $\mu_A(u) = r_A(u) \cos \theta_A(u)$ and $\nu_A(u) = r_A(u) \sin \theta_A(u)$.

To alleviate the notations, in the following, when there is no ambiguity, $r_A(u)$ and $\theta_A(u)$ will be noted r and θ respectively.

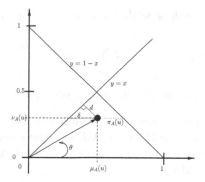

Fig. 1. Geometrical representation of an intuitionistic fuzzy set.

Using trigonometric identities, we have:

$$\mu_A(u) + \nu_A(u) = r(\cos\theta + \sin\theta)$$
$$= r\sqrt{2}\,\left(\frac{\sqrt{2}}{2}\cos\theta + \frac{\sqrt{2}}{2}\sin\theta\right)$$
$$= r\sqrt{2}\,\sin\!\left(\theta + \frac{\pi}{4}\right)$$

The intuitionistic index can thus be rewritten as $\pi_A(u) = 1 - r\sqrt{2}\sin(\theta + \frac{\pi}{4})$.

Moreover, let d be the distance from $(\mu_A(u), \nu_A(u))$ to the straight line $y = x$ and δ be the projection according to the U axis of $(\mu_A(u), \nu_A(u))$ on $y = x$ (ie. $\delta = |\mu_A(u) - \nu_A(u)|$).

It is easy to see that $d = r|\sin(\frac{\pi}{4} - \theta)|$ and $\delta = \frac{d}{\sin\frac{\pi}{4}}$. Thus, we have $\delta = \sqrt{2}\,r|\sin(\frac{\pi}{4} - \theta)|$.

In the following, for the sake of simplicity, when there is no ambiguity, $\mu_A(u_i)$, $\nu_A(u_i)$, $r_A(u_i)$ and $\theta_A(u_i)$ will be respectively noted μ_i, ν_i, r_i and θ_i.

3.2 Rewriting AIFS Entropies

With the notations introduced in the previous paragraph, entropy E_1 can be rewritten as:

$$E_1(A) = 1 - \frac{\sqrt{2}}{2n}\sum_{i=1}^{n} r_i|sin(\frac{\pi}{4} - \theta_i)|.$$

With this representation of the AIFS, it is easy to see that:

- if θ_i is given, E_1 decreases when r_i increases (ie. when the AIFS gets closer to $y = 1 - x$, and thus, when it tends to be a classical fuzzy set): the nearer an AIFS is from the straight line $y = 1 - x$, the lower its entropy.
- if r_i is given, E_1 increases when θ_i tends to $\frac{\pi}{4}$ (ie. when the knowledge on the non-membership decreases): the closer to $y = x$ it is, the higher its entropy.

A similar study can be done with E_2. In our setting, it can be rewritten as:

$$E_2(A) = n - \sqrt{2} \sum_{i=1}^{n} r_i \sin(\theta_i + \frac{\pi}{4}).$$

With this representation of the AIFS, we can see that:

- if θ_i is given, then E_2 decreases when r_i increases: the closer to $(0,0)$ (ie. the more "completely intuitionistic") the AIFS is, the lower its entropy.
- if r_i is given, E_2 decreases when θ_i increases: the farther an AIFS is from $y = x$, the higher its entropy.

A similar study can be done with E_3 that can be rewritten as:

$$E_3(A) = \frac{1}{2n} \sum_{i=1}^{n} \left(1 - r_i\sqrt{2}|\sin(\frac{\pi}{4} - \theta_i)|\right)\left(2 - r_i\sqrt{2}\sin(\frac{\pi}{4} + \theta_i)\right).$$

It is interesting to highlight here two elements of comparison: on one hand a behavioural difference between E_1 (resp. E_3) and E_2: if they vary similarly according to r_i, they vary in an opposite way according to θ_i; on the other hand, a similar behaviour between E_1 and E_3.

To illustrate these similarities and differences, a set of experiments have been conducted and results are provided in Sect. 4.

4   Experimental Study

In this section, we present some results related to experiments conducted to compare AIFS entropies.

4.1   Correlations Between AIFS Entropies

The first experiment has been conducted to see if some correlations could be highlighted between each of the three presented AIFS entropies.

First of all, an AIFS A is randomly generated. It is composed of n points, n also randomly generated and selected from 1 to $n_{max} = 20$. Afterwards, the AIFS entropy of A is valued for each of the three AIFS entropies presented in Sect. 2: E_1, E_2 and E_3. Then, a set of $n_{AIFS} = 5000$ such random AIFS is built.

A correlogram to highlight possible correlations between the values of $E_1(A)$, $E_2(A)$ and $E_3(A)$ is thus plot and presented in Fig. 2.

In this figure, each of these 9 spots (i, j) should be read as follows. The spot line i and column j corresponds to:

- if $i = j$: the distribution of the values of $E_i(A)$ for all A;
- if $i \neq j$: the distribution of $(E_j(A), E_i(A))$ for each random AIFS A.

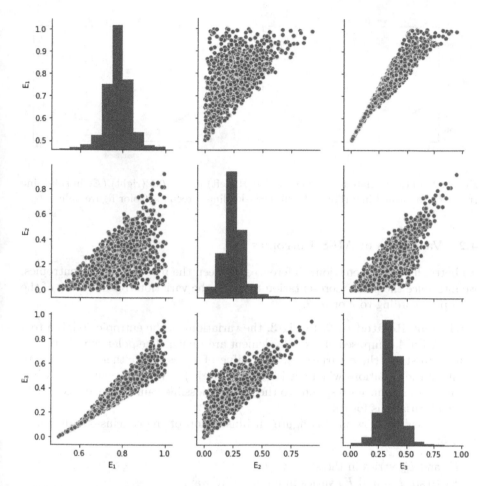

Fig. 2. Correlations between $E_1(A)$, $E_2(A)$ and $E_3(A)$ for 5000 random AIFS.

This process has been conducted several times, with different values for n_{max} and for n_{AIFS}, with similar results.

It is clear that there is no correlation between the entropies. It is noticeable that E_1 and E_2 could yield to very different values for the same AIFS A. For instance, if $E_1(A)$ equals 1, the value of $E_2(A)$ could be either close to 0 or equals to 1 too.

As a consequence, it is clear that these entropies are highly different but no conclusion can be drawn about the elements that bring out this difference.

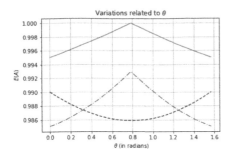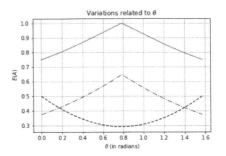

Fig. 3. Variations related to θ when $r = 0.01$ (left) or $r = 0.5$ (right) (E_1 in solid line (red), E_2 in dashed line (blue), E_3 in dash-dot line (green)). (Color figure online)

4.2   Variations of AIFS Entropies

To better highlight behaviour differences between the presented AIFS entropies, we introduce the polar representation to study the variations of the values of the entropy according to r or to θ.

Variations Related to θ. In Fig. 3, the variations of the entropies related to θ for an AIFS A composed of a single element are shown. The polar representation is used to study these variations. The value of r is set to either 0.01 (left) to highlight the variations when r is low, or 0.5 (right) to highlight variations when r is high (this value corresponds to the highest possible value to have a complete range of variations for θ).

It is easy to show in this figure an illustration of the conclusions drawn in Sect. 3:

- E_1 and E_3 varies in the same way;
- E_1 (resp. E_3) and E_2 varies in an opposite way;
- all of these AIFS entropies reach an optimum for $\theta = \frac{\pi}{4}$. It is a maximum for E_1 and E_3 and a minimum for E_2.
- the optimum is always 1 for E_1 for any r, but it depends on r for E_2 and E_3.
- for all entropies, the value when $\theta = 0$ (resp. $\theta = \frac{\pi}{2}$) depends on r.

Variations Related to r. In Fig. 4, the variations of the entropies related to r for an AIFS A composed of a single element are shown. Here again, the polar representation is used to study these variations.

According to the polar representations of the AIFS entropies, it is easy to see that E_1 and E_2 vary linearly with r, and in a quadratic form for E_3.

We provided here the variations when $\theta = 0$ (ie. the AIFS is on the horizontal axis), $\theta = \frac{\pi}{8}$ (ie. the AIFS is under $y = x$), $\theta = \frac{\pi}{4}$ (ie. the AIFS is on $y = x$) and when $\theta = \frac{\pi}{2}$ (ie. the AIFS is on the vertical axis). We don't provide results when the AIFS is below $y = x$ as it is similar to the results when the AIFS is under with a symmetry related to $y = x$ (as it can be seen with the variations when $\theta = \frac{\pi}{2}$ which are similar to the ones when $\theta = 0$).

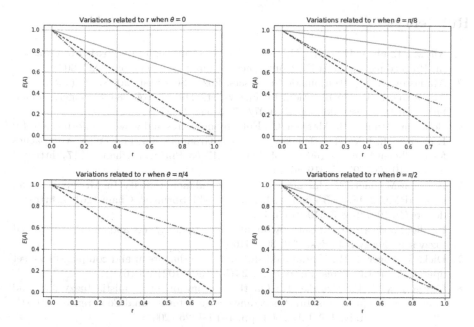

Fig. 4. Variations related to r when θ varies (E_1 in solid line (red), E_2 in dashed line (blue), E_3 in dash-dot line (green)). (Color figure online)

For each experiment, r varies from 0 to $r = (\sqrt{2}\cos(\frac{\pi}{4} - \theta))^{-1}$ when $\theta \neq \frac{\pi}{2}$.

It is easy to highlight from these results some interesting behaviour of the AIFS entropies when r varies:

- they all varies in the same way but not with the same amplitude;
- all of these AIFS entropies reach an optimum when $r = 0$ (ie. the AIFS is completely intuitionistic;
- E_1 takes the optimum value for any r when $\theta = \frac{\pi}{4}$.

5   Conclusion

In this paper, we introduce a new approach to compare measures of entropy in the setting of intuitionistic fuzzy sets. We introduce the use of a polar representation to study the three main AIFS entropies of the literature.

This approach is very promising as it enables us to highlight the main differences in behaviour that can exist between measures. Beyond this study on the AIFS, such a polar representation could thus be an interesting way to study bipolar information-based measures.

In future work, our aim is to develop this approach and apply it to other AIFS entropies, for instance [8], and other bipolar representations of information.

References

1. Ali, M., Tamir, D.E., Rishe, N.D., Kandel, A.: Complex intuitionistic fuzzy classes. In: Proceedings of the IEEE Conference on Fuzzy Systems, pp. 2027–2034 (2016)
2. Atanassov, K.T.: Intuitionistic fuzzy sets. Fuzzy Sets Syst. **20**, 87–96 (1986)
3. Atanassov, K.T.: On Intuitionistic Fuzzy Sets Theory, vol. 283. Springer, Heidelberg (2012). https://doi.org/10.1007/978-3-642-29127-2
4. Bouchon-Meunier, B., Marsala, C.: Entropy measures and views of information. In: Kacprzyk, J., Filev, D., Beliakov, G. (eds.) Granular, Soft and Fuzzy Approaches for Intelligent Systems. SFSC, vol. 344, pp. 47–63. Springer, Cham (2017). https://doi.org/10.1007/978-3-319-40314-4_3
5. Bouchon-Meunier, B., Marsala, C.: Entropy and monotonicity. In: Medina, J., et al. (eds.) IPMU 2018. CCIS, vol. 854, pp. 332–343. Springer, Cham (2018). https://doi.org/10.1007/978-3-319-91476-3_28
6. Burillo, P., Bustince, H.: Entropy on intuitionistic fuzzy sets and on interval-valued fuzzy sets. Fuzzy Sets Syst. **78**, 305–316 (1996)
7. Dick, S., Yager, R.R., Yazdanbakhsh, O.: On Pythagorean and complex fuzzy set operations. IEEE Trans. Fuzzy Syst. **24**(5), 1009–1021 (2016)
8. Grzegorzewski, P., Mrowka, E.: On the entropy of intuitionistic fuzzy sets and interval-valued fuzzy sets. In: Proceedings of the International Conference on IPMU 2004, Perugia, Italy, 4–9 July 2004, pp. 1419–1426 (2004)
9. Guo, K., Song, Q.: On the entropy for Atanassov's intuitionistic fuzzy sets: an interpretation from the perspective of amount of knowledge. Appl. Soft Comput. **24**, 328–340 (2014)
10. Kampé de Fériet, J.: Mesures de l'information par un ensemble d'observateurs. In: Gauthier-Villars (ed.) Comptes Rendus des Séances de l'Académie des Sciences, volume 269 of série A, Paris, pp. 1081–1085, Décembre 1969
11. Kampé de Fériet, J.: Mesure de l'information fournie par un événement. Séminaire sur les questionnaires (1971)
12. Montes, I., Pal, N.R., Montes, S.: Entropy measures for Atanassov intuitionistic fuzzy sets based on divergence. Soft. Comput. **22**(15), 5051–5071 (2018). https://doi.org/10.1007/s00500-018-3318-3
13. Szmidt, E., Kacprzyk, J.: New measures of entropy for intuitionistic fuzzy sets. In Proceedings of the Ninth International Conference on Intuitionistic Fuzzy Sets (NIFS), vol. 11, Sofia, Bulgaria, pp. 12–20, May 2005
14. Tamir, D.E., Ali, M., Rishe, N.D., Kandel, A.: Complex number representation of intuitionistic fuzzy sets. In: Proceedings of the World Conference on Soft Computing, Berkeley, CA, USA (2016)
15. Zadeh, L.A.: The information principle. Inf. Sci. **294**, 540–549 (2015)

Decision Making, Preferences and Votes

Generalized Weak Transitivity
of Preference

Thomas A. Runkler[✉]

Siemens AG, Corporate Technology, Otto–Hahn–Ring 6, 81739 Munich, Germany
thomas.runkler@siemens.com

Abstract. Decision making processes are often based on (pairwise) preference relations. An important property of preference relations is transitivity. Many types of transitivity have been proposed in the literature, such as max–min and max–max transitivity, restricted max–min and max–max transitivity, additive and multiplicative transitivity, or Łukasiewicz transitivity. This paper focuses on weak transitivity. Weak transitivity has been defined for additive preference relations. We extend this definition to multiplicative preference relations and further introduce a generalized version called generalized weak transitivity. We show that for reciprocal additive and multiplicative preference relations weak transitivity is equivalent to generalized weak transitivity, and we also illustrate generalized weak transitivity for preference relations that are neither additive nor multiplicative. Finally, we show how a total order (ranking of the options) can be constructed for any generalized weak transitive preference relation.

Keywords: Preference relations · Weak transitivity · Decision making

1 Introduction

Decision making processes are often based on preference relations [2,6,7,9,14, 17,18]. Given a set of n options, a (pairwise) preference relation is specified by an $n \times n$ preference matrix

$$P = \begin{pmatrix} p_{11} & \cdots & p_{1n} \\ \vdots & \ddots & \vdots \\ p_{n1} & \cdots & p_{nn} \end{pmatrix} \qquad (1)$$

where each matrix element $p_{ij} \geq 0$ quantifies the degree of preference of option i over option j, where $i, j = 1, \ldots, n$. We distinguish additive (or fuzzy) preference [21] and multiplicative preference [16]. An important property of preference relations is transitivity [8,11]. Given three options i, j, and k, transitivity specifies the relation between the preference of i over j, j over k, and i over k. Many types of transitivity have been defined for preference relations, such as max–min and max–max transitivity [4,22], restricted max–min and max–max transitivity [18], additive and multiplicative transitivity [18], or Łukasiewicz transitivity [5,13].

© Springer Nature Switzerland AG 2020
M.-J. Lesot et al. (Eds.): IPMU 2020, CCIS 1237, pp. 119–128, 2020.
https://doi.org/10.1007/978-3-030-50146-4_10

This paper focuses on weak transitivity [19]. We extend the definition of weak transitivity for additive preference relations to multiplicative preference relations (Definition 2). We derive an equivalent formulation of weak transitivity for reciprocal additive and multiplicative preference relations and use this to define a more general version of weak transitivity called generalized weak transitivity (Definition 3). We show that for reciprocal additive and multiplicative preference relations weak transitivity is equivalent to generalized weak transitivity (Theorems 1 and 2), and we also illustrate generalized weak transitivity for preference relations that are neither of additive nor multiplicative type. Finally, we show how a total order (ranking of the options) can be constructed for any generalized weak transitive preference relation (Theorem 3).

This paper is structured as follows: Sect. 2 briefly reviews additive preference relations and weak transitivity. Section 3 moves on to multiplicative preference relations and introduces a definition of weak transitivity for multiplicative preference relations. Section 4 develops a joint formula of weak transitivity for reciprocal additive and multiplicative preference relations and uses this to define generalized weak transitivity. Section 5 shows how a total order of elements can be constructed from any generalized weakly transitive preference relation. Finally, Sect. 6 summarizes our conclusions and outlines some directions for future research.

2   Weakly Transitive Additive Preference

Consider a preference matrix P. We call P an *additive* preference matrix if and only if the following two conditions hold:

1. $p_{ij} \in [0, 1]$ for all $i, j = 1, \ldots, n$, and
2. $p_{ij} = 0.5$ if and only if the options i and j are equivalent.

This implies that all elements on the main diagonal of P are

$$p_{ii} = 0.5 \tag{2}$$

for all $i = 1, \ldots, n$. We call 0.5 the *neutral* additive preference. An additive preference matrix P is called *reciprocal* if and only if

$$p_{ij} + p_{ji} = 1 \tag{3}$$

for all $i, j = 1, \ldots, n$. This always holds for $i = j$ because of (2), so it is sufficient to check this condition only for $i \neq j$. Several different types of transitivity have been defined for preference relations. In this paper we consider weak transitivity as defined by Tanino [19]:

Definition 1 (weakly transitive additive preference). *An $n \times n$ additive preference matrix P is called* weakly transitive *if and only if*

$$p_{ij} \geq 0.5 \wedge \quad p_{jk} \geq 0.5 \Rightarrow p_{ik} \geq 0.5 \tag{4}$$

for all $i, j, k = 1, \ldots, n$,

This can be interpreted as follows: if i is preferred over j and j is preferred over k, then i is preferred over k, where in all cases preference is not less than neutral. For reciprocal additive preference relations it is sufficient for weak transitivity to check (4) only for $i \neq j$, $j \neq k$, and $i \neq k$, because all other cases are trivial.

Consider for example the preference matrix

$$P^A = \begin{pmatrix} 0.5 & 0.7 & 0.9 \\ 0.3 & 0.5 & 0.8 \\ 0.1 & 0.2 & 0.5 \end{pmatrix} \tag{5}$$

All elements of P^A are in the unit interval and all elements on the main diagonal are equal to the neutral additive preference 0.5, so P^A is an additive preference matrix. For the sums of preferences and reverse preferences we obtain

$$p_{12}^A + p_{21}^A = 0.7 + 0.3 = 1 \tag{6}$$
$$p_{13}^A + p_{31}^A = 0.9 + 0.1 = 1 \tag{7}$$
$$p_{23}^A + p_{32}^A = 0.8 + 0.2 = 1 \tag{8}$$

so P^A is reciprocal. For the preferences not smaller than neutral we obtain

$$p_{12}^A \geq 0.5, \; p_{23}^A \geq 0.5, \; p_{13}^A \geq 0.5 \tag{9}$$

so (4) holds and P^A is weakly transitive. To summarize, P^A is a weakly transitive reciprocal additive preference matrix.

Next consider the preference matrix

$$P^B = \begin{pmatrix} 0.5 & 0.2 & 0.7 \\ 0.8 & 0.5 & 0.4 \\ 0.3 & 0.6 & 0.5 \end{pmatrix} \tag{10}$$

All elements of P^B are in the unit interval and all elements on the main diagonal are 0.5, so P^B is an additive preference matrix. The sums of preferences and reverse preferences are

$$p_{12}^B + p_{21}^B = 0.2 + 0.8 = 1 \tag{11}$$
$$p_{13}^B + p_{31}^B = 0.7 + 0.3 = 1 \tag{12}$$
$$p_{23}^B + p_{32}^B = 0.4 + 0.6 = 1 \tag{13}$$

so P^B is reciprocal. However, for the preferences not smaller than neutral we obtain

$$p_{13}^B \geq 0.5, \; p_{32}^B \geq 0.5, \; p_{12}^B \not\geq 0.5 \tag{14}$$

so (4) does not hold and P^B is not weakly transitive. Hence, P^B is a not weakly transitive reciprocal additive preference matrix.

3 Weakly Transitive Multiplicative Preference

Consider again a preference matrix P. We call P a *multiplicative* preference matrix if $p_{ij} = 1$ if and only if the options i and j are equivalent. This implies that all elements on the main diagonal of P are

$$p_{ii} = 1 \tag{15}$$

for all $i = 1, \ldots, n$. We call 1 the *neutral* multiplicative preference. A multiplicative preference matrix P is called *reciprocal* if and only if

$$p_{ij} \cdot p_{ji} = 1 \tag{16}$$

for all $i, j = 1, \ldots, n$. This always holds for $i = j$ because of (15), so it is sufficient to check this condition only for $i \neq j$. We modify the condition (4) for weak transitivity of additive preference relations by replacing the neutral additive preference 0.5 by the neutral multiplicative preference 1, and obtain

Definition 2 (weakly transitive multiplicative preference). *An $n \times n$ multiplicative preference matrix P is called* weakly transitive *if and only if*

$$p_{ij} \geq 1 \land \quad p_{jk} \geq 1 \Rightarrow \quad p_{ik} \geq 1 \tag{17}$$

for all $i, j, k = 1, \ldots, n$.

For reciprocal multiplicative preference relations it is sufficient for weak transitivity to check (17) only for $i \neq j$, $j \neq k$, and $i \neq k$, because (just as for additive preference relations) all other cases are trivial.

Now consider the preference matrix

$$P^C = \begin{pmatrix} 1 & 2 & 4 \\ 1/2 & 1 & 3 \\ 1/4 & 1/3 & 1 \end{pmatrix} \tag{18}$$

All elements on the main diagonal of P^C are one, so P^C is a multiplicative preference matrix. For the products of preferences and reverse preferences we obtain

$$p_{12}^C \cdot p_{21}^C = 2 \cdot 1/2 = 1 \tag{19}$$
$$p_{13}^C \cdot p_{31}^C = 4 \cdot 1/4 = 1 \tag{20}$$
$$p_{23}^C \cdot p_{32}^C = 3 \cdot 1/3 = 1 \tag{21}$$

so P^C is reciprocal. For the preferences not smaller than neutral (≥ 1 for multiplicative preferences) we obtain

$$p_{12}^C \geq 1, \, p_{23}^C \geq 1, \, p_{13}^C \geq 1 \tag{22}$$

so (17) holds and P^C is weakly transitive. To summarize, P^C is a weakly transitive reciprocal multiplicative preference matrix.

As another example consider

$$P^D = \begin{pmatrix} 1 & 1/3 & 4 \\ 3 & 1 & 1/2 \\ 1/4 & 2 & 1 \end{pmatrix} \tag{23}$$

which has ones on the main diagonal, so P^D is a multiplicative preference matrix. The products of preferences and reverse preferences are

$$p_{12}^D \cdot p_{21}^D = 1/3 \cdot 3 = 1 \tag{24}$$
$$p_{13}^D \cdot p_{31}^D = 4 \cdot 1/4 = 1 \tag{25}$$
$$p_{23}^D \cdot p_{32}^D = 1/2 \cdot 2 = 1 \tag{26}$$

so P^D is reciprocal. However, the preferences not smaller than neutral are

$$p_{13}^D \geq 1, \; p_{32}^D \geq 1, \; p_{12}^D \not\geq 1 \tag{27}$$

so (17) is violated and P^D is not weakly transitive. To summarize, P^D is a not weakly transitive reciprocal multiplicative preference matrix.

4   Generalized Weak Transitivity

Let us revisit additive preference. An additive preference matrix P is reciprocal if and only if (3)

$$p_{ij} + p_{ji} = 1$$

for all $i, j = 1, \ldots, n$, which implies

$$p_{ij} \geq 0.5 \Rightarrow p_{ji} = 1 - p_{ij} \leq 0.5, \Rightarrow p_{ij} \geq p_{ji} \tag{28}$$

and

$$p_{ij} \geq p_{ji} \Rightarrow p_{ij} \geq 1 - p_{ij} \Rightarrow 2p_{ij} \geq 1 \Rightarrow p_{ij} \geq 0.5 \tag{29}$$

and so

$$p_{ij} \geq 0.5 \Leftrightarrow p_{ij} \geq p_{ji} \tag{30}$$

For reciprocal additive preference relations we can therefore rewrite the condition for weak transitivity (4)

$$p_{ij} \geq 0.5 \wedge p_{jk} \geq 0.5 \Rightarrow p_{ik} \geq 0.5$$

to the equivalent condition

$$p_{ij} \geq p_{ji} \wedge p_{jk} \geq p_{kj} \Rightarrow p_{ik} \geq p_{ki} \tag{31}$$

for all $i, j, k = 1, \ldots, n$.

Now let us revisit multiplicative preference in the same way. A multiplicative preference matrix P is reciprocal if and only if (16)

$$p_{ij} \cdot p_{ji} = 1$$

for all $i, j = 1, \ldots, n$, which implies

$$p_{ij} \geq 1 \Rightarrow p_{ji} = 1/p_{ij} \leq 1 \Rightarrow p_{ij} \geq p_{ji} \tag{32}$$

and

$$p_{ij} \geq p_{ji} \Rightarrow p_{ij} \geq 1/p_{ij} \Rightarrow p_{ij}^2 \geq 1 \Rightarrow p_{ij} \geq 1 \tag{33}$$

and so

$$p_{ij} \geq 1 \Leftrightarrow p_{ij} \geq p_{ji} \tag{34}$$

For reciprocal multiplicative preference relations we can therefore rewrite the condition for weak transitivity (17)

$$p_{ij} \geq 1 \wedge p_{jk} \geq 1 \Rightarrow p_{ik} \geq 1$$

to the equivalent condition

$$p_{ij} \geq p_{ji} \wedge p_{jk} \geq p_{kj} \Rightarrow p_{ik} \geq p_{ki} \tag{35}$$

for all $i, j, k = 1, \ldots, n$, which is equivalent to the condition (31) that we obtained for additive preference. So, for reciprocal additive preference and for reciprocal multiplicative preference we obtain the same condition (31) = (35) for weak transitivity. This leads us to

Definition 3 (generalized weakly transitive preference). *An $n \times n$ preference matrix P is called* generalized weakly transitive *if and only if*

$$p_{ij} \geq p_{ji} \wedge p_{jk} \geq p_{kj} \Rightarrow p_{ik} \geq p_{ki} \tag{36}$$

for all $i, j, k = 1, \ldots, n$, $i \neq j$, $j \neq k$, $i \neq k$.

Notice that we exclude the cases $i = j$, $j = k$, and $i = k$ here, because these are trivial for reciprocal additive and multiplicative preference relations, and so we also want to exclude them for arbitrary preference relations. This means that elements on the main diagonal of P are irrelevant for generalized weak transitivity.

It is easy to check that P_A (5) and P_C (18) satisfy generalized weak transitivity, and P_B (10) and P_D (23) do not. In general, the following two theorems relate weakly transitive additive and multiplicative preference to generalized weakly transitive preference.

Theorem 1 (additive generalized weak transitivity). *A reciprocal additive preference matrix is weakly transitive if and only if it is generalized weakly transitive.*

Theorem 2 (multiplicative generalized weak transitivity). *A reciprocal multiplicative preference matrix is weakly transitive if and only if it is generalized weakly transitive.*

Proof. The proof for both Theorems follows immediately from the equivalence of Eqs. (31), (35), and (36). □

So, for reciprocal additive and multiplicative preference matrices weak transitivity and generalized weak transitivity are equivalent. Therefore, let us now look at preference matrices which are neither additive nor multiplicative. Consider for example the preference matrix

$$P^E = \begin{pmatrix} 1 \ 6 \ 9 \\ 2 \ 4 \ 8 \\ 5 \ 3 \ 7 \end{pmatrix} \qquad (37)$$

The elements on the main diagonal of P^E are different from 0.5 and 1, so P^E is neither an additive preference matrix nor a multiplicative preference matrix. However, it may seem reasonable to interpret P^E as a preference matrix. Imagine for example that the rows and columns of this matrix correspond to soccer teams, and each element p_{ij}^E corresponds to the number of times that team i has won over team j. The elements on the main diagonal are chosen arbitrarily and may be ignored for generalized weak transitivity, as pointed out above. For the off–diagonal preference pairs we have

$$p_{12}^E \geq p_{21}^E, \ p_{23}^E \geq p_{32}^E, \ p_{13}^E \geq p_{31}^E \qquad (38)$$

so (36) holds and therefore P^E is generalized weakly transitive.

As another example consider

$$P^F = \begin{pmatrix} 1 \ 5 \ 8 \\ 9 \ 4 \ 2 \\ 3 \ 6 \ 7 \end{pmatrix} \qquad (39)$$

Again, the main diagonal is different from 0.5 and 1, so this is neither an additive preference matrix nor a multiplicative preference matrix. For the off–diagonal preference pairs of P_F we obtain

$$p_{13}^F \geq p_{31}^F, \ p_{32}^F \geq p_{23}^F, \ p_{12}^F \not\geq p_{21}^F \qquad (40)$$

so (36) is violated and therefore P^F is not generalized weakly transitive.

5   Total Order Induced by Generalized Weak Transitivity

Generalized weak transitivity is an important property of preference relations because it allows to construct a total order (ranking) of the elements.

Theorem 3 (total order for generalized weak transitivity). *If an $n \times n$ preference matrix is generalized weakly transitive, then we can construct a total order o of the n elements so that $o_1 \geq o_2 \geq \ldots \geq o_n$.*

Proof. If a preference matrix P is generalized weakly transitive, then from (36) follows that for each $i, j, k = 1, \ldots, n$ with $i \neq j$, $j \neq k$, $i \neq k$. we will have at least one of the following six cases:

$$p_{ij} \geq p_{ji}, p_{jk} \geq p_{kj}, p_{ik} \geq p_{ki} \Rightarrow o_i \geq o_j \geq o_k \tag{41}$$

$$p_{ik} \geq p_{ki}, p_{kj} \geq p_{jk} \, p_{ij} \geq p_{ji} \Rightarrow o_i \geq o_k \geq o_j \tag{42}$$

$$p_{ji} \geq p_{ij}, p_{ik} \geq p_{ki}, p_{jk} \geq p_{kj} \Rightarrow o_j \geq o_i \geq o_k \tag{43}$$

$$p_{jk} \geq p_{kj}, p_{ki} \geq p_{ii}, p_{ji} \geq p_{ij} \Rightarrow o_j \geq o_k \geq o_i \tag{44}$$

$$p_{ki} \geq p_{ik}, p_{ij} \geq p_{ji}, p_{kj} \geq p_{jk} \Rightarrow o_k \geq o_i \geq o_j \tag{45}$$

$$p_{kj} \geq p_{jk}, p_{ji} \geq p_{ij}, p_{ki} \geq p_{ik} \Rightarrow o_k \geq o_j \geq o_i \tag{46}$$

If more than one of these cases is satisfied, then we have ties $p_{ij} = p_{ji}$ or $p_{jk} = p_{kj}$ or $p_{ik} = p_{ki}$, which implies $o_i = o_j$ or $o_j = o_k$ or $o_i = o_k$, respectively, so the order may not be strict. For an arbitrary generalized weakly transitive preference matrix we can therefore pick arbitrary three elements i, j, k and construct a total order for these. If we have a total order for $p \geq 3$ elements $i_1, i_2, \ldots, i_p$, then for any additional element k from (36) follows that we will have one at least of the following three cases

$$p_{ki_1} \geq p_{i_1 k}, p_{i_1 i_2} \geq p_{i_2 i_1}, p_{ki_2} \geq p_{i_2 k} \Rightarrow o_k \geq o_{i_1} \geq o_{i_2} \tag{47}$$

$$p_{i_{p-1} i_p} \geq p_{i_p i_{p-1}}, p_{i_p k} \geq p_{ki_p}, p_{i_{p-1} k} \geq p_{ki_{p-1}} \Rightarrow o_{i_{p-1}} \geq o_{i_p} \geq o_k \tag{48}$$

or we can find an index $j \in \{1, 2, \ldots, p-1\}$ for which

$$p_{i_j k} \geq p_{ki_j}, p_{ki_{j+1}} \geq p_{i_{j+1} k}, p_{i_j i_{j+1}} \geq p_{i_{j+1} i_j} \Rightarrow o_{i_j} \geq o_k \geq o_{i_{j+1}} \tag{49}$$

In the first case, we can insert element k before element i_1 and obtain the new total order $k, i_1, i_2, \ldots, i_p$. In the second case, we can insert element k after element i_p and obtain the new total order $i_1, i_2, \ldots, i_p, k$. And in the third case we can insert element k between elements i_j and i_{j+1} and obtain the new total order $i_1 \ldots, i_j, k, i_{j+1} \ldots, i_p$. Again, several of these three cases may be satisfied if we have ties $p_{ki_1} = p_{i_1 k}$ or $p_{ki_{p-1}} = p_{i_{p-1} k}$ or $p_{ki_j} = p_{i_j k}$ or $p_{ki_{j+1}} = p_{i_{j+1} k}$, which lead to equal ranks $o_k = o_{i_1}$ or $o_k = o_{i_{p-1}}$ or $o_k = o_{i_j}$ or $o_k = o_{i_{j+1}}$, respectively, so the order may be not strict. Using this scheme we can construct a total order for any generalized weakly transitive preference matrix. □

For example, the preference matrix P^E (37) is generalized weakly transitive, so we have the relation (38) for the off–diagonal preference pairs which yields the total order $o_1 \geq o_2 \geq o_3$. In the semantic context of the application we can interpret this as a ranking of soccer teams, where team 1 is ranked on top, then team 2, and finally team 3. Notice again that the resulting total order may be not strict, so the ranking can contain ties. As another example, the preference matrix P^F (39) is *not* generalized weakly transitive, so the relation (40) for the off–diagonal preference pairs does *not* yield a total order, since $o_1 \geq o_3 \geq o_2$ contradicts $o_1 \not\geq o_2$. Here, it is not possible to construct a ranking of the soccer teams.

6   Conclusions

We have extended the concept of weak transitivity from additive preference relations to multiplicative preference relations and further in a generalized version to arbitrary preference relations. We have shown that generalized weak transitivity is an important property of preference relations because it allows to construct a strict order of elements (ranking of options) which is very useful in decision making processes.

Many questions have been left open for future research, such as:

- What are efficient algorithms to test a preference matrix for generalized weak transitivity and to construct the corresponding total order?
- How does generalized weak transitivity relate to other mathematical properties of preference relations such as monotonicity or positivity?
- How can weakly transitive preference relations be constructed from utilities [10] or from rank orders [12]?
- How do these types of transitivity extend to interval valued preferences [1,3, 15,20]?

References

1. Bilgiç, T.: Interval-valued preference structures. Eur. J. Oper. Res. **105**(1), 162–183 (1998)
2. Brans, J.-P., Vincke, P., Mareschal, B.: How to select and how to rank projects: the PROMETHEE method. Eur. J. Oper. Res. **24**(2), 228–238 (1986)
3. Cavallo, B., Brunelli, M.: A general unified framework for interval pairwise comparison matrices. Int. J. Approximate Reasoning **93**, 178–198 (2018)
4. Dubois, D., Prade, H.: Fuzzy Sets and Systems. Academic Press, London (1980)
5. Duddy, C., Piggins, A.: On some oligarchy results when social preference is fuzzy. Soc. Choice Welf. **51**(4), 717–735 (2018). https://doi.org/10.1007/s00355-018-1134-4
6. Fürnkranz, J., Hüllermeier, E.: Preference learning. In: Sammut, C., Webb, G.I. (eds.) Encyclopedia of Machine Learning, pp. 789–795. Springer, Boston (2010). https://doi.org/10.1007/978-0-387-30164-8_662
7. Herrera, F., Herrera-Viedma, E., Chiclana, F.: Multiperson decision-making based on multiplicative preference relations. Eur. J. Oper. Res. **129**(2), 372–385 (2001)
8. Herrera-Viedma, E., Herrera, F., Chiclana, F., Luque, M.: Some issues on consistency of fuzzy preference relations. Eur. J. Oper. Res. **154**(1), 98–109 (2004)
9. Orlovsky, S.A.: Decision-making with a fuzzy preference relation. Fuzzy Sets Syst. **1**(3), 155–167 (1978)
10. Runkler, T.A.: Constructing preference relations from utilities and vice versa. In: Carvalho, J.P., Lesot, M.-J., Kaymak, U., Vieira, S., Bouchon-Meunier, B., Yager, R.R. (eds.) IPMU 2016. CCIS, vol. 611, pp. 547–558. Springer, Cham (2016). https://doi.org/10.1007/978-3-319-40581-0_44
11. Runkler, T.A.: Mapping utilities to transitive preferences. In: Medina, J., et al. (eds.) IPMU 2018. CCIS, vol. 853, pp. 127–139. Springer, Cham (2018). https://doi.org/10.1007/978-3-319-91473-2_11

12. Runkler, T.A.: Canonical fuzzy preference relations. In: Kearfott, R.B., Batyrshin, I., Reformat, M., Ceberio, M., Kreinovich, V. (eds.) IFSA/NAFIPS 2019 2019. AISC, vol. 1000, pp. 542–555. Springer, Cham (2019). https://doi.org/10.1007/978-3-030-21920-8_48

13. Runkler, T.A.: Generating preference relation matrices from utility vectors using Łukasiewicz transitivity. In: Kóczy, L.T., Medina-Moreno, J., Ramírez-Poussa, E., Šostak, A. (eds.) Computational Intelligence and Mathematics for Tackling Complex Problems. SCI, vol. 819, pp. 123–130. Springer, Cham (2020). https://doi.org/10.1007/978-3-030-16024-1_16

14. Runkler, T.A., Chen, C., Coupland, S., John, R.: Just-in-time supply chain management using interval type-2 fuzzy decision making. In: IEEE International Conference on Fuzzy Systems, New Orleans, Louisiana, USA, pp. 1149–1154, June 2019

15. Runkler, T.A., Coupland, S., John, R.: Interval type-2 fuzzy decision making. Int. J. Approximate Reasoning 80, 217–224 (2017)

16. Saaty, T.L.: Analytic hierarchy process. In: Gass, S.I., Fu, M.C. (eds.) Encyclopedia of Operations Research and Management Science, pp. 52–64. Springer, Boston (2013). https://doi.org/10.1007/978-1-4419-1153-7_31

17. Sousa, J.M., Palm, R.H., Silva, C.A., Runkler, T.A.: Optimizing logistic processes using a fuzzy decision making approach. IEEE Trans. Syst. Man Cybern. A 33(2), 245–256 (2003)

18. Tanino, T.: Fuzzy preference orderings in group decision making. Fuzzy Sets Syst. 12(2), 117–131 (1984)

19. Tanino, T.: Fuzzy preference relations in group decision making. In: Kacprzyk, J., Roubens, M. (eds.) Non-Conventional Preference Relations in Decision Making. Lecture Notes in Economics and Mathematical Systems, vol. 301, pp. 54–71. Springer, Heidelberg (1988). https://doi.org/10.1007/978-3-642-51711-2_4

20. Türkşen, İ.B., Bilgiç, T.: Interval valued strict preference with Zadeh triples. Fuzzy Sets Syst. 78(2), 183–195 (1996)

21. Zadeh, L.A.: Similarity relations and fuzzy orderings. Inf. Sci. 3(2), 177–200 (1971)

22. Zimmermann, H.J.: Fuzzy Set Theory and Its Applications. Kluwer Academic Publishers, Boston (1985)

Investigation of Ranking Methods Within the Military Value of Information (VoI) Problem Domain

Behrooz Etesamipour[(⊠)] [iD] and Robert J. Hammell II

Department of Computer and Information Sciences, Towson University,
Towson, MD, USA
{betesamipour, rhammell}@towson.edu

Abstract. Determining the relative importance among vast amounts of individual pieces of information is a challenge in the military environment. By aggregating various military intelligence experts' knowledge, decision support tools can be created. A next step in the continuing research in this area is to investigate the use of three prominent ranking methods for aggregating opinions of military intelligence analysts with respect to the Value of Information (VoI) problem domain. This paper offers discussion about ongoing VoI research and demonstrates outcomes from a military-related experiment using Borda count, Condorcet voting, and Instant-runoff voting (IRV) methods as ranking aggregation models. These ranking methods are compared to the "ground truth" as generated by the current fuzzy-based VoI prototype system. The results by incorporating the ranking models on the experiment's data demonstrate the efficacy of these methods in aggregating Subject Matter Expert (SME) opinions and clearly demonstrate the "wisdom of the crowd" effect. Implications related to ongoing VoI research are discussed along with future research plans.

Keywords: Value of Information · Decision support · Information aggregation · Borda count · Condorcet voting · Instant-runoff voting · Rank aggregation

1 Introduction

The enormous volume of data generated everyday by computer systems and internet activities around the world cannot be easily processed and prioritized. It is an overwhelming challenge to analyze all pieces of data. The concept of Big Data Analytics introduces new challenges with its characteristics of enormous growth in data size, volume, and velocity as well as variability in data scope, data structure, data format, and data variety [1]. There are limitations in technological processing resources and human analytical expertise that do not allow the examination of all data generated every day. In a time-constraint environment such as the military, prioritizing the data can help to converge attention on the most important data first.

A "ranking" challenge in everyday human life can be defined as when someone is trying to rank a set of items based on some criterion with the goal in-mind to order those items from "best" to "worst" within some context. Essentially, the human

© Springer Nature Switzerland AG 2020
M.-J. Lesot et al. (Eds.): IPMU 2020, CCIS 1237, pp. 129–142, 2020.
https://doi.org/10.1007/978-3-030-50146-4_11

decision making process subconsciously reasons over multiple pieces of information and orders them in some way to achieve the "best" possible decision. Accordingly, an entirely new challenge can be identified when there are multiple individuals with multiple opinions trying to reason over multiple pieces of information.

Knowledge elicitation and aggregation from multiple individuals can sometimes provide a better outcome than the answer of any individual. This demonstrates the "wisdom of the crowd" phenomenon in which the aggregated crowd's outcome is closer to correct answer than all or most of the individual answers [2]. This work considers the matter of aggregating information by first starting with the task of ranking multiple independent judgments from multiple independent individuals. The emphasis of this paper is towards military decision making where vast amounts of data gathered from collective intelligence undertakings need to be prioritized.

Information assessment to judge and analyze the high value information, termed as Value of Information (VoI) [3], is very critical for military operations. In recent work to automate the VoI determinations, a Fuzzy Associative Memory (FAM) architecture was used to develop a decision support system for military intelligence analysts [3]. A fuzzy-based prototype system was constructed to provide VoI ratings for individual pieces of information considering the characteristics of information content, source reliability, timeliness, and mission context [3]. Later research was done that included additional knowledge elicitation efforts with subject matter experts that resulted in a complex, multi-FAM system [4].

The approach for this research is to explore the use of the ranking and voting methods of Borda count, Condorcet voting, and Instant-runoff voting (IRV) with respect to the VoI problem domain. Initially, all methods will be compared to the VoI determinations produced by the original fuzzy-based VoI system. The results from the comparisons will illustrate how well these ranking systems match the fuzzy-based approach. Using these aggregation models can also provide a way to quantitatively assess the efficacy of the recently developed VoI prototype since there are no other VoI-producing systems to compare with the fuzzy method.

The remainder of this paper is organized as follows: first, background information is presented pertaining to aggregation and some of the more popular approaches, the military value of information challenge, and a brief description of the current VoI prototype. Following that are sections that describe an experiment using Borda count, Condorcet voting, and Instant-runoff voting models and then the methodology for moving forward to accomplish the above aim of this work. Finally, conclusions and future work are discussed.

2   Background

Information judged on source reliability and content importance can be prioritized in different levels to be addressed and taken into actionable decisions on a given timeline. Information about the enemy, the battlefield environment, and the situation allow the commander and staff to develop a plan, seize and retain the initiative, build and maintain momentum, and exploit success [5]. Information aggregation and prioritization are key components in this process.

2.1   Aggregation

Based on a phenomenon called the "wisdom of crowd" [2], the aggregated rank of the crowd's choices in a voting poll has been generally identified to provide an estimate very close to the true answer; this is very helpful in many estimation tasks. This aggregated rank of the crowd's choices is usually represented in an order from the best value to the worst value. The "wisdom of the crowd" effect was first demonstrated by Francis Galton, who showed that averaging estimates from individuals regarding the weight of an ox produced a close approximation of the actual weight [6]. Consequently, many researchers believe "Crowdsourcing" [7] is driving the future of businesses by obtaining opinions on different subjects from a large and rapidly growing group of Internet users. The "wisdom of the crowd" approach is used in multiple real-world applications ranging from prediction applications aimed at markets and consumer preferences to web-based applications such as spam filtering, and others [8].

This research investigates aggregation within the military VoI problem domain by first considering aggregation related to ranking problems. The topic of aggregation and its need for, and approaches to, combining rankings from multiple sources is certainly not a new challenge; in fact, it has been studied for hundreds of years. The earliest examples of combining rankings relate to the area of "voting" and go back to the 18th century. The French mathematician and astronomer Jean-Charles de Borda in the 1700s proposed a voting method wherein voters would rank all candidates rather than selecting only one [9].

Rank aggregation has long been used in the social choice domain as well as in such fields as applied psychology, information retrieval, marketing, and others; a classification of rank aggregation methods is offered in [10]. Consistent with the overall goal of this research the focus is currently on Borda Count, Condorcet, and Instant-runoff ranking approaches. These three ranking models were chosen as representations to investigate the use of rank ordering techniques within the VoI problem domain.

2.2   Methods of Data Aggregation

Ranking is an example of obtaining assessments from a group of people to achieve one single decision on a winner. Voting on the top movie, electing the president, passing a bill, and many more examples are efforts to produce one single decision. In addition to single winner determination in balloting, voting is used to produce a ranked list. The three ranking methods compared in this paper are discussed below.

Borda Count

The Borda count model is a simple, statistical heuristic method that is widely used in voting theory [8]. The basic idea is that each item in an individual ranking is assigned points based on the position in which it is placed; then, the total points for each item is computed. The resultant totals for each item are used to sort the items and provide the aggregate ranking. One of the primary advantages of the Borda count method is that it is simple to implement and understand. Additionally, Borda count performs well with respect to rank performance measurements.

In this approach, for i items to be ranked by p participants, the total for the ith item, τ_i, can be written as:

$$\tau_i = \sum_{j=1}^{p} y_{ij}$$

where y_{ij} is the point value given to item i by participant j. The final, aggregated ranking is then found by ordering the item totals with any ties broken at random.

This traditional model is used in many applications, especially in sports. It is often used to choose the winners of sports awards such as the Heisman trophy in college football, selecting the Most Valuable Player in professional baseball, and ranking sports teams by the Associated Press and United Press International [11]. The Borda count model is considered as a baseline aggregation method for winner selection in voting. It is also used for aggregating rankings of data for decision making to produce the "wisdom of the crowd" phenomena.

The Borda count procedure is explained in [12] with an example for ranking 4 states in order from high to low by population with 5 participants. The model produces combined rankings that typically perform well relative to individual rankings.

Condorcet Method

The Condorcet method as an election method selects the winning candidate as the one who has gained the majority of the votes in an election process against all of the candidates in a head-to-head election comparison [13]. The Condorcet method requires making a pairwise comparison between every candidate. When there is single ranking item that beats every item in an election, it is called Condorcet Winner. Technically, the Condorcet winner candidate is the one candidate that wins every two-way contest against every other alternative candidate. The winning candidate should beat every other candidate in a head to head election which means the winning candidate should win a runoff election regardless of who it is competing against [14]. The Condorcet voting technique was first advocated by the 18th-century French mathematician and philosopher Marie Jean Antoine Nicolas Caritat, the Marquis de Condorcet [15].

A number of Condorcet-Compliant algorithms exist such that if there is a Condorcet winner, it would be elected; otherwise, they have different behavior. The proper Condorcet method is chosen based on how appropriate it is for a given context. A relatively new single-winner Condorcet election method called Schulze Voting or Beatpath is used in this paper and is described next. The Schulze method is recognized as a common means of solving a Condorcet's Paradox, which is a situation wherein the voters have cyclic preferences such that there is no Condorcet winner.

Condorcet/Schulze Voting (Beatpath)

As one of the Condorcet methods, Schulze Voting (or Schwartz Sequential dropping (SSD) or Beatpath) is a relatively new single-winner election method proposed by Markus Schulze in 1997 [16]. The Schulze method is an effective ranking method which many organizations and some governments are starting to apply. Debian Linux distribution has incorporated the Schulze algorithm into their constitution which can be found under the Appendix on the Debian Linux constitution [17].

While comparable to Borda count as a highly recognizable method for ranking, the Condorcet methods are more problematic to implement as they require pairwise comparisons between all candidates. Therefore, the Condorcet cycle of pairwise comparisons grows as the number of candidates grows.

In this method, after each voter ranks the candidates based on the order of preferences, a head-to-head comparison of all pairs of candidates is conducted to determine the winner of each pair. If there is one candidate that wins in all its pair comparisons, the candidate is Condorcet Winner. If there is no winner, the next step is to determine the pairwise preferences for all pair candidates in a matrix. For each head-to-head pairwise comparison, the number of voters who preferred candidate A over candidate B and vice versa is counted and noted. Once this is done, all the strongest paths for each pairwise comparison are identified; this is the most difficult and computationally intensive step. Finally, the items are ranked by their strongest path computations, producing the winner (and second place, third place, and so on). The full details of the algorithm, along with examples, can be found in [16].

Instant-runoff Voting
Similar to the Condorcet voting method, Instant-runoff voting (IRV) is a preferential ranking method which is used in single-seat elections; this method is useful when there are more than two competing candidates. Basically, voters rank the candidates or items in order of preference rather than showing support for only one candidate. There are countries that use Instant-runoff voting in their election systems such as selecting members of the Australian House of Representatives and the house of Australian state parliaments [18]. The IRV method establishes more fairness in an election when there are multiple candidates dividing votes from the more popular point of the political spectrum such that an unopposed candidate from the unpopular base might win simply by being unopposed.

In this method, once the ballots are counted for each voter's top choice, the candidate with the fewest votes will be eliminated if there is no candidate winning a simple majority (more than half) of the votes. The votes from the voters who have voted for the defeated candidate will be allocated to the total of their next preferred choice. This step will be repeated until the process produces a simple majority for one candidate; this candidate becomes the winner. At some point throughout this process, the race gets narrowed down to only two candidates; this is called an "instant runoff" competition which leads to a comparison of the top two candidates head-to-head. IRV can establish fairness and save the votes from like-minded voters supporting multiple candidates when it comes to a vote-splitting situation in an election.

In 2002, Senator John McCain from Arizona in his campaign supported instant-runoff voting and said that Instant-runoff voting "will lead to good government because voters will elect leaders who have the support of a majority." [19]. The Instant-runoff can avoid the chaos of the US 2000 presidential election and guarantee the elected candidates to have the broadest amount of support. Based on the "Record of Illinois 92nd General Assembly Bills", in 2002, Illinois Senator Barack Obama introduced SB 1789 in the Senate that created Instant-runoff voting for Congress in state primaries and local elections [20].

2.3   Rank Evaluation Metric

While none of the models discussed above require a known ground truth to produce a ranking, the existence of a ground truth will be assumed to allow the resultant rankings to be evaluated. The particular ranking metric will be discussed next.

Kendall's tau distance is an ordinal association measure between two random variables. Kendall's tau coefficient measures the rank correlation of similarities between two sets of ranks and it was developed in 1938 by Maurice Kendall [21]. This correlation coefficient based on Kendall's tau is used to compare the similarity of the derived rankings with the "ground truth" ranking. Kendall's tau is a non-parametric measure that is used to quantify the relationships between columns of ranked data. The computed value ranges from 1 to −1, inclusive [22]. A value of 1 means the rankings are identical while a value of −1 means the rankings are in the exact opposite order. Kendall's tau is calculated using the formula:

$$\frac{(C - D)}{(C + D)}$$

where C represents the number of concordant pairs in the ranking and D represents the number of discordant pairs. Concordant pairs are how many larger ranks are below a specific rank in the list; discordant pairs are how many smaller ranks are below a specific rank. This Kendall's tau value is explained in detail in [12].

Basically, Kendall tau distance measures the number of pairwise swaps needed to bring two orderings into alignment. The Kendall tau distance can be represented in a chance agreement distribution curve as shown in Fig. 1. Higher Kendall tau values indicate a greater disagreement between the ground truth and some resultant aggregated ranking. The lowest possible Tau value is 0, which indicates that the two rankings are identical. The highest possible Tau value can be calculated as:

$$Tau = n(n - 1)/2$$

where n equals the number of items being ranked. When comparing the aggregate ranking with the ground truth, Kendall tau distance can help to measure how closely rankings inferred by a ranking model match the latent ground truth. It is a way to quantify the relationships between columns of ranked data as well as the rankings provided by the participants.

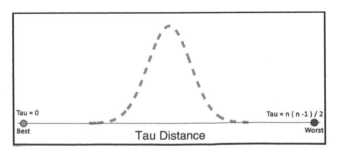

Fig. 1. Kendall tau chance agreement distribution

2.4   Value of Information Challenge

Decision making in military environments involves massive cognitive and temporal resources to examine options and decide on the best alternative. Monitoring and analyzing real-time data to identify the most important information in the context of the mission at hand is challenging for intelligence groups. Information assessment to select and prioritize the high value information is very critical for military operations.

The Value of Information (VoI) metric is used to grade the importance of individual pieces of information. The process of making a VoI determination for a piece of information is a multi-step, human-intensive process. Intelligence analysts and collectors must make these decisions based on the characteristics of the information and also within different operational situations.

U.S. military doctrinal guidance for determining VoI is imprecise at best [5, 23]. The guidance provides two tables for judging the "reliability" and "content" of a piece of data, with each characteristic broken into six categories. Reliability relates to the information source, and is ranked from A to F (reliable, usually reliable, fairly reliable, not usually reliable, unreliable, and cannot judge). Information content is ranked from 1 to 6 (confirmed, probably true, possibly true, doubtfully true, improbable, and cannot judge).

This U.S. military doctrinal guidance does not clearly provide a method to associate these categories for information value determination. Moreover, there is no instruction on how to combine other attributes that may contribute to information value. Two other potential data characteristics include mission context (the operational tempo and decision cycle for a specific mission) and timeliness (time since a piece of information was obtained).

2.5   VoI Prototype Architecture

A prototype system using a Fuzzy Associative Memory (FAM) model has been developed to offer an effective framework for determining the VoI based on the information's content, source reliability, latency, and mission context [24]. For the prototype system, three inputs are used to make the VoI decision: source reliability, information content, and timeliness. The overall architecture of the fuzzy system is shown in Fig. 2. Instead of using one 3-dimensional FAM, two 2-dimensional FAMs were used; the reasoning behind this decision was presented in detail in [24]. The VoI metric is defined as the second FAM output and the overall system output. Note that mission context is handled by using three separate VoI FAMs. The correct VoI FAM is automatically selected based on the indicated mission context, which ranges from 'tactical' (high-tempo) to 'operational' (moderate-tempo), to 'strategic' (slow-tempo).

Fuzzy rules are used in the FAMs to capture the relationships between the input and output domains. Knowledge elicitation from military intelligence Subject Matter Experts (SMEs) was used to build the fuzzy rules [25]. More detailed descriptions of the FAMs, the fuzzy rules bases, the domain decompositions, and other implementation aspects of the prototype system can be found in [3] and [24]. The series of surveys and interviews with SMEs that were used to integrate cognitive requirements, collect functional requirements, and elicit the fuzzy rules is presented in [25].

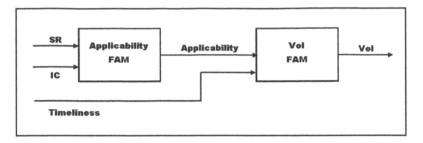

Fig. 2. VoI prototype system architecture

Note that there is no current system against which the results can be compared. As such, the system has not been tested comprehensively due to the human-centric, context-based nature of the problem and usage of the system. Approaches to validate (or partially validate) the system that do not require an extensive, expensive experiment are desired; this research seeks to assist in that effort (as explained further later).

3   Ranking Aggregation Experiment

This section describes an investigative experiment using the Borda count, Condorcet, and Instant-runoff voting methods to aggregate rankings from multiple participants. The experiment was devised not only to gather data for comparing the ranking methods to the VoI prototype system, but also to aid in understanding what data might be needed to relate the ranking models to future continued study in the VoI domain.

The VoI system details provided above in the Background section mention the use of SMEs and a knowledge acquisition process to provide a basis for constructing fuzzy rules. It should be clear that the involvement of multiple experts provides the potential for disagreement in deciding how to combine information characteristics, and multiple pieces of information, to arrive at a "value of information" determination. Another goal of this experiment was to provide first-hand familiarity regarding the efficacy of the Borda count, Condorcet, and IRV rank aggregation models with respect to their potential contribution to the VoI research.

3.1   Experiment Details and Implementation

During the summer of 2019, a team of researchers from the U.S. Army Research Laboratory (part of the U.S. Army Combat Capabilities Development Command) conducted an experiment with 34 military participants as SMEs. Each participant provided rankings for 10 different card decks, where each deck consisted of 5 or 7 cards (5 decks had 5 cards; 5 decks had 7 cards). Each card represented a piece of military information; each participant ranked each card (within each deck) based upon the attributes of source reliability, information content, and latency. An example card is depicted in Fig. 3.

Source Reliable	Information Content	Latency
B	**3**	**Somewhat**
Usually Reliable	Possibly True	**Recent**

Fig. 3. Value of Information card attributes

The experiment was conducted via a computerized online interface. The resultant "Ranked Data Set" reflects the SME's determination of how each piece of information ranks with respect to information value (VoI). The highest ranking card represents the information with the greatest perceived "value" (highest VoI determination), while the lowest ranking card represents the piece of information with the lowest perceived value (lowest VoI determination). As mentioned before, each SME was charged with ranking 10 card decks in this manner. At the completion of the experiment, the Borda count, Condorcet voting, and IRV methods were used to aggregate the rankings of the 34 SMEs within each deck.

3.2 Borda Count, Condorcet, and IRV Implementation for Experiment

For the Borda count implementation, the calculations were performed in Microsoft Excel to convert the alphabetic reviewer designations to numeric rankings lists. The spreadsheet data were then imported into RStudio, an integrated development environment for the R programming language, where the aggregate Borda ranking was derived and compared to the ground truth, producing the Kendall tau distance.

For the Condorcet method, the algorithm presented in [16] was implemented in the "R" programing language and executed in RStudio. This function performs the ranking, starting with the full ballot and finding a pure Condorcet winner if one exists; otherwise, the strongest path computations are done and a winner is computed based on the Schulze algorithm as described earlier.

In the Instant-runoff voting method, a Python module was used which determines the winner card of this experiment using Instant-runoff voting rules. The algorithm executed separately for each of the 10 card deck sessions. The algorithm initially counts all the first-place choices. If there is a card in deck that has a true majority, meaning over half of the counts, the winner card with the first place is recorded. But after the initial count, if none of the cards have over 50% of participants' choices, then the card that is in last place with least number of votes is eliminated. The participants who had selected that defeated card now have their second-place cards counted. The total counts are recorded again, and this process repeats until a true majority winner card is found. Once the winner is determined, the first preferred or winner card is recorded and removed from card deck to re-run this procedure for producing a ranking of cards for second place and-so-forth.

3.3 Methodology and Results

The Borda count, Condorcet voting, and IRV methods descried above were applied to the data sets of the 10 card decks independently; again, each deck had 34 rankings – one from each SME. For each aggregated ranking, the tau distance was computed by counting the number of bubble sort swaps required to make it match the ground truth. The ground truth was derived by computing the VoI for each card using the prototype Fuzzy system, and then ranking the cards based on this value (highest to lowest).

The result for one of the 10 card sets is visually represented in Fig. 4. The performance of the rankings inferred by all three methods as compared to the individual participant rankings is shown; note that all three ranking methods had identical performance in this case. The value of the tau distance for each participant is approximated by the person symbols. The x-axis represents the range of tau distance values that are possible. Tau distance is the number of swaps needed to make some given ranking equal to the ground truth ranking; thus, it measures the "performance" of an SME's ranking relative to the "correct" answer [21].

The light gray circle on the left side indicates the best possible ranking which is equal to the ground truth (tau distance of 0). The dark circle to the right indicates the worst possible case in which the rank order is the total reverse of the ground truth (for 5 cards that value is 10). The dotted curve depicts the chance tau distance distribution which would correspond to rankings being generated at random. The tau distance for the rankings produced by all three methods is indicated by the light-dotted circle with tau distance of 1. In this representation, the small tau distance indicates that the models produce a combined ranking that performs well relative to the individual performances, and that the aggregated rankings are very close to the ground truth.

A summary of the experimental results is shown in Table 1. The ground truth ranking for each of the 10 decks is given along with the aggregated ranking produced by each of the ranking methods. The "B" rows represent the result from the Borda count method, while "C" represents Condorcet and "I" represents IRV. The "Tau Dist" column gives the Kendall Tau Distance for the associated aggregated ranking compared to the worst case number of swaps that would be possible. The Tau distance is used to measure the ranking accuracy relative to the ground truth.

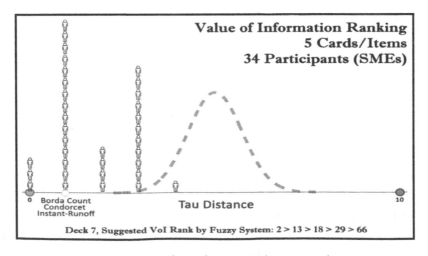

Fig. 4. Example ranking and Borda count result

The somewhat random card numbers in each deck have been mapped to an ordered set of integers reflecting the amount of cards used in the specific deck. This was done so that it is easier to understand how the aggregated rankings match the ground truth. For example, for Deck 7, the actual order of the cards for the ground truth is <2, 13, 18, 29, 66>; 2 is mapped to 1, 13 is mapped to 2, etc. So that <1, 2, 3, 4, 5> represents the ground truth order of the cards. The aggregated ranking produced by all 3 ranking models gave the actual order of <2, 18, 13, 29, 66>, which maps to <1, 3, 2, 4, 5>. It is easy to observe that the aggregated order produced by the ranking methods differs in the 2nd and 3rd cards, such that 1 swap (tau distance of 1) out of a possible 10 (worst case ranking) is required to achieve the ground truth order.

The Kendall tau distance values demonstrate that the aggregated rankings from the SMEs are relatively close to the ground truth. Note that the three methods are almost identical in their resultant aggregated rankings. The only differences are where the IRV method is 1 swap worse for Deck 8 and 1 swap better for Deck 2. Given that the Borda count model is much easier to implement, this model is deemed the "best".

Table 1. Experimental data ranking results

Deck	Ground Truth Ranking	Aggregated Ranking		Tau Dist
8	1 2 3 4 5 6 7	B	1 2 3 4 5 6 7	0/21
		C	1 2 3 4 5 6 7	0/21
		I	1 2 4 3 5 6 7	1/21
10	1 2 3 4 5 6 7	B	1 3 2 4 5 6 7	1/21
		C	1 3 2 4 5 6 7	1/21
		I	1 3 2 4 5 6 7	1/21
4	1 2 3 4 5 6 7	B	1 4 2 6 5 3 7	5/21
		C	1 4 2 6 5 3 7	5/21
		I	1 4 2 6 5 3 7	5/21
6	1 2 3 4 5 6 7	B	1 5 2 6 7 3 4	7/21
		C	1 5 2 6 7 3 4	7/21
		I	1 5 2 6 7 3 4	7/21
2	1 2 3 4 5 6 7	B	1 4 5 6 2 7 3	7/21
		C	1 4 5 6 2 7 3	7/21
		I	1 4 5 6 2 3 7	6/21
1	1 2 3 4 5	B	1 2 3 4 5	0/10
		C	1 2 3 4 5	0/10
		I	1 2 3 4 5	0/10
7	1 2 3 4 5	B	1 3 2 4 5	1/10
		C	1 3 2 4 5	1/10
		I	1 3 2 4 5	1/10
5	1 2 3 4 5	B	1 3 2 4 5	1/10
		C	1 3 2 4 5	1/10
		I	1 3 2 4 5	1/10
9	1 2 3 4 5	B	1 4 3 2 5	2/10
		C	1 4 3 2 5	2/10
		I	1 4 3 2 5	2/10
3	1 2 3 4 5	B	1 4 2 5 3	3/10
		C	1 4 2 5 3	3/10
		I	1 4 2 5 3	3/10

There are two cases where the Tau distance is 6 or 7 (out of a possible 21); the remainder of the results show much greater accuracy. These results, from Decks 2 and 6, have caused the ARL researchers to go back and look at their data and the fuzzy rules used to produce the ground truth ranking.

The experimental results demonstrate the efficacy of the ranking models in aggregating SME opinions, and highlight the "wisdom of the crowd" effect. As a corollary, it can also be said that the suggested rankings produced by the fuzzy VoI system (used as the ground truth) are reasonable given the aggregated SME rankings. As final highlight to this experiment, note that the ground truth and the aggregated rankings always agree on the most important piece of information (the first position in all rankings match) and they almost always agree on the least important piece of information.

3.4   Significance of the Experiment

While the Borda count, Condorcet, and IRV methods did not match the ground truth exactly in all instances, the methodology has shown promise to be a viable aggregation method in this particular domain. As previously mentioned, there is no current system with which to compare the VoI system results. Only a subjective validation of the efficacy of the systems by the SMEs has been possible. By providing a different approach for arriving at the basis for the rules and architectures of the systems, these methods are useful in providing the missing quantitative support of the systems.

As mentioned earlier, the fuzzy rules used in the VoI prototype systems were constructed based on knowledge elicitation processes with military intelligence SMEs. Notably, the experts did not always agree in their opinions and answers. The differences occurred in providing an applicability rating for a given SR/IC combination as well as in determining the VoI output for a given Timeliness/Applicability pattern. Further, it was not unusual for the SMEs to have varying interpretations of the linguistic terms (e.g. "fairly reliable", etc.). Based on the experimental results, it is hoped that one or more of the ranking methods can perhaps aid in better aggregating the expert's opinions in the knowledge elicitation process to create more accurate fuzzy rules. The observed differences between the fuzzy rankings and the aggregated SME rankings have already motivated a reexamination of the original fuzzy rule construction to ensure the systems are accurately representing the "ground truth".

The results herein display only the aggregated SME rankings and do not examine the spread of the 34 individual rankings (as shown in the Fig. 4 example). It is possible that extenuating circumstances could have influenced the aggregations in the instances where they did not closely match the ground truth. Factors such as the experts not fully understanding the nuances in a particular card deck (notably decks 2 and 6), or missing an understanding of the mission context in which the information would be used, or others could have come into play. Examination of variables that may have influenced the rankings is underway at the U.S. Army Research Laboratory.

Finally, while the current VoI systems assign a specific value determination to a piece or pieces of information, there may be times when military analysts simply need to rank information according to its perceived importance without the need for a numerical value. The three ranking models could certainly be useful in this regard.

4 Conclusion

In a military environment, the process of analyzing the relative value of vast amounts of varying information is currently carried out by human experts. These experts, known as military intelligence analysts, have the significant challenge of evaluating and prioritizing available information, in a time-constrained environment, to enable commanders and their staffs to make critical mission decisions. Within the domain of military intelligence analyst decision support, this work considers the matter of aggregating information. One goal of this research was to investigate the use of the Borda count, Condorcet voting, and Instant-runoff voting (IRV) aggregation models with respect to the Value of Information (VoI) problem. Another goal was to provide a way to quantitatively assess the efficacy of the recently developed VoI prototypes.

This paper presented discussion about ongoing VoI research and a preliminary experiment using the Borda count, Condorcet voting, and IRV models. The results demonstrated the usefulness of these ranking models in aggregating SME opinions and clearly highlighted the "wisdom of the crowd" effect. Additionally, this work offered some quantitative validation of the current VoI prototypes by providing results for comparison to those from the fuzzy-based systems. These efforts will help to optimize the current fuzzy rules, motivate additional knowledge elicitation efforts, and influence the development of other VoI decision support architectures altogether.

Future research will include the comparison of one or more of the models used here to the well-known Bayesian Thurstonian ranking model. As the U.S. Army continues to move toward improving its ability to create situational awareness, the ability to successfully aggregate information will be a critical factor.

Acknowledgment. The researchers would like to thank the U.S. Army Research Laboratory, and Dr. Timothy Hanratty, Eric Heilman, and Justine Caylor in particular, for their efforts and insights that made this research possible.

References

1. Couldry, N., Powell, A.: Big data from the bottom up. Big Data Soc. 1(2), 5 (2014). https://doi.org/10.1177/2053951714539277
2. Surowiecki, J.: The Wisdom of Crowds. Anchor, New York (2005). Reprint edition
3. Hanratty, T., Newcomb, E.A., Hammell II, R.J., Richardson, J., Mittrick, M.: A fuzzy-based approach to support decision making in complex military environments. Int. J. Intell. Inf. Technol. 12(1), 1–30 (2016). https://doi.org/10.4018/IJIIT.2016010101
4. Hanratty, T., Heilman, E., Richardson, J., Caylor, J.: A Fuzzy-logic approach to information amalgamation. In: Proceedings of the 2017 IEEE International Conference on Fuzzy Systems, Naples, Italy (2017). https://doi.org/10.1109/FUZZ-IEEE.2017.8015667
5. US Army, US Army Field Manual (FM) 2-22.3, Human Intelligence Collection Operations, Washington, DC, Washington, DC: US Army, p. 384 (2006)
6. Galton, F.: Vox populi. Nature 75, 450–451 (1970)
7. Estellés-Arolas, E., González-Ladrón-de-Guevara, F.: Towards an integrated crowdsourcing definition. J. Inf. Sci. 38(2), 189–200 (2012). https://doi.org/10.1177/0165551512437638

8. Steyvers, M., Lee, M., Miller, B., Hemmer, P.: The wisdom of crowds in the recollection of order information. In: NIPS2009 Proceedings of the 22nd International Conference on Neural Information Processing Systems, Vancouver, British Columbia (2009). https://dl. acm.org/doi/10.5555/2984093.2984293
9. Borda, J.-C.: On elections by ballot. In: Histoire de l'Academie Royale des Sciences for the Year 1781 (1781)
10. Li, X., Wang, X., Xiao, G.: A comparative study of rank aggregation methods for partial and top ranked lists in genomic applications. Briefings Bioinform. **20**(1), 178–189 (2017)
11. Easley, D., Kleinberg, J.: Networks, Crowds, and Markets: Reasoning About a Highly Connected World, pp. 736–746. Cambridge University Press, Cambridge (2010). https://doi. org/10.1017/CBO9780511761942
12. Etesamipour, B., Hammell II, R.J.: Aggregating information toward military decision making. In: 17th IEEE/ACIS International Conference on Software Engineering Research, Management and Applications, Honolulu, Hawaii, (2019). https://doi.org/10.1109/SERA. 2019.8886807
13. Pivato, M.: Condorcet meets Bentham. J. Math. Econ. **59**, 58–65 (2015). https://doi.org/10. 1016/j.jmateco.2015.04.006
14. Newenhizen, J.V.: The Borda method is most likely to respect the Condorcet principle. Econ. Theory **2**(1), 69–83 (1992). https://doi.org/10.1007/BF01213253
15. Condorcet, M.de.: Essay on the Application of Analysis to the Probability of Majority Decisions (1785)
16. Schulze, M.: The Schulze Method of Voting. eprint arXiv:1804.02973 (2018)
17. Constitution, D.: Constitution for the Debian Project (v1.7) (2016)
18. Bennett, S., Lundie, R.: Australian Electoral Systems. Parliament of Australia Department of Parliamentary Services (2007)
19. Lavin, N., Robinson, R.: John McCain understood how ranked choice voting strengthens our democracy. FairVote (2018)
20. Richie, R., Penrose, D.: When Barack Obama Was a Leader in Seeking Fair Voting Systems. FairVote (2012)
21. Kendall, M.: A new measure of rank correlation. Biometrika **30**, 81–89 (1938)
22. Abdi, H.: The Kendall Rank Correlation Coefficient. Encyclopedia of Measurement Statistics. Sage, Thousand Oaks (2007)
23. North Atlantic Treaty Organization (NATO) Standard Agreement 2022. STANAG magazine, STANAG 2022 Intelligence Reports. (Edition 8) Annex
24. Hammell II, R.J., Hanratty, T., Heilman, E.: Capturing the value of information in complex military environments: a fuzzy-based approach. In: Proceedings of 2012 IEEE International Conference on Fuzzy Systems (FUZZ-IEEE 2012), Brisbane, Australia, (2012). https://doi. org/10.1109/FUZZ-IEEE.2012.6250786
25. Hanratty, T., Heilman, E., Dumer, J., Hammell, R.J.: Knowledge elicitation to prototype the value of information. In: Proceedings of the 23rd Midwest Artificial Intelligence and Cognitive Sciences Conference (MAICS 2012), Cincinnati, OH (2012)

Combining Multi-Agent Systems and Subjective Logic to Develop Decision Support Systems

César González-Fernández$^{(\boxtimes)}$, Javier Cabezas, Alberto Fernández-Isabel, and Isaac Martín de Diego

Data Science Laboratory, Rey Juan Carlos University,
c/ Tulipán, s/n, 28933 Móstoles, Spain
{cesar.gonzalezf,javier.cabezas,alberto.fernandez.isabel,
isaac.martin}@urjc.es
http://www.datasciencelab.es

Abstract. Nowadays, the rise of the interconnected computer networks and the increase of processed data have led to producing distributed systems. These systems usually separate multiple tasks into other simpler with the goal of maintaining efficiency. This paradigm has been observed for a long time in different animal organisations as insect colonies and fish shoals. For this reason, distributed systems that emulate the biological rules that govern their collective behaviour have been developed. *Multi-Agent Systems (MAS)* have shown their ability to address this issue. This paper proposes *Ant Colony based Architecture with Subjective Logic (ACA-SL)*. It is a bio-inspired model based on ant colony structures. It makes use of *MAS* to distribute tasks and *Subjective Logic (SL)* to produce *Decision Support Systems (DSS)* according to the combination of individual opinions. A system implementation based on the proposed architecture has been generated to illustrate the viability of the proposal. The proposed architecture is intended to be the starting point for developing systems that solve a variety of problems.

Keywords: Multi-Agent system · Subjective Logic · Bio-inspired system · Distributed organisation · Decision Support System

1 Introduction

In recent times, the heyday of the Internet and the advance of technology have produced tons of data which are processed by several systems [1]. These systems apply the strategy of separating data into simpler and smaller pieces in order to process them efficiently. Thus, the information extraction task and the generation of knowledge are simplified. This issue has led to the resurface of distributed systems.

Multi-Agent Systems (MAS) [2] are a specific case of distributed systems. They use agents that are software abstractions able to perform tasks and to satisfy the associated goals interacting with the environment around. These agents

© Springer Nature Switzerland AG 2020
M.-J. Lesot et al. (Eds.): IPMU 2020, CCIS 1237, pp. 143–157, 2020.
https://doi.org/10.1007/978-3-030-50146-4_12

present desirable features as: autonomy [3], flexible behaviour to react to changes in the environment in a timely fashion [4], and dynamic interaction between them [5].

Nevertheless, these systems have some shortcomings. The organisation to solve specific situations is one of the most typical challenges [6]. In this regard, bio-inspired mechanisms are one of the most used self-organisation solutions [7]. Thus, they can adapt some social animal behaviour to solve specific situations. Typical instances of these mechanisms are insect colonies [8], fish shoals [9] and mammals packs [10], where the solution of a complex problem is achieved by the individuals solving simpler issues.

This paper proposes *Ant Colony based Architecture with Subjective Logic* (*ACA-SL*), a novel architecture based on bio-inspired *MAS* and *Subjective Logic* (*SL*) [11] to develop distributed *Decision Support Systems(DSS)* [12]. These latter are able to produce evaluation according to a specific topic or domain according to a previously obtained knowledge. *ACA-SL* has been developed as part of the SABERMED project, which is funded by the Spanish Ministry of Economy and Competitiveness. *ACA-SL* emulates the behaviour of ant colonies to execute distributed jobs. The way in which jobs are defined exhibits a high degree of flexibility for multiple application scenarios. For this purpose, there will be several agents assuming the same role as workers in ant colonies. Jobs are defined as the combination of very diverse tasks which may hold some dependencies among them. This fact provides agents with the capability to work on the same job at the same time. The architecture combines the opinions generated by following the rules, and methodologies defined in the *SL*.

A system based on *ACA-SL* has been implemented to show the viability of the proposal. It has been used to analyse websites and generate an opinion concerning the degree of trust applicable to them. The created opinions are the result of processing related information extracted from the websites under analysis (e.g. domain and Whois).

The remainder of the paper is structured as follows. Section 2 situates the proposal in the domain and make comparisons with previous approaches. Section 3 details *ACA-SL* and its components. Section 4 presents the experiments. Finally, Sect. 5 concludes and proposes future guidelines.

2 Background

This section introduces the foundations of *ACA-SL*. It overviews the concept of *MAS* (see Sect. 2.1), both by defining it and also by providing some details on the current state of art. Secondly, insect colonies and their internal organisational procedures are introduced (see Sect. 2.2). Finally, *SL* foundations and most common applications are presented (see Sect. 2.3).

2.1 Multi-Agent Systems

Agents can be defined as intelligent autonomous entities able to act, partially perceive the environment they live in, interact with it and communicate with other agents [13]. They take part in an organised activity in order to satisfy the particular goals they were designed to, both by executing a set of skills and by interacting with other agents. These goals are evaluated by a mental state. The mental state acts as the brain of the agents, containing steps and rules. Therefore, agents show pro-activeness (they exhibit goal-directed behaviour by taking initiative), reactivity (they perceive their environment and respond in a timely way to changes that may occur in the environment) and social awareness (they cooperate with other agents in order to accomplish their tasks) [14].

MAS [15] are a loosely coupled set of agents situated in a common environment that interact with each other to solve complex problems that are beyond the individual capacities or knowledge of each agent [16]. These systems are found in a wide spectrum of heterogeneous applications such as simulations [17], optimisation problems [18] and computers games [19]. *MAS* have been also used in the literature with the purpose of distributing very demanding tasks [20]. They are able to use agents that perform simple tasks in order to generate a more complex and demanding one. Fields of application where these kind of approaches are considered are road traffic [21] and communication networks [22].

There are multiple frameworks available to implement software based on *MAS*. Many of them respond to the restless evolution and unremitting development occurring both in the industry and in the scientific community. JADE (Java Agent Development framework) [23], FIPA-OS (Foundation for Intelligent Physical Agents Operating System) [24] and SPADE (Smart Python Agent Development Environment) [25] exemplify some of the existing options at disposal of the user.

MAS can be designed by using Agent-Based Modelling (ABM) [26] and Agent-Oriented Software Engineering (AOSE) techniques [27]. These ones are considered by solid methodologies to simulate relationships and communication channels between agents. Instances of well-known agent methodologies are INGENIAS [28] and Tropos [29].

ACA-SL models a *MAS* that identifies agents as workers belonging to an ant colony. Notice that at this point, these workers are only modelled through a finite state machine, instead of defining explicitly goals and mental states. The implementation achieved to validate the proposal has been developed using the SPADE framework.

2.2 Insect Colonies

Many species of social insects exhibit the division of labour among their members. This behaviour can be observed in bumblebee colonies [30], termites colonies [31] and wasp colonies [32]. The specific task allocation can be determined by multiple features. The age of the individual [33], the body size [34] or the position held in the nest [35] are some instances of these features. Several works concentrate on these behaviours in order to propose new task allocation strategies in artificial systems [36].

Regarding the task allocation method used by individuals, it can be modelled by using response thresholds [37]. These thresholds refer to individual tendency to react to task-associated stimuli. For the specific case of ants, it is considered that every task can exert certain level of influence over them. Thus, if the stimulus that a task exerts on an ant is higher than its response threshold, the ant engages to this task. This leads to considering the existence of castes in which individuals may have different response thresholds. In the case of artificial systems, the use of these response thresholds to solve labour division have been used to enhance response times and load balancing issues [38].

ACA-SL uses a model based on two different types of ants according to a specific response threshold. The architecture provides a specific definition to the measure of the stimuli and the response threshold level.

2.3   Subjective Logic

SL [11] is a type of logic that allows playing with subjective beliefs. These ones are modelled as *opinions*. The *opinions* represent the probabilities of a proposition or an event with a certain degree of uncertainty. *SL* defines a set of operations that can be applied to the *opinions*. Typical instances of these operations are: *addition, subtraction, cumulative fusion* and *transitivity*.

SL extends the traditional belief function model [39]. This logic is also different from Fuzzy logic [40]. While Fuzzy logic uses vague propositions but provides accurate measurements, *SL* works with clear propositions and uncertain measures.

Given a binomial variable (*true* or *false*) representing a proposition x, and a source of *opinions* A, an *opinion* provided by A about x, w_x^A, is represented by a quadruple of values as follows:

$$w_x^A = \{b_x, d_x, u_x, a_x\}, \tag{1}$$

where b_x is the mass belief (belief supporting that x is *true*), d_x is the disbelief mass (belief supporting that x is *false*), u_x is the uncertainty mass and a_x is the atomicity rate.

Regarding the features of *SL*, they have made this logic suitable for applying it to multiple projects covering different knowledge areas. Thus, in general, it can be used to build frameworks for *DSS* [41]. More specifically, *SL* can be used in Trust Network Analysis to calculate the trust between different parts of the network where trust measures can be expressed as beliefs [42]. Analogously, in mobile networks, *SL* can be used to calculate the reputation of the communication nodes [43]. *SL* can be also used in applications independent from the technology (for instance, legal reasoning [44]).

The proposed architecture makes use of *SL* to handle the beliefs that can be generated as a result of the different tasks processed. These beliefs are modelled as *opinions* using only a specific subset of operations.

3   Proposed Architecture

This section details *ACA-SL*. The aim of this architecture is to produce a design to develop *DSS* able to make evaluations. It combines bio-inspired *MAS*

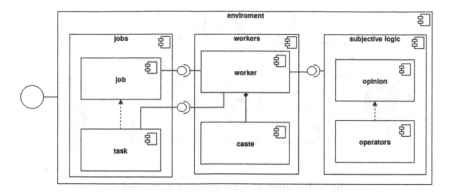

Fig. 1. Components defined by *ACA-SL*.

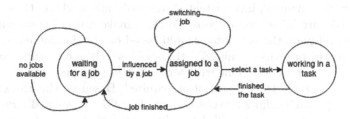

Fig. 2. Life cycle of a worker as a finite state machine.

approaches with *SL* to achieve this issue. Overall jobs are decomposed into multiple tasks which are assigned to the different agents. Agents take full responsibility on a successful accomplishment of the assigned tasks. Notice that *ACA-SL* defines the baselines on how the jobs should be divided into atomic tasks. Individual results arising from their internal processes are then combined to obtain a solution for the global problems.

Figure 1 shows an overview of the proposed architecture. A system based on this architecture takes responsibility on executing the jobs. These ones correspond to needs that the external systems may require to satisfy.

Next sections address the internal procedures followed by agents, detailing jobs and their inner structure. They also describe how *SL* is implemented in the proposal.

3.1   Multi-Agent System Based on Ant Colonies

Analogous to nature, the proposed architecture presents an environment where workers live in. The behaviour of workers is represented by a finite state machine with three states (see Fig. 2).

Delving into the behaviour of workers, the registering of a new job represents a change happening in the environment. These changes (i.e. new jobs) exert stimuli that are perceived by workers, which can be influenced by it. To prevent

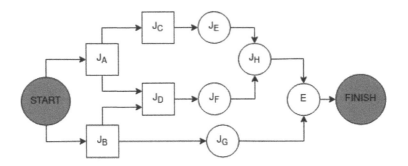

Fig. 3. Instance of a job graph and its tasks.

that some task remains in inconsistent states, only idle workers (those not running any task) are influenced by jobs. Workers make a decision on whether to take responsibility on the new jobs created based on this job influence.

Every worker presents an internal threshold which is compared to the job influence value itself to determine whether the latter presents a higher value and consequently, a switch to the new job is required. Based on this threshold, the bio-inspired approach defines two castes of workers [37]: *major* and *minor*. Those workers simulating *major* ants will show a higher threshold than the one assigned to the workers representing *minor* ants. This feature allows reserving workers to carry out specific jobs. For example, if the influence of a job is calculated based on its priority, the major workers only perform high-priority jobs. This feature plays a crucial role in systems where resources availability, response times and load balancing are critical and very demanding [45].

3.2   The Job Workflow

A job gets represented by directed graphs (see, for instance, Fig. 3). Its component tasks can be interpreted as the multiple possible road-maps linking the *start* node (i.e. starting point) with the *finish* (i.e. finishing point). The workers assigned to a job that are not running any task are placed at the *start* node. These workers are continuously checking the status of all tasks connected to the *start* node via directed edges. If all connected tasks present are being run by other worker, then workers wait at the *start* node. Those tasks connected to the *start* node that are not being executed, present themselves as potential candidates to be selected by workers.

Workers are oriented towards the task selection issue. Thus, there are measures which provide cost values to the different edges between nodes.

A job is considered to be successfully completed when the worker handling the last task represented by the *finish* node completes its duties. Notice that jobs are independent from each other.

Regarding the intermediate nodes of this graph, they represent the different tasks in which the job is divided, giving shape to multiple possible paths linking

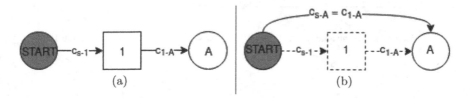

Fig. 4. a.- State of the path from (start) to (A) before task (1) is completed. b.- State of the path from (start) to (A) after task (1) is completed.

the *start* node with the *finish* node. It may be the case when there is not any path from the *start* node to the *finish* node (e.g. the intermediate nodes have raised errors). This situation translates into a job finishing with failures during execution.

Notice that the next tasks available in the path are the next related to the last completed one (i.e. a completed task is no longer visible as available tasks for workers). Figure 4 illustrates this point with an example. Let c_{s-1} be the cost associated to the edge connecting nodes *start* and (1). Let c_{1-A} be the cost between (1) and (A). Figure 4(a) shows one path and one task connected to the *start* node along with the cost involved in the different edges. When the task (1) is completed, it is hidden and the *start* node gets virtually connected to the node (A) by establishing a new edge with cost value c_{s-A} equal to the c_{1-A}.

Regarding the tasks, they are considered as atomic. Every worker assigned to a job is responsible for carrying out just one of the component tasks at a time. Hence, a one-to-one relationship between workers and tasks is established. Tasks assigned to workers contain specific prerequisites to be fulfilled. These preconditions are addressed to ensure correct alignment of workers.

According to these preconditions, tasks are organised into two main groups. The first group considers the tasks that require the fulfillment of all the prerequisites to be executed (labelled as *strict*), while the second group includes tasks executed every time a requirement is satisfied (labelled as *soft*).

In Fig. 3, the *strict* tasks are represented by squares, while the *soft* tasks are pictured by circles. In this example, task represented by node (J_D) cannot be performed until task (J_A) and (J_B) are completed. On the other hand, task (J_H) is executed when (J_E) or (J_F) are completed. On this way, the requirements can be only satisfied with the result of an individual previous task.

Tasks follow a specific workflow to manage their own internal state. Five states are defined in this regard: *waiting, running, completed, error* and *blocked*. Figure 5 shows the dynamic flow and possible relationships between states.

When a new job is created, all component tasks are in *waiting* state. Once a worker is in a position to start with a task (i.e. fulfillment of its particular requirements), the task changes its internal *waiting* state to *running*. A successful completion of the task results in a *completed* state for it. However, if errors were found during the process, the task changes to *error* state. Notice that any other worker can take responsibility for a task in the *error* state to seek its completion

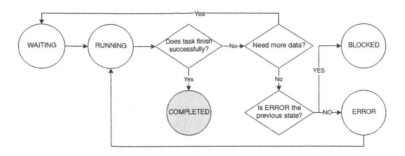

Fig. 5. Flow diagram of the states of a job.

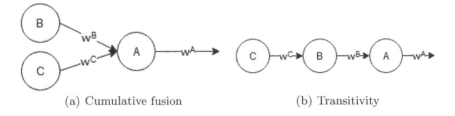

(a) Cumulative fusion   (b) Transitivity

Fig. 6. Graphical representation of the *SL* operator.

(even the same original worker). However, tasks can also enter into a *blocked* state when, after being in *error* and proceed with retrial, *completed* state is not reached. All tasks in *blocked* state are removed from the pool of available tasks for workers, which results in not considering neither their nodes nor the edges connected to them in the graph.

3.3 Combining Opinions with Subjective Logic

The proposed architecture lies in its ability to deal with beliefs. In pursue of that feature, *SL* is considered as a methodology to manage these beliefs. The belief can be the results of the execution of a task.

Considering the fact that *ACA-SL* currently finds itself at a very early stage, just binomial opinions are taken into account in the remaining of this section. Likewise, a reduced subset formed by two operators is contemplated in the proposed architecture (see Fig. 6) : *cumulative fusion operator* ($w_x^{A \diamond B} = w_x^A \oplus w_x^B$) and *transitivity operator* ($w_x^{A;B} = w_x^A \otimes w_x^B$).

The use of *SL* enables to manage opinions given by multiples sources. These opinions can be combined. Furthermore, the sources can have different robustness levels based on the confidence in each of them. The confidence in a source can be assigned by manual setup, using rules defined by human experts, or can be dynamically defined by training the system (e.g. using a previously evaluated dataset).

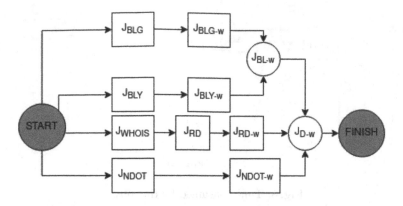

Fig. 7. Job graph produced for the experiment.

Table 1. System configurations for the proposed experiments.

	#major	#minor	major th.	minor th.
Configuration 1	0	1	–	0
Configuration 2	5	20	5	0
Configuration 3	5	20	10	0
Configuration 4	20	50	5	0

4   Experiments

A *DSS* based on *ACA-SL* has been implemented to evaluate the validity of the proposed architecture. The system purpose is to identify malicious web domains. Thus, given a domain, the system is capable of generating an opinion about it. This opinion is based on specific methods gathered from the literature of the domain. These methods are to query well-known blacklists [46], to check both the number of dots in the domain [47] and the registration date of the domain [48].

A job that includes the specific methods has been created to evaluate a domain. Figure 7 shows the graph of the implemented job. This job is divided into multiple tasks. (J_{BLG}) and (J_{BLY}) tasks query the blacklists of *Google* and *Yandex* respectively. (J_{WHOIS}) obtains the Whois, while (J_{RD}) extracts the registration date from the Whois data, and (J_{NDOT}) obtains the number of dots in domain. The objective of these tasks is to retrieve information about the domain. This information is then used by the following tasks to generate opinions. (J_{BLG-w}) and (J_{BLY-w}) give an opinion based on blacklists responses, (J_{BL-w}) combines preceding opinions, (J_{RD-w}) generates an opinion about the registration date, (J_{NDOT-w}) use the count of dots in the domain to give the opinion and, finally, (J_{D-w}) combine all of these opinions to provide a final resulting opinion about the domain.

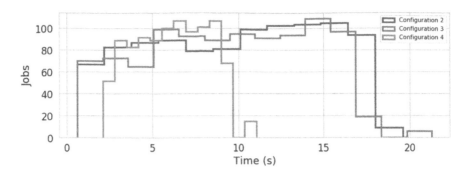

Fig. 8. Time consumed by the jobs.

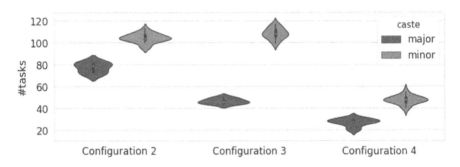

Fig. 9. Number of tasks carried out by each worker.

Four configurations of workers have been tested. Table 1 shows the parameters for each experiment. *#major* and *#minor* indicate the number of ants belonging to each category, while *major th.* and *minor th.* reflect their respective thresholds.

The system has also been customised according to specific settings. First, when the influence of multiple jobs exceeds the threshold of a worker, the worker selects the job with the greatest influence. Secondly, Eq. 2 is used to calculate the influence exerted by a job (I_j). This influence is proportional to the age of the job ($T_j(s)$) (i.e. the current time subtracting the initial time the job is registered in the system) and inversely proportional to the square of the number of workers (W_j^2) assigned to it. To avoid a potential division by zero, one is added to the denominator:

$$I_j = \frac{T_j(s)}{W_j^2 + 1} \tag{2}$$

In this configuration, the two blacklist methods are preferred over the rest ones. To indicate this preference, the cost of the edges used to form the paths passing through these tasks is set up with lower values. Finally, given a domain, not appearing in a blacklist is not sufficient to consider it as trustworthy.

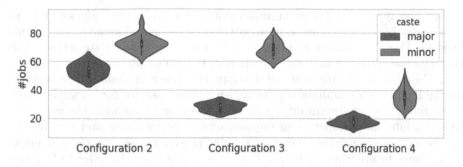

Fig. 10. Number of times each worker change its assigned job.

To test the application, 1000 domains have been processed. For each domain, a job is registered in the system. Figure 8 shows the number of jobs and their time consumed until completion according to the selected configurations. Configuration 1 is not depicted because the time consumed is one order of magnitude higher. Times spent in Configurations 2 and 3 are similar (their jobs take between 1 and 22 seconds to finish). Configuration 4 obtains shorter times. This result illustrates that the use of several agents had a positive effect in the performance of the system.

Figures 9 and 10 depict the number of tasks and the number of job changes each worker performs in both castes respectively. In all configurations, the major workers perform less tasks than minor workers. Also, the Configuration 3 shows a larger gap between castes than the rest of the configurations. This fact is a consequence of the configured thresholds.

This experiment shows how a good selection of the configuration parameters improves the performance. The time consumed by the jobs varies drastically depending on the number of workers available. This fact indicates that the job division and the use of multi-workers are suitable. Finally, the use of different thresholds has provoked that the number of tasks performed by a major worker decreased. If needed, the resources associated to these workers can be reserved by adjusting these thresholds.

5 Conclusions

This paper introduces *ACA-SL*, a bio-inspired architecture based on ant colony structures. It combines *MAS* and *SL* to correctly manage high distributed *DSS*.

The architecture makes use of agents taking the role of ant workers. They deal with registered jobs which are external requests placed to the colony. To facilitate the parallel processing, these jobs are defined as a set of tasks which are individually assigned to the different workers. These tasks make up a graph defining the job. Finally, *SL* is used to handle the opinions generated during the process.

A basic application has been implemented to validate the proposal. Experiments have been carried out to illustrate that the proposed architecture is truly feasible. They have shown the importance of defining appropriate settings (i.e. the edge costs, the job influence equation or the job graph shape).

$ACA\text{-}SL$ is in an early stage of development. However, foundations followed during its design are addressed to establish a good basis for future implementations. Some instances exemplifying these aspects can be found in the capability to setup internal parameters to improve offered performance and the division of jobs in tasks to guarantee correct parallel processing and resources management. Some future works will arise from this contribution. In order to facilitate future implementations based on the proposed architecture, a complete framework development is being considered. This framework will follow the ABM methodology and the Model Driven Architecture (MDA) guidelines. There is a plan to extend some of the features already defined, and to increase the number of castes with the purpose of improving the flexibility of the system. New SL operators will be also considered in future projects.

Acknowledgments. Research supported by grant from the Spanish Ministry of Economy and Competitiveness, under the Retos-Colaboración program: SABERMED (Ref: RTC-2017-6253-1) and the support of NVIDIA Corporation with the donation of the Titan V GPU.

References

1. Pratap, A.: Analysis of big data technology and its challenges. Int. Res. J. Eng. Technol. (IRJET) **6**, 5094–5098 (2019)
2. Zeghida, D., Meslati, D., Bounour, N.: Bio-IR-M: a multi-paradigm modelling for bio-inspired multi-agent systems. Informatica **42**(3) (2018)
3. Wooldridge, M., Jennings, N.R.: Intelligent agents:theory and practice. Knowl. Eng. Rev. **10**(2), 115–152 (1995)
4. Weyns, D., Omicini, A., Odell, J.: Environment as a first class abstraction in multiagent systems. Auton. Agent. Multi-Agent Syst. **14**(1), 5–30 (2007). https://doi.org/10.1007/s10458-006-0012-0
5. Sun, R., et al.: Cognition and Multi-agent Interaction: From Cognitive Modeling to Social Simulation. Cambridge University Press, Cambridge (2006)
6. Horling, B., Lesser, V.: A survey of multi-agent organizational paradigms. Knowl. Eng. Rev. **19**(4), 281–316 (2004)
7. Jean-Pierre, M., Christine, B., Gabriel, L., Pierre, G.: Bio-inspired mechanisms for artificial self-organised systems. Informatica **30**(1) (2006)
8. Fewell, J.H., Harrison, J.F.: Scaling of work and energy use in social insect colonies. Behav. Ecol. Sociobiol. **70**(7), 1047–1061 (2016). https://doi.org/10.1007/s00265-016-2097-z
9. Ward, A.J., Herbert-Read, J.E., Sumpter, D.J., Krause, J.: Fast and accurate decisions through collective vigilance in fish shoals. Proc. Natl. Acad. Sci. **108**(6), 2312–2315 (2011)
10. Muro, C., Escobedo, R., Spector, L., Coppinger, R.: Wolf-pack (canis lupus) hunting strategies emerge from simple rules in computational simulations. Behav. Process. **88**(3), 192–197 (2011)

11. Jøsang, A.: Subjective Logic. Springer, Heidelberg (2016). https://doi.org/10. 1007/978-3-319-42337-1
12. He, C., Li, Y.: A survey of intelligent decision support system. In: 2017 7th International Conference on Applied Science, Engineering and Technology (ICASET 2017), pp. 201–206. Atlantis Press (2017)
13. Garro, A., Mühlhäuser, M., Tundis, A., Mariani, S., Omicini, A., Vizzari, G.: Intelligent agents and environment. In: Encyclopedia of Bioinformatics and Computational Biology: ABC of Bioinformatics, p. 309 (2018)
14. Railsback, S.F., Grimm, V.: Agent-Based and Individual-based Modeling: A Practical Introduction. Princeton University Press, Princeton (2019)
15. Michel, F., Ferber, J., Drogoul, A.: Multi-agent systems and simulation: a survey from the agent community's perspective. In: Multi-Agent Systems, pp. 17–66. CRC Press (2018)
16. Pipattanasomporn, M., Feroze, H., Rahman, S.: Multi-agent systems in a distributed smart grid: design and implementation. In: 2009 IEEE/PES Power Systems Conference and Exposition. PSCE2009, pp. 1–8. IEEE (2009)
17. Fernández-Isabel, A., Fuentes-Fernández, R.: An agent-based platform for traffic simulation. In: Corchado, E., Snášel, V., Sedano, J., Hassanien, A.E., Calvo, J.L., Ślęzak, D. (eds.) Soft Computing Models in Industrial and Environmental Applications, 6th International Conference SOCO 2011. Advances in Intelligent and Soft Computing, vol. 87, pp. 505–514. Springer, Heidelberg (2011). https://doi.org/10. 1007/978-3-642-19644-7_53
18. González-Briones, A., De La Prieta, F., Mohamad, M.S., Omatu, S., Corchado, J.M.: Multi-agent systems applications in energy optimization problems: a state-of-the-art review. Energies 11(8), 1928 (2018)
19. Conati, C., Klawe, M.: Socially intelligent agents in educational games. In: Dautenhahn, K., Bond, A., Cañamero, L., Edmonds, B. (eds.) Socially Intelligent Agents. Multiagent Systems, Artificial Societies, and Simulated Organizations, vol. 3, pp. 213–220. Springer, Boston (2002). https://doi.org/10.1007/0-306-47373-9_26
20. Fernández-Isabel, A., Fuentes-Fernández, R., de Diego, I.M.: Modeling multi-agent systems to simulate sensor-based smart roads. Simul. Model. Pract. Theory 99, 101994 (2020)
21. Salehinejad, H., Talebi, S.: Dynamic fuzzy logic-ant colony system-based route selection system. Appl. Comput. Intell. Soft Comput. 2010, 13 (2010)
22. Yan, X., Li, L.: Ant agent-based QoS multicast routing in networks with imprecise state information. In: Shi, Z.-Z., Sadananda, R. (eds.) PRIMA 2006. LNCS (LNAI), vol. 4088, pp. 374–385. Springer, Heidelberg (2006). https://doi.org/10. 1007/11802372_36
23. Bellifemine, F., Poggi, A., Rimassa, G.: JADE: a FIPA2000 compliant agent development environment. In: Proceedings of the Fifth International Conference on Autonomous Agents, pp. 216–217 (2001)
24. Yang, Y.J., Sung, T.-W., Wu, C., Chen, H.-Y.: An agent-based workflow system for enterprise based on FIPA-OS framework. Expert Syst. Appl. 37(1), 393–400 (2010)
25. Spade: scheduler for parallel and distributed execution from mobile devices
26. Crooks, A.T., Heppenstall, A.J.: Introduction to agent-based modelling. In: Heppenstall, A., Crooks, A., See, L., Batty, M. (eds.) Agent-Based Models of Geographical Systems, pp. 85–105. Springer, Dordrecht (2012). https://doi.org/ 10.1007/978-90-481-8927-4_5

27. Shehory, O., Sturm, A. (eds.): Agent-Oriented Software Engineering: Reflections on Architectures, Methodologies, Languages, and Frameworks. Springer, Heidelberg (2014). https://doi.org/10.1007/978-3-642-54432-3

28. Pavón, J., Gómez-Sanz, J.J., Fuentes, R.: The INGENIAS methodology and tools. In: Agent-Oriented Methodologies, no. 9, pp. 236–276 (2005)

29. Bresciani, P., Perini, A., Giorgini, P., Giunchiglia, F., Mylopoulos, J.: Tropos: An agent-oriented software development methodology. Auton. Agent. Multi-Agent Syst. 8(3), 203–236 (2004). https://doi.org/10.1023/B:AGNT.0000018806.20944. ef

30. Goulson, D.: Bumblebees: Behaviour, Ecology, and Conservation. Oxford University Press on Demand, Oxford (2010)

31. Korb, J., Thorne, B.: Sociality in termites. In: Comparative Social Evolution, pp. 124–153 (2017)

32. MacDonald, J., Deyrup, M.: The social wasps (hymenoptera: Vespidae) of Indiana. Great Lakes Entomol. 22(3), 7 (2017)

33. Seid, M.A., Traniello, J.F.: Age-related repertoire expansion and division of labor in pheidole dentata (hymenoptera: Formicidae): a new perspective on temporal polyethism and behavioral plasticity in ants. Behav. Ecol. Sociobiol. 60(5), 631–644 (2006). https://doi.org/10.1007/s00265-006-0207-z

34. Jandt, J.M., Dornhaus, A.: Spatial organization and division of labour in the bumblebee Bombus impatiens. Anim. Behav. 77(3), 641–651 (2009)

35. Tschinkel, W.R.: The nest architecture of the florida harvester ant, pogonomyrmex badius. J. Insect Sci. 4(1), 21 (2004)

36. Cicirello, V.A., Smith, S.F.: Wasp nests for self-configurable factories. In: Proceedings of the Fifth International Conference on Autonomous Agents, pp. 473–480 (2001)

37. de Oliveira, V.M., Campos, P.R.: The emergence of division of labor in a structured response threshold model. Phys. A: Stat. Mech. Appl. 517, 153–162 (2019)

38. Duarte, A., Pen, I., Keller, L., Weissing, F.J.: Evolution of self-organized division of labor in a response threshold model. Behav. Ecol. Sociobiol. 66(6), 947–957 (2012). https://doi.org/10.1007/s00265-012-1343-2

39. Shafer, G.: A Mathematical Theory of Evidence, vol. 42. Princeton University Press, Princeton (1976)

40. Zadeh, L.A.: Fuzzy logic. Computer 21(4), 83–93 (1988)

41. Sidhu, A.S.: Recommendation framework based on subjective logic in decision support systems, Ph.D. thesis, University of Windsor (2014)

42. Jøsang, A., Hayward, Pope, S.: Trust network analysis with subjective logic. In: Proceedings of the 29th Australasian Computer Science Conference (ACSW 2006), pp. 885–894. Australian Computer Society (2006)

43. Liu, Y., Li, K., Jin, Y., Zhang, Y., Qu, W.: A novel reputation computation model based on subjective logic for mobile ad hoc networks. Future Gener. Comput. Syst. 27(5), 547–554 (2011)

44. Jøsang, A., Bondi, V.A.: Legal reasoning with subjective logic. Artif. Intell. Law 8(4), 289–315 (2000). https://doi.org/10.1023/A:1011219731903

45. Khan, M.W., Wang, J., Ma, M., Xiong, L., Li, P., Wu, F.: Optimal energy management and control aspects of distributed microgrid using multi-agent systems. Sustain. Cities Soc. 44, 855–870 (2019)

46. Gerbet, T., Kumar, A., Lauradoux, C.: A privacy analysis of Google and Yandex safe browsing. In: 2016 46th Annual IEEE/IFIP International Conference on Dependable Systems and Networks (DSN), pp. 347–358. IEEE (2016)

47. Ma, J., Saul, L.K., Savage, S., Voelker, G.M.: Beyond blacklists: learning to detect malicious web sites from suspicious URLs. In: Proceedings of the 15th ACM SIGKDD International Conference on Knowledge Discovery and Data Mining, pp. 1245–1254 (2009)
48. McGrath, D.K., Gupta, M.: Behind phishing: an examination of phisher modi operandi. In: LEET, no. 4, p. 8 (2008)

Decision Under Ignorance: A Comparison of Existing Criteria

Zoé Krug[1,2], Romain Guillaume[1(✉)], and Olga Battaïa[3]

[1] Université de Toulouse-IRIT, Toulouse, France
`Romain.Guillaume@irit.fr`
[2] ISAE-SUPAERO, Université de Toulouse, Toulouse, France
`zoe.krug@isae.fr`
[3] Kedge Business School, 680 cours de la Liberation, 33405 Talence, France
`olga.battaia@kedgebs.com`

Abstract. In this study, we compare the behavior of classic Hurwicz criterion with three more recent criteria τ-anchor, R^* and R_*. This evaluation is realized on linear optimization problems with uncertain costs coefficients taking into account the risk aversion of the decision maker. The uncertainty is represented by a scenario set.

1 Introduction

Decision or optimization problems often arise in an uncertain context. Depending on available information, several approaches have been proposed to model this uncertainty (e.g. possibility theory [17], evidence theory [18], etc.). In this paper, we focus on the case of low knowledge on possible states, namely decision under ignorance. In this case the decision-maker is able to give the set of possible values of optimization problem parameters but she/he is not able to differentiate them. In other words, all possible parameter values are all possible (this is a particular case in possibility theory when all scenarios have possibility equal to 1). Hurwicz-Arrow proposed a decision under ignorance theory [3] that specifies the properties that a criterion must satisfy. One of the most popular criteria in this context is the Wald criterion (*maxmin* criterion). Recently, a considerable amount of literature on robust optimization has studied this *maxmin* criterion [1,4], and [10]. This criterion is very pessimistic since it focuses on the worst case scenario. Moreover, it is necessary to meet the underlying condition that all scenarios are almost possible (ignorance context). Otherwise, other criteria are more relevant, see [10]. Other criteria have been proposed to take decision-making under ignorance behavior into account. The oldest one is the Hurwciz criterion which consists in modeling optimism by making a linear aggregation with the best and the worst evaluation. This criterion has been used to model the behavior of a decision-maker in different contexts (see [5,14,15,19] ...etc) and has been spread to include imprecise probability theory [12]. This criterion has been criticized in a sequential decision context since it does not satisfy the desired properties in this decision context (for more details see [6,11]).

In order to satisfy the properties of the sequential decision pointed out in [11] and that of the decision under ignorance, two criteria have been recently

© Springer Nature Switzerland AG 2020
M.-J. Lesot et al. (Eds.): IPMU 2020, CCIS 1237, pp. 158–171, 2020.
https://doi.org/10.1007/978-3-030-50146-4_13

proposed namely R^* and R_* [8]. On the other hand, Giang [9] proposes a new criterion namely τ-anchor which satisfies the decision under ignorance property [3] and with Anscombe-Aumann's [2] ideas of reversibility and monotonicity that had been used to characterize subjective probability.

The aim of this paper is to discuss those four criteria in the context of a linear programming problem. We tackle the problem of optimization under ignorance by taking the optimism of the decision-maker into account as a bi-objective optimization problem where the first criterion is the pessimistic point of view and second one is the optimistic point of view. So, we study some properties such as the Pareto optimality of the optimal solution to those criteria. The paper is organized as follows. Firstly, we set out the problem being studied, then we recall the decision under ignorance and we present the four criteria that will be studied. Then, we compare the properties of the optimal solutions to the linear programming problem for all those criteria. Then, the computational aspects of R_* are discussed. Finally, we propose a new criterion which generalizes Hurwicz, R_* and R^* which satisfy the decision under ignorance properties and under some conditions the Pareto optimality.

2 Problem Under Study

In this paper we focus on a Linear Program[1] (Eq. 1) where profit coefficients are uncertain.

Notations

- N: the set of decisions,
- M: the set of constraints,
- x_i: the value of decision $i \in N$,
- $a_{i,j}$: the coefficient of decision variable $i \in N$ for constraints $j \in M$
- p_i: the profit of decision variable $i \in N$,
- b_j: the coefficient of constraints $j \in M$,
- $\mathcal{X}$: the set of feasible solutions (defined by constraints 1.(a) and 1.(b))

$$\max \quad \sum_{i \in N} p_i x_i \tag{1}$$
$$\text{s.t.}$$
$$(a) \ \sum_{i \in N} a_{i,j} x_i \le b_j \ \forall j \in M$$
$$(b) \quad x_i \ge 0 \quad \forall i \in N$$

To model the uncertainty we are given a *scenario set* **S**, which contains all possible vectors of the profit coefficients, called *scenarios*. We thus only know that one profit scenario $s \in \mathbf{S}$ will occur, but we do not know which one until a

---

[1] Throughout this paper we assume that the feasible set of solutions is not empty and is bounded.

solution is computed. The profit of decision variable $i \in N$ under scenario $s \in \mathbf{S}$ is denoted p_i^s and we assume that $p_i^s \geq 0$. No additional information for the scenario set $\mathbf{S}$, such as a probability distribution, is provided. Two methods of defining scenario sets are popular in the existing literature (see, e.g., [4,13] and [16]). The first one is discrete uncertainty representation, $\mathbf{S}^D = \{s_1, \cdots, s_K\}$ contains $K > 1$ explicitly listed scenarios. The second one is interval uncertainty set $\mathbf{S}^I = \prod_{i \in N}[\underline{p}_i, \overline{p}_i]$.

The profit of solution $X = (x_i)_{i \in N}$ depends now on scenario $s \in \mathbf{S}, \mathbf{S} \in \{\mathbf{S}^D, \mathbf{S}^I\}$, and will be denoted as $f(X, s) = \sum_{i \in N} p_i^s x_i$. So the profit of solution $X = (x_i)_{i \in N}$ is a set $F(X) = \{f(X, s), \forall s \in \mathbf{S}\}$. In order to choose a solution which takes into account the optimism of the decision maker, different criteria aggregating minimal and maximal possible values of the profit could be used. In this paper, we will study four different criteria, namely the Hurwicz, τ-anchor, R^* and R_* criterion.

3   Background

In this section, we recall the main results of the decision under ignorance and define the criteria we will consider.

3.1   Decision Under Ignorance

Firstly, we recall the main results of the decision under ignorance theory developed by Hurwicz and Arrow [3]. Two solutions X_1 and X_2 are isomorphic if there is one-to-one mapping h from the set of scenarios such that $\forall s \in \mathbf{S}, f(X_1, s) = f(X_2, h(s))$. Solution X_2 is said to be derived from solution X_1 by deleting duplicate if $F(X_2) \subset F(X_1)$ and for each $w \in F(X_1) \backslash F(X_2)$, there exists $w' \in F(X_2)$ such that $w = w'$. The decision under ignorance is based on 4 axioms (called HA axioms):

A) (Weak order): $\succeq_I$ is a weak order.
B) (Invariance under relabeling axiom (symmetry)). If two solution are isomorphic then they are indifferent.
C) (Invariance under deletion of duplicate states). If X_2 is derived from X_1 by deleting duplicates then X_1 and X_2 are indifferent.
D) (Weak dominance axiom). If X_1 , X_2 are solutions on the same scenario set $\mathbf{S}$ and $\forall s \in \mathbf{S}$, $f(X_1, s) \geq f(X_2, s)$ then $X_1 \succeq_I X_2$.

Theorem 1 *(Hurwicz-Arrow). The necessary and sufficient condition for preference I on the set of solutions $\mathcal{X}$ to satisfy properties A through D is that*

$$X_1 \succeq_I X_2 \ if \ \min_{s \in \mathbf{S}} f(X_1, s) \geq \min_{s \in \mathbf{S}} f(X_2, s) \ and \ \max_{s \in \mathbf{S}} f(X_1, s) \geq \max_{s \in \mathbf{S}} f(X_2, s).$$

The HA theorem says that the comparison between two sets of prizes corresponds to comparing their extremes. If both extremes of one set are greater than or equal to their counterparts in another set then the former is preferred to the latter. The intermediate members of the set do not matter. The criteria presented below satisfy the HA axioms.[2]

3.2   The Hurwicz Criterion [3]

The Hurwicz criterion seeks for a solution that minimizes the convex combination of the best and worst performances (the total profit) across all scenarios. In this case, we solve the following problem:

$$\max_{X \in \mathcal{X}} ((1 - \alpha) \min_{s \in \mathbf{S}} f(X, s) + \alpha \max_{s \in \mathbf{S}} f(X, s)) \tag{2}$$

where $\alpha \in [0, 1]$ is called *optimism-pessimism index*. Clearly, if $\alpha = 1$ then we solve the problem with criterion *max-max*; if $\alpha = 0$ then we solve the problem with criterion *max-min*. Hence, $\alpha \in [0, 1]$ controls the relative importance of two extremes min and max.

3.3   The τ-Anchor Criterion [9]

Recently Giang proposed a new criterion called τ-anchor where *max-min* and *max-max* are special cases as the Hurwicz criterion does. Initially those criteria were defined on $[0, 1]$ but in a linear program context we define that criterion on $]-\infty, +\infty[$. $\tau \in]-\infty, +\infty[$ is called the *tolerance for ignorance* because it characterizes the behavior under ignorance of an individual decision-maker.

$$\max_{X \in \mathcal{X}} C\varepsilon(F(X)) = \begin{cases} \max_{s \in \mathbf{S}} f(X, s) & \text{if } \max_{s \in \mathbf{S}} f(X, s) < \tau \\ \tau & \text{if } \min_{s \in \mathbf{S}} f(X, s) \leq \tau \leq \max_{s \in \mathbf{S}} f(X, s) \\ \min_{s \in \mathbf{S}} f(X, s) & \text{if } \min_{s \in \mathbf{S}} f(X, s) > \tau \end{cases}$$
$$\tag{3}$$

The behavior of a decision-maker is: if *tolerance for ignorance* value of the decision-maker (τ) is not possible she/he evaluates solution (X) using the closest possible profit to her/his characteristic value. Otherwise all solutions containing her/his characteristic value as possible profit are considered equivalent and equal to τ.

3.4   The R_* and R^* Criteria [8]

More recently criteria R_* and R^* have been proposed to take into account the optimism of the decision-maker in the context of a sequential decision problem under total ignorance [8] since they satisfy the properties desired for sequential decision problems. Like τ-anchor those criteria have been defined on $[0, 1]$ but in a linear program context we define those criteria on $]-\infty, +\infty[$. $e \in]-\infty, +\infty[$

---

[2] Note that the average or Ordered weighted average [20] (which Hurwicz generalizes) does not satisfy the properties A through D.

is called the neutral value. Both of these criteria are also known as uni-norm aggregation functions [21].

$$\max_{X \in \mathcal{X}} R_*(F(X)) = \begin{cases} \min_{s \in \mathbf{S}} f(X,s) & \text{if } \min_{s \in \mathbf{S}} f(X,s) < e \\ \max_{s \in \mathbf{S}} f(X,s) & \text{otherwise} \end{cases} \quad (4)$$

$$\max_{X \in \mathcal{X}} R^*(F(X)) = \begin{cases} \min_{s \in \mathbf{S}} f(X,s) & \text{if } \max_{s \in \mathbf{S}} f(X,s) < e \\ \max_{s \in \mathbf{S}} f(X,s) & \text{otherwise} \end{cases} \quad (5)$$

R_* specifies that if one of the $f(X,s)$'s is lower than e then the min operator is applied, otherwise max is applied. R^* specifies that if one of the $f(X,s)$'s is greater than e then the max operator is applied, otherwise min is applied. One can see that these two uni-norms (R_*, R^*) generalize the min and max, as Hurwicz does (max is recovered when $e = -\infty$, min when $e = +\infty$). The identity element e can represent the optimism threshold (like α for Hurwicz).

4   Discussion on Hurwicz, τ-Anchor, R^*, and R_*

To compare those criteria we formulate the problem of taking into account the optimism of the decision-maker as a bi-objective optimization problem:

$$\max_{X \in \mathcal{X}} \{g^{max}(X) = \max_{s \in \mathbf{S}} f(X,s), g^{min}(X) = \min_{s \in \mathbf{S}} f(X,s)\} \quad (6)$$

We will call the *robust solution* the solution optimal for objective function g^{max} and *opportunistic solution* the solution optimal for objective g^{min}. To perform the analysis we need to recall the notions of Pareto optimality[3].

Definition 1. *A solution X_1 is called **Pareto optimal** if there is no $X_2 \neq X_1$ for which $g^i(X_2) \geq g^i(X_1) \ \forall \ i \in \{max, min\}$ and $\exists i \in \{max, min\} \ g^i(X_2) > g^i(X_1)$.*

Definition 2. *A solution X_1 is called **weakly Pareto optimal** if there is no $X_2 \neq X_1$ for which $g^i(X_2) > g^i(X_1) \ \forall \ i \in \{max, min\}$.*

Hence, in this section we firstly study the general properties (namely: *Pareto optimality* and *weak Pareto optimality*) of the optimal solution for all those criteria without taking the property of the linear programming problem into account. From the results of this study we will focus on two criteria, namely Hurwicz and R_*. First, we need introduce some additional notations.

---

[3] The Pareto preferences of problem 6 satisfies the HA axioms.

Notations

- $\mathcal{G} = G(\mathcal{X})$: the feasible set in objective space of problem 6.
- $\mathcal{F} = F(\mathcal{X})$: the feasible set in scenario space.
- $\mathcal{P}$: the set of Pareto optimal solutions of problem 6.
- $\mathcal{P}^w$: the set of weak Pareto optimal solutions of problem 6.
- **M**: the set of optimal solutions of problem 6 with only criterion g^{max},
- **m**: the set of optimal solutions of problem 6 with only criterion g^{min},
- **H**: the set of optimal solutions of problem 6 when Hurwicz criterion is used $\forall \alpha \in [0,1]$,
- **Cε**: the set of optimal solutions of problem 6 when τ-anchor criterion is used $\forall \tau \in]-\infty, +\infty[$
- **R$_*$**: the set of optimal solutions for criterion $R_*, \forall e \in]-\infty, +\infty[$,
- **R***: the set of optimal solutions for criterion $R^*, \forall e \in]-\infty, +\infty[$,

4.1   General Comparison

From Theorem 3.3 and Theorem 3.4 [7] and the fact that Hurwicz criterion is a convex combination of $g^{max}(X)$ and $g^{min}(X)$ we have the following proposition:

Proposition 1. *For scenario set* $\mathbf{S} \in \{\mathbf{S}^D, \mathbf{S}^I\}$*, we have* $\mathbf{H} = \{\mathbf{M} \cup \mathbf{m} \cup (cov(\mathcal{G}) \cap \mathcal{P})\}$[4].

From Proposition 1, we can see that using the Hurwicz criterion a decision-maker with $\alpha \in]0,1[$ can access to compromised solutions which are optimal in the Pareto sense. Nevertheless, if the Pareto front is strictly concave, the optimal solution $\forall \alpha \in [0,1]$ are the optimal solution for maxmin or maxmax criteria. In the next section, we will discuss this point in details.

Let us now study the set of possible optimal solutions for the τ-anchor criterion. τ-anchor breaks down the evaluation space into three areas. We will call those areas: the *min* area when min aggregator is applied, the *equivalent* area when all solutions in this area have the same evaluation $C\varepsilon(F(X)) = \tau$ and finally the *max* area when the max aggregator is applied.

To better understand the behavior of the decision-maker applying the τ-anchor criterion, we look at four possible cases of localization of feasible profit set $\mathcal{F}$ on these three areas. Figure 1 illustrates those cases for a problem with 2 discrete scenarios $\{s_1, s_2\}$ where the *min*, *equivalent* and *max* areas are respectively represented by a red, white and green area. The case (a) shows that all feasible solutions can be considered as equivalent if the solution is good enough for one scenario but too bad for another. In other words, in the case where there is no feasible solution having as maximal evaluation a value greater than τ on both scenarios. The case (b) is close to the case (a) with the exception that the optimal solution has constraints on the maximal possible profit. In case (c), the optimal solutions are the optimal solutions for the maxmax criteria. In the last case, case (d), the optimal solution is the solution for the maxmin criterion. The Proposition 2 sums up the discussion above.

---

[4] $cov(\mathcal{G})$ is the convex hull of $\mathcal{G}$.

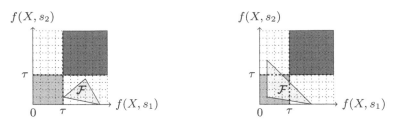

(a) All feasible solutions are in *equivalent* area (b) they exist feasible solutions in *equivalent* area

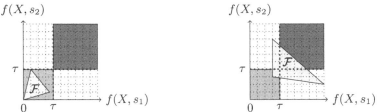

(c) All feasible solutions are in the *max* area   (d) they exist feasible solutions in *min* area

Fig. 1. τ-Anchor

Proposition 2. *Whatever the scenario set* $\mathbf{S} \in \{\mathbf{S}^D, \mathbf{S}^I\}$, *We have:*

$$\mathbf{C}\varepsilon = \begin{cases} \mathbf{M} & if\ \forall X \in \mathcal{X}\ \max_{s \in \mathbf{S}} f(X,s) < \tau \\ \mathbf{m} & if\ \exists X \in \mathcal{X}\ such\,that\ \min_{s \in \mathbf{S}} f(X,s) \geq \tau \\ \{\mathcal{X} \mid \max_{s \in \mathbf{S}} f(X,s) \geq \tau\} & else \end{cases}$$

(7)

In addition to the fact that all solutions may be considered as equivalent, the τ-anchor does not look very interesting from the point of view of bi-objective optimization compared to the Hurwicz criterion since it cannot prefer the Pareto optimal solution which is a compromise between robust and opportunistic solutions.

R^*, as τ-anchor, cuts the evaluation space into areas with the difference being that there is no *equivalent* area. Figure 2 illustrates two interesting situations: (a) all feasible solutions are in the *min* area and (b) there exists a feasible solution in the *max* area. One can see that in case (a) the best solution is a robust solution since we maximize the minimal value without constraints. In case (b), the best solution is the opportunistic solution since we maximize the maximal solution with constraints on the maximal value (greater than e) which is always true if we are in this case. The Proposition 3 sums up the discussion above.

Proposition 3. *For the scenario set* $\mathbf{S} \in \{\mathbf{S}^D, \mathbf{S}^I\}$ *we have* $\mathbf{R}^* = \{\mathbf{M} \cup \mathbf{m}\}$.

This proposition shows that the uni-norm R^* does not look interesting compared to the Hurwicz criterion since only the extreme (robust or opportunistic)

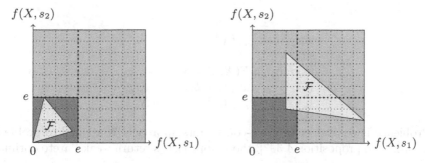

(a) All feasible solutions are in the *min* area   (b) they exist feasible solutions in the *max* area

Fig. 2. R^*

solutions can be preferred. Thus, R^* and τ-anchor have almost opposite behavior since one prefers the robust one and the other prefers the opportunistic one. A decision-maker who is consistent with one of them only needs to know the robust and opportunistic solutions in order to choose one compliant with his/her behaviour without being given any other information, i.e. there is no need to explicitly define the value of τ or e.

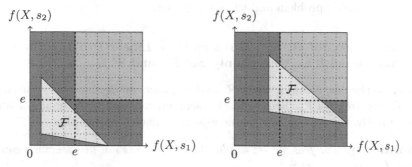

(a) All feasible solutions are in *min* area   (b) they exist feasible solutions in *max* area

Fig. 3. R_*

Let us now focus on the optimization with R_* criterion. From the definition, we can distinguish 2 cases: the case without any feasible solution in the *max* area (Fig. 3.(a)) and the case with (Fig. 3.(b)). In the former case, the optimal solution is the robust one, in the latter case, the optimal solution is the solution which is optimal for the following optimization problem:

$$\begin{aligned}
\max \quad & \max_{s \in \mathbf{S}} F(X, s) \\
\text{s.t.} \quad & \\
(a) \quad & \min_{s \in \mathbf{S}} F(X, s) \geq e \\
(b) \quad & \sum_{i \in N} a_{i,j} x_i \leq b_i \quad \forall j \in M, \\
(c) \quad & x_i \geq 0 \qquad\qquad \forall i \in N
\end{aligned}$$

(8)

Problem 8 is equivalent to the ϵ-constraints approach [7] applied to problem 6. According to proposition 4.3 [7], those approaches return weak Pareto optimal solutions.

Proposition 4. *For scenario set* $\mathbf{S} \in \{\mathbf{S}^D, \mathbf{S}^I\}$ *we have* $\mathbf{R}_* = \mathcal{P}^w$.

From Proposition 4 and 3, we can see that the two uni-norm R_* and R^* differ fundamentally. R^* is a rule to choose between the robust or opportunistic solution while R_* can prefer a compromise solution.

The conclusion of this section is that on one hand we have τ-anchor and R^* which are criteria encoding a rule to choose between the robust and opportunistic solution. On the other hand the Hurwicz and R_* criteria may give the possibility to the decision-maker to prefer a solution which is a compromise between the robust and the opportunistic solution. To study the difference between both criteria in greater depth we need to take into account the characteristics of the linear programming problem and the scenario set.

4.2   Comparison of R_* and Hurwicz for a Linear Programming Problem with an Uncertainty Set $\mathbf{S}^D$ and $\mathbf{S}^I$

To continue the discussion between R_* and Hurwicz, we need to study the shape of the Pareto front of Problem 6. In this section, we start with the uncertainty set $\mathbf{S}^D$. Firstly, the feasible set in the objective space $\mathcal{G}$ is investigated.

Proposition 5. *The feasible set in objective space* $\mathcal{G}$ *is not necessarily a convex polytope for scenario set* $\mathbf{S}^D$.

Corollary 1. *The set of Pareto optimal solution* $\mathcal{P}$ *is not necessarily convex.*

From Propositions 1, 4, and 5, we have the following theorem:

Theorem 2. *For scenario set* $\mathbf{S}^D$ *we have* $\mathbf{H} \subseteq \mathbf{R}_*$.

Proposition 5, Corollary 1 and Theorem 2 are illustrated by the example below:

Example 1. Let us consider two scenarios $s_1 = (p_1^1 = 1, p_2^1 = 0)$ and $s_2 = (p_1^2 = 0, p_2^2 = 1)$ and the following constraints:

$$\begin{aligned}
\text{s.t. } & x_1 + 0.45 \cdot x_2 \leq 8, \\
& x_2 \leq 6, \\
& x_1, x_2 \geq 0
\end{aligned}$$

Figure 4 represent the set of solutions where the x-axis is $\max_{s \in \mathbf{S}} F(X, s)$ and y-axis is $\min_{s \in \mathbf{S}} F(X, s)$. The set of solutions $\mathbf{H}$ and $\mathbf{R}_*$ are represented in red. Since $\mathbf{H}$ is a linear combination of the min and max criteria, it behaves as a straight line. Thus, it will never reach the solutions that are in the concave part of the Pareto front. Because of this phenomenon $\mathbf{H}$ is too restrictive. More precisely, the solutions that offer a good guarantee but have a good opportunity (for instance the coordinates point [2.2,7]) will never be considered. Conversely, $\mathbf{R}_*$ is too permissive and accepts solutions dominated in the Pareto sense.

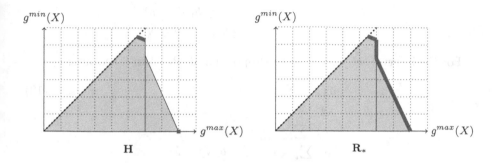

Fig. 4. Comparison of $\mathbf{H}$ and $\mathbf{R}_*$

Let us focus on the case of interval uncertainty set.

Proposition 6. $(\underline{p}_i)_{i \in N} = argmin_{s \in \mathbf{S}^I} F(X^*, s), \forall X^* \in \mathcal{X}$ and $(\overline{p}_i)_{i \in N} = argmax_{s \in \mathbf{S}^I} F(X^*, s), \forall X^* \in \mathcal{X}$.

Proposition 7. *The feasible set in objective space $\mathcal{G}$ is a convex polytope for the scenario set $\mathbf{S}^I$.*

Corollary 2. *The set of Pareto optimal solutions $\mathcal{P}$ is convex.*

From Propositions 1, 4, and 7 we have the following theorem:

Theorem 3. *For the scenario set $\mathbf{S}^I$, we have $\mathbf{H} = \mathbf{R}_*$.*

However, it should be noted that R_* criterion is less unstable for low variation of e than Hurwicz for low variation of α. To our opinion, it handles better the notion of optimism than Hurwicz.

5 Resolution of LP with R_*

We consider R_* criterion since the resolution for Hurwicz, R^* and τ-anchor also requires the resolution of the maxmin and maxmax problems, therefore the conclusion is similar to that for R_*. From a computational point of view, the problem of optimizing an LP under ignorance with a discrete scenario set or interval set using R_* is not harder than an LP problem. For the discrete set, we can easily build an algorithm to solve $|\mathbf{S}^D| + 1$ LP in the worst case. We first need to solve the robust problem and if the solution is better than e then to solve Model 9 for all $s' \in \mathbf{S}^D$ scenarios and choose the best one.

$$\forall s' \in \mathbf{S} \quad \max \quad \sum_{i \in N} p_i^{s'} x_i \tag{9}$$

$$s.t.$$

$$(a) \quad \sum_{i \in N} p_i^s x_i \geq e \quad \forall s \in \mathbf{S},$$
$$(b) \quad \sum_{i \in N} a_{i,j} x_i \leq b_j \quad \forall j \in M,$$
$$(c) \qquad x_i \geq 0 \qquad \forall i \in N$$

For the interval set, according to Proposition 6, Problem 6 becomes:

$$\max \quad \sum_{i \in N} \overline{p}_i x_i \tag{10}$$

$$s.t.$$

$$(a) \quad \sum_{i \in N} \underline{p}_i x_i \geq e$$
$$(b) \quad \sum_{i \in N} a_{i,j} x_i \leq b_i \quad \forall j \in M,$$
$$(c) \qquad x_i, \geq 0 \qquad \forall i \in N$$

6   Generalization of R_*, R^* and Hurwicz Criteria

To propose a generalization of R_*, R^* and Hurwicz, we introduce a new aggregation function I that depends only on the possible maximal and minimal values. It is a parametric aggregation function with 4 parameters $e \in] - \infty, +\infty[$, $a \in [0, 1]$, $b \in [0, 1]$ and $c \in [0, 1]$. To have nondecreasing function (on $\min(F(X))$ and on $\max(F(X))$), we need to add constraints to the parameters: $a \leq b \leq c$.[5] This criterion replaces the min and max function in R_* with the Hurwicz criterion with a different value of α (H^a is the value of the Hurwicz criterion with value $\alpha = a$):

$$\max_{X \in \mathcal{X}} I^{e,a,b,c}(F(X)) = \begin{cases} H^a(F(X)) & \text{if } H^b < e \\ H^c(F(X)) & \text{else} \end{cases} \tag{11}$$

One can see that if $a = b = 0$ and $c = 1$, we obtain R_*. With $a = 0$ and $b = c = 1$, we obtain R^*. There exist more than one parameter which makes the equivalence to Hurwicz, e.g. $a = b = c = \alpha$, $e \in] - \infty, +\infty[$.

As we have shown in the previous section, R_* have the advantage of making more solutions accessible. From some point of view, it enables greater finesse in taking the optimism of the DM into account. However, it can return a dominated solution in the Pareto sense. Some values of parameters I will combine the benefits of R_* and Hurwicz in the sense that the returned optimal solution will

---

[5] Note that the value returned with the use of Hurwicz increases when α increases.

be always Pareto optimal and will possibly include a solution from the concave part of the Pareto front (see Proposition 8, where $\mathcal{I}$ is the set of possible optimal solutions for parameters I).

Proposition 8. $\{\mathbf{H}|\alpha \in]0,1[\} \subseteq \{\mathcal{I}|a,c \in]0,1[and\, a < c\} \subseteq \mathcal{P} \subseteq \mathbf{R}_*.$

Proposition 9. *Criterion I satisfies the HA axioms.*

Example 2. Let us illustrate on Example 1 the use of parameters I. Figure 5.(a) and Fig. 5.(b) illustrate the case where e is a non-compensatory border and the decision-maker is not fully optimistic even if the minimal value of e is guaranteed. Figure 5.(c) and Fig. 5.(d) illustrate the case where e is a compensatory border and the maximal value greater or equal to 8.5 compensates a minimal value equal to 0. The decision-maker is optimistic in the case Fig. 5.(b) or less optimistic Fig. 5.(d). This generalized criterion can be used to specify finely the preferences of the decision-maker.

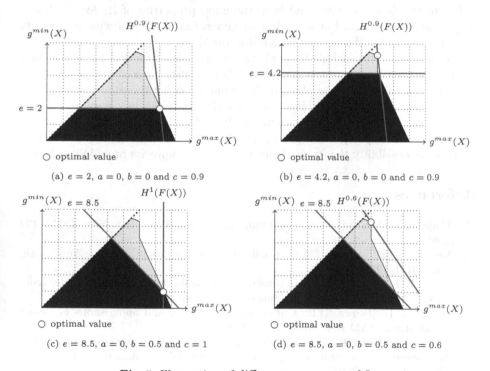

(a) $e = 2$, $a = 0$, $b = 0$ and $c = 0.9$

(b) $e = 4.2$, $a = 0$, $b = 0$ and $c = 0.9$

(c) $e = 8.5$, $a = 0$, $b = 0.5$ and $c = 1$

(d) $e = 8.5$, $a = 0$, $b = 0.5$ and $c = 0.6$

Fig. 5. Illustration of different parameters of I

7   Conclusion

In this paper, we compare four criteria capable of taking the optimism of a decision-maker into account in the context of decision under ignorance, namely

Hurwicz criterion, τ-anchor, R^* and R_*. We show that they can be categorized into two different classes. The first class includes τ-anchor, R^*. They are the criteria which always lead to the extreme solutions: the robust one if the decision maker is pessimistic and to the opportunistic solution if the decision maker is optimistic. The second class includes R_* and Hurwicz. The use of these criteria may lead to a compromised solution which is not completely robust nor completely opportunistic.

We show that R_* and Hurwicz do not lead to the same solution of a linear programming problem. More precisely, R_* solution can be on the concave part of Pareto front while Hurwicz solution can be only on the convex part of Pareto front. Moreover, R_* is more stable to small changes in optimistic/pessimistic parameter value (namely e) than Hurwicz. Previously it was shown that R_* has similar good mathematical properties for sequential decision problems in comparison with Hurwicz. This leads us to conclude that R_* is a good criterion to take the decision-maker's optimism into account in a context of ignorance.

We also develop a generalization of R_* and Hurwicz which gives more flexibility to the decision-maker and keep the good properties of R_* for the linear programming problem, but since it is a generalization of Hurwicz it loses its good properties for the sequential decision problem.

We have also to conclude that taking the optimism of a decision-maker into account in the case where the uncertainty is described by a convex polytope, is a computationally more complex problem than choosing the robust solution. In further research, we are planning to deepen the study on the complexity of the problem for different types of uncertainty including that of a convex polytope. Another research perspective is to generalize R_* and I for other uncertainty contexts as possibility theory, evidence theory, and imprecise probability theory.

References

1. Ahuja, R.K.: Minimax linear programming problem. Oper. Res. Lett. **4**, 131–134 (1985)
2. Anscombe, F.J., Aumann, R.J.: A definition of subjective probability. Ann. Math. Stat. **34**(1), 199–205 (1963)
3. Arrow, K.J., Hurwicz, L.: An optimality criterion for decision-making under ignorance. In: Uncertainty and Expectations in Economics, pp. 1–11 (1972)
4. Bertsimas, D., Brown, D.B., Caramanis, C.: Theory and applications of robust optimization. SIAM Rev. **53**(3), 464–501 (2011)
5. Chassein, A., Goerigk, M., Kasperski, A., Zieliński, P.: Approximating multiobjective combinatorial optimization problems with the OWA criterion. arXiv preprint arXiv:1804.03594 (2018)
6. Dubois, D., Fargier, H., Guillaume, R., Thierry, C.: Deciding under ignorance: in search of meaningful extensions of the Hurwicz criterion to decision trees. In: Grzegorzewski, P., Gagolewski, M., Hryniewicz, O., Gil, M.Á. (eds.) Strengthening Links Between Data Analysis and Soft Computing. AISC, vol. 315, pp. 3–11. Springer, Cham (2015). https://doi.org/10.1007/978-3-319-10765-3_1
7. Ehrgott, M.: Multicriteria Optimization, vol. 491. Springer, Heidelberg (2005). https://doi.org/10.1007/3-540-27659-9

8. Fargier, H., Guillaume, R.: Sequential decision making under uncertainty: ordinal uninorms vs. the Hurwicz criterion. In: Medina, J., Ojeda-Aciego, M., Verdegay, J.L., Perfilieva, I., Bouchon-Meunier, B., Yager, R.R. (eds.) IPMU 2018. CCIS, vol. 855, pp. 578–590. Springer, Cham (2018). https://doi.org/10.1007/978-3-319-91479-4_48

9. Giang, P.H.: Decision making under uncertainty comprising complete ignorance and probability. Int. J. Approximate Reasoning **62**, 27–45 (2015)

10. Gorissen, B.L., Yanıkoğlu, İ., den Hertog, D.: A practical guide to robust optimization. Omega **53**, 124–137 (2015)

11. Jaffray, J.-Y.: Linear utility theory for belief functions. Oper. Res. Lett. **8**(2), 107–112 (1989)

12. Jeantet, G., Spanjaard, O.: Optimizing the Hurwicz criterion in decision trees with imprecise probabilities. In: Rossi, F., Tsoukias, A. (eds.) ADT 2009. LNCS (LNAI), vol. 5783, pp. 340–352. Springer, Heidelberg (2009). https://doi.org/10.1007/978-3-642-04428-1_30

13. Kouvelis, P., Yu, G.: Robust Discrete Optimization and Its Applications, vol. 14. Springer, Heidelberg (2013)

14. Kuhn, K.D., Madanat, S.M.: Model uncertainty and the management of a system of infrastructure facilities. Transp. Res. Part C: Emerg. Technol. **13**(5–6), 391–404 (2005)

15. Lau, H.C., Jiang, Z.-Z., Ip, W.H., Wang, D.: A credibility-based fuzzy location model with Hurwicz criteria for the design of distribution systems in B2C e-commerce. Comput. Ind. Eng. **59**(4), 873–886 (2010)

16. Minoux, M.: Robust network optimization under polyhedral demand uncertainty is NP-hard. Discrete Appl. Math. **158**(5), 597–603 (2010)

17. Dubois, D., Prade, H.: Possibility Theory: An Approach to Computerized Processing of Uncertainty. Springer, Heidelberg (1988). https://doi.org/10.1007/978-1-4684-5287-7

18. Shafer, G.: A Mathematical Theory of Evidence, vol. 42. Princeton University Press, Princeton (1976)

19. Sheng, L., Zhu, Y., Wang, K.: Uncertain dynamical system-based decision making with application to production-inventory problems. Appl. Math. Model. **56**, 275–288 (2018)

20. Yager, R.R.: Generalized OWA aggregation operators. Fuzzy Optim. Decis. Making **3**(1), 93–107 (2004). https://doi.org/10.1023/B:FODM.0000013074.68765.97

21. Yager, R.R., Rybalov, A.: Uninorm aggregation operators. Fuzzy Sets Syst. **80**(1), 111–120 (1996)

Multi-agent Systems and Voting: How Similar Are Voting Procedures

Janusz Kacprzyk[1,4]([✉]) [iD], José M. Merigó[2], Hannu Nurmi[3],
and Sławomir Zadrożny[1,4] [iD]

[1] Systems Research Institute, Polish Academy of Sciences,
ul. Newelska 6, 01-447 Warsaw, Poland
kacprzyk@ibspan.waw.pl
[2] Department of Management Control and Information Systems,
University of Chile, Av. Diagonal Paraguay 257, 8330015 Santiago, Chile
jmerigo@fen.uchile.cl
[3] Department of Political Science, University of Turku, 20014 Turku, Finland
hnurmi@utu.fi
[4] Warsaw School of Information Technology (WIT), 01-447 Warsaw, Poland

Abstract. We consider the problem of the evaluation of similarity of voting procedures which are crucial in voting, social choice and related fields. We extend our approach proposed in our former works and compare the voting procedures against some well established and intuitively appealing criteria, and using the number of criteria satisfied as a point of departure for analysis. We also indicate potential of this approach for extending the setting to a fuzzy setting in which the criteria can be satisfied to a degree, and to include a distance based analysis. A possibility to use elements of computational social choice is also indicated.

Keywords: Voting · Social choice · Voting procedure · Similarity ·
Binary pattern

1 Introduction

This paper is basically concerned with some aspects of a highly advocated trend of considering voting theory to be an important part of the broadly perceived area of multiagent systems (cf. Endriss [4], Pitt, Kamara, Sergot and Artikis [29], Dodevska [3], to just cite a few). In many multiagent systems we need a mechanism that can be used by the agents to make collective decisions, and an effective and efficient way of doing this can be by *voting*, that is by employing a voting procedure. There are very many voting procedures and there has been much research in this area, both related to theoretical analyses and development of voting protocols or procedures. Voting procedures are getting more complex, notably in view of a rapidly increasing importance of all kind of voting in large (maybe huge) sets of agents, for instance because of a proliferation of e-voting, and a rapidly increasing use of computers for this purpose. There is therefore

© Springer Nature Switzerland AG 2020
M.-J. Lesot et al. (Eds.): IPMU 2020, CCIS 1237, pp. 172–184, 2020.
https://doi.org/10.1007/978-3-030-50146-4_14

an urgent need to use in the context of voting – which has traditionally been discussed within economics, decision theory or political science – new concepts, and tools and techniques of computer science and numerical analyses. This is basically offered by the new and rapidly developing area of *computational social choice* (cf. the handbook Brandt, Conitzer, Endriss, Lang and Procaccia [2]).

Multiagent systems are composed of agents, real or virtual entities exemplified by people (individuals, groups or organizations), software, etc. The agents act, interact, cooperate, collaborate, exchange information, etc. In multiagent systems decision processes, including voting, are usually highly uncertain and changeable (dynamic), and proceed in a distributed way, electronic voting is often advocated. In general, the winning option should be in line with opinions or preferences of as many agents as possible. Preferences, which are also an important topic in artificial intelligence (AI), are often used in multiagent models though it is not easy to elicit and deal with multiagent preferences.

We propose here to consider one of interesting problems in voting, both in the traditional social choice setting and a new multiagent setting, which can be stated as follows: there is a multitude of possible voting procedures proposed in the literature and it would be expedient to have a tool to determine if and to what extent they are similar to each other. This could help us, for instance, to use instead of a good but computationally demanding voting procedure a computationally simpler one if it is sufficiently similar to the former one.

In this paper we will discuss the similarity meant as how many requirements (conditions) which are usually assumed in voting are fulfilled jointly by a particular pair of voting procedures, and then to consider as similar those procedures which have similar values. We will use here the idea of our former, more qualitative approach to the comparison of voting procedures, cf. Kacprzyk, Nurmi and Zadrożny [11], and a more qualitative one by Fedrizzi, Kacprzyk and Nurmi [6]. It should be noted that an intuitively justified comparison of voting procedure might be with respect to the results obtained. This might however be too much related to a particular voting problem considered, and not general enough. One can argue, as we do, that the satisfaction (or not) of some important and widely accepted requirements for the voting procedure is possibly related to which solutions they can yield, that is, such a requirement related analysis can be implicitly equivalent to a result related analysis.

Moreover, in this paper we extend the above proposal by a new, fuzzy logic based one in which a "soft" measure of the number of requirements jointly satisfied is used which can be exemplified by a degree to which "about n, many, most, etc. requirements are fulfilled". This can be extended to the case of to which degree the requirements for the voting procedures are satisfied but this will not be considered here as it is not obvious and needs a deeper analysis of what social choice theorists and practitioners think about the intensity of satisfaction.

In our setting there are $n, n \geq 2$ individuals who present their testimonies over the set of $m, m \geq 2$, options. These testimonies can be, for instance, *individual preference relations*, orderings over the set of options, etc. In our approach we focus in principle on *social choice functions*, a class of social choice procedures that select a single social outcome, i.e. an option that best reflects the opinions of the individuals. *Voting procedures* are here perhaps the best known and most intuitively appealing examples. A voting procedure is meant to determine the winner of an election as a function of the votes cast by the voters.

Of a major concern in voting theory is the extent to which voting procedures satisfy some plausible and reasonable axioms, and more foundational results are of a rather negative type, that is, their essence is that plausible and intuitively obvious assumptions are usually incompatible, cf. the Arrow impossibility theorem. For more information, cf. Kelly [15], Nurmi [21], Riker [33], etc.

Much less attention has been paid to the problem of how similar/dissimilar are the voting procedures the number of which is very high. Except for a foundational reference book by Nurmi [21], and a new book by Teixeira de Almeida, Costa Morais and Nurmi [40], one can cite here: Elkind, Faliszewski and Slinko [5], McCabe-Dansted and Slinko [17], Richelson [32], etc.

In this paper we deal with this problem. We take into account a set of popular and well established criteria against which the voting procedures are usually evaluated. To reduce the size of this set, and the size of the set of the voting procedures, we use first the idea of Fedrizzi, Kacprzyk and Nurmi [6] in which these sets are reduced using a qualitative type analysis based on elements of Pawlak's [27], Pawlak and Skowron [28] rough sets theory to obtain the most specific non-redundant characterization of the particular voting procedures with respect to the criteria assumed.

Then, using this reduced representation, we consider the problem of how similar/dissimilar the particular voting procedures are, that is, of how to measure the degree of their similarity/dissimilarity. Our measure is derived from the number of criteria satisfied by a particular procedure, and in our view two voting procedures are similar if they jointly satisfy a similar number of criteria as proposed by Kacprzyk et al. [11]. As mentioned, this could be a good indicator of possibly similar results in terms of functioning, i.e. voting results yielded.

Next, we propose to extend this simple measure by using a fuzzy linguistic quantifier based aggregation to obtain the degree of satisfaction of, e.g., a few, many, most, much more than a half, etc. criteria. We also mention a possibility of using for the above fuzzy linguistic quantifier based aggregation the OWA operators, notably their recent extensions, cf. Kacprzyk, Yager and Merigó [12].

2   A Comparison of Voting Procedures

We are concerned with *social choice functions* which may be, for our purposes, equated with *voting procedures*. The literature on social choice is very rich, and a multitude of various social choice functions (voting procedures) has been proposed which can be simple and sophisticated, intuitively appealing and not, widely employed and not, etc. and there are little or no indications as to which one to use in a particular problem. For information on the comparison and evaluation of voting procedures, cf. the classic sources, e.g., Richelson [32],

Straffin [37], Nurmi [21]; the recent book by Teixeira de Almeida et al., Costa Morais and Nurmi [40] provides much information on new approaches too.

In our context a simple and intuitive approach for the comparison of voting procedures using rough sets as a point of departure, has been proposed by Fedrizzi, Kacprzyk and Nurmi [6]. For a lack of space we will not present it here and refer the reader to that paper. We will just use as the point of departure the reduced problem representation obtained by using the rough sets.

We consider here the following 13 popular voting procedures:

1. Amendment: an option is proposed as a preliminary solution and then motions may be submitted to change it with another option; if such a motion gets required support then the proposed option is considered as a preliminary solution; if there are no more motions then the final vote for the current preliminary option is carried out,
2. Copeland: selects the option for which the number of times it beats other options minus the number of times it looses to other options in pairwise comparisons is the highest,
3. Dodgson: each voter provides a rank ordered list of all options, from the best to worst, and the option wins for which we need to perform the minimum number of pairwise swaps (summed over all candidate options) before they become a Condorcet winner,
4. Schwartz: selects the set of options over which the collective preferences are cyclic and the entire cycle is preferred over the other options; when a Condorcet winner exists this is the single element in such a set otherwise there may be many options,
5. Max-min: selects the option for which the greatest pairwise score for another option against it is the least one of score among all options,
6. Plurality: each voter selects one option (or none if abstains), and the options with the most selection votes win,
7. Borda: each voter provides a linear ordering of the options to which the so-called Borda score is assigned: in case of n candidates, $n - 1$ points is given to the first ranked option, $n - 2$ to the second ranked, etc., these numbers are added up for the options over all orderings which yields the Borda count, and the option(s) with the highest Borda count wins.
8. Approval: each voter selects (approves) a subset of the candidate options and the option(s) with the most votes is/are the winner(s).
9. Black: selects the Condorcet winner, i.e. an option that beats or ties all other options in pairwise comparisons, when one exists or, otherwise, the Borda count winner (as described above),
10. Runoff: plurality vote is used first to find the top two options (or more if there are ties), and then there is a runoff between these options with the one with the most votes to win.
11. Nanson: the Borda count is used, at each step dropping the candidate with the smallest score (majority),
12. Hare: the ballots are linear orders over the set of options, and repeatedly the options are deleted which receive the lowest number of first places in the votes, and the option(s) that remain(s) are the winner(s),

13. Coombs: each voter rank orders all of the options, and if one option is ranked first (among non-eliminated options) by an absolute majority of the voters, then this is the winner, otherwise, the option which is ranked last by a plurality of the voters is eliminated, and this is repeated.

Notice that these voting procedures are well known and popular but, clearly, are just examples of a multitude of possible procedures known in the literature and even employed (cf. the book by Teixeira de Almeida et al. [40]). For example, some recent promising procedures are not used in this paper as Schulze's method [35], Tideman's ranked pairs method [38], and many other ones.

These voting procedures used are based on highly reasonable, desirable and intuitively appealing properties but it is difficult to say if and how similar or different they are. Such an analysis of similarity/dissimilarity can proceed by comparing the voting procedures against some well founded and reasonable criteria (requirements). A multitude of various criteria are possible, and no voting procedure will satisfy all of them. The comparison of voting procedures is therefore a non-trivial task and is to a large extent subjective. We will try to use some formal tools to make it more objective.

The satisfaction of the following criteria is often advocated (cf. Nurmi [21]):

1. Majority winner criterion: if there exists a majority (at least 50%) of voters who rank a single option at the top of the ranking, higher than all other options, then this option should win,
2. Mutual majority criterion: if there exists a majority of voters ranking a group of options higher than all other options, one of the options from that group should win.
3. Majority loser criterion: if a majority of voters prefers every other option over a given one, the latter option should not win,
4. Monotonicity criterion: it is impossible to make a winning option lose by ranking it higher, or to cause a losing option to win by ranking it lower,
5. Consistency criterion: if the electorate is divided in two groups and an option wins in both groups, then it should win in general,
6. Weak Pareto criterion: whenever all voters rank an option higher than another option, the latter option should never be chosen,
7. Participation criterion: it should always be better to vote honestly than not to vote at all,
8. Condorcet winner criterion: if an option beats every other option in pairwise comparisons, then it should always win,
9. Condorcet loser criterion: if an option loses to every other option in pairwise comparisons, it should always loose,
10. Independence of irrelevant alternatives: if an option is added or removed, the relative rankings of the remaining options should remain the same,
11. Independence of clones: the outcome of voting should be the same if we add options identical to the existing ones (clones),
12. Reversal symmetry: if individual preferences of each voter are inverted, the original winner should never win,

13. Heritage criterion: if an option is chosen from the entire set of options using a particular voting procedure, then it should also be chosen from all subsets of the set of options (to which it belongs) using the same voting procedure and under the same preferences.
14. Polynomial time: it should be possible to find the winner in polynomial time with respect to the number of options and voters.

In general, a criterion can be said to be "stronger" (more widely adopted/adhered to) than another one when it is satisfied by more voting procedures. This will be of relevance for our consideration.

For clarity and simplicity of presentation and interpretation, we will only use the following 7 "strong" criteria (the letters A, ..., G correspond to the labels of columns in the tables to be shown), which can be claimed to be especially important (cf. Fedrizzi, Kacprzyk and Nurmi [6]), in the analysis and comparison of voting procedures: A – Condorcet winner, B – Condorcet loser, C – majority winner, D – monotonicity, E – weak Pareto winner, F – consistency, and G – heritage, and a similar analysis can be extended to all 13 criteria listed before, as well as many other ones which can be found in the literature.

In the tables showing results of the subsequent steps of our approach, the rows will correspond to the 13 voting procedures analyzed in the paper: Amendment, Copeland, Dodgson, Schwartz, Max-min, Plurality, Borda, Approval, Black, Runoff, Nanson, Hare, and Coombs.

The columns correspond to 7 above mentioned criteria: Condorcet winner, Condorcet loser, majority winner, monotonicity, weak Pareto winner, consistency, and heritage.

The point of departure is presented in Table 1 which shows which voting procedure satisfies which criterion: "0" stands for "does not satisfy", and "1" stands for "satisfies".

It should be noticed that the data set given in Table 1 can be directly used for the comparison of the 13 voting procedures considered with respect to the 7 criteria assumed. Basically, such a comparison can be accomplished by comparing the consecutive pairs of the binary rows corresponding to the voting procedures with each other using some methods for the determination of similarity and dissimilarity (cf. Kacprzyk et al. [11]). However, this would not provide any deeper insight into the differences between the voting procedures as the comparison would concern just particular voting procedures and not their more or less homogeneous classes. This problem is closely related to Kacprzyk and Zadrożny's [13,14] OWA operator based approach to the classification of voting procedures into a number of more general classes that are related, first, to the order in which the aggregation via an OWA operator proceeds and, second, to specific sets of weights of the respective OWA operators. In this paper we use another way of comparing the voting procedures based on an analysis of how many criteria are jointly fulfilled, and on the related distance between the voting procedures, and then on some structural analyses using human consistent natural language summaries.

Table 1. Satisfaction of 7 criteria by 13 voting procedures

Voting procedure	Criteria						
	A	B	C	D	E	F	G
Amendment	1	1	1	1	0	0	0
Copeland	1	1	1	1	1	0	0
Dodgson	1	0	1	0	1	0	0
Schwartz	1	1	1	1	0	0	0
Max-min	1	0	1	1	1	0	0
Plurality	0	0	1	1	1	1	0
Borda	0	1	0	1	1	1	0
Approval	0	0	0	1	0	1	1
Black	1	1	1	1	1	0	0
Runoff	0	1	1	0	1	0	0
Nanson	1	1	1	0	1	0	0
Hare	0	1	1	0	1	0	0
Coombs	0	1	1	0	1	0	0

Table 2. Satisfaction of 7 criteria by a reduced number (9 families) of voting procedures

Voting procedure	Criteria						
	A	B	C	D	E	F	G
Amendment	1	1	1	1	0	0	0
Copeland	1	1	1	1	1	0	0
Dodgson	1	0	1	0	1	0	0
Max-min	1	0	1	1	1	0	0
Plurality	0	0	1	1	1	1	0
Borda	0	1	0	1	1	1	0
Approval	0	0	0	1	0	1	1
Runoff	0	1	1	0	1	0	0
Nanson	1	1	1	0	1	0	0

The first step of the approach proposed in Fedrizzi et al. [6] and Kacprzyk et al. [11] is the simplification of the problem in the sense of the reduction of the number of voting procedures using elements of Pawlak's rough sets theory (cf. Pawlak [27], Pawlak and Skowron [28]). Basically, first, we merge those voting procedures which satisfy the same properties, i.e. under the set of criteria assumed they may be considered to be equivalent. We obtain therefore the following 9 voting procedures (Table 2): Amendment (which stands now for Amendment and Schwartz), Copeland (which stands now for Copeland and Black), Dodgson, Max-min, Plurality, Borda, Approval, Runoff (which stands now for Runoff, Hare and Coombs), and Nanson. These are equivalence classes of the indiscernibility relation which may be defined as usual, in line with the rough sets theory.

This step may be followed by another one aiming at reducing also the number of criteria. One may identify the so-called *indispensable criteria* (the core) which are meant as those whose omission will make at least one pair of voting procedures indistinguishable. That is, such criteria are necessary to differentiate between the voting procedures. Then, we finally obtain the reduced representation of the voting procedures versus the criteria as shown in Table 3 which expresses the most crucial properties or criteria of the voting procedures in the sense that the information it conveys would be sufficient to restore all information given in the source Table 1. For details we refer the reader to cf. Fedrizzi, Kacprzyk and Nurmi [6] and Kacprzyk, Nurmi and Zadrożny [11]. However, in this paper we will assume that only the first step is executed, i.e., the number of voting procedures is reduced but all the criteria are preserved.

Table 3. Satisfaction of the criteria belonging to the core by the particular voting procedures

Voting procedure	Criteria			
	A	B	D	E
Amendment	1	1	1	0
Copeland	1	1	1	1
Dodgson	1	0	0	1
Max-min	1	0	1	1
Plurality	0	0	1	1
Borda	0	1	1	1
Approval	0	0	1	0
Runoff	0	1	0	1
Nanson	1	1	0	1

3 Similarity and Distances Between Voting Procedures: An Indiscernibility Based Analysis

We operate here on the characterization of the voting procedures shown in the Table 2. This will better serve the purpose of presenting a new approach to the comparison of voting procedures, and also provide a point of departure for further works in which similarity analyses will be performed on reduced representations.

For each pair of voting procedures, $(x, y) \in V^2$, where V is the set of voting procedures (9 in our case, as in Table 2), and for each criterion z, $z \in Z$, where Z is the set of criteria assumed (7 in our case, as in Table 1), we define the following function $v_z : V \times V \longrightarrow \{0, 1\}$, such that

$$v_z(x, y) = \begin{cases} 1 \text{ if } x \text{ and } y \text{ take on the same values for criterion } z \\ 0 \text{ otherwise} \end{cases} \quad (1)$$

For example, for the data given in Table 1:

$$v_A(Amendment, Copeland) = 1$$
$$v_E(Amendment, Copeland) = 0$$

In the simplest way the agreement between two voting procedures, $x, y \in V$, denoted by $A(x, y)$, $A : V \times V \longrightarrow \{0, \ldots, \text{card } Z\}$, can be defined in terms of $v_z(x, y)$ given by (1) as follows:

$$A(x, y) = \sum_{z \in Z} v_z(x, y) \quad (2)$$

that is as the number of simultaneous satisfaction/dissatisfaction of the criteria.

Therefore, we get the following matrix of agreements (cf. Table 4). In Table 4, the agreement between the same voting procedures does not matter, so that "-" is put, and since the agreement function is symmetric, we only define the upper half of the matrix.

Table 4. Values of agreements between the particular voting procedures due to (2)

Voting procedure	Voting procedure								
	Amendment	Copeland	Dodgson	Max-min	Plurality	Borda	Approval	Runoff	Nanson
Amendment	–	6	4	5	3	3	1	4	5
Copeland		–	5	6	4	4	1	5	6
Dodgson			–	6	4	2	2	5	6
Max-min				–	5	3	2	4	5
Plurality					–	5	4	4	3
Borda						–	4	4	3
Approval							–	1	0
Runoff								–	6
Nanson									–

It can be illustrative to present the results in the form of some summarizing statements. One can notice that the Copeland, Max-Min, Dodgson and Nanson form a group of voting procedures which are no more than two criteria away of each other. Quite closely related to that group are Runoff and Amendment. The so-called positional methods, that is, Plurality, Borda and Approval, seem to be rather far away from the rest of the procedures in terms of the number of criteria they differ by. This holds particularly for Approval.

It is easy to see that this indiscernibility analysis based on the sets of criteria satisfied jointly by pairs of voting procedures yields here sets of cardinality 0, 1, 2, 3, 4, 5, 6. Of course, in practice it is usually not important if two voting procedures differ by 2 or 3, 4 or 5, or 5 or 6 criteria so that one can use here aggregated values, for instance, by merging these values, which would yield a more compact representation.

However, such a merging of numbers of criteria may be difficult because it needs a deep insight into how important the particular criteria are, and the satisfaction of which combination of them is relevant.

Moreover, it may often be convenient to use a natural language description of the similarity, for instance similar in terms of: a low, medium or high, a few and many, about n, etc. number of criteria satisfied.

Technically, this can easily be done by using tools and techniques of fuzzy logic, to be more specific the well known fuzzy logic based calculus of linguistically quantified propositions by Zadeh.

The method presented in this section, and the results obtained, which is based on some indiscernibility analyses, may be viewed to be somewhat qualitative. To proceed to a more quantitative analysis, we can use the normalized

Table 5. Normalized distance between the particular voting procedures due to (3)

Voting procedure	Voting procedure								
	Amendment	Copeland	Dodgson	Max-min	Plurality	Borda	Approval	Runoff	Nanson
Amendment	—	$\frac{1}{7}$	$\frac{3}{7}$	$\frac{2}{7}$	$\frac{4}{7}$	$\frac{4}{7}$	$\frac{6}{7}$	$\frac{3}{7}$	$\frac{2}{7}$
Copeland		—	$\frac{2}{7}$	$\frac{1}{7}$	$\frac{3}{7}$	$\frac{3}{7}$	$\frac{6}{7}$	$\frac{2}{7}$	$\frac{1}{7}$
Dodgson			—	$\frac{1}{7}$	$\frac{3}{7}$	$\frac{5}{7}$	$\frac{5}{7}$	$\frac{2}{7}$	$\frac{1}{7}$
Max-min				—	$\frac{2}{7}$	$\frac{4}{7}$	$\frac{5}{7}$	$\frac{3}{7}$	$\frac{2}{7}$
Plurality					—	$\frac{2}{7}$	$\frac{3}{7}$	$\frac{3}{7}$	$\frac{4}{7}$
Borda						—	$\frac{3}{7}$	$\frac{3}{7}$	$\frac{4}{7}$
Approval							—	$\frac{6}{7}$	$\frac{4}{7}$
Runoff								—	$\frac{1}{7}$
Nanson									—

distance between two voting procedures $x, y \in V$ which can be defined in a straightforward way as

$$D(x, y) = 1 - \frac{A(x, y)}{\operatorname{card} Z} \tag{3}$$

where $A(x, y)$ is given by (2) and card Z is the number of criteria.

Therefore, using (3), we obtain the matrix of normalized distances between the voting procedures given by Table 5.

Then, a distance based analysis can be performed along the lines of Kacprzyk, Nurmi and Zadrożny [11] but this is outside of the scope of this paper that is focused on an indiscernibility analysis.

4   Concluding Remarks

We have presented a new approach to the evaluation of similarity/dissimilarity of voting procedures. We followed the approach proposed in our former works and compared the voting procedures against some well established and intuitively appealing criteria, and using the number of criteria satisfied as a point of departure. We have indicated some further research directions, notably using elements of fuzzy logic to describe the delicacy of the comparison of voting procedures and also a possibility to extend the analysis to a distance based reasoning.

We hope that this work will help solve one of problems that exists in the use of voting procedures in multiagent systems (cf. Dodevska [3] or Endriss [4]). Moreover, for large scale voting problems which are more and more important in practice, computational social sciences can provide a rich set of tools and techniques which will help solve our problem (cf. Elkind, Faliszewski and Slink [5]).

References

1. Arrow, K.J., Sen, A.K., Suzumura, K. (eds.): Handbook of Social Choice and Welfare, 1st edn. Elsevier, Amsterdam (2002)
2. Brandt, F., Conitzer, V., Endriss, U., Lang, J., Procaccia, A.D. (eds.): Handbook of Computational Social Choice. Cambridge University Press, Cambridge (2016)
3. Dodevska, Z.: Voting in multi-agent systems. Tehnika-Menadžment **69**(5), 724–740 (2019)
4. Endriss, U.: Social choice theory as a foundation for multiagent systems. In: Müller, J.P., Weyrich, M., Bazzan, A.L.C. (eds.) MATES 2014. LNCS (LNAI), vol. 8732, pp. 1–6. Springer, Cham (2014). https://doi.org/10.1007/978-3-319-11584-9_1
5. Elkind, E., Faliszewski, P., Slinko, A.: On the role of distances in defining voting rules. In: van der Hoek, W., Kaminka, G.A., Lespérance, Y., Luck, M., Sen, S. (eds.) Proceedings of 9th International Conference on Autonomous Agents and Multiagent Systems (AAMAS 2010), pp. 375–382 (2010)
6. Fedrizzi, M., Kacprzyk, J., Nurmi, H.: How different are social choice functions: a rough sets approach. Qual. Quant. **30**, 87–99 (1996)
7. Gibbard, A.: Manipulation of voting schemes: a general result. Econometrica **41**(4), 587–601 (1973)

8. Kacprzyk, J.: Group decision making with a fuzzy majority. Fuzzy Sets Syst. **18**, 105–118 (1986)
9. Kacprzyk, J., Fedrizzi, M.: A 'human consistent' degree of consensus based on fuzzy logic with linguistic quantifiers. Math. Soc. Sci. **18**, 275–290 (1989)
10. Kacprzyk, J., Fedrizzi, M., Nurmi, H.: Group decision making and consensus under fuzzy preferences and fuzzy majority. Fuzzy Sets Syst. **49**, 21–31 (1992)
11. Kacprzyk, J., Nurmi, H., Zadrożny, S.: Using similarity and dissimilarity measures of binary patterns for the comparison of voting procedures. In: Kacprzyk, J., Filev, D., Beliakov, G. (eds.) Granular, Soft and Fuzzy Approaches for Intelligent Systems. SFSC, vol. 344, pp. 141–169. Springer, Cham (2017). https://doi.org/10.1007/978-3-319-40314-4_8
12. Kacprzyk, J., Yager, R.R., Merigó, J.M.: Towards human-centric aggregation via ordered weighted aggregation operators and linguistic data summaries: a new perspective on Zadeh's inspirations. IEEE Comput. Intell. Mag. **14**(1), 16–30 (2019). https://doi.org/10.1109/MCI.2018.2881641
13. Kacprzyk, J., Zadrożny, S.: Towards a general and unified characterization of individual and collective choice functions under fuzzy and nonfuzzy preferences and majority via the ordered weighted average operators. Int. J. Intell. Syst. **24**(1), 4–26 (2009)
14. Kacprzyk, J., Zadrożny, S.: Towards human consistent data driven decision support systems using verbalization of data mining results via linguistic data summaries. Bull. Pol. Acad. Sci.: Tech. Sci. **58**(3), 359–370 (2010)
15. Kelly, J.S.: Arrow Impossibility Theorems. Academic Press, New York (1978)
16. Kelly, J.S.: Social Choice Theory. Springer, Berlin (1988). https://doi.org/10.1007/978-3-662-09925-4
17. McCabe-Dansted, J.C., Slinko, A.: Exploratory analysis of similarities between social choice rules. Group Decis. Negot. **15**(1), 77–107 (2006)
18. Moulin, H.: The strategy of social choice. Advanced Textbooks in Economics. North-Holland, Amsterdam (1983)
19. Moulin, H.: Axioms of Cooperative Decision Making. Cambridge University Press, Cambridge (1991)
20. Merrill, S.: Making Multicandidate Elections More Democratic. Princeton University Press, Princeton (1988)
21. Nurmi, H.: Comparing Voting Systems. D. Reidel, Dordrecht (1987)
22. Nurmi, H.: Discrepancies in the outcomes resulting from different voting schemes. Theory Decis. **25**, 193–208 (1988a)
23. Nurmi, H.: Inferential modes in applying social choice theory. In: Munier, B.R., Shakun, M.F. (eds.) Compromise, Negotiation and Group Decision. D. Reidel, Dordrecht (1988b)
24. Nurmi, H.: An assessment of voting system simulations. Publ. Choice **73**, 459–487 (1992)
25. Nurmi, H., Kacprzyk, J.: On fuzzy tournaments and their solution concepts in group decision making. Eur. J. Oper. Res. **51**, 223–232 (1991)
26. Nurmi, H., Kacprzyk, J., Fedrizzi, M.: Probabilistic, fuzzy and rough concepts in social choice. Eur. J. Oper. Res. **95**, 264–277 (1996)
27. Pawlak, Z.: Rough Sets: Theoretical Aspects of Reasoning About Data. Kluwer, Dordrecht (1991)
28. Pawlak, Z., Skowron, A.: Rudiments of rough sets. Inf. Sci. **177**(1), 3–27 (2007)
29. Pitt, J., Kamara, L., Sergot, M., Artikis, A.: Voting in multi-agent systems. Comput. J. **49**(2), 156–170 (2006)

30. Plott, C.R.: Axiomatic social choice theory: an overview and interpretation. Am. J. Polit. Sci. **20**, 511–596 (1976)
31. Shoham, Y., Leyton-Brown, K.: Multiagent Systems: Algorithmic, Game Theoretic, and Logical Foundations. Cambridge University Press, Cambridge (2009)
32. Richelson, J.: A comparative analysis of social choice functions I, II, III: a summary. Behav. Sci. **24**, 355 (1979)
33. Riker, W.H.: Liberalism Against Populism. W. H. Freeman, San Francisco (1982)
34. Satterthwaite, M.A.: Strategy-proofness and Arrow's conditions: existence and correspondence theorems for voting procedures and social welfare functions. J. Econ. Theory **10**, 187–217 (1975)
35. Schulze, M.: A new monotonic, clone-independent, reversal symmetric, and Condorcet-consistent single-winner election method. Soc. Choice Welfare **36**(2), 267–303 (2011)
36. Schwartz, T.: The Logic of Collective Choice. Columbia University Press, New York (1986)
37. Straffin, P.D.: Topics in the Theory of Voting. Birkhäuser, Boston (1980)
38. Tideman, N.T.: Collective Decisions and Voting: The Potential for Public Choice. Ashgate Publishing, Farnham (2006)
39. Yager, R.R., Kacprzyk, J., Beliakov, G. (eds.): Recent Developments in the Ordered Weighted Averaging Operators: Theory and Practice. Springer, Heidelberg (2011). https://doi.org/10.1007/978-3-642-17910-5
40. de Almeida, A.T., Morais, D.C., Nurmi, H.: Systems, Procedures and Voting Rules in Context. AGDN, vol. 9. Springer, Cham (2019). https://doi.org/10.1007/978-3-030-30955-8

Optimization and Uncertainty

Softening the Robustness of Optimization Problems: A New Budgeted Uncertainty Approach

Romain Guillaume[1], Adam Kasperski[2], and Paweł Zieliński[2]($\boxtimes$)

[1] Université de Toulouse-IRIT, Toulouse, France
`Romain.Guillaume@irit.fr`
[2] Wrocław University of Science and Technology, Wrocław, Poland
{`Adam.Kasperski,Pawel.Zielinski`}`@pwr.edu.pl`

Abstract. In this paper an optimization problem with uncertain parameters is discussed. In the traditional robust approach a pessimistic point of view is assumed. Namely, a solution is computed under the worst possible parameter realizations, which can lead to large deterioration of the objective function value. In this paper a new approach is proposed, which assumes a less pessimistic point of view. The complexity of the resulting problem is explored and some methods of solving its special cases are presented.

Keywords: Robustness · Uncertainty · Optimization

1 Introduction

In this paper we wish to investigate the following optimization problem with uncertain parameters:

$$
\begin{aligned}
\max(\min) \ & \boldsymbol{c}^T \boldsymbol{x}, \\
\text{s.t.} \quad & \widetilde{\boldsymbol{a}}_i^T \boldsymbol{x} \le b_i \quad i \in [m], \\
& \boldsymbol{x} \in \mathbb{X} \subseteq \mathbb{R}_+^n.
\end{aligned}
\tag{1}
$$

In formulation (1), $\boldsymbol{x}$ is n-vector of nonnegative decision variables, $\boldsymbol{c}$ is n-vector of deterministic objective function coefficients, $\widetilde{\boldsymbol{a}}_i = (\widetilde{a}_{i1}, \dots, \widetilde{a}_{in})$ is n-vector of uncertain constraint coefficients, $i \in [m]$ ($[m]$ denotes the set $\{1, \dots, m\}$), and $\mathbb{X}$ is a bounded subset of $\mathbb{R}_+^n$, where $\mathbb{R}_+$ is the set of nonnegative reals. For example, if $\mathbb{X}$ is a bounded polyhedron, then (1) is an uncertain linear programming problem. If $\mathbb{X} \subseteq \{0,1\}^n$, then (1) is an uncertain combinatorial optimization problem. We will first assume that the right-hand sides b_i, $i \in [m]$, of the constraints are deterministic. Later, we will also discuss the case with uncertain b_i. We can assume w.l.o.g. that the objective function coefficients are

---

Romain Guillaume was partially supported by the project caasc ANR-18-CE10-0012 of the French National Agency for Research. Adam Kasperski and Paweł Zieliński were supported by the National Science Centre, Poland, grant 2017/25/B/ST6/00486.

© Springer Nature Switzerland AG 2020
M.-J. Lesot et al. (Eds.): IPMU 2020, CCIS 1237, pp. 187–200, 2020.
https://doi.org/10.1007/978-3-030-50146-4_15

precise. Otherwise, in a minimization problem, we can replace the objective function with minimization of a new variable t and add one additional uncertain constraint $\widetilde{\boldsymbol{c}}^T \boldsymbol{x} - t \leq 0$. The transformation for maximization problems is similar.

A method of solving (1) depends on the information available. If $\widetilde{\boldsymbol{a}}_i$ is a vector of random variables with known probability distribution, then the ith imprecise constraint can be replaced with a chance constraint of the form $\Pr(\widetilde{\boldsymbol{a}}_i^T \boldsymbol{x} \leq b_i) \geq 1 - \epsilon$, where $\epsilon \in [0, 1)$ is a given risk level [8]. Assume that we only know that $\widetilde{\boldsymbol{a}}_i \in \mathcal{U}_i \subseteq \mathbb{R}^n$, where $\mathcal{U}_i$ is a given *uncertainty* (*scenario*) *set*. In this paper we use the following interval model of uncertainty [3]. For each uncertain coefficient $\widetilde{a}_{ij}$ an interval $[\widehat{a}_{ij} - \Delta_{ij}, \widehat{a}_{ij} + \Delta_{ij}]$ is provided, where $\widehat{a}_{ij}$ is the *nominal value* of $\widetilde{a}_{ij}$ and Δ_{ij} is the *maximal deviation* of the value of $\widetilde{a}_{ij}$ from its nominal one. The interval can be interpreted as a support of random variable $\widetilde{a}_{ij}$, symmetrically distributed around its nominal (expected) value [3]. Set $\mathcal{U}_i$ is the Cartesian product of the uncertainty intervals $[\widehat{a}_{ij} - \Delta_{ij}, \widehat{a}_{ij} + \Delta_{ij}]$, $j \in [n]$. Let $\widehat{\boldsymbol{a}}_i \in \mathbb{R}^n$ be a vector of the nominal constraint coefficients. After replacing the uncertain vectors $\widetilde{\boldsymbol{a}}_i$ with their nominal counterparts $\widehat{\boldsymbol{a}}_i$ for each $i \in [m]$, we get a deterministic *nominal problem* with the optimal objective value equal to $\widehat{c}$. Using the robust optimization framework [2,10,13], the ith imprecise constraint can be replaced with

$$\max_{\boldsymbol{a}_i \in \mathcal{U}_i} \boldsymbol{a}_i^T \boldsymbol{x} \leq b_i, \tag{2}$$

which ensures that $\boldsymbol{x}$ is feasible for all scenarios $\boldsymbol{a}_i \in \mathcal{U}_i$.

The application of *strict robustness* concept (2) results in a very conservative constraint, in which we assume by the non-negativity of $\boldsymbol{x}$, that the true realization of all the coefficients will be at $\widehat{a}_{ij} + \Delta_{ij}$, $j \in [n]$. Hence the objective value of the strict robust solution can be much less than $\widehat{c}$. This phenomenon is called a *price of robustness* [3] and large price of robustness is often regarded as the main drawback of the strict robust optimization. However, in many practical situations, the true realization of $\widetilde{\boldsymbol{a}}_i$ will be rather closer to $\widehat{\boldsymbol{a}}_i$, as the extreme values of the coefficients are less probable to occur, especially when everything goes smoothly without any perturbations [5].

Several approaches have been proposed in the literature to soften the strict robustness. One of the most popular was introduced in [3]. The key idea is to assume that at most Γ_i coefficients in the ith constraint will take the values different than their nominal ones. To simplify presentation, we will assume that Γ_i is an integer in $\{0, \ldots, n\}$. Accordingly, the ith constraint becomes then

$$\max_{\{\boldsymbol{a}_i \in \mathcal{U}_i : |\{a_{ij} : a_{ij} \neq \widehat{a}_{ij}, j \in [n]\}| \leq \Gamma_i\}} \boldsymbol{a}_i^T \boldsymbol{x} \leq b_i. \tag{3}$$

Notice that the case when $\Gamma_i = 0$ only ensures that $\boldsymbol{x}$ is feasible under the nominal scenario (we get the nominal problem). On the other hand, $\Gamma_i = n$ ensures that $\boldsymbol{x}$ is feasible under all scenarios and, in this case, (3) is equivalent to (2). The parameter Γ_i allows decision makers to control the robustness of the constraint. By changing Γ_i, we get a family of solutions with different levels of robustness. However, it is still assumed that Γ_i constraint coefficients may take their worst values, which represents a pessimistic point of view. In [5,12] an approach to soften the robustness of (2) was proposed. The idea is to assume that the solution cost should be of some predefined distance to $\hat{c}$, which can be achieved by allowing additional constraint violations (see [12] for details). Another method consists in replacing very conservative minmax criterion with the minmax regret one (see [9,10] for more details and the references given there). However, the minmax regret problems are typically hard to solve, even for linear programming problems [1].

In this paper we propose a new approach to soften the strict robustness. The idea will be to modify the approach proposed in [3], by replacing the pessimistic point of view with a more optimistic one. We will still assume that Γ_i constraint coefficients can take the worst values. However, contrary to (3), we will assume that this will happen in the best possible case. Consequently, the objective function will be optimized over the larger set of feasible solutions and the optimal objective value will be closer to $\hat{c}$.

2   New Concept of Choosing Robust Solutions

In this section we propose a new concept to soften the conservatism of the strict robust approach. We will use the same model of uncertainty as the one described in the previous section. Namely, for each uncertain coefficient $\tilde{a}_{ij}$ we define the uncertainty interval $[\hat{a}_{ij} - \Delta_{ij}, \hat{a}_{ij} + \Delta_{ij}]$. Also $\Gamma_i \in \{0, \ldots, n\}$ specifies the number of coefficients in the ith constraint, whose values can be different from their nominal ones. Let $\Phi_i = \{\boldsymbol{\delta}_i \in \{0,1\}^n : \sum_{j \in [n]} \delta_{ij} = \Gamma_i\}$. A fixed vector $\boldsymbol{\delta}_i \in \Phi_i$ induces the following convex uncertainty set:

$$\mathcal{U}_{\boldsymbol{\delta}_i} = \{\boldsymbol{a}_i \in \mathbb{R}^n : a_{ij} \in [\hat{a}_{ij} + \delta_{ij}\Delta_{ij}], \ j \in [n]\}.$$

Since $\boldsymbol{x} \in \mathbb{R}^n_+$, the constraint (3) can be rewritten equivalently as

$$\max_{\boldsymbol{\delta}_i \in \Phi_i} \max_{\boldsymbol{a}_i \in \mathcal{U}_{\boldsymbol{\delta}_i}} \boldsymbol{a}_i^T \boldsymbol{x} \leq b_i. \tag{4}$$

We can now provide the following interpretation of (4). Given a solution $\boldsymbol{x}$, we first choose the worst uncertainty set $\mathcal{U}_{\boldsymbol{\delta}_i}$ and then the worst scenario $\boldsymbol{a}_i$ in this set. This represents a pessimistic point of view. From an optimistic point of view, we can assume that the best uncertainty set is chosen in the first step, which leads to the following constraint:

$$\min_{\boldsymbol{\delta}_i \in \Phi_i} \max_{\boldsymbol{a}_i \in \mathcal{U}_{\boldsymbol{\delta}_i}} \boldsymbol{a}_i^T \boldsymbol{x} \leq b_i. \tag{5}$$

If $\mathbb{X}$ is a polyhedron, then the set of feasible solutions to (4) is convex, because it can be represented by an intersection of polyhedral sets. Indeed, $\max_{\boldsymbol{a}_i \in \mathcal{U}_{\boldsymbol{\delta}_i}} \boldsymbol{a}_i^T \boldsymbol{x} = \sum_{j \in [n]} (\hat{a}_{ij} + \delta_{ij}\Delta_{ij})x_j$ and (4) is equivalent to the family (conjunction) of the linear constraints $\sum_{j \in [n]} (\hat{a}_{ij} + \delta_{ij}\Delta_{ij})x_j \leq b_i$ for all $\boldsymbol{\delta}_i \in \Phi_{\boldsymbol{\delta}_i}$. On the other hand, the set of feasible solutions to (5) need not to be convex, because it is a union of polyhedral sets. Constraint (5) is equivalent to

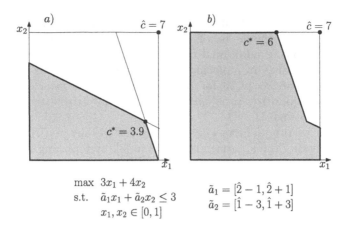

$$\max\ 3x_1 + 4x_2$$
$$\text{s.t.}\quad \tilde{a}_1 x_1 + \tilde{a}_2 x_2 \le 3$$
$$x_1, x_2 \in [0, 1]$$

$$\tilde{a}_1 = [\hat{2} - 1, \hat{2} + 1]$$
$$\tilde{a}_2 = [\hat{1} - 3, \hat{1} + 3]$$

Fig. 1. A sample problem with one uncertain constraint and $\Gamma_1 = 1$. In a) the set of feasible solutions using (4) and in b) the set of feasible solutions using (5) are shown.

$\sum_{j \in [n]} (\widehat{a}_{ij} + \delta_{ij} \Delta_{ij}) x_j \le b_i$ for at least one $\boldsymbol{\delta}_i \in \Phi_{\boldsymbol{\delta}_i}$, so it is a disjunction of a family of linear constraints.

Let us illustrate this by the example shown in Fig. 1. The set of feasible solutions to the nominal problem with $\tilde{a}_1 = 2$ and $\tilde{a}_2 = 1$ is $[0, 1] \times [0, 1]$, which gives the optimal solution $x_1 = x_2 = 1$ with $\widehat{c} = 7$. Using the concept (4) we get the set of feasible solutions shown in Fig. 1a, which is conjunction of the constraints $3x_1 + x_2 \le 3$ and $2x_1 + 4x_2 \le 3$. The optimal solution is $x_1 = 0.9$, $x_2 = 0.3$ with the optimal objective value equal 3.9. Using the concept (5) we get the set of feasible solutions depicted in Fig. 1b, which is disjunction of the constraints $3x_1 + x_2 \le 3$ and $2x_1 + 4x_2 \le 3$. The optimal solution is then $x_1 = 0.67$, $x_2 = 1$ with the objective value equal to 6. This solution has lower price of robustness. Observe, however, that the resulting set of feasible solutions is not convex.

3   Solving the Problem

Using the concept (5) we can rewrite the uncertain problem (1) as follows:

$$\max\ \boldsymbol{c}^T \boldsymbol{x}$$
$$\text{s.t.}\ \sum_{j \in [n]} (\widehat{a}_{ij} + \delta_{ij} \Delta_{ij}) x_j \le b_i\ i \in [m],$$
$$\sum_{j \in [n]} \delta_{ij} = \Gamma_i \qquad\qquad i \in [m], \qquad\qquad (6)$$
$$\delta_{ij} \in \{0, 1\} \qquad\qquad i \in [m], j \in [n],$$
$$\boldsymbol{x} \in \mathbb{X}.$$

Binary variables δ_{ij} select the uncertainty set $\mathcal{U}_{\boldsymbol{\delta}_i}$ in the ith constraint. The nonlinear terms $\delta_{ij} x_j$ can be linearized by applying standard techniques. In consequence, if $\mathbb{X}$ is described by a system of linear constraints, then the resulting problem is a mixed integer linear one. In Sect. 5 we will investigate the complexity of (6) and two its special cases.

4   Illustrative Example

In this section we will evaluate our concept by computational experiments. We will perform experiments for the continuous 0-1 knapsack problem with uncertain constraint coefficients (weights). The following model is the counterpart of (6) for the uncertain continuous 0-1 knapsack problem:

$$
\begin{aligned}
&\max \ c^T x \\
&\text{s.t.} \ \sum_{j \in [n]} (\widehat{a}_j + \delta_j \Delta_j) x_j \leq b, \\
&\qquad \sum_{j \in [n]} \delta_j = \Gamma, \\
&\qquad \delta \in \{0, 1\}^n, \\
&\qquad x \in [0, 1]^n.
\end{aligned}
\tag{7}
$$

An instance of the problem is generated as follows. We fix $n = 100$, c_j is a random integer, uniformly distributed in $[10, 100]$, $\widehat{a}_j$ is a random integer, uniformly distributed in $[20, 60]$ and $\Delta_j = \sigma \widehat{a}_j$, where σ is a random real from the interval $[0, 1]$. We also fix $b = 0.4 \sum_{j \in [n]} \widehat{a}_i$. By changing Γ from 0 to 100, we obtained a family of solutions to the *pessimistic* problem with the constraint (4) and to the *optimistic* problem with the constraint (5), i.e. to problem (7). Let x be a feasible solution for some fixed Γ. We define $dev(x) = (\widehat{c} - c^T x)/\widehat{c}$, which is the price of robustness of x, expressing a relative distance of $c^T x$ to the optimal objective function value of the nominal problem. The quantity $viol(x)$ is an empirical estimation of the probability of the constraint violation, which is computed as follows. We generated 10 000 random scenarios (constraint coefficient values) by choosing uniformly at random the value of $\tilde{a}_j$ from $[\widehat{a}_j - \Delta_j, \widehat{a}_j + \Delta_j]$, $j \in [n]$. Then $viol(x)$ is the fraction of scenarios under which x is infeasible.

The obtained results are shown in Fig. 2. As one can expect, the optimistic approach results in lower price of robustness, but also in larger risk of the constraint violation. Both approaches are equivalent for the boundary values of Γ equal to 0 or 100. Furthermore, the pessimistic problem quickly decreases the constraint violation $viol(x)$ and increases the price of robustness $dev(x)$ as Γ increases and a compromise, between $viol(x)$ and $dev(x)$, is reached for $\Gamma \approx 10$. While for the optimistic problem a similar compromise is reached for $\Gamma \approx 75$. Accordingly, combining the two approaches, i.e. the pessimistic and optimistic points of view, one can provide a larger family of solutions whose profile is shown in Fig. 2. One of them can be ultimately chosen by the decision maker, who can take a risk-aversion or some other factors into account.

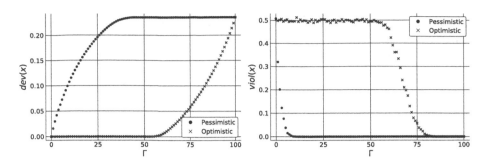

Fig. 2. The values of $dev(\boldsymbol{x})$ and $viol(\boldsymbol{x})$, where $\boldsymbol{x}$ is an optimal solution to the pessimistic (4) or optimistic (5) problem for $\Gamma \in \{0, \ldots, 100\}$.

5   Uncertain Constraint Coefficients

In this section we proceed with the study of problem (6). We provide a negative complexity result for it and some positive results for its two special cases. The following theorem characterizes the complexity of problem (6):

Theorem 1. *Problem (6) is strongly NP-hard and not at all approximable even if* $\mathbb{X} = [0, 1]^n$ *and* $\Gamma_i \in \{0, 1\}$ *for each* $i \in [m]$.

Proof. Consider the strongly NP-complete 3-SAT problem [6], in which we are given a set of boolean variables $\{x_1, \ldots, x_n\}$ and a set of clauses $C_1, \ldots, C_m$. Each clause C_i contains three literals $\{p_i, q_i, r_i\}$, where $p_i, q_i, r_i \in \{x_1, \overline{x}_1, \ldots, x_n, \overline{x}_n\}$. We ask if there is a $0 - 1$ assignment to the variables which satisfies all the clauses. Given an instance of 3-SAT we build the following program:

$$
\begin{aligned}
\max\ & (1 - t) \\
\text{s.t.}\ & (-1 + 2\delta_{1j})x_j + (-1 + 2\delta_{2j})\overline{x}_j \leq -1 & j \in [n], \\
& \delta_{1j} + \delta_{2j} = 1 & j \in [n], \\
& p_i + q_i + r_i \geq 1 - t & \forall C_i = \{p_i, q_i, r_i\}, \\
& \delta_{1j}, \delta_{2j} \in \{0, 1\} & j \in [n], \\
& x_j, \overline{x}_j, t \in [0, 1] & j \in [n].
\end{aligned}
\tag{8}
$$

Observe that (8) is a special case of (6), where $\Gamma_i \in \{0, 1\}$ for each constraint and $\mathbb{X} = [0, 1]^{n+1}$. Notice that the clause constrains can be equivalently rewritten as $-p_i - q_i - r_i - t \leq -1$ with $\Gamma_i = 0$. In any feasible solution to (8), we must have $x_j = 1, \overline{x}_j = 0$ or $x_j = 1, \overline{x}_j = 0$ for each $j \in [n]$. Indeed, the constraint $\delta_{1j} + \delta_{2j} = 1$ forces $x_j - \overline{x}_j \leq -1$ or $\overline{x}_j - x_j \leq -1$. Since $x_j, \overline{x}_j \in [0, 1]$ the property is true. Also, (8) is feasible, because by setting $t = 1$, we can satisfy all the constraints associated with the clauses. We will show that the answer to 3-SAT is yes if the optimal objective value to (8) is 1 and 0, if the answer is no.

   Assume that the answer to 3-SAT is yes. Then, there is a $0 - 1$ assignment to the variables $x_1, \ldots, x_n$, which satisfies all the clauses. We construct a feasible solution to (8) as follows. The values of x_j are the same as in the truth assignment and $\overline{x}_j = 1 - x_j, j \in [n]$. Also, $\delta_{1j} = 1 - x_j$ and $\delta_{2j} = 1 - \overline{x}_j$ for each $j \in [n]$. Finally $t = 0$. The clause constraints are satisfied by the assumption that $x_1, \ldots, x_n$ satisfies all the clauses. The objective value for this feasible solution equals 1.

Assume that the answer to 3-SAT is no, but the optimal objective value to (8) is greater than 0, so $t < 1$. Since $x_j, \overline{x}_j \in \{0, 1\}$ in every feasible solution to (8), there must be at least one variable with the value of 1 in each clause constraint. But, as the answer to 3-SAT is no, there must be j such that $x_j = \overline{x}_j$, which contradicts the feasibility of x_j and $\overline{x}_j$ for $t < 1$. Hence $t = 1$ and the optimal objective value of (8) is 0. $\qquad\square$

5.1   0-1 Knapsack Problem

In this section we study the 0-1 knapsack problem with uncertain weights that is a special case of problem (6) in which $\mathbb{X} = \{0, 1\}^n$ and $m = 1$, i.e. we investigate the following problem:

$$
\begin{aligned}
\max \quad & \sum_{j \in [n]} c_j x_i \\
\text{s.t.} \quad & \sum_{j \in [n]} (\widehat{a}_j + \delta_j \Delta_j) x_j \le b, \\
& \sum_{j \in [n]} \delta_j = \Gamma, \\
& \delta_j \in \{0, 1\} && j \in [n], \\
& x_j \in \{0, 1\} && j \in [n].
\end{aligned}
\tag{9}
$$

Problem (9) is NP-hard, because the deterministic 0-1 knapsack problem, obtained by fixing $\Gamma = 0$, is already NP-hard [6]. We will show that (9) can be reduced the following *constrained shortest path problem* in which: we are given a network $G = (V, A)$ with a cost $c_a \ge 0$ and a weight $w_a \ge 0$ specified for each arc $a \in A$. We seek a shortest $s - t$ path in G whose total weight does not exceed b. This problem is NP-hard [6]. However, it can be solved in pseu-dopolynomial time $O(|A|b)$ in acyclic networks, assuming that $b \in \mathbb{Z}_+$, by using dynamic programming and it admits a fully polynomial approximation scheme (FPTAS) [7]. Consequently the problem (9) can be solved in pseudopolynomial time and has a FPTAS as well.

The method of constructing the corresponding network $G = (V, A)$ for $n = 6$ and $\Gamma = 4$ is shown in Fig. 3 (the idea for arbitrary n and $\Gamma \in \{0, \ldots, n\}$ is the same). Network G is composed of n layers. In the jth layer we consider all possible four cases for the variable x_j. Namely, $x_j = 1$ and $\delta_j = 0$ (solid *horizontal* arc); $x_j = 0$ and $\delta_j = 0$ (dashed *horizontal* arc); $x_j = 1$ and $\delta_j = 1$ (solid *diagonal* arc); $x_j = 0$ and $\delta_j = 1$ (dashed *diagonal* arc). We seek a longest $s - t$ path in G whose weight is not greater than b. Observe that this problem can be easily reduced to the constrained shortest path problem by replacing c_j with $c_{\max} - c_j$ for each $j \in [n]$, where $c_{\max} = \max_{j \in [n]} c_j$. Since each $s - t$ path has the same number of arcs, the longest constrained shortest path in G is the same as the shortest constrained path in the modified network.

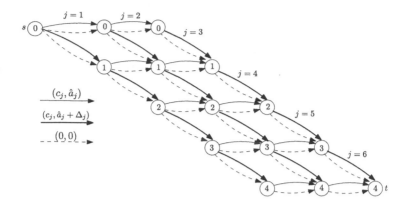

Fig. 3. Network for $n = 6$ and $\Gamma = 4$.

It is easy to see that each optimal solution to the constrained longest path problem in G corresponds to an optimal solution to (9). Each feasible $s - t$ path in G contains exactly Γ diagonal arcs, which correspond to $\delta_j = 1$, and exactly $n - \Gamma$ horizontal arcs which correspond to $\delta_j = 0$. For each diagonal and horizontal arcs the path indicates whether $x_j = 1$ or $x_j = 0$, which provides a feasible solution to (9). Since the computed path is the longest one, it corresponds to an optimal solution to (9).

5.2 Continuous 0-1 Knapsack Problem

In this section we examine the continuous version of the 0-1 knapsack problem with uncertain weights, i.e. the model (7) discussed in Sect. 4 (a special case of problem (6)). This model can be linearized in a standard way by introducing additional n variables $y_j \geq 0$, $j \in [n]$, which express $y_j = \delta_j x_j$, and $2n$ constraints of the form $y_j \leq \delta_j$, $y_j \geq x_j - (1 - \delta_j)$ for $j \in [n]$. In this section we will transform (7) into a mixed integer linear program having only $n + 1$ variables (including n binary variables) and at most $2n + 1$ additional linear constraints. We also propose an upper bound, which can be computed in $O(n^2)$ time.

Fix δ_j, $j \in [n]$, and consider the following linear programming problem (with dual variables β, α_j, in brackets):

$$
\begin{aligned}
\max \ & \sum_{j \in [n]} c_j x_j \\
\text{s.t.} \ & \sum_{j \in [n]} (\widehat{a}_j + \delta_j \Delta_j) x_j \leq b && [\beta], \\
& x_j \leq 1 && j \in [n] \ [\alpha_j], \\
& x_j \geq 0 && j \in [n].
\end{aligned}
\tag{10}
$$

The dual to (10) is

$$\min \beta b + \sum_{j \in [n]} \alpha_j$$
$$\text{s.t.} \ \beta(\widehat{a}_j + \delta_j \Delta_j) + \alpha_j \geq c_j \ j \in [n], \tag{11}$$
$$\alpha_j \geq 0 \qquad\qquad j \in [n],$$
$$\beta \geq 0.$$

In an optimal solution to (11), we can fix $\alpha_j = [c_j - \beta(\widehat{a}_j + \delta_j \Delta_j)]_+$, where $[y]_+ = \max\{0, y\}$. Hence (11) can be rewritten as

$$\min_{\beta \geq 0} g(\boldsymbol{\delta}, \beta) = \beta b + \sum_{j \in [n]} [c_j - \beta(\widehat{a}_j + \delta_j \Delta_j)]_+. \tag{12}$$

Proposition 1. *For any $\boldsymbol{\delta}$, the function $g(\boldsymbol{\delta}, \beta)$ attains minimum at $\beta = 0$ or $\beta = \frac{c_k}{\widehat{a}_k}$ or $\beta_k = \frac{c_k}{\widehat{a}_k + \Delta_k}$ for some $k \in [n]$.*

Proof. Let us reorder the variables so that

$$\frac{c_1}{\widehat{a}_1 + \delta_1 \Delta_1} \geq \frac{c_2}{\widehat{a}_2 + \delta_2 \Delta_2} \geq \cdots \geq \frac{c_n}{\widehat{a}_n + \delta_n \Delta_n}.$$

Let $k \in [n]$ be the smallest index in $[n]$ such that $\sum_{j \in [k]} (\widehat{a}_j + \delta_j \Delta_j) > b$. If there is no such k, then we fix $x_j = 1$ for each $j \in [n]$, obtaining a feasible solution to (10) with the objective value $\sum_{j \in [n]} c_j$. The objective value of (11) for $\beta^* = 0$ is also $\sum_{j \in [n]} c_j$ so, by strong duality $\beta^* = 0$ is optimal.

Assume that $k < n$. Fix $b' = \sum_{j \in [k-1]} (\widehat{a}_j + \delta_j \Delta_j) \leq b$ and $b' = 0$ if $k = 1$. Let us construct a feasible solution to (10) by setting $x_j = 1$ for $j \in [k-1]$ and $x_k = (b - b')/(\widehat{a}_k + \delta_k \Delta_k)$. The objective value of (10) is $\sum_{j \in [k-1]} c_j + c_k x_k$. Let us now construct a feasible solution to (11) by fixing $\beta^* = \frac{c_k}{\widehat{a}_k + \delta_k \Delta_k}$. Using the fact that the optimal $\alpha_j^* = [c_j - \beta(\widehat{a}_j + \delta_j \Delta_j)]_+$, $j \in [n]$, we conclude that $\alpha_j^* = 0$ for $j \geq k$, and we get the objective value of (11)

$$\frac{c_k b}{\widehat{a}_k + \delta_k \Delta_k} + \sum_{j \in [k-1]} c_j - \frac{c_k}{\widehat{a}_k + \delta_k \Delta_k} b' = \sum_{j \in [k-1]} c_j + c_k x_k.$$

Hence β^* is optimal according to the strong duality. Since $\beta^* = \frac{c_k}{\widehat{a}_k + \delta_k \Delta_k}$ and $\delta_k \in \{0, 1\}$ the proposition follows. $\qquad\qquad\qquad\qquad\qquad\qquad\qquad \square$

Using the fact that $\boldsymbol{\delta} \in \{0, 1\}^n$, let us rewrite $g(\boldsymbol{\delta}, \beta)$ as follows:

$$g(\boldsymbol{\delta}, \beta) = \beta b + \sum_{j \in [n]} \left([c_j - \beta \widehat{a}_j]_+ + \delta_j([c_j - \beta(\widehat{a}_j + \Delta_j)]_+ - [c_j - \beta \widehat{a}_j]_+)\right).$$

Setting $\phi(\beta) = \beta b + \sum_{j \in [n]} [c_j - \beta \widehat{a}_j]_+$ and $\psi_j(\beta) = [c_j - \beta(\widehat{a}_j + \Delta_j)]_+ - [c_j - \beta \widehat{a}_j]_+$ yields

$$g(\boldsymbol{\delta}, \beta) = \phi(\beta) + \sum_{j \in [n]} \delta_j \psi_j(\beta).$$

Fix $\mathcal{B} = \{\frac{c_k}{\widehat{a}_k} : k \in [n]\} \cup \{\frac{c_k}{\widehat{a}_k + \Delta_k} : k \in [n]\} \cup \{0\}$. Proposition 1 now shows that the optimal values of $\boldsymbol{\delta}$ can be found by solving the following max-min problem:

$$\max_{\boldsymbol{\delta} \in \Phi} \min_{\beta \in \mathcal{B}} g(\boldsymbol{\delta}, \beta), \tag{13}$$

which can be represented as the following program:

$$\begin{aligned}
\max \; & t \\
\text{s.t.} \; & t \le \phi(\beta) + \sum_{j \in [n]} \psi_j(\beta)\delta_j \; \beta \in \mathcal{B}, \\
& \sum_{j \in [n]} \delta_j = \Gamma, \\
& \delta_j \in \{0, 1\} \qquad\qquad j \in [n].
\end{aligned} \tag{14}$$

Model (14) has n binary variables and one continuous variable. Since $|\mathcal{B}| \le 2n+1$, the number of constraints is at most $2n + 2$. Observe that the size of $\mathcal{B}$ can be smaller, since some ratios in $\mathcal{B}$ can be repeated. Unfortunately, no polynomial time algorithm for solving (14) is known, so the complexity of the problem remains open. Observe that we can use (13) to compute an upper bound for (10). Namely, by exchanging the min-max operators we get

$$\max_{\boldsymbol{\delta} \in \Phi} \min_{\beta \in \mathcal{B}} g(\boldsymbol{\delta}, \beta) \le \min_{\beta \in \mathcal{B}} \max_{\boldsymbol{\delta} \in \Phi} g(\boldsymbol{\delta}, \beta) := UB.$$

For a fixed $\beta \in \mathcal{B}$, the optimal values of $\boldsymbol{\delta}$ can be found in $O(n)$ time by solving a selection problem (see, e.g., [4]). Hence UB can be computed in $O(n^2)$ time.

6   Uncertain Right Hand Sides

In this section we will show how to cope with uncertain right hand sides of the constraints. To simplify the presentation, we will assume that the constraint coefficients are deterministic. We thus study the following problem

$$\begin{aligned}
\max z = & \boldsymbol{c}^T \boldsymbol{x} \\
& \boldsymbol{A}\boldsymbol{x} \le \widetilde{\boldsymbol{b}}, \\
& \boldsymbol{x} \in \mathbb{X}.
\end{aligned} \tag{15}$$

where $\boldsymbol{A}$ is $m \times n$ matrix of precise constraint coefficients and $\widetilde{\boldsymbol{b}}$ is an m vector of uncertain right hand sides. The meaning of $\mathbb{X}$ is the same is in the previous sections. Assume that $\widetilde{b}_i$ is only known to belong to the interval $[\widehat{b}_i - \Delta_i, \widehat{b}_i + \Delta_i]$, $i \in [m]$. Let $\Phi = \{\boldsymbol{\delta}_i \in \{0, 1\}^m : \sum_{i \in [m]} \delta_i = \Gamma\}$. A fixed vector $\boldsymbol{\delta} \in \Phi$ induces the uncertainty set $\mathcal{U}_{\boldsymbol{\delta}} = \{\boldsymbol{b} \in \mathbb{R}^m : b_i \in [\widehat{b}_i - \delta_i \Delta_i, \widehat{b}_i], i \in [m]\}$. Using the optimistic approach (see Sect. 2), we can transform (15) into the following problem:

$$\begin{aligned}
\max z = & \boldsymbol{c}^T \boldsymbol{x} \\
& \min_{\boldsymbol{\delta} \in \Phi} \max_{\boldsymbol{b} \in \mathcal{U}_{\boldsymbol{\delta}}} \boldsymbol{A}\boldsymbol{x} \le \boldsymbol{b}, \\
& \boldsymbol{x} \in \mathbb{X}.
\end{aligned} \tag{16}$$

Problem (16) can be rewritten as follows:

$$\max \boldsymbol{c}^T \boldsymbol{x}$$
$$\sum_{j \in [n]} \widehat{a}_{ij} x_j \leq \widehat{b}_i - \delta_i \Delta_i \quad i \in [m],$$
$$\sum_{i \in [m]} \delta_i = \Gamma \qquad\qquad i \in [m], \tag{17}$$
$$\delta_i \in \{0, 1\} \qquad\qquad i \in [m].$$
$$\boldsymbol{x} \in \mathbb{X}.$$

Observe that (17) can be solved by trying all possible vectors of $\boldsymbol{\delta}$, which can be done in reasonable time if the number of constraints is not large. In particular, if $\mathbb{X}$ is a polyhedron and m is constant, then (17) can be solved in polynomial time. The next theorem characterizes the problem complexity when m is a part of input.

Theorem 2. *Problem (17) is strongly NP-hard and not at all approximable even if $\mathbb{X}$ is a bounded polyhedron.*

Proof. Consider the strongly NP-complete 3-SAT problem [6], in which we are given a set of boolean variables $\{x_1, \dots, x_n\}$ and a set of clauses $C_1, \dots, C_m$. Each clause C_i contains three literals $\{p_i, q_i, r_i\}$, where $p_i, q_i, r_i \in \{x_1, \overline{x}_1, \dots, x_n, \overline{x}_n\}$. We ask if there is a $0-1$ assignment to the variables which satisfies all the clauses. Given an instance of 3-SAT we build the following program:

$$\max (1 - t)$$
$$-x_j \leq 0 - \delta_j \cdot 1 \qquad j \in [n],$$
$$-\overline{x}_j \leq 0 - \delta'_j \cdot 1 \qquad j \in [n],$$
$$x_j + \overline{x}_j \leq 1 \qquad j \in [n],$$
$$\textstyle\sum_{j \in [n]} (\delta_j + \delta'_j) = n, \tag{18}$$
$$p_i + q_i + r_i \geq 1 - t \quad \forall C_i = \{p_i, q_i, r_i\},$$
$$\delta_j, \delta'_j \in \{0, 1\} \qquad j \in [n],$$
$$x_j, \overline{x}_j \in [0, 1] \qquad j \in [n],$$
$$t \in [0, 1].$$

Notice that in any feasible solution to (18) we must have $\delta_j = 1$ and $\delta'_j = 0$; or $\delta_j = 0$ and $\delta'_j = 1$. Indeed, if $\delta_j = \delta'_j = 1$, then $x_j \geq 1$, $\overline{x}_j \geq 1$ which contradicts $x_j + \overline{x}_j \leq 1$. If $\delta_j = \delta'_j = 0$, then there must be some $k \in [n]$ such that $\delta_k = \delta'_k = 1$ and we again get contradiction. Now $\delta_j = 1$ and $\delta'_j = 0$ implies $x_j = 1$ and $\overline{x}_j = 0$, and $\delta_j = 0$ and $\delta'_j = 1$ implies $x_j = 0$ and $\overline{x}_j = 1$. The rest of the proof is the same as in the proof of Theorem 1. □

7 The Shortest Path Problem with Uncertain Costs

Let $\mathbb{X}$ be the set of characteristic vectors of all $s - t$ paths in a given network $G = (V, A)$, $\mathbb{X} \subseteq \{0, 1\}^{|A|}$. Suppose that the arc costs are uncertain, and they are specified as intervals $[\hat{c}_{ij} - \Delta_{ij}, \hat{c}_{ij} + \Delta_{ij}]$, $(i,j) \in A$. Applying the approach from Sect. 2, we get the following model:

$$
\begin{aligned}
\min \quad & \sum_{(i,j) \in A} (c_{ij} + \delta_{ij}\Delta_{ij})x_{ij} \\
\text{s.t} \quad & \sum_{(i,j) \in A} \delta_{ij} = \Gamma \\
& \delta_{ij} \in \{0,1\} \qquad (i,j) \in A, \\
& \boldsymbol{x} \in \mathbb{X}.
\end{aligned}
\tag{19}
$$

We now show that (19) can be solved in polynomial time. Given network $G = (V, A)$, we form network $G' = (V, A')$ having the same set of nodes with the same s and t. For each arc $(i, j) \in A$ we create two parallel arcs, namely the *solid arc* $(i, j) \in A'$ with cost $\hat{c}_{ij}$ and weight 1 and the *dashed arc* $(i, j) \in A'$ with cost $\hat{c}_{ij} + \Delta_{ij}$ and weight 0. An example is shown in Fig. 4. We solve the constrained shortest path problem in G' with $b = |A| - \Gamma$, i.e. we seek a shortest $s - t$ path in G' whose weight does not exceed $|A| - \Gamma$. This problem can be solved in $O(|A|b)$-time [7] which is $O(|A|^2)$, by the definition of b.

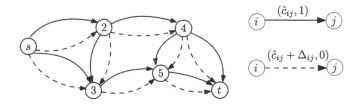

Fig. 4. Network G' for $G = (V, A)$. We seek a shortest $s - t$ path with weight at most $|A| - \Gamma$, i.e. which uses at most $|A| - \Gamma$ solid arcs.

To see that the transformation is correct, let P be a path in G' with cost $c(P)$. Path P is of the form $P_S \cup P_D$, where P_S is the set of solid arcs in P and P_D is the set of dashed arcs is P. Since $|P_S| \leq |A| - \Gamma$, we get $|A \backslash P_S| \geq \Gamma$. We form a feasible solution to (19) with the cost at most $c(P)$ as follows. If $(i,j) \in P$ and (i,j) is a solid arc, we fix $x_{ij} = 1$ and $\delta_{ij} = 0$; if $(i,j) \in P$ and (i,j) is a dashed arc, we fix $x_{ij} = 1$. We also fix $\delta_{ij} = 1$ for any subset of Γ arcs in $A \backslash P_S$. It is easy to see that we get a feasible solution to (19) with the objective value at most $c(P)$. Conversely, let (x_{ij}, δ_{ij}) be a feasible solution to (19) with the objective value c^*. We construct a corresponding path P in G' as follows. If $x_{ij} = 1$ and $\delta_{ij} = 0$, then we add the solid arc (i,j) to P; if $x_{ij} = 1$ and $\delta_{ij} = 1$, then we add the dashed arc (i,j) to P. Since x_{ij} describe an $s - t$ path in G, the set of arcs P is an $s - t$ path in G'. Suppose that we chose more than $|A| - \Gamma$ solid arcs to construct P. Then $|\{(i,j) \in A : \delta_{ij} = 0\}| > |A| - \Gamma$, and $|\{(i,j) \in A : \delta_{ij} = 1\}| < \Gamma$, a contradiction with the feasibility of δ_{ij}. Hence P is a feasible path in G' and $c(P)$ is equal to c^*. So, the cost of an optimal path in G' is at most c^*.

The proposed technique, consisting in arc duplication, can be used to solve other network problems. For example, when $\mathbb{X}$ is the set of characteristic vectors of spanning trees in G, then we can use an algorithm for the constrained spanning tree problem described in [11].

8 Conclusion

In this paper we have proposed a new approach to deal with uncertainty in optimization problems. Our idea is to soften the assumption that the worst scenario will occur for a given solution. We can thus use the pessimistic and the optimistic approaches to provide a broader family of solutions, one of which can be ultimately chosen by decision maker. Unfortunately, the proposed approach may lead to computationally harder problems. In particular, even the case of linear programming problems is NP-hard. However, we have shown in this paper some examples of optimization problems with uncertain parameters for which, after applying the approach, effective solution methods can be constructed.

The proposed approach can be too optimistic. Namely, the computed solution can be infeasible with large probability. Hence, and interesting research direction is to combine the pessimistic and the optimistic approaches, by using some aggregation methods.

References

1. Averbakh, I., Lebedev, V.: On the complexity of minmax regret linear programming. Eur. J. Oper. Res. **160**, 227–231 (2005). https://doi.org/10.1016/j.ejor.2003.07.007
2. Ben-Tal, A., El Ghaoui, L., Nemirovski, A.: Robust Optimization. Princeton Series in Applied Mathematics. Princeton University Press, Princeton (2009)
3. Bertsimas, D., Sim, M.: The price of robustness. Oper. Res. **52**, 35–53 (2004). https://doi.org/10.1287/opre.1030.0065
4. Cormen, T., Leiserson, C., Rivest, R.: Introduction to Algorithms. MIT Press, Cambridge (1990)
5. Fischetti, M., Monaci, M.: Light robustness. In: Ahuja, R.K., Möhring, R.H., Zaroliagis, C.D. (eds.) Robust and Online Large-Scale Optimization. LNCS, vol. 5868, pp. 61–84. Springer, Heidelberg (2009). https://doi.org/10.1007/978-3-642-05465-5_3
6. Garey, M.R., Johnson, D.S.: Computers and Intractability. A Guide to the Theory of NP-Completeness. W. H Freeman and Company, New York (1979)
7. Hassin, R.: Approximation schemes for the restricted shortest path problem. Math. Oper. Res. **17**, 36–42 (1992). https://doi.org/10.1287/moor.17.1.36
8. Kall, P., Mayer, J.: Stochastic Linear Programming. Models, Theory, and Computation. Springer, Heidelberg (2011). https://doi.org/10.1007/978-1-4419-7729-8
9. Kasperski, A., Zieliński, P.: Robust discrete optimization under discrete and interval uncertainty: a survey. In: Doumpos, M., Zopounidis, C., Grigoroudis, E. (eds.) Robustness Analysis in Decision Aiding, Optimization, and Analytics. ISORMS, vol. 241, pp. 113–143. Springer, Cham (2016). https://doi.org/10.1007/978-3-319-33121-8_6

10. Kouvelis, P., Yu, G.: Robust Discrete Optimization and its Applications. Kluwer Academic Publishers, Boston (1997). https://doi.org/10.1007/978-1-4757-2620-6
11. Ravi, R., Goemans, M.X.: The constrained minimum spanning tree problem. In: Karlsson, R., Lingas, A. (eds.) SWAT 1996. LNCS, vol. 1097, pp. 66–75. Springer, Heidelberg (1996). https://doi.org/10.1007/3-540-61422-2_121
12. Schöbel, A.: Generalized light robustness and the trade-off between robustness and nominal quality. Math. Methods Oper. Res. **80**(2), 161–191 (2014). https://doi.org/10.1007/s00186-014-0474-9
13. Soyster, A.L.: Convex programming with set-inclusive constraints and applications to inexact linear programming. Oper. Res. **21**, 1154–1157 (1973). https://doi.org/10.1287/opre.21.5.1154

Hierarchical Reasoning and Knapsack Problem Modelling to Design the Ideal Assortment in Retail

Jocelyn Poncelet[1,2(✉)], Pierre-Antoine Jean[1(✉)], Michel Vasquez[1(✉)], and Jacky Montmain[1(✉)]

[1] EuroMov Digital Health in Motion, Univ Montpellier,
IMT Mines Alès, Alès, France
{jocelyn.poncelet,pierre-antoine.jean,michel.vasquez,
jacky.montmain}@mines-ales.fr
[2] TRF Retail, 116 allée Norbert Wiener, 30000 Nîmes, France

Abstract. The survival of a supermarket chain is heavily dependent on its capacity to maintain the loyalty of its customers. Proposing adequate products to customers is the issue of the store's assortment. With tens thousands of products on shelves, designing the ideal assortment is theoretically a thorny combinatorial optimization problem. The approach we propose includes prior knowledge on the hierarchical organization of products by family to formalize the ideal assortment problem into a knapsack problem. The main difficulty of the optimization problem remains the estimation of the expected benefits associated to changes in the product range of products' families. This estimate is based on the accounting results of similar stores. The definition of the similarity between two stores is then crucial. It is based on the prior knowledge on the hierarchical organization of products that allows approximate reasoning to compare any two stores and constitutes the major contribution of this paper.

Keywords: Optimal assortment in mass distribution · Semantic similarity measures · Knapsack problem

1 Introduction

Competition in large retailers is becoming increasingly intense; therefore, in order to satisfy fluctuating demand and customers' increasing expectations, deal with the competition and remain or become market leaders, retailers must focus on searching for sustainable advantages. The survival of a supermarket chain is heavily dependent on its capacity to maintain the loyalty of its customers [11,12]. Proposing adequate products to its customers is the issue of the store's assortment, *i.e.*, products offered for sale on shelves [10,13]. Moreover, retailers are faced to manage high stores' networks. This way, they use a common assortment shared in the stores' network to allow an easier management [14].

© Springer Nature Switzerland AG 2020
M.-J. Lesot et al. (Eds.): IPMU 2020, CCIS 1237, pp. 201–214, 2020.
https://doi.org/10.1007/978-3-030-50146-4_16

Therefore, stores share a common and centralised assortment [18] with some tolerated exceptions to take into account specific characteristics of stores in the network [16]. To improve their global performance, retailers aim to increase their knowledge on stores' specific characteristics in the network to suggest the optimal assortment for each store, *e.g.*, they try to identify the products that perform remarkably in some stores of the network to recommend them to other *similar* stores. However, the definition of *similar* stores is not so obvious: it can be related to the localisation of stores, their assortments on shelves, their revenues, their format, *e.g.*, Hypermarket, Supermarket... [11]. This concept of similarity plays a central role in this contribution.

In this article we address the question of the ideal assortment in supermarkets. To better understand the complexity of the task, it must be remembered that some hypermarkets offer up to 100,000 products [16]. Defining the ideal assortment in a department store consists in selecting this set of products. More formally, this thorny problem corresponds to an insoluble combinatorial optimization problem. In practice, decisions are made locally by a category manager, while the problem of the ideal assortment should correspond to a global decision at the store level. To tackle this question, retailers have prior knowledge available [17]. Indeed, department store are organized into categories, *e.g.*, food, household products, textiles, etc. These categories are themselves divided into families or units of need (*e.g.* textiles category is derived into woman, man and child sections and so on). This hierarchical organization of products makes it possible to reason about families of products, structure the decision and thus avoid combinatorial explosion. Most of the time, a hypermarket cannot choose a single product to increase its offer. Indeed, this additional item necessarily belongs to a level of assortment or product line, generally in adequacy with the size or the location of the store: choosing a product requires to take all products associated to the same level of assortment [16,18]. For example, if a store offers a soda section, it can be satisfied with a minimum offer, *e.g.*, Coca-Cola 1.5L; but it can also claim a product range more consistent: for example, it would like to offer Lipton 2L, nevertheless, increase cannot be realized product by product, but by subset of products and the final offer should be Coca-Cola 1.5L + Lipton 2L + Orangina 1.5L + Schweppes 1.5L.

The proposed approach includes prior knowledge on the hierarchical organization of products by family and constraints on levels of assortment for each family. It proposes to calculate the ideal assortment from the overall point of view of store' managers. The ideal assortment thus appears as a combinatorial optimization problem that can be solved thanks to approximate reasoning based on the products' hierarchy of abstraction. The main difficulty of the optimization problem remains the estimation of the expected benefits associated to any increase of the product range in a given family. This estimate is based on the accounting results of similar stores. The definition of the *similarity* between two stores is then crucial. It is based on the prior knowledge on the hierarchical organization of products that allows approximate reasoning and constitutes the major contribution of this paper.

2   Modelling the Ideal Assortment as a Combinatorial Optimization Problem

Let Ω be the department store.

F_i is the ith family of products, *i.e.* a set of products that are related to a same use category or consumption unit (*e.g.*, soft-drinks, household electrical products, etc.).

Recursively, any family F_i is a specialization of a super family: *e.g.*, Coca-cola $\in F_{Soda} \subset F_{SoftDrinks} \subset F_{Drinks}$.

Products can thus be organized within a taxonomic partial order defining an abstraction hierarchy (Fig. 1). Products are the most specific classes of this partial order, the leaves of the taxonomy.

Let us distinguish the particular case of families of products, *i.e.* the families the lower in the hierarchy, the less abstracted ones because their descendants are concrete products (direct parents of products). For each of these families of products F_i, a product range or level of assortment $s(F_i)$ is defined: for each family of products, the department store may choose the wideness of $s(F_i)$ in a finite set of opportunities imposed by the direction of the stores' network. Formally, for each family, a hierarchy of subsets of products $s^{k_i}(F_i) = 1..n$ in the sense of the inclusion relationship (*i.e.* $s^{k_i}(F_i) \subset s^{k_i+1}(F_i)$) is defined and the department store can only choose among the subsets $s^{k_i}(F_i)$ as product range for F_i (*e.g.* imagine the minimal product range of the Soda family would be Coca-Cola 1.5L, the second one Coca-Cola 1.5L + Lipton 2L + Orangina 1.5L + Schweppes 1.5L, and so on). Thus, $s(F_i)$ can only be a subset of products that belongs to this finite set of product ranges $s^{k_i}(F_i) = 1..n$ defined a priori by retailer. The size of $s(F_i)$ is then the level k_i of assortment such that $s(F_i) = s^{k_i}(F_i)$. In practice, k_i is a natural number that may vary from 1 to 9. $k_i = 1$ when the product range for the family of products F_i is minimal and $k_i = 9$ when it is maximal.

Therefore, we can write: $\Omega \triangleq \bigcup\limits_{i=1}^{n} s^{k_i}(F_i)$. An expected turnover $p(s^{k_i}(F_i))$ and a storage cost $c(s^{k_i}(F_i))$ can be associated to each $s^{k_i}(F_i)$. $c(s^{k_i}(F_i))$ represents the storage cost the department store allocates for the family F_i.

For any super family in the hierarchical organization of products, its expected turnover and its storage capacity are simply computed recursively as the sum of the expected turnovers and storage capacities of the product it covers.

Designing the assortment of a department store then consists in choosing the rank k_i for each family of products (see Fig. 1). Obviously, the higher $p(\Omega) \triangleq \bigcup\limits_{i=1}^{n} p(s^{k_i}(F_i))$, the better the assortment of Ω. Nevertheless, without further constraints, $p(\Omega)$ should be necessarily maximal when $k_i = 9 \ \forall \ i = 1..n$. In practice, $\sum\limits_{i=1}^{n} c(s^{k_i}(F_i))$ is generally far below $\sum\limits_{i=1}^{n} c(s^9(F_i))$ for obvious storage or cost constraints $\mathcal{C}$. Let us consider $\mathcal{I}$ a subset of families. It can be necessary to model constraints related to this super family. For example:

$$c(s^{k_{s.drinks}}(F_{s.drinks})) + c(s^{k_{beers}}(F_{beers})) + c(s^{k_{waters}}(F_{waters})) \leq \mathcal{C}_{\mathcal{I}=Beverages}$$

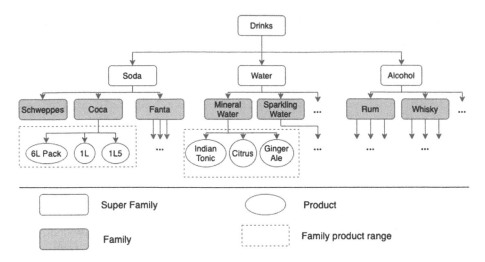

Fig. 1. Products organization and department store assortment as the union of families' product ranges

means that the storage capacity (or the cash flow) related to Beverages (superfamily $F_{\mathcal{I}}$) is limited to $\mathcal{C}_{\mathcal{I}}$. A lower $c_{\mathcal{I}}$ bound can also be introduced: in our example, $c_{\mathcal{I}}$ represents the minimal level of investment for the superfamily family Drinks. For any superfamily, such local constraints can be added to the optimization problem.

$$\text{Arg}\max_{k_i, i=1..n} \sum_{i=1}^{n} p(s^{k_i}(F_i))$$

Under:

$$\sum_{i=1}^{n} c(s^{k_i}(F_i)) \leq C - global\ constraint$$

$$\text{For some } \mathcal{I} \text{ in } 2^{\{1..n\}}, c_{\mathcal{I}} \leq \sum_{i=1}^{|\mathcal{I}|} c(s^{k_i}(F_i)) \leq C_{\mathcal{I}} - global\ constraint$$

This combinatorial optimization problem is known as the knapsack problem with mono dimensional constraints and bounded natural number variables.

3   Estimate of the Expected Turnover in the Knapsack Problem

Let consider that one of the assortments to be assessed in the optimization problem includes the increase of the product range of the product family F_i: $s^{k_i}(F_i)$ is upgraded as $s^{k_i+1}(F_i)$. The storage cost (or purchase price) $c(s^{k_i+1}(F_i))$ can be easily completed by the store because it is a basic notion in the retail segment. It is thus easy to inform this point in the optimization problem. On the other hand, it is thornier to estimate $p(s^{k_i+1}(F_i))$ that is however essential to assess the expected performance of the new product range. When the level of assortment of the store is k_i, it is easy to fill in its turnover $p(s^{k_i}(F_i))$ in the knapsack problem but $p(s^{k_i+1}(F_i))$ cannot directly assessed.

We have to design an estimator of $p(s^{k_i+1}(F_i))$. It can only be estimated from other reference measurements encountered in other similar stores. The basic idea is that the more similar these "reference" stores are to the store of concern, the more reliable the estimation. The most difficult problem is to define what "reference" means. Intuitively, the "reference" stores are departments that are "close" to the department store of concern and offer $s^{k_i+1}(F_i)$ to their customers. $p(s^{k_i^{\Omega}+1}(F_i))$ can then be computed for example as the weighted mean or the max of the $p(s^{k_i^{\Omega'}+1}(F_i))$ values, where Ω' are the reference stores neighbours of Ω. For sake of simplicity, the neighborhood is restricted to the nearest reference store in our experiments. The next issue is now to define what "close to" means.

This concept of distance between any two department store is the crucial issue. Roughly speaking, Ω should be similar to Ω' when the turnovers of Ω and Ω' are distributed in the same way over the hierarchical organization of products. It implies they have approximately the same types of customers.

Intuitively, the distance between any two stores should be based on a classical metrics space where the n dimensions would correspond to all the products that are proposed by the department store of a given chain; the value of each coordinate would be the turnover of the product for example, and would be null if the department store does not propose this product. Because some hypermarkets offer up to 100,000 products, the clustering process on such a metrics space would be based on a sparse matrix and then suffer from the space dimension. Furthermore, such a distance would not capture the hierarchical organization of products in the concept of similarity. Indeed, let's go back to the hierarchical organization of products in families. We can note that a `fruits` and `vegetables` specialist department store is obviously closer to a large grocery store than to a hardware store because the first two are `food` superstores whereas the last one is a speciality store: the first two propose the same super family F_{Food}. This intuitive similarity cannot be assessed with classical distances. The hierarchical products organization in families of products and super families is prior knowledge to be considered when assessing how similar two departments stores are. It is necessary to introduce more appropriate measures that take advantage of this organization to assess the similarity of any two departments store. This notion of similarity measures is detailed in Sect. 4.

In previous sections, we have introduced the levels of assortment $s^{k_i}(F_i), k = 1..n$ for any product family F_i. Note that the increase from $s^{k_i}(F_i)$ to $s^{k_i+1}(F_i)$ must generate an improved turnover for the product family F_i to be worthwhile. By contrast, it requires a higher storage cost $c(s^{k_i+1}(F_i))$ than $c(s^{k_i}(F_i))$. Therefore, the storage cost of at least one product family $F_{j,j\neq i}$ must be reduced to keep the overall storage cost of the department store constant. Then, the reduced turnover $p(s^{k_j-1}(F_j))$ of the family F_j has yet to be estimated to complete the optimization problem. However, this estimation can easily be processed. Indeed, because $s^{k_j-1}(F_j) \subset s^{k_j}(F_j)$, $p(s^{k_j-1}(F_j))$ can simply be deduced from $p(s^{k_j}(F_j))$: it is the sum of the turnovers of all products that belong to $s^{k_j}(F_j) \cap s^{k_j-1}(F_j)$. There is clearly an assumption behind this estimation: the disappearance of a product will not change drastically the turnover of other products of the same family. At this stage, for any store Ω, $\forall (k_1, k_2, \ldots, k_n) \in [1..9]^n$, we can estimate any $p(s^{k_i+1}(F_i))$ as the corresponding turnover of the closest reference store to Ω.

Because designing the assortment of a department store consists in choosing the rank k_i for each family of products, we could now naively enumerate and evaluate any potential assortment in $[1..9]^n$ to select the best one that will be the solution of the optimization problem.

4   Taxonomy and Abstraction Reasoning

The similarity measure that meets our expectations relies on the taxonomical structure that organises products and product families in the department store since Ω should be similar to Ω' when the turnovers of Ω and Ω' are distributed in the same way over the hierarchical organization of products. Generally, in the literature, the elements of the taxonomical structure are named concepts (or classes). A taxonomical structure defines a partial order of the key concepts of a domain by generalizing and specializing relationships between concepts (e.g. Soft drinks generalizes Soda that in turn generalizes Coca or Schweppes). Taxonomies give access to consensual abstraction of concepts with hierarchical relationships, e.g. Vegetables defines a class or concept that includes beans, leeks, carrots and so on, that are more specific concepts. Taxonomies are central components of a large variety of applications that rely on computer-processable domain expert knowledge, e.g. medical information and clinical decision support systems [1]. They are largely used in Artificial Intelligence systems, Information Retrieval, Computational Linguistics... [2].

In our study, using products taxonomy allows synthetizing and comparing the sales of department store through abstraction reasoning. In retail world, product taxonomy can be achieved by different means. Retailers or other experts can build this commodity structure. Most approaches usually introduce the Stock Keeping Unit (SKU) per item [3] or product categories (e.g. Meat, Vegetables, Drinks, etc.). Some researchers adopt the cross-category level indicated by domain experts and/or marketers [4].

More formally, we consider a concept taxonomy $T = (\preceq, C)$ where (C) stands for the set of concepts (i.e. class of products in our case) and ($\preceq$) the partial ordering. We denote $A(c) = \{x \in C/c \preceq x\}$ and $D(c) = \{x \in C/x \preceq c\}$ respectively the ancestors and descendants of the concept $c \in C$. The root is the unique concept without ancestors (except itself) $(A(root) = \{root\})$ and a concept without descendant (except itself) is denoted a leaf (in our case a leaf is a product) and $D(leaf) = \{leaf\}$. We also denote $leaves\text{-}c$ the set of leaves (*i.e.* products in our study) that are included in the concept (or class) c, *i.e.*, *leaves-c*= $D(c) \cap$ *leaves*.

4.1   Informativeness Based on Taxonomy

An important aspect of taxonomies is that they give the opportunity to analyse intrinsic and contextual properties of concepts. Indeed, by analysing their topologies and additional information about concept usage, several authors have proposed models, which take advantage of taxonomies in order to estimate the Information Content (IC) of concepts [5]. IC models are designed to mimic human, generally consensual and intuitive, appreciation of concept informativeness. As an example, most people will agree that the concept Cucumber is more informative than the concept Vegetables in the sense that knowing the fact that a customer buys Cucumber is more informative than knowing that he buys Vegetables. Indeed, various taxonomy-driven analyses, such as computing the similarity of concepts, extensively depend on accurate IC computational models. Initially, Semantic Similarity Measure (SSM) were designed in an "ad-hoc" manner for few specific domains [6]. Research have been done in order to get a theoretical unifying framework of SSMs and to be able to compare them [1,7].

More formally, we denote I the set of instances, and $I^*(c) \subseteq I$ the instances that are explicitly associated to the concept c. We consider that no annotation associated to an instance can be inferred, *i.e.*, $\forall\, c, c' \in C$, with $c \preceq c', I^*(c) \cap I^*(c') = \emptyset$. We denote $I(c) = I$ the instances that are associated to the concept c considering the transitivity of the taxonomic relationship and concept partial ordering $\preceq$, *e.g.* $I(\text{Vegetables}) \subseteq I(\text{Food})$. We therefore obtain $\forall\, c \in C, |I(c)| = \sum_{x \in\, D(c)} |I^*(x)|$.

In our approach, we use sales receipt to count the instances of concept: obviously, only products appear on sales receipt, and then only instances of products can occur in practice. The information is only carried by the leaves of the taxonomy (products in our case), $\forall c \notin$ *leaves*, $|I^*(c)| = 0$.

Due to the transitivity of the taxonomic relationship the instances of a concept $c \in C$ are also instances of any concept subsuming c, *i.e.*, Vegetables $\preceq$ Food $\Rightarrow I(\text{Vegetables}) \subseteq (\text{Food})$. This central notion is generally used to discuss the specificity of a concept, *i.e.* how restrictive a concept is with regard to I. The more restrictive a concept, the more specific it is considered to be. In the literature, the specificity of a concept is also regarded as the Information Content (IC). In this paper we will refer to the notion of IC defined through a function $IC : C \longrightarrow \mathbb{R}^+$. In accordance to knowledge modelling constraints, any IC function must monotonically decrease from the leaves to the root of the taxonomy such as $c \preceq c' \Rightarrow IC(c) \geq IC(c')$.

In this paper, extrinsic evidence has been used to estimate concept informativeness (*i.e.* that can be found outside the taxonomy). This is an extrinsic approach, based on Shannon's Information Theory and proposes to assess the informativeness of a concept by analysing a collection of items. Originally defined by Resnik [5], the IC of a concept c is defined to be inversely proportional to

$pro(c)$, the probability that c occurs in a collection. Considering that evidence of concept usage can be obtained by studying a collection of items (here, products) associated to concepts, the probability that an instance of I belongs to $I(c)$ can be defined such as $pro : 2^c \rightarrow [0,1]$ with $pro(c) = |I(c)|/|I|$. The informativeness of a concept is next assessed by defining: $IC(c) = -log(pro(c))$.

We will then use extrinsic IC in our proposal to capture concept usage in our specific application context. Let us note T the taxonomy of products, F the set of families (or classes). The leaves of T are the products (*e.g.*, Coca-cola 1.5L). Classes that directly subsume leaves are product families (*e.g.* Soda) with which assortment levels are associated ($(k_1, k_2, \ldots, k_n) \in [1..9]^n$); other classes are super family of products (*e.g.*, Soft-drinks). Let us derive these notions in our modeling. The above "collection of items" corresponds to the products that a network of department store within a same chain sails. $\forall x \in leaves(T), p^{\Omega}(x)$ (*i.e.* x is a product) is the turnover related to the product x in the store Ω. We define the probability mass pro as:

$$
\left|
\begin{aligned}
&\forall x \in leaves(T), |I^*(x)| \triangleq |I(x)| = \sum_{\Omega} p^{\Omega}(x) \text{ then} \\[1em]
&pro(x) = \frac{\sum_{\Omega} p^{\Omega}(x)}{\sum_x \sum_{\Omega} p^{\Omega}(x)} \text{ and } IC(x) = -log(pro(x)) \\[1em]
&\forall f \notin leaves, |I^*(f)| = 0, |I(f)| = \sum_{x \in leaves\text{-}f} |I(x)|, \\[1em]
&pro(f) = \frac{\sum_{x \in leaves\text{-}f} \sum_{\Omega} p^{\Omega}(x)}{\sum_x \sum_{\Omega} p^{\Omega}(x)} \text{ and } IC(f) = -log(pro(f))
\end{aligned}
\right.
$$

4.2   Similarity Measures Based on Taxonomy

After the informativeness of a concept is computed, we can now explain how to compute the similarity of any two concepts using concepts' informativeness. We recall some famous Semantic Similarity Measure (SSM) based on the informativeness of concepts and usually used in Information Retrieval. One common SSM is based on the Most Informative Common Ancestor (MICA) also named the Nearest Common Ancestor (NCA). For example, in Fig. 1, the MICA of Coca-cola 1.5L and Schweppes 1.5L is Soda while the MICA of Soda and Water is Drinks (the root in Fig. 1). Resnik [5] is the first to implicitly define the MICA: this is the concept that subsumes two concepts c_1 and c_2 that has the higher IC (*i.e.*, the most specific ancestor):

$$sim_{\text{Resnik}}(c_1, c_2) = IC(MICA(c_1, c_2))$$

Such SSMs allow comparing any two concepts. However, as stores are associated to subsets of concepts, we still have to introduce group similarities to compare sub-sets of concepts. Indirect SSMs have been proposed [8,9]. The Best Match Average (BMA) [8] is a composite average between two sets of concepts, here A and B:

$$sim_{\mathrm{BMA}}(A, B) = \frac{1}{2|B|} \sum_{c \in B} sim_m(c, A) + \frac{1}{2|A|} \sum_{c \in A} sim_m(c, B)$$

where $sim_m(c, X) = \max_{c' \in X} sim(c, c')$ and $sim(c, c')$ is any IC-based pairwise SSM. It is thus the average of all maximum similarities of concepts in A regarding B and vice-versa. This is the most common group similarity. See [8,9] for a complete review.

Pairwise and groupwise SSMs allow comparing any two subsets of concepts (products in our case) when a taxonomical structure defines a partial order of the key concepts of a domain. In our study, they allow to capture the idea that two stores Ω and Ω' are similar when their turnovers are distributed in the same way over the hierarchical organization of products.

5 Illustration and Experiments

This section aims to illustrate the modelling and the data processing chain described in the preceding sections. It is illustrated how designing the ideal assortment in retail thanks to reasoning on an abstraction hierarchy of products, semantic similarity measures and knapsack formalization. The required parameters and variables for this modelling are:

1. A taxonomy of products shared in the store network.
2. A product range (or level of assortment $s^{k_i}(F_i)$) defined for all families F_i of products.
3. A storage cost associated to each product range for each family: for each family, a hierarchy of subsets of products $s^{k_i}(F_i), k_i = 1..n$ is defined, and the higher k_i, the higher the corresponding cost $c(s^{k_i}(F_i))$.
4. For each store, a turnover $p(s^{k_i}(F_i))$ is associated to each product range of each family $s^{k_i}(F_i)$.
5. Storage capacity thresholds are introduced to manage storage constraints (see local constraints in Sect. 2).

Figure 2 illustrates the required data. The example in Fig. 2 takes into account three stores (M1, M2 and M3):

1. We only consider two products' families denoted F for Fruits and V for Vegetables. There are two product ranges for the Fruits family (*i.e.*, k_F is 1 or 2) and three for the Vegetables family (*i.e.*, k_V is 1, 2 or 3). We have $S^1(F) \subset S^2(F)$ and $S^1(V) \subset S^2(V) \subset S^3(V)$.
2. Each product range has its own storage cost: $c(S^1(F)) = 346; c(S^2(F)) = 1191; c(S^1(V)) = 204; c(S^2(V)) = 866; c(S^3(V)) = 2400$.
3. From the given product range associated with each store, in this example (k_F, k_V), their turnover can be computed that is $p(S^{k_F}(F)) + p(S^{k_V}(V))$.
4. Each store has the following storage capacity: $SCM_1 = 1670; SCM_2 = 2700; SCM_3 = 5540$ which implies that $c(S^{k_F}(F)) + c(S^{k_V}(V)) \leq SCM$ for each store with given values of a couple of variables (k_F, k_V).

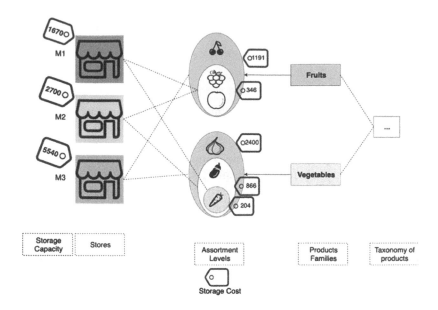

Fig. 2. Required parameters and variables for the knapsack model

Any change in $S^{k_F}(F)$ or $S^{k_V}(V)$ entails turnovers variations. The optimal assortment problem consists in identifying the best couple of values for k_F and k_V. This result is achieved by solving the knapsack problem formalized in this paper. The main difficulty is the assessment of the turnovers when k_F and k_V are changed into $k_F + j$ and $k_V + j'$. An estimation of these turnovers has to be computed in order to evaluate the performance of the candidate values $k_F + j$ and $k_V + j'$ in the knapsack problem. As explained above this estimation is based on the turnovers of similar stores that propose $k_F + j$ and $k_V + j'$ for families Fruits and Vegetables. To this end, we apply semantic similarity measures on the product taxonomy to compute a similarity matrix between stores (*cf.* Sect. 4). The unknown turnovers are then assessed from those of the most similar stores. The stores' similarity matrix is based on semantic similarity measures (in this experiment, the Resnik's measure for the semantic similarity measure and the BMA for the groupwise measure using the semantic library tools[1]). Note that, this step allows defining similarities between stores and can be used to define semantic clusters of stores [19]. Once the matrix is defined, it is used to estimate the turnovers of increased candidate product ranges ($k_F + j$ and $k_V + j'$) estimated as the corresponding turnovers of the nearest stores that propose $k_F + j$ and $k_V + j'$ for F and V. An example of estimation of product range turnovers is proposed in Fig. 3.

---

[1] https://www.semantic-measures-library.org/sml/index.php?.

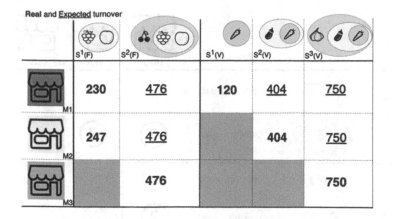

Fig. 3. Estimation of the turnovers of increased product ranges

The last step consists in exploiting the previous estimation in the knapsack problem. As explained above, the knapsack problem aims identifying the ideal product range for each family in order to find: $\text{Arg} \max_{k_i, i=1..n} \sum_{i=1}^{n} p(s^{k_i}(F_i))$ while respecting (at least) the overall storage cost constraint $\sum_{i=1}^{n} c(s^{k_i}(F_i)) \leq C$ defined in Sect. 2. This step involves assessing any combination of $s^{k_i}(F_i)$ for all categories of products F_i. Constraints regarding the storage costs $c(s^{k_i}(F_i))$ can be applied on any category of products which allow reducing complexity of the *knapsack problem* thanks to local reduction of possible solutions (see local constraints in Sect. 2). An illustration on how local constraints reduce the set of solutions is available in the Fig. 4.

For example in Fig. 4, the highest level of assortment for vegetables $s^3(V)$ is greater than the total storage capacity of stores M1 and M2. This information allows eliminating the upgrade $s^3(V)$ for the Vegetables family in stores M1 and M2. Finally, in this example three upgrades can be envisaged:

1. Store M1 can improve its Fruits assortment from $S^1(F)$ to $S^2(F)$:

$$c(S^2(F)) + c(S^1(V)) \leq \text{SCM}_1$$

2. Store M1 can improve its Vegetables assortment from $S^1(V)$ to $S^2(V)$:

$$c(S^1(F)) + c(S^2(V)) \leq \text{SCM}_1$$

3. Store M2 can improve its Fruits assortment from $S^1(F)$ to $S^2(F)$:

$$c(S^2(F)) + c(S^2(V)) \leq \text{SCM}_2$$

Then, store M3 owns already all products, so no upgrade is feasible. Due to its storage capacity, store M2 can only improve its Fruits assortment. Store

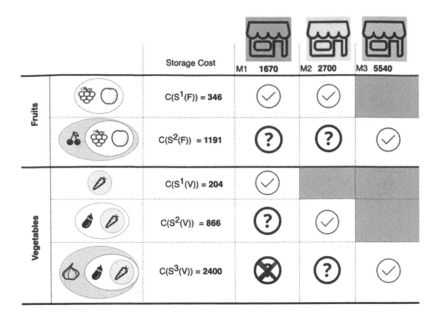

Fig. 4. Example of local constraints

M1 is available to improve either its `Fruits` or its `Vegetables` assortments. The Fig. 3 provides the turnovers' estimations for any feasible product range upgrade. The optimal upgrades can now be deduced from it. Therefore, store M1 should upgrade its `Vegetables` assortment from $S^1(V)$ to $S^2(V)$ to improve its turnover. This trivial example allows highlighting how *knapsack problem* can be simplified thanks to local restrictions and taxonomical reasoning.

   This example was a mere illustration. The naive optimization of the assortment would consist in trying all possible subsets of products without considering constraints (from stores or from range products). In other words, it requires to try all possible combinations of products whatever their category. In our example, without taxonomy, we should basically reason on the set of products: `apple`, `grapefruit`, `cherry`, `carrot`, `eggplant` and `onion`. With only 6 products, we have 63 possibilities $[2^n - 1]$ which have to be tried for each store. In our toy example, reasoning on the taxonomy of products and managing storage cost constraints significantly reduce the research space. We have shown in other articles referring to the biomedical field the interest of semantic similarities when the dimensions of space are organized by a domain taxonomy [15].

   To ensure that this process is scalable with a real dataset from retail, we have built three benchmarks based on the Google Taxonomy[2] that we report in this paper. Experiments have been processed on 1 CPU from an Intel Core I7-2620M 2.7GHz 8Go RAM. We exploited the CPLEX library (IBM CPLEX 1.25) and each benchmark requires less than one second. These benchmarks simply allow

---

[2] https://www.google.com/basepages/producttype/taxonomy.fr-FR.txt.

us to claim that our complete data processing to compute the ideal assortment of stores of a network can be achieved even for significantly large problems as referred in Table 1. Semantic interpretations of our work are yet to be done and require the intervention of domain experts and evaluations over large periods of time. This assessment is outside the scope of this article and will be carried out as part of the commercial activity of TRF Retail.

Table 1. Benchmarks' details

	Benchmark 1	Benchmark 2	Benchmark 3
Number of stores	15	30	50
Number of levels of range product	4	16	20
Number of families of products	12	80	200
Number of variables	180	2 400	10 000

6   Conclusion

The aim of the paper is to propose a methodology allowing improvement of retailers' assortments. Indeed, the ultimate goal consists in proposing adequate products to stores depending on their specific constraints. To achieve this goal semantic approaches are used not only to improve knowledge on stores but also to make the estimations of the consequences of assortment changes more reliable. As a matter of fact, the proposed approach includes prior knowledge from the taxonomy of products used to formalize the ideal assortment problem into a knapsack problem. The estimation of the expected benefits associated to changes in the product range of products' families is based on results of similar stores. Those similarities are identified by means of semantic similarity measures we previously studied in the field of biomedical information retrieval. The use of semantic approaches brings more appropriate results to retailers because it includes part of their knowledge on the organization of products sold.

The management of the products' taxonomy notably reduces the search space for the knapsack problem. It also allows defining an appropriate similarity matrix between the stores of a network that takes into account the way the turnovers of the stores are distributed. It implies they have approximately the same types of customers. The definition of this similarity is crucial for the estimation of turnovers required in the knapsack problem. This process should be computed repetitively to allow continuous improvement which is a key factor in retail sector. Actually, we are working on the integration of more sophisticated constraints in our optimization problem in order to capture more complex behaviors of retailers.

214   J. Poncelet et al.

References

1. Harispe, S., Sanchez, D., Ranwez, S., Janaqi, S., Montmain, J.: A framework for unifying ontology-based semantic similarity measures: a study in the biomedical domain. J. Biomed. Inform. **48**, 38–53 (2014)
2. Harispe, S., Imoussaten, A., Trousset, F., Montmain, J.: On the consideration of a bring-to-mind model for computing the information content of concepts defined into ontologies, pp. 1–8 (2015)
3. Kim, H.K., Kim, J.K., Chen, Q.Y.: A product network analysis for extending the market basket analysis. Expert Syst. Appl. **39**(8), 7403–7410 (2012)
4. Ibadvi, A., Shahbazi, M.: A hybrid recommendation technique based on product category attributes. Expert Syst. Appl. **36**(9), 11480–11488 (2009)
5. Resnik, P.: Using information content to evaluate semantic similarity in a taxonomy. In: Proceedings of IJCAI-95, pp. 448–453 (1995)
6. Sanchez, D., Batet, M.: Semantic similarity estimation in the biomedical domain: an ontology-based information-theoretic perspective. J. Biomed. Inform. **44**(5), 749–759 (2011)
7. Janaqi, S., Harispe, S., Ranwez, S., Montmain, J.: Robust selection of domain-specific semantic similarity measures from uncertain expertise. In: Laurent, A., Strauss, O., Bouchon-Meunier, B., Yager, R.R. (eds.) IPMU 2014. CCIS, vol. 444, pp. 1–10. Springer, Cham (2014). https://doi.org/10.1007/978-3-319-08852-5_1
8. Schlicker, A., Domingues, F.S., Rahnenfhrer, J., Lengauer, T.: A new measure for functional similarity of gene products based on gene ontology. BMC Bioinform. **7**, 302 (2006). https://doi.org/10.1186/1471-2105-7-302
9. Pesquita, C., Faria, D., Bastos, H., Falcao, A., Couto, F.: Evaluating go-based semantic similarity measures. In Proceedings of the 10th Annual Bio-Ontologies Meeting, vol. 37, p. 38 (2007)
10. Yucel, E., Karaesmen, F., Salman, F.S., Türkay, M.: Optimizing product assortment under customer-driven demand substitution. Eur. J. Oper. Res. **199**(3), 759–768 (2009)
11. Kok, A.G., Fisher, M.L., Vaidyanathan, R.: Assortment planning: review of literature and industry practice. In: Retail Supply Chain Management, pp. 1-46 (2006)
12. Agrawal, N., Smith, S.A.: Optimal retail assortments for substitutable items purchased in sets. Naval Res. Logist. **50**(7), 793–822 (2003)
13. Alptekinoglu, A.: Mass customization vs. mass production: variety and price competition. Manuf. Serv. Oper. Manag. **6**(1), 98–103 (2004)
14. Boatwright, P., Nunes, J.C.: Reducing assortment: an attribute-based approach. J. Mark. **65**(3), 50–63 (2001)
15. Poncelet, J., Jean, P.A., Trousset, F., Montmain, J.: Impact des mesures de similarité sémantique dans un algorithme de partitionnement : d'un cas biomédical à la détection de comportements de consommation. SFC, pp. 1–6 (2019)
16. Huffman, C., Kahn, B.E.: Variety for sale: mass customization or mass confusion? J. Retail. **74**, 491–513 (1998)
17. Pal, K.: Ontology-based web service architecture for retail supply chain management. Eur. J. Oper. Res. **199**, 759–768 (2009)
18. Netessine, S., Rudi, N.: Centralized and competitive inventory model with demand substitution. Oper. Res. **51**, 329–335 (2003)
19. Poncelet, J., Jean, P.-A., Trousset, F., Montmain, J.: Semantic customers' segmentation. In: El Yacoubi, S., Bagnoli, F., Pacini, G. (eds.) INSCI 2019. LNCS, vol. 11938, pp. 318–325. Springer, Cham (2019). https://doi.org/10.1007/978-3-030-34770-3_26

Towards Multi-perspective Conformance Checking with Aggregation Operations

Sicui Zhang[1,2(✉)], Laura Genga[2], Lukas Dekker[3], Hongchao Nie[4],
Xudong Lu[1,2], Huilong Duan[1], and Uzay Kaymak[2,1]

[1] School of Biomedical Engineering and Instrumental Science, Zhejiang University,
Hangzhou, People's Republic of China
zhangsicui@zju.edu.cn
[2] School of Industrial Engineering,
Eindhoven University of Technology, Eindhoven, The Netherlands
[3] Cardiology Department, Catharina Hospital, Eindhoven, The Netherlands
[4] Philips Research, Eindhoven, The Netherlands

Abstract. Conformance checking techniques are widely adopted to validate process executions against a set of constraints describing the expected behavior. However, most approaches adopt a crisp evaluation of deviations, with the result that small violations are considered at the same level of significant ones. Furthermore, in the presence of multiple data constraints the overall deviation severity is assessed by summing up each single deviation. This approach easily leads to misleading diagnostics; furthermore, it does not take into account user's needs, that are likely to differ depending on the context of the analysis. We propose a novel methodology based on the use of aggregation functions, to assess the level of deviation severity for a set of constraints, and to customize the tolerance to deviations of multiple constraints.

Keywords: Conformance checking · Fuzzy aggregation · Data perspective

1 Introduction

Nowadays organizations often define procedures describing how their processes should be performed to satisfy a set of constraints, e.g., to minimize the throughput time or to comply with rules and regulations. A widely used formalism to represent these procedures consists in so-called *process models*, that are graphic or logic formalism representing constraints defined on organization processes, e.g., by the order of execution of the activities. However, it is well documented in literature that real process behavior often deviates from the expected process, which often leads to performance issues or opens the way to costly frauds [12].

© Springer Nature Switzerland AG 2020
M.-J. Lesot et al. (Eds.): IPMU 2020, CCIS 1237, pp. 215–229, 2020.
https://doi.org/10.1007/978-3-030-50146-4_17

In recent years, the increasing use by organizations of information systems (e.g., ERP, SAP, MRP and so on) to support and track the execution of their processes enabled the development of automatic, data-driven techniques to assess the compliance level of the real process behavior. Among them, *Conformance checking* techniques have been gaining increasing attention both from practitioners and academics [1,2,5,6,23]. Given an *event log*, i.e., a log file tracking data related to activities performed during process executions, conformance checking techniques are able to pinpoint discrepancies (aka, deviations) between the log and the corresponding model. While classic conformance checking techniques only deal with the *control-flow* of the process, i.e., the activities execution order, in recent years also some *multi-perspective conformance checking*, aimed to deal also with data constraints, have become more and more relevant [23,25].

Nevertheless, there are still several open challenges to implement multi-perspective conformance checking. Among them, here we focus on the lack of appropriate modeling mechanisms for dealing with the *uncertainty* and *graduality* often characterizing human-decisions in real-world processes. State of the art techniques implement a *crisp* approach: every execution of an activity is considered as either *completely wrong* or *completely correct*. [13,23,25].

While this assumption is well grounded to deal with the control-flow (indeed, each activity is either executed at the right moment, or it is not), when addressing data constraints it can easily lead to misleading results. A well-known example of this issue can be found in the healthcare domain. Let us assume that a surgery department implements a guideline stating that the systolic blood pressure (SBP) of a patient has to be lower than 140 to proceed with a surgery. It is reasonable to expect that sometimes clinicians will not refuse to operate patients whose SBP is 141, since this is quite a small deviation and delaying the surgery could be more dangerous for the patient. Clearly, surgeries performed with this value of SBP are likely to be much less problematic than surgeries performed with a SBP equal to, e.g., 160. However, conformance checking techniques would simply mark both these cases as 'not compliant', without allowing for any distinction. This behavior is undesirable, since it is likely to return in output a plethora of not-interesting deviations, at the same time hiding those which could deserve further investigation. We investigated this issue in our previous work [29], where we proposed to use fuzzy sets, which are used to present the flexibility in the constraints and the goals in fuzzy optimization [20], to determine the severity of violations of a single soft constraint per activity.

However, the previous work used basic strategy of standard conformance checking techniques for dealing with *multiple constraints deviations*; namely, the total degree of data deviations of that activity is computed by summing up the costs for all the violated constraints. This strategy poses some important limitations when investigating the data compliance. First, it introduces an asymmetry in the assessment of control-flow and data deviations. While control-flow deviations for each activity express the level of compliance of the activity to control-flow constraints (either fully compliant or wrong), in the presence of multiple data constraints the obtained value does not give an indication of the overall level of compliance to the constraints set. Furthermore, no customization to the user's needs is provided. First, in this setting data violations tend to be considered more severe than control-flow ones, even if this might not fit with user's intention. Furthermore, different contexts might require tailored functions to assess multiple data deviations severity.

In this paper, we address this issue by proposing a novel fuzzy conformance checking methodology based on the use of aggregation functions, which have been proved feasible for modeling simultaneous satisfaction of aggregated criteria [20]. With respect to previous work, the approach brings two main contributions: a) it applies fuzzy aggregation operators to assess the level of deviation severity for a set of constraints, and b) it allows to customize the tolerance to deviations of multiple constraints. As a proof-of-concept, we tested the approach over synthetic data.

The remainder of this paper is organized as follows. Section 2 introduces a running example to discuss the motivation of this work. Section 3 introduces basic formal notions. Section 4 illustrates our approach, and Sect. 5 presents results obtained by a set of synthetic experiments. Section 6 discusses related work. Finally, Sect. 7 draws some conclusions and presents future work.

2   Motivation

Consider, as a running example, a loan management process derived from previous work on the event log of a financial institute made available for the BPI2012 challenge [3,15]. Figure 1 shows the process in BPMN notation. The process starts with the submission of a loan application. Then, the application passes through a first assessment of the applicant's requirements and, if the requested amount is greater than 10000 euros, also through a more thorough fraud detection analysis. If the application is not eligible, the process ends. Otherwise, the application is accepted, an offer to be sent to the customer is selected and the details of the application are finalized. After the offer has been created and sent to the customer, the latter is contacted to discuss the offer with her. At the end of the negotiation, the agreed application is registered on the system. At this point, further checks can be performed on the application, if the overall duration is still below 30 days and the Amount is larger than 10000, before approving it.

Let us assume that this process is supported by some system able to track the execution of its activities in a so-called event log. In practice, this is a collection of *traces*, i.e., sequences of activities performed within the same process execution, each storing information like the execution timestamp of the execution, or other data element [1]. As an example, let us consider the following traces[1] showing two executions of the process in Fig. 1 (note that we use acronyms rather than complete activity names) : $\sigma_1 = \langle (A\_S, \{Amount =$

---

[1] We use the notation $(act, \{att_1 = v_1, \ldots, att_n = v_n\})$ to denote the occurrence of activity act in which variables $att_1 \ldots att_n$ are assigned to values $v_1, \ldots v_n$. The symbol $\perp$ means that no variable values are changed when executing the activity.

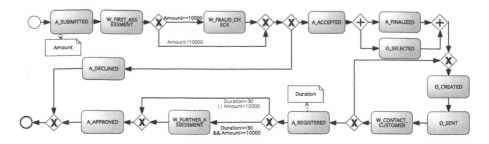

Fig. 1. The loan management process.

$8400\}), (W\_FIRST\_A, \perp), (W\_F\_C, \perp), (A\_A, \perp), (A\_F, \perp), (O\_S, \perp), (O\_C, \perp),$
$(O\_S, \perp), (W\_C, \perp), (A\_R, \{Duration = 34\}), (W\_F\_A, \perp), (A\_AP, \perp), \rangle; \sigma_2 =$
$\langle (A\_S, \{Amount = 1400\}), (W\_FIRST\_A, \perp), (W\_F\_C, \perp), (A\_A, \perp), (A\_F, \perp),$
$(O\_S, \perp), (O\_C, \perp), (O\_S, \perp), (W\_C, \perp), (A\_R, \{Duration = 24\}), (W\_F\_A, \perp),$
$(A\_AP, \perp), \rangle.$ Both executions violate the constraints defined on the duration and
the amount of the loan, according to which the activity W_F_A should have been
anyway skipped.

Conformance checking techniques also attempt to support the user in investigating the *interpretations* of a deviation. In our case, the occurrence of the
activity W_F_A could be considered either as a 1) control-flow deviation (i.e.,
data are corrected but the activity should not have been executed) or as a 2)
data-flow deviation (i.e., the execution of the activity is correct but data have not
been properly recorded on the system). In absence of domain knowledge in determining what is the real explanation, conformance checking techniques assess the
severity (aka, cost) of the possible interpretations and select the least severe one,
assuming that this is the one closest to the reality. In our example, conformance
checking would consider σ_1 as a control-flow deviation, since the cost would be
equal to 1, while data-flow deviation would correspond to 2, having two violated
constraints; for σ_2, instead, the two interpretations would be equivalent, since
only one data constraint is violated. In previous work [29] we investigated how
to use fuzzy membership function to assess severity of data deviations taking
into account the magnitude of the deviations. However, the approach still comes
with some limitations when considering multiple constraints. Indeed, with this
approach the overall severity of the data deviation for an activity is assessed by
a simple sum operation. For example, let us suppose that with the method in
[29] we obtained a cost of 0.3, 0.8 for the violations of *Amount* and *Duration* in
W_F_A in σ_1, thus obtaining a total cost of 1.1, and 0.8 and 0 in σ_2, thus obtaining, a total cost of 0.8. In this setting, activities involving multiple constraints
will tend to have an interpretation biased towards control-flow deviations, since
the higher the number of constraints, the higher the the data-deviation cost.
Furthermore, it is worth noting that the comparison between the two traces can
be misleading; in one case, constraints are violated, even if one only slightly
deviated; while in the second case only one constraint is violated, even if with
quite a strong deviation. However, the final numerical results are quite similar,

thus hiding the differences. This example shows how the use of the simple sum function can impact the results significantly, without the user realizing it and, above all, without providing the user with any customization mechanism. For example, the user might want to assess the data-compliance level in terms of the percentage of satisfied constraints, or by considering only the maximum cost, and so on. However, current techniques do not allow for this kind of customization.

3 Preliminaries

This section introduces a set of concepts that will be used through the paper.

3.1 Conformance Checking: Aligning Event Logs and Models

Conformance checking techniques detect discrepancies between a process model and the real process execution. Here we define the notion of process model using the notation from [2], enriched with data-related notions explained in [13].

Definition 1 (Process model). *A process model $M = (P, P_I, P_F, A_M, V, W, U, T, G, Values)$ is a transition system defined over a set of activities A_M and a set of variables V, with states P, initial states $P_I \subseteq P$, final states $P_F \subseteq P$ and transitions $T \subseteq P \times (A_M \times 2^V) \times P$. $U(V_i)$ represents the domain of V_i for each $V_i \in V$. The function $G : A_M \rightarrow Formulas(V \cup \{V_i' \mid V_i \in V\})$ is a guard function, i.e., a boolean formula expressing a condition on the values of the data variables. $W : A_M \rightarrow 2^V$ is a write function, that associates an activity with the set of variables which are written by the activity. Finally, $Values : P \rightarrow \{V_i = v_i, i = 1..|V| \mid v_i \in U(V_i) \cup \{\bot\}\}$ is a function that associates each state with the corresponding pairs variable=value.*

The firing of an activity $s = (a, w) \in A_M \times (V \nrightarrow U)$ in a state p' is *valid* if: 1) a is enabled in p'; 2) a writes all and only the variables in $W(a)$; 3) $G(a)$ is *true* when evaluate over $Values(p')$. To access the components of s we introduce the following notation: $vars(s) = w$, $act(s) = a$. Function $vars$ is also overloaded such that $vars(V_i) = w(V_i)$ if $V_i \in dom(vars(s))$ and $vars(s, V_i) = \bot$ if $V_i \notin dom(vars(s))$. The set of valid process traces of a model M is denoted with $\rho(M)$ and consists of all the valid firing sequences $\sigma \in (A_M \times (V \nrightarrow U))^*$ that, from an initial state P_I lead to a final state P_F. Figure 1 provides an example of a process model in BPMN notation.

Process executions are often recorded by means of an information system in event logs. Formally, let S_N be the set of (valid and invalid) firing of activities of a process model M; an **event log** is a multiset of traces $\mathbb{L} \in \mathbb{B}(S_N^*)$. Given an event log L, conformance checking builds an *alignment* between L and M, mapping "moves" occurring in the event log to possible "moves" in the model. A "no move" symbol "$\gg$" is used to represent moves which cannot be mimicked. For convenience, we introduce the set $S_N^{\gg} = S_N \cup \{\gg\}$. Formally, we set s_L to be a transition of the events in the log, s_M to be a transition of the activities in the model. A move is represented by a pair $(s_L, s_M) \in S_N^{\gg} \times S_N^{\gg}$ such that:

- (s_L, s_M) is a *move in log* if $s_L \in S_N$ and $s_M = \gg$
- (s_L, s_M) is a *move in model* if $s_M \in S_N$ and $s_L = \gg$
- (s_L, s_M) is a *move in both without incorrect data* if $s_L \in S_N$, $s_M \in S_N$ and $act(s_L) = act(s_M)$ and $\forall V_i \in V(vars(s_L, V_i) = vars(s_M, V_i)))$
- (s_L, s_M) is a *move in both with incorrect data* if $s_L \in S_N$, $s_M \in S_N$ and $act(s_L) = act(s_M)$ and $\exists V_i \in V \mid vars(s_L, V_i) \neq vars(s_M, V_i))$.

Let $A_{LM} = \{(s_L, s_M) \in S_N^{\gg} \times S_N^{\gg} \mid s_L \in S_N \vee s_M \in S_N\}$ be the set of all legal moves. The *alignment* between two process executions σ_L, $\sigma_M \in S_N^*$ is $\gamma \in A_{LM}^*$ such that the projection of the first element (ignoring $\gg$) yields σ_L, and the projection on the second element (ignoring $\gg$) yields σ_M.

Example 1. Let us consider the model in Fig. 1 and the trace σ_1 in Sect. 2.

Table 1 shows two possible alignments γ_1 and γ_2 for activity W_F_A. For Alignment γ_1, the pair (W_F_A, W_F_A) is a move in both with incorrect data, while in γ_2 the move $(W\_F\_A, \bot)$ is matched with a $\gg$, i.e., it is a move on log. (In remaining part, *Amount* and *Duration* are abbreviated to A and D).

Table 1. Two possible alignments between σ_M and σ_L

Alignment γ_1		Alignment γ_2	
Log	Model	Log	Model
...	...	...	...
$(W\_F\_A, \{8000, 34\})$	(W_F_A)	$(W\_F\_A, \{8000, 34\})$	$(\gg)$
...	...	...	...

As shown in Example 1, there can be multiple possible alignments for a given log trace and process model. Our goal is to find the *optimal alignment*, i.e., the alignment with minimum cost. To this end, the severity of deviations is assessed by means of a *cost function*.

Definition 2 (Cost function, Optimal Alignment). *Let σ_L, σ_M be a log trace and a trace, respectively. Given the set of all legal moves A_N, a cost function k assigns a non-negative cost to each legal move: $A_N \to \mathbb{R}_0^+$. The cost of an alignment γ between σ_L and σ_M is computed as the sum of the cost of all the related moves: $K(\gamma) = \sum_{(S_L, S_M) \in \gamma} k(S_L, S_M)$. An **optimal** alignment of a log trace and a process trace is one of the alignments with the lowest cost according to the provided cost function.*

3.2   Fuzzy Set Aggregation Operators

Aggregation operations (AOs) are mathematical functions that satisfy minimal boundary and monotonicity conditions, and are often used for modeling decision making processes, since they allow to specify how to combine the different criteria that are relevant when making a decision [17,27].

In literature, many AOs have been defined (see [18,19,22] for an overview), with different level of complexity and different interpretations. A commonly used class of aggregation operators are the t-norms, which are used to model conjunction of fuzzy sets. In compliance analysis, one often tries to satisfy all constraints on the data, and so t-norms are suitable operators for modeling soft constraints in compliance analysis. Widely used t-norms are the minimum, product and the Yager operators [21].

In addition to the t-norms, other aggregation operators could also be used, depending on the goals of the compliance analysis. We do not consider other types of aggregation operators in this paper, but, in general, one could use the full flexibility of different classes of fuzzy set aggregation operators that have been used in decision making (see, e.g. [11]).

4   Proposed Compliance Analysis Method

We introduce a compliance checking approach tailored to dealing with decision tasks under multiple guards, to enhance the flexibility of the compliance assessing procedure. To this end, we investigate the use of AOs.

4.1   Aggregated Cost Function

Compliance checking in process analysis is based on the concept of alignment between a process model and a process trace that minimizes a cost of misalignment. The computation of an optimal alignment relies on the definition of a proper cost function for the possible kind of moves (see Sect. 3). Most of state-of-the art approaches adopt (variants of) the standard distance function defined in [2], which sets a cost of 1 for every move on log/model (excluding invisible transitions), and a cost of 0 for synchronous moves. Multi-perspective approaches extend the standard cost function to include data costs. Elaborating upon these approaches, in previous work [29] we defined our fuzzy cost function as follows.

Definition 3 (Data-aware fuzzy cost function). *Let* (S_L, S_M) *be a move between a process trace and a model execution,* $W(S_M)$ *be the set of variables written by the activity related to* S_M*, and let* $\mu_i(var(S_L, V_i))$ *be a fuzzy membership function returning the compliance degree of single variable* $var(S_L, V_i)$*. For the sake of simplicity, we write it as* μ_i *in the following. Then we define* $(1 - \mu_i)$ *as the data cost of this deviation. The cost* $k(S_L, S_M)$ *is defined as:*

$$
k(S_L, S_M) = \begin{cases} 1 & \text{if } (S_L, S_M) \text{ is a move in log} \\ 1 + |W(S_M)| & \text{if } (S_L, S_M) \text{ is a move in model} \\ \sum_{\forall V_i \in V} (1 - \mu_i) & \text{if } (S_L, S_M) \text{ is a move in both} \end{cases} \tag{1}
$$

This cost function assigns a cost equal to 1 for a move in log; 1 plus the number of variables that should have been written by the activity for a move in model; finally, the sum of the cost of the deviations $(1-\mu_i)$ for the data variables if it's a move in both. Note that the latter consider both the case of move with incorrect and incorrect data. As discussed in Sect. 2, summing up all the data cost presents important limitations to assess the conformance of multiple constraints. Therefore, in the present work, we propose a new version of our fuzzy cost function with the goal of standardize every move within the range (0,1) and allow the user to customize the cost function to her needs.

Definition 4 (AOs based cost function). *Let $\pi(\mu_1, \mu_2, ..., \mu_n)$ be an user-defined aggregated membership function of multiple variables. Then $(1 - \pi)$ is the overall deviation cost of a set of variables. The cost $k(S_L, S_M)$ is defined as:*

$$k(S_L, S_M) = \begin{cases} 1 & if\,(S_L, S_M) \text{ is a move in log} \\ 1 + |W(S_M)| & if\,(S_L, S_M) \text{ is a move in model} \\ 1 - \pi(\mu_1, \mu_2, ..., \mu_n) & if\,(S_L, S_M) \text{ is a move in both.} \end{cases} \quad (2)$$

4.2   Using A* to Find the Optimal Alignment

The problem of finding an optimal alignment is usually formulated as a search problem in a directed graph [14]. Let $Z = (Z_V, Z_E)$ be a directed graph with edges weighted according to some cost structure. The A* algorithm finds the path with the lowest cost from a given source node $v_0 \in Z_v$ to a node of a given goals set $Z_G \subseteq Z_V$. The cost from each node is determined by an evaluation function $f(v) = g(v) + h(v)$, where:

- $g : Z_V \rightarrow \mathbb{R}^+$ gives the smallest path cost from v_0 to v;
- $h : Z_V \rightarrow \mathbb{R}_0^+$ gives an estimate of the smallest path cost from v to any of the target nodes.

If h is *admissible*,i.e. it underestimates the real distance of a path to any target node v_g, then A* finds a path that is guaranteed to have the overall lowest cost.

The algorithm works iteratively: at each step, the node v with lowest cost is taken from a priority queue. If v belongs to the target set, the algorithm ends returning node v. Otherwise, v is expanded: every successor v_0 is added to the priority queue with a cost $f(v_0)$.

Given a log trace and a process model, to employ A* to determine an optimal alignment we associate every node of the search space with a prefix of some complete alignments. The source node is an empty alignment $\gamma_0 = \langle \rangle$, while the set of target nodes includes every complete alignment of σ_L and M. For every pair of nodes (γ_1, γ_2), γ_2 is obtained by adding one move to γ_1.

The cost associated with a path leading to a graph node γ is then defined as $g(\gamma) = K(\gamma) + \epsilon|\gamma|$, where $K(\gamma) = \sum_{s_L, s_M \in \gamma} k(s_L, s_M)$, with $k(s_L, s_M)$ defined as in (1), $|\gamma|$ is the number of moves in the alignment, and ϵ is a negligible cost, added to guarantee termination. Note that the cost g has to be strictly increasing. While we do not give a formal proof for the sake of space, it is straight to see that g is obtained in our approach by the sum of all non negative elements. Therefore, while moving from an alignment prefix to a longer one, the cost can never decrease. For the definition of the heuristic cost function $h(v)$ different strategies can be adopted. Informally, the idea is computing, from a given alignment, the minimum number of moves (i.e., the minimum cost) that would lead to a complete alignment. Different strategies have been defined in literature, e.g., the one in [2], which exploits Petri-net marking equations, or the one in [28], which generates possible states space of a BPMN model.

Example 2. Let us analyze possible moves to assign to the activity W_F_A in σ_1. Let us assume that the memberships of the variables are $\mu_A = 0.4$ and $\mu_D = 0.2$. According to (2) and *Product* t-norm we get the fuzzy cost function $k(S_L, S_M)$.

$$k(S_L, S_M) = \begin{cases} 1 & , \textit{move in log} \\ 1 & , \textit{move in model} \\ 1 - \mu_A \cdot \mu_D & , \textit{move in both} \end{cases} \quad (3)$$

Figure 2 shows the portion of the space states for the alignment building of σ_1. At node #11, $f = 0$, since no deviations occurred so far. From here, there are two possible moves that could be selected, one representing a move on log (on the left), one a move on model (on the right) and finally a move in both (in the middle). Since using the *Product* aggregation the data cost is equal to 0.92, the algorithm selects the move in both, being the one with the lowest cost.

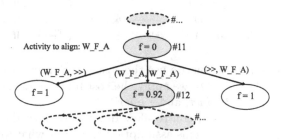

Fig. 2. The alignment with the new aggregated function.

5   Experiment and Result

This section describes a set of experiments we performed to obtain a proof-of-concept of the approach. We compared the diagnostics returned by an existing approach [29] and our new cost functions with three $t - norm$ aggregations. More precisely, we aimed to get the answer to the question: *what is the impact of different aggregation operations on the obtained alignments?* In particular, we assess the impact of the aggregation function in terms of a) differences in the overall deviation cost, and b) difference in terms of the interpretation, i.e., the moves selected by the alignment algorithm as the best explanation for the deviation.

5.1   Settings

In order to get meaningful insights on the behavior we can reasonably expect by applying the approach in the real world, we employ a realistic synthetic event log, consisting of 50000, introduced in a former paper [16], obtained starting from one real-life logs, i.e., the event log of the BPI2012 challenge [2]. We evaluated the compliance of this log against a simplified version of the process model in [16], to which we added few data constraints (see Fig. 1). The approach has been implemented as an extension to the tool developed by [28], designed to deal with BPMN models. Our process model involves two constraints for the activity W_F_A, i.e., $Amount \geq 10000$ and $Duration \leq 30$.

Here we assume that $Amount \in (3050, 10000)$ and $Duration \in (30, 70)$ represent a tolerable violation range for the variables. Since we cannot refer to experts' knowledge, we derived these values from simple descriptive statistics. In particular, we considered values falling within the third quartile as acceptable. The underlying logic is that values which tend to occur repeatedly are likely to indicate acceptable situations. Regarding the shape of the membership functions for the variables, here we apply the linear function μ, as reported below.

$$\mu_1(A) = \begin{cases} 1 & \text{, if } A \geq 10000 \\ 0 & \text{, if } A \leq 2650 \\ \frac{A-2650}{7350} & \text{, if } 2650 < A < 10000; \end{cases} \qquad \mu_2(D) = \begin{cases} 1 & \text{, if } D \leq 30 \\ 0 & \text{, if } D \geq 69 \\ \frac{69-D}{39} & \text{, if } 30 < D < 69 \end{cases} \qquad (4)$$

For the classic sum function, we use the cost function provided by (1); while for the new approach with AOs, we apply the cost function in (2). We tested the $t-norms$: $Minimum$, $Product$, and $Yager$.

When data deviations and control-flow deviations show the same cost, we picked the control-flow move. This assumption simulates what we would do in a real-world context. Indeed, without a-priori knowledge on the right explanation, it is reasonable to assume that it is more likely that the error was executing the activity, rather than accepting out-of-range data deviations.

5.2   Results

Note that here we focus on the activity W_F_A, since, in our log, is the only one involving multiple data constraints. Table 2 shows differences in terms of number and type of moves, as well as in terms of costs. The columns $\#move\ in$ log, $\#move\ in\ data$ show the number of traces in which the alignment has selected

---

[2]   https://www.win.tue.nl/bpi/doku.php?id=2012:challenge.

for the activity W_F_A a move in log or a move in data, respectively. The column "Average costs" shows the average alignment cost. The conformance checking algorithms selects for each activity the move corresponding to the minimum cost. Therefore, the differences among the chosen move depend on the different costs obtained on W_F_A when applying different operators. To provide a practical example of the impact of the aggregated cost on the obtained diagnostics, below we discuss the results obtained for one trace.

Table 2. Number of different moves of the activity W_F_A.

	#move in log	#move in data	Average cost
Sum	707	350	0.823
Min	660	397	0.804
Product	660	397	0.814
Yager	678	379	0.811

Table 3. The cost of possible moves

	#move in log	#move in data
Sum	1	1.003
Min	1	0.513
Product	1	0.751
Yager	1	0.709

Table 4. The optimal alignments

	Log	Model	Move type	Cost
S	W_F_A	$\gg$	#move in log	1
M	W_F_A	W_F_A	#move in data	0.513
P	W_F_A	W_F_A	#move in data	0.751
Y	W_F_A	W_F_A	#move in data	0.709

Example 3. Let us consider the *trace* $\sigma_{\#2859} = \langle (A\_S, \{Amount = 6400\})$, $(W\_FIRST\_A, \perp), (A\_A, \perp), (A\_F, \perp), (O\_S, \perp), (O\_C, \perp), (O\_S, \perp), (W\_C, \perp)$, $(W\_F, \perp), (O\_C, \perp), (O\_S, \perp), (W\_C, \perp), (A\_R, \{Duration = 50\}), (W\_F\_A, \perp)$ $, (A\_AP, \perp), \rangle$. According to their membership functions (4), $\mu_1(A = 6400) = 0.5102$ and $\mu_2(D = 50) = 0.4872$. Therefore, the corresponding costs are 0.4898 and 0.5128. Table 3 shows the cost of possible moves for W_F_A according to the aggregation functions. Table 4 shows the move picked by each function to build the alignment. Using the Sum function, the data cost is 1.003, so that a move-in-log is chosen as an optimal alignment. In the other cases, instead, the move in data is the one with the lowest cost. Since both the deviations fall in the acceptable range, this interpretation is likely to be more in line with the user's expectations.

The observations made for the example can be generalized to the overall results of Table 2, which shows a set of traces whose interpretation is heavily affected by the chosen cost function. As expected, the Sum function is the most biased towards the choice of move in log interpretation. It selects 40 moves in log more that Product and Min, and 29 more than Yager. One can argue that this choice is likely not one the human analyst would have expected. Indeed, we are using Yager with $\omega = 2$ [11], that means that when both the variables show severe deviations, we expect the data cost to be 1 and move-in-log to be picked. This means that at least 29 of the aligned traces were marked as move-in-log also if both the constraints did not show severe deviations. We argue that this behavior can be misleading for the analyst or, anyway, not being in line with her needs. The Product function marks other 18 traces as move-in-data, in addition to the ones marked by the Yager. This was expected, since the Product function relaxes the requirements on the full satisfaction of the set of constraints. Nevertheless, this implies that in all these 18 traces the deviations always fell in the tolerance range. Therefore, also these situations might have been better represented as data deviations, depending on the analysts' needs. As regards the Min function, it returns a full data deviation in the presence of at least one deviation outside the deviation range, which explains why it returned the same alignments of the Product function. The overall alignments costs are in line with the expectations. The Sum function returns the highest average cost, as expected, the Min the lowest, while the Yager and the Product behave similarly, and the difference can likely be explained with the 18 traces of difference discussed above. While the absolute difference among the costs is not very relevant, these results show that both the alignments and the assessment of the deviations are impacted by the choice of the cost function, thus highlighting once again the need for a more flexible approach to compliance assessment allowing the user to tailor the cost function to her context.

6 Related Work

During the last decades, several conformance checking techniques have been proposed. Some approaches [9,10,26] propose to check whether event traces satisfy a set of compliance rules, typically represented using declarative modeling. Rozinat and van der Aalst [24] propose a token-based technique to replay event traces over a process model to detect deviations, which, however, has been shown to provide misleading diagnostics in some contexts [4]. Recently, alignments have been proposed as a robust approach to conformance checking based on the use of a cost function [2]. While most of alignment-based approaches use the standard distance cost function as defined by [2], some variants have been proposed to enhance the provided diagnostics, e.g., the work of Alizadeh et al. [8], which computes the cost function by analyzing historical logging data. Besides the control flow, there are also other perspectives like data, or resources, that are often crucial for compliance checking analysis. Few approaches have investigated how to include these perspectives in the analysis: [7] extends the approach in [8] by taking into account data describing the contexts in which the activities occurred. Some approaches proposed to compute the control-flow first then assessing the compliance with respect to the data perspective, e.g. [13]. These methods gives priority to check the control flow, with the result that some important deviations can be missed. [23] introduces a cost function balancing different perspectives, thus obtaining more precise diagnostics. The approaches mentioned so far assume a crisp evaluation of deviations. To the best of our knowledge, the only work which explored the use of a fuzzy cost function is our previous work [29] which, however, did not consider multiple constraints violation.

7 Conclusion

In this work, we investigated the use of fuzzy aggregation operations in conformance checking of process executions to deal with multiple data constraints for an activity. The proposed approach enhances significantly the flexibility of compliance checking, allowing the human analyst to customize the compliance diagnostic according to her needs. We elaborated upon the relevance of this aspect both theoretically and with some examples.

As a proof of concept, we implemented the approach and tested it over a synthetic dataset, comparing results obtained by cost functions with classic sum function and three different aggregations. The experiments confirmed that the approach generates more "balanced" diagnostics, and introduces the capability of personalizing the acceptance of deviations for multiple guards.

Nevertheless, there are several research directions still to be explored. In future work, first we plan to test our approach with real-world data. Furthermore, we intend to investigate the usage of different aggregation functions, as well as the possibility of extending the notion of aggregation to take into account also other kinds of deviations. Finally, we intend to investigate potential applications, for example in terms of on-line process monitoring and support, with the aim of enhancing the system resilience to exceptions and unforeseen events.

Acknowledgements. The research leading to these results has received funding from the Brain Bridge Project sponsored by Philips Research.

References

1. van der Aalst, W., et al.: Process mining manifesto. In: Daniel, F., Barkaoui, K., Dustdar, S. (eds.) BPM 2011. LNBIP, vol. 99, pp. 169–194. Springer, Heidelberg (2012). https://doi.org/10.1007/978-3-642-28108-2_19
2. Van der Aalst, W., Adriansyah, A., van Dongen, B.: Replaying history on process models for conformance checking and performance analysis. Wiley Interdisc. Rev.: Data Min. Knowl. Discovery **2**(2), 182–192 (2012)
3. Adriansyah, A., Buijs, J.C.: Mining process performance from event logs. In: La Rosa, M., Soffer, P. (eds.) BPM 2012. LNBIP, vol. 132, pp. 217–218. Springer, Heidelberg (2013). https://doi.org/10.1007/978-3-642-36285-9_23
4. Adriansyah, A., van Dongen, B.F., van der Aalst, W.M.P.: Towards robust conformance checking. In: zur Muehlen, M., Su, J. (eds.) BPM 2010. LNBIP, vol. 66, pp. 122–133. Springer, Heidelberg (2011). https://doi.org/10.1007/978-3-642-20511-8_11

5. Adriansyah, A., van Dongen, B.F., van der Aalst, W.M.: Memory-efficient alignment of observed and modeled behavior. BPM Center Report **3**, 1–44 (2013)
6. Adriansyah, A., Munoz-Gama, J., Carmona, J., van Dongen, B.F., van der Aalst, W.M.P.: Alignment based precision checking. In: La Rosa, M., Soffer, P. (eds.) BPM 2012. LNBIP, vol. 132, pp. 137–149. Springer, Heidelberg (2013). https://doi.org/10.1007/978-3-642-36285-9_15
7. Alizadeh, M., De Leoni, M., Zannone, N.: Constructing probable explanations of nonconformity: a data-aware and history-based approach. In: 2015 IEEE Symposium Series on Computational Intelligence, pp. 1358–1365. IEEE (2015)
8. Alizadeh, M., de Leoni, M., Zannone, N.: History-based construction of alignments for conformance checking: formalization and implementation. In: Ceravolo, P., Russo, B., Accorsi, R. (eds.) SIMPDA 2014. LNBIP, vol. 237, pp. 58–78. Springer, Cham (2015). https://doi.org/10.1007/978-3-319-27243-6_3
9. Borrego, D., Barba, I.: Conformance checking and diagnosis for declarative business process models in data-aware scenarios. Expert Syst. Appl. **41**(11), 5340–5352 (2014)
10. Caron, F., Vanthienen, J., Baesens, B.: Comprehensive rule-based compliance checking and risk management with process mining. Decis. Support Syst. **54**(3), 1357–1369 (2013)
11. da Costa Sousa, J.M., Kaymak, U.: Model predictive control using fuzzy decision functions. IEEE Trans. Syst. Man. Cybern. Part B (Cybern.) **31**(1), 54–65 (2001)
12. de Leoni, M., van der Aalst, W.M.P., van Dongen, B.F.: Data- and resource-aware conformance checking of business processes. In: Abramowicz, W., Kriksciuniene, D., Sakalauskas, V. (eds.) BIS 2012. LNBIP, vol. 117, pp. 48–59. Springer, Heidelberg (2012). https://doi.org/10.1007/978-3-642-30359-3_5
13. de Leoni, M., van der Aalst, W.M.P.: Aligning event logs and process models for multi-perspective conformance checking: an approach based on integer linear programming. In: Daniel, F., Wang, J., Weber, B. (eds.) BPM 2013. LNCS, vol. 8094, pp. 113–129. Springer, Heidelberg (2013). https://doi.org/10.1007/978-3-642-40176-3_10
14. Dechter, R., Pearl, J.: Generalized best-first search strategies and the optimality of A. J. ACM (JACM) **32**(3), 505–536 (1985)
15. Genga, L., Alizadeh, M., Potena, D., Diamantini, C., Zannone, N.: Discovering anomalous frequent patterns from partially ordered event logs. J. Intell. Inf. Syst. **51**(2), 257–300 (2018). https://doi.org/10.1007/s10844-018-0501-z
16. Genga, L., Di Francescomarino, C., Ghidini, C., Zannone, N.: Predicting critical behaviors in business process executions: when evidence counts. In: Hildebrandt, T., van Dongen, B.F., Röglinger, M., Mendling, J. (eds.) BPM 2019. LNBIP, vol. 360, pp. 72–90. Springer, Cham (2019). https://doi.org/10.1007/978-3-030-26643-1_5
17. Grabisch, M., Labreuche, C.: Fuzzy measures and integrals in MCDA. In: Greco, S., Ehrgott, M., Figueira, J.R. (eds.) Multiple Criteria Decision Analysis. ISORMS, vol. 233, pp. 553–603. Springer, New York (2016). https://doi.org/10.1007/978-1-4939-3094-4_14
18. Grabisch, M., Marichal, J.L., Mesiar, R., Pap, E.: Aggregation functions: construction methods, conjunctive, disjunctive and mixed classes. Inf. Sci. **181**(1), 23–43 (2011)
19. Grabisch, M., Marichal, J.L., Mesiar, R., Pap, E.: Aggregation functions: means. Inf. Sci. **181**(1), 1–22 (2011)
20. Kaymak, U., Sousa, C., João, M.: Weighted constraints in fuzzy optimization (2001)

21. Keresztfalvi, T.: Operations on fuzzy numbers extended by yager's family of t-norms. Math. Res. **68**, 163 (1993)
22. Klir, G.J., Yuan, B.: Fuzzy Sets and Fuzzy Logic: Theory and Applications. Prentice-Hall, Inc., Upper Saddle River (1995)
23. Mannhardt, F., de Leoni, M., Reijers, H.A., van der Aalst, W.M.P.: Balanced multi-perspective checking of process conformance. Computing **98**(4), 407–437 (2015). https://doi.org/10.1007/s00607-015-0441-1
24. Rozinat, A., Van der Aalst, W.M.: Conformance checking of processes based on monitoring real behavior. Inf. Syst. **33**(1), 64–95 (2008)
25. Song, W., Jacobsen, H.A., Zhang, C., Ma, X.: Dependence-based data-aware process conformance checking. IEEE Trans. Serv. Comput. (2018). https://doi.org/10.1109/TSC.2018.2821685
26. Taghiabadi, E.R., Gromov, V., Fahland, D., van der Aalst, W.M.P.: Compliance checking of data-aware and resource-aware compliance requirements. In: Meersman, R., et al. (eds.) OTM 2014. LNCS, vol. 8841, pp. 237–257. Springer, Heidelberg (2014). https://doi.org/10.1007/978-3-662-45563-0_14
27. Torra, V., Narukawa, Y.: Modeling decisions: Information Fusion and Aggregation Operators. Springer Science & Business Media, Berlin (2007). https://doi.org/10.1007/978-3-540-68791-7
28. Yan, H., et al.: Aligning event logs to task-time matrix clinical pathways in BPMN for variance analysis. IEEE J. Biomed. Health Inform. **22**(2), 311–317 (2017)
29. Zhang, S., Genga, L., Yan, H., Lu, X., Duan, H., Kaymak, U.: Towards multi-perspective conformance checking with fuzzy sets. arXiv:2001.10730 (2020)

On the Impact of Fuzzy Constraints in the Variable Size and Cost Bin Packing Problem

Jorge Herrera-Franklin[1] , Alejandro Rosete[2(✉)] , Milton García-Borroto[2] ,
Carlos Cruz-Corona[3] , and David A. Pelta[3]

[1] Maritime Transportation Division, Center of Research and Environmental
Management of Transport, La Habana, Cuba
`franklin@cimab.transnet.cu`
[2] Universidad Tecnológica de La Habana José Antonio Echeverría (CUJAE),
La Habana, Cuba
`{rosete,mgarciab}@ceis.cujae.edu.cu`
[3] Department of Computer Science and AI, Universidad de Granada, Granada, Spain
`{carloscruz,dpelta}@decsai.ugr.es`
`http://www.cimab.transnet.cu`

Abstract. The Variable Size and Cost Bin Packing Problem (VSCBPP)
consists of minimizing the cost of all bins used to pack a set of items without exceeding the bins capacities. It is a well known NP-Hard problem
with many practical applications.

In this contribution we assume that the capacity of a bin can be understood in a flexible way (so it may allow some overload) thus leading to
a fuzzy version of the VSCBPP with fuzzy constraints.

We solve the proposed fuzzy VSCBPP by using the parametric approach based on α-cuts, thus defining a set of related crisp problems.

By using three different solving algorithms and several instances, we
explore the impact of different degrees of relaxation not only in terms of
cost, but also in the structure of the solutions.

Keywords: Combinatorial optimization · Variable Size and Cost Bin
Packing Problem · Fuzzy constraint · Parametric approach

1 Introduction

The Variable Sized Bin Packing Problem (VSBPP) is a generalization of the
Bin Packing Problem that was first formalized by Friesen and Langston [5]. It
consists in packing a set of items in a set of heterogeneous bins with different
sizes or capacities. The objective is to minimize the number of bins that are
used. For each size, it is assumed an inexhaustible supply of bins. Crainic *et. al.*
[2] states that by minimizing the cost of all used bins, the problem became the
Variable Size and Cost Bin Packing Problem (VSCBPP).

© Springer Nature Switzerland AG 2020
M.-J. Lesot et al. (Eds.): IPMU 2020, CCIS 1237, pp. 230–240, 2020.
https://doi.org/10.1007/978-3-030-50146-4_18

Some variants of the problem were defined to allow the management of imprecise and/or uncertain information in the problem data. One of the pioneers treating this subject is Crainic *et. al.* [3] studying a real-life application in logistics with uncertainty on the characteristics of the items. Also Wang *et. al.* [16] describes a chance-constrained model where the item sizes are uncertain, while Peng and Zhang [10] introduce the uncertainty on item volumes and bin capacities.

Here, we consider that the capacity of a bin is associated with the maximum weight it can hold. So, for example bins with 25, 50 or 75 kg. capacity may exist. In the standard formulation of the problem, such values are used as crisp constraints. However, it has perfect sense to consider such capacity values as approximate ones and thus using fuzzy constraints instead of crisp ones. In other words, some overloading will be allowed in the bins. To the best of our knowledge, there are not variants of the Variable Size and Cost Bin Packing Problem (VSCBPP) with fuzzy constraints.

The aim of the paper is twofold. Firstly, we introduce a fuzzy version of the VSCBPP that allows some overloading of the bins, which means to relax the satisfaction of capacity constraints; and secondly, we explore the impact that the fuzzy constraints (and the associated relaxations) have, not only in terms of cost, but also in the structure of the solutions. In order to do this, some randomly generated instances of the proposed fuzzy VSCBPP (FVSCBPP) are solved following the parametric approach that transforms the fuzzy problem into a set of crisp problem based on α-cuts [4,14]. Then, each of these instances is solved by an exact solver (SCIP) [15] and two problem-specific heuristics.

The paper is organized as follows. Section 2 presents the VSCBPP fuzzy model. Section 3 explains how the proposed FVSCBPP may be solved by using a parametric approach. This is illustrated in several instances presented in Sect. 4. Finally, Sect. 5, is devoted to conclusions.

2   Fuzzy Variable Size and Cost Bin Packing Problem (FVSCBPP)

In this section we firstly present the basic Variable Size and Cost Bin Packing Problem (VSCBPP) before introducing the proposed fuzzy extension. Problem parameters and the standard VSCBPP formulation [1,7,8] are described next. Being

$I = \{1\ldots,i,\ldots,n\}$	set of items
w_i	weight of the item $i \in I$
$J = \{1\ldots,j,\ldots,m\}$	set of bins
W_j	capacity (or size) of the bin $j \in J$
C_j	cost of the bin $j \in J$
x_{ij}	binary variable: 1 if item i is packed in bin j; 0 otherwise
y_j	binary variable: 1 if bin j is used; 0 otherwise

Then, the VSCBPP is then formulated as follows:

$$Min \sum_{j \in J} C_j y_j \tag{1}$$

$$s.t. \sum_{j \in J} x_{ij} = 1, \quad i \in I \tag{2}$$

$$\sum_{i \in I} w_i x_{ij} \leqslant W_j y_j, \quad j \in J \tag{3}$$

$$x_{ij} \in \{0, 1\}, i \in I, \ j \in J \tag{4}$$

$$y_j \in \{0, 1\}, \quad j \in J \tag{5}$$

The objective function (1) minimizes the cost of the bins used for packing all the items. Constraint (2) ensures that each item i is packed in one and only one bin (items are not divisible). Inequality (3) is the capacity constraint: for each used bin j, the sum of weights of packed items can not exceed its capacity; (4) and (5) are domain constraints. This formulation involves every single bin regardless its type, i.e., a list of bins is one of the problem inputs and it may have more bins than items since there must be enough bins to pack all the items that fulfill the constraint for every type (3). Here, the term type is referred to the capacity of the bin, i.e., its size.

Here we consider that the capacity constraint (3) can be understood in "flexible" (fuzzy) terms:

$$\sum_{i \in I} w_i x_{ij} \leqslant^f W_j y_j, j \in J \tag{6}$$

where $\leqslant^f$ stands for "approximately smaller than or equal to".

This implies that solutions may be either feasible or infeasible depending on the interpretation of the fuzzy relation (6). Indeed, all solutions may be considered feasible with different degrees of membership.

A decision maker may clearly states that the solutions that do not exceed the bin capacity W_j are definitely feasible. In addition, some small overloads may be also considered feasible. Let's suppose there is a bin j with capacity $W_j = 10$. A solution where the items in j weights 10.01 units will have a higher degree of feasibility than another one with weight 12. In turn, a solution that try to pack items with weight 20 in such bin with 10 units of capacity, may be consider unfeasible. In terms of decision making, this relaxation that may imply a small overload in the bins may be preferable if it allows to reduce the total cost in the objective function (1).

To model this situation, we consider that a decision maker must define the tolerance T_j that defines the maximum admissible relaxation for each bin j. Figure 1 shows the function to measure the degree of accomplishment (for a given solution) of constraint (6) in terms of the bin capacity W_j and tolerance T_j.

To understand such function, let's call tw_j to the sum of the weights of the items placed in each bin j. Then, if $tw_j \leq W_j$, then such assignment of items

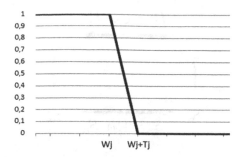

Fig. 1. Membership function for the accomplishment of capacity constraint (y axis) in terms of the sum of the items' weights (x axis).

to the bin is feasible with degree 1. In turn, if $tw_j > W_j + T_j$ feasibility is zero. When $tw_j \in [W_j, W_j + T_j]$ then such solution will feasible with different degrees between $[0, 1]$.

In order to solve the problem, the fuzzy constraint will be managed using the parametric approach [14] and the concept of α−cuts. In very simple terms, this allows to obtain several crisp instances based on different values of α. If S is the whole set of solutions we may define S^α as the crisp set of solutions that satisfy the constraints with, at least, a given degree α.

$$S^\alpha = \{s \in S \mid \mu(s_j) \geq \alpha\} \tag{7}$$

According to (7), a solution $s \in S$ is considered α-feasible if it is feasible with a degree above α. This implies that there are different sets S^α of feasible solutions for different values of α. As different sets S^α may include different set of solutions, the optimal solution for each S^α may be different. So to solve the fuzzy version of VSCBPP we will use the constraint (8) instead of (3).

$$\sum_{i \in I} w_i x_{ij} \leq (W_j + (1 - \alpha)T_j)y_j, j \in J \tag{8}$$

When $\alpha = 1$ the most restricted definition of capacity is taken into account, thus having the original crisp problem (no flexibility). When $\alpha = 0$, the most flexible situation is reached. The allowed bin overload is maximum. Consequently, the value α define the degree of relaxation that is admitted.

3   Solving the Fuzzy VSCBPP

The parametric approach [14], illustrated in Fig. 2, is used to solve the problem. The main idea is to transform the fuzzy problem into a family of crisp problems. Initially, a set of different α values is defined and for each value, a crisp problem is obtained and then solved. Finally, it should be remarked that the final result consists of a *set* of solutions related to each value of α.

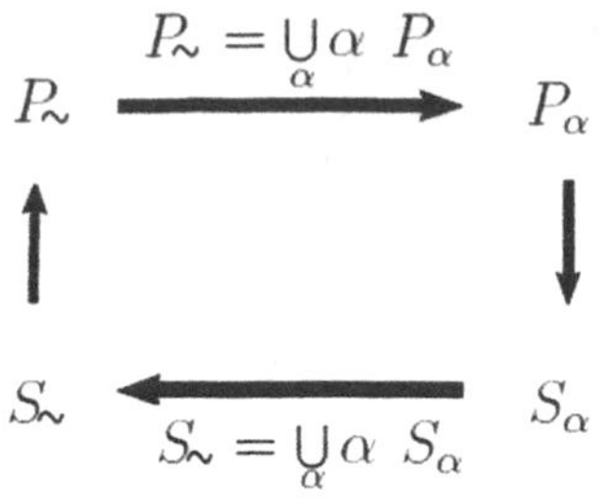

$$P_{\sim} \xrightarrow{\;\; P_{\sim} = \bigcup_{\alpha} \alpha\, P_{\alpha}\;\;} P_{\alpha}$$

$$S_{\sim} \xleftarrow{\;\; S_{\sim} = \bigcup_{\alpha} \alpha\, S_{\alpha}\;\;} S_{\alpha}$$

Fig. 2. General scheme of the parametric approach

In our case, we consider 11 values of $\alpha \in \{0.0, 0.1, \ldots, 0.9, 1.0\}$ [1]. For solving the different crisp problems we consider in this paper three solution methods. An exact solver based on Integer Linear Programming (SCIP solver), and two heuristics: First Fit Decreasing (FFD) and Best Fit Decreasing (BFD) [6,11]. FFD is a deterministic heuristic that place each item in the first bin where it is possible to place it, where BFD chooses the bin where the item best fits. Both heuristics repeat this process item by item, taking them in descending order. It is worth noting that heuristics methods do not guarantee optimality.

In the case of the SCIP solver, the optimality is only guaranteed after a considerable amount of time. In controlled conditions (for example, a maximum execution time of one hour for each instance) we observed that optimality is not guaranteed for all cases. It is import to remark that our focus here is on analyzing each fuzzy solution (i.e., the set of solutions obtained for the base instance with different values of α).

3.1 Test Instances

In this experiment we take three base instances following previous works in VSCBPP [1,8]. These base instances are used as the crisp original instances, associated with $\alpha = 1$. These three base instances of the FVSCBPP result in 33 crisp ones that need to be solved (one for each value of the 11 α values considered). Each base instance contains 25 items and 3 bin types with $W \in \{50, 100, 150\}$, and the tolerance was set to $T \in \{6, 5, 7\}$, respectively. The weight of each item is randomly assigned using a uniform distribution in $[20, 120]$.

It must be remarked that the three base instances differ in the relation between the cost C_j of each bin and its capacity W_j, as it was previously conceived in other works [1,7]. This functional relation is the origin of the name used to identify each instance: Concave (Cc) where $C_j = 15\sqrt{W_j}$, Linear (Ln) where $C_j = 10W_j + 32$, and Proportional (Pr) where $C_j = 0.1W_j$).

The three cost functions (Cc, Ln, and Pr) produce different behaviors. For example, according to the function Pr is the same to use three bins with $W_j = 50$ than using one bin with $W_j = 150$, and both are the same than using a bin with

---

[1] Alternative schemes for exploring the values of α were recently presented in [13].

$W_j = 50$ plus another with $W_j = 100$. In the other cost functions, it is better to use a bin with $W_j = 150$ than using three bins with $W_j = 50$ (in the case of the function Ln this implies 5% additional cost, while in Cc this value is 73%). The same occurs in the comparison of using a bin of $W_j = 150$ with respect to use a bin $W_j = 50$ and a bin with $W_j = 100$ (in the case of Ln this implies 4% additional cost, while in Cc this value is 38%). These are just some examples of the implications of the cost function.

The test instances are available in www.cimab.transnet.cu/files/iFSCBPP. zip in order to allow replication of the presented results.

4   Results and Discussion

As stated before, for each test instance we define 11 α values: $\alpha \in [0.0, 0.1, 0.2, \ldots, 1.0]$, leading to 33 crisp problems. Each one of these problems is solved in two ways. Firstly using a Mixed Integer Programming solver SCIP [12], with the problem model coded using the ZIMPL format [9]. SCIP is expected to return the optimal solution. Secondly, two problem-specific heuristics for the Bin Packing Problem are used: First Fit Decreasing (FFD) and Best Fit Decreasing (BFD) [6,11].

Runs were performed in an Intel Core i5 processor with 2.4 GHz of clock speed and 8 Gb of RAM. The running time of both heuristics is less than one second (after the items are sorted). But for running the SCIP solver, a maximum number of 4 parallel threads and 60 min (one hour) of maximal execution time were set.

The analysis of the results is divided in two parts. In the first one we consider the impact of α in the solutions' costs. Then, we analyze such impact in terms of the solutions' structure.

Figure 3 shows the impact of the α values on the solutions costs for every solver and test instances. A clear tendency is observed: as the problem is more relaxed ($\alpha \to 0$), the cost of the solution diminished. When α is near to 0 (fully relaxed case), the available space in the bins is higher and less cost is needed to store all the items. As the relaxation decreases ($\alpha \to 1$) the cost increases.

Figure 3 shows an additional interesting feature. When a heuristic is used, there is no guarantee to obtain a better solution if the problem is more relaxed. This is not the case with the SCIP solver, where if $\alpha_1 < \alpha_2$ then the corresponding solutions s_1, s_2 satisfies that $f(s_1) \leq f(s_2)$.

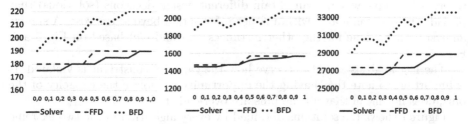

Fig. 3. Impact of the constraint violation in the cost (α in x-axis vs. cost in y-axis) for the test instances: Pr(left), Cc(center), Ln(right).

Finally, we can observe that the heuristic FFD obtains very similar results to those of the SCIP solver but using a very simple approach and an extremely reduced time. Although for most of the cases the SCIP solver required less than a minute, for some cases (mainly in the Ln instance) it did not finish within an hour. In such a case, the best solution found up to that point is reported.

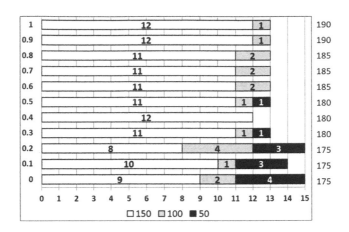

Fig. 4. Structure of the solutions (SCIP Solver) for the Pr instance in terms of α (y-axis). The number of bins of each type is shown.

In order to analyze the impact of α in terms of solutions' structure, Fig. 4 shows the solutions obtained by the SCIP solver for the Pr instance for each α value. Every row displays the number of bins of each type (150: white, 100: grey, 50: black) used together with the cost.

If the capacity constraint is very strict, 12 big bins and 1 middle sized are needed. As the relaxation increases, a better cost can be achieved with 11 big bins and 2 middle sized. The cases of $\alpha = 0.5$ and $\alpha = 0.4$ are also interesting. In the former, the three types of bins are used in a solution with cost 180. But an additional relaxation of the constraint allows to pack all the items using 12 big bins. A similar situation happens for the most relaxed cases $\alpha \geq 0.2$ where three different solutions with the same cost are displayed. So we have here another benefit of using fuzzy constraints and the parametric approach, where in a simple way we can obtain different design decisions (solutions) that provide the decision maker with a richer information beyond the cost. A similar analysis can be done for the other instances (Ln and Cc) based on Fig. 5 and Fig. 6.

The last analysis aims to observe how the capacity constraint is violated. It is important to note that having the opportunity to violate the capacity of the bins, does not mean that all the bins are overloaded.

Figure 7 shows the solutions obtained by every algorithm in the two extreme situations: $\alpha = 1$ and $\alpha = 0$. While bars correspond to big bins (capacity

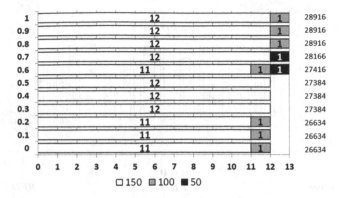

Fig. 5. Structure of the solutions (SCIP Solver) for the Ln instance in terms of α (y-axis). The number of bins of each type is shown.

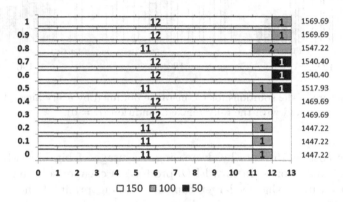

Fig. 6. Structure of the solutions (SCIP Solver) for the Cc instance in terms of α (y-axis). The number of bins of each type is shown.

150), grey ones to mid-size bins (capacity 100) and black ones to small size bins (capacity 50). If a bar is taller than its capacity, then such bin are overloaded (makes use of the relaxation). The horizontal lines within the bars identify the items packed.

Figure 7 (on top) displays the crisp case (no relaxation is allowed). Both SCIP solver and FFD heuristic achieved a solution with the same cost and the same structure. They used 12 big bins and 1 middle sized but, as it can be observed the items are packed differently. Again, this kind of visualization allows a user to take a more informed decision. One may argue that aspects like the level of occupancy of the bins should be taken into account within the objective function. However, in our opinion, that would complicate the solution of the problem.

If we consider the fully relaxed version ($\alpha = 0$), we observe that the capacity violation is small. Nevertheless, it allows for a relevant cost reduction. If the comparison is made in a column wise manner, we can compare the none relaxed vs. the fully relaxed solutions obtained by every algorithm. The SCIP solver

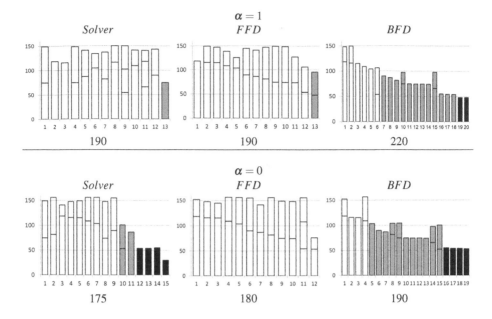

Fig. 7. Structure of the solutions for Cc instance obtained by every algorithm with no relaxation ($\alpha = 1$) and with the fully relaxed condition ($\alpha = 0$).

reduces the cost in 15 units by decreasing the usage of big bins while adding more of the smaller ones. In turn FFD, produces a cheaper solution differently: use big bins with a slightly larger capacity (taking profit of the relaxation). Finally, BFD obtained the greatest decrease (30 units) but using 30 bins. It can be noticed that such solution has the same cost of those obtained by SCIP and FFD when $\alpha = 1$. Despite the slight violation of the capacity, it is clear that the BFD solution may be harder to "manage": larger number of bins and use the three available types.

5   Conclusions

This paper presents a fuzzy version of the Variable Size and Cost Bin Packing Problem (VSCBPP), where the bins capacity is considered flexible. Allowing such flexibility is relevant in many practical situations because it may allow to obtain cheaper solutions.

The proposed fuzzy version of VCSBPP is expressed in terms of fuzzy constraints that allow to respect the limitations regarding the capacities with a certain tolerance. Based on the parametric approach, the solution of the fuzzy problem consists of a set of solutions that may show a trade-off between relaxation (violation of the original condition) and benefit (cost reduction).

Our experimental study shows first that all the algorithms tested can achieve cheaper solutions as the relaxation increases. The analysis of the solution revealed that different algorithms manage the flexibility in different ways, thus allowing to obtain a diverse set of solutions.

This is a crucial aspect for a decision maker, for whom, different solutions with similar cost can be provided. Then, such solutions can be analyzed from other points of view beyond costs like how easy/hard is to manage the selected bins, or how easy/hard is to transport them. Although such features may be included in the cost function (this is far from trivial), this will lead to a more complex and harder to solve model.

Acknowledgements. This work is supported by the project *"Optimization of the distribution and composition of the fleet in midterm considering acquisition mode"* of the Maritime Transport Division of the Center of Research and Environmental Management of Transport (Cimab) and the PhD Academic Program of the Informatics Engineering Faculty of the Universidad Tecnológica de La Habana "José Antonio Echeverría" (CUJAE) as part of PhD research of the first author. D. Pelta and C. Cruz acknowledge the support of Project- TIN2017-86647-P from Spanish Ministry of Economy and Competitiveness, (including FEDER funds, from the European Union).

References

1. Correia, I., Gouveia, L., da Gama, F.S.: Solving the variable size bin packing problem with discretized formulations. Comput. Oper. Res. **35**(6), 2103–2113 (2008). https://doi.org/10.1016/j.cor.2006.10.014. ISSN 0305-0548
2. Crainic, T.G., Perboli, G., Rei, W., Tadei, R.: Efficient lower bounds and heuristics for the variable cost and size bin packing problem. Comput. OR **38**, 1474–1482 (2011)
3. Crainic, T.G., Gobbato, L., Perboli, G., Rei, W., Watson, J.P., Woodruff, D.L.: Bin packing problems with uncertainty on item characteristics: an application to capacity planning in logistics. Procedia - Soc. Behav. Sci. **111**(5), 654–662 (2014). https://doi.org/10.1016/j.sbspro.2014.01.099
4. Ebrahimnejad, A., Verdegay, J.L.: Fuzzy Sets-Based Methods and Techniques for Modern Analytics. SFSC, vol. 364. Springer, Cham (2018). https://doi.org/10.1007/978-3-319-73903-8
5. Friesen, D.K., Langston, M.A.: Variable sized bin-packing. SIAM J. Comput. **15**, 1 (1986). https://doi.org/10.1137/0215016. ISSN 1095-7111
6. Haouari, M., Serairi, M.: Heuristics for the variable sized bin-packing problem. Comput. Oper. Res. **36**, 2877–2884 (2009). https://doi.org/10.1016/j.cor.2008.12.016. ISSN 0305-0548
7. Hemmelmayr, V., Schmid, V., Blum, C.: Variable neighbourhood search for the variable sized bin packing problem. Comput. Oper. Res. **39**, 1097–1108 (2012). https://doi.org/10.1016/j.cor.2011.07.003. ISSN 0305-0548
8. Kang, J., Park, S.: Algorithms for the variable sized bin packing problem. Eur. J. Oper. Res. **147**, 365–372 (2003). 0377-2217/03, ISSN 0377-2217
9. Koch, T.: Zimpl User Guide (Zuse Institute Mathematical Programming Language)). PH.D. thesis, Technische Universität Berlin, February 2004. http://www.zib.de/Publications/abstracts/ZR-04-58/. zIB-Report 04-58

10. Peng, J., Zhang, B.: Bin packing problem with uncertain volumes and capacities (2012). http://www.researchgate.net/profile/Jin_Peng13/publication/265533462_ Bin_Packing_Problem_with_Uncertain_Volumes_and_Capacities/links/55657a3808 aec4b0f4859d5d.pdf
11. Pillay, N., Qu, R.: Hyper-heuristics: theory and applications. In: Packing Problems, Natural Computing Series, pp. 67–73. Springer, Cham (2018). https://doi.org/10. 1007/978-3-319-96514-7
12. Schwarz, C.: An Introduction to SCIP. University of Bayreuth, September 2010
13. Torres, M., Pelta, D.A., Lamata, M.T.: A new approach for solving personalized routing problems with fuzzy constraints. In: IEEE International Conference on Fuzzy Systems (FUZZ) (2018)
14. Verdegay, J.L.: Fuzzy mathematical programming. In: Fuzzy Information and Decision Processes, pp. 231–237 (1982)
15. Vigerske, S., Gleixner, A.: SCIP: global optimization of mixed-integer nonlinear programs in a branch-and-cut framework. Optim. Methods Soft. **33**(3), 563–593 (2018). https://doi.org/10.1080/10556788.2017.1335312
16. Wang, S., Li, J., Mehrotra, S.: Chance-constrained bin packing problem with an application to operating room planning (2019). http://www.optimization-online. org/DB_HTML/2019/02/7053.html

Artificial Bee Colony Algorithm Applied to Dynamic Flexible Job Shop Problems

Inês C. Ferreira[1], Bernardo Firme[1(✉)] ⓘ, Miguel S.E. Martins[1] ⓘ,
Tiago Coito[1] ⓘ, Joaquim Viegas[1] ⓘ, João Figueiredo[2] ⓘ, Susana M. Vieira[1] ⓘ,
and João M.C. Sousa[1] ⓘ

[1] IDMEC - Instituto Superior Técnico, Universidade de Lisboa, Lisboa, Portugal
{ines.azevedo.ferreira,bernardo.firme,miguelsemartins,
tiagoascoito,susana.vieira,jmcsousa}@tecnico.ulisboa.pt
[2] Universidade de Évora, Évora, Portugal
jfig@uevora.com

Abstract. This work introduces a scheduling technique using the Artificial Bee Colony (ABC) algorithm for static and dynamic environments. The ABC algorithm combines different initial populations and generation of new food source methods, including a moving operations technique and a local search method increasing the variable neighbourhood search that, as a result, improves the solution quality. The algorithm is validated and its performance is tested in a static environment in 9 instances of Flexible Job Shop Problem (FJSP) from Brandimarte dataset obtaining in 5 instances the best known for the instance under study and a new best known in instance mk05. The work also focus in developing tools to process the information on the factory through the development of solutions when facing disruptions and dynamic events. Three real-time events are considered on the dynamic environment: jobs cancellation, operations cancellation and new jobs arrival. Two scenarios are studied for each real-time event: the first situation considers the minimization of the disruption between the previous schedule and the new one and the second situation generates a completely new schedule after the occurrence. Summarizing, six adaptations of ABC algorithm are created to solve dynamic environment scenarios and their performances are compared with the benchmarks of two case studies outperforming both.

Keywords: Dynamic environment · New jobs arrival · Operations cancellation · Jobs cancellation · Flexible job shop rescheduling

1 Introduction

Factories and governments are launching the fourth industrial revolution called Industry 4.0, due to the dynamic nature of the manufacturing environments and

This work was supported by FCT, through IDMEC, under LAETA, project UIDB/50022/2020.

M.-J. Lesot et al. (Eds.): IPMU 2020, CCIS 1237, pp. 241–254, 2020.
https://doi.org/10.1007/978-3-030-50146-4_19

the growing of the virtual world. Industrial production will be highly flexible in production volume and suppliers, and above all sustainable and efficient [10]. Smart factories using Industry 4.0 based on collaborative systems represent the future of industrial networks.

According to a PWC survey from 2013 [4], 50% of German enterprises plan their new industrial network and 20% are already involved in smart factories. A survey by American Society for Quality (ASQ), from 2014 [1], states that 82% of organizations that implemented smart manufacturing experienced increased efficiency, 49% experienced fewer product defects and 45% experienced increased customer satisfaction [8]. Hence, companies can highly benefit from the implementation of Industry 4.0 concepts.

Industry 4.0 represents a smart manufacturing network concept where machines and products interact with each other without human intervention. Supply chains in such networks have dynamic structures which evolve over time. In these settings, short-term supply chain scheduling is challenged by sources of uncertainty. Manufacturing environments are dynamic by nature and there are several events which can occur and change the system status affecting the performance, known as real-time events. Exchanging data and information between different parties in real time is the key element of smart factories; such data could represent production status, energy consumption behaviour, material movements, customer orders and feedback, suppliers' data, etc. The next generation of smart factories, therefore, will have to be able to adapt, almost in real time, to the continuously changing market demands, technology options and regulations [2].

2   Flexible Job Shop Problem

The Flexible Job Shop Scheduling Problem (FJSSP) is a generalization of the classical Job Shop Scheduling Problem (JSSP). The JSSP follows the idea that a set of jobs ($J = \{1, 2, ..., m\}$) is processed by a set of machines ($M_k = \{1, 2, ..., m\}$). Every job consists of a finite set of operations and the processing of the operation has to be performed on a preassigned machine. The $i-th$ operation of job j, denoted by O_{ji}, is processed on machine $k \in M$ and the JSSP solves the assignment of jobs to the machines. The order of operations of each job is fixed, meaning the JSSP doesn't need to solve the operation sequence. The aim of the classical static n-by-m JSSP is to find a schedule for processing n jobs on m machines fulfilling an objective function. The FJSSP has one more condition than the JSSP, it is imposed job variability which creates the need for an operation sequence solution, besides the assignment machine solution.

The FJSSP follows some ideas, rules and assumptions. No machine is able to process more than one job at the same time and no job may be processed by more than one machine. Each job and each operation must be processed exactly one time. There is independence between machines and jobs. The sequence of machines a job visits is specified, having a linear precedence structure. If the precedent operation is still being processed, the remaining operations commenced until the processing is completed. The processing time of the O_{ji} operation using a specific machine takes $PT_{jik} > 0$ time unities and is known. Machines must always be available at the usage time zero.

The ideas, rules and assumptions can be formulated as follows. There are m machines defined as $M_k = \{1, 2, ..., m\}$. There are a group of n jobs independent of each other defined as $J = \{1, 2, ..., n\}$. Each job has a set of h operations defined as $O_{ji} = \{O_{j1}, O_{j2}, ..., O_{jh}\}$. For each operation O_{ji}, there is a set of machines capable of performing it. The set is denoted by $M_{ji} \subset M_k$. If $M_{ji} = M_k$ for all i and k, the problem becomes a complete flexible job shop problem. The processing time of operations O_{ji} on machine k is stated as PT_{jik}. The start time for every operation O_{ji} on machine k is presented as ST_{jik}. The finishing time of operation O_{ji} on machine k is presented as FT_{jik}.

This problem has also some constraints. The technique constraint describes that the operation must be processed after all precedent operations have been processed. The operations should not be overlapped and the machine will be available to other operations, if the previous operations are completed. The resource constraint demands that one machine can only handle exactly one operation at a time. There is also a precedence constraint for operations of the same job. The objective of the FJSSP is to determine a feasible schedule minimizing the makespan that is the maximum completion time of the jobs. In other words, the total elapsed time between the beginning of the first task and the completion of the last task.

3   Artificial Bee Colony (ABC) Algorithm

The implementation of the ABC algorithm developed to solve the FJSSP for static environments is described and it was based on the work of [11]. The ABC Algorithm is inspired by the intelligent foraging behaviour of a honeybee swarm. The model that leads to the emergence of the collective intelligence of honey bee swarms consists of three essential components: food sources, employed foragers and unemployed foragers.

3.1   ABC Procedure to Static Scheduling

The objective value of the solution is $f(FS_i)$ that represents the selected food source, FS_i. The total number of food sources is characterized by t_{FS} and p_i is the probability of selecting a food source. The algorithm for ABC is as follows:

1. Initialize parameters.
2. Initialize a population of food sources, FS_i.
3. Calculate the objective value $f(FS_i)$ of each food source FS_i and then determine the best food source g_{best}.
4. Employed bees phase
   (a) For every employed bee generate the new food source FS_i^1 from FS_i.

244 I. C. Ferreira et al.

(b) Onlooker bees phase
  i. Calculate the probability of selecting the food source FS_i^1 according to equation $p_i = \dfrac{[f(FS_i)]}{\sum_{j=1}^{t_{FS}}[f(FS_i)]}$.
  ii. Calculate the number of onlooker bees to be sent to the food source FS_i^1 according to $NE_i = p_i \times n_{ob}$.
  iii. For every onlooker bee generate N_i new food sources FS_i^2 from FS_i^1 using local search according to its termination criteria.
  iv. Choose the best from all the N_i food sources generated and set it as FS_i^{best}
  v. If $f(FS_i^{best}) \leq f(FS_i^1)$, then $FS_i^1 = FS_i^{best}$
  vi. For all the employed bees, if $f(FS_i^{best}) \leq f(gbest)$, then $gbest = FS_i$
5. Scout bees phase
  (a) Initialize scout bees with random solutions and update g_{best}, if possible.
  (b) Determine the worst employed bees and replace them with the best scout bees, if they are better.
6. If the stopping criteria is met the output is g_{best} and its objective value; otherwise, go back to point 4.

Following the steps of ABC algorithm, it is important to define some concepts and techniques utilized:

- **Solution Representation:** The solutions are a combination of two vectors: machine assignment and operation sequence. The first one codes the assignment of operations to the machines and the second one codes the processing sequence of operations for each job. This dual coding vector is a representation of a feasible solution.
- **Population Initialization:** To guarantee an initial population with quality, diversity and capability of avoiding falling in a local optimal, a hybrid way to generate the food sources was utilized. The machine assignment initialization uses three rules: random rule, local minimum processing time rule and global minimum processing time rule with a probability of occurence of $\{0.6, 0.2, 0.2\}$ respectively. The operation sequence initialization uses: random rule, most work remaining rule (MWR) and most number of operations remaining rule (MOR) with a probability of occurrence of $\{0.8, 0.1, 0.1\}$
- **Crossover Operators:** To evolve the machine assignment vector two crossover operators, the two-point crossover and the uniform crossover, are applied. Also, a method of a crossover called the Precedence Preserving Order-Based Crossover (POX) is implemented to evolve the operation sequence vector.
- **Mutation for Machine Assignment:** To enhance the exploration capability in the employed bee search phase, a mutation operator for the machine assignment is embedded in the ABC algorithm.
- **Local Search Based on Critical Path:** The local search strategy based on the critical path is proposed and embedded in the searching framework to enhance the local intensification capability for the onlooker bees.

– **Termination Criteria:** There are two termination conditions set to terminate the algorithm: a number of trials ter_{max} to improve the solution and the pre-defined number of iterations gen_{max}.

Critical Path Theory. The earliest starting time of the operation O_{ji} is denoted as ST_{ji}^{E} and the latest starting time is ST_{ji}^{L}. If the operation O_{ji} is processed on the machine k, then the operation processed previously is PM_{ji}^{k} and operation processed next is NM_{ji}^{k}. $PJ_{ji} = O_{j-1i}$ is the operation of the job i that precedes O_{ji}. $NJ_{ji} = O_{j+1i}$ is the operation of the job j that is next to O_{ji}. The starting time of the dummy starting node $ST_{E}(0) = 0$. If the node has no job predecessor, the earliest completion time is $c^{E}(PJ_{ji}) = 0$. If it has no machine predecessor, the earliest completion time is $c^{E}(PM_{ji}^{k}) = 0$. The latest completion time of the ending node is equal to the makespan of the schedule, $c^{L}(N+1) = c_{M}(G)$.

The earliest completion time of the operation is described by Eq. 1 and the latest completion time by Eq. 2.

$$c_{ji}^{E} = ST_{ji}^{E} + PT_{jik} \qquad (1)$$

$$c_{ji}^{L} = ST_{ji}^{L} + PT_{jik} \qquad (2)$$

The earliest starting time is calculated by Eq. 3 and latest starting time of each node by Eq. 4.

$$ST_{ji}^{E} = max\{c^{E}(PJ_{ji}), c^{E}(PM_{ji}^{k})\} \qquad (3)$$

$$ST_{ji}^{L} = min\{S^{L}(NJ_{ji}), S^{L}(NM_{ji}^{k}\}) \qquad (4)$$

The total slack of operation is the amount of time that an activity can be delayed or anticipated without increasing makespan. The total slack of each node is calculated using Eq. 5.

$$TS_{ji} = ST_{ji}^{L} - ST_{ji}^{E} \qquad (5)$$

The makespan of a schedule is defined by the length of its critical path, implying that any delay in the start of the critical operation will delay the schedule. The idea behind the local search of the critical path is to analyze all critical operations to verify the possibility of scheduling them earlier.

Moving Operations. In order to simplify the notation, the operation to be moved O_{ji} is called r and the candidate operation O_{lk} to have O_{ji} assigned before is called v.

$$TS_{v} \geq PT_{r} \qquad (6)$$

$$ST_{v}^{L} \geq max\{c_{r-1}^{E}, c_{v-1}^{E}\} + PT_{r} \qquad (7)$$

$$ST_{r+1}^{L} \geq max\{c_{r-1}^{E}, c_{v-1}^{E}\} + PT_{r} \qquad (8)$$

The above moving operations process is repeated until all the critical operations of the present food source are moved or until the termination criteria for the local search is met. If the food source being searched has no more critical operations the search is terminated, otherwise the procedure is applied a certain number of times $move_{max}$ to have better improvement of the final schedule. To accept the new solution generated one of the following conditions is satisfied: the new solution has a smaller makespan or the new solution has the same makespan but has fewer critical paths.

4   Dynamic Scheduling

In the industrial world, scheduling systems often operate under dynamic and stochastic circumstances and it is inevitable to encounter some disruptions which are inherently stochastic and non-optimal. Therefore, algorithms which guarantee quick and good solutions for the scheduling are strongly needed.

In this work, heuristics were made in order to be possible to solve dynamic scheduling cases, through rescheduling. Not only the adaptations created in this work are able to solve unpredictable scenarios, but also reoptimize the solution.

4.1   Earl Job Cancelation (ABC-R1)

The first scenario (R1) under study is the rescheduling when a job is cancelled early enough making possible the acceptance and feasibility in the factory to adjust to the significant changes. When the dynamic event arrives, a new schedule will be constructed. The early job cancellation algorithm, ABC-R1, has several mechanisms included to improve the solution as much as possible.

Early Job Cancellation Procedure

1. Load the static scheduling;
2. Initialize parameters: the deleted job (J^{delete});
3. Initialize a population of food sources (FS_i) from the loaded static scheduling. In the machine assignment vector and in the operation sequence vector delete the operations belonging to the deleted job;
4. Calculate the objective value $f(FS_i)$ of the food source FS_i and then establish the best food source g_{best};
5. Onlooker bees phase
   (a) For every onlooker bee N_i generate new food sources $F_{S_i}^2$ from $F_{S_i}^1$ using local search according to its termination criteria;
   (b) Choose the best from all the N_i food sources generated and set it as the best;
   (c) If the stopping criteria are met go to the next point; otherwise, go back to point 5;
6. Initialize a population of food sources (FS_i^3) from the loaded static scheduling. In the machine assignment vector and in the operation sequence vector delete the operations belonging to the deleted job. The process will be done in a cycle, first deleting the operations one at each time, and then trying to anticipate the sequenced operations belonging to the same machine the deleted operation belongs to;

7. Onlooker bees phase
   (a) For every onlooker bee generate N_i new food sources FS_i^2 from FS_i^1 using local search according to its termination criteria;
   (b) Choose the best from all the N_i food sources generated and set it as FS_i^{best};
   (c) If $f(FS_i^{best}) \leq f(FS_i^1)$, then $FS_i^1 = FS_i^{best}$;
   (d) For all the employed bees, if $f(FS_i^{best}) \leq f(gbest)$, then $gbest = FS_i$;
   (e) If the stopping criteria is met go to the employed bee phase; otherwise, go back to point 7;
8. Employed bees phase
   (a) For every employed bee generate the new food source FS_i^1 from FS_i;
       i. Applying crossover operators to the machine assignment vector;
       ii. Applying crossover operators to the operation sequence vector;
       iii. Applying mutation operator to the machine assignment vector;
       iv. Local search for the critical path according to the termination criteria;
       v. If $f(FS_i^{best}) \leq f(FS_i^1)$, then $FS_i^1 = FS_i^{best}$;
       vi. For all the employed bees, if $f(FS_i^{best}) \leq f(gbest)$, then $gbest = FS_i$;
9. Scout bees phase
   (a) Initialize scout bees with random solutions and update g_{best}, if possible;
   (b) Determine the worst employed bees and replace them with the best scout bees if those are better;
10. If the stopping criteria are met the output is g_{best} and its objective value; otherwise, go back to point 4.

4.2   Late Job Cancellation (ABC-R2)

The second scenario (R2) appear when the job cancellation order arrives and the static scheduling is already being used on the factory, so it is important to erase it without altering the scheduling previously done, with the goal of having the less disruption and disturbance as possible on the factory.

Late Cancellation Procedure

1. Load the static scheduling;
2. Initialize parameters: the arrival time of the cancellation order (time of job cancellation) and the job which was cancelled (J^{delete});
3. Initialize the population of food source (FS_i) from the loaded static scheduling;
4. Using the machine assignment vector and the operation sequence vector, calculate the search space containing all the possible positions to anticipate the operations, according to the *time delete job*;
5. Each one of the operations which have the possibility to be anticipated will be introduced in the best position possible of the search space, respecting to the precedence constraints;
6. The output is the g_{best}.

4.3 Early Operation Cancelation (ABC-R3)

The third scenario (R3) is similar to the procedure described for ABC-R1. The main difference is that ABC-R1 implies the cancellation of all the operations belonging to the job and ABC-R3 implies the cancellation of part of the job operations.

Early Operation Cancellation Procedure The main differences are in the first, second and fourth steps. In the first step, a parameter describing which operation will be deleted is additionally initialized, namely ($delete_{operation}$). In the second step, the procedure to initialize the population is similar but operations processed before the operation cancelled are kept on the machine assignment and on the operation sequence vectors. The last difference is in the fourth step: the operations processed before the cancelled operation are kept on the vectors.

4.4 Late Operation Cancelation (ABC-R4)

The fourth scenario (R4) is a solution created when it is important to generate a solution similar to the previous one. This scenario has a late operation cancellation order at a time defined as the time of operation cancellation. The precedence constraints imply the cancellation of certain operations from the schedule, not only the one which was cancelled. The main differences to ABC-R2 are in the first and second steps. In the first step, the parameter setting which operation will be deleted is additionally initialized, called (O_{ji}^{delete}), and the variable of time initialized is the time of operation cancellation. In the second step, the procedure to initialize the population is similar but operations processed before the operation cancelled are kept.

Late Operation Cancellation Procedure

1. Load the static scheduling;
2. Initialize parameters: the arrival time of the cancellation order (time of operation cancellation) and the operation which was cancelled (O_{ji}^{delete});
3. Initialize the population of food source (FS_i) from the loaded static scheduling deleting the operations from the machine assignment vector and the operation sequence vector;
4. Using the new machine assignment vector and the operation sequence vector, the search space containing all the possible positions to anticipate the operations will be calculated, according to the *time of operation cancellation*;
5. Each one of the operations which have the possibility to be anticipated will be introduced in the best position possible of the search space, with respect to the precedence constraint;
6. The output is g_{best}.

4.5   Early New Job Arrival (ABC-R5)

The fifth scenario (R5) is characterized by the unexpected arrival of a new job . The ABC-R5 was created as a more reactive model and has several mechanisms included to improve the solution as much as possible.

Early New Job Arrival Procedure

1. Load the static scheduling;
2. Initialize parameters: global parameters + new Job (J^{new});
3. Initialize a population of food sources (FS_i);
   (a) The machine assignment vector from static scheduling is called u_{old}. The machine assignment for the new job is done independently of the remaining jobs, meaning a vector containing the machine assignment information of only the new job is initialized. This vector is called u_{new}. Both vectors are joined in one vector u', creating the machine assignment vector for the new situation of $n + 1$ jobs; The operation sequence vector is initialized from scratch.
4. Calculate the objective value $f(FS_i)$ of the food source FS_i and then establish the best food source g_{best};
5. Employed bees phase
   (a) For every employed bee generate the new food source FS_i^1 from FS_i;
      i. Applying crossover operators to the machine assignment vector;
      ii. Applying crossover operators to the operation sequence vector;
      iii. Applying mutation operator to the machine assignment vector;
      iv. Using the local search for the critical path according to the termination criteria;
      v. If $f(FS_i^{best}) \leq f(FS_i^1)$, then $FS_i^1 = FS_i^{best}$;
      vi. For all the employed bees, if $f(FS_i^{best}) \leq f(gbest)$, then $gbest = FS_i$;
6. Onlooker bees phase
   (a) For every onlooker bee generate N_i new food sources FS_i^2 from FS_i^1 using local search according to its termination criteria;
   (b) Choose the best from all the N_i food sources generated and set it as FS_i^{best};
   (c) If $f(FS_i^{best}) \leq f(FS_i^1)$, then $FS_i^1 = FS_i^{best}$;
   (d) For all the employed bees, if $f(FS_i^{best}) \leq f(gbest)$, then $gbest = FS_i$;
   (e) If the local search stopping criteria is met go to the employed bee phase; otherwise, go back to point 5;
7. Scout bees phase
   (a) Initialize scout bees with random solutions and update g_{best}, if possible;
   (b) Determine the worst employed bees and replace them with the best scout bees if those are better;
8. If the stopping criteria are met the output is g_{best} and its objective value; otherwise, go back to point 6;

4.6   Late New Job Arrival (ABC-R6)

The sixth scenario (R6) simulates a new order arrival and the need to introduce it on the system having the lowest disruption possible. It is considered that the static scheduling was already in production until the arrival time of the order, making it impossible to introduce the new operations before the time new job appears.

Late New Job Arrival Procedure

1. Load the static scheduling;
2. Initialize parameters: global parameters + the arrival time of the new job (*time new job appears*) and the new number of jobs (n_{new});
3. Initialize the population of food source (FS_i) from the loaded static scheduling;
4. Using the machine assignment vector and the operation sequence vector, calculate the search space containing all the possible positions to introduce the operations of the new job, according to the *time new job appears*;
5. Each one of the new operations will be introduced in the best position possible of the search space, respecting the precedence constraints;
6. The output is g_{best}.

5   Results

To verify and validate the implementation of the ABC algorithm for FJSSP in static and dynamic environments, the algorithm was tested in benchmark datasets and compared to other algorithms.

– **Static scheduling**
  1. Brandimarte dataset [3] and compared to algorithms [5,6,9,11–13]
– **Dynamic scheduling**
  1. **Benchmark 1** - The problem is formulated in [7] where there are 6 machines, 13 different jobs and for each job the number of the operations is {3, 2, 3, 4, 3, 3, 2, 3, 2, 3, 3, 3, 3}, respectively. There is a total of 37 operations. The job 11 is cancelled as a dynamic occurrence.
  2. **Benchmark 2** - In [14], the benchmark treats the arrival of three new jobs ($J^{new} = \{14, 15, 16\}$) and each job has {3, 2, 3} operations, respectively.

5.1   Parameters

To perform the initialization of the population, the determination of several parameters is needed. The values for termination criteria (ter_{max} and gen_{max}), local search termination criteria ($move_{max}$), the number of the employed bees (n_{eb}), the number of onlooker bees (n_{ob})and the number of scout bees (n_{sb}) are presented in Table 1.

Table 1. Global parameters for static and dynamic scheduling

Global parameters						
	gen_{max}	ter_{max}	$move_{max}$	n_{eb}	n_{ob}	n_{sb}
Static	$2 \times n \times m$	$1,5 \times n \times m$	$n \times m$	$3 \times n$	$11 \times n$	$0,2 \times n$
Dynamic	$n \times m$	$0.4 \times n \times m$	n	$3 \times n$	$2 \times n$	n

5.2   Static Scheduling

The results are compared in Table 2 and all the results were obtained after twenty runs selecting the best individual. The proposed algorithm is within the best performing algorithms and it produces good results when compared to PVNS and ABC. PVNS achieves four optimal solutions, ABC achieves six and the proposed ABC also achieves six optimal solutions and one is a new reached lower bound. Comparing hGAJobs, it was better in six instances and equal in other three. Comparing to LEGA, it was equal in one instances and better in seven. From the comparation with KBACO, the proposed ABC is better in six instances and equal in three. When comparing TSPCB, it performed better in five datasest, and equally good in four of them. The good performance of the proposed implementation is guaranteed by the combination of different initial populations, including a moving operations technique and a local search method increasing the neighbourhood search that, as a result, improves the solution quality. It is very likely that with more time and maybe better tuned parameters, the proposed implementation would reach the optimum solutions in more instances.

Valuable to note, a new best known lower bound was reached for the mk05 benchmark of the Brandimarte dataset in static environment. The lower bound for mk05 is 169 and the previously last known lower bound found was 172 in [9] and [11]. The new reached lower bound is 169, three units of time smaller than the previous one.

Table 2. The results of the Brandimarte instances when solved using different algorithms.

Instances	hGAJobs [5]	LEGA[6]	PVNS[13]	KBACO [12]	TSPCB [9]	ABC [11]	Proposed
Mk01	40	40	40	39	40	40	39
Mk02	27	29	**26**	29	**26**	**26**	**26**
Mk03	**204**	N/	**204**	**204**	**204**	**204**	**204**
Mk04	**60**	67	**60**	65	62	**60**	**60**
Mk05	173	176	173	173	172	172	**169***
Mk06	64	67	67	67	65	60	**58**
Mk07	141	147	144	144	140	**139**	140
Mk08	**523**	**523**	**523**	**523**	**523**	**523**	**523**
Mk09	312	320	**307**	311	310	**307**	308
Mk10	211	229	208	229	214	208	-

5.3 Dynamic Scheduling

Early Job Cancellation (ABC-R1)

Comparing Benchmark 1 and ABC-R1 : In the case study using benchmark 1, the makespan of the initial scheduling is 66 and the makespan after the cancellation of the job 11 is 62. The maximum makespan obtained is 52, which is considerably smaller than the makespan obtained in study 1, and an important fact is that the worst solution obtained using the implemented algorithm is 11,29% better. The maximum improvement of the solution obtained using ABC-R1 is 19,4%. It is possible to conclude that ABC-R1 is highly competitive compared to the solution proposed in the case study using benchmark 1 and it always obtains a substantial lower makespan.

Late Job Cancellation (ABC-R2)

Comparing Benchmark 1 and ABC-R2 : Only one result for ABC-R2 is presented because the algorithm has no stochastic behaviour and, as a consequence, the results obtained are the same for each run. The original makespan of the scheduling in the case study using benchmark 1, before the cancellation of the job 11 at 8 units of time, was 66 and after the cancellation is 62. Using the ABC-R2 a makespan of 55 is obtain in 1, 7 seconds. The makespan obtained with ABC-R2 is 7 units of time smaller and it has an improvement of 11, 29%.

Early Operation Cancellation (ABC-R3). To evaluate the performance of ABC-R3, mk04 was used. All the operations, from 2 to 5, were set one at each time as the cancellation of the operation. All the results were obtained after three runs, selecting the best individual. The makespan of the static scheduling of mk04 is 60. The makespan obtained using ABC-R3 is 38, 54, 42 and 38, respectively, and is always smaller than the original of the static scheduling. Another important note is that the maximum run time was 99,2 s and the minimum was 0,3 s. A good solution was achieved in a short period of time to solve the problem.

Late Operation Cancellation (ABC-R4). To test the performance of the ABC-R4 solving a late cancellation of the operation, mk01 was used. For job 7 and job 10, one at each time were set as the dynamic event. All the results were obtained after three runs, selecting the best individual. The makespan of the static scheduling of mk04 is 60. The makespan of these dynamic events is always smaller, being 37 for O_{72} and 33 for $O_{10,2}$ and the run time was 1, 5 seconds. A good solution was achieved to solve the problem of a late order to cancel one operation.

Early New Job Arrival (ABC-R5)

Comparing Benchmark 2 and ABC-R5 : In the case using benchmark 2, the makespan of the initial scheduling is 66 and the one after the three orders arrival is 78. It is possible to conclude that ABC-R5 is highly competitive, when compared to the results obtained using benchmark 2, because even the maximum makespan obtained of 71 using the ABC-R5 is smaller than the makespan of 78 obtained using benchmark 2. Other reason to be considered a highly competitive solution is the significative improvement of the solution obtained using the ABC-R5 and it is important to notice that the worst improvement was 9%, which still a relevant improvement.

Late New Job Arrival (ABC-R6)

Comparing Benchmark 2 and ABC-R6 Only one result of the ABC-R6 algorithm is presented because the algorithm has no stochastic behaviour. The original makespan of the scheduling in the case study using benchmark 2, before the new orders arrive, was 66. The time new job appears is 8 units of time and after the arrival of the three orders the makespan became 78. Using the ABC-R6, for three orders arrival at 8 units of time, a makespan of 66 is obtained in 1 second. Comparing this result with the result obtained in benchmark 2, there was an improvement of 15, 38% in the solution. The cases of just one new order and two new orders arrival were also studied and, in both cases, a makespan of 66 was obtained. In fact, the developed solution is substantially better than the case study solution using benchmark 2.

6   Conclusions

The main objective of this work was to develop tools capable of creating a schedule solution in a dynamic environment. To achieve this objective, firstly a static scheduling algorithm was implemented using an Artificial Bee Colony algorithm and then, as a response to unpredicted or disruptive events, such as jobs cancellation, operations cancellation, and new job arrivals, six heuristics were created and implemented to solve the dynamic problem. Therefore, the Artifical Bee Colony algorithm was extended to innovative solutions for the dynamic environment.

After testing the algorithm in benchmark problems and comparing it to other published algorithms, the implemented solution was verified to be a good solution and achieved the optimal solution in six of the ten instances. Valuable to note, a new optimal solution for one of the instances was found, which it is three units of time smaller than the last one known and only one more than the lower bound.

Notwithstanding, this work's primary goal was to create the six adaptations of the ABC algorithm to dynamic scenarios. The scenarios were tested against the static scheduling obtained using the benchmark problems in the static environment, to fulfil the objective of evaluating the performance of the adapted algorithms and create instances for dynamic testing. All the implementations achieve good solutions in a very short time. Additionally, the solution obtained using ABC-R1, ABC-R2, ABC-R5 and ABC-R6 were compared to the solution obtained using benchmarks belonging to case studies. The makespan was always smaller, while compared to the benchmarks, and it was always several units of time smaller. It is important to refer, that the worst improvement of a solution obtained using one of the adaptations was 9, 0%, which is still a relevant improvement comparatively to the benchmarks results and all the results have an improvement. All in all, very promising solutions are shown for dynamic scheduling.

References

1. American Society for Quality: Manufacturing outlook survey. Technical report (2013)
2. Azevedo, A., Almeida, A.: Factory templates for digital factories framework. Rob. Comput. Integr. Manuf. **27**(4), 755–771 (2011)
3. Brandimarte, P.: Routing and scheduling in a flexible job shop by tabu search. Ann. Oper. Res. **41**(3), 157–183 (1993). https://doi.org/10.1007/BF02023073
4. Company - PWC: Deutschland hinkt bei industrie 4.0 hinterher – smart factory (2013). http://www.pwc.de/de/pressemitteilungen/2013/deutschland-hinkt-bei-industrie-4-0-hinterher-smart-factory-etabliert-sichnur-langsam.jhtml. Accessed 28 Aug 2018
5. Cunha, M.: Scheduling of Flexible Job Shop Problem in Dynamic Environments. Master's thesis, Instituto Superior Técnico (2017)
6. Ho, N.B., Tay, J.C., Lai, E.M.K.: An effective architecture for learning and evolving flexible job-shop schedules. Eur. J. Oper. Res. **179**(2), 316–333 (2007)
7. Honghong, Y., Zhiming, W.: The application of adaptive genetic algorithms in FMS dynamic rescheduling. Int. J. Comput. Integr. Manuf. **16**(6), 382–397 (2003)
8. Ivanov, D., Dolgui, A., Sokolov, B., Werner, F., Ivanova, M.: A dynamic model and an algorithm for short-term supply chain scheduling in the smart factory industry 4.0. Int. J. Prod. Res. **54**(2), 386–402 (2016)
9. Li, J.Q., Pan, Q.K., Suganthan, P., Chua, T.: A hybrid tabu search algorithm with an efficient neighborhood structure for the flexible job shop scheduling problem. Int. J. Adv. Manuf. Technol. **52**(5–8), 683–697 (2011). https://doi.org/10.1007/s00170-010-2743-y
10. Shrouf, F., Ordieres, J., Miragliotta, G.: Smart factories in industry 4.0: a review of the concept and of energy management approached in production based on the internet of things paradigm. In: International Conference on Industrial Engineering and Engineering Management (IEEM), 2014 IEEE, pp. 697–701. IEEE (2014)
11. Wang, L., Zhou, G., Xu, Y., Wang, S., Liu, M.: An effective artificial bee colony algorithm for the flexible job-shop scheduling problem. Int. J. Adv. Manuf. Technol. **60**(1–4), 303–315 (2012)
12. Xing, L.N., Chen, Y.W., Wang, P., Zhao, Q.S., Xiong, J.: A knowledge-based ant colony optimization for flexible job shop scheduling problems. Appl. Soft Comput. **10**(3), 888–896 (2010)
13. Yazdani, M., Amiri, M., Zandieh, M.: Flexible job-shop scheduling with parallel variable neighborhood search algorithm. Expert Syst. Appl. **37**(1), 678–687 (2010)
14. Zakaria, Z., Petrovic, S.: Genetic algorithms for match-up rescheduling of the flexible manufacturing systems. Comput. Ind. Eng. **62**(2), 670–686 (2012)

Games

From Truth Degree Comparison Games to Sequents-of-Relations Calculi for Gödel Logic

Christian Fermüller, Timo Lang$^{(\boxtimes)}$, and Alexandra Pavlova$^{(\boxtimes)}$

Technische Universität Wien, Vienna, Austria
{timo,alexandra}@logic.at

Abstract. We introduce a game for (extended) Gödel logic where the players' interaction stepwise reduces claims about the relative order of truth degrees of complex formulas to atomic truth comparison claims. Using the concept of disjunctive game states this semantic game is lifted to a provability game, where winning strategies correspond to proofs in a sequents-of-relations calculus.

1 Introduction

Fuzzy logics, by which we mean logics where the connectives are interpreted as functions of the unit interval $[0,1]$, come in many variants. Even if we restrict attention to t-norm based logics, where a left continuous t-norm $\circ$ serves as truth function for conjunction and the (unique) residuum of $\circ$ models implication, there are still infinitely many different fuzzy logics to choose from. Almost all of these logics feature truth functions that yield values that are in general different from 0 and 1, but also different from each argument value. E.g., the function $f(x) = 1 - x$ often serves as truth function for negation. However, if we take the minimum, $\min(x, y)$, as t-norm modeling conjunction $\wedge$, the corresponding residuum as truth function for implication $\rightarrow$, and define the negation by $\neg A = A \rightarrow \bot$ [1], we arrive at *Gödel logic*, where every formula evaluates to either 0, 1, or to the value of one of the propositional variables occurring in it. Moreover, Gödel logic is the only t-norm based fuzzy logic, where whether a formula is true (i.e., evaluates to 1) does not depend on the particular values in $[0,1]$ that interpret the propositional variables, but only on the *order*[2] of these values.

In this paper, we look at Gödel logic from a game semantic point of view. After explaining, in Sect. 2, for the simple case of classical logic restricted to negation, conjunction, and disjunction, how a semantic game may be turned into a calculus for proving validity, we turn to Gödel logic G (and its extension

---

[1] $\bot$ denotes *falsum* and always evaluates to 0.
[2] An order of n values in $[0,1]$ is given here by $0 \sharp_0 x_1 \sharp_1 \ldots x_n \sharp_n 1$, where $\sharp_i \in \{<, \leq, =\}$.

C. Fermüller—Research supported by FWF project P 32684.
T. Lang and A. Pavlova—Research supported by FWF project W1255-N23.

M.-J. Lesot et al. (Eds.): IPMU 2020, CCIS 1237, pp. 257–270, 2020.
https://doi.org/10.1007/978-3-030-50146-4_20

$G^\triangle$ with the $\triangle$-operator) in Sect. 3. We introduce a truth degree comparison game, where a player **P** seeks to uphold, against attacks by opponent **O**, a claim of the form $F < G$ or $F \leq G$, expressing that the truth value of F is smaller (or equal) to that of G under a given interpretation. The interaction of **P** and **O** stepwise reduces the initial truth comparison claim to an atomic claim that can be immediately checked. In Sect. 5, we lift the game from truth degree comparison claims for concrete interpretations to the level of validity, i.e., to comparison claims that hold under every interpretation. Following the general clue given in Sect. 2, the key ingredient is the notion of disjunctive states, triggering disjunctive strategies. It turns out that disjunctive winning strategies for **P** correspond to proofs in an analytic proof system, called sequents-of-relations calculus, introduced in [6]. We conclude in Sect. 6 by a brief summary of our results, followed by suggestions for future research in this area.

2   From Classical Semantic Games to Sequent Calculus

Before focusing on Gödel logic, let us illustrate how to turn a semantic game into an analytic proof system in its simplest case: classical propositional logic CL with $\wedge$, $\vee$, and $\neg$ as the only connectives.

Following Hintikka [14], a semantic game for CL can be described as follows. There are two players, say *You* and *I*, who, at any state of the game, can either be in the role of a proponent **P** or in the role of an opponent **O** with respect to the claim that a current formula F is true in a given interpretation $\mathcal{J}$. The game starts with *You* in role **O** and me (player I) in role **P**. It proceeds in accordance with the following rules, which refer to the players only via their current roles.

$(R_\wedge)$: If the current formula is of the form $A \wedge B$ then **O** chooses whether to continue with A or with B as the new current formula.

$(R_\vee)$: If the current formula is of the form $A \vee B$ then **P** chooses whether to continue with A or with B as the new current formula.

$(R_\neg)$: If the current formula is of the form $\neg A$ then the roles of the players are switched and the game continues with A as the new current formula.

(R_{at}): If the current formula A is atomic, the game ends with **P** winning iff A is true in the given interpretation.

It is straightforward to show that I, the initial **P**, have a winning strategy in the game for formula F and interpretation $\mathcal{J}$ iff F is true under $\mathcal{J}$. The game thus characterizes the fundamental notion of truth in a model (interpretation).

The just described semantic game can be turned into a provability game by lifting its states to *disjunctive states*. By this we mean that any state of the provability game consists of a disjunction of states of the semantic game. At any disjunctive state I pick one disjunct where the current formula is non-atomic. If all formulas are atomic, we have reached a final disjunctive state of the provability game. We call such a disjunctive state *winning* (for me, i.e., player I) if for every interpretation there is at least one disjunct (state) where I win.

For states of the semantic game (and thus disjunctive components of the provability game) and each formula F in these states, let us write $I\!:\!F$ if I am in the role of $\mathbf{P}$, and $You\!:\!F$ if You are in the role of $\mathbf{P}$ (and thus I am in the role $\mathbf{O}$). The rules of the provability game may then be denoted as follows, where $\mathcal{D}$ denotes a, possible empty, disjunction of component states.

$$\frac{(I\!:\!A)\bigvee\mathcal{D}\quad(I\!:\!B)\bigvee\mathcal{D}}{(I\!:\!A\wedge B)\bigvee\mathcal{D}}\qquad\frac{(You\!:\!A)\bigvee(You\!:\!B)\bigvee\mathcal{D}}{(You\!:\!A\wedge B)\bigvee\mathcal{D}}$$

$$\frac{(I\!:\!A)\bigvee(I\!:\!B)\bigvee\mathcal{D}}{(I\!:\!A\vee B)\bigvee\mathcal{D}}\qquad\frac{(You\!:\!A)\bigvee\mathcal{D}\quad(You\!:\!B)\bigvee\mathcal{D}}{(You\!:\!A\vee B)\bigvee\mathcal{D}}$$

$$\frac{(You\!:\!A)\bigvee\mathcal{D}}{(I\!:\!\neg A)\bigvee\mathcal{D}}\qquad\frac{(I\!:\!A)\bigvee\mathcal{D}}{(You\!:\!\neg A)\bigvee\mathcal{D}}$$

In these rules, the component state exhibited in the lower disjunctive state is the one picked by me. Notice that a branching into two disjunctive successor states (premises of the rule) only occurs if You has to move in the underlying semantic game. In contrast, if I am to move, the component state picked by me splits into two states, i.e., into two components (disjuncts) of the given disjunctive state.

Again, it is straightforward to check that I have a winning strategy for the provability game starting in state $I\!:\!F$ iff F is valid in CL. Actually, the above rules can be seen as classical sequent (or, equivalently, as tableau) rules in disguise. If one translates the labels 'I'/'You' as 'to the right/left of the sequent arrow', respectively, one indeed arrives at the rules introducing conjunction, disjunction, and negation in the classical sequent calculus $\mathbf{LK}$ (or more precisely, its variant $\mathbf{G3}$ without structural rules [18]). For example:

$$\frac{(You\!:\!A)\bigvee(I\!:\!C)\quad(You\!:\!B)\bigvee(I\!:\!C)}{(You\!:\!A\vee B)\bigvee(I\!:\!C)}\qquad\text{corresponds to}\qquad\frac{A\Rightarrow C\quad B\Rightarrow C}{A\vee B\Rightarrow C}$$

Winning disjunctive states turn into initial sequents $\Gamma,p\Rightarrow p,\Delta$ such that only variables occur in $\Gamma\cup\Delta\cup\{p\}$. Clearly the structural rules of $\mathbf{LK}$, namely permutation, weakening, and contraction, remain sound in the interpretation of sequents as disjunctive game states. Winning strategies in the provability game thus translate into $\mathbf{LK}$ proofs.

We suggest that the sketched transformation of a semantic game into a provability game via moving from single states (referring to particular interpretations) into disjunctive states (referring to all possible interpretations) can be seen as a general principle, rather than a trick that works only for (a fragment of) propositional CL. An arguably more interesting case of this transformation has been worked out in [10] for (infinite-valued) Łukasiewicz logic Ł: Taking Giles' game for Ł [11,12] as a starting point on the semantic level, we arrive at disjunctive states that can be interpreted as hypersequents. Indeed, as shown in [10], one can systemically derive the logical rules of the hypersequent calculus $\mathsf{HŁ}$, originally introduced in [16], in this manner.

In the following, we will apply the transformation of a semantic game into a provability game, and thus a corresponding analytic proof system, to (a somewhat extended version of) Gödel logic.

3   Extended Gödel Logic

Gödel logic occured for the first time in an article by Kurt Gödel [13] where he proved that intuitionistic logic is not a finite-valued logic. It was axiomatized and further investigated by Michael Dummett [9]. As a fuzzy logic, it is characterized by the following truth functions for conjunction, disjunction, and implication:

$$\|A \wedge B\|_{\mathcal{J}} = \min(\|A\|_{\mathcal{J}}, \|B\|_{\mathcal{J}}), \qquad \|A \vee B\|_{\mathcal{J}} = \max(\|A\|_{\mathcal{J}}, \|B\|_{\mathcal{J}}),$$

$$\|A \rightarrow B\|_{\mathcal{J}} = \begin{cases} 1 & \text{if } \|A\|_{\mathcal{J}} \leq \|B\|_{\mathcal{J}} \\ \|B\|_{\mathcal{J}} & \text{otherwise.} \end{cases}$$

These truth functions extend any *interpretation*, i.e., any assignment of *truth values* to propositional variables to compound formulas. In principle, any set V, where $\{0, 1\} \subseteq V \subseteq [0, 1]$ can be taken here as set of truth values. We are mostly interested in infinite-valued Gödel logic G, which is a t-norm based fuzzy logic, where $V = [0, 1]$, min is the t-norm modeling conjunction, and the corresponding residuum modeling implication. We include the propositional constants $\bot$ and $\top$ in G, interpreted by $\|\bot\|_{\mathcal{J}} = 0$ and $\|\top\|_{\mathcal{J}} = 1$. The *atomic formulas* of G are the propositional variables and the propositional constants.

Negation in G is a defined connective, given by $\neg A = A \rightarrow \bot$. We moreover extend G to $\mathsf{G}^{\triangle}$ by including the following *projection operator* [2]:

$$\|\triangle A\|_{\mathcal{J}} = \begin{cases} 1 & \text{if } \|A\|_{\mathcal{J}} = 1 \\ 0 & \text{otherwise.} \end{cases}$$

The set of all $[0, 1]$-valued interpretations is denoted $\mathbf{Int}^{[0,1]}$. An interpretation $\mathcal{J} \in \mathbf{Int}^{[0,1]}$ *satisfies* a formula F and is called a *model* of F (written $\mathcal{J} \models F$) if $\|F\|_{\mathcal{J}} = 1$. F is *valid* if all interpretations are models of F.

4   Truth Degree Comparison Games

Below, we will focus on *truth degree comparison claims*, or just *claims*, of the form $F \leq G$ or $F < G$, where F and G are $\mathsf{G}^{\triangle}$-formulas. An interpretation $\mathcal{J}$ *satisfies* such a claim if $\|F\|_{\mathcal{J}} \leq \|G\|_{\mathcal{J}}$ or $\|F\|_{\mathcal{J}} < \|G\|_{\mathcal{J}}$, respectively.

Note that truth comparison claims can be reduced to single $\mathsf{G}^{\triangle}$-formulas in the following sense: $\mathcal{J}$ satisfies $F \leq G$ iff $\mathcal{J}$ satisfies $F \rightarrow G$ and $\mathcal{J}$ satisfies $F < G$ iff $\mathcal{J}$ satisfies $\neg \triangle (G \rightarrow F)$.

We introduce a semantic game for the stepwise reduction of arbitrary truth degree comparison claims to atomic ones. Game states consist of truth degree comparison claims $F \triangleleft G$, where $\triangleleft$ is either $\leq$ or $<$. Furthermore each non-atomic claim carries a *marking* which points to a non-atomic formula in the claim (either F or G). In the Hintikka-style game of Sect. 2 for CL we had to distinguish between the players identities (I and *You*) and their current roles $\mathbf{P}$ or $\mathbf{O}$. The truth degree comparison game for $\mathsf{G}^{\triangle}$ does not feature role switches; therefore we can identify the two players with $\mathbf{P}$ and $\mathbf{O}$, respectively. Given an interpretation $\mathcal{J}$, at any state $F \lhd G$ player **P** seeks to defend and **O** to refute the claim that $\mathcal{J}$ satisfies $F \lhd G$. If F and G are atomic formulas the game is in an *atomic state*, where **P** wins (and **O** loses) if $\|F\|_{\mathcal{J}} \lhd \|G\|_{\mathcal{J}}$.

At each state of the game, **P** and **O** make moves according to the rules below resulting in a successor claim where the marked formula has been decomposed. If the successor claim is not atomic, then in a final (implicit) move, a *regulation function* ρ marks one of the non-atomic formulas in the successor claim. The resulting claim is the *successor state* of the game.

For each connective there are four rules, according to whether the connective appears in a marked formula on the left or on the right, and whether the truth degree comparison is strict or non-strict, i.e., of the form $F < G$ or $F \leq G$. Some of the rules can be represented in a uniform manner using $\lhd$ to stand for either $<$ or $\leq$ (consistently within the rule). In the following, the exhibited compound formula is the marked formula of the state[3].

$A \wedge B \lhd C$: **P** chooses whether the game continues with $A \lhd C$ or with $B \lhd C$.
$C \lhd A \wedge B$: **O** chooses whether the game continues with $C \lhd A$ or with $C \lhd B$.

$A \vee B \lhd C$: **O** chooses whether the game continues with $A \lhd C$ or with $B \lhd C$.
$C \lhd A \vee B$: **P** chooses whether the game continues with $C \lhd A$ or with $C \lhd B$.

$A \to B \leq C$: **P** chooses one of the following intermediary states, where it is **O**'s turn to choose:
   (1): the game continues with $\top \leq C$;
   (2): **O** chooses whether the game continues with $B < A$ or with $B \leq C$.
$C \leq A \to B$: **P** chooses whether the game continues with $A \leq B$ or with $C \leq B$.
$A \to B < C$: **O** chooses whether the game continues with $B < A$ or with $B < C$.
$C < A \to B$: **P** chooses between
   (1): the game continues with $C < B$;
   (2): **O** chooses whether the game continues with $A \leq B$ or with $C < \top$.

$\triangle A \leq C$: **P** chooses whether to continue with $A < \top$ or with $\top \leq C$.
$C \leq \triangle A$: **P** chooses whether to continue with $\top \leq A$ or with $C \leq \bot$.
$\triangle A < C$: **O** chooses whether to continue with $A < \top$ or with $\bot < C$.
$C < \triangle A$: **O** chooses whether to continue with $\top \leq A$ or with $C < \top$.

We can picture these game rules as decision trees. E.g, the rule for the game state $A \to B \leq C$ corresponds to the tree in Fig. 1. The leaves of this tree, i.e., $\top \leq C, B < A$ and $B \leq C$, are the possible successor claims of $A \to B \leq C$. Given the regulation ρ, we can then further expand the successor claims into decision trees according to the game rules.

We therefore see that each game can be viewed as a finite tree $\tau_{\rho}^{\mathcal{J}}[F \lhd G]$ of (marked) truth comparison claims, rooted in the initial claim $F \lhd G$ and branching according to the rules of the truth degree comparison game and the regulation ρ until all leaves are atomic states, i.e., states where the compared formulas are either variables or propositional constants $\bot$ or $\top$. If the interpretation

---

[3] This convention will be followed often throughout the article.

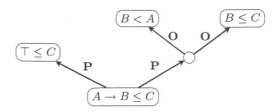

Fig. 1. Decision tree of a game rule

$\mathcal{J}$ satisfies the truth comparison claim at an atomic state ν then ν is a *winning state* of $\tau_\rho^{\mathcal{J}}[F \lhd G]$ for **P**.

Example 1. Below is the tree $\tau_\rho^{\mathcal{J}}[p \wedge (p \to q) \le p \wedge q]$. The formulas marked by the regulation ρ are underlined.

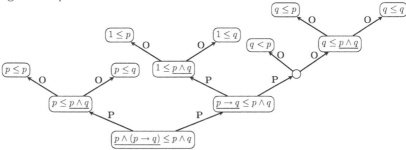

In the case that $\|q\|_{\mathcal{J}} < \|p\|_{\mathcal{J}} < 1$, the winning states are $p \le p$, $q < p$, $q \le p$ and $q \le q$.

A *strategy* σ for **P** in $\tau_\rho^{\mathcal{J}}[F \lhd G]$ is a subtree of $\tau_\rho^{\mathcal{J}}[F \lhd G]$ which is obtained from pruning all but one **P**-labelled outgoing branches from every node in the tree, while keeping **O**-labelled branches intact. Clearly, the remaining tree specifies how **P** is to move at the given state, while all possible choices of **O** are still recorded. σ is a *winning strategy*, hereinafter referred to as *ws*, for **P** if all leaf nodes of σ are winning states for **P**.

Example 2. To the right is a strategy for **P** in the game $\tau_\rho^{\mathcal{J}}[p \wedge (p \to q) \le p \wedge q]$, which is obtained from the tree in Example 1 by pruning the right branch stemming from the root. It is a winning strategy if and only if $\|p\|_{\mathcal{J}} \le \|q\|_{\mathcal{J}}$.

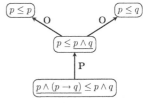

Each non-atomic game state $F \lhd G$ corresponds to exactly one of the 12 game rules described above, and we describe its set $\mathrm{Pow}(F \lhd G)$ of **P**-*powers*[4] as follows:

---

[4] This notion is similar to the general definition of a power in game theory, cf. [19].

A set X is a **P**-power of $F \lhd G$ if it is a subset-minimal set of claims such that in the game state $F \lhd G$, **P** can enforce that the successor claim is among the claims in X.

For example, in the game state $A \to B \leq C$ with $A \to B$ marked (cf. Fig. 1), **P** can make a move so that the successor claim is $\top \leq C$. Alternatively, she can make a move ensuring that the successor claim is one of $B < A$ or $B \leq C$, but she does not know which one since this depends on a move by **O**. Hence we have

$$\text{Pow}(A \to B \leq C) = \{\{\top \leq C\}, \{B < A, B \leq C\}\}.$$

As further examples,

$$\text{Pow}(A \wedge B \lhd C) = \{\{A \lhd C\}, \{B \lhd C\}\},$$
$$\text{and} \quad \text{Pow}(C \lhd A \wedge B) = \{\{C \lhd A, C \lhd B\}\}.$$

For an atomic state $F \lhd G$, we formally set $\text{Pow}(F \lhd G) = \{\{F \lhd G\}\}$.

Proposition 1 (Soundness of game rules). *For any game state $F \lhd G$ and any $\mathcal{J} \in \mathbf{Int}^{[0,1]}$, $\mathcal{J} \models F \lhd G$ iff for some $X \in \text{Pow}(F \lhd G)$, $\mathcal{J}$ satisfies all formulas in X.*

Proof For an atomic state $F \lhd G$, this holds by definition. For non-atomic $F \lhd G$, this is proved for all 12 types of game states seperately. Consider for example the state $A \to B \leq C$ and its **P**-power

$$\{\{\top \leq C\}, \{B < A, B \leq C\}\}.$$

Let $\mathcal{J} \in \mathbf{Int}^{[0,1]}$. Then either $\|A\|_{\mathcal{J}} \leq \|B\|_{\mathcal{J}}$, in which case $\|A \to B\|_{\mathcal{J}} = 1$ and so $\mathcal{J}$ satisfies the claim $A \to B \leq C$ iff $\mathcal{J}$ satisfies $\top \leq C$. Or $\|A\|_{\mathcal{J}} > \|B\|_{\mathcal{J}}$: Then $\|A \to B\|_{\mathcal{J}} = \|B\|_{\mathcal{J}}$ and so $\mathcal{J}$ satisfies the claim $A \to B \leq C$ iff $\mathcal{J}$ satisfies $B \leq C$.

We prove the equivalence for the other two examples given above. For the game state $A \wedge B \lhd C$ with

$$\text{Pow}(A \wedge B \lhd C) = \{\{A \lhd C\}, \{B \lhd C\}\}$$

we observe that an interpretation $\mathcal{J}$ satisfies $A \wedge B \lhd C$ if and only if we either have $\|A\|_{\mathcal{J}} \leq \|B\|_{\mathcal{J}}$ and $\|A\|_{\mathcal{J}} \lhd \|C\|_{\mathcal{J}}$, or alternatively $\|A\|_{\mathcal{J}} > \|B\|_{\mathcal{J}}$ and $\|B\|_{\mathcal{J}} \lhd \|C\|_{\mathcal{J}}$.

Finally, for the game state $C \lhd A \wedge B$ with

$$\text{Pow}(C \lhd A \wedge B) = \{\{C \lhd A, C \lhd B\}\}$$

we observe that an interpretation $\mathcal{J}$ satisfies $C \lhd A \wedge B$ if and only if $\mathcal{J}$ satisfies both $C \lhd A$ and $C \lhd B$.

The remaining 9 cases can be shown similarly. □

Proposition 2. *For any interpretation $\mathcal{J}$ and regulation ρ, if **P** has a winning strategy in $\tau_{\rho}^{\mathcal{J}}[F \lhd G]$ then $\mathcal{J}$ satisfies $F \lhd G$, where $\lhd \in \{<, \leq\}$.*

Proof. By induction on the tree height of a ws $\sigma \subseteq \tau_\rho^{\mathcal{J}}[F \lhd G]$. If the height of σ is 1, then $F \lhd G$ is atomic, and since σ is a ws, $F \lhd G$ must therefore be a winning state of $\tau_\rho^{\mathcal{J}}[F \lhd G]$. Hence $\mathcal{J} \models F \lhd G$.

Now assume that the height of σ is at least 2, and let $S_1, \ldots, S_n$ be the successor claims of $F \lhd G$ in $\tau_\rho^{\mathcal{J}}[F \lhd G]$ which are contained in σ. Since σ is a strategy for **P**, the set $\{S_1, \ldots, S_n\}$ is a **P**-power of $\tau_\rho^{\mathcal{J}}[F \lhd G]$. Now for each $i \leq n$, let σ_i be the subtree of σ with root S_i. Then each σ_i is a ws for **P** in $\tau_\rho^{\mathcal{J}}[S_i]$, and so by induction hypothesis $\mathcal{J} \models S_i$ for every $i \leq n$. We have thus shown that all claims in a **P**-power of $\tau_\rho^{\mathcal{J}}[F \lhd G]$ are satisfied by $\mathcal{J}$, and so by Proposition 1 it follows that $\mathcal{J}$ satisfies $F \lhd G$. □

Proposition 3. *If an interpretation $\mathcal{J}$ satisfies $F \lhd G$, where $\lhd \in \{<, \leq\}$, then **P** has a winning strategy in $\tau_\rho^{\mathcal{J}}[F \lhd G]$ for any regulation ρ.*

Proof. If $\mathcal{J}$ satisfies $F \lhd G$, then by Proposition 1 there is a power $X \in$ Pow$(F \lhd G)$ (where the marking in $F \lhd G$ is set according to ρ) such that $\mathcal{J}$ satisfies all claims in X. So **P** can enforce that the successor state of $F \lhd G$ in the game $\tau_\rho^{\mathcal{J}}[F \lhd G]$ is contained in X.

Repeating the same kind of reasoning, we see that **P** can always move ensuring that the resulting game state is satisfied by $\mathcal{J}$, and in particular, any atomic state ν reached using this strategy will be a winning state in $\tau_\rho^{\mathcal{J}}[F \lhd G]$ for **P**. □

5   Disjunctive Game States as Sequents-of-relations

As an immediate consequence of Propositions 2 and 3 we have:

Theorem 1. *The following are equivalent:*

1. *$F \lhd G$ is valid in $\mathsf{G}^{\triangle}$*
2. *For some regulation ρ, **P** has a ws in $\tau_\rho^{\mathcal{J}}[F \lhd G]$ for every $\mathcal{J} \in \mathbf{Int}^{[0,1]}$*
3. *For any regulation ρ, **P** has a ws in $\tau_\rho^{\mathcal{J}}[F \lhd G]$ for every $\mathcal{J} \in \mathbf{Int}^{[0,1]}$.*

In particular, although different regulations ρ lead to different games, the choice of the regulation does not matter if one is only interested in the winnability of a game.

A family $(\sigma_{\mathcal{J}})_{\mathcal{J} \in \mathbf{Int}^{[0,1]}}$ of ws for the games $\tau_\rho^{\mathcal{J}}[F \lhd G]$ witnesses that the claim $F \lhd G$ is valid. We may think of $(\sigma_{\mathcal{J}})_{\mathcal{J} \in \mathbf{Int}^{[0,1]}}$ as a *proof* of $F \lhd G$, but this notion of provability would not be efficient since $(\sigma_{\mathcal{J}})_{\mathcal{J} \in \mathbf{Int}^{[0,1]}}$ is an infinite object. However, we now show that an infinite family of strategies such as $(\sigma_{\mathcal{J}})_{\mathcal{J} \in \mathbf{Int}^{[0,1]}}$ can be encoded into a single *disjunctive winning strategy*. In doing so, we follow the approach sketched in Sect. 2 for classical logic.

First, define a *disjunctive state* D to be a finite nonempty multiset of claims written $D = S_1 \bigvee \ldots \bigvee S_n$. A disjunctive state is called *atomic* if all of its disjuncts are atomic claims. We say that an interpretation $\mathcal{J}$ satisfies a disjunctive state D, and write $\mathcal{J} \models D$, if $\mathcal{J}$ satisfies at least one of the disjuncts of D. A disjunctive state D is called *winning* if it is an atomic state satisfied by every interpretation.

For a set $\mathcal{P} = \{X_1, \ldots, X_n\}$ where each X_i is a set of claims, we define $\bigvee \mathcal{P}$ as the set of all disjunctive states

$$S_1 \bigvee \ldots \bigvee S_n$$

where for each $i \leq n$, $S_i \in X_i$.

Definition 1 (disjunctive rule). *Let S be a non-atomic claim and D a disjunctive state. A disjunctive rule is a rule of the form*

$$\frac{D \bigvee D_1 \quad \ldots \quad D \bigvee D_k}{D \bigvee S}$$

where for a game state S' obtained from marking a formula in S, the sequence $D_1, \ldots, D_k$ is an enumeration of $\bigvee \mathrm{Pow}(S')$.

As an example, let S be the claim $A \to B \leq C$ and S' the corresponding game state where $A \to B$ is marked. Recall that

$$\mathrm{Pow}(A \to B \leq C) = \{\{\top \leq C\}, \{B < A, B \leq C\}\}$$

and so

$$\bigvee \mathrm{Pow}(A \to B \leq C) = \{(\top \leq C \bigvee B < A), (\top \leq C \bigvee B \leq C)\}.$$

The corresponding disjunctive rule is thus:

$$\frac{D \bigvee (\top \leq C) \bigvee (B < A) \quad D \bigvee (\top \leq C) \bigvee (B \leq C)}{D \bigvee (A \to B \leq C)}$$

Figure 2 contains the disjunctive rules corresponding to all 12 types of game states.

Definition 2 (Disjunctive strategy). *Let D be a disjunctive state. A disjunctive strategy for $\mathbf{P}$ in D is a tree of disjunctive states built using disjunctive rules, and with root D. A disjunctive strategy is called* winning strategy *if all its leaves are disjunctive winning states.*

For the time being, disjunctive strategies will just be syntactic objects rather than strategies in some game. We will however discuss later on how to interpret disjunctive strategies in a game theoretic sense.

Proposition 4 (Soundness of disjunctive rules). *Let $\mathcal{J} \in \mathbf{Int}^{[0,1]}$. Then $\mathcal{J}$ satisfies the conclusion of a disjunctive rule iff $\mathcal{J}$ satisfies all of its premises.*

Proof Let the disjunctive rule be presented as in Definition 1. Assume first that $\mathcal{J} \vDash D \bigvee S$. If $\mathcal{J} \vDash D$, then clearly $\mathcal{J}$ satisfies all premises of the disjunctive rule as well. On the other hand, if $\mathcal{J} \vDash S$, then by Proposition 1 there exists a power $X \in \mathrm{Pow}(S)$ such that all claims in X are satisfied by $\mathcal{J}$. It follows that $\mathcal{J} \vDash D_i$ for every $i \leq n$ because D_i contains a disjunct from X.

For the other direction, assume that $\mathcal{J} \nvDash D \bigvee S$. Then $\mathcal{J} \nvDash D$ and $\mathcal{J} \nvDash S$. The latter implies, again by Proposition 1, that every power $X \in \mathrm{Pow}(S)$ contains a state not satisfied by $\mathcal{J}$. The disjunctive combination of all these failing states is one of the D_i's, and so $\mathcal{J}$ does not satisfy the premise $D \bigvee D_i$. $\qquad \square$

$$\frac{D\bigvee(A \lhd C)\bigvee(B \lhd C)}{D\bigvee(A \wedge B \lhd C)} {\scriptstyle\wedge\lhd} \qquad \frac{D\bigvee(C \lhd A)\quad D\bigvee(C \lhd B)}{D\bigvee(C \lhd A \wedge B)} {\scriptstyle\lhd\wedge} \qquad \frac{D\bigvee(A \lhd C)\quad D\bigvee(B \lhd C)}{D\bigvee(A \vee B \lhd C)} {\scriptstyle\vee\lhd}$$

$$\frac{D\bigvee(C \lhd A)\bigvee(C \lhd B)}{D\bigvee(C \lhd A \vee B)} {\scriptstyle\lhd\vee} \qquad \frac{D\bigvee(\top \le C)\bigvee(B < A)\quad D\bigvee(\top \le C)\bigvee(B \le C)}{D\bigvee(A \to B \le C)} {\scriptstyle\to\le}$$

$$\frac{D\bigvee(B < A)\quad D\bigvee(B < C)}{D\bigvee(A \to B < C)} {\scriptstyle\to<} \qquad \frac{D\bigvee(C < B)\bigvee(A \le B)\quad D\bigvee(C < B)\bigvee(C < \top)}{D\bigvee(C < A \to B)} {\scriptstyle<\to}$$

$$\frac{D\bigvee(A \le B)\bigvee(C \le B)}{D\bigvee(C \le A \to B)} {\scriptstyle\le\to} \qquad \frac{D\bigvee(A < \top)\bigvee(\top \le C)}{D\bigvee(\triangle A \le C)} {\scriptstyle\triangle\le} \qquad \frac{D\bigvee(\top \le A)\bigvee(C \le \bot)}{D\bigvee(C \le \triangle A)} {\scriptstyle\le\triangle}$$

$$\frac{D\bigvee(A < \top)\quad D\bigvee(\bot < C)}{D\bigvee(\triangle A < C)} {\scriptstyle\triangle<} \qquad \frac{D\bigvee(\top \le A)\quad D\bigvee(C < \top)}{D\bigvee(C < \triangle A)} {\scriptstyle<\triangle}$$

Fig. 2. Disjunctive rules

Theorem 2. $F \lhd G$ *is valid in* $\mathsf{G}^\triangle$ *iff there is a disjunctive ws for* **P** *in* $F \lhd G$.

Proof. Given the claim $F \lhd G$ (seen as a disjunctive state with one component), we can exhaustively apply disjunctive rules to it in any order, and eventually obtain a disjunctive strategy with atomic leaves. By Proposition 4 (and a simple induction on the height of the tree), all leaves of this tree will be disjunctive winning states because $F \lhd G$ is valid by assumption.

Conversely, if there is a disjunctive ws for **P** in $F \lhd G$, then by definition all of its leaves are winning states. Again by Proposition 4 and a simple induction on the tree height, it follows that all disjunctive states in the ws are valid. Hence in particular, the claim $F \lhd G$ is valid.   □

To use disjunctive ws as a proof system for $\mathsf{G}^\triangle$, the only thing left to establish is that we can efficiently check whether the leaves of a disjunctive strategy are winning. Indeed, this holds true:

Lemma 1. *It is decidable in PTIME whether an atomic disjunctive state is winning.*

Proof. See Theorem 4 in [7].   □

From this and Theorem 2 it follows that the disjunctive ws form a *propositional proof system* for $\mathsf{G}^\triangle$ in the sense of Cook-Reckhow (cf. the survey [17]).

The disjunctive strategies can be seen as strategies in the usual game-theoretic sense, with respect to a game that we are going to define now. The $\mathsf{G}^\triangle$-*provability game on* D starts with a (disjunctive) state D. At each turn, **P** picks one disjunctive rule whose conclusion matches the current disjunctive state. Then, **O** chooses one of the premises of this disjunctive rule as the successor state. If an atomic disjunctive state is reached, **P** wins if the state is a winning state (in the earlier sense). Clearly, disjunctive ws are the same as ws in the provability game, and so we have:

Theorem 3. $F \lhd G$ *is valid in* $\mathsf{G}^{\triangle}$ *iff there is a ws for* **P** *in the* $\mathsf{G}^{\triangle}$-*provability game on* $F \lhd G$.

We can think of the provability game as a game where multiple instances of a truth degree comparison game $\tau_\rho^{\mathcal{J}}[F \lhd G]$ are played simultaneously, for varying interpretations $\mathcal{J}$. Or alternatively, we may imagine that **P** plays $\tau_\rho^{\mathcal{J}}[F \lhd G]$ without knowing the interpretation $\mathcal{J}$. Now whenever **P** faces a choice in one of the degree comparison games, she simply encodes *all* possible moves she could make into the strategy in the provability game. The claim that **P** then defends is that for every $\mathcal{J}$, at least one of the subgames she plays necessarily leads to a winning state.

More compact representations of disjunctive strategies are sometimes possible. For example, consider the following disjunctive rule:

$$\frac{D_1 \bigvee \ldots \bigvee D_n}{D_1 \bigvee \ldots \bigvee D_n \bigvee D_{n+1}} \; ew$$

It is easy to see that whenever there is a disjunctive ws for **P** in $D_1 \bigvee \ldots \bigvee D_n$, then there is also a disjunctive ws for **P** in $D_1 \bigvee \ldots \bigvee D_n \bigvee D_{n+1}$. So if we allow the rule ew in the construction of disjunctive ws, we still characterize validity in $\mathsf{G}^{\triangle}$. However, disjunctive ws with ew might be smaller. The intuitive (bottom-up) reading of ew is the following: If during the construction of a disjunctive ws for $D_1 \bigvee \ldots \bigvee D_n \bigvee D_{n+1}$ player **P** finds out that already the disjuncts $D_1 \bigvee \ldots \bigvee D_n$ lead to a winning state, then she can discard the redundant disjunct D_n.

Example 3. Below is a disjunctive ws for the claim in Example 1, which uses the rule ew:

$$\cfrac{\cfrac{p \leq p}{(p \leq p) \bigvee (p \to q \leq p \wedge q)} \; ew \quad \cfrac{\cfrac{\cfrac{(p \leq q) \bigvee (q < p)}{(p \leq q) \bigvee (\top \leq p \wedge q) \bigvee (q < p)} \; ew \quad \cfrac{\cfrac{(p \leq q) \bigvee (q \leq p) \quad (p \leq q) \bigvee (q \leq q)}{(p \leq q) \bigvee (q \leq p \wedge q)} \leq \wedge}{(p \leq q) \bigvee (\top \leq p \wedge q) \bigvee (q \leq p \wedge q)} \; ew}{\cfrac{(p \leq q) \bigvee (p \to q \leq p \wedge q)}{} } \to \leq}{(p \leq p \wedge q) \bigvee (p \to q \leq p \wedge q)} \leq \wedge}{p \wedge (p \to q) \leq p \wedge q} \wedge \leq$$

The disjunctive ws are very close to proofs in the *sequents-of-relations* calculus $\mathbf{RG}_\infty$, and its extension $\mathbf{RG}_\infty^{\triangle}$ capturing the $\triangle$ projection operator, as developed in [3,4,6]. The approach there is algebraic rather then game-theoretic.

On a purely notational level, the sequents-of-relations calculus differs from the disjunctive ws by the use of the symbol $\mid$ instead of $\bigvee$, making it fit into the framework of *hypersequent calculi* as developed independently by Mints, Pottinger and Avron (cf. the survey [5]).

The other differences are: $\mathbf{RG}_\infty^{\triangle}$ includes the structural rules

$$\frac{D}{D \bigvee S} \; ew \qquad \text{and} \qquad \frac{D \bigvee S \bigvee S}{D \bigvee S} \; ec$$

of *external weakening* (see the discussion above) and *external contraction*, and it features the logical rules

$$\frac{D\bigvee(\top \le C)\bigvee(B < A)\quad D\bigvee(B \le C)}{D\bigvee(A \to B \le C)}\to\le^* \qquad \frac{D\bigvee(C < B)\bigvee(A \le B)\quad D\bigvee(C < \top)}{D\bigvee(C < A \to B)}<\to^*$$

instead of our rules $\to\le$ and $<\to$ (cf. Fig. 2). All other rules are the same.

To show the equivalence of both calculi, we can proceed as follows. First, for the rule variants $\to\le^*$ and $<\to^*$ the analogue of Proposition 4 can be shown:

Lemma 2. *Let* $\mathcal{J} \in \mathbf{Int}^{[0,1]}$. *Then* $\mathcal{J}$ *satisfies the conclusion of the rule* $\to\le^*$ *(resp.* $<\to^*$*) iff* $\mathcal{J}$ *satisfies all of the premises of* $\to\le^*$ *(resp.* $<\to^*$*).*

Proof. Assume $\mathcal{J} \nvDash D$ (otherwise the statement is obvious).

If $\|A\|_{\mathcal{J}} \le \|B\|_{\mathcal{J}}$, then $\mathcal{J}$ satisfies the conclusion of $\to\le^*$ iff $\|C\|_{\mathcal{J}} = 1$, and this is equivalent to the statement that $\mathcal{J}$ satisfies the premises of $\to\le^*$, since $\mathcal{J} \nvDash D$ and $\mathcal{J} \nvDash (B < A)$. If on the other hand $\|A\|_{\mathcal{J}} > \|B\|_{\mathcal{J}}$, then $\mathcal{J}$ satisfies the conclusion of $\to\le^*$ iff $\|B\|_{\mathcal{J}} \le \|C\|_{\mathcal{J}}$. This in turn is equivalent to saying that $\mathcal{J}$ satisfies the premises of $\to\le^*$ since it satisfies the left premise by assumption, and the right premise reduces to $B \le C$ since $\mathcal{J} \nvDash D$.

The argument for the rule $<\to^*$ is similar. ☐

It follows that the proof of Theorem 2 goes through if we use $\to\le^*$ and $\to\le^*$ as disjunctive rules instead of their non-starred versions. The additional structural rules *ew* and *ec* are in fact redundant, since already the system without them is complete for $\mathsf{G}^\triangle$. Note however that the inclusion of redundant rules might lead to shorter proofs. More such rules for the sequents-of-relations calculus are discussed in [6].

6   Summary and Conclusion

We have investigated Gödel logic, one of the fundamental fuzzy logics, from a game semantic perspective. In Sect. 4, we presented a game for reducing truth degree comparison claims $F < G$ or $F \le G$, i.e., claims about the relative order of arbitrary $\mathsf{G}^\triangle$-formulas, to atomic comparison claims. This amounts to a generalization of Hintikka's well known semantic game for classical logic. As illustrated in Sect. 2 for the simple case of classical propositional logic, semantic games can be systematically lifted to provability games. The latter operate on the level of validity rather than the level of truth in a model and thus correspond to analytic proof systems. Indeed, Gentzen's sequent system for classical logic can be interpreted from a game perspective in this manner. In Sect. 5, we have applied this general scheme to the more involved case of the truth degree comparison game and demonstrated that moving from single states to disjunctions of states yields a characterization of validity in $\mathsf{G}^\triangle$ in terms of 'disjunctive winning strategies'. Moreover, disjunctions of states can be viewed as sequents-of-relations in the sense of [3,6]. Hence, disjunctive winning strategies provide an interpretation of proofs in this calculus.

A number of topics for further research arise from our game based take on Gödel logic. While it has already been shown in [10] that a similar approach relates Giles's game for Łukasiewicz logic to a corresponding hypersequent calculus, it remains open whether and how this method can be extended to yet further fuzzy logics. Even for Gödel logic itself, one may ask whether not only sequents-of-relations but also the arguably better known hypersequent calculus **HLC** of Avron [1] can be systematically related to a truth degree comparison game. This might also open the way to generalize to the first order level, since in contrast to the sequents-of-relations calculus, **HLC** can straightforwardly be extended to include quantifier rules. Due to its attractiveness for certain applications, an extension of Gödel logic featuring an involutative negation, in addition to standard Gödel-negation, has received some attention [8]. In future work we plan to extend our truth degree comparison games to include also this type of negation. Finally, we like to point out that a game based approach to fuzzy logics may open the route to more sophisticated models of reasoning under vagueness than can be achieved by sticking with truth functional logics. It is natural to ask what happens if the players of a game have only imperfect information about their opponent's moves. For classical logic this leads to Independence Friendly (IF) logic of Hintikka and Sandu [15]. Given the fact that vagueness may be seen as a phenomenon involving a lack of full share of (precise) information between speaker and hearer of vague statements, it seems attractive to explore the impact of imperfect information on truth degree comparison games.

References

1. Avron, A.: Hypersequents, logical consequence and intermediate logics for concurrency. Ann. Math. Artif. Intell. **4**, 225–248 (1991)
2. Baaz, M.: Infinite-valued Gödel logics with 0-1-projections and relativizations. In: Gödel 1996: Logical Foundations of Mathematics, Computer Science and Physics–Kurt Gödel's Legacy, Proceedings, Brno, Czech Republic, August 1996, pp. 23–33. Association for Symbolic Logic (1996)
3. Baaz M., Ciabattoni A., Fermüller, C.: Sequent of relations calculi: a framework for analytic deduction in many-valued logics. In: Fitting M., Orlowska E. (eds.) Beyond Two: Theory and Applications of Multiple-Valued Logic. Studies in Fuzziness and Soft Computing, vol. 114, pp. 157–180. Springer, Heidelberg (2003). https://doi.org/10.1007/978-3-7908-1769-0_6
4. Baaz, M., Ciabattoni, A., Fermüller, C.G.: Cut-elimination in a sequents-of-relations calculus for Gödel logic. In: 31st IEEE International Symposium on Multiple-Valued Logic, ISMVL 2001, Proceedings, Warsaw, Poland, 22–24 May 2001, pp. 181–186. IEEE Computer Society (2001)
5. Baaz, M., Ciabattoni, A., Fermüller, C.G.: Hypersequent calculi for Gödel logics - a survey. J. Log. Comput. **13**(6), 835–861 (2003)
6. Baaz, M., Fermüller, C.G.: Analytic calculi for projective logics. In: Murray, N.V. (ed.) TABLEAUX 1999. LNCS (LNAI), vol. 1617, pp. 36–51. Springer, Heidelberg (1999). https://doi.org/10.1007/3-540-48754-9_8

7. Ciabattoni, A., Fermüller, C.G., Metcalfe, G.: Uniform rules and dialogue games for fuzzy logics. In: Baader, F., Voronkov, A. (eds.) LPAR 2005. LNCS (LNAI), vol. 3452, pp. 496–510. Springer, Heidelberg (2005). https://doi.org/10.1007/978-3-540-32275-7_33

8. Ciabattoni, A., Vetterlein, T.: On the (fuzzy) logical content of CADIAG-2. Fuzzy Sets Syst. **161**(14), 1941–1958 (2010)

9. Dummett, M.: A propositional calculus with denumerable matrix. J. Symb. Logic **24**(2), 97–106 (1959)

10. Fermüller, C.G., Metcalfe, G.: Giles's game and the proof theory of Łukasiewicz logic. Studia Logica **92**(1), 27–61 (2009)

11. Giles, R.: A non-classical logic for physics. Studia Logica **33**(4), 397–415 (1974)

12. Giles, R.: A non-classical logic for physics. In: Wojcicki, R., Malinowski, G. (eds.) Selected Papers on Łukasiewicz Sentential Calculi, pp. 13–51. Polish Academy of Sciences (1977)

13. Gödel, K.: Zum Intuitionistischen Aussagenkalkül. J. Symb. Logic **55**(1), 344–344 (1990). (1932) (Reprint)

14. Hintikka, J.: Logic, Language-Games and Information: Kantian Themes in the Philosophy of Logic. Clarendon Press, Oxford (1973)

15. Mann, A.L., Sandu, G., Sevenster, M.: Independence-Friendly Logic. A Game-Theoretic Approach. Cambridge University Press, Cambridge (2011)

16. Metcalfe, G., Olivetti, N., Gabbay, D.: Sequent and hypersequent calculi for Abelian and Łukasiewicz logics. ACM Trans. Comput. Logic (TOCL) **6**(3), 578–613 (2005)

17. Segerlind, N.: The complexity of propositional proofs. Bull. Symb. Logic **13**(4), 417–481 (2007)

18. Troelstra, A.S., Schwichtenberg, H.: Basic Proof Theory. Cambridge Tracts in Theoretical Computer Science, 2nd edn. Cambridge University Press, Cambridge (2000)

19. Van Benthem, J.: Logic in Games. MIT press, Cambridge (2014)

Ordinal Graph-Based Games

Arij Azzabi[1,3(✉)], Nahla Ben Amor[1], Hélène Fargier[2], and Régis Sabbadin[3]

[1] LARODEC, ISG-Tunis, Université de Tunis, Tunis, Tunisia
arij.azzabi@gmx.fr
[2] IRIT-CNRS, Université de Toulouse, Toulouse, France
[3] INRAE-MIAT, Université de Toulouse, Toulouse, France

Abstract. The graphical, hypergraphical and polymatrix games frameworks provide concise representations of non-cooperative normal-form games involving many agents. In these *graph-based* games, agents interact in simultaneous local subgames with the agents which are their neighbors in a graph. Recently, ordinal normal form games have been proposed as a framework for game theory where agents' utilities are ordinal. This paper presents the first definition of *Ordinal Graphical Games* (OGG), *Ordinal Hypergraphical Games* (OHG), and *Ordinal Polymatrix Games* (OPG). We show that, as for classical graph-based games, determining whether a pure NE exists is also NP-hard. We propose an original CSP model to decide their existence and compute them. Then, a polynomial-time algorithm to compute possibilistic mixed equilibria for graph-based games is proposed. Finally, the experimental study is dedicated to test our proposed solution concepts for ordinal graph-based games.

Keywords: Possibility theory · Ordinal game theory · Algorithms · Complexity

1 Introduction

Game theory is a natural framework to consider when modeling complex multi-agent systems. The larger the number of agents in these systems, the more computational issues arise. However, there exist situations where the utility of players only depends on a small subset of other players' strategies. Accordingly, researchers in AI proposed compact representations for games, pursuing the seminal work on graphical games [11]. *Polymatrix games* [20], *graphical games* [11] and *hypergraphical games* [16] have been proposed as a convenient way to represent games with multiple players and local interactions. These models offer the possibility to exploit local interactions among players and can require exponentially less space than usual normal-form games to represent. In hypergraphical games, agents' interactions are represented by a hypergraph where each agent (vertex) can be involved in several normal-form subgames (hyperedges). If the utility of each agent depends on exactly one subgame, then the game is a graphical game. If all subgames involve only two players then the game is a polymatrix game. In this work, we are interested in these three classes of games.

As for standard representations, the overall aim for players with compact representations of games is to compute a Nash equilibrium (NE) [13]. Significant work has

© Springer Nature Switzerland AG 2020
M.-J. Lesot et al. (Eds.): IPMU 2020, CCIS 1237, pp. 271–285, 2020.
https://doi.org/10.1007/978-3-030-50146-4_21

been devoted to finding pure or mixed NE for polymatrix, graphical and hypergraphical games. [11] proposed a *message passing* type algorithm (*TreeNash*) for computing NE on *tree structured* graphical games. [10,15] extended the *TreeNash* algorithm to arbitrary graphical games, by defining *NashProp*, a heuristic *Loopy Belief Propagation*-type algorithm. Concerning polymatrix games, [12] have demonstrated that a mixed NE could be found by a reduction to a *Linear Complementarity Problem (LCP)*. In a different line of works, [17] have studied *constrained* pure NE in different subclasses of polymatrix games. They have shown that the problem of finding pure NE is tractable in these subclasses. [2] proposed *Valued Nash Propagation (VNP)*, an algorithm for finding a pure NE in hypergraphical games and showed that VNP works efficiently when the hypertree-width is bounded. [19] proposed an algorithm for solving Asymmetric Distributed Constraint Satisfaction problems (ADisCSP), in order to find approximate NE for hypergraphical games. When it comes to reflecting realistic games situations, local interactions between players is only one aspect. Another important feature of games is that preferences of players may not always be easily quantified. Sometimes, only an ordinal ranking of joint strategies can be reasonably expressed by "players". Pure NE are hopefully invariant to the quantitative embedding of ordinal preference scales. However, mixed-equilibria are sensitive to non-linear transformations of the preference scales of players, which makes usual game theory unable to easily tackle ordinal preferences over joint strategies. Therefore, *Ordinal games* [3] have been studied as a framework to tackle games with ordinal preferences. However, until recently there has been little advancement in the analysis of equilibria in ordinal games. [3] studied only pure strategies in ordinal games. Then, a definition of mixed strategies has been recently proposed in the possibility theory framework [1]. In the same line, [8] have proposed the definition of randomization over actions using possibilistic approaches to study and compare both qualitative and quantitative equilibrium concepts based on the Sugeno integral and Choquet integral [7]. However, to our knowledge, all works dedicated to the study of ordinal games are limited to normal-form games, while the two aspects of compactness and ordinal preferences occur naturally in human elicited games situations. Our goal is to overcome the lack of solution concepts and algorithms for compactly represented ordinal games.

The contributions of the present paper are fourfold: (i) We give the first definition of *Ordinal Graphical Games* (OGG), *Ordinal Hypergraphical Games* (OHG) and *Ordinal Polymatrix Games* (OPG). These definitions allow, in some cases, an exponentially more compact representation of ordinal games than in [1], for example. We also study both pure and possibilistic mixed NE in these games. (ii) We show that, as for cardinal graph-based games, deciding whether a pure NE exists is NP-complete. (iii) We propose and implement solution approaches for finding pure and mixed NE for graph-based ordinal games, the algorithm computing mixed-NE being polytime in the game description. (iv) We end the paper with an experimental study.

2 Background

2.1 Extensive and Compact Representations of Normal Form Games

A *normal form game* represents strategic interactions between players with conflicting objectives. Extensive normal form games representations are exploited to compute equilibrium strategies between players. A normal form game is defined as follows [18]:

Definition 1 (Normal form game). *A normal form game is a triple* $G = \langle N, \mathcal{A}, U \rangle$*:*

- $N = \{1, ..., n\}$ *is a set of n players.*
- $\mathcal{A} = \mathcal{A}_1 \times \ldots \times \mathcal{A}_n$*: $\mathcal{A}_i$ is a set of strategies available to player i. $a = (a_1, \ldots, a_n)$ denotes a joint strategy.*
- $U = \{u_i : \mathcal{A} \to \mathbf{R}\}_{i \in N}$ *is a set of real-valued utility functions.*

The classical definition of pure NE in a normal-form game is the following:

Definition 2 (Pure Nash equilibrium). *Let $G = \langle N, \mathcal{A}, U \rangle$ be a normal form game. A pure NE is a strategy $a^* \in \mathcal{A}$ such that $u_i(a^*) \geq u_i(a_i, a^*_{-i})$, $\forall i \in \{1, ..., n\}$, $\forall a_i \in \mathcal{A}_i$, where $a^*_{-i} =_{def} (a^*_1, \ldots, a^*_{i-1}, a^*_{i+1}, \ldots, a^*_n)$.*

Extensive normal form games expressions are unable to model games with more than dozen of players (utility tables representations are exponential in the number of players). Fortunately, in realistic games situations with many players, interactions are often only "local". The utility of players only depends on the strategies chosen by few other players. Compact representations of games have thus been largely studied.

In this paper, we are particularly interested in three models of compactly-represented normal-form games, based on graph theory: *graphical games* [11], *polymatrix games* [20] and *hypergraphical games* [16].

These three frameworks represent normal form games $\langle N, \mathcal{A}, \bar{U} \rangle$, where the utility functions of players $\bar{U} = \{\bar{u}_i : \mathcal{A} \to \mathbf{R}\}_{i \in N}$ have some particular structure:

- In a graphical game the local utility functions of players are defined by: $\bar{U} = \{\bar{u}_i : \mathcal{A}_{M_i} \to \mathbf{R}\}_{i \in N}$, where $i \in M_i \subseteq N, \forall i \in N$. These local utility functions concisely represent (when $|M_i| < n$) the utility functions of players in the corresponding normal form game. These are defined by $U = \{u_i : \mathcal{A} \to \mathbf{R}\}$, where

$$u_i(a) = \bar{u}_i(a_{M_i}), \forall i \in N, \forall a \in \mathcal{A}. \tag{1}$$

- In hypergraphical games the utility function of any player is a sum of local utility functions over subgames involving only few players. There are K subgames and $N^k \subseteq N, \forall k = 1..K$, is the set of players involved in subgame k. The local utility functions of player i are defined as: $\bar{U}_i = \{\bar{u}_i^k : \mathcal{A}_{N^k} \to \mathbf{R}\}_{i \in N^k}$. In the corresponding normal form game, global utility functions are defined as $U = \{u_i : \mathcal{A} \to \mathbf{R}\}$, where

$$u_i(a) = \sum_{\substack{k \in \{1,\ldots,K\} \\ i \in N^k}} \bar{u}_i^k(a_{N^k}), \forall i \in N, \forall a \in \mathcal{A}. \tag{2}$$

– In a polymatrix game, the local utility functions of players are defined by: $\bar{U} = \{\bar{u}_{ij} : \mathcal{A}_{\{i,j\}} \to \mathbf{R}\}_{(i,j)\in E\subseteq N^2}$ where E is a set of pairs of players involved in 2-player games. In the corresponding normal form game, global utility functions are defined as $U = \{u_i : \mathcal{A} \to \mathbf{R}\}$, where

$$u_i(a) = \sum_{j,(i,j)\in E} \bar{u}_{ij}(a_{\{i,j\}}), \forall i \in N, \forall a \in \mathcal{A}. \tag{3}$$

2.2 Ordinal Games Within the Possibility Theory Framework

[1] have introduced the definition of possibilistic mixed strategies in ordinal games. These definitions are based on the possibilistic decision theory framework. First, we give an overview of the possibility theory framework. Possibility theory [4] can be seen as a qualitative counterpart to probability theory. The basic concept in possibility theory is the notion of *possibility distribution* π. It is a mapping from a set of states S to a finite ordered scale $L = \{0_L < \ldots < 1_L\}$, equipped with the order-reversing function $\nu : L \to L$. π gives some knowledge about state $s \in S$: $\pi(s) = 1_L$ indicates that s is totally plausible, $\pi(s) = 0_L$ means that s is impossible and $\pi(s) > \pi(s')$ implies that s is more plausible than s'. π is assumed to be normalized: there is at least one completely possible state (s^* such that $\pi(s^*) = 1_L$). Assuming π, the possibility $\Pi(E)$ and the necessity $N(E)$ of any event $E \subseteq S$ can be computed: $\Pi(E) = \sup_{s\in E} \pi(s)$ determines to what extent E is *consistent* with the knowledge expressed by π whereas $N(E) = \nu\left(\Pi(\bar{E})\right) = \nu\left(\sup_{s\notin E} \pi(s)\right)$ evaluates to what extent $\neg E$ is inconsistent, hence, it determines the certitude level of E implied by knowledge π.

In light of qualitative (possibilistic) decision problems under uncertainty, where each result is assessed by an ordinal utility function $\mu : S \mapsto \Delta$, [4,5] have introduced qualitative pessimistic utility (U^{pes}), which is a counterpart to von Neumann and Morgenstern's [18] expected utility:

$$U^{pes}(\pi) = \min_{s\in S} \; max(\nu(\pi(s)), \mu(s)) \tag{4}$$

U^{pes} generalizes the Wald criterion and determines to what degree it is certain (i.e., according to measure N) that μ achieves a good utility. While pure NEs are similar in ordinal and cardinal games, ordinal games do not admit stochastic mixed strategies, since one cannot compute the mathematical expectation of a probability distribution over ordinal rewards. However, possibilistic mixed strategies can be considered as a qualitative counterpart to probabilistic mixed strategies in cardinal games and have been justified in terms of equilibria in ordinal games, in [1].

Definition 3 (Ordinal game). *An ordinal game $\mathcal{OG}$ is a tuple $\langle N, (\mathcal{A}_i)_{i\in N}, (\mu_i)_{i\in N}\rangle$:*

– $N = \{1, ..., n\}$ *is a set of n players.*
– $\mathcal{A} = \mathcal{A}_1 \times ... \times \mathcal{A}_n$: $\mathcal{A}_i$ *is a set of actions available to player i.*
– L *is a finite ordinal scale.*
– $\mu = \{\mu_i : \mathcal{A} \mapsto L\}_{i\in N}$ *is a set of ordinal utility functions.* $\mu_i(a)$ *is the ordinal utility of player i in the ordinal game when the joint strategy of players is a.*

Example 1 (Ordinal game). Assume two farmers own neighbour fields. Each farmer decides what to sow in her field. The set of possible crops is: *Wheat (W)* or *Organic Wheat (OW)*. Each farmer has to be cautious in her choice of crop because sowing organic crop near to non-organic crops reduces the profit of the organic crop. Throughout the examples we consider the following unique ordinal satisfaction and uncertainty scale: scale $L = \{0, ..., 4\}$. Emojis (😨, 😟, 😐, 🙂, 😃) are used to distinguish satisfaction from uncertainty levels, but both sets are in bijection. The ordinal utilities of farmers are given in the following table:

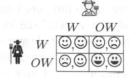

Since the concept of pure NE is ordinal in nature, its definition is the same in the possibilistic framework as in the classical framework:

Definition 4 (Pure NE in ordinal games). *Let us consider an ordinal game $\mathcal{OG} = \langle N, (\mathcal{A}_i)_{i \in N}, (\mu_i)_{i \in N} \rangle$. $a^* \in \mathcal{A}$ is a pure NE of $\mathcal{OG}$, iff: $\mu_i(a_i^*, a_{-i}^*) \geq \mu_i(a_i, a_{-i}^*)$, $\forall i \in N, \forall a_i \in \mathcal{A}_i$.*

Note that, in the qualitative case, a pure NE verifies: $\exists a^* \in \mathcal{A}$ s.t. $\pi(a_i^*) = \max(L), \forall i \in N$ and $\pi(a_i) = \min(L), \forall a_i \neq a_i^*, \forall i \in N$.

Example 2 (Cont. Example 1). One can check that the ordinal game has two pure NEs: (W,W) and (OW,OW). In these NE, both farmers are somehow satisfied (levels 3 and 4 $\in L$) and have no incentive to deviate.

The concept of mixed strategy in the possibility theory framework [1] is described as a *possibility distribution* over the alternatives of player i, i.e., $\pi_i : \mathcal{A}_i \mapsto L$. Hence, π_i is a ranking over the options included in $\mathcal{A}_i$, showing a player's preferences. π_i can also be usefully interpreted by other players as a likelihood of play of player i, i.e., a ranking of the options that player i is likely to play. As usual, π_i is normalized, i.e., $\max_{a_i \in \mathcal{A}_i} \pi_i(a_i) = 1_L$. A joint possibilistic mixed strategy verifies:

$$\pi(a) = \min_{i \in N} \pi_i(a_i), \forall a = (a_1, ..., a_n) \in \mathcal{A}. \tag{5}$$

The pessimistic possibilistic decision criterion [4] is used to evaluate the utility of π to player i:

$$\mu_i^{PES}(\pi) = \min_{a \in \mathcal{A}} \max(\nu(\pi(a)), \mu_i(a)). \tag{6}$$

where $\nu : L \to L$ is the order-reversing function of L.

A (least specific) Possibilistic Mixed Equilibrium (ΠME) is defined as a set $\pi^* = (\pi_1^*, ..., \pi_n^*)$ of normalized possibility distributions expressing individual preferences, where no player has incentive to deviate unilaterally from her strategy.

Definition 5 (Possibilistic Mixed Equilibrium (ΠME)). *For a given ordinal game $\mathcal{OG} = \langle N, (\mathcal{A}_i)_{i \in N}, (\mu_i)_{i \in N} \rangle$, $\pi^* = (\pi_1^*, ..., \pi_n^*)$ is a ΠME iff it satisfies, for any possibilistic mixed strategy π, $\mu_i^{PES}(\pi^*) \geq \mu_i^{PES}(\pi_i, \pi_{-i}^*)$, $\forall i \in N, \forall \pi_i : \mathcal{A}_i \to L$, where $\pi_{-i}^* =_{def} (\pi_1^*, ..., \pi_{i-1}^*, \pi_{i+1}^*, ..., \pi_n^*)$.*

Example 3 (Cont. Example 1). Let us consider possibilistic mixed strategy (π_1^*, π_2^*), where $\pi_1^*(W) = \pi_2^*(W) = 4$ and $\pi_1^*(OW) = \pi_2^*(OW) = 1$.

One can check that $\mu_1^{PES}(\pi^*) = \mu_2^{PES}(\pi^*) = 3$ and that this is a possibilistic equilibrium in the sense of [1]. No player can improve her pessimistic utility by changing her mixed strategy.

Mixed strategies in cardinal games are evaluated by their expected utility, reflecting the assumption that games are repeated and that utility compensate. In the ordinal framework, mixed strategies can be seen as refining "worst-case" strategies (i.e. minimax strategies). [1] have proposed a different interpretation of these: A player's own strategy is a form of "commitment to play" she announces to other players (I will preferably play actions with highest possibility degree but I may play different actions as well). Then an equilibrium results from different rounds of discussions during which players successively lower the plausibility of playing actions, until no one feels better of changing her current strategy.

A ΠME is generally not unique. A *least specific* ΠME is one where the utility of any player can only decrease when it unilaterally transforms its possibility distribution into a less specific one. The interest of a *least specific* ΠME is that it sets only the lightest possible constraints on every players' strategies. One can check that in the previous example, π^* is a least specific ΠME. By replacing π_1^* with $\pi_1(W) = 4$, $\pi_1(OW) = 2$, we get $\mu_1^{PES}(\pi_1, \pi_2^*) = 2 < \mu_1^{PES}(\pi^*)$. [1] have proved that a least specific ΠME for an ordinal game could be computed in polynomial time, and have provided a polynomial time algorithm to compute a ΠME through successive improvement of strategies.

3   Ordinal Graphical, Hypergraphical and Polymatrix Games

In the previous section, we have recalled the framework of possibilistic ordinal game theory, which has been proposed to model, in particular, human elicited game situations, where preferences between strategies are usually best modelled in "ordinal" ways. Another important features of human-elicited games is a need for compactness of expression. One cannot easily rank joint strategies where actions of many players are involved. In particular, the notion of local interactions is worth exploring in the context of ordinal games as well. This section introduces the ordinal counterparts to graphical, hypergraphical and polymatrix games.

3.1   Motivating 'Farmers' Example

We briefly present a toy problem which illustrates both ordinal and graphical aspects of the game. Let assume that we have n farmers each with a unique field arranged in the form of a grid. M_i denotes the set of farmers (including farmer i), which actions may influence the utility of i. Typically, M_i will include the (at most four) nearest neighbours of i. Each year and according to her subjective preferences, each farmer decides what to sow in her field. The set of possible crops is $\{Meadow(M), Wheat(W), Canola(C), Organic\ Wheat(OW), Organic\ Canola(OC)\}$. The utility of any farmer i aggregates *production, biodiversity* and *pollination* ordinal utility functions. If there are n players, this game requires $O\left(n|A|^5\right)$ space to represent as an (ordinal) graphical game or $O\left(n|A|^2\right)$ space as an ordinal polymatrix game, instead of $O\left(n|A|^n\right)$ space as a normal form ordinal game.

3.2 Ordinal Graph-Based Games: Definitions

In this section, we provide definitions of three new ordinal games classes: graphical (OGG), hypergraphical (OHG) and polymatrix (OPG). These are defined in the framework of possibilistic game theory, by considering local ordinal utility functions $\bar{\mu}_i$.

Definition 6 (OGG utility functions). *In an OGG the local utility functions of players are defined as:*

$$\bar{\mu} = \{\bar{\mu}_i : \mathcal{A}_{M_i} \to L\}_{i \in N} ,$$

where $M_i \subseteq N, \forall i \in N$ is a subset of players. In the corresponding ordinal normal form game, global utility functions are defined as $\mu = \{\mu_i : \mathcal{A} \to L\}_{i \in N}$, where

$$\mu_i(a) = \bar{\mu}_i(a_{M_i}), \forall i \in N, \forall a \in \mathcal{A}.$$

In the case of ordinal graphical games, the analogy with cardinal games is direct, since utility functions μ_i require no aggregations of local utilities.

In an OHG, local utilities are combined using an ordinal aggregator. In the following, the local utilities are aggregated through a *minimum* operator, which is coherent with preference aggregation in an adversarial framework.

Definition 7 (OHG utility functions). *In an ordinal hypergraphical game the local utility functions of players are defined as:*

$$\bar{\mu}_i = \left\{\bar{\mu}_i^k : \mathcal{A}_{N^k} \to L\right\}_{i \in N^k} ,$$

where $N^k \subseteq N, \forall k = 1..K$ (K is the number of subgames).

In the corresponding ordinal normal form game, global utility functions are defined as $\mu = \{\mu_i : \mathcal{A} \to L\}$, where

$$\mu_i(a) = \min_{\substack{k \in \{1,\dots,K\} \\ i \in N^k}} \bar{\mu}_i^k(a_{N^k}), \forall a \in \mathcal{A}. \tag{7}$$

Note that a given OHG can be easily cast as an OGG, by defining $M_i = \cup_{k,i \in N^k} N^k, \forall i \in N$ and

$$\bar{\mu}_i(a_{M_i}) = \min_{\substack{k \in \{1,\dots,K\} \\ i \in N^k}} \bar{\mu}_i^k(a_{N^k}), \forall i, a_{M_i} \in \mathcal{A}_{M_i}. \tag{8}$$

Finally, OPG can be defined as specific cases of OHG where each agent is involved in simultaneous 2-player games. Formally, once again we consider a specific ordinal utility function:

Definition 8 (OPG utility functions). *In an OHG, the local utility functions of players are defined as:*

$$\bar{\mu} = \{\bar{\mu}_{ij} : \mathcal{A}_{\{i,j\}} \to L\}_{(i,j) \in E \subseteq N^2} ,$$

where E is a set of edges defining the 2-player games.

In the corresponding ordinal normal form game, global utility functions are defined as:

$$\mu_i(a) = \min_{j,(i,j) \in E} \bar{\mu}_{ij}(a_i, a_j), \forall a \in \mathcal{A}.$$

4   Computing Pure NE in Graph-Based Ordinal Games

4.1   Hardness

Informally, a pure NE in a graph-based ordinal game is a joint action $a^* \in \mathcal{A}$ from which no player has an incentive to deviate unilaterally.

In the case of OGG, the definition of pure NE may exploit the *locality* of possibilistic utility functions:

Proposition 1 (Pure NE in OGG).   *Let $\mathcal{G} = \langle N, \mathcal{A}, \{M_i\}_{i \in N}, \bar{\mu} \rangle$ be an ordinal graphical game. $a^* \in A$ is a pure NE of $\mathcal{G}$, iff: $\forall i \in N, \forall a_i \in \mathcal{A}_i$.*

$$\bar{\mu}_i(a_i^*, a_{M_i - \{i\}}^*) \geq \bar{\mu}_i(a_i', a_{M_i - \{i\}}^*).$$

Proof sketch. The proposition results from Definition 4 and Definition 6.   □

Recall that an OHG can be cast as an OGG. According to Proposition 1 and Eq. 8, we can prove the following corollary:

Proposition 2. a^* *is a pure NE of an OHG iff* $\forall i \in N, \forall a_i \in \mathcal{A}_i$,

$$\min_{\substack{k \in \{1, \ldots, K\} \\ i \in N^k}} \mu_i^k(a_i^*, a_{N^k \setminus \{i\}}^*) \geq \min_{\substack{k \in \{1, \ldots, K\} \\ i \in N^k}} \mu_i^k(a_i, a_{N^k \setminus \{i\}}^*).$$

In the same way, an OPG is an OHG where all subgames contain exactly two players.

Proposition 3. a^* *is a pure NE of OPG iff* $\forall i \in N, \forall a_i \in \mathcal{A}_i$,

$$\min_{j, (i,j) \in E} \bar{\mu}_{ij}(a_i^*, a_j^*) \geq \min_{j, (i,j) \in E} \bar{\mu}_{ij}(a_i, a_j^*).$$

We now show that deciding the existence of a pure NE in ordinal graphical games is a difficult problem, even in a very restricted setting.

Proposition 4.   *Deciding whether an ordinal graphical game has a pure NE is NP-complete. Hardness holds even if $\mathcal{G}$ has 3-bounded neighborhood, and the number of actions is fixed.*

Proof sketch.   **Membership.** We can decide the membership by guessing a joint action a and verifying that a is a NE. Clearly the latter task takes time polynomial in the size of the game.

Hardness. [6] have shown that deciding the existence of a pure NE in (usual) graphical games is NP-complete even for 3-bounded neighborhood games, i.e., where each player has 3 neighbors at most. Now, just note that pure NE in ordinal and cardinal graphical games are the same notion when no utility functions aggregations are performed (i.e. when there is a single ordinal utility function for each player). Thus, if one is given a graphical game as input, it can be transformed (in polynomial time) into an ordinal graphical game, by plunging the utilities into an ordinal scale. The pure NE are then equal in both cardinal and ordinal problems.   □

We then prove that deciding the existence of pure NE in OPG and OHG is also NP-complete, which is less obvious at first glance:

Proposition 5. *Deciding whether an OPG or an OHG admits a pure NE is NP-complete.*

Proof sketch. The membership part is easy for both OPG and OHG. The hardness part, for OPG, relies on a reduction of the K-INDEPENDENT SET problem[1]. Hardness for OHG results from the hardness result for OPG.   □

4.2 CSP Modeling in Graph-Based Ordinal Games

A Constraint Satisfaction Problem (CSP) is a triple $(\mathcal{X}; D; \mathcal{C})$ where $\mathcal{X}$ is a set of variables, D is the set of domains of these variables and $\mathcal{C}$ is a set of constraints over variables values. Modeling a graph-based ordinal game as a CSP is useful in the sense that we can take advantage of existing CSP solvers in order to find pure NE(s) within reasonable time. In this section, we show how to model Ordinal Graphical Games (and OHG and OPG) as a CSP and show that the solutions of the induced CSP are pure NE for the original game. Note that [6] proposed a CSP modeling of (cardinal) graphical games in order to find pure NE. The concepts of pure NE are identical in cardinal and ordinal graphical games since utilities in each games are not combined, unlike in polymatrix and hypergraphical games. Still, our CSP[2] model is different from that of [6]. Indeed, the fact that local utilities are aggregated by a minimum operator and not a *sum* leads to a different reduction (a simpler one, in fact), to a different problem. This different form allows it to be extended to OHG and OPG.

Definition 9 (CSP modeling). *Let $\mathcal{G} = \langle N, \mathcal{A}, \{M_i\}_{i \in N}, \bar{\mu} \rangle$ be an ordinal graphical game. We define the CSP model $(\mathcal{X}; D; \mathcal{C})$ of $\mathcal{G}$ as follows:*

- $\mathcal{X} = \{A_1, ..., A_n\}$; *each variable A_i represents the action of player i ($N = \{1, ..., n\}$).*
- $D = \mathcal{A}_1 \times ... \times \mathcal{A}_n$; $\mathcal{A}_i$ *is the domain of variable A_i, that is the set of allowed strategies of player i.*
- $\mathcal{C} = \{C_{i,a_i'}, i \in \{1, ..., n\}, a_i' \in \mathcal{A}_i\}$, *can be seen as binary-valued functions $C_{i,a_i'} : \mathcal{A}_{M_i} \to \{0, 1\}, \forall i, a_i'$, satisfying:*
  $C_{i,a_i'}(a_{M_i}) = 1$ *iff* $\bar{\mu}_i(a_{M_i}) \geq \bar{\mu}_i(a_i', a_{M_i - i}), \forall i, a_i', a_{M_i}$.

Note that there are $\sum_{i=1}^{n} |A_i|$ constraints $C_{i,a_i'}$ (of arity $|M_i|$). Remark also that $C_{i,a_i'}(a_{M_i})$ is satisfied if and only if a_i' is a non-dominated action of player i. So, obviously, the following proposition holds:

---

[1] The proofs are omitted for the reason of brevity; they can be found here (anonymous address): Proofs.

[2] Our CSP model uses integer-valued variables. In our actual implementation, we used binary variables x_{i,a_i}, where $x_{i,a_i} = 1$ iff $A_i = a_i$, for any pair (i, a_i) and the constraints were changed accordingly. Still, the two problems are equivalent and we describe here the "non-binary" model, which is more "readable".

Proposition 6. $a^* = (a_1^*, ..., a_n^*) \in \mathcal{A}$ is a pure NE of ordinal graphical game $\mathcal{G}$ if and only if it is a solution of CSP $(\mathcal{X}; D; \mathcal{C})$.

Proof sketch. The proof directly results from the definition of the constraints in terms of non-dominated strategies.   □

Since ordinal hypergraphical and polymatrix games can be represented as ordinal graphical games, they can also be modelled as CSP. However, note that in the conversion to a graphical game, conciseness may be lost. In the extreme case of a polymatrix game with relations between every pairs of players, the representation of the resulting ordinal graphical game is exponentially larger than that of the original game. Fortunately, in the case of ordinal hypergraphical/polymatrix games, each constraint $C_{i,a_i'}$ of the corresponding ordinal graphical game can be equivalently replaced in the CSP with an equivalent set of constraints (of reasonable sizes):

$$\mathcal{C}_{i,a_i'} = \left\{ C_{i,a_i'}^k (a_{N^k}) \right\}_{i \in N^k}, \text{ where } C_{i,a_i'}^k (a_{N^k}) = 1 \text{ iff } \mu_i^k (a_{N^k}) \geq \mu_i^k (a_i', a_{N^{k-i}}).$$
(9)

Indeed, recall that $\bar{\mu}_i(a_{M_i}) = \min_{\substack{k \in \{1,...,K\} \\ i \in N^k}} \mu_i^k (a_{N^k})$. Then it directly follows that:

$$C_{i,a_i'}(a_{M_i}) = 1 \text{ iff } \left(C_{i,a_i'}^k (a_{N^k}) = 1, \forall k \text{ s.t. } i \in N^k \right).$$

Thus, for both hypergraphical and polymatrix games, the search for pure NE can be performed through modelling as a CSP of similar size as that of the original problem.

5 Possibilistic Mixed Equilibria in Ordinal Graph-Based Games

In this section, we show that computing a possibilistic mixed equilibrium in OGG (and OPG and OHG) takes polynomial time in the size of the game. To start with, let an OGG $\mathcal{G} = \langle N, \mathcal{A}, \{M_i\}_{i \in N}, \bar{\mu} \rangle$ be given. Let us assume that $\pi = \{\pi_i\}_{i=1..n}$ is a mixed possibilistic strategy over $\mathcal{A} = \mathcal{A}_1 \times ... \times \mathcal{A}_n$. As for ordinal games [1], the utility of π in an OGG is measured using the pessimistic criterion μ^{pes}. It can be shown that the expression of μ^{pes} decomposes according to the structure of the graphical game.

Proposition 7. *The pessimistic utility for player i of a joint mixed possibilistic strategy in an OGG, OPG or OHG is:*

$$\mu_i^{pes}(\pi) = \min_{a_{M_i}} \max \left(\max_{j \in M_i} \nu(\pi(a_j)), \bar{\mu}_i(a_{M_i}) \right).$$
(10)

Proof sketch: The proposition results from the expression of $\mu_i(a) = \bar{\mu}_i(a_{M_i}), \forall i \in N, \forall a \in \mathcal{A}$ as well as from the decomposability of $\pi(a_{M_i}) = \min_{j \in M_i} \pi_j(a_j)$, through elementary computations.   □

The subset of *dominated actions* for player i, $D_i \subseteq \mathcal{A}_i$ can be defined as follows:

$$D_i = \{a_i \in \mathcal{A}_i \text{ s.t. } \mu_i^{pes}(a_i, \pi_{-i}) \leq \mu_i^{pes}(\pi)\}.$$

Now, we can prove that,

Proposition 8. *The computation of a mixed ΠME in OGG, OHG and OPG is polynomial in the size of the game.*

Proof. First, note that $\mu_i^{pes}(\pi)$ in Proposition 7 takes polynomial time to compute for OGG, OHG and OPG, given that the expression of $\bar{\mu}_i(a_{M_i})$ decomposes in OHG and OPG. The computation of a possibilistic mixed equilibrium in an ordinal game requires iterative calls to an IMPROVE procedure (Algorithm 1), as shown in [1]. Basically, the solution algorithm proposed in [1] in order to compute a mixed equilibrium consists in starting with uniform possibilistic strategies for every players $(\pi_i(a_i) = 1_L, \forall i, a_i)$ and then "improving" the current mixed strategy of a single player of the game, by applying Algorithm 1. At every time steps, a new player is chosen, which strategy is improved. The algorithm stops when no player can see her strategy improved. It is shown in [1] that the algorithm converges to a possibilistic mixed strategy in time polynomial in the expression of the ordinal game. Let us show that the same result holds in the case of ordinal graph-based games, First, note that one call to the IMPROVE procedure takes polynomial time, since $\mu_i^{pes}(\pi)$ takes polynomial time to compute[3]. Now, we need to

---

Algorithm 1. IMPROVE procedure

1: Input: $(\mathcal{G}, \pi_{loc}, i)$
2: Output: π_{loc}
3: $\pi \leftarrow \pi_{loc}$
4: **if** $(\Pi_i(\mathcal{A}_i \setminus D_i) = 1_L)$ **and** $(\mu_i^{pes}(\pi) < 1_L)$ **then**
5:    **for** $a_i \in D_i$
6:       $\pi_i(a_i) \leftarrow \min\left(\pi_i(a_i), \underline{\nu(\mu_i^{pes}(\pi))}\right)$
7:    **end for**
8: **end if**
9: $\pi_{loc} \leftarrow \pi$

---

prove that a possibilistic mixed equilibrium is reached within a polynomial number of calls to IMPROVE. This results directly from the observation made in [1], that each improvement reduces strictly the possibility $\pi_i(a_i)$ of one alternative of one player i. Since at least one alternative for each player should keep possibility 1_L, the number of iterations of the algorithm is bounded by:

$$N_{iter} = \sum_{i=1..n} |L|(|\mathcal{A}_i| - 1) \leq n|L|na_{max}; na_{max} = \max_{i=1..n} |\mathcal{A}_i|. \tag{11}$$

$\square$

6 Experimental Study

We empirically evaluated the time execution of pure and mixed NE computation in various ordinal games. To this end, we built and solved CSP models using the CHOCO solver [9], providing a single pure NE or a proof of non-existence. The mixed NE

---

[3] $\underline{\nu(\mu_i^{pes}(\pi))}$ is, by definition, the degree immediately below $\nu(\mu_i^{pes}(\pi))$ in L.

computation algorithm (PME) was implemented in MATLAB. All experiments were performed on an Intel(R) Core(TM) i5-7200U CPU, 2.5 Ghz processor, with 8 Gb RAM memory, 64 bits architecture and under Windows 10 OS. Both algorithms were tested on a dataset of problems, including randomly generated problems and "Farmers games":

- **Randomly generated Games.** The structure of OHGs were generated randomly, by controlling the number of players n, the number of actions per player, m, the size of hyperedges, s ($s = 2$ in the case of OPG). The number of local games, K, was computed from a connectivity parameter, $c = \frac{K}{NPG}$, where $NPG = \frac{n!}{s!(n-s)!}$ is the maximal number of distinct subgames. The local games correspond to four types of games included in the Gamut suite [14]: "Chicken Games (CG)", "Compound Games (COG)", "Random Games (RG)" and "Dispersion Games (DG)". Local games where generated using Gamut, then their utilities were made ordinal. Every local games of a game are of the same type. The following combinations of parameters were considered: (i) $m = 2$ and $(n, c) \in \{3, 4, 5..., 15\} \times \{0.4, 0.8\}$ and (ii) $n = 8$ and $(m, c) \in \{2, ..., 7\} \times \{0.4, 0.8\}$. For every combinations of parameters, we solved 100 randomly generated games and we computed the average solution times.
- **Farmers games.** (Defined in Sect. 3.1). In these games, we vary the dimension of the grid by considering grids of dimensions 2×2, 2×3, 2×4, 3×3, 4×3 and 4×4. Results on our tested games are shown in Figs. 1, 2, 3 and 4, respectively.

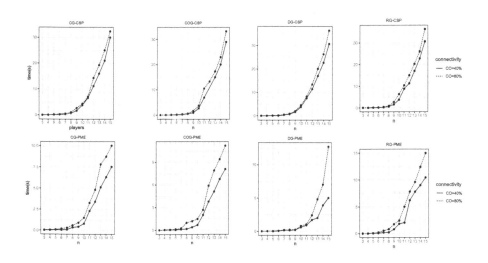

Fig. 1. Avg. runtime on ordinal hypergraphical games

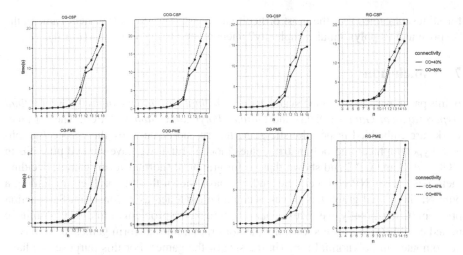

Fig. 2. Avg. runtime on ordinal polymatrix games

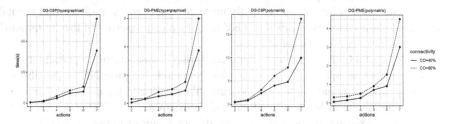

Fig. 3. Avg. runtime on Dispersion games with $n = 8$

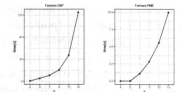

Fig. 4. Avg. runtime on Farmers games

From Figs. 1 and 2, we notice that, for all games, the CSP algorithm is able to return a proof of existence or non existence of pure NE. Besides, the PME algorithm always returns a possibilistic mixed equilibrium efficiently. As theoretically expected, pure NE existence is experimentally harder to prove than mixed NE computation, especially for "difficult" games (more players, more actions per player). Another result of the experimental study is that ordinal polymatrix games seem easier to solve than ordinal hypergraphical games. In addition, the connectivity of games seems to have only a second-order impact on experimental time-complexity. Figure 3 shows that the number of actions impacts the execution time for both algorithms. This conclusion holds for all tested games. For the most difficult games we considered (Farmers games), the "exponential vs polynomial complexity" phenomenon really shows up (Fig. 4).

7 Conclusion

In this paper, we have introduced and defined *Ordinal Graphical Games* (OGG) *Ordinal Hypergraphical Games* (OHG) and *Ordinal Polymatrix Games* (OPG). These frameworks are embedded in possibilistic game theory and inspired by the classical graphical, hypergraphical and polymatrix games models. First, we have studied pure NE in OGG, OHG and OPG and shown that, as for graphical (normal-form) games, deciding their existence is NP-complete. Second, we have shown that the problem of finding a pure NE in ordinal graph-based games could be modelled as a Constraint Satisfaction problem (CSP). Finally, we focused our attention on the problem of finding possibilistic mixed equilibria. We have shown that, as for ordinal normal-form games, a ΠME can be computed in polynomial time (in the size of the game). For this purpose, we have proposed an adapted version of the current algorithm for ordinal games and we have shown that it runs in polynomial time. This result is surprising at first glance, since OGG, OHG and OPG admit exponentially more compact representations than normal form ordinal games. However, this result is due to the nice properties of the "minimum" aggregator used to combine local utilities.

The choice of a CSP modelling to compute pure NE was natural, in particular since it provides a natural and easy way to model the search for pure NE in possibilistic games. It is even more natural than in the cardinal case, due to the use of the *minimum* operator to aggregate utilities in local games. Furthermore, the CSP approach allows to make use of existing efficient solvers and do not require to develop specific solution algorithms. However, in the context of (cardinal) graphical games, the family of TreeNash/Nashprop algorithms [10] has been advocated to compute exact/approximate mixed NE, in particular for graphical games where the underlying graphical structure is a tree. These algorithms require, from a conceptual point of view, to propagate messages between players in the form of continuous multivariate functions $T : [0, 1]^k \rightarrow \{0, 1\}$, expressing "best responses" to mixed strategies. Since this is not possible in practice, approximate (discretized) or exact (exponential size) representations of these functions are propagated to compute equilibria. In ordinal graphical games, since the set of possibility distributions is a finite set, we may define similar message propagation algorithms where the messages are finite tables. This is an interesting avenue for further research. However, it remains to compare the efficiency of these message-passing algorithms to that of the ones we propose.

References

1. Ben Amor, N., Fargier, H., Sabbadin, R.: Equilibria in ordinal games: a framework based on possibility theory. In: International Joint Conferences on Artificial Intelligence (IJCAI 2017), pp. 105–111 (2017)
2. Chapman, A.C., Farinelli, A., de Cote, E.M., Rogers, A., Jennings, N.R.: A distributed algorithm for optimising over pure strategy Nash equilibria. In: Association for the Advancement of Artificial Intelligence (AAAI 2010), pp. 749–755 (2010)

3. Cruz, J.B., Simaan, M.A.: Ordinal games and generalized Nash and Stackelberg solutions. J. Optim. Theory Appl. **107**, 205–222 (2000). https://doi.org/10.1023/A:1026476425031

4. Dubois, D., Prade, H.: Possibility theory and its applications: a retrospective and prospective view. In: Della Riccia, G., Dubois, D., Kruse, R., Lenz, H.-J. (eds.) Decision Theory and Multi-Agent Planning. CICMS, vol. 482, pp. 89–109. Springer, Vienna (2006). https://doi.org/10.1007/3-211-38167-8_6

5. Dubois, D., Prade, H., Sabbadin, R.: Decision-theoretic foundations of qualitative possibility theory. Eur. J. Oper. Res. **128**, 459–478 (2001)

6. Gottlob, G., Greco, G., Scarcello, F.: Pure Nash equilibria: hard and easy games. J. Artif. Intell. Res. **24**, 357–406 (2005)

7. Grabisch, M., Labreuche, C.: A decade of application of the Choquet and Sugeno integrals in multi-criteria decision aid. Q. J. Oper. Res. **6**, 1–44 (2008). https://doi.org/10.1007/s10288-007-0064-2

8. Hosni, H., Marchioni, E.: Possibilistic randomisation in strategic-form games. Int. J. Approx. Reas. (IJAR 2019) **114**, 204–225 (2019)

9. Jussien, N., Rochart, G., Lorca, X.: Choco: an open source Java constraint programming library (2008)

10. Kearns, M.: Algorithmic game theory, chap. 7. In: Graphical Games, pp. 159–180. Cambridge University Press (2007)

11. Kearns, M., Littman, M.L., Singh, S.: Graphical models for game theory. In: Uncertainty in Artificial Intelligence (UAI 2001), pp. 253–260 (2001)

12. Miller, D.A., Zucker, S.W.: Copositive-plus Lemke algorithm solves polymatrix games. Oper. Res. Lett. **10**, 285–290 (1991)

13. Nash, J.: Non-cooperative games. Ann. Math. 286–295 (1951)

14. Nudelman, E., Wortman, J., Shoham, Y., Leyton-Brown, K.: Run the GAMUT: a comprehensive approach to evaluating game-theoretic algorithms. In: International Joint Conference on Autonomous Agents and Multiagent Systems (AAMAS 2004), pp. 880–887 (2004)

15. Ortiz, L.E., Kearns, M.: Nash propagation for loopy graphical games. In: Advances in Neural Information Processing Systems, pp. 817–824 (2003)

16. Papadimitriou, C.H., Roughgarden, T.: Computing correlated equilibria in multi-player games. J. ACM **55**, 14–46 (2008)

17. Simon, S., Wojtczak, D.: Constrained pure Nash equilibria in polymatrix games. In: Association for the Advancement of Artificial Intelligence (AAAI 2017), pp. 691–697 (2017)

18. Von Neumann, J., Morgenstern, O.: Theory of Games and Economic Behavior (1948)

19. Wahbi, M., Brown, K.N.: A distributed asynchronous solver for Nash equilibria in hyper-graphical games. In: European Conference on Artificial Intelligence (ECAI 2016), pp. 1291–1299 (2016)

20. Yanovskaya, E.B.: Equilibrium points in polymatrix games. Litovskii Matematicheskii Sbornik **8**, 381–384 (1968)

Real World Applications

On Relevance of Linguistic Summaries – A Case Study from the Agro-Food Domain

Anna Wilbik[1]($\boxtimes$) (iD), Diego Barreto[1], and Ge Backus[2]

[1] Information Systems, Eindhoven University of Technology,
Eindhoven, The Netherlands
a.m.wilbik@tue.nl
[2] Connecting Agri and Food, Uden, The Netherlands
g.backus@connectingagriandfood.nl

Abstract. We present an application of linguistic summaries in the agro-food domain. We focus on the relevance aspect. Using the interviews we determine which linguistic summaries are useful and appropriate for target users (farmers). The user evaluation with a TAM survey indicates that linguistic summaries allow farmers to understand quickly the past performance of their pig barns.

Keywords: Linguistic summaries · Relevance · Case study · Computing with words

1 Introduction

Recent developments in IoT technology allow more and more data to be collected. For instance, in agro-food domain sensor data can improve the crop yield and quality prediction [29,31] and optimization of food supply [34]. Those data can be analyzed through statistical analysis [13,15], machine learning methods [12] or visual analytics [32,40]. Visual analytics technology is often used for the descriptive analytics purposes and understanding the data. However, many dashboards fail to communicate efficiently and effectively to the user [11]. Also, novice professionals struggles to understand simple graphs, especially if the information is not clear-cut and visually prominent [33]. Therefore, presenting the information in another way than visual, such as verbal, may be beneficial for the user.

One of the ways to obtain verbalization of data is to employ the linguistic summarization approach. Linguistic summaries can automatically generate natural language like sentences with aim of capturing the essence of data [16,24]. Linguistic summaries have proven to be useful in several applications, e.g. retail [17] and eldercare [39].

Linguistic summaries are being developed and matured as the method by many researchers, for instance by proposing new quality measures, e.g., [8,17], or more efficient generation methods [10,18,28].

© Springer Nature Switzerland AG 2020
M.-J. Lesot et al. (Eds.): IPMU 2020, CCIS 1237, pp. 289–300, 2020.
https://doi.org/10.1007/978-3-030-50146-4_22

Yet, one of the remaining challenges is that often too many linguistic summaries are being generated [10]. Thus, it is important to determine which linguistic summaries are useful and appropriate for target users achieving a balanced trade-off between complexity and precision of information [28]. Naturally usefulness or relevance is very much context dependent. In this paper we tackle this issue, by proposing a method to detect relevant linguistic summaries in a case from the agro-food domain. We consider a case of a pig barn and a climate sensor that captures the conditions in the barn. We want to present to the farmers only the linguistic summaries that they consider useful and allowing them to take actions that improve their operations.

This paper is structured as follows. Next section describes the background on linguistic summarization, Sect. 3 describes the context of the case study and Sect. 4 the proposed the method to detect relevant summaries. Section 5 presents the results of the case study and is followed by the concluding remarks.

2   Background

In this paper we follow the approach of Yager [41], which was considerably advanced and then implemented by Kacprzyk [16], Kacprzyk and Yager [21], Kacprzyk et al. [22,23].

In this approach linguistic data summaries are quantified propositions with two possible protoforms (or templates):

- simple protoform:
$$Qy\text{'s are } P; \tag{1}$$

    e.g. *Most* cars are *new*
- extended protoform:
$$Q\ Ry\text{'s are } P; \tag{2}$$

    e.g. *Most new* cars are *fast*

where Q is the quantifier, P is the summarizer, and R is an optional qualifier, which are all modeled as fuzzy sets over appropriate domains.

The truth value, describing the validity of the summary, is the basic measure of the quality of the summary. Many methods for calculating the truth value have been proposed [8]. But the truth value is not the only quality measure of a linguistic summary. Kacprzyk et al. [17,23] proposed four additional measures, namely the degree of specificity, the degree of appropriateness, the degree of covering, and the length of the summary. Bugarin et al. [5] were differentiating between evaluating a single summary sentence and a set of summaries. They proposed several measures that capture aspects, such as coverage, length, and specificity. An overview of different quality criteria can be found in [8,28].

Those quality measures can be used to select a smaller set of the true summaries to be shown to the user. Three different approaches of employing the other quality measures can distinguished, namely using the thresholds [4,39], dominance of linguistic summaries [6] and aggregating the quality measures [9,17]. Yet, despite those methods, still too many linguistic summaries are obtained [10].

Linguistic summaries has been applied to different types of data: numerical [7,14,36], time series [19,20,30], sensor data [35,39], texts [37], videos [1–3] and processes [10,38] in different application domains, such as retail [17], eldercare [39] or finance [19].

3 Case Study in the Agro-Food Domain

In this section we describe the context of the case study as the context is very important for the relevance. We consider a case of a climate sensor installed in a pig barn. This sensor measures the temperature, air humidity and carbon dioxide (CO_2) inside the barn every 10 min. A dashboard was created to present the data collected by the sensor and some additional readings coming from the nearby weather station like outside air temperature and outside air humidity.

The dashboard initially showed the live gauges displaying the information about current values of inside temperature, humidity and CO_2. Also three time series plots were introduced to display the data (temperature, humidity and CO_2) over a longer period of time, typically 28 days. However, many farmers found this dashboard difficult to understand and deduce appropriate actions. To overcome this difficulty, it was decided to extend the dashboard with linguistic summaries.

Through discussions with the domain experts and a selected group of farmers (users) a few decisions were made. Firstly, it was decided to use only simple type protoforms, as they can provide the overview on the past. Moreover it was decided to create linguistic summaries about inside humidity, CO_2, and daily range of inside temperatures (maximum-minimum), as the variations in the temperature in the barn is more important that the actual values. Thirdly, three linguistic labels for each variable and five quantifiers were defined and validated. Their membership functions are depicted in Figs. 1 and 2, respectively. The quantifiers has been designed in such a way, for a given argument only one of them has a value of membership function higher than 0.7. This means that at most only one summary will be generated for a summarizer. By this, we will not present to the user both summaries as e.g. "Almost all CO_2 levels are too high" and "Most CO_2 levels are too high", as the second sentence is obsolete. Also each of the variables has desired values and undesired values. Temperature range and CO_2 has undesired values only on one end, while humidity has undesired values on both ends. At last it was decided also to display at most two summaries per variable for comprehensibility purposes [27].

Also, the farmers indicated that they want to see information that triggers them for an action. Therefore linguistic summaries like "almost all CO_2 values are OK" they don't find relevant and useful.

4 Method for Selecting Relevant Linguistic Summaries

The input from the domain experts and the farmers was used to propose a method that determined which of the true summaries can be considered as relevant and presented to the users. The most important feature of this case is

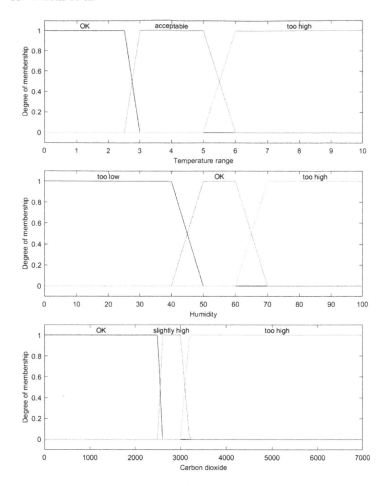

Fig. 1. Membership functions for the linguistic variables of temperature range, humidity and CO_2.

that there is a desired state and users are only interested when things are not as expected. When considering those two requirements we can distinguish two cases:

– desired case - when most (if not all) data are in the desired (ok) range
– undesired case - when majority of the data is not in the desired (ok) range

In order to develop the method that selects relevant summaries, some additional information from the users is needed. One of them is to rank of the linguistic labels for each variables. For the one tailed variables, like CO_2 levels, this is rather trivial task, the more undesired the value, the higher priority has the linguistic label. For the two tailed variables, like the humidity, where desired values are in the middle of the domain, setting the priority can be more challenging, as both extreme values can have same priorities or one of them is more

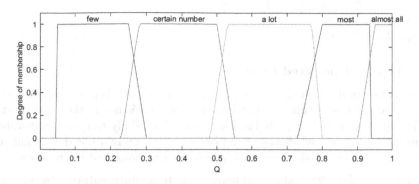

Fig. 2. Membership functions for quantifiers.

important than the other. We assume that priority of the qualifiers is positively correlated with the amount they describe, i.e. the bigger amount the qualifier describes, the higher the priority.

Second question is about what is more important to the user: summarizer or qualifier. For instance, consider the CO_2 level. The question is which sentence is more useful to the user:

- certain number of the CO_2 values is slightly high
- a few of the CO_2 values is too high

In the considered case all farmers have indicated the second summary as more useful, giving the higher weight to the summarizer than qualifier.

The proposed method consist of 5 steps:

- Step 1: Calculate all summaries with truth value $T > \theta_T$. We assumed θ_T of 0.7. For each linguistic value at most three summaries will be created.
- Step 2: Classify case as desired or undesired. This is done by looking at the summary with the biggest quantifier, in our case *a lot*, **most** or *almost all*. If the summarizer's label is a "desired" one, we consider a case as desired, otherwise it is undesired.
- Step 3: Prune all summaries with "desired" linguistic label. In our case, we used "OK" as "desired" linguistic label for all linguistic variables.
- Step 4: Order the summaries according to the priority. In our case this is according to the importance of the summarizer label.
- Step 5: Display summaries: maximally one for desired case and maximally two for undesired case.

5   Results

We will show now several examples of data with linguistic summaries obtained. Most of those examples are real data obtained from farms in the Netherlands from February 2018 to February 2019. A few dummy examples are shown to demonstrate extreme cases. We divided those examples in two groups:

- type 1: undesired case
- type 2: desired case

5.1 Type 1: Undesired Case

In this subsection we show three examples that we classified as undesired case. We defined the case as undesired, when most of data is not in the desired range.

The first example is depicted in Fig. 3 and shows daily range of temperature. Please note that desired values are on one extreme. Only a few data points can be described as ok. In this case two summaries are presented to the user:

- *certain number* (30%-50%) of differences in temperature during the day were *too high* (6 °C or more),
- *certain number* (30%-50%) of differences in temperature during the day were *slightly high* (3–5 °C).

The first of them is about *too high* differences in temperature, since this linguistic label has the highest priority than the others, as indicated by the farmers.

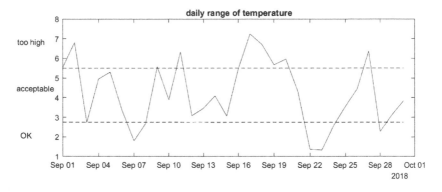

Fig. 3. Example: daily range of temperature – an undesired case

Another example (Fig. 4) depicts the levels of CO_2 in a barn. Here again two summaries were obtained:

- *Few* (5%-35%) of the CO_2 values were *too high* (3200 ppm or more),
- *A lot* (55%-75%) of the CO_2 values were *slightly high* (2600–3200 ppm).

Even more that there are many more observations with only *slightly high* CO_2 values, farmers indicated that the ordering should be done based on the importance of the summarizer, rather than qualifier.

The last example (Fig. 5) for this type of undesired cases is when only one summary is found relevant to be shown to the user.

- *Most* (80%-95%) of the humidity values were *too low* (below 40%).

Please note, that this data are dummy data, created for the completeness purpose.

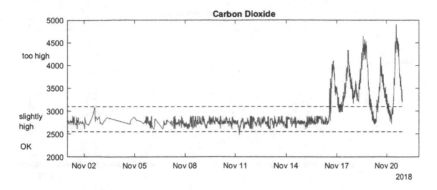

Fig. 4. Example: CO_2 values – an undesired case

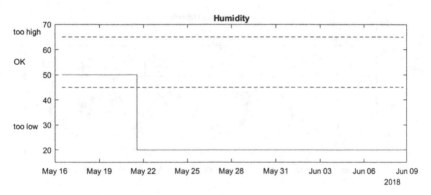

Fig. 5. Example: humidity values – an undesired case (dummy data)

5.2 Type 2: Desired Case

In this subsection we show two examples that we classified as desired case, that is when most of data is in the desired range.

The first example (Fig. 6) shows the shows daily range of temperature, with majority of values being ok.

Hence, in this case only a summary shown below is presented to the user.

– *few* (5%–25%) of differences in temperature during the day were *slightly high* (3–5 °C).

The other example (Fig. 7) depicts CO_2 levels. All the values were in the *OK* range, hence in this case no summary is presented to the user.

5.3 Evaluation with the Users

We have evaluated this method with farmers. First the farmers were using the dashboard for two weeks, next they were asked to fill the questionnaire. Only 12 farmers has filled the questionnaire. The questionnaire was based on Technology

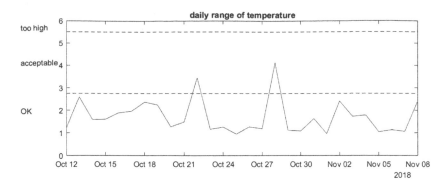

Fig. 6. Example: daily range of temperature – an desired case

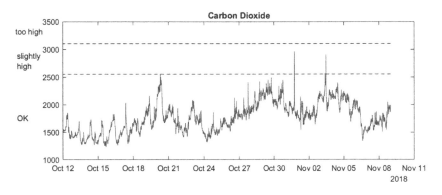

Fig. 7. Example: CO$_2$ levels – an desired case

Acceptance Model (TAM) [25]. TAM assess four aspects: perceived ease of use, perceived usefulness, satisfaction and intention to use [26]. We have used the following 14 statements evaluated on the 7 point Likert scale:

Q1: Using LSs enable farmers to get information from the barn quickly.
Q2: Using LSs saves time of the farmers.
Q3: Using LSs improves my tracking efficiency of the barn.
Q4: Using LSs improves performance of the barn.
Q5: It is clear and understandable what are the LSs.
Q6: I find easy to read the LSs.
Q7: The words used in the LSs are easy and adequate.
Q8: I find good that the system calculates automatically the LSs.
Q9: I am completely satisfied with having sentences in the dashboard.
Q10: I am completely satisfied with the LSs that were presented.
Q11: I feel confident on using the LSs presented.
Q12: I can accomplish adjustments to barn quickly using LSs.
Q13: I will use LSs for tracking variables of the barn.
Q14: I would recommend to use LSs to other farmers.

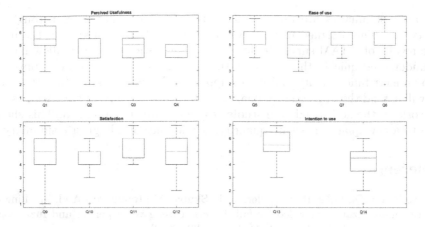

Fig. 8. Average scores for the results of TAM survey

The results of the survey are shown in Fig. 8.

The statements with the lowest score were "using LSs improves performance of the barn" from perceived usefulness aspect and "I would recommend to use LSs to other farmers" from intention to use. Analyzing the different scores of each farmer and the comments they did, it can be explained that these scores were low because according to some of them, it was expected that the linguistic summaries showed an alert instead of showing a summary of the past. Moreover farmers are reluctant to tell other farmers what to do.

On the other hand, the highest scores were for "using LSs enable farmers to get information about the barn quickly" from perceived usefulness; "I feel confident on using the LSs presented" from satisfaction; and "I will use LSs for tracking the variables of the barn" from intention to use aspect. This shows that, in general, farmers trust in the system and believe that linguistic messages are a good tool for making decisions in the barns.

When analyzing the factors in general, ease of use was the best characteristic of LSs, whereas perceived usefulness was the lowest score among the factors, but not with a significant difference. These results showed that linguistic messages are an appropriate solution for the farmers as they find it easy to use, but there is still more room for improvement on usefulness for the information provided.

To sum up, linguistic summaries are a good tool for the farmers; however, the linguistic texts that farmers expected were more about an alerting system, rather than a summarization system, which means that for the future, there is still some potential for extending the linguistic summaries with messages that not only describe the past, but support also decision making.

6   Concluding Remarks

We have analyzed and described an application of linguistic summaries in the agro-food domain. We have focused on the relevance aspect. We determine which linguistic summaries are useful and appropriate for target users through the interviews. The designed method was implemented and tested with the users. The results of a TAM survey indicated that linguistic summaries allow farmers to understand quickly the past performance, yet the farmers would appreciate also an alert functionality and more support in decision making. In future work we will extend the linguistic summaries with the suggestions of the users, as well as work on the additional case studies to generalize the approach. This should lead to a new quality measure quantifying relevance of a linguistic summary.

References

1. Anderson, D., Luke, R.H., Keller, J.M., Skubic, M., Rantz, M., Aud, M.: Linguistic summarization of video for fall detection using voxel person and fuzzy logic. Comput. Vis. Image Underst. **1**(113), 80–89 (2009)
2. Anderson, D., Luke, R.H., Keller, J.M., Skubic, M., Rantz, M., Aud, M.: Modeling human activity from voxel person using fuzzy logic. IEEE Trans. Fuzzy Syst. **1**(17), 39–49 (2009)
3. Anderson, D., Luke, R.H., Stone, E., Keller, J.M.: Segmentation and linguistic summarization of voxel environments using stereo vision and genetic algorithms. In: Proceedings IEEE International Conference on Fuzzy Systems, World Congress on Computational Intelligence, pp. 2756–2763 (2010)
4. Baczko, T., Kacprzyk, J., Zadrozny, S.: Towards knowledge driven individual integrated indicators of innovativeness. In: Knowledge-Based Intelligent System Advancements: Systemic and Cybernetic Approaches, pp. 129–140. IGI Global (2011)
5. Bugarín, A., Marín, N., Sánchez, D., Trivino, G.: Aspects of quality evaluation in linguistic descriptions of data. In: 2015 IEEE International Conference on Fuzzy Systems (FUZZ-IEEE), pp. 1–8 (2015)
6. Castillo-Ortega, R., Marín, N., Sánchez, D., Tettamanzi, A.G.B.: Quality assessment in linguistic summaries of data. In: Greco, S., Bouchon-Meunier, B., Coletti, G., Fedrizzi, M., Matarazzo, B., Yager, R.R. (eds.) IPMU 2012. CCIS, vol. 298, pp. 285–294. Springer, Heidelberg (2012). https://doi.org/10.1007/978-3-642-31715-6_31
7. Castillo-Ortega, R., Marìn, N., Sànchez, D.: A fuzzy approach to the linguistic summarization of time series. Multiple-Valued Log. Soft Comput. **17**(2–3), 157–182 (2011)
8. Delgado, M., Ruiz, M.D., Sanchez, D., Vila, M.A.: Fuzzy quantification: a state of the art. Fuzzy Sets Syst. **242**, 1–30 (2014). https://doi.org/10.1016/j.fss.2013.10.012
9. Díaz, C.A.D., Pérez, R.B., Morales, E.V.: Using linguistic data summarization in the study of creep data for the design of new steels. In: 2011 11th International Conference on Intelligent Systems Design and Applications, pp. 160–165. IEEE (2011)
10. Dijkman, R.M., Wilbik, A.: Linguistic summarization of event logs - a practical approach. Inf. Syst. **67**, 114–125 (2017)
11. Few, S.: Information Dashboard Design: The Effective Visual Communication of Data. O'Reilly (2006)
12. Flach, P.: Machine Learning: The Art and Science of Algorithms That Make Sense of Data. Cambridge University Press, Cambridge (2012)

13. Hastie, T., Tibshirani, R., Friedman, J.: The Elements of Statistical Learning: Data Mining, Inference, and Prediction. Springer, Heidelberg (2009)
14. Jain, A., Keller, J.M.: On the computation of semantically ordered truth values of linguistic protoform summaries. In: 2015 IEEE International Conference on Fuzzy Systems, FUZZ-IEEE 2015, Istanbul, Turkey, 2–5 August 2015, pp. 1–8 (2015)
15. James, G., Witten, D., Hastie, T., Tibshirani, R.: An Introduction to Statistical Learning, vol. 112. Springer, Heidelberg (2013). https://doi.org/10.1007/978-1-4614-7138-7
16. Kacprzyk, J.: Intelligent data analysis via linguistic data summaries: a fuzzy logic approach. In: Decker, R., Gaul, W. (eds.) Classification and Information Processing at the Turn of Millennium, pp. 153–161. Springer, Heidelberg (2000). https://doi.org/10.1007/978-3-642-57280-7_17
17. Kacprzyk, J., Strykowski, P.: Linguistic summaries of sales data at a computer retailer: a case study. In: Proceedings of IFSA 1999, vol. 1, pp. 29–33 (1999)
18. Kacprzyk, J., Wilbik, A.: Towards an efficient generation of linguistic summaries of time series using a degree of focus. In: Proceedings of the 28th North American Fuzzy Information Processing Society Annual Conference - NAFIPS 2009 (2009)
19. Kacprzyk, J., Wilbik, A., Zadrożny, S.: Linguistic summarization of time series using a fuzzy quantifier driven aggregation. Fuzzy Sets Syst. **159**(12), 1485–1499 (2008)
20. Kacprzyk, J., Wilbik, A., Zadrożny, S.: An approach to the linguistic summarization of time series using a fuzzy quantifier driven aggregation. Int. J. Intell. Syst. **25**(5), 411–439 (2010)
21. Kacprzyk, J., Yager, R.R.: Linguistic summaries of data using fuzzy logic. Int. J. Gen. Syst. **30**, 33–154 (2001)
22. Kacprzyk, J., Yager, R.R., Zadrożny, S.: A fuzzy logic based approach to linguistic summaries of databases. Int. J. Appl. Math. Comput. Sci. **10**, 813–834 (2000)
23. Kacprzyk, J., Zadrożny, S.: Fuzzy linguistic data summaries as a human consistent, user adaptable solution to data mining. In: Gabrys, B., Leiviska, K., Strackeljan, J. (eds.) Do Smart Adaptive Systems Exist?, vol. 173, pp. 321–339. Springer, Heidelberg (2005). https://doi.org/10.1007/3-540-32374-0_16
24. Kacprzyk, J., Zadrożny, S.: Linguistic database summaries and their protoforms: toward natural language based knowledge discovery tools. Inf. Sci. **173**, 281–304 (2005)
25. Kwon, H.S., Chidambaram, L.: A test of the technology acceptance model: The case of cellular telephone adoption. In: Proceedings of the 33rd Annual Hawaii International Conference on System Sciences, pp. 7-pp. IEEE (2000)
26. Lai, P.: The literature review of technology adoption models and theories for the novelty technology. JISTEM-J. Inf. Syst. Technol. Manag. **14**(1), 21–38 (2017)
27. Lewis, R.L.: Interference in short-term memory: the magical number two (or three) in sentence processing. J. Psycholinguist. Res. **25**(1), 93–115 (1996)
28. Marín, N., Sánchez, D.: On generating linguistic descriptions of time series. Fuzzy Sets Syst. **285**, 6–30 (2016). https://doi.org/10.1016/j.fss.2015.04.014. Special Issue on Linguistic Description of Time Series
29. Mkhabela, M., Bullock, P., Raj, S., Wang, S., Yang, Y.: Crop yield forecasting on the Canadian prairies using MODIS NDVI data. Agric. Forest Meteorol. **151**(3), 385–393 (2011)
30. Moyse, G., Lesot, M.J., Bouchon-Meunier, B.: Linguistic summaries for periodicity detection based on mathematical morphology. In: 2013 IEEE Symposium on Foundations of Computational Intelligence (FOCI), pp. 106–113 (2013)

31. Panda, S.S., Ames, D.P., Panigrahi, S.: Application of vegetation indices for agricultural crop yield prediction using neural network techniques. Remote Sens. **2**(3), 673–696 (2010)
32. Pretorius, A.J., Van Wijk, J.J.: What does the user want to see? What do the data want to be? Inf. Vis. **8**(3), 153–166 (2009)
33. Reiter, E.: Non-experts struggle with information graphics (2017). https://ehudreiter.com/2017/10/02/non-experts-struggle-graphs/
34. Rong, A., Akkerman, R., Grunow, M.: An optimization approach for managing fresh food quality throughout the supply chain. Int. J. Prod. Econ. **131**(1), 421–429 (2011)
35. Ros, M., et al.: Linguistic summarization of long-term trends for understanding change in human behavior. Proceedings of the IEEE International Conference on Fuzzy Systems, FUZZ-IEEE 2011, pp. 2080–2087 (2011)
36. Smits, G., Nerzic, P., Pivert, O., Lesot, M.: Efficient generation of reliable estimated linguistic summaries. In: 2018 IEEE International Conference on Fuzzy Systems, FUZZ-IEEE 2018, Rio de Janeiro, Brazil, 8–13 July 2018. pp. 1–8 (2018)
37. Szczepaniak, P.S., Ochelska, J.: Linguistic summaries of standardized documents. In: Last, M., Szczepaniak, P.S., Volkovich, Z., Kandel, A. (eds.) Advances in Web Intelligence and Data Mining, vol. 23, pp. 221–232. Springer, Heidelberg (2006). https://doi.org/10.1007/3-540-33880-2_23
38. Wilbik, A., Dijkman, R.: Linguistic summaries of process data. In: Proceedings of IEEE International Conference on Fuzzy Systems (FUZZ-IEEE 2015) (2015)
39. Wilbik, A., Keller, J.M., Alexander, G.L.: Linguistic summarization of sensor data for eldercare. In: Proceedings of the IEEE International Conference on Systems, Man, and Cybernetics (SMC 2011), pp. 2595–2599 (2011)
40. Wong, P.C., Thomas, J.: Visual analytics. IEEE Comput. Graph. Appl. **5**, 20–21 (2004)
41. Yager, R.R.: A new approach to the summarization of data. Inf. Sci. **28**, 69–86 (1982)

Data-Driven Classifiers for Predicting Grass Growth in Northern Ireland: A Case Study

Orla McHugh[4(✉)] [iD], Jun Liu[4] [iD], Fiona Browne[2] [iD],
Philip Jordan[1] [iD], and Deborah McConnell[3]

[1] Ulster University, Coleraine, Northern Ireland, UK
[2] Datactics, Belfast, Northern Ireland, UK
[3] AFBI, Belfast, Northern Ireland, UK
[4] Ulster University, Jordanstown, Northern Ireland, UK
McHugh-03@ulster.ac.uk

Abstract. There are increasing pressures to combat climate change and improve sustainable land management. The agriculture industry is one of the most challenging areas for these changes, especially in Northern Ireland, as agriculture is one of the larger industries. Research has been carried out across the island of Ireland into methods of improving farm efficiency in multiple areas of farming, including livestock health, machinery improvements, and crop growth. Research has been carried out in this study into grass growth in the dairy farming sector, specifically within Northern Ireland. Grass growth prediction aims to inform farmers and policy makers in their decision-making process regarding sustainable land management in agriculture. The present work focuses on analysing and evaluating how data-driven classifiers can be used for grass growth prediction using the data related to soil content, weather, grass quality components *etc.* Four classifiers, namely Decision Trees, Random Forest, Naïve Bayes, and Neural Networks, are chosen for this purpose. Classification results based on a real-world data set are analysed and compared to evaluate and illustrate the performance and robustness of the classifiers. The results indicate that it is difficult to declare a single classifier with the highest performance and robustness. Nevertheless, it indicates that tree classification methods are better suited to the data to be studied, as opposed to probabilistic methods and weighted methods, *e.g.,* the naïve Bayes classifier obtained a predictive performance of 78% when classifying spring seasonal grass growth data.

Keywords: Climate change · Grass growth prediction · Data-driven classifier

1 Introduction

Climate change is a global issue that has become more pressing in recent years. Countries around the world are adopting strategies to combat the effects of climate change and reduce their greenhouse gas (GHG) output. This includes the United Kingdom (UK) and Ireland, which currently have plans in place to reduce GHG emissions from the 1990 baseline [1]. However, these strategies have not been enough to significantly reduce the output, therefore, more needs to be done. Each region within the UK has specific targets to achieve including Northern Ireland (NI), which must

© Springer Nature Switzerland AG 2020
M.-J. Lesot et al. (Eds.): IPMU 2020, CCIS 1237, pp. 301–312, 2020.
https://doi.org/10.1007/978-3-030-50146-4_23

reduce its carbon emission by 35% by 2030, to meet UK targets [2]. In NI, one of the main contributors to GHG emissions is the agricultural sector, which produces almost 30% of the total NI output [2]. Dairy farming is one of the largest agricultural industries in NI. According to the Committee on Climate Change, agricultural emissions in NI have continuously increased since 2009 despite efforts to improve the efficiency in dairy farming [2]. Therefore, it is vital that tools and support are provided to farmers and stakeholders within the industry to inform them on solutions and actions that can improve farming efficiency and reduce emissions.

This study relates to the improvement of dairy farming efficiency by focusing on sustainable land management and examining grass growth which is one of the cheapest feed sources for livestock in NI [3]. Grass growth rates are variable across the year and depend on various factors, with some of the most influential factors being meteorological *e.g.*, rainfall, solar radiation, and temperature. NI has a temperate climate that allows for a long growing season. Soil conditions such as temperature and moisture also have an influence on grass growth, curtailing growth particularly when soils are oversaturated or excessively dry. Other factors relating to management also impact grass growth such as fertiliser application, grazing intensity, and grazing rotation length.

Grass related data have been collected by the Agri-Food and Biosciences Institute (AFBI) across NI in their GrassCheck project. AFBI is a research and development organisation that supports the Department for Agriculture, Environment, and Rural Affairs (DAERA) and other UK government bodies and public organisations. The GrassCheck project consists of farmer research gathered across 50 locations in NI including beef, sheep, dairy, and crop plot farming. The project will run for three years from 2018 to 2020 collecting grass growth data, grass quality data, grazing event data, and meteorological data.

The authors of this research have performed an exploratory statistical analysis of the GrassCheck dataset detailed in [4]. In this study, the R programming language was used to provide a statistical overview, correlation analysis, and linear regression analysis of the GrassCheck data to identify the grass growth predictive features. A boxplot visualising the variance in the grass growth features (including pre-grazing cover, utilisation, and soil moisture) illustrated the variability of grass growth over an 8-month period in which data was recorded. A correlation analysis identified strong positive relations between offtake, pre-grazing cover and grass growth and strong negative relations between post-grazing cover and grass growth. Linear regression was performed on the GrassCheck dataset to determine which features had the greatest influence on grass growth. Using this method, pre-grazing cover and the available amount of grass to livestock (known as available) features were shown to be the best fit models when used as the explanatory variables. Other statistically significant features include offtake, utilisation, and month [4]. However, this study is still limited in finding the in-depth pattern for grass growth prediction. Advanced data analytics are expected to further enhance predictions by using, for example, data-driven classification models such as neural networks, naïve Bayes, and decision trees to expand on the exploratory statistics used to analyse grass growth data.

Therefore, the aim of this research was to aid in understanding how various grassland features contribute to the prediction of grass growth, and to analyse and evaluate various classification models to deduce which are the most suitable for grass growth prediction.

This paper is organised as follows; Sect. 2 provides an overview of related research in the area. Section 3 provides the methodology underpinning the research, with results and discussion presented in Sect. 4. Conclusions and future work are discussed in Sect. 5.

2 Related Work

In Ireland, research has developed a grass growth prediction model for dairy based farming [3, 5]. The Moorepark St. Gilles Grass Growth Model, known as the MoSt GG model, is a descriptive model providing insight into grass growth at paddock levels in Ireland. There are various inputs into this model including forecasted meteorological data, management strategy information, and fertiliser application, specifically nitrogen (N). The outputs of this model include daily grass growth, N information such as the soil content, grass content, grass uptake, and nitrate leaching. The output from the model was compared to the output from an experimental farm in Cork, Ireland, for a period of two years. It was observed that, while the model was successful in improving some areas of prediction from a previous model, $i.e.,$ better prediction of production per cutting date and per plot, it was not always accurate in others. For example, the N prediction in grass content and nitrate leaching was underestimated, potentially since the model does not consider previous years management techniques. Although this model was not designed for NI, the same principles can be applied to aid constructing a decision support system to support sustainability in NI. The decision support system could be expanded to make predictions to support farmers and policy makers in their decisions regarding sustainability.

Classification approaches, such as decision trees, artificial neural networks, and support vector machines have been used in multiple research studies for different classification problems. These include agricultural issues such as crop disease prediction [6], crop yield prediction [7], and grassland biomass estimation [8]. This research discusses the various classification methods used in agricultural prediction. When predicting crop disease, multiple classification methods were used including, neural networks, naïve Bayesian, random forest, decision trees, support vector machines, k-nearest neighbor, and ensemble models [6]. In this study, it was found that random forest and Gaussian naïve Bayes classifiers performed better than other classifiers when predicting binary data, while neural networks and random forest were better when predicting the original dataset. Multiple linear regression and density-based clustering classification methods have been used in this research area [7]. Multiple linear regression, neural networks, and adaptive neuro-fuzzy inference systems were also used in the area of grassland biomass estimation [8]. This research highlighted the use of the neuro-fuzzy system as it performed better when estimating biomass than the artificial neural networks and the multiple linear regression.

3 Materials and Methods

3.1 Data

The data used in this research is grassland data provided by AFBI, from the Grass-Check project. Different features have been collected including grass growth, grass quality, grazing events, and meteorological data. The features within this dataset have been outlined in Table 1 below.

Table 1. List of features in the datasets

Feature	Description	Type	Unit	Example
ID	The farm identifier assigned to the farms	Numeric		1–12, 41–48
Month	The month that the grass growth took place	Numeric		3 (March)
Week	The week that the grass growth took place	Numeric		12
County	The county in NI in which the farm is located	Categorical		Down
Grass Growth	The daily grass growth averaged across all paddocks on the farm	Numeric	Kg DM/Ha	Range: 0–145.3
Field	Farmer description of field where grass quality measurements are recorded	Qualitative		"Yard Field"
Conditions	Farmer description of weather conditions on day of measurement	Qualitative		"Bright and Sunny"
Dry Matter (DM)	The proportion of total grass components (fibers, proteins *etc.*) remaining in the grass after water is removed	Numeric	%	Range: 9.6–40.4
Crude Protein	The protein content in the grass, minus effluent losses	Numeric	%	Range: 9.7–26.8
Acid Detergent Fiber (ADF)	Measurement of digestibility, via the cellulose, lignin, and lignified nitrogen content of the grass	Numeric	%	Range: 2.6–37
WSC	The soluble sugars released from the grass in the animal	Numeric	%	Range: 0–23.5
Metabolisable Energy (ME)	The energy content of the grass measured in megajoules of energy per kilogram of dry matter (MJ/Kg/DM)	Numeric	MJ/Kg/DM	Range: 9.9–13.4
Paddock	Farmer description of paddock where grazing event took place	Qualitative		"MC2B"
Event	Type of grazing event *i.e.*, grazed, or part grazed by animals or cut by machinery	Quantitative		Grazed
Date	The date of the grazing event	Date		30/03/18
Notes	Any notes made by farmer relevant to the event	Qualitative		"part grazed – value changed"

(*continued*)

Table 1. (*continued*)

Feature	Description	Type	Unit	Example
Pre-grazing Cover	The amount of grass in the paddock immediately before animal grazing	Numeric	Kg DM/Ha	Range: −2200–8100
Post-grazing Cover	The amount of grass in the paddock after animals have grazed	Numeric	Kg DM/Ha	Range: 50–4920
Available	The amount of grass available to livestock in the paddock, calculated as pre-grazing cover − 1500	Numeric	Kg DM/Ha	Range: −3700–6600
Offtake	The amount of grass removed by animals at that grazing event, calculated as the pre-grazing − post-grazing cover	Numeric	Kg DM/Ha	Range: −4012.5–6889
Utilisation	The amount of grass consumed by animals (*i.e.*, the offtake) as a proportion of that available	Numeric	Kg DM/Ha	Range: -174–6.33
Total Rainfall	The total rainfall fallen on the day of grazing	Numeric	mm	Range: 0–36.4
Air Temperature	The average air temperature on the day of grazing	Numeric	°C	Range: −2.03–25.33
Solar Radiation	The average solar radiation on the day of grazing	Numeric	W/m^2	Range: 0–610.375
Soil Moisture	The average moisture levels in the soil on the day of grazing	Numeric	cb	Range: −1.09–200

Note: Kg DM/Ha: Kilogram Dry Matter per Hectare; WSC: Water Soluble Carbohydrate

There is a total of 4917 records that have been labelled using the classifications of High, Medium, and Low. The data have been binned into these labels based on calculating interquartile ranges on the grass growth value, where the lower quartile is 32.5 kg DM/Ha and the upper quartile is 75.9 kg DM/Ha. Approximately 25% of the records are low, 50% are medium, and 25% are high. In numerical terms, this equates to 1146 low records, 2336 medium records, and 1153 high records. The prediction of grass growth can help farmers make management decisions about grazing, cutting, and other areas of farming decisions in order to improve farm efficiency. For instance, knowing that there will be a low grass growth rate in the next month can allow farmers to ensure they have adequate stocks of feed concentrates to ensure the wellbeing of their livestock. There are some grass growth entries that are missing, which results in 282 records with an unknown classification category. These unknown variables were removed from analysis during this study. The missing variables have been introduced through the method of data collection used to collect the data which relied on individual farmer input. Missing data was also introduced via the grass growth dataset being measured daily, while the grass quality and grazing events were not measured daily, but measured more sporadically. This meant when the datasets were joined on the Farm ID and the Date, there were empty variables where there was no recordings in the grass quality and grazing event datasets.

3.2   Data-Driven Classifiers Overview

The following prediction models were chosen due to their ease of use, and popularity within the predictive analytics domain and application in the agricultural industry.

Decision Tree (DT). DT classifies instances via a tree structure, where individual attributes are represented by nodes, and there are links between nodes. The DT calculates the information gained from the attribute and makes decisions based on which attribute has the most information gain [6].

Random Forest (RF). The RF can be described as a collection of individual decision trees that work together as an ensemble [6]. This classification model is useful as each individual decision tree is unlikely to make the same mistakes as the others, and therefore, the classification is safer from error.

Naïve Bayes (NB). NB is a probabilistic classifier that assumes attributes are independent of each other and they carry the same weight when making predictions [9].

Neural Network (NN). A NN is a classifier that, like a decision tree, uses nodes and links to make predictions. However, each node is assigned a weight, with priorities at each node split being given to the feature that has a larger weight [9].

3.3   Case Studies for Grass Growth Prediction

Two case studies were carried out in the analysis of the GrassCheck dataset. Firstly, analysis of the 2018 dataset was performed where missing data were included (4917 instances, 19 features). Secondly, the same dataset was analysed where instances containing missing data were removed (107 instances, 19 features).

The datasets were divided into seasonal data (winter data were excluded as there is no grass growth during these months and, therefore, no recordings take place). The dataset was divided into Spring (March-May), Summer (June-August), and Autumn (September-October). This resulted in imbalanced datasets as there is more likely to be high growth rates in summer months and lower growth rates towards the cooler times of year. To resolve this imbalance, the larger sets could have been reduced to the approximate size of the smallest set, resulting in even divisions. This method was not applied as it would result in a dataset that is too small to perform classification models. Therefore, the Synthetic Minority Oversampling Technique (SMOTE) [10] was applied to the smallest set in the dataset in order to make synthetic data that resembled actual data in the dataset. In the Spring dataset, SMOTE was applied at 250%, which means the smallest dataset is increased by 250%. SMOTE was applied again at 35%, resulting in approximately balanced numbers of Low, Medium, and High labelled data (384, 376, 273, respectively). In the Summer dataset, SMOTE was applied at 150%, and 90%, to result in approximately balanced numbers of Low, Medium, and High data (642, 675, 665, respectively). In the Autumn dataset, SMOTE was applied at 100%, 250%, and 60%, to result in approximately balanced numbers of Low, Medium, and High data (299, 292, 291, respectively).

For each case study, ten-fold cross validation was carried out to evaluate each classifier. The evaluation metrics in this research are Kappa statistics, Mean Absolute

Error (MAE), Root Mean Square Error (RMSE), Precision, Recall, F-Measure, and ROC Area. Several features in the dataset were determined to not have informational values, including Farm ID, Field, Conditions, Paddock, and Notes. As discussed in the statistical analysis performed on the data [4], some features have a stronger correlation with grass growth. This was further analysed using feature selection methods including the Pearson's Correlation Coefficient, Information Gain Attribute Evaluation, and Wrapper Subset Evaluation.

4 Results and Discussions

This section summarises the results and discusses the outcome of the experiments. The analysis performed consists of four classifiers used on multiple variations of the dataset. This includes yearly divisions and period divisions of spring, summer and autumn. Classification analysis was performed on the data using techniques including DT, RF, NN, and NB. The tables below displays the outcome of the analysis, by showing the percentage of correct predictions, Kappa statistics, MAE, RMSE, precision (P), recall, F-measure (FM), and ROC, respectively.

4.1 Yearly Analysis

The tables below (Table 2 and Table 3) display information from the classification of data over the year of 2018. Table 2 shows the evaluation metrics on the whole dataset without handling the missing data, which contains 4917 instances. Table 3 displays the evaluation metrics on the same dataset, but with all instances including the missing variable removed, resulting in classification being performed on 107 instances.

Table 2. Evaluation metrics on the whole dataset.

Class	Acc (%)	Kappa	MAE	RMSE	P	Recall	FM	ROC
DT	73	**0.5624**	**0.2511**	0.3567	0.737	**0.735**	**0.731**	0.845
RF	73	0.546	0.2784	**0.3523**	**0.753**	0.734	0.725	**0.871**
NB	61	0.3497	0.2998	0.4306	0.628	0.610	0.599	0.738
NN	59	0.3303	0.2995	0.4386	0.613	0.599	0.596	0.731

Table 3. Evaluation metrics on the dataset excluding the variables with the missing data.

Class	Acc (%)	Kappa	MAE	RMSE	P	Recall	FM	ROC
DT	65	0.405	0.2914	0.4188	0.678	0.654	0.631	0.695
RF	80	**0.6676**	0.2376	**0.3211**	**0.809**	**0.804**	**0.794**	**0.917**
NB	62	0.3816	0.2504	0.4385	0.622	0.617	0.619	0.772
NN	76	0.5934	**0.1732**	0.3753	0.757	0.757	0.752	0.863

Table 2 shows that the tree classifiers, *i.e.*, DT and RF, have the best metrics out of the four classifiers. DT has the greatest Kappa statistic of 0.56, lowest MAE of 0.25, and highest recall and F-measure of 0.74 and 0.73, respectively. RF shows the best performance in RMSE with 0.35, precision with 0.75, and ROC area with 0.87. The results show that, overall, the RF classification method performed the best over the two datasets as the ROC curve is the highest in both sets at 0.871 and 0.917. It also has the lowest RSME across the classifiers at 0.3523 and 0.3211. However, RF is susceptible to influence by an imbalanced dataset, *i.e.*, the Medium category is a larger set than the other two growth rates, resulting in a skewed output.

The NN classifier showed the greatest improvement when missing data were removed from the dataset in terms of all metrics, *e.g.*, it increased from 0.33 to 0.59 in Kappa statistics, while reducing the MAE from 0.30 to 0.17, from Table 2 to Table 3, as a networks performance will increase when all features are available, *i.e.*, when there are no missing data. The NB classifier showed little difference in terms of performance when comparing the analysis using the full dataset, to the dataset where missing values are with no strong improvement observed. This is due to the assumption of independence of the attributes, as not all the attributes in these data are independent, *e.g.*, offtake, available, and utilisation depend on the pre-grazing and post-grazing cover.

4.2 Seasonal Analysis

The tables below (Table 4, Table 6, and Table 8) display information from the classification of data pertaining to spring, summer, and winter. Table 5, Table 7, and Table 9 show the evaluation metrics on the spring, summer, and winter dataset, respectively, where SMOTE has been applied to balance the classes in the datasets, resulting in balanced categories of low, medium, and high.

Table 4. Evaluation metrics for Spring dataset (normal).

Class	Acc (%)	Kappa	MAE	RMSE	P	Recall	FM	ROC
DT	79	**0.6468**	0.1944	**0.3252**	**0.794**	**0.790**	**0.790**	0.874
RF	79	0.637	0.2268	0.3268	0.787	0.786	0.786	**0.888**
NB	63	0.3743	0.2458	0.4393	0.635	0.627	0.616	0.801
NN	78	0.6304	**0.1565**	0.3547	0.782	0.782	0.782	0.865

Table 5. Evaluation metrics for Spring dataset (SMOTE).

Class	Acc (%)	Kappa	MAE	RMSE	P	Recall	FM	ROC
DT	84	0.7594	0.1523	0.2889	0.842	0.840	0.839	0.921
RF	85	**0.7726**	0.1519	**0.2653**	**0.850**	**0.848**	**0.849**	**0.958**
NB	78	0.666	0.1605	0.3399	0.784	0.777	0.777	0.915
NN	84	0.7567	**0.1154**	0.3058	0.838	0.838	0.838	0.934

All the classifiers were improved from Table 4 to Table 5, in terms of correctly predicting instances when the dataset classes have been balanced, *e.g.*, NB showed the largest increase of 15%, from 63 to 78%. However, this does not mean that NB is a good classification method for these data. Although it has shown the most improvement in all features, including MAE and RMSE (reduction of 0.09 and 0.10 respectively), it is the overall least successful when predicting the level of grass growth, again due to the assumption made of independence. RF could be considered a good performer as second to NB, as it improved the most across most of the metrics. For instance, the Kappa statistic increased by 0.14, and the precision and recall have increased by 0.062 and 0.063, respectively. Overall, there are minor differences between DT, RF, and NN as they have similar evaluation outputs across all of the metrics.

Table 6. Evaluation metrics for Summer dataset (normal).

Class	Acc (%)	Kappa	MAE	RMSE	P	Recall	FM	ROC
DT	77	**0.6047**	0.2112	0.345	**0.769**	**0.768**	**0.764**	0.847
RF	75	0.5655	0.236	**0.3403**	0.754	0.748	0.741	**0.868**
NB	64	0.4402	0.2686	0.4153	0.671	0.638	0.640	0.797
NN	74	0.5496	**0.1967**	0.3943	0.739	0.737	0.731	0.810

Table 7. Evaluation metrics for Summer dataset (SMOTE).

Class	Acc (%)	Kappa	MAE	RMSE	P	Recall	FM	ROC
DT	80	0.6926	0.1622	0.3252	0.798	0.795	0.796	0.907
RF	84	**0.7607**	0.1619	**0.2768**	**0.848**	**0.841**	**0.841**	**0.952**
NB	71	0.572	0.1949	0.3655	0.721	0.714	0.702	0.903
NN	82	0.7365	**0.1242**	0.321	0.829	0.824	0.825	0.923

In Table 6, DT has the largest percentage of correctly predicted instances of 77%. It also has the best performance in Kappa statistic, precision, recall, and F-measure (0.60, 0.77, 0.77, 0.76, respectively). However, when SMOTE is applied in Table 7 to balance the classes in this dataset, RF becomes the better classification method as it has the best metric value in Kappa, RMSE, precision, recall, F-measure, and ROC area, (0.76, 0.28, 0.85, 0.84, 0.84, and 0.95 respectively). RF shows the greatest improvement in six of the eight metrics, including precision (0.75 to 0.85), recall (0.75 to 0.84), and F-measure (0.74 to 0.84). This means the classifier is returned more accurate results. NN shows the greatest improvement in RMSE, where it reduced from 0.39 to 0.32, and in the ROC area, where it increased from 0.91 to 0.92. Overall, all of the classification methods have improved in all of the evaluation metrics with the addition of synthetic data.

Table 8. Evaluation metrics for Autumn dataset (normal).

Class	Acc (%)	Kappa	MAE	RMSE	P	Recall	FM	ROC
DT	80	**0.627**	0.1861	0.3296	**0.799**	**0.804**	**0.794**	0.834
RF	78	0.5827	0.2111	**0.323**	0.783	0.780	0.776	**0.875**
NB	48	0.2083	0.3471	0.5412	0.665	0.483	0.533	0.693
NN	75	0.5382	**0.1685**	0.369	0.749	0.752	0.750	0.846

Table 9. Evaluation metrics for on Autumn dataset (SMOTE).

Class	Acc (%)	Kappa	MAE	RMSE	P	Recall	FM	ROC
DT	84	0.7534	0.1513	0.3043	0.836	0.836	0.836	0.904
RF	84	**0.7567**	0.1726	**0.2822**	**0.839**	**0.838**	**0.838**	**0.946**
NB	67	0.5129	0.2185	0.4351	0.712	0.675	0.664	0.874
NN	81	0.7177	**0.1307**	0.3233	0.810	0.812	0.810	0.918

Table 8 shows the evaluation metrics of the classifiers before SMOTE has been applied, and indicates that the DT classifier could be considered the most appropriate method due to its performance across the metrics. For example, it has the greatest Kappa statistic of 0.63, the highest precision and recall rates of 0.80 each, and F-measure which is 0.79. However, when the synthetic data were added to the dataset, RF became the most accurate classifier due to its performance in Kappa, RMSE, precision, recall, F-measure, and ROC area (0.76, 0.28, 0.84, 0.84, 0.84, and 0.95 respectively). Again, NB showed the greatest improvement out of the four classifiers, with improvements across all of the features including Kappa statistic (increase of 0.30), MAE (reduction of 0.13), and RMSE (reduction of 0.11). However, NB was the least accurate classification method for this dataset as DT, RF, and NB, performed better across all of the metrics, *e.g.,* NB has an MAE of 0.22 when SMOTE was applied, while the other classifiers have an MAE of 0.17 or below. NN showed the greatest improvement in precision as it increased from 0.67 to 0.71, although the increases in each of the classifiers in this metric were very minor.

4.3   Further Study and Discussions

Three methods of feature selection were carried out as well on the dataset with no missing values including the Correlation Attribute Evaluation. This method is also known as the Pearson's Correlation Coefficient, in which attributes are ranked on how much information they provide to the prediction of the target class. The results of this method show there is more of a correlation between Dry Matter, Soil Moisture, Offtake, Pre-Grazing Cover, Available and the target category class. Attributes such as Total Rainfall, Week, Month, and Crude Protein, have less of a correlation as the values are closer to 0 than 1.

Information Gain Attribute Evaluation with a ranker filter was also used for feature selection. In this method, attributes such as Date, Offtake, Dry Matter, Available, and Pre-Grazing Cover, provide more information to the prediction of the target class. Attributes which provide less information for predicting included Acid Detergent Fiber, Dry Matter, and Post Grazing Cover, while attributes such as Month, WSC, Utilisation, Crude Protein, Air Temperature, Total Rainfall, and Post-Grazing Cover, had no information gain with a value of 0.

Another method of feature selection used on the data was the Wrapper Subset Evaluation, using a Decision Tree with the Best First Ranker method. This method uses a decision tree to evaluate numerous subsets to determine the best subset. In this method, the merit of the best subset was 0.748, and found that the optimal number of folds for this dataset is 5 folds. Attributes identified as having the most significance included Week, ADF, ME, Utilisation and Total Rainfall.

The feature selection methods described above have selected different attributes as being the most informative. Each of the classification methods were run again using the information provided by the feature selection methods. However, the results proved to be poorer when features were removed. A Python script was developed to perform the same feature selection methods, and difficulties were found due to the text fields (County, Event, Date) and negative numbers (Utilisation, Offtake, Available) in the dataset, which are problematic in the feature selection methods chosen. The negative numbers were normalised and the text fields were categorised, and the classification was run again. As it had been before, each of the methods chose different features as important and there was no strong similarity between features. As well as this, features which would be designated important in real life (*e.g.*, rainfall) were not classed very highly and vice versa. Therefore, some further investigation on the feature selection methods to better suit the available dataset needs to be done, along with the more elaborate data pre-processing method to be used to classify and clean the data in order for the above feature selection methods become feasible.

5 Conclusions and Future Work

The present work focused on analysing and evaluating four data-driven classifiers for grass growth prediction using some real grass data collected related to soil content, weather, grass quality components *etc*. From the above study, it was found that tree classifiers were better methods of classification, namely the DT and RF methods. DT performed better in datasets which contained imbalance, such as in the seasonal divisions of spring, summer, and autumn. It consistently performed the best in Kappa statistics, precision, recall and F-measure across all seasonal data. RF performed consistently in RMSE and ROC area in both imbalanced and balanced datasets, with its best performance values as low as 0.27 (RMSE), and as high as 0.96 (ROC) on the spring dataset with synthetic data. This was the highest ROC value across all the classifiers, while the lowest value was 0.69, produced by NB on the autumn dataset. Once synthetic data were applied, and the imbalance was eradicated, RF became the overall best classifier in each of the experiments that was carried out. It consistently performed the best in Kappa, RMSE, precision, recall, F-measure, and ROC area. NB could be considered the least successful classification method, as its accuracy and evaluation metrics were well below that of the other three methods. The best performance from this classifier was on the spring dataset with SMOTE applied, in which its accuracy was 78%, while the other classifiers were 84% and 85% accurate in the same dataset. Other measures including MAE (0.16) were good in the NB classifier, however, this was the highest error rate in the dataset. Other interesting results were produced by the DT classifier, as it reduced in performance, over all of the evaluation metrics, from the whole year to the dataset when instances containing missing data were removed. This is due to the size of the dataset, as a small dataset of 107 instances does not contain enough information for the DT to make accurate decisions. The above study has demonstrated the good potential of using data analytics for grass growth prediction, although the overall performance of those four classifiers are not exceptional considering, for example, the accuracy rate, which is partially due to the quality of data (missing data, imbalance and uncertainty inside), and partially due to limiting to only four classifiers. More elaborate data preprocessing and cleaning methods can be used, and other types of classifiers can be also explored further in future work. One limitation of this study is the inability to explain how the classifiers came to the conclusion of their prediction. At present the classifiers are assigned greater weights when there is higher information gain, and lower weights when there is little information gain. Future work will consider the use of expert knowledge to assign weights to attributes, which will allow the conclusion to be better explained to users, and to give definitive reasoning for the prediction.

This study underpins research for aiding farmers and policy makers in their decisions regarding sustainable land management. The agriculture and farming industry of NI requires tools and strategies to encourage sustainable land management, especially due to its greater contribution to gaseous emissions in NI. The study has highlighted the need for a system that can handle missing data and uncertainty. The data-driven approach is expected to be combined with expert knowledge from the industry and models must be integrated to enhance the overall performance and create a multilayer decision support system, to support farmers and policymakers when making land sustainability decisions.

References

1. Parliament of the United Kingdom: Climate Change Act 2008 (2008)
2. Committee on Climate Change: Reducing emissions in Northern Ireland (2019)
3. McDonnell, J., Brophy, C., Ruelle, E., Shalloo, L., Lambkin, K., Hennessy, D.: Weather forecasts to enhance an Irish grass growth model. Eur. J. Agron. **105**, 168–175 (2019)
4. McHugh, O., Browne, F., Liu, J., Jordan, P.: A decision analytic framework and exploratory statistical case study analysis of grass growth in Northern Ireland. J. Adv. Inf. Technol. **11**(1), 15–20 (2020)
5. Ruelle, E., Hennessy, D., Delaby, L.: Development of the Moorepark St Gilles grass growth model (MoSt GG model): A predictive model for grass growth for pasture based systems. Eur. J. Agron. **99**, 80–91 (2018)
6. Ayub, U., Moqurrab, S.A.: Predicting crop diseases using data mining approaches: classification. In: 1st International Conference on Power, Energy and Smart Grid (ICPESG) (2018)
7. Ramesh, D., Vardhan, B.V.: Analysis of crop yield prediction using data mining techniques. IJRET: Int. J. Res. Eng. Technol. **4**(01), 470–473 (2015)
8. Ali, I., Cawkwell, F., Dwyer, E., Green, S.: Modeling managed Grassland biomass estimation by using multitemporal remote sensing. IEEE J. Sel. Top. Appl. Earth Observ. Remote Sensing **10**(07), 3254–3264 (2017)
9. Corrales, D.C., Corrales, J.C., Figueroa-Casas, A.: Towards detecting crop diseases and pest by supervised learning. Ingeniería y Universidad **19**(01), 207–228 (2015)
10. Chawla, N.V., Bowyer, K.W., Hall, L.O., Kegelmeyer, W.P.: SMOTE: synthetic minority over-sampling technique. J. Artif. Intell. Res. **16**, 321–357 (2002)

Forecasting Electricity Consumption in Residential Buildings for Home Energy Management Systems

Karol Bot[1] , Antonio Ruano[1,2(✉)] ,
and Maria da Graça Ruano[1,3]

[1] Faculty of Science and Technology, University of Algarve,
8005 Faro, Portugal
aruano@ualg.pt
[2] IDMEC, Instituto Superior Técnico, Universidade de Lisboa, Lisbon, Portugal
[3] CISUC, University of Coimbra, Coimbra, Portugal

Abstract. Prediction of the energy consumption is a key aspect of home energy management systems, whose aim is to increase the occupant's comfort while reducing the energy consumption. This work, employing three years measured data, uses radial basis function neural networks, designed using a multi-objective genetic algorithm (MOGA) framework, for the prediction of total electric power consumption, HVAC demand and other loads demand. The prediction horizon desired is 12 h, using 15 min step ahead model, in a multi-step ahead fashion. To reduce the uncertainty, making use of the preferred set MOGA output, a model ensemble technique is proposed which achieves excellent forecast results, comparing additionally very favorably with existing approaches.

Keywords: Home consumption forecasting · HVAC consumption forecasting · Prediction methods · Neural networks · Multi-objective optimization · Home energy management systems · Ensemble modelling

1 Introduction

The consumption of energy has increased substantially in the building sector in the past years, fueled primarily by the growth in population, households and commercial floor space. For this reason, Home Energy Management Systems (HEMS) are becoming increasingly important to invert this continuously increasing trend. HEMS offer advantages to both building occupants and electricity suppliers. For the former, they are a means to reduce energy consumption in a household, or, perhaps more important to the occupants, by reducing their electricity bill. For suppliers, making use of smart grid

The authors would like to acknowledge the support of Programa Operacional Portugal 2020 and Operational Program CRESC Algarve 2020 grant 01/SAICT/2018. Antonio Ruano acknowledges the support of Fundação para a Ciência e Tecnologia, through IDMEC, under LAETA, grant UIDB/50022/2020.

M.-J. Lesot et al. (Eds.): IPMU 2020, CCIS 1237, pp. 313–326, 2020.
https://doi.org/10.1007/978-3-030-50146-4_24

technology, HEMS enable the implementation of several Demand Response (DR) mechanisms [1].

If the HEMS is able to control the operation of devices in a home, it is necessary to separate the consumption of non-deferrable (or non-schedulable) appliances, from deferrable (or schedulable) devices. As the efficiency of DR techniques can be improved making use of forecasts of electricity consumption (and electricity generation if renewables are employed), then both the consumption of schedulable and non-schedulable appliances must be predicted [2].

Methodologies based on computation intelligence are the ones that are most used for short-term load forecasting. However, a certain degree of uncertainty is typically found in those forecasts [3], which typically can be reduced using ensembles of models [4].

1.1 Literature Review

Computational intelligence models are developed by measuring the inputs and outputs of the system and fitting a linear or non-linear mathematical model to approximate the operation of the building [5]. These models are based on the implementation of a function deduced only from samples of training data describing the behavior of a specific system, being this way well suited when physical relations are not known [6, 7]. For buildings, the advantage of computational intelligence models over physical methods is that the former do not require knowledge of the building geometry and physical phenomena to deduce an accurate prediction model. However, the lack of proper data can become an issue for the use of computational learning methods [7], because the accuracy is strongly depending on the quality and amount of available data.

Reviews of prediction of energy consumption in buildings with computational intelligence methods can be found, for instance, in [8, 9]. According to the mentioned works, in which more than 100 cases were analyzed, these techniques are proven to be very effective. Among these methods, Artificial Neural Networks (ANN) are the primary models employed to evaluate and predict energy consumption [10–12]. The main input data used to feed this technique may be segmented in two main categories: weather-related parameters and building-related parameters. Concerning the weather-related parameters, atmospheric temperature is the parameter most used as an exogenous variable, but also solar radiation availability and relative humidity are employed. Considering the building related parameters, the total building energy consumption data is the most used variable (as endogenous variable), followed by parameters as occupancy, usage of devices, indoor temperatures and fenestration characteristics.

According to the partition of electricity considered, the prediction of the electric energy consumption may have different focus. Most studies deal with the whole-building energy consumption [13–15]; other focus only on heating demand [12, 16], only on cooling demand [17], on both heating and cooling [18], and also on the detailed segmentation considering devices and other uses as water heating [19].

The prediction horizon of reviewed studies was segmented in hourly fractions, hour, day, month and year, with varying prediction time steps (most hourly for one-day as a prediction horizon, and daily for the one-month horizon).

The validation methods of the prediction models also varied between the use of analytical proofs, experimental analysis, model comparison, reference comparison and simulation comparison, being the first two the most used. Additionally, an extensive review concerning categorization of forecasting parameters may be found in [20, 21].

1.2 Objectives and Work Organization

This work proposes the use of an ensemble of models to be used for producing forecasts of home electric consumption data, considering total consumption, of schedulable equipment, and of non-schedulable devices, to be employed in HMES schemes. The models employed are Radial Basis Function (RBFs) Neural Networks models, designed using a Multi-Objective Genetic (MOGA) algorithm.

MOGA has been employed successfully in a variety of applications (please see [22–26] to name just a few). In all these works, results obtained with MOGA have been compared with other available methods, relevant to the application at hand. The objective of this paper is not to compare MOGA with other methods for forecasting energy, but to verify if MOGA results could be improved with ensemble averaging of the models in the non-dominated set.

Experimental data obtained from the Honda Smart Home US, located in Davis, United States, are employed as a case study.

The paper is divided in five sections. Section 1 introduces the scope of the work, objectives and work organization, and a brief literature review. Section 2 presents the description of the case study. Section 3 introduces the MOGA methodology, and its use for ensemble averaging. Section 4 presents and discusses the results. Conclusions are drawn in Sect. 5.

2 Honda Smart Home US – Case Study Description

This work uses data obtained in the Honda Smart Home (HSM) US [22]. This building is located on the West Village campus of the University of California, Davis. The building is classified as a Net Zero Energy Building, used sustainable construction materials, has a radiant floor and night ventilation. Electric appliances and lighting have high efficiency, and the HVAC system employs a ground-source heat pump. The household has a complex home energy management system to control the electric systems. Details about the construction, electric appliances and data acquisition system details can be found in its website [22].

The group responsible for the HSM makes available experimental data every six months. Based on the public available data, some studies were developed, focused mainly on the integration between electric vehicles and the smart home, and the home management systems of the HVAC solutions, as well as construction practices. The present work uses the HSM data to design the prediction models and test their accuracies (Fig. 2).

To develop the present study, four variables are used from the HSM data set. They are the total average electric power demand, the HVAC power demand, all the "other" electric loads except the HVAC (equipment, lighting, energy management system

Fig. 1. Architecture of the Honda Smart Home US. Source: [27]

equipment, and other miscellaneous loads), as well as the outdoor temperature. Figure 2 presents the data from sensors, with a 15-min average of 1-min sampled data, from January 1st 2016 to December 31st 2018. During the considered three-year period, the power hits a maximum of 8,57 kW, a minimum of 0 kW and a mean value of 0,99 kW. During the same period, the outside temperature has a maximum of 42,82 ° C, a minimum of −2,00 °C and an average temperature value of 17 °C.

3   Methodology of Predictive Model Design

The data set is composed by 15 min averages of power consumption of the HSM (total, HVAC, and others) and outdoor temperature, during the three years (2016, 2017 and 2018). Additionally, a codification of each day, within a week, considering holidays and their position within the week [28], was employed to associate the patterns of consumption to the calendar days. The model intends to predict the power consumption for a prediction horizon of 12 h, using steps of 15 min, in a multi-step fashion.

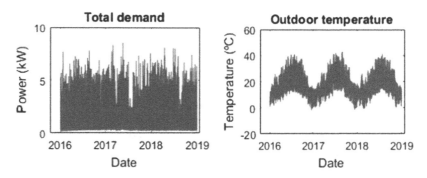

Fig. 2. HSM average data of power and temperature.

RBF models are employed, in a Nonlinear AutoRegressive with eXogenous inputs (NARX) configuration. Two exogenous variables (v2 – outdoor temperature, v3 – day code) and their delays are used as inputs, together with delays of the modelled variable (v1 - electric power).

Three problems are considered, aiming to model the total demand (P1), HVAC demand (P2) and "other" demand (P3). As it will be explained later, two different models will be designed for each problem. As those models will be subsequently used in a predictive control scheme, the main goal of the models is to obtain a small Root Mean Square Error (RMSE) over the chosen prediction horizon. Notice that a 15 min time-step is employed to meet the technical requirements for interchanging energy information between prosumer and the energy suppliers [29] in the Portuguese market.

3.1 Construction of Data Sets

This work uses the ApproxHull algorithm, proposed in [30], to select data for training, testing and validation sets used in model design. ApproxHull is an incremental randomized approximate convex hull (CH) algorithm, applicable to high dimension data, that treats memory and computation time efficiently The convex hull vertices obtained are compulsory introduced in the training set, so that the model can be designed with data covering the whole operational range.

Very briefly ApproxHull starts with an initial convex hull and subsequently the current convex hull grows by adding the new vertices into it. A pre-processing phase is performed on the original data set before applying the convex hull, scaling all data in the range of $[-1,1]$. The maximum and minimum of each dimension form the initial convex hull vertices. Then, it generates a population of k facets based on the current convex hull, selects the furthest points in the current facets population as new vertices of the convex hull, and integrates them in the current convex hull. A detailed explanation of ApproxHull may be found in [30].

The data set for the P1 problem is composed of three full years of data, from January 2016 to December 2018, while the data sets for P2 and P3 start in April 2016 due to the lack of HVAC data in the first three months of 2016. To each variable [v1, v2, v3], the admissible lags are associated to three periods: period 1 (lags immediately before the current sample), period 2 (lags around one day before), and period 3 (a week before). The admissible lags employed are: P1 - [1, 20], P2 - [4, 4, 0], P3 - [4, 0, 0].

A training set (S_{tr}) and a testing set (S_{te}) are used in MOGA execution (please see the next Section). When MOGA stops its execution, the non-dominated or preferable (if restrictions are employed) set of models is evaluated on a third data set, the validation data set (S_{va}). The size of S_{tr} is 60% of the whole set, and S_{te} and S_{va} have a size of 20% each. All convex hull points are incorporated in the training set. These sets are supplied to MOGA.

3.2 Multi-objective Genetic Algorithm Design

The model design is considered as a multi-objective optimization problem, with possible restrictions and priorities associated to the objectives. Genetic algorithms can evolve trained model structures that meet pre-specified design criteria in acceptable computing time. Globally, the ANN structure optimization problem can be viewed as sequence of actions undertaken by the model designer, which should be repeated until pre-specified design goals are achieved. These actions can be grouped into three major categories: problem definition, solutions generation and analysis of results (for a detailed explanation of the design framework used, MOGA, please consult [6]).

In this problem, for the former category, the objectives to minimize are the RMSEs of the training set, of the testing set, the model complexity ($O(\mu)$) and the forecasting error (ε_p). This last criterion is obtained as described in Eq. 1, where D is an additional simulation set, with p data points, and E is an error matrix (Eq. 2):

$$\varepsilon_p(D, PH) = \sum_{i=1}^{PH} RMSE(E(D, PH), i) \tag{1}$$

$$E(D, PH) = \begin{bmatrix} e[1,1] & e[1,2] & \cdots & e[1,PH] \\ e[2,1] & e[2,2] & \cdots & e[2,PH] \\ \vdots & \vdots & \ddots & \vdots \\ e[p-PH,1] & e[p-PH,2] & \cdots & e[p-PH,PH] \end{bmatrix} \tag{2}$$

MOGA is executed with 100 generations, population size of 100, proportion of random emigrants of 0.10 and crossover rate of 0.70. The admissible range of neurons vary from 2 to 10, while the possible inputs vary from 2 to 20, out of the possible 68. As MOGA employs a multi-objective formulation, its results are a set of non-dominated models or preferable models, if restrictions are employed. From this set, the final selection of one model is then performed based on the objective values obtained, ε_{va} and ε_p.

Taking into consideration the unconstrained results, a second design is performed, with constraints on some of the objectives. These are called Problems b, in contrast with the unconstrained ones, denoted as a. The restrictions considered for P1b, P2b and P3b were, respectively: RMSE(ε_{tr}). = [0,13; 0,25; 0,12]; RMSE(ε_{te}) = [0,13; 0,25; 0,12] and $O(\mu)$ = [200; 110; 120].

3.3 Ensemble Averaging

The output of MOGA is not a single solution, but a set of non-dominated models (or preferable models, if restrictions are used). This last set of models can be employed for ensemble averaging. As the forecasting criterion (1) is not used as a MOGA objective, in a few situations, models within the set can deliver a bad prediction performance. This can be solved if the median of the results obtained in the dominant (or the preferable) set, and not their mean value, is used as output of the ensemble.

4 Results and Discussion

4.1 ApproxHull

The results obtained by the ApproxHull algorithm, are presented in Table 1. As previously explained, the number of samples correspond to the available data for each problem description. The number of features is equal for the three problems, and the ratio used for sets distribution is constant as well.

4.2 Non-dominated Sets

Considering scaled data within the range $[-1,1]$, the minimum results of ε_{tr}, ε_{te} and ε_{va}, for the non-dominated models (P$_{*-a}$), or preferable sets (P$_{*-b}$) are presented in Table 2. There, the mean value is used for $O(\mu)$. It is possible to conclude that larger RMSE errors are obtained for the P2 problem, reflecting the modelling difficulty of the HVAC operation. The smallest values are obtained for P3; however, it should also be noted that the values of P1 are only slightly higher than the P3.

Table 1. ApproxHull results.

Problem	Samples	Features	CH Vertices	S_{tr}	S_{te}	S_{va}
P$_1$	104519	69	1711	62711	20903	20905
P$_2$	95780	69	10288	57468	19156	19156
P$_3$	95780	69	3145	57468	19156	19156

Table 2. Non-dominated/preferable sets statistics

Problem	ε_{tr}	ε_{te}	ε_{va}	$O(\mu)$
P$_{1-a}$	0,1218	0,1147	0,1149	70,8
P$_{1-b}$	0,1284	0,1147	0,1148	28,0
P$_{2-a}$	0,1974	0,1770	0,1770	57,0
P$_{2-b}$	0,1979	0,1774	0,1771	33,0
P$_{3-a}$	0,1134	0,1120	0,1139	68,6
P$_{3-b}$	0,1140	0,1123	0,1137	50,2

4.3 Selected Models

Equation (3) to (8) present the selected models for P$_{1-a}$, P$_{1-b}$, P$_{2-a}$, P$_{2-b}$, P$_{3-a}$ and P$_{3-b}$, respectively. Further details and performance values obtained with the selected models are presented in Table 3. In this table $\|w\|_2$ denotes the 2-norm of the linear parameters, which is related with the model condition.

Table 3. Selected models results.

Problem	Features	Neurons	$O(\mu)$	$\|w\|_2$	ε_{te}	ε_{tr}	ε_{va}	ε_p
P$_{1-a}$	18	8	152	56,89	0,1246	0,1170	0,1174	9,46
P$_{1-b}$	19	9	180	29,53	0,1604	0,1524	0,1545	9,34
P$_{2-a}$	3	2	8	4,61	0,2296	0,2044	0,2071	19,29
P$_{2-b}$	5	2	12	5,17	0,2289	0,2038	0,2066	18,74
P$_{3-a}$	18	9	171	21,82	0,1149	0,1131	0,1153	2,02
P$_{3-b}$	6	5	35	2,33	0,1652	0,1629	0,1673	1,39

$$v1(k) = f(v1(k-1), v1(k-2), v1(k-3), v1(k-4), v1(k-18), v1(k-96),$$
$$v1(k-669), v2(k-3), v2(k-6), v2(k-10)) \tag{3}$$

$$v1(k) = f(v1(k-1), v1(k-12), v1(k-13), v1(k-17), v1(k-18),$$
$$v1(k-92), v1(k-96), v1(k-97), v1(k-672), v1(k-673),$$
$$v1(k-675), v2(k-2), v2(k-5), v1(k-11), v2(k-12), v2(k-16),$$
$$v2(k-20), v2(k-99)) \tag{4}$$

$$v1(k) = f(v1(k-1), v1(k-14), v3(k-1)) \tag{5}$$

$$v1(k) = f(v1(k-1), v1(k-2), v1(k-97), v1(k-676), v2(k-5),$$
$$v2(k-6), v2(k-11), v2(k-13), v2(k-14)) \tag{6}$$

$$v1(k) = f(v1(k-1), v1(k-3), v1(k-4), v1(k-5), v1(k-95),$$
$$v1(k-100), v1(k-671), v2(k-11), v2(k-15), v2(k-20), v2(k-94)) \tag{7}$$

$$v1(k) = f(v1(k-1), v1(k-5), v1(k-92), v1(k-94),$$
$$v1(k-95), v1(k-98), v1(k-668), v1(k-676), v2(k-3),$$
$$v2(k-11), v2(k-20), v2(k-93)) \tag{8}$$

It should be noted that, apart from model (5), all models use samples of the modelled variable around 1 day and 1 week before. Most models use the outside temperature, but only model (5) uses the day code.

4.4   Accuracy of Prediction

To analyze the prediction results a one-month period, the month of October 2017, was employed. A prediction horizon of 12 h was considered, which means that 48 steps-ahead forecasts were employed. Figure 3 present the plots of real measured data (denoted as Target), and one-step ahead predictions for P_{3-b} (best model selected considering the prediction error), considering just one week of the prediction period.

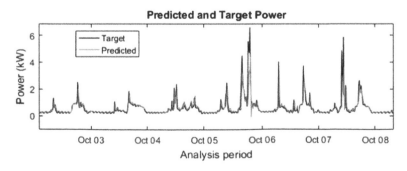

Fig. 3. Target and predicted values result of the best model of P_{3-b}.

In order to better graphically represent the comparison between target values and prediction values for all the problems, the 1-step prediction errors for all problems, for that particular week, are shown in Figs. 4, 5 and 6.

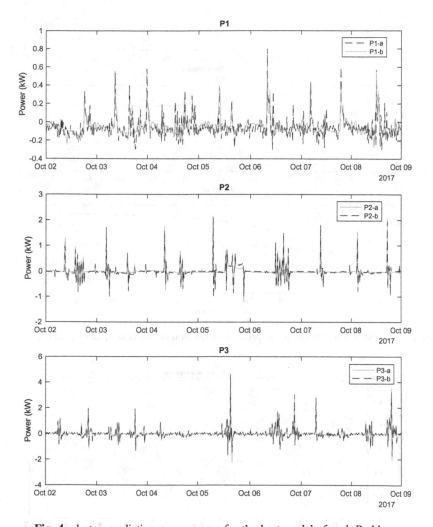

Fig. 4. 1-step prediction power errors for the best model of each Problem

The scaled prediction RMSE evolutions along the prediction horizon are presented in Fig. 5, for the 6 problems. With the exception of P2, the models obtained with a constrained formulation present smaller RMSE values.

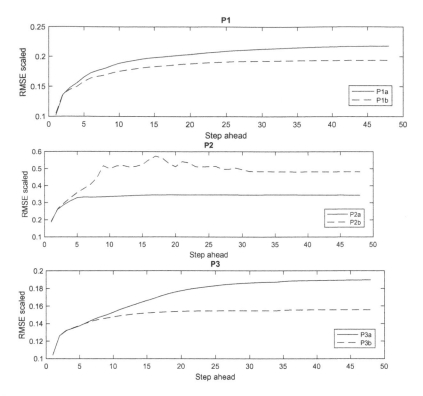

Fig. 5. RMSE evolution along the prediction horizon of the best model of each Problem.

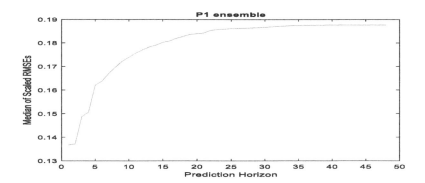

Fig. 6. RMSE median evolution along the prediction horizon for P1

4.5   Model Ensemble

The results presented before are obtained for the single model that has been selected for each of the six different cases. As explained in Sect. 4.3, the ensemble results are obtained by taking the median of the results obtained with each model in the preferable set. They are presented in Figs. 6, 7 and 8, for problems P1 to P3. If these figures are compared with Figs. 3, 4 and 5, it is clear that better forecasting performance is obtained with the model ensemble.

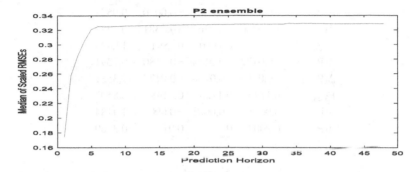

Fig. 7. RMSE median evolution along the prediction horizon for P2

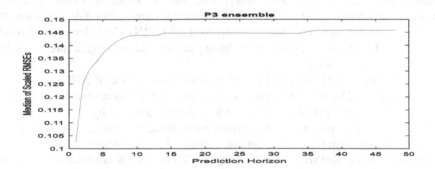

Fig. 8. RMSE median evolution along the prediction horizon for P3

Table 4 presents the ensemble RMSEs ($P_{\text{-ensemble}}$) for the training, testing and validation sets, as well as the respective differences ($\Delta(P)$) with the results obtained with the selected models, shown in Table 3. The last column shows additionally the prediction error obtained for the whole month of October 2017. It can be seen that for the RMSEs, the majority of the ensemble performs better than the selected models. In terms of the RMSE evolution over PH, which is the most important goal, all ensemble values are significantly better that the selected models.

Besides the analysis made in this work, it is important to compare the obtained results with the results of related studies for the prediction of energy demand in buildings. It is however quite tricky to perform a quantitative assessment of the proposed techniques, since their performances will depend on the training data used as input [7]. Additionally, it is not so common to find results of forecasting load demand within a prediction horizon, and this is much more difficult for individual households. If we narrow this analysis to the forecasting of different load classes (total, schedulable and non-schedulable), to the best of our knowledge, there are no available results.

Table 4. εtr, ε_{te} ε_{va} and ε_p – Ensemble (P$_{-ensemble}$) and best models.

Problem	ε_{tr}	ε_{te}	ε_{va}	ε_p
P$_{1\text{-ensemble}}$	0,1244	0,1163	0,1164	8,6203
Δ(P$_{1\text{-a}}$)	−0,0002	−0,0007	−0,0010	−0,9524
Δ(P$_{1\text{-b}}$)	−0,0360	−0,0361	−0,0381	−0,1473
P$_{2\text{-ensemble}}$	0,2124	0,1890	0,1891	15,429
Δ(P$_{2-a}$)	−0.0172	−0,0154	−0,0180	−0,7391
Δ(P$_{2\text{-b}}$)	−0,0165	−0,0148	−0,0175	−7,3581
P$_{3\text{-ensemble}}$	0,1143	0,1123	0,1148	6,8507
Δ(P$_{3\text{-a}}$)	0,0011	−0,0001	−0,0003	−1,4184
Δ(P$_{3\text{-b}}$)	0,0003	0	0.011	−0,3669

In [21], different prediction models (ANN-NAR, Hidden Markov Models, Support Vector Machines (SVM), MultiLayer Perceptrons and Deep Belief Networks) were designed for one-step daily and weekly forecasts. 8 weeks of 1-hour data were extracted from Pecan Street database, in 4 different scenarios. For daily forecasts, the RMSEs varied between 4.02 (ANN-NAR) to 1.48 (DBN) kW. Much better results were obtained in the present work, using three years of data, although a forecast ceiling of a half day is considered.

The authors of [31] compared the forecasting performance of ANNs, SVMs and Least-Squares SVMs, with different data resolutions and forecasting horizons, with several models, each applied to a different load profile, obtained by clustering the load profiles. In the same way as in the previous work, these are one-step-ahead forecasts, although with different forecasting horizons. The best results obtained for a house with similar load profile, RMSEs within the range of 0.8 to 1.6 kW are obtained for a time resolution of 30 min and a 12-hours forecast. Again, the results presented in this paper compare very favorably with these values.

5   Conclusions

This work focused on improving the accuracy of predictive models for the energy demand in buildings, using ensembles of RBF models designed with a MOGA framework. Real data, obtained from the Honda Smart Home US for three years, were used in this work. Three problems were analyzed, each one in two design versions (unconstrained and constrained).

For a common prediction horizon of 12 h, it was shown that the best results were for the problem in which the predicted variable is the power consumption of "other" loads (not considering the HVAC), followed by the Problem where the total demand is the predicted variable. The Problem where HVAC demand is the modelled variable obtains the lowest accuracy, due to the higher volatility of the time series.

Comparing MOGA designs, the best forecasting results were obtained with a constrained formulation, expect for the HVAC modelling. The model ensemble approach obtained, for all cases considered, the best prediction results. This scheme is obviously applicable to all classification, prediction and forecasting problems. For the case at hand, although a quantitative comparison is impossible, the prediction accuracy obtained in this work compares favorably with other existing approaches.

Future work will employ these forecasting models for model predictive scheduling of a real household in the South of Portugal, with PV energy production and electricity storage.

References

1. Pau, G., Collotta, M., Ruano, A., Qin, J.: Smart home energy management. Energies **10**, 382 (2017)
2. Ruano, A., Hernandez, A., Ureña, J., et al.: NILM techniques for intelligent home energy management and ambient assisted living: a review. Energies **12**, 2203 (2019)
3. Makridakis, S., Bakas, N.: Forecasting and uncertainty: a survey. In: Risk and Decision Analysis, pp 37–64 (2016)
4. Shamshirband, S., Jafari Nodoushan, E., Adolf, J.E., et al.: Ensemble models with uncertainty analysis for multi-day ahead forecasting of chlorophyll a concentration in coastal waters. Eng. Appl. Comput. Fluid Mech. **13**, 91–101 (2019)
5. Killian, M., Kozek, M.: Ten questions concerning model predictive control for energy efficient buildings. Build. Environ. **105**, 403–412 (2016)
6. Ferreira, P.M., Ruano, A.E.: Evolutionary multiobjective neural network models identification: evolving task-optimised models. In: Ruano, A.E., Várkonyi-Kóczy, A.R. (eds.) New Advances in Intelligent Signal Processing. SCI, vol. 372, pp. 21–53. Springer, Heidelberg (2011). https://doi.org/10.1007/978-3-642-11739-8_2
7. Foucquier, A., Robert, S., Suard, F., et al.: State of the art in building modelling and energy performances prediction: a review. Renew. Sustain. Energy Rev. **23**, 272–288 (2013)
8. Loyola, M.: Big data in building design: a review. ITcon **23**, 259–284 (2018)
9. Wei, Y., Zhang, X., Shi, Y., et al.: A review of data-driven approaches for prediction and classification of building energy consumption. Renew. Sustain. Energy Rev. **82**, 1027–1047 (2018)
10. Ahmad, T., Chen, H., Guo, Y., Wang, J.: A comprehensive overview on the data driven and large scale based approaches for forecasting of building energy demand: a review. Energy Build. **165**, 301–320 (2018)
11. Rahman, A., Srikumar, V., Smith, A.D.: Predicting electricity consumption for commercial and residential buildings using deep recurrent neural networks. Appl. Energy **212**, 372–385 (2018). https://doi.org/10.1016/j.apenergy.2017.12.051
12. Ai, S., Chakravorty, A., Rong, C.: Household power demand prediction using evolutionary ensemble neural network pool with multiple network structures. Sensors **19** (2019). https://doi.org/10.3390/s19030721
13. Wakui, T., Sawada, K., Kawayoshi, H., et al.: Optimal operations management of residential energy supply networks with power and heat interchanges. Energy Build. **151**, 167–186 (2017)
14. Do, H., Cetin, K.S.: Evaluation of the causes and impact of outliers on residential building energy use prediction using inverse modeling. Build. Environ. **138**, 194–206 (2018)
15. Fayaz, M., Kim, D.: A prediction methodology of energy consumption based on deep extreme learning machine and comparative analysis in residential buildings. Electronics **7**, 222 (2018)

16. Arabzadeh, V., Alimohammadisagvand, B., Jokisalo, J., Siren, K.: A novel cost-optimizing demand response control for a heat pump heated residential building. Build. Simul. **11**, 533–547 (2018). https://doi.org/10.1007/s12273-017-0425-5

17. Moon, J.W., Jung, S.K.: Development of a thermal control algorithm using artificial neural network models for improved thermal comfort and energy efficiency in accommodation buildings. Appl. Therm. Eng. **103**, 1135–1144 (2016)

18. Geysen, D., De Somer, O., Johansson, C., et al.: Operational thermal load forecasting in district heating networks using machine learning and expert advice. Energy Build. **162**, 144–153 (2018)

19. Babaei, T., Abdi, H., Lim, C.P., Nahavandi, S.: A study and a directory of energy consumption data sets of buildings. Energy Build. **94**, 91–99 (2015)

20. Amasyali, K., El-Gohary, N.M.: A review of data-driven building energy consumption prediction studies. Renew. Sustain. Energy Rev. **81**, 1192–1205 (2018). https://doi.org/10.1016/j.rser.2017.04.095

21. Mynhoff, P., Mocanu, E., Gibescu, M.: Statistical learning versus deep learning: performance comparison for building energy prediction methods. In: 8th IEEE PES Innovative Smart Grid Technology Conference Europe (2018)

22. Harkat, H., Ruano, A.E., Ruano, M.G., Bennani, S.D.: GPR target detection using a neural network classifier designed by a multi-objective genetic algorithm. Appl. Soft Comput. **79**, 310–325 (2019)

23. Hajimani, E., Ruano, M.G., Ruano, A.E.: An intelligent support system for automatic detection of cerebral vascular accidents from brain CT images. Comput. Methods Programs Biomed. **146**, 109–123 (2017)

24. Teixeira, C.A., Pereira, W.C.A., Ruano, A.E., Ruano, M.G.: On the possibility of non-invasive multilayer temperature estimation using soft-computing methods. Ultrasonics **50**, 32–43 (2010)

25. Mestre, G., et al.: An intelligent weather station. Sensors **15**, 31005–31022 (2015)

26. Khosravani, H., Ruano, A., Ferreira, P.: A comparison of four data selection methods for artificial neural networks and support vector machines. IFAC-PapersOnLine **50**, 11227–11232 (2017)

27. Honda: Honda Smart Home US (2019). https://www.hondasmarthome.com/

28. Ferreira, P.M., Ruano, A.E., Pestana, R., Kóczy, L.T.: Evolving RBF predictive models to forecast the Portuguese electricity consumption. IFAC Proc. Vol. 2 (2009). https://doi.org/10.3182/20090921-3-TR-3005.00073

29. Presidência do Conselho de Ministros: Decreto-Lei n.º 162/2019 de 25 de outubro (2019)

30. Khosravani, H.R., Ruano, A.E., Ferreira, P.M.: A convex hull-based data selection method for data driven models. Appl. Soft Comput. J. **47**, 515–533 (2016). https://doi.org/10.1016/j.asoc.2016.06.014

31. Yildiz, B., Bilbao, J.I., Dore, J., Sproul, A.B.: Short-term forecasting of individual household electricity loads with investigating impact of data resolution and forecast horizon. Renew. Energy Environ. Sustain. **3**, 3 (2018)

Solving Dynamic Delivery Services Using Ant Colony Optimization

Miguel S. E. Martins[1]([✉])(ID), Tiago Coito[1](ID), Bernardo Firme[1](ID),
Joaquim Viegas[1](ID), João M. C. Sousa[1](ID), João Figueiredo[2](ID),
and Susana M. Vieira[1](ID)

[1] IDMEC, Instituto Superior Técnico, Universidade de Lisboa, Lisbon, Portugal
{miguelsemartins,tiagoascoito,bernardo.firme,
joaquim.viegas,jmsousa,susana.vieira}@tecnico.ulisboa.pt
[2] IDMEC, Universidade de Évora, Évora, Portugal
jfig@uevora.pt

Abstract. This article presents a model for courier services designed to guide a fleet of vehicles over a dynamic set of requests. Motivation for this problem comes from a real-world scenario in an ever-changing environment, where the time to solve such optimization problem is constrained instead of endlessly searching for the optimal solution. First, a hybrid method combining Ant Colony Optimization with Local Search is proposed, which is used to solve a given static instance. Then, a framework to handle and adapt to dynamic changes over time is defined. A new method pairing nearest neighbourhood search with subtractive clustering is proposed to improve initial solutions and accelerate the convergence of the optimization algorithm. Overall, the proposed strategy presents good results for the dynamic environment and is suitable to be applied on real-world scenarios.

Keywords: Pickup delivery problem · Ant Colony Optimization · Local Search · Time windows · Dynamic requests

1 Introduction

According to data on *The World's Cities in 2018* United Nations (2018), it is clear that big cities are bound to grow both in size and number. This brings many concerns regarding the already problematic vehicle saturation on urban settlements. Small efficiency increases can have a big impact all-around when applied at a larger scale. This is especially relevant for transportation companies, whose main activity often involves driving and thus requires careful planning not to travel on heavy traffic situations.

This work was supported by FCT, through IDMEC, under LAETA, project UIDB/50022/2020.

M.-J. Lesot et al. (Eds.): IPMU 2020, CCIS 1237, pp. 327–341, 2020.
https://doi.org/10.1007/978-3-030-50146-4_25

The results presented in this article culminate from the development of a model for courier services based on the general demands of a real world scenario. At its core, the problem at hand focuses on transporting goods between two locations while efficiently handling the routing of a vehicle fleet, this is, a Vehicle Routing Problem (VRP). Since it operates on a dynamic environment, it is constructed to handle both changing traffic conditions and insertion of new customer requests over time.

2   Vehicle Routing Problem

The VRP category is a combinatorial optimization and integer programming set of problems. It deals on how to direct a fleet to serve interest points in the most profitable way. The most common problem on combinatorial problems is the large number of possible solutions. Many of these problems are in fact NP-Hard, i.e., there is no guarantee the optimal result can be reached in polynomial computation time Steiglitz (1982).

The origin of this problem can be traced to the Travelling Salesman Problem (TSP), in Flood (1956). Here the goal is to make a single salesman find the round-trip through each and all the cities only once. From here, a broader generalization was made in 1959, the Vehicle Routing Problem (VRP), credited to Dantzig (1959). The focus of the VRP shifts to finding the optimal set of routes for a fleet of vehicles to service a given set of customers. An example comparison between the two problems can be visualized in Fig. 1.

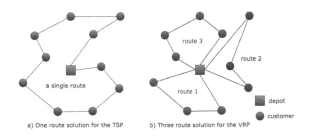

Fig. 1. TSP and VRP example for the same set of points

Many other constraints exist to model different real-life scenarios, many often approached as specialized sub-problems of the VRP Montoya-Torres (2015):

- **Capacitated VRP (CVRP)**: customers have specific demands for amount of goods and vehicles have finite capacity, which can be volume or weight;
- **Time Windows**: if locations must be always be visited within a time interval, these are called *hard time windows*. If the visit can be outside the time interval but the solution incurs into a penalty, they are *soft time windows*.

- **VRP with Pickup and Delivery (PDVRP)**: when there is a need to *pickup* an order from a specific location and *deliver* it to another. A *request* is then characterized by the pair pickup-delivery.

If the conditions of a problem instance remain unchanged during the whole problem, it is called *static*. Otherwise it is a *dynamic* problem, which can entail changes in orders, number of vehicles, travel costs, etc. The simplest approach to a dynamic problem is to solve it as static, and restart the optimization if the conditions change. When considering real world applicability, there might not be enough processing time to achieve good solutions between updates. An interesting modification is presented by Ferruci (2014), where the working span is separated into successive time intervals of fixed length, each is solved as a static instance. During this interval, any new requests are buffered for insertion at the end of the time span.

2.1 Metaheuristics

Metaheuristics are a common approach when exact methods are inapplicable to large search spaces. At their core, they are a set of general directives that can be adapted into a big variety of problems with little changes to the inner workings of the algorithm. Metaheuristics also have the advantage of exploring large search spaces without getting trapped in local minima by allowing temporary deterioration of the solution. However, these strategies often require long computational times and a careful parameter tuning to provide good solutions. An example of a widely used metaheuristic, and the one used for this paper, is the Ant Colony Optimization (ACO). As many other, this metaheuristic is based a natural behaviour, the foraging process of a colony of ants.

The ACO formulation is convenient to the VRP formulation since ants are assumed to travel on a weighted node graph, visiting every node once and stopping only when returning to the starting point. Each vertex $i \in V$ in the graph represents an interest point (be it pickup, delivery or depot). Each edge $ij \in C$ has associated two things. The first is *pheromone trail*, τ_{ij}, information left by previous ants (the attractiveness of past visits). The second characteristic of each edge is a cost η_{ij}, translating the effort to transverse that arc (usually distance between nodes), called the *heuristic information*.

Ants *construct solutions* by sequentially deciding which node to visit next, weighting *pheromone trail* information left by previous ants (the attractiveness of past visits) and the *heuristic information* (translating the effort to transverse that arc, usually distance between nodes). The probability of travelling trough an edge is then given by Eq. 1.

$$p_{ij}^k = \frac{\tau_{ij}^\alpha \times \eta_{ij}^\beta}{\sum_{j \in feasible\ set} \tau_{ij}^\alpha \times \eta_{ij}^\beta} \tag{1}$$

with τ_{ij} being the pheromone trail value and η_{ij} being the heuristic information, both associated with c_{ij}. Parameters α and β are ACO model parameters intended to balance relative weight importance.

Daemon actions is an optional step that includes other actions not based on real ant behaviour, and is further explained in Sect. 2.2. The last step of the algorithm is to *update pheromones*, computed using Eq. 2.

$$\eta_{ij} = (1 - \rho) \times \eta_{ij} + \rho \times f(s_{best}) \tag{2}$$

where ρ is the evaporation coefficient (rate at which pheromone trail values wear off) and $f(s_{best})$ is a value derived from the quality of the best solution s_{best}, commonly called the fitness value.

The decision process is probabilistic since moves with low attractiveness can still be selected, even if less often. To provide a better balance between choosing good moves and exploring less common moves, the *pseudorandom proportional rule* is used, which can be defined as:

- if $q > q_O$, Eq. 1 is used (biased exploration)
- if $q \leq q_O$, next node is dictated by $arg\ max_{u \in J_r^k}\{[\tau_{ij}]^\alpha \times [\eta_{ij}]^\beta\}$ (exploitation)

2.2   Local Search

Local search (LS) is intended to exploit the current solution s in search of improvements in the immediate neighbourhood solutions. LS procedures for VRP usually fall into the *edge-exchange* category. Starting from a feasible solution, new solutions are generated by deleting k edges and replacing them with new edges. These connect the same nodes in a different ways, completing the cycle, performing what is known as a *k-exchange*.

A noteworthy local search strategy for the VRP is the *or-exchange* or *or-opt*. Instead of deleting arcs as in *edge-exchange* strategies, a certain chunk of size s from the route, or *slice*, is separated from the rest. The *neighbourhood search space* is composed by all feasible solutions after moving the slice to each position around the original one, up to L steps in every direction. A schema for this can be seen on Fig. 2. Note that when exchanging more than one node, the original travel order of a slice is maintained.

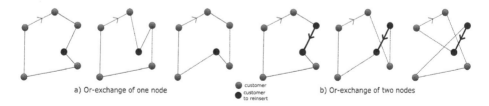

a) Or-exchange of one node   ● customer   b) Or-exchange of two nodes
● customer to reinsert

Fig. 2. Or-opt local search strategy for VRP, for k = 1 and k = 2

3   Proposed Approach

The motivation for this paper is to solve a real world delivery problem. Thus, considering the complex combinatorial challenge it represents, a metaheuristic algorithm is used to tackle the big solution search space. Ant Colony Optimization is a graph based method, which is an intuitive way to formulate a VRP. To make the algorithm more efficient, the ACO is to be paired with a Local Search to further exploit good solutions by trying small changes and that can lead to big improvements.

The central piece of the approach is the *static solver*, which optimizes an existing solution using a hybrid Ant Colony System paired with Local Search. The initial solution is given by an *initial solution constructor* module, as detailed in Sect. 3.2.

The approach taken to solve dynamic instances is based on the presented solution for static problems. Starting from an initial feasible solution, the *static solver* is applied for a limited period of time. This intends to represent the precomputation of requests already known beforehand. At this point we are at the beginning of the working span and will next repeat the same set of directives until all requests have been serviced:

- Insert new requests buffered during previous interval into the best routes;
- Deploy the best obtained solution after the insertion to the physical vehicles;
- Predict changes to deployed route at current interval's end-state, namely new vehicle positions and serviced requests;
- While current interval's end isn't reached, optimize the end-state prediction;
- Output from the *static solver* the best and the latest found solutions;
- Group these two solutions with the state of the deployed route to form the best selected routes;
- Update vehicle positions, serviced requests and distance matrix.

3.1 Mathematical Formulation

An adaptation from the well-known mathematical formulation of the VRPTW Hasle (2007) is presented, where the goal is to service as efficiently as possible a set of customers requests $\mathcal{R} = \{1,...,n\}$. Every request is defined by a pickup and a delivery location, each represented by a unique graph node out of a total $2n$ nodes. The full set of customers to service is given by $\mathcal{C} = \{p_1, d_1, p_2, d_2, ..., p_n, d_n\}$ where p_r is the pickup node of request r from the subset $\mathcal{P} = \{1, 3..., 2n-1\} \subset \mathcal{C}$ and d_r is the delivery node of request r from subset $\mathcal{D} = \{2, 4..., 2n\} \subset \mathcal{C}$.

In order to service each customer request we have available k vehicles. The depot location is split into k nodes forming the set of depot nodes $\mathcal{W} = \{1, ..., k\}$. The mathematical formulation is dimensioned for a graph $\mathcal{G}(\mathcal{N}, \mathcal{A})$, where $\mathcal{A} \subseteq \mathcal{N} \times \mathcal{N}$ is the set of graph edges representing all travel possibilities between nodes and $\mathcal{N} = \mathcal{W} \cup \mathcal{C}$ represent the graph nodes. $\mathcal{V} = \{1, ..., k\}$ is the set of homogeneous vehicles. Each vehicle has a fixed hire cost of e_k and a maximum capacity given by $q \geq l_i$, $i \in \{1, ..., n\}$ where l_i is the load capacity demand for customer i, i.e. how much of a vehicle's available capacity a request will occupy.

The variable x_{ij}^k is a binary parameter that expresses if a vehicle travels directly from node i to node j. For each arc ij, it takes the value 1 if vehicle

k travels directly from i to j and 0 otherwise. x_{kj}^k represents an arc between a depot node and node j, serviced by vehicle k. Similarly, x_{ik}^k expresses if an arc between a customer node i and the depot is serviced by vehicle k Each arc is also defined in terms of travel time, t_{ij} specific positive travel cost, c_{ij}, for each arc in $\mathcal{A}$. The variable s_i^k is the exact time of service at each point i by vehicle k and $[a_i, b_i]$ is the time window specified for node i. Finally, $\mathcal{M} = [m_1, m_2, m_3, m_4]$ is a vector composed by scaling factors, which define the priority of each term in the objective function.

minimize

$$
m_1 \sum_{k \in \mathcal{V}} \sum_{(i,j) \in \mathcal{A}} c_{ij} x_{ij}^k + m_2 \sum_{k \in \mathcal{V}} \sum_{j \in \mathcal{C}} e_k x_{kj}^k +
$$
$$
m_3 \sum_{k \in \mathcal{V}} \sum_{(i,j) \in \mathcal{A}} \max(s_j^k - b_j, 0) x_{ij}^k + m_4 \sum_{k \in \mathcal{V}} \sum_{(i,j) \in \mathcal{A}} \max(a_j - s_j^k, 0) x_{ij}^k \tag{3}
$$

subject to

$$
\sum_{k \in \mathcal{V}} \sum_{j \in \mathcal{C}} x_{ij}^k = 1, \ \forall \ i \in \mathcal{C} \tag{4}
$$

$$
\sum_{(i,j) \in \mathcal{A}} l_i x_{ij}^k \leq q, \ \forall \ k \in \mathcal{V} \tag{5}
$$

$$
\sum_{p \in \mathcal{P}} x_{hp_n}^k - \sum_{d \in \mathcal{D}} x_{gd_n}^k = 0, \ \forall \ h \in \mathcal{N}, \ \forall \ g \in \mathcal{N}, \ \forall \ k \in \mathcal{V}, \ \forall \ n \in \mathcal{R} \tag{6}
$$

$$
\sum_{j \in \mathcal{C}} x_{kj}^k = 1, \ \forall \ k \in \mathcal{V} \tag{7}
$$

$$
\sum_{i \in \mathcal{V}} x_{ih}^k - \sum_{j \in \mathcal{N}} x_{hj}^k = 0, \ \forall \ h \in \mathcal{C}, \ \forall \ k \in \mathcal{V} \tag{8}
$$

$$
\sum_{i \in \mathcal{V}} x_{ik}^k = 1, \ \forall \ k \in \mathcal{V} \tag{9}
$$

$$
x_{ij}^k (s_i^k + t_{i,j} - s_j^k) \leq 0, \ \forall \ (i,j) \in \mathcal{A}, \forall \ k \in \mathcal{V} \tag{10}
$$

$$
a_i \leq s_i^k, \ \forall \ i \in \mathcal{N}, \ \forall \ k \in \mathcal{V} \tag{11}
$$

$$
x_{ij}^k \in \{0,1\}, \ \forall \ (i,j) \in \mathcal{A}, \ \forall \ k \in \mathcal{V} \tag{12}
$$

Equation 3 defines the objective function. It is the sum of four different terms, each multiplied a scaling factor from the vector $\mathcal{M}$. The first term, $m_1 \sum_{k \in \mathcal{V}} \sum_{(i,j) \in \mathcal{A}} c_{ij} x_{ij}^k$, represents the route specific travel costs. It takes into account the sum of each cost, c_{ij}, for all arcs travelled by each vehicle of the fleet, this is where $x_{ij}^k = 1$. The second term, $m_2 \sum_{k \in \mathcal{V}} \sum_{j \in \mathcal{C}} e_k x_{kj}^k$, gives the cost of hiring vehicles. This value is independent of vehicle travelled distance since such costs are already covered in the first term. It sums the one-off cost of hiring each vehicle in the solution by multiplying each vehicle cost c_k by x_{kj}^k, which represents leaving the depot. This means that for all unused vehicles there is no travel out of the depot, making $x_{kj}^k = 0$ and thus not considering the unused vehicle cost in the objective function.

The third term, $m_3 \sum_{k \in \mathcal{V}} \sum_{(i,j) \in \mathcal{A}} \max(s_j^k - b_j, 0) x_{ij}^k$, considers how late each node visit is. Deliveries have a specified time window. The start of this time window is hard, but the ending of it is soft. This means that a location cannot be visited before the time window starts, but can be visited after it ends. The third term is used to penalize late arrivals, which is visiting a node after its time window has ended. This does not make the solution unfeasible but has a negative impact on the objective function. The third term sums how late all the deliveries were, and due to the max operator each contribution to total lateness is always equal or greater than zero. While this term handles the end time of a time window, the last term handles the start.

Since deliveries can't be made before the specified time window, if a vehicle arrives earlier it must wait. This wait time is penalized in the forth term, $m_4 \sum_{k \in \mathcal{V}} \sum_{(i,j) \in \mathcal{A}} \max(a_j - s_j^k, 0) x_{ij}^k$. This is done similarly to the previous term, but now looking at the difference between the arrival time and the time window start. Both of these terms use the max operator, making this objective function non-linear.

The constraint represented in Eq. 4 assures that all customers are serviced only once. Vehicle capacity constraint is represented by Eq. 5. Equation 6 assures the same vehicle services both pickup and delivery nodes of the respective request. The expression 7 defines that there is only one tour per vehicle, which can be empty. Equation at 8 ensures that if a vehicle arrives at a costumer location it also departs from the mentioned customer location. Equation 9 define all tours' ending location as the depot. For a vehicle to travel directly from i to j Eq. 10 states the arrival time at customer j is such that it allows travelling between i and j. Expression 11 specifies that the early limit of a time window a_i is hard and Eq. 12 denotes the x_{ij}^k variable as binary. Thus, the model is non-linear due to the non-linear max operations in Eq. 3, quadratic terms in Eq. 10 and integrality constraints at Eq. 12.

3.2   Initial Solution Constructor

To the presented method is given as input info on the orders to solve and on the fleet to manage. Orders are given as a list of pickup-delivery pairs, their respective locations and time-windows. Regarding the fleet, to the method is given the total number of vehicles, their capacity, travel speed and cost.

The *initial solution constructor* is then used to generate a feasible solution from scratch, to be used as a starting point for the *static solver* module. Since good starting solutions lead to faster convergence, a proposed new strategy based on Nearest Neighbourhood Search (NNS) and subtractive clustering is presented. Customer locations nodes are clustered, and to each clusters a single vehicle is assigned following the NNS heuristic. A comparison between using only NNS or clustered NNS is showed later in Table 1, which supports the decision to use pre-clustering.

To mitigate the negative effects of clustering locations, and not time availability, the maximum number of allowed vehicles is always used. This helps reducing lateness and excessive waiting times in the initial solutions, even if at the cost of using more vehicles. To do so, an heuristic starts with a very low *cluster influence range* parameter and subtractive clustering is applied on the midpoint of each pickup-delivery pair. Having an extremely low *cluster influence range* results in each midpoint being a cluster centre, i.e, it asks for as many vehicles as there are pickup-delivery pairs. If this parameter creates a solution with more vehicles than allowed by the problem, the parameter is increased and the process repeated until the maximum number of vehicles is in use.

3.3   Static Solver

The *static solver* combines ACO and Local Search. The overall logic can be seen on Fig. 3. Starting from a feasible solution, module specific variables are initialized. The *pheromone trail matrix* is also here generated. It is a $Q \times Q$ matrix, where Q is the length of $\mathcal{N}$, and it is initialized uniformly at the value $\tau_0 = 0.5$.

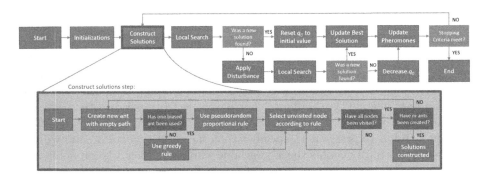

Fig. 3. Static solver, with the module *Construct Solutions* further detailed

Pheromone limits are dynamic and depend on the quality of the current best solution, meaning they will be updated every time a new global best solution, s_{best}, is found. Pheromone trail matrix limits are computed according to the following equations Gambardella (2015):

$$\tau_{max} = \frac{1}{\rho \times f(s_{best})} \tag{13}$$

$$\tau_{min} = \frac{\tau_{max}}{2Q} \tag{14}$$

The ant behaviour is guided by a pseudorandom proportional rule, as explained at Sect. 2.1. One *biased ant* is also used each cycle, where the value of q_0 is set to 1 making this ant greedily select the most attractive move.

At each cycle, the *static solver* can work either with or without the *local search* module. If no new solutions have been found for more than LS_{limit} cycles, as defined in Eq. 15, LS is used. This is done to let the algorithm use ACO to explore the search space without the overhead of the LS as long as solution improvements are being regularly found.

$$LS_{\text{limit}} = \frac{\log(\frac{1}{2Q})}{\log(1-\rho)} \tag{15}$$

When the iterative cycle starts, m new ants are generated using the *new ant* generator module. If an improving solution is found, both s_{best} and pheromone matrix limits are updated. After all ants are computed, the *local search* method is applied to the most promising ants on the top ant fitness list from current iteration (size specified by *top ants_number*) until no improvement is found, and are compared with the global best ant, s_{best}.

If no best solution was found up to this point on the current cycle, a disturbance is induced on s_{best} ant, and to the disturbed solution s_{new} the *local search* method is applied. The number of new solutions generated by the disturbance method is given by *perturbed_ants_number* times. Every s_{new} is saved on *exchange_memory* to avoid applying a local search on equal solutions and save valuable computational time. The disturbance introduced can either be a *shift*, where a pickup delivery pair is moved from on vehicle to another, or a *switch*, where one pair of each vehicle switch places.

When eventually the time limit is surpassed for the *static solver* module, it stops running iteratively and outputs the best found solution as result.

3.4 Dynamic Solver

To handle the dynamic changes over time, the working horizon is divided into successive intervals of length $time_{ss}$. During each interval, the working conditions (time and distances matrix) will remain unchanged and any new requests appearing during this interval will be buffered, i.e., saved for later insertion. When the end of the interval is reached, time and distance matrix are updated and new requests inserted in the already existing routes. This way, during the length of the interval, the problem instance does not change and the *static solver* method can be applied.

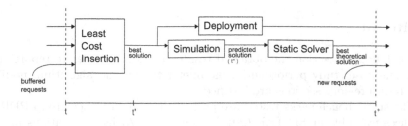

Fig. 4. Cycle of the dynamic solver

After running the static solver to generate a solution from scratch, with the currently known locations, an iterative cycle as represented on Fig. 4 is entered. Denoting the start of the interval as T, the best solutions obtained in the previous optimization interval serve as input to the *least cost insertion* module, which adds any new requests buffered during the previous cycle and outputs the best found solution. This is the solution to be deployed to the physical vehicles, $\dot{s}_{deployed}$, at time T', which would start to travel immediately accordingly to this route, ignoring any previous orders. Simultaneously to this vehicle deployment, another module called *end-state simulator* predicts where the vehicles will be at the end of the current time interval. This predicted route, $\dot{s}^*$, is used as input to the *static solver*, which will try to find improvements to this solution for $t_{ss} - (T' - T)$ minutes.

When the end of the time interval T^+ is reached, the *static solver* module is stopped. The best solution from the *static solver*, now $\dot{s}_{best}$, is grouped with the previously deployed solution, $\dot{s}_{best}$, and together they serve as input to next cycle's *least cost insertion model*, at T^+. All requests made available between T and T^+ are now introduced into the routes. The cycle is repeated until no more requests need to be serviced.

When solving dynamic instances, some changes are needed in the *static solver* module to account for requests that have already been partially serviced, this is, whose pickup has already been serviced in the real world and thus is irreversibly tied to a vehicle. This means that the *new ant* module must also account for vehicle history when constructing the feasible node list. When generating new solutions with the local search strategy, any time a feasibility check is done it needs to also take into account nodes outside the current planned route but which are on the vehicle history.

To account for changes between the predicted end-state environment and the real environment at T', another module is needed to check if at the interval's end the predicted state matches the real state and fix anything needed accordingly. The module *vehicle position update* is responsible for predicting where the state of the system will be at the end of the interval. It works by letting the vehicles follow their current routes until the time of end of interval, using new generated distance and times matrices. For the time being, to these new distance and times matrices random noise is added using the module *vehicle noise*. The *vehicle position update* is also used to re-calculate a predicted route at the end of the time interval, this time with the correct distance and times matrices.

4 Results

On the following section the obtained results are detailed. Before reporting on static and case study performance, parameter tuning and algorithm modifications from previous sections are justified.

The static benchmarks tested are part of the 100 customer group PDPTW instances available at SINTEF (2016). How pickup/delivery locations are distributed geographically can be seen in the file name prefix: *lc* files are clustered geographically; *lr* are randomly distributed; *lcr* mix previous two. Also, file prefixes ending in 1 have short and overlapped time windows and have many vehicles available, while files ending in 2 have long and spaced time windows and few available vehicles.

4.1   Initial Solution Construction

To compare implementing the construction method with and without pre-clustering the customer nodes, Table 1 was generated. Average values of 20 min runs are presented for both options, showing the following values with and without clustering: fitness of initial solution; time until solution with no lateness; best solution fitness after 20 min. The best values in each case are highlighted in bold.

Table 1. Comparing initial solution construction using Nearest Neighbourhood Search with and without pre-clustering with the best values for each comparison in bold

File name	Strategy	Initial Fitness	Time (min)	Final Fitness
lc101	Normal	**188.58**	0	**182.89**
	Clustering	190.49	0	**182.89**
lc201	Normal	9.97E+09	8.04	437.11
	Clustering	**5.53E+09**	**2.41**	**310.02**
lr101	Normal	**7.08E+09**	**2.54**	**348.73**
	Clustering	9.67E+09	2.82	355.91
lr202	Normal	**182.89**	0	**182.90**
	Clustering	190.49	0	**182.90**
lcr101	Normal	8.66E+09	6.42	421.25
	Clustering	**3.98E+09**	**2.19**	**309.89**
lcr201	Normal	**7.33E+09**	3.57	367.03
	Clustering	1.06E+10	**2.60**	**359.92**

Comparing the approach with and without clustering, no consistent improvement can be found for initial fitness and time until a solution with no lateness. However there is almost always a clear fitness improvement after running for 20 min. Pre-clustering the requests and then applying a NNS to each cluster will be the strategy used in all other sections.

4.2   Static Solver Modifications

On Table 2 we can see different runs for various modifications of the parameter q_0. It was noted that a high q_0 value is especially valuable at the start of the algorithm, but might negatively impact the search for better solutions once we are closer to the optimal solution. With this in mind, the idea of iteratively decreasing q_0 value with each iteration where a new solution isn't found is tested

in the problem at hand. For this formulation, the best combination found was
using $q_0 = 0.9$ initially and decreasing it by 20% each iteration with no better
solution found.

Next, in Table 3, we see a comparison between having or not a biased ant
in each cycle, this is, single ant with $q_0 = 1$ independently of previous q_0 values of decreasing factor. Since this change has a positive impact on the used
formulation, it was used for all the following tests.

4.3   Static Instances

Table 4 presents the average values for the error tables per type of file. Clustered
data behaves differently from the others files as it manages to always reach the
optimal number of vehicles for type 2 files. While for type 1 it does not reach
the optimal value for all of them, it reaches a lower distance than the given by
the optimal. For the other two the conclusions are similar, with average vehicle
number error of 2 or less. Distance does fall below the optimal value as with the
clustered files, but instead has an average error around 20%.

Table 2. Different tests done on the pseudorandom proportional rule parameter, q_0

File ID	q_0	Multiplier	Vehicles	Total Distance
18	0	-	24	1960.55
	0.9	-	24	1849.61
	0.9	0.9	**23**	1995.27
	0.9	0.8	**23**	**1878.41**
	0.8	0.5	24	1895.69
23	0	-	15	1440.08
	0.9	-	15	1450.43
	0.9	0.9	16	1555.86
	0.9	0.8	**14**	**1338.17**
	0.8	0.5	**14**	1417.95

Table 3. Test runs with and without biased ant

File	Vehicles		Distance	
	Normal	Biased ant	Normal	Biased ant
lr101	23.33	**20.70**	1875.33	**1775.71**
lr109	13.33	**13.20**	1463.15	**1376.57**
lr205	4.00	**3.60**	1224.59	**1190.37**
lrc101	14.33	**13.20**	1728.43	**1679.72**
lrc102	6.00	**5.20**	**1756.58**	1763.81

Table 4. Average distance from optimal solutions for static solver

File	Vehicle				Distance			
	Mean	%	Best	%	Mean	%	Best	%
lc1	0.31	3.46	0.22	1.39	−41.49	−4.06	−23.79	−2.00
lc2	0	0	0	0	5.38	0.91	0.96	0.16
lr	0.18	1.83	0.12	1.31	−19.44	−1.72	−12.14	−0.98
lr1	1.39	12.99	0.67	6.57	131.87	11.33	58.50	4.99
lr2	1.05	40.54	0.55	21.96	329.41	35.31	168.63	18.15
lr	1.23	26.17	0.61	13.93	226.34	22.80	111.17	11.29
lcr1	2	16.62	1.5	12.42	161.70	7.34	98.89	2.84
lcr2	1.35	38.51	1	29.16	371.76	28.86	226.77	15.66
lc	1.68	28.93	1.25	21.83	266.73	17.86	162.83	9.07

4.4 Dynamic Case Study

The presented case study is based on food service distribution centres and its most common design constraints. For this example, real customer requests from a typical day of a distribution company are detailed regarding real order hour and location. This specific example models the distribution services of one restaurant.

All pickups happen at the depot location, the restaurant, and it is assumed that after a customer makes an order, the delivery time window starts in 45 min and lasts for 15 min. For this implementation, the distance between real world locations is computed using the Haversine formula. However, the created model can work with any matrix giving the distances and travel times between nodes, for example given by the *Google Maps Distance Matrix API*.

First, the data is processed by the *static solver* module for 30 min, similarly to the approach on the static benchmarks. This will give a solution to be considered as the optimal when performing any dynamic tests. The case study data is similar to the benchmarks in size, with 47 requests to service. In terms of scheduling horizon, the case study matches the type 2 files. Visualization of a solution found can be seen on Fig. 5 a).

For the static run, 2 vehicles were able to service all requests without lateness for a total of 51.4 km travelled. For the dynamic solution, with a $t_{sim} = 15$ min and a $t_{look} = 45$ min the solution obtained is represented in Fig. 5 b). It uses 4 vehicles instead of 2 and has a total travelled distance of 68.6 km.

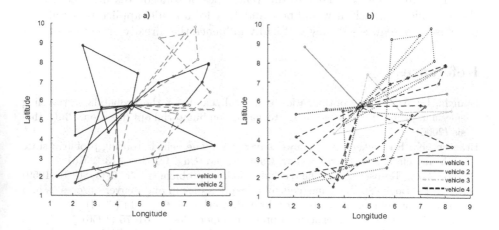

Fig. 5. Comparison of static and dynamic solutions for the case study

5 Conclusions

The main objective was accomplished by defining an algorithm able to solve the original problem. Further, the proposed approach is suitable for implementation in a real world environment, being able to deal with the tight time windows available to solve such heavily constrained problems. Considering the benchmark problems where the data is not clustered, the proposed approach does not match the competition when comparing with other multi-vehicle pickup delivery problems, except on a few instances. However under the case study presented, the proposed approach is able to give satisfactory solutions within the given time constraints, generating solutions without any delays in the delivery time windows.

In general, the proposed approach shows a good performance in the validation benchmarks. The introduction of the initial clustering step improved the overall results of the proposed approach, only little improvements are needed to consider them competitive with other approaches from the state of the art. The developed strategy for initial solution construction seems very promising and worth exploring further. The hybrid approach improves the solution quality, with the inevitable cost of extra computational time mainly due to the LS module.

5.1   Future Work

The proposed model can be improved by adding the ability to handle different types of vehicles and more advanced waiting strategies. A first attempt to improve the algorithm would be to further improve the Local Search method, namely adding more diversity to different types of solution disturbance. As for the dynamic approach, it would be mandatory for a model applied to a real case to refuse new requests if they will badly influence the already accepted routes.

References

Bianchi, L.: Ant colony optimization and local search for the probabilistic traveling salesman problem: a case study in stochastic combinatorial optimization. Ph.D. thesis (2006)

Braekers, K., Ramaekers, K., Nieuwenhuyse, I.V.: The vehicle routing problem: state of the art classification and review. Comput. Ind. Eng. **99**, 300–313 (2016)

Dantzig, G.B., Ramser, J.H.: The truck dispatching problem stable. **6**(1), 80–91 (1959)

Ferrucci, F., Bock, S.: Real-time control of express pickup and delivery processes in a dynamic environment. Transp. Res. Part B: Methodol. **63**, 1–14 (2014)

Flood, M.M.: The traveling-salesman problem. Oper. Res. **4**, 61–75 (1956)

Gambardella, L.M.: Coupling ant colony system with local search. Ph.D. thesis (2015)

Hasle, G., Lie, K.-A., Quak, E.: Geometric modelling, numerical simulation, and optimization: applied mathematics. SINTEF (2007)

Johnson, D.S., Mcgeoch, L.A.: The traveling salesman problem: a case study in local optimization. In: Local Search in Combinatorial Optimization, pp. 215–310 (1997)

Li, H., Lim, A.: A Metaheuristic for the Pickup and Delivery Problem with Time Windows (2001)

Mitrović-Minić, S., Krishnamurti, R., Laporte, G.: Double-horizon based heuristics for the dynamic pickup and delivery problem with time windows. Transp. Res. Part B: Methodol. **38**(8), 669–685 (2004)

Montoya-Torres, J.R., López, J., Nieto, S., Felizzola, H., Herazo-Padilla, N.: A literature review on the vehicle routing problem with multiple depots. Comput. Ind. Eng. **79**, 115–129 (2015)

SINTEF Applied Mathematics: Transportation Optimization Portal - TOP (2008). https://www.sintef.no/projectweb/top/pdptw/li-lim-benchmark/

Steiglitz, K., Papadimitrou, C.H.: Combinatorial Optimization: Algorithms and Complexity. Dover Publications, Mineola (1982)

United Nations: The World's Cities in 2018 - Data Booklet. Department of Economics and Special Affaris, Population Division (2018)

Acoustic Feature Selection with Fuzzy Clustering, Self Organizing Maps and Psychiatric Assessments

Olga Kamińska$^{(\boxtimes)}$ ⓘ, Katarzyna Kaczmarek-Majer$^{(\boxtimes)}$ ⓘ,
and Olgierd Hryniewicz$^{(\boxtimes)}$

Systems Research Institute, Polish Academy of Sciences,
Newelska 6, 01-447 Warsaw, Poland
{o.kaminska,k.kaczmarek,olgierd.hryniewicz}@ibspan.waw.pl
http://www.ibspan.waw.pl/

Abstract. Acoustic features about phone calls are promising markers for prediction of bipolar disorder episodes. Smartphones enable collection of voice signal on a daily basis, and thus, the amount of data available for analysis is quickly growing. At the same time, even though the collected data are crisp, there is a lot of imprecision related to the extraction of acoustic features, as well as to the assessment of patients' mental state. In this paper, we address this problem and perform an advanced approach to feature selection. We start from the recursive feature elimination, then two alternative approaches to clustering (fuzzy clustering and self organizing maps) are performed. Finally, taking advantage of the partially assumed labels about the state of a patient derived from psychiatric assessments, we calculate the degree of agreement between clusters and labels aiming at selection of most adequate subset of acoustic parameters. The proposed method is preliminary validated on the real-life data gathered from smartphones of bipolar disorder patients.

Keywords: Self organizing maps · Fuzzy C-Means · Recursive feature selection · Cluster agreement · Bipolar disorder episode prediction

1 Introduction

Bipolar disorder (BD) is a chronic mental illness characterized with changing episodes from euthymia (state of health) through depression to mania (euphoric state) and the mixed states (depressive and manic symptoms present). BD affects more than 2% of the world's population [1]. The risk of a new episode can be reduced significantly by an early detection and an appropriate treatment. However, the frequency of visits with the psychiatrist is usually insufficient to provide early intervention, and patients by themselves are usually not aware of the need of treatment if a new episode starts. Therefore, in the recent years,

© Springer Nature Switzerland AG 2020
M.-J. Lesot et al. (Eds.): IPMU 2020, CCIS 1237, pp. 342–355, 2020.
https://doi.org/10.1007/978-3-030-50146-4_26

smartphone becomes an increasingly important tool in the early prediction of a starting episode and smartphone-based objective data become a valid markers in predicting BD episode recurrence [2].

Although, the acoustic data collection can be performed during the everyday life of a patient, labeled data are limited only to days around the psychiatric assessments. The mental state of a patient between psychiatric assessments is often unknown. In majority of the related work, see e., [2,3], the problem is stated as a supervised learning task. Recently, to alleviate the problems of uncertainty about patients state and limited data, Kamińska et al. [4] applied unsupervised learning technique (self-organizing maps) to find groups (clusters) in acoustic data for each patient without taking into account the psychiatric assessments. As a consequence, the whole dataset was used for learning rather than constraining it only to a few days before and after the visit to the psychiatrist. Then, the relation between the learned clusters and the labels from psychiatric assessment was investigated. Kamińska et al. [4] have noted that the degree of agreement between the results of unsupervised learning (clusters learned on acoustic data) and the results of the psychiatric assessments is related to the type of the BD phases recognized during the psychiatric evaluation.

The extraction of acoustic features is accompanied by several uncertainties. First of all, the device used by a patient and the quality of its microphone are unknown. Secondly, voices in the background have some influence on the quality of the collected parameters. Also, due to unexpected technical issues, some phone calls were simply not recorded without clear understanding of this situation. Other phone calls are not recorded because BDapp on patient's smartphone was off due to unknown reason (patients declared that they will use it). Finally, due to memory shortages, not all frames of a phone call could are processed and are simply omitted. Only some of them were selected. At the same time, the process of assigning labels under the psychiatric evaluations is also accompanied by several uncertainties. Its outcomes are subject to the condition during the visit. The mental state and mood of a patient could change quickly after it. Nonetheless, the BD phase of a patient assessed by a doctor during the interview is usually assigned as labels to the surrounding days assuming a specific ground-truth for the analyses. Often authors apply 7 days before the psychiatric assessment and 2 days after [3]. However, there might not be one common ground-truth that describes adequately all patients. Therefore, it is essential to explore the structure of the acoustic data and investigate what is the quality of learned clusters.

Due to all these uncertainties, in this research we incorporate fuzzy clustering for alternative subsets of acoustic features. This paper is a continuation of our previous works [4]. However, now we aggregate data to a single phone call, whereas in citech26ref4 the aggregation has been done to one day. Aggregation process relay on collecting all acoustic parameters [5] for each phonecall for each patient and then calculate the quartiles for received values. We perform an extensive comparative analysis aiming at selection of: (1) smaller subset of acoustic parameters that will require less computational efforts; (2) fuzzy clustering algorithm for the considered smartphone-based acoustic data to reflect the related imprecision. This research is a step forward the superior goal that is an adequate prediction of BD episode recurrence using smartphone-based acoustic features.

The main novelty of this paper consists in application of Fuzzy C-means and Self organizing map algorithms for truncated datasets from the RFE algorithm. The unsupervised algorithm is selected over the supervised ones to alleviate the problem of limited labeled data and aiming at exploration of the whole data and investigating whether they can be grouped into clusters. The proposed approach is validated on the real-life dataset coming from the voice calls of patients suffering from bipolar disorder and the degree of agreement between learned clusters and psychiatric labels is evaluated.

That paper is organized as follows. In Sect. 2, methodology applied in this research is described, starting from the observational study on bipolar disorder to the brief description of the unsupervised approaches. Then, results of experiments are presented in Sect. 3. In the last Section, main conclusions are discussed.

2   Methodology

2.1   Observational Study and Acoustic Feature Extraction

Motivation for this research comes from analyzing real-world data collected in a recent observation study[1]. The study included patients diagnosed with bipolar disorder (F31 according to ICD-10 classification). In total, 33 patients were enrolled and used a dedicated smartphone application in everyday life for up to 15 months (starting in September 2017 and ending in December 2018). The study was conducted in the Department of Affective Disorders, Institute of Psychiatry and Neurology in Warsaw, Poland. Each patient was associated to a psychiatrist and control visits were scheduled. The evaluation of the mental state was performed by psychiatrists using both - the standardized measures of depressive and manic symptoms: Hamilton Depression Rating Scale (HAMD) and Young Mania Rating Scale (YMRS), as well as clinician's own assessment based on his experience with BD patients. The interviews were performed with various frequency depending on the need identified by the doctor or a patient.

Participants of the study received a dedicated mobile application, called *BDMon* able to collected acoustic features about phone calls. Patient's voice signal was divided into short 10–20 ms frames (withing a frame it is approximately stationary). With the use of an adopted version of a common library: openSMILE [5], the extended Geneva Minimalistic Acoustic Parameter Set (eGeMAPS) for voice research was extracted from each frame. The considered set contains 73 parameters connected with energy, spectral parameters and cepstral

---
[1] Data considered in this paper come from CHAD project − entitled "Smartphone-based diagnostics of phase changes in the course of bipolar disorder" (RPMA.01.02 00-14-5706/16-00) that was financed from EU funds (Regional Operational Program for Mazovia) in 2017–2018.

parameters (e.g., bandwidth energy, energy ration in different bands, relative volume) and 13 parameters connected with sound source (e.g., intonation contour). The meaning of that parameters is mostly connected with loudness in particular bands, voice energy, pitch etc.

The current state of the art lacks a clear indication which of the acoustic features are the best predictors of BD phase. In [4], the authors use the following 12 parameters: spectral slope in the ranges 0–500 Hz and 500–1500 Hz, energies in bands 0–650 Hz and 1000–4000 Hz, alpha ratio, ratio of the energy in band 50–100 Hz to the energy in band 1000–5000 Hz, spectral roll-off point the frequency below which 25% of the spectrum energy is concentrated, harmonicity of the spectrum, maximal position of the FFT spectrum, Hammarberg index, entropy of the spectrum, modulated loudness (RASTA), and zero-crossing rate. Different subset of parameters was selected with filter feature selection by [3], who use the following parameters: kurtosis energy, mean second and mean third MFCC, mean fourth delta MFCC, max ZCR and mean HNR, std and range F0. Within this research, recursive feature elimination are used as first step to select a subset of predictors.

2.2 Recursive Feature Elimination (RFE)

To obtain significant voice parameters we apply one of the automatic feature selection methods called Recursive Feature Elimination (RFE) [6]. The idea of the RFE technique is to build a model with all variables and after that the algorithm removes one by one the weakest variables until there will be achieved established number of variables. To find the optimal number of features cross-validation is used with RFE algorithm to obtain the best scoring collection of features.

2.3 Self Organizing Maps (SOM)

Results of the fuzzy c-mean clustering are compared to the clustering using self-organizing map algorithm known also as Kohonen network [7]. For each patient, we performed clustering of the call aggregates (quartiles) of theirs voice features using the *kohonen* package from the CRAN repository for R language, [8]. Important feature of the self-organizing maps is the preservation of neighborhood between the clusters in the two-dimensional space. Similarly, as in [4], we apply rectangular map topology of dimensions 3×1, which is intended to identify the two most distant affective states mania and depression and euthymia in between. The fourth affective state, mixed, is in its nature a combination of both, manic and depressive symptoms, and as confirmed with our preliminary experiments, it is more adequately represented as a mixture of depression and mania within the 3×1 Kohonen network, than as an additional dimension of a map, e.g., 4×1.

2.4 Fuzzy C-Means Algorithm (FCM)

Fuzzy C-means [9] another cluster algorithm is applied for the acoustic data and compared with the SOM algorithm. The specificity of this algorithm is that one value could be clustered as a cluster A with some membership, and the same value could be clustered as a cluster B with another membership. It might seem as thougest examples to identify mixed phase. In mixed phase patients could be for some time in depression and for some time mania. For the comparative purposes, the number of clusters was predefined and assumed as 3. Package e1071 from CRAN repository has been used.

2.5 Evaluation Metric

To compare the degree of agreement between the learned clusters in acoustic data and the labels from the psychiatric assessment, we apply the clustering agreement metric as applied in [4] according to the following formulas, and similar to the Rand Index.

$$f_{t_v,i} = \frac{\sum_{t=t_v-7}^{t_v+2} I\{c_t = i\}}{\sum_{t=t_v-7}^{t_v+2} I\{c_t \text{ is not missing}\}} \tag{1}$$

$f_{t_v,i}$ frequencies of each cluster
I indicator function taking value:
   1 - if the predicates are true
   0 - otherwise
c_t denote a cluster which was assigned to on day
    to a particular patient

where I is the indicator function taking value one if the predicate in curly brackets is true and zero otherwise.

We extract the data around every pair of visits and assign them to clusters trained on the remaining data. Therefore, we aim at comparing two groupings of the same data, one done by the clustering algorithm and the other by psychiatric assessments extrapolated to 7 days before and 2 after the visit.

Then, we compare the distributions f_{t_A} and f_{t_B} for two visits A and B with the normalized absolute difference

$$a_{t_A,t_B} = 1 - \frac{1}{2}\sum_{i=1}^{3} |f_{t_A,i} - f_{t_B,i}| \tag{2}$$

a_{t_A,t_B} normalized absolute difference

2.6 Diagram Representation

In order to easily visualize the received results, they were presented using a heatmaps diagrams for each pair of patient visits. On the X axis there are labels (received by psychiatrifor the first visit and on the Y axis there are labels for the second visit. The value presented in the graph is the degree of agreement (2) described above and calculated for each pair of visits to available patients and then averaged for the same pair of visits in reverse order. Values close to 0 mean that the clusters received on two different visits are different from each other, while values close to 1 mean that the clusters obtained on two different visits are similar to each other. The expected values for this chart are as follows. We strive for the highest possible values on the diagonal of the matrix - which means that the received clusters for visits with the same label are similar to each other. However, we strive to keep the remaining values as close to 0 as possible, which means that during two visits with different labels, the received clusters are different.

3   Experimental Results

Two set of experiments have been conducted. In the first one, the RFE method has been applied with various parameters for each patient individually due to the high variability between patients. In the second set of experiments, we apply fuzzy clustering vs. self organizing maps and evaluate the degree of agreement between learned clusters and psychiatric assessments.

3.1   RFE on Acoustic Data

RFE calculation has been prepared using *caret* package coming from CRAN repository and 10-fold cross validation. We present and discuss detailed RFE results for 2 exemplary patients. Both patients have 3 visits for which the patient used BDmon application in the surrounding days. At each visit the mental state of a patient was assessed by the doctor (e.g., euthymia, depression). The ground-truth for the analysis is considered as in [4] and [3] and all phone calls conducted in period starting from 7 days before visit, the day of visit and 2 days after visit received the label (which was given during that visit). The number of total labeled phone calls for considered patients is summarized in Table 1. For each phone call, 86 acoustic parameters are extracted for all its frames (frame length: 10–20 ms), so usually there are thousands of frames used as training data for the RFE algorithm for one phone call.

As observed in Table 1, data are incomplete and for 3 out of 6 visits records from some days are missing (2 days for visit from 20.06, 6 days for 07.08 and 1 day for 19.06).

Results obtained by the RFE for both patients are presented in Table 2 and Table 3. It turned out that for patient 1472 the best results are received when all 86 variables are taken into account and then accuracy of that model is slightly above 80%. Similarly, for patient 2582 the best model is the one that uses 86 parameters and its accuracy equals to 65%. However, the difference in accuracy for smaller number of parameters is relatively small and for example, the accuracy with 8 parameters (reduction by over 90%) amounts to 78.2% and 61.4%, respectively. These results are very promising.

Table 1. Summary of considered available data: psychiatric assessments and recorded phones calls for 2 exemplary patients in the days surrounding the visit to the doctor (labeled data).

Patient	Visit date	Psychiatric assignments	Nb. of phone calls	Nb of surrounding days with active BDmon app
1472	28.03	Mixed	188	10
1472	20.06	Euthymia	142	8
1472	07.08	Mixed	73	4
2582	19.06	Euthymia	57	9
2582	17.07	Mixed	75	10
2582	09.10	Depression	69	10

Another coefficient called Kappa presented in Table 2 and Table 3 points to classification accuracy because is useful during class imbalance. Classification is normalized at the baseline of random chance on dataset. Received values oscillate around 0.3 which is interpreted as fair agreement.

It is also important to mention that th RFE method is rather time-consuming. Calculations for one patient lasted more than $17\,h^2$ when it was conducted for 5% of randomly selected frames from each voice call.

Table 2. Results received by RFE methods for patient 1472

Patient 1472 (binary classification)				
Variables	Accuracy	Kappa	AccuracySD	KappaSD
4	0.761	0.192	0.003	0.011
8	**0.782**	0.250	0.006	0.027
16	0.793	0.291	0.006	0.010
86	**0.802**	**0.323**	**0.002**	**0.007**

Time calculation: 17.34 h

The applied RFE methods returned the subset of ordered variables, that achieved the best results for classification (according to the random forest algorithm).

The final subsets of first 10 parameters learned separately on data of both patients are presented in Table 4. Selected first 10 most relevant parameters in received order because of that received accuracy between 86 parameters and 8 parameters published in Table 2 and Table 3 are small.

Analysis of the received parameters shows that majority of the parameters coming from Mel-Frequency Cepstral Coefficinet Fourier transformate (group of

---

[2] 3,1 GHz Intel Core i7 500GB SSD, 16 GB Ram.

Table 3. Results received by RFE methods for patient 2582

Patient 2582 (3-class classification)				
Variables	Accuracy	Kappa	AccuracySD	KappaSD
4	0.538	0.138	0.011	0.026
8	**0.614**	0.293	0.005	0.010
16	0.636	0.328	0.004	0.008
86	**0.650**	**0.345**	**0.003**	**0.006**
Time calculation: 17.17 h				

Table 4. Parameters selected by RFE for patients 1472 (left) and 2582 (right)

No.	Parameter
1	**f0env_sma**
2	**slope0500_sma3**
3.	**pcm_fftMag_mfcc_1**
4	pcm_fftMag_mfcc_3_
5	**pcm_fftMag_mfcc_4**
6	loudness_sma3
7	**pcm_fftMag_mfcc_6**
8	pcm_fftMag_mfcc_9_
9	slope5001500_sma3
10	pcm_fftMag_mfcc_8_

No.	Parameter
1	**pcm_fftMag_mfcc_4_**
2	**slope0500_sma3**
3.	**f0env_sma**
4	pcm_fftMag_mfcc_2_
5	**pcm_fftMag_mfcc_1_**
6	pcm_fftMag_fband0250_sma
7	pcm_fftmag_spectralentropy-_sma_compare
8	pcm_fftMag_mfcc_0_
9	**pcm_fftMag_mfcc_6_**
10	pcm_zcr_sma

variables: pcm_fftMag_mfcc_n) which indicates range of pitch. We conclude that 5 (out of 10) parameters (marked in bold in Table 4) are present in both of the subsets and these 5 parameters are considered as *RFE subset* of parameters for the clustering algorithms in next Sections.

3.2 Fuzzy C-Means vs. Self Organizing Maps

The second set of experiments consists of application of fuzzy clustering and self organizing maps to two alternative subsets variables:

- *RFE subset* (described in Sect. 3.1.);
- 12 features subjectively selected by medical experts and data analysts as introduced in [4] denoted as *Kam20*.

For every phone call, the selected acoustic features were extracted from the 10–20 ms frames and aggregated by five-number summary consisting of quartiles (0, 0.25, 0.5, 0.75, 1). The resulting two datasets were used as input for the unsupervised learning algorithms. This experiment was performed for all patients available for the *BDmon* study who have at least one pair of assessments with at least 2 data points available during the surrounding days of the assumed ground-truth (-7 to $+2$ days). As a result, 17 patients and 62 pairs were considered in this experiment. Degree of agreement (2) was calculated for each pair of assessments. Next, we averaged results for the same (concordant) and different (incompatible) types of BD episodes assessed during visits.

It also needs to be noted, that in [4], which was the inspiration for the *Kam20* subset, different level of aggregation was applied and all of frames coming form mobile calls from the last 3 days were aggregated into quartiles. In this research, the aggregates are calculated for all individual frames from one phone calls and this procedure is implemented for each voice parameter.

Self Organizing Maps. Results received from SOMs for both subsets of parameters are depicted in Fig. 1. As observed, there are notable differences between the two heatmaps.

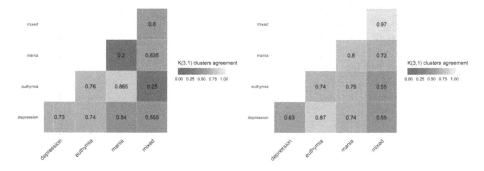

Fig. 1. Degree of agreement for SOM for (left) Kam20 parameters (right RFE subset of acoustic parameters.

On the left diagram, there are results of the degree of agreement where SOM algorithm is used on *Kam20* parameters. On the diagonal where we strive to achive values aiming to 1, we received following values: 0.73 for agreement between depression-depression and 0.76 for agreement between euthymia-euthymia which are quite satisfying, 0.2 for agreement between mania and mania seems insufficient due to specificity of that phase, and 0.6 for agreement between mixed and mixed - that value is rather high considering the overall difficulty to identify the mixed state (depressive and manic symptoms are present).

The results obtained on the remaining positions strive for the lowest possible values. As observed, the degree of agreement between states euthymia-mania is high and equal to 0.865, and this result is contrary to the knowledge of medical experts and their intuitions. Euthymia is the state of health and mania is the state of BD disease. We'd rather expect that clusters learned for data around these two types of labeled visits does not agree to a high degree. The remaining results are quite satisfactory like in case mixed-euthymia where receive 0.25 and between mania-depression where receive 0.54.

On the right diagram of Fig. 1, there are results of the degree of agreement where SOM algorithm used parameters selected by RFE methods.

On the diagonal we received the following values: 0.63 for agreement between depression-depression which is worse than using *Kam20*; 0.74 for the agreement between euthymia-euthymia which is quite satisfying; 0.8 for the agreement between mania and mania is an increase (compared to the previous heatmap) in a positive way, and 0.97 for agreement between mixed and mixed - that value is very impressive.

Results obtained on remaining position strive for the lowest possible values. All of the remaining values are above 0.5 (which could be a border) which is satisfactory only to some extent.

Summarizing this result, overall the *RFE subset* delivered better degrees of agreement than *Kam20*. It is surprising that when we aim to as low value as possible, got the highest degree (comaprison of mania and euthymia on the first heatmap).

Fuzzy C-Means. Similarly to SOM, results received from fuzzy clustering differ between alternative subsets of acoustic features. Figure 2 summarizes the degree of agreement for both subsets.

On the left diagram there are results of the degree of agreement where algorithm used *Kam20* parameters. On the diagonal where we strive to achieve values aiming to 1, we received the following values: 0.53 for agreement between depression-depression and 0.57 for agreement between euthymia-euthymia which are sufficient only to some extent; 0.2 for the agreement between mania and mania which is definitely insufficient due to the specificity of phase, and 0.6 for the agreement between the mixed and mixed.

Results obtained on the remaining position strive for the lowest possible values. The worst results were received again between the states of euthymia-mania where the degree of agreement is high and equal to 0.835 which is very high. In case of euthymia-mixed that could be sufficient because degree of agreement is only 0.25

On the right diagram there are results of the degree of agreement where algorithm used parameters selected by RFE methods.

On the diagonal we received the following values: 0.5 for the agreement between depression-depression which is moderately high compared to previous heatmaps; 0.65 for the agreement between euthymia-euthymia which is quite satisfying; 0.8 for the agreement between mania and mania where is increase in a positive way; and 0.97 for agreement between mixed and mixed - that value is very impressive.

Results obtained on remaining position strive for the lowest possible values. Two of the remaining values are below 0.5 which is satisfactory in comparison to the previous heatmaps. Overall others values are lowest then in previous examples.

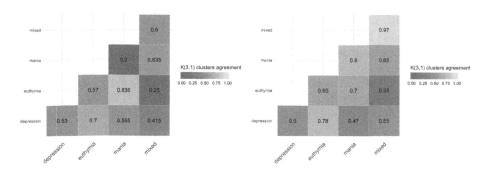

Fig. 2. Degree of agreement for Fuzzy C-Means algorithm (left) Kam20 parameters (right) RFE subset of acoustic parameters

Detailed results of the experiments for the Fuzzy C-mean algorithm using parameters coming from RFE methods are presented in Table 5. Meaning of columns is as follow:
Visit A & Visit B - means visit order number, Assessment A & Assessment B - contain received labels from psychiatrists, Grouping agreement - is the coefficient calculated for each case using (2), Valid days A & Valid days B - contain the number of days that have any of voice parameters for that day.

Comparative Analysis. To compare all of the received results, the average degree of agreement for particular pairs of visits were calculated and are presented in Table 6.

We distinguish pairs of visits with the concordant labels of psychiatric assessment, namely: euthymia-euthymia, depression-depression, mania-mania, mixed-mixed. The average degree of agreement for these concordant labels is the highest for the FCM applied in the *RFE* subset of with parameters and amounts to 0.71. For the incompatible labels (e.g. euthymia-depression), the average degree of agreement is expected to be the lowest, and again, the FCM on *RFE* approach outperforms other variants (0.51).

At the same time, it needs to be noted that the fact that for SOM we receive in general clusters that are not that well corresponding to the psychiatric assessments does not necessarily mean that the applied clustering approach is making a mistake. It needs to be noted that there is a lot of uncertainty related to the psychiatric assessments itself, including the fact that the episodes are determined depending on the total number of points using to the Hamilton Scale of Depression (HAMD), and e.g., 8 points are classified as depressive episode whereas 7 points are regarded still as a healthy episode (euthymia).

Table 5. The degree of agreement between clusters learned by the fuzzy clustering with RFE features vs. labels from psychiatric assessment based on pairs of visits

Patient ID	Visit A	Visit B	Assessment A	Assessment B	Grouping agreement	Valid days A	Valid days B
837	1	2	euthymia	euthymia	0.80	2	10
1472	2	3	mania	mixed	0.70	9	10
1472	2	4	mania	mixed	0.61	9	6
1472	3	4	mixed	mixed	0.96	10	6
2004	1	2	euthymia	euthymia	0.83	3	10
2004	1	3	euthymia	depression	0.96	3	10
2004	2	3	euthymia	depression	0.70	10	10
2582	2	3	mixed	mania	0.80	10	10
2582	2	4	mixed	euthymia	0.50	10	6
2582	2	5	mixed	depression	0.60	10	10
2582	3	4	mania	euthymia	0.40	10	6
2582	3	5	mania	depression	0.40	10	10
2582	4	5	euthymia	depression	0.70	6	10
4248	1	2	depression	depression	0.00	3	3
4248	1	3	depression	depression	0.60	3	10
4248	1	4	depression	depression	1.00	3	10
4248	1	5	depression	mania	0.33	3	3
4248	2	3	depression	depression	0.33	3	10
4248	2	4	depression	depression	0.33	3	10
4248	2	5	depression	mania	0.33	3	3
4248	3	4	depression	depression	0.80	10	10
4248	3	5	depression	mania	0.60	10	3
4248	4	5	depression	mania	1.00	10	3
4953	2	3	mania	mixed	0.70	10	3
4953	2	4	mania	depression	0.70	10	10
4953	3	4	mixed	depression	0.80	3	10
5656	1	2	euthymia	euthymia	0.55	7	9
5656	1	3	euthymia	depression	0.85	7	7
5656	1	4	euthymia	depression	1.00	7	3
5656	2	3	euthymia	depression	0.69	9	7
5656	2	4	euthymia	depression	0.55	9	3
5656	3	4	depression	depression	0.85	7	3
5659	1	2	euthymia	euthymia	0.40	3	10
5736	2	3	depression	mixed	0.40	10	10
5768	2	3	euthymia	euthymia	0.60	10	10
5768	2	4	euthymia	euthymia	0.80	10	10
5768	3	4	euthymia	euthymia	0.70	10	10
6139	3	4	mania	mixed	0.00	3	6
6139	3	5	mania	depression	0.00	3	2
6139	4	5	mixed	depression	0.00	6	2
6601	1	3	euthymia	euthymia	0.50	2	6
6601	1	4	euthymia	depression	1.00	2	6
6601	3	4	euthymia	depression	0.50	6	6
8866	1	2	depression	mixed	0.88	3	9
8866	1	3	depression	depression	0.91	3	8
8866	2	3	mixed	depression	0.36	9	8
9341	2	3	mania	mania	0.80	1	5
9341	2	4	mania	mixed	1.00	1	3
9341	2	5	mania	euthymia	0.20	1	5
9341	3	4	mania	mixed	0.00	5	3
9341	3	5	mania	euthymia	0.60	5	5
9341	4	5	mixed	euthymia	0.20	3	5
9829	1	2	depression	depression	0.10	2	10
9829	1	3	depression	depression	0.50	2	10
9829	1	4	depression	depression	0.00	2	10
9829	1	5	depression	depression	1.00	2	10
9829	2	3	depression	depression	0.50	10	10
9829	2	4	depression	depression	0.60	10	10
9829	2	5	depression	depression	0.20	10	10
9829	3	4	depression	depression	0.50	10	10
9829	3	5	depression	depression	0.80	10	10
9829	4	5	depression	depression	0.10	10	10

Table 6. The average degree of agreement for concordant, semi-concordant and incompatible labels. Concordant labels are as follows: E-E, D-D, M-M, X-X; incompatible labels: E-M, E-D, E-X; and semi-concordant labels: M-X, D-X, where E denoted euthymia (healthy state), D stands for depression, M for the mania and X denotes the mixed state

Avg degree of agreement	FCM-RFE	FCM-Kam20	SOM-RFE	SOM-Kam20
Concordant	**0.71**	0.47	0.79	0.57
Incompatible	**0.51**	0.65	0.73	0.68
Semi-concordant	**0.52**	0.54	0.63	0.59

3.3 Conclusions

Recursive feature elimination enabled to significantly reduce the number of important acoustic features (from 86 to 5 parameters). Futhermore, we conclude that the degree of agreement between clustering results and psychiatric labels vary between the applied fuzzy and SOM clustering methods and the subsets of acoustic features. The highest degree of agreement for concordant BD episodes has been achieved using RFE subset of 5 parameters and the fuzzy clustering algorithm (Fuzzy C-means). The lowest degree of agreement for incompatible BD episodes has been achieved also for the fuzzy clustering algorithm. Thus, the most satisfactory results for the degree of agreement has been achieved by this fuzzy clustering and the subset of acoustic features selected with the RFE method.

In future work, we consider representation of the smartphone data as fuzzy numbers instead of vectors with crips quartiles. Also, we plan to futher examine the characteristic of clusters obtained with Fuzzy C-means algorithm and interpret them from the medical perspective. It seems that RFE methods is very promising and should be tested for higher number of patients, to obtain more accurate results. However, that is time consuming, so it needs to be tested in more efficient environment.

Acknowledgment. Datasets considered in this paper were collected in the CHAD project – entitled "Smartphone-based diagnostics of phase changes in the course of bipolar disorder" (RPMA.01.02.00-14-5706/16-00) that was financed from EU funds (Regional Operational Program for Mazovia) in 2017–2018. The authors thank psychiatrists and patients that participated in the observational study for their commitment. The authors thank the researchers Karol Opara and Weronika Radziszewska from Systems Research Institute, Polish Academy of Sciences for their support in data preparation and analysis, as well as the researchers Monika Dominiak, Anna Wójcińska and Łukasz Święcicki from Institute of Psychiatry and Neurology for their advice and comments.

References

1. Grande, I., et al.: Bipolar disorder. In: The Lancet, vol. 387, no. 10027, pp. 1561–1572 (2016). https://doi.org/10.1016/S0140-6736(15)00241-X, http://www.sciencedirect.com/science/article/pii/S014067361500241X, ISSN: 0140-6736

2. Faurholt-Jepsen, M., et al.: Objective smartphone data as a potential diagnostic marker of bipolar disorder. Aust. New Zealand J. Psychiatry **53**(2), 119–128 (2019). https://doi.org/10.1177/0004867418808900, PMID: 30387368

3. Gruüerbl, A., Muaremi, A., Osmani, V.: Smartphone-based recognition of states and state changes in bipolar disorder patients. IEEE J. Biomed. Health Inform. **19**(1), 140–148 (2015)

4. Kamińska, O., et al.: Self-organizing maps using acoustic features for prediction of state change in bipolar disorder. In: Marcos, M., et al. (eds.) KR4HC/TEAAM - 2019. LNCS (LNAI), vol. 11979, pp. 148–160. Springer, Cham (2019). https://doi.org/10.1007/978 3-030-37446-4_12

5. Wollmer, M., Eyben, F., Schuller, B.: openSMILE - The Munich Versatile and Fast Open-Source Audio Feature Extractor (2010)

6. Guyon, I., et al.: Gene selection for cancer classification using support vector machines. Mach. Learn. **46**(1–3), 389–422 (2002). https://doi.org/10.1023/A:1012487302797

7. Kohonen, T.: Self-organizing maps (1995). https://doi.org/10.1007/978-3-642-97610-0

8. Wehrens, R., Kruisselbrink, J.: Flexible self-organizing maps in kohonen 3.0. J. Stat. Softw. **87**(7), 1–18 (2018). ISSN: 1548-7660, https://doi.org/10.18637/jss.v087.i07, https://www.jstatsoft.org/v087/i07

9. Bezdeck, J.C., Ehrlich, R., Full, W.: FCM: fuzzy C-means algorithm. Comput. Geosci. **10**(2–3), 191–203 (1984)

Knowledge Processing and Creation

Concept Membership Modeling
Using a Choquet Integral

Grégory Smits[1]([✉]), Ronald R. Yager[2], Marie-Jeanne Lesot[3],
and Olivier Pivert[1]

[1] University of Rennes – IRISA, 6074 Lannion, France
{gregory.smits,olivier.pivert}@irisa.fr
[2] Machine Intelligence Institute - IONA College, New Rochelle, NY, USA
yager@panix.com
[3] Sorbonne Université, CNRS, LIP6, 75005 Paris, France
marie-jeanne.lesot@lip6.fr

Abstract. Imprecise and subjective concepts, as e.g. *promising students*, may be used within data mining tasks or database queries to faithfully describe data properties of interest. However, defining these concepts is a demanding task for the end-user. We thus provide a strategy, called CHOCOLATE, that only requires the user to give a tiny subset of data points that are representative of the concept he/she has in mind, and that infers a membership function from them. This function may then be used to retrieve, from the whole dataset, a ranked list of points that satisfy the concept of interest. CHOCOLATE relies on a Choquet integral to aggregate the relevance of individual attribute values among all the representative points as well as the representativity of sets of such attribute values. As a consequence, a valuable property of the proposed approach is that it is able to both capture properties shared by most of the user-selected representative data points as well as specific properties possessed by only one specific representative data point.

Keywords: Fuzzy concept · Fuzzy measure · Choquet integral

1 Introduction

Datasets generally contain points described by precise numerical and categorical values whereas users, when they express properties about data, often use imprecise, complex, context-dependent and subjective concepts. As an illustrative example, consider a set of students described by their marks, prepared diploma, number of repeated years, having a scholarship grant, etc. These values are all precise ones. However, to describe some properties a student may possess, vague concepts as *promising*, *dynamic* or *weak* to name a few, whose definitions are subjective and context-dependent, are more naturally used. The definition of such complex properties is not an easy task for an end-user as data points may satisfy a given concept for very different reasons. A student may for instance

M.-J. Lesot et al. (Eds.): IPMU 2020, CCIS 1237, pp. 359–372, 2020.
https://doi.org/10.1007/978-3-030-50146-4_27

be considered *promising* because of his/her marks in a selective major, or for having restarted studies after a long break, among others.

This paper proposes the CHOCOLATE strategy, which stands for CHOquet-based COncept LeArning from a Tiny set of Examples, for inferring the membership function of a fuzzy concept from a small set of representative data points provided by a user. From these examples, CHOCOLATE infers a fuzzy measure that quantifies the extent to which combinations of attribute values match the underlying concept, as well as a measure quantifying the relevance of each attribute value individually. These two quantities are then aggregated using a Choquet integral, that gives the method its name, to determine the degree to which a point satisfies the concept.

The paper is structured as follows. Section 2 discusses the polysemous notion of *concept*, as well as some related works about concept learning. Section 3 presents the principles of the proposed approach, then detailed in Sect. 4, that infers a membership function from a small set of representative data points. Section 5 describes an illustrative example on a 2D toy dataset that serves to emphasize the characteristics of the approach. Section 6 shows the usefulness of CHOCOLATE to *fuzzy query by example* and recommendation systems.

2 Related Works on Concept Definition and Learning

This section discusses the notion of a concept, as well as some methods proposed in the literature for learning concepts. Among other things, these methods vary according to the types of concepts they can extract as well as their inputs.

Concept Extent and Intent. The notion of a concept has been widely used in artificial intelligence and data management, mainly following the twofold definition introduced by R. Wille [13]: a concept has an extensional definition, that consists of the set of its members (i.e. data points), and an intentional one that corresponds to the set of data properties (i.e. attribute values) shared by the members of its extent. It is possible to build, through so-called derivation functions, the extensional set of a concept from its intentional definition, and *vice versa*. Formal Concept Analysis, FCA [4], then identifies concepts as fixed points of the combination of these two functions: concepts are the sets of points that are exactly those that possess all the properties in the set, that in turn are shared by exactly the points in the set.

Another example is the fuzzy prototype approach [6,9] inspired from cognitive definitions of class representatives [10]: classes instances, defining the class extent, can be used to identify attribute values that are shared by the class members and that differ from the values observed by the members of other classes, so as to determine the class intent.

This concept definition, and its implementation in FCA and fuzzy prototypes, rely on a conjunctive definition of concepts, that in particular imposes that all members of the extent have common attribute values. It thus does not allow data to satisfy a concept for different reasons, i.e. to have disjunctive concept intents.

The proposed CHOCOLATE approach relaxes this constraint, e.g. making it possible to handle the case of *promising student* mentioned in the introduction.

Crisp and Fuzzy Concepts. Existing approaches to define and learn concepts can be structured depending on whether they consider crisp or fuzzy definitions, i.e. whether the point assignment to the concepts is binary or weighted.

The initial concept definition [13] mentioned above as well as Formal Concept Analysis are binary. They have been extended in several ways to Fuzzy Formal Concept Analysis (e.g. see the comparative study [2]), allowing for more flexibility: they rely on assessing the truth degree of the statement "the data in the extent all have the properties of the intent", where both the extent and the intent can be fuzzy sets (or only one of them). They thus still consider a conjunctive definition of concepts. The implementation of prototypes proposed in [6,9] also relies on fuzzy subsets. The CHOCOLATE approach proposed in this paper also relies on fuzzy concept definition, and outputs a membership function that associates each data point with its membership degree to the concept.

Discriminative Concepts and Counterexamples. A third property concerns the relations between concepts. In agreement with cognitive principles [10], fuzzy prototypes define a concept in opposition to others. They take into account distinctive and shared features between concepts. On the contrary, FCA deals with each concept independently from the others.

This distinction has consequences on the approach requirements: in order to take other concepts into account, it is necessary to dispose of examples of the considered concept, but also counterexamples, i.e. instances of the other concepts. This point of view opens the way to formalizing the concept learning task as a classification task, distinguishing the data points that satisfy a considered concept from those that do not satisfy it. Dealing with each concept individually means that only representatives of the current concept are required. This paper considers the case where a concept is learned independently from the others, only requiring a few representative points. To alleviate the requirements on the user, it also considers the case where only very few representative examples are available. When the number of available observations is very small, most classification approaches based on the observation of regularities cannot be applied.

Other Related Approaches. It may be argued that the considered definition and constraints on concept learning is related to the subspace clustering task [1, 5,12]: the latter is an unsupervised task that aims at decomposing a set of points into homogeneous subsets and simultaneously determining the subspaces they are situated in. As a consequence, it provides characterizations of the clusters, that may be interpreted as concepts extracted from the data. However, it does not take as input any example of a concept of interest and it would lead to identify several concepts at once, defining them in opposition to one another. As a consequence, it requires more data than the framework considered in this paper. The proposed CHOCOLATE approach can also be related to preference inference from partial definition [3], but it differs by the fact that, instead of ranking the items, it aims at inferring a measure from it to evaluate how much other points are related to the ones provided by the user.

3   Overview of the Proposed CHOCOLATE Approach

After presenting the notations and the toy example used throughout the paper, this section gives an overview of the proposed CHOCOLATE approach and presents its underlying principle, making clear the benefits of using a Choquet integral.

3.1   Notations

The paper relies on the following notations: we consider a dataset $\mathcal{D}$ containing n data points $\{x_1, x_2, \ldots, x_n\}$ from a universe $\mathcal{X}$. Each data point is described by the values it takes on m attributes $A_1, A_2, \ldots, A_m$. A data point x is thus represented as $x = \langle x.A_1, x.A_2, \ldots, x.A_m \rangle$, where $x.A_i$ denotes the value taken by x on the definition domain D_i of attribute A_i. The attributes can be numerical or categorical ones.

Throughout the paper, a *property* is defined as a couple made of an attribute and a value, denoted (A_i, p). A point x is said to possess a property (A_i, p) if $x.A_i = p$ and a set of properties s if $\forall (A_i, p) \in s$, $x.A_i = p$. Reciprocally, the properties of x are all the couples $(A_i, x.A_i)$ for $i = 1 \ldots m$.

As an example throughout the paper, we consider a dataset describing students on $m = 5$ attributes: *level* with domain $D_{level} = \{$Bsc.1, Bsc.2, Bsc.3, Mast., PhD$\}$, *major* with domain $D_{major} = \{$architecture, biology, literature, computer science, maths, physics$\}$, *grant* with domain $\{0, 1\}$ indicating whether the student receives a grant or not, *rep. years* with domain $[\![1, 10]\!]$ indicating the number of repeated years and *mark* with domain $\{A^+, A, A^-, \ldots D^-, F\}$.

3.2   Considered Concept Definition

Based on the discussion from previous section, the concepts considered in this paper are fuzzy ones, they possibly have a disjunctive definition and they are derived from very few instances that the user considers as representative, without disposing of counterexamples. The user-provided representative examples of concept C, denoted by $\mathcal{E}_C$, constitutes a partial extent for C, seen as an initial definition of the extent that has to be completed with appropriate points from the rest of the data with soft assignments.

Definition 1. *A concept C is a fuzzy subset of the universe $\mathcal{X}$, described by its membership function $S_c : \mathcal{X} \to [0, 1]$. It is induced from a user-defined partial extent denoted by $\mathcal{E}_C \subseteq \mathcal{D}$ and associated with a linguistic label (generally an adjective from the natural language).*

Table 1. Partial extension of the concept *promising student*

	Level	Major	Grant	Rep. years	Mark
x_1	Bsc.1	Physics	1	0	A^+
x_2	PhD	Maths	1	0	B^+
x_3	Bsc.3	Architecture	0	0	A^+
x_4	Bsc.3	Computer sciences	1	0	A^-
x_5	Mast.	Biology	1	0	A^+

Example 1. Throughout the paper, we consider that a user wants to define the concept *promising student* and initiates the concept definition by giving the partial extension given in Table 1.

This example illustrates the significant difference between the definition of a concept in [13] and the one used here. Following the discussion given in Sect. 2, we consider that different data points may satisfy the same concept without necessarily sharing all their attribute values. The proposed approach is thus able to capture both the specificities of each member from the partial concept extent as well as properties shared by most of them.

3.3 Underlying Principles of CHOCOLATE

CHOCOLATE initiates the definition of a concept with the few user-given representative points. The central question addressed in this paper is to define a measure to identify other points satisfying the concept of interest, in a fuzzy framework, i.e., quantifying the extent to which they satisfy it.

For a given point, CHOCOLATE first determines if its properties individually match the user-given partial concept extent. We propose to consider that the more frequent a property among the representative elements, the higher its individual matching degree. This step makes it possible to identify properties that appear important individually to define the considered concept.

The complementary step is to identify combinations of properties that are both possessed by the point to be evaluated and by representative elements of the concepts. It aims at preventing from combining properties that are individually frequent but actually are never observed together, thus leading to a non-additive way of evaluating the importance of a set of values.

Finally, the Choquet integral is used to aggregate the individual and combined assessments of the point properties: a high membership degree is assigned to points possessing a large set of properties that are individually representative. In addition, the proposed aggregation strategy makes it possible to identify a specific subset of properties possessed by only one (or a few) of the representative elements of the concept, and to give some importance to this subset even if it is composed of properties that are individually not shared by most of the other representative elements.

4   Proposed Approach to Concept Learning

This section details the proposed functions used in CHOCOLATE to determine the importance of each property individually and of subsets of these properties possessed by a point to evaluate.

4.1   The δ_i Function: Property Wrt. Partial Concept Extent

The function that quantifies the importance of a property (A_i, p) wrt. a partial concept extension $\mathcal{E}_C$ is denoted by δ_i: it serves to determine whether a value p taken by attribute A_i is representative of the given partial concept extent. The more p is shared, for attribute A_i, by representative elements of the concept, the more important it is.

As a consequence, the δ_i function checks how often p appears in $\mathcal{E}_C$ for attribute A_i: the degree $\delta_i(p)$ is the highest $(\delta_i(p) = 1)$ when p is observed for attribute A_i among all data points in $\mathcal{E}_C$. A first binary matching approach consists in defining:

$$\delta_i(p) = \frac{|\{x \in \mathcal{E}_C / x.A_i = p\}|}{|\mathcal{E}_C|}. \tag{1}$$

Example 2. Using the concept extension given in Table 1, we obtain the following matching degrees for the properties (*rep. years, 0*) and (*mark, A^+*): $\delta_{rep.\ years}(0) = 1$ and $\delta_{mark}(A^+) = \frac{3}{5}$.

Instead of performing a binary matching, based on strict equality, between the candidate value p and the A_i values of the data points in $\mathcal{E}_C$, a less drastic comparison can be performed allowing some approximation, as

$$\delta_i(p) = \frac{|\{x \in \mathcal{E}_C / sim_i\,(p,\ x.A_i) \geq \eta_i\}|}{|\mathcal{E}_C|}. \tag{2}$$

This relaxed definition for δ_i implies the use of an appropriate similarity measure sim_i on the domain of attribute A_i and the use of a similarity threshold η_i. For numerical attributes, this similarity measure can for instance be the absolute value of the difference, but many other possibilities exist, see e.g. [7]. It can also be indirectly defined by specifying a fuzzy partition on the concerned domain D_i, which makes it possible to take into account the indistinguishability of some values wrt. the satisfaction of the fuzzy terms that form the partition, see e.g. [11].

4.2   The μ Measure: Set of Properties Wrt Partial Concept Extent

Once the properties have been individually evaluated, the importance of *sets* of such properties is quantified. Whereas an individual value is considered important if it frequently appears in the partial concept extent, the importance of a subset of values depends on its size and whether it appears at least once in the partial concept extent. The fuzzy measure μ thus serves to quantify the extent to which the subset of attribute values matches one of the representative data points in $\mathcal{E}_C$. The μ score is maximal if the assessed set of properties exactly corresponds to one of the representative data points in $\mathcal{E}_C$. Denoting $s = \{(A_i, p_i), i = 1 \ldots |s|\}$ such a set of properties, the binary approach defines μ as:

$$\mu(s) = \max_{x \in \mathcal{E}_C} \frac{1}{m} |\{(A_i, p_i) \in s / x.A_i = p_i\}|, \tag{3}$$

where m denotes the number of attributes. It is straightforward to check that μ is a fuzzy measure: $\mu(\emptyset) = 0$, and if $s \subseteq s'$, then $\mu(s) \leq \mu(s')$ as, for any $x \in \mathcal{E}_C$, $\{(A_i, p_i) \in s / x.A_i = p_i\} \subseteq \{(A_i, p_i) \in s' / x.A_i = p_i\}$.

Example 3. Let us consider the two following examples: $s = \{(level,$ PhD$),$ $(rep.\ years, 0), (mark, B^-)\}$ and $s' = \{(level,$ PhD$), (major,$ maths$), (grant, 1),$ $(rep.\ years, 2), (mark, B^+)\}$, and again the concept extent from Table 1. Then, $\mu(s) = \frac{2}{5} = 0.4$ (data point x_2) and $\mu(s') = \frac{4}{5} = 0.8$ (data point x_2 again): s' is closer to a representative point than s that is only observed to the level $2/5$ in the best case.

As the δ_i's, the μ measure can be turned into a more gradual version by considering a similarity measure instead of a strict equality:

$$\mu(s) = \max_{x \in \mathcal{E}_C} \frac{1}{m} |\{(A_i, p_i) \in s / sim_i(p_i, x.A_i) \geq \eta_i\}|. \tag{4}$$

4.3 The S_C Function: Data Point Wrt. Partial Concept Extent

The final step combines the evaluations of atomic properties and set of properties. We propose to use the Choquet integral to perform this aggregation of the δ_i and μ evaluations. It especially takes into account, when comparing a set of properties wrt. representative data points (using the μ function), if these evaluated properties are individually specific to one data point or shared by many. This makes it possible to differentiate between a set of properties possessed by only a single representative data point and another set of properties of the same size but shared by several representative data points.

The candidate sets of properties are defined as the set of the most promising ones, according to their individual δ values: let H_j denote the subset of the j properties that best match the representative data points from $\mathcal{E}_C$, i.e. the j properties with maximal δ values. Formally, let σ be a ranking function such that $\sigma(i) = j$ iff. j is the rank of δ_i in decreasing order. H_j is then defined as $H_j = \{(A_i, x.A_i)/\sigma_i(i) \leq j\}$ for $j = 1 \ldots m$ and $H_0 = \emptyset$. In addition, $\kappa_j(x)$ denotes the jth value among the δ_i, i.e. the matching degree of the jth most representative property possessed by x wrt. concept C (formally $\kappa_j(x) = \delta_{\sigma^{-1}(j)}(x.A_{\sigma^{-1}(j)})$).

We thus propose to define the membership degree of a data point x to concept C based on the set of representative examples $\mathcal{E}_C$ as

$$S_C(x) = \sum_{j=1}^{m} (\mu(H_j) - \mu(H_{j-1})) \kappa_j(x), \tag{5}$$

Table 2. Computation details of $S_C(x)$ for $x = \langle \text{PhD, literature, 1, 0, B}^- \rangle$ and $C = promising\ student$ using $\mathcal{E}_C$ from Table 1.

$\delta_{level}(\text{PhD}) =$	$\dfrac{1}{4} = \kappa_3(x)$	$H_1 = \{(rep.\ years, 0)\}$	$\mu(H_1) = \dfrac{1}{5}$
$\delta_{major}(\text{literature}) = 0$	$= \kappa_4(x)$	$H_2 = H_1 \cup \{(grant, 0)\}$	$\mu(H_2) = \dfrac{2}{5}$
$\delta_{grant}(1) =$	$\dfrac{3}{4} = \kappa_2(x)$	$H_3 = H_2 \cup \{(level, \text{PhD})\}$	$\mu(H_2) = \dfrac{3}{5}$
$\delta_{rep.\ years}(0) =$	$1 = \kappa_1(x)$	$H_4 = H_3 \cup \{(major, \text{literature})\}$	$\mu(H_2) = \dfrac{3}{5}$
$\delta_{mark}(\text{B}^-) =$	$0 = \kappa_5(x)$	$H_5 = H_4 \cup \{(mark, B)\}$	$\mu(H_2) = \dfrac{3}{5}$

Let us notice that the implication $x \in \mathcal{E}_C \to S_C(x) = 1$ holds only if $\mathcal{E}_C = \{x\}$ which means that C is defined by only one prototypical example. This is because the S_c measure makes it possible to capture a similarity with atypical examples in $\mathcal{E}_C$, here understood as an example with possibly low internal resemblance to the other examples from C.

Example 4. Consider the data point $x = \langle \text{PhD, literature, 1, 0, B}^- \rangle$ and the partial extent of the concept *promising student* from Table 1. Its membership degree can be computed as detailed in Table 2: it shows the individual δ_i values, their respective rank κ_j, and the derived H_j together with their μ values. The satisfaction of x wrt. concept C is:

$$S_C(x) = \frac{1}{5} \times 1 + \frac{1}{5} \times \frac{3}{4} + \frac{1}{5} \times \frac{1}{4} + 0 \times 0 + 0 \times 0 = \frac{2}{5}$$

This means that the considered student is rather not on the side of *promising*, with a membership degree slightly lower than 0.5, based on his/her comparison with the user-provided examples of *promising students*. Indeed, he/she does not share enough set of properties with these examples, even if he/she received a grant and did not repeat years.

5   Empirical Study of CHOCOLATE's Behavior

This section illustrates the behavior of the approach on 2D toy examples. We first consider the example represented on Fig. 1, for which the user provides $\mathcal{E}_C$ of size 4, represented by the five diamond points. CHOCOLATE is applied using the relaxed version of the δ_i and μ functions, respectively defined in Eq. (2) and (4) setting sim to the complement to 1 of the Euclidean distance normalized by its maximal observed value and with $\eta_i = 0.7$. The inferred membership degrees for all points on a regular grid in the considered universe are shown by the grey levels, the points whose membership degree is lower than 0.2 are not shown.

Results. A non-uniform effect around the points in $\mathcal{E}_C$ can be observed, as well as a grouping effect: the membership degrees are higher around the three

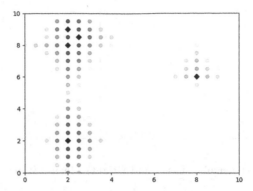

Fig. 1. Membership degrees inferred by CHOCOLATE for the concept with partial extent $\mathcal{E}_C$ defined by the diamond points.

grouped representative points in the region around (2, 8) than for the isolated point with coordinates (2, 2) and even more so for (8, 6). The latter however still has a non-symmetrical impact on the concept definition. The results also show that CHOCOLATE identifies that $A_1 = 2$ is a very important value: all points with this value have a high membership degree. The second important characteristic is $A_2 \approx 8.5$ although to a lesser extent. This explains the absence of symmetry around the representative point with coordinates (8, 6).

This toy dataset clearly shows the main characteristics of our approach: the δ_i functions quantify the importance of each property individually. This allows for instance to capture the fact that the four representative points on the left side of the domain share close A_1 values. Then the fuzzy measure μ gives more importance to the points located close to the three top-left points, because they share close values on the two dimensions. However contrary to a purely distance-based approach (see below), the proximity with at least one of the representatives is also taken into account, this is for instance the case of the points close to the right hand side representative.

This strategy thus enforces an interesting compromise between the importance of each property, the number of properties shared between a data point and a representative, and the number of representative elements of the concepts possessing these properties. This leads to a strategy that is both i) able to handle a comparison with very different representative elements of a concept taking into account their diversity and ii) able to identify properties shared by several of these representative elements.

Comparison to Aggregated Distance Approaches. As baseline approaches, we propose to compare CHOCOLATE with membership degrees derived solely from the (Euclidean) distances to the representative points in $\mathcal{E}_C$. Two aggregation operators are considered, defining the membership degrees as:

$$\mathcal{S}^1_C(x) = \min_{z \in \mathcal{E}_C} d(x, z) \qquad \mathcal{S}^2_C = \text{avg}_{z \in \mathcal{E}_C} d(x, z), \qquad (6)$$

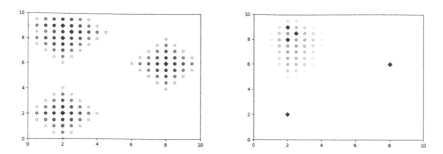

Fig. 2. (Left) Nearest neighbor-based and (Right) Mean-distance-based membership computations.

where $d(x, z) \in [0, 1]$ is the Euclidean distance normalized by the maximum value. The obtained results are shown on Fig. 2: $\mathcal{S}_C^1$ is a nearest-neighbour approach that leads to membership degrees uniformly distributed around the given representative points and does not make it possible to characterize the underlying concept. On the other hand, the mean-distance approach is very sensitive to a grouping effect, even other normalisation strategies would not allow for a fair integration of the isolated representative points.

Case of Disjunctive Concepts. As a complement to illustrate CHOCO-LATE's ability to both capture and give importance to properties shared by several user-selected representative data points, and to be able to take into account an isolated representative data point, another situation is depicted in Fig. 3, which makes the isolated points more isolated: the user-selected representative data points are composed of a quite compact cluster of data points (in the top left corner) and two isolated data points, one being close to that cluster on the x-axis property.

CHOCOLATE's behavior is interesting as the generated membership function is "attracted" by this cluster of representative data points but still takes into account the isolated representative data points and does not discard them. Again, the presence of two distant representative points, one on the bottom left and one on the right, makes the assignment around the cluster asymmetrical but with an emphasis on the x-values close to 2.

Applied on this same representative data points setting, the minimal and mean distance-based strategies are illustrated in Fig. 3. It shows that for the nearest neighbor strategy the presence of a compact group of representative data points has no impact on the assignment around the atypical data points. As for the mean distance-based approach, this larger compact group of representative data points simply gives more weight to the area they are located in.

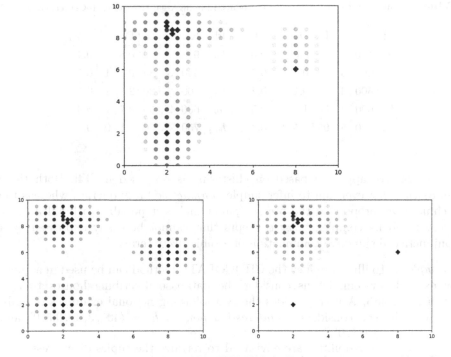

Fig. 3. Membership degrees inferred by CHOCOLATE (Top), nearest-neighbor (Bottom left) and mean-distance (Bottom right), for a partial extent containing a cluster of representative data points and two isolated ones

6   Examples of Possible Applications

In many applicative contexts where user interaction is considered, a crucial issue is to capture the user's intent requiring a minimum of knowledge from him/her. In such contexts, the proposed CHOCOLATE approach may be applied with the aim of making the most of only a few examples representing the user's intent. Two examples of such applicative contexts are described here.

Fuzzy Query by Example. Query by example is a database and information retrieval paradigm [15] that is used to acquire results based either on: one (or several) input tuple(s) provided by the user, or the evaluation by the user of a set of examples (positively, negatively, ...) reflecting the content of the database. The expected output contains elements that are similar to the input tuple(s) provided as example(s), or that reflect the choices of the user if prototypical examples are evaluated.

In [8], we proposed an approach that infers a fuzzy query from user-assessed prototypical examples, based on an algorithm determining the fuzzy modalities (from a fuzzy vocabulary defined on the attribute domains) that best represent the positive examples (and at the same time discards the negative ones).

370   G. Smits et al.

Table 3. Houses dataset (R: rent, N: nbRooms, L: livingArea, D: distanceToCenter, G: garden)

Id	R	N	L	D	G	$\mathcal{S}(h_i)$	Id	R	N	L	D	G	$\mathcal{S}(h_i)$
h_1	450	4	83	0.6	0	0.5	h_6	900	7	99	0.8	1	0.4
h_2	700	5	82	0.5	1	0.3	h_7	450	3	35	0.1	1	0.5
h_3	500	3	45	0.1	1	0.2	h_8	300	2	20	2.1	1	0.1
h_4	600	3	50	1.7	1	0.5	h_9	600	2	50	0.7	1	0.3
h_5	400	5	95	3.5	0	0.3	h_{10}	750	4	50	1.1	0	0.5

An alternative approach based on clustering is proposed in [14]. Both these approaches, however aim to infer simple *conjunctive* fuzzy queries, whereas the technique we propose in the present paper makes it possible to build a fuzzy query involving complex fuzzy concepts that cannot be easily expressed by a conjunctive/disjunctive combination of atomic fuzzy terms.

Example 5. To illustrate how the CHOCOLATE method can be used in a query by example system, let us consider the database describing houses to rent given in Table 3. A user provides the two following fictional examples describing what he/she considers as *interesting houses*: $h = \langle 400, 3, 40, 0.5, 0 \rangle$ and $h' = \langle 800, 6, 99, 0.5, 0 \rangle$.

Then, two possibilities are envisaged to retrieve the tuples of interest from the database. If a strict version of the δ_i and μ functions is used (Eqs. 1 and 3), then a query of the following form is executed to return the tuples whose values on the different attributes are present in the partial concept extent:

Q = SELECT * FROM houses
    WHERE rent IN (400,800) AND nbRooms IN (3,6) AND...;

When relaxed versions of the δ_i and μ functions are used (Eqs. 2 and 4), the submitted query returns all the tuples whose values are close enough to those present in the partial concept extent:

Q_r = SELECT * FROM houses
    WHERE (sim(rent, 400) <= ETA1 OR sim(rent, 800) <= ETA1) AND ...;

CHOCOLATE is then applied to the tuples returned by the query to rank them according to their satisfaction of the concept partially defined by the user-provided representative tuples. Column $\mathcal{S}(h_i)$ in Table 3 gives the results obtained with the relaxed version of δ_i and μ (Eqs. 2 and 4) and similarity

$$sim_i(v, v') = 1 - \frac{|v - v'|}{\max_h h.A_i - \min_h h.A_i},$$

where A is the considered attribute and the threshold values are 0.7 and 0.9 for ETA1 and ETA2 respectively.

These results show that the membership function built using CHOCOLATE takes into account the specificity of each of the two provided representative elements: h_6 receives a high score because of its similarity with h', and h_7 because it is rather close to h. It also shows that the distance to the city center and the absence of a garden are important properties that counterbalance other less desired properties, as e.g. in the case of h_{10}.

Data Point Recommendation. A second possible applicative context where CHOCOLATE can be useful is that of *recommendation systems*. Most recommendation strategies, especially those based on collaborative filtering, are not able to suggest meaningful recommendations for new users, which is known as the *cold start problem*. A particular interesting aspect of the proposed approach is to be able to find data points that best match the user history, even for a very small available history.

Consider for instance a user who has shown an interest for the manga *Dreamland* by Reno Lemaire, the novel *L'Étranger* by Albert Camus, the science-fiction novel *Ravage* by René Barjavel and the philosophical essay *Discours de la méthode* by René Descartes. Thematically speaking, these books are very different but they all have a French author and half of them are novels. The Choquet integral-based approach would favor French novels by Camus or Barjavel and then French philosophical essays or French comics, etc.

7 Conclusion and Perspectives

When knowledge has to be acquired from user-defined data, one cannot expect to have large labelled training sets. In this work, we face the problem of trying to build, from a small set of data points that illustrate the possible meanings of a concept, a function that can then be used to quantify the extent to which a new data point matches the underlying concept. With such a small set of representative data points and without any negative examples, classical statistical approaches cannot be applied. This is why we propose CHOCOLATE, an alternative strategy based on a generic aggregation framework, namely the Choquet integral involving a fuzzy measure. This method can both identify subsets of properties shared by representative data points of the concept and take into account their specificities, properties possessed by only one or a few representative elements. This approach can be used to recommend data points that best match a user's interest expressed by a few selected examples only.

In this paper, the proposed approach has been applied on toy examples to show its behavior. The interesting obtained results open several perspectives for future works. The first one, despite the topical aspect of the approach, is to assess the relevance of the generated results in a concrete applicative context of a recommendation system, so as to show that relevant recommendations are provided for users having a scarce history. At a more theoretical level, another perspective concerns the extension of CHOCOLATE to handle imprecise descriptions of typical points that form the partial concept extent. Another improvement of the approach would be to help users select the prototypical elements that define the concept they have in mind. To do so, several strategies from the query by examples field may be envisaged.

References

1. Agrawal, R., Gehrke, J., Gunopulos, D., Raghavan, P.: Automatic subspace clustering of high dimensional data for data mining applications. In: Proceedings of the ACM SIGMOD International Conference on Management of Data, pp. 94–105 (1998)
2. Bělohlávek, R., Vychodil, V.: What is a fuzzy concept lattice? In: Proceedings of the 3rd International Conference on Concept Lattices and Their Applications, CLA05, pp. 34–45 (2005)
3. Fürnkranz, J., Hüllermeier, E. (eds.): Preference Learning. Springer, Heidelberg (2010). https://doi.org/10.1007/978-3-642-14125-6
4. Ganter, B., Wille, R.: Formal Concept Analysis: Mathematical Foundations. Springer, Heidelberg (2012). https://doi.org/10.1007/978-3-642-29892-9
5. Lesot, M.-J.: Subspace clustering and some soft variants. In: Ben Amor, N., Quost, B., Theobald, M. (eds.) SUM 2019. LNCS (LNAI), vol. 11940, pp. 433–443. Springer, Cham (2019). https://doi.org/10.1007/978-3-030-35514-2_33
6. Lesot, M.J., Rifqi, M., Bouchon-Meunier, B.: Fuzzy prototypes: from a cognitive view to a machine learning principle. In: Bustince, H., Herrera, F., Montero, J. (eds.) Fuzzy Sets and Their Extensions: Representation, Aggregation and Models, pp. 431–452. Springer, Heidelberg (2008). https://doi.org/10.1007/978-3-540-73723-0_22
7. Lesot, M.-J., Rifqi, M., Benhadda, H.: Similarity measures for binary and numerical data: a survey. J. Knowl. Eng. Soft Data Paradigms (KESDP) 1(1), 63–84 (2009)
8. Moreau, A., Pivert, O., Smits, G.: Fuzzy query by example. In: Proceedings of ACM SAC 2018, pp. 688–695 (2018)
9. Rifqi, M.: Constructing prototypes from large databases. In: Proceedings of the International Conference on Information Processing and Management of Uncertainty in Knowledge-Based Systems (IPMU 1996), pp. 300–306 (1996)
10. Rosch, E.: Principles of categorization. In: Rosch, E., Lloyd, B. (eds.) Cognition and Categorization, pp. 27–48. Lawrence Erlbaum Associates (1978)
11. Smits, G., Pivert, O., Duong, T.N.: On dissimilarity measures at the fuzzy partition level. In: Medina, J., et al. (eds.) IPMU 2018. CCIS, vol. 854, pp. 301–312. Springer, Cham (2018). https://doi.org/10.1007/978-3-319-91476-3_25
12. Vidal, R.: A tutorial on subspace clustering. IEEE Sig. Process. Mag. 28(2), 52–68 (2010)
13. Wille, R.: Restructuring lattice theory: an approach based on hierarchies of concepts. In: Rival, I. (ed.) Ordered Sets, vol. 83, pp. 445–470. Springer, Dordrecht (1982). https://doi.org/10.1007/978-94-009-7798-3_15
14. Zadrozny, S., Kacprzyk, J., Wysocki, M.: On a novice-user-focused approach to flexible querying: the case of initially unavailable explicit user preferences. In: Proceedings of the 10th International Conference on Intelligent Systems Design and Applications, ISDA 2010, pp. 696–701 (2010)
15. Zloof, M.M.: Query-by-example: a data base language. IBM Syst. J. 16(4), 324–343 (1977)

Using Topic Information to Improve Non-exact Keyword-Based Search for Mobile Applications

Eugénio Ribeiro[1,2,3]($\boxtimes$) , Ricardo Ribeiro[1,3] , Fernando Batista[1,3] ,
and João Oliveira[3]

[1] INESC-ID, Lisbon, Portugal
eugenio.ribeiro@inesc-id.pt
[2] Instituto Superior Técnico, Universidade de Lisboa, Lisbon, Portugal
[3] Instituto Universitário de Lisboa (ISCTE-IUL), Lisbon, Portugal

Abstract. Considering the wide offer of mobile applications available nowadays, effective search engines are imperative for an user to find applications that provide a specific desired functionality. Retrieval approaches that leverage topic similarity between queries and applications have shown promising results in previous studies. However, the search engines used by most app stores are based on keyword-matching and boosting. In this paper, we explore means to include topic information in such approaches, in order to improve their ability to retrieve relevant applications for non-exact queries, without impairing their computational performance. More specifically, we create topic models specialized on application descriptions and explore how the most relevant terms for each topic covered by an application can be used to complement the information provided by its description. Our experiments show that, although these topic keywords are not able to provide all the information of the topic model, they provide a sufficiently informative summary of the topics covered by the descriptions, leading to improved performance.

Keywords: Application search · Topic information · Non-exact queries

1 Introduction

Nowadays, the offer of mobile applications with different functionality in app stores is constantly increasing. Thus, although users spend most of their time inside the applications, they also spend a significant amount of time searching for and installing new applications. This reveals the need for effective search and recommendation systems. However, most queries in app store search engines contain just the name of the application that the user is looking for. This means

This work was supported by Portuguese national funds through Fundação para a Ciência e a Tecnologia (FCT), with reference UIDB/50021/2020, and PT2020, project number 39703 (AppRecommender).

© Springer Nature Switzerland AG 2020
M.-J. Lesot et al. (Eds.): IPMU 2020, CCIS 1237, pp. 373–386, 2020.
https://doi.org/10.1007/978-3-030-50146-4_28

that users target specific applications, either because they were suggested to them by acquaintances or they found them using other approaches, such as web search. Word of mouth has always been an important form of marketing. Thus, searching for applications suggested by acquaintances is normal. On the other hand, searching for applications on the web is somewhat of a countersense, since app stores have specialized search engines. However, those engines are typically unable to semantically interpret the queries, considering their characteristics and context. Thus, they lose to web search engines, which are able to process more complex queries by crawling large amounts of data. Overall, data is the defining factor, since queries in app store search engines are typically short and the amount of data available to search on is reduced, especially in comparison to the whole web. Thus, in order to deliver better search results, app store search engines must overcome the data problem, either by semantically interpreting the queries or by inferring additional information from the existing data to improve the match ratio between the queries and relevant applications.

Topic information has been proved important in the context of information retrieval [23], including in search for applications [14, 24], since it enables matching when similar contexts are referred to using different words. However, while the existing approaches to topic-based retrieval are based on similarity between topic distributions, the highly distributed search approaches used in most app stores are based on keyword-matching and boosting according to popularity factors. In this paper, we explore means to include topic information in such approaches, in order to improve their ability to retrieve relevant applications for non-exact queries, without impairing their computational performance. More specifically, we start by creating topic models specialized on application descriptions. Then, we identify the most relevant and distinctive terms to represent each topic. Finally, we explore how the relevant terms for each topic covered by an application can be used to complement the information provided by the words of its description in the context of non-exact keyword-based search.

In the remainder of the paper, we start by providing an overview on related work on search for mobile applications, in Sect. 2. Then, in Sect. 3, we present our approach for including topic information in keyword-based search. Section 4 describes our experimental setup, including the dataset, evaluation approach, and implementation details that allow future reproduction of our experiments. The results of those experiments are presented and discussed in Sect. 5. Finally, Sect. 6 summarizes the contributions of this paper and provides pointers for future work.

2   Related Work

The algorithms behind the search engines of the two major mobile app stores, Google Play [9] and Apple's App Store [1], are constantly evolving and, since they are proprietary, not all the details are disclosed. However, it is known that they are mostly based on keyword-matching with multiple fields regarding the applications and boosting based on popularity factors or for business purposes. Most alternative app stores are also proprietary and use similar search approaches. Among these, many are based on the Lucene search engine [5] or one of the highly distributed search engines built on top of it, such as Solr [19] or Elasticsearch [2], which focus on speed and availability.

For instance, Aptoide's search engine [21] is based on Elasticsearch and performs keyword-matching between the terms present in the query and fields containing application information regarding its name, its package, and its description. Furthermore, in order to improve the match ratio, it includes alternatives of the name, such as abbreviations, lemmatized words, and split and merged versions of multi-word names. Matches with each of these fields contribute to the relevance score with different weights. Furthermore, information regarding the number of downloads of the application, its rating, the number of users that rated the application, and whether it should be promoted for business purposes is used to boost the score.

Mobilewalla [6, 7] uses an application search engine based on Lucene. The keyword-matching fields include the application name, description, and its categories, while boosting fields include the rating and rank of the application, its age and the frequency of releases, the number of users that commented and rated the application, the number of applications in the same categories, and information about the developer. The main difference from Aptoide's approach is that the computation of alternatives is not on the application side, but rather on the query side. That is, the knowledge base does not include alternative application names, but multiple versions of the query are generated by stemming and lemmatizing its words. Furthermore, if using all the terms in the query does not lead to the retrieval of enough results, alternative queries are generated by dropping part of the terms. Alternatively, the query can be expanded by replacing terms with corresponding synonyms or hyponyms.

To reduce the number of mismatches in keyword search caused by the use of different terms by the users and developers, Tencent's MyApp [24] extends the queries performed in its search engine with topic and tag information. The set of more than a thousand topics was obtained by applying Latent Dirichlet Allocation (LDA) [3] to the title and descriptions of a million applications. Using this model, each application can then be represented as a topic distribution. Since the queries are typically too short for performing an accurate inference of their topic distribution, they are extended with information from the applications which have been clicked on after similar queries. By computing the similarity between the topic distribution of an extended query and those of the applications, the search engine is able to identify the most relevant applications for the query in terms of topic. Tag information is used to add fine-grained semantics to the query. The set of tags of an application is a filtered combination of human labels and tags obtained by crawling web and usage data regarding that application. A query is extended with tags using a template-based method which uses information from clicked applications to select the templates. Finally, the LambdaMART algorithm [4] is applied to aggregate the applications obtained through term, topic, and tag matching and order them for presentation to the user. Although considering topic and tag information leads to a higher match ratio, the query extensions and the computation of its topic distribution introduce a high computational overhead during search.

Park et al. [14] explored the use of user reviews to improve the match ratio by bridging the gap between the vocabulary used by users and developers. Furthermore, in their study, they compared the performance of multiple retrieval approaches – BM25(F) [18], Query Likelihood (QL) [16], and LDA-Based Document Model (LBDM) [22]. The first is based on keyword-matching, the second on language modeling, and the last on the combination of keyword- and topic-matching. While relying solely on application descriptions, the highest performance on a set of more than 50 non-exact queries was achieved using LBDM with a topic model with 300 topics trained on the descriptions of 40,000 applications. This confirms that topic information is able to complement the information explicitly present in descriptions by providing associations with words that refer to similar topics. Furthermore, using the information provided by user reviews significantly improved the performance of every retrieval approach. The best results were achieved using an approach that combines language modeling with topic-based retrieval. Separate topic models are trained on descriptions and reviews, but the review-level model is conditioned by the description-level one. This allows the identification of review topics that do not match any description topic and, thus, are not relevant for application retrieval. In a later study, Park et al. [13] also relied on language and topic modeling to induce queries from users' social media text and recommend relevant applications.

3   Topic Information for Keyword-Based Search

When topic information is used for retrieval, documents are typically ranked according to the similarity between their topic distribution and that of the query. The approach we describe below enables the representation of topic information as keywords that can be used by keyword-based retrieval approaches. These keywords correspond to a set of terms that are sufficiently relevant and distinctive to identify a topic and, thus, can function as its summary. In addition to how the the keywords are generated from the topic models and included in the retrieval approaches, we also describe preprocessing and topic model training approaches that allow the generated models and the corresponding keywords to focus on relevant aspects for mobile application search. Since our intent is to show that topic information can be represented as keywords, we focus on obtaining that information from application descriptions. However, the approach can be generalized to other textual information sources, such as user reviews.

3.1   Preprocessing

In the preprocessing phase, each description is split into sentences and dependency parsed and its tokens are Part-of-Speech (POS) tagged and lemmatized. By splitting into sentences, we are able to train both generic models based on whole descriptions and more specific ones based on the sentences. POS tagging allows filtering by the word classes that are more relevant for application search, such as nouns, adjectives, and verbs. While the first reveal the concepts focused by the description, the second reveal their characteristics. Furthermore, in this context, non-auxiliary verbs typically reveal functionality. By combining the POS tags and the dependency parse of each sentence, we can identify adjectives and verbs that are negated. This is important to avoid grouping descriptions or sentences that have opposite meanings. Finally, lemmatization simplifies matching and leads to the generation of more constrained models.

Additionally, while terms that occur in a small set of descriptions or sentences are unrelated to the most relevant topics covered by the whole collection, terms that occur in a large portion of the collection are typically not discriminative. Thus, we discard tokens that are commonly classified as stopwords, as well as those which have a document frequency below a threshold df_{min} or above a threshold df_{max}. The most appropriate values for these thresholds vary according to the model. Those used in our experiments are detailed in Sect. 4.3.

Since the descriptions are lemmatized, we also lemmatize queries, in order to enable matching. No additional preprocessing is performed on the queries.

3.2 Topic Models

We obtain our topic models using a classical LDA approach [3]. In an LDA model, topics are seen as term distributions while documents are seen as mixtures of topics. Thus, the definitions of term and document have a wide impact on the aspects that are actually modeled. In typical applications of LDA, documents are relatively large pieces of text, such as news articles or reports, and the terms are the words in the documents, excluding stopwords. However, as referred in the previous section, we can split the descriptions in different ways and filter the tokens by specific word classes, in order to identify topics that are more informative for application retrieval.

Regarding terms, after the preprocessing described in the previous section, when training the topic models, we discard tokens that are not nouns, adjectives, or non-auxiliary verbs. Furthermore, negated verbs and adjectives are distinguished from their positive counterparts.

In terms of documents, the most straightforward approach is to consider each description a document. However, since the LDA model uses a Bag of Words (BoW) approach, it is not aware of the dependency relations between nouns and adjectives nor between verbs and their arguments. Thus, it assumes that all the terms that occur in the document are related in the same manner. This leads to the identification of more generic topics that may group terms that are not directly related in the descriptions. On the other hand, each individual sentence in a description typically contains terms that are directly related. Thus, training a sentence-level model leads to the identification of more constrained topics. Since both kinds of topic may provide relevant information for application retrieval, we train both a description-level model and a sentence-level model.

Finally, similarly to any application of LDA, the number of topics, N_t, must be defined a priori. The selection of an appropriate value for this parameter reduces the probability of identifying topics that are either too generic to be useful or so specific that capture irrelevant aspects. However, this categorization depends on the intended use for the topics. Furthermore, the best value typically depends on the dimensionality of the collection and the number of terms in the vocabulary. Thus, the most appropriate number of topics is expected to differ between the description- and sentence-level models.

3.3   Topic Keywords

Having trained the topic models, in order to use the information that they capture in the context of keyword-based retrieval approach, it must be transformed into keywords. A straightforward approach is to represent each topic by the top n terms in its distribution. However, using a fixed number of terms may lead either to the inclusion of non-relevant terms or the discarding of terms that are relevant for a topic. Thus, we use the approach described in Algorithm 1 to identify the set of relevant terms for each topic. The idea behind it is to approximate the term distribution of a topic by a negative exponential function and select the terms that appear before the inflection point as relevant. Thus, given a term distribution, T, the algorithm starts by sorting it in decreasing weight order. Then, only the n terms with highest weight in the distribution are considered, as long as their weight is above a residual threshold, r. To account for noisy distributions, the weights of the terms are then smoothed using a weighted running average that further approximates the distribution to a negative exponential one. The remainder of the algorithm identifies the inflection point by analyzing the weight differences between consecutive terms.

Algorithm 1. Relevant Terms

Input: T // The term weight distribution
Input: r // The residual weight threshold
Input: n // The maximum number of terms
Output: R // The relevant terms
1: $T \leftarrow \text{SORT}(\{(t, w) \in T\}, (t_i, w_i) < (t_j, w_j) := w_i > w_j)$
2: $W \leftarrow \{w_i : (t_i, w_i) \in T, w_i > r, 0 < i \leq n\}$
3: $W \leftarrow \text{WEIGHTEDRUNNINGAVERAGE}(W)$
4: $d \leftarrow \textbf{false}$
5: **for** $i = 1 : |W|$ **do**
6:      $m \leftarrow (W_i - W_{i-1}) \times |W|$
7:      **if** d **and** $m > -1$ **then**
8:           **break**
9:      **else if** $m < -1$ **and not** d **then**
10:          $d \leftarrow \textbf{true}$
11:     **end if**
12: **end for**
13: $R \leftarrow T_{1:i}$
14: **return** R

3.4 Application Retrieval

The approach described in the previous section identifies the set of relevant terms for a topic. To identify the set of description-level topic keywords for an application, a, we use the corresponding topic model to compute the topic mixture of its description. Then, we discard topics with weight below a residual threshold, r. The keywords are then given by the aggregate of the relevant terms for the remaining topics. The set of sentence-level topic keywords is computed in a similar fashion. However, the topic mixture is computed for each sentence in the application's description and the keywords of the application are given by the aggregate of the topic keywords of its sentences.

For retrieval purposes, each application is represented by a set of three textual fields for keyword matching, $\{a_d, a_{st}, a_{dt}\}$, corresponding to its textual description, the set of sentence-level topic keywords, and the set of document-level topic keywords, respectively. In our experiments, we explore two keyword-based retrieval approaches – BM25F [18] and Elasticsearch [2]. While the first is a widely used information retrieval approach for semi-structured textual data, the latter is a highly distributed search engine focused on speed and availability. Given a query, q, both return a list of applications ordered by relevance score. However, the scoring function differs. The adaptation of the two scoring functions to our problem is presented below.

BM25F. We use the same formulation of the base BM25F scoring function found in several previous studies (e.g. [8,14,15]):

$$\text{score}(q, a) = \sum_{t \in q \cap a} \left(\text{idf}(t) \times \frac{(k_3 + 1)c(t, q)}{k_3 + c(t, q)} \times \frac{(k_1 + 1)c'(t, a)}{k_1 + c'(t, a)} \right) \quad (1)$$

where $\text{idf}(t)$ is the inverse document frequency of term t in the set of descriptions, k_1 and k_3 are parameters that can be tuned according to the problem, $c(t, q)$ is t's count in q and $c'(t, a)$ is t's normalized count in a, weighted by field:

$$c'(t, a) = \frac{w_d \cdot c(t, a_d)}{1 - b + b\frac{|a_d|}{\bar{n}}} + w_{st} \cdot c(t, a_{st}) + w_{dt} \cdot c(t, a_{dt}) \quad (2)$$

where w_d, w_{st}, and w_{dt} are the weights given to textual descriptions, sentence-level topic information, and description-level topic information, respectively, and b is a parameter that controls the strength of the normalization according to the mean description length, $\bar{n}$. We do not include normalization factors for topic information, since the number of topic keywords is not relevant for the problem.

Elasticsearch. Scoring in Elasticsearch is based on Lucene's Practical Scoring Function, which computes individual scores for each field, f, as

$$\text{score}(q, f) = \frac{1}{\sqrt{\sum_{t \in q} \text{idf}(t)^2}} \times \frac{|q \cap f|}{|q|} \times \sum_{t \in q} \left(\frac{\text{tf}(t, f) \cdot \text{idf}(t)^2 \cdot w_f}{\sqrt{|f|}} \right) \quad (3)$$

where the first factor is a cross-query normalization factor, the second factor boosts according to the number of matching terms, $\mathrm{tf}(t, f)$ is the term frequency of t in f, $\mathrm{idf}(t)$ is the inverse document frequency of t in the field f of all applications, and w_f is the weight of the field.

The relevance score of an application for a query is then given by

$$\mathrm{score}(q, a) = (1 - \mathrm{tb}) \cdot \max_{f \in a} \mathrm{score}(q, f) + \mathrm{tb} \cdot \sum_{f \in a} \mathrm{score}(q, f) \qquad (4)$$

where tb is a parameter that controls the extent to which the non-top scoring fields contribute for the overall relevance score of the application.

4 Experimental Setup

In this section, we describe our experimental setup, including the dataset, the evaluation approach, and implementation details that enable the reproduction of our experiments in future studies.

4.1 Dataset

In our experiments, we use the dataset crawled by Park et al. [14], which features information regarding 43,041 mobile applications. Among other less relevant information, for each application, it includes the name, category, description, developer, date of publication, price, and number of downloads. Furthermore, it includes review information in the form of the number of reviews, the average rating, and textual data of up to 50 reviews per application, with a total of 1,385,607 reviews. Additionally, the dataset features 56 non-exact queries generated from forum posts that targeted an application with a specific functionality. Each of these queries is paired with relevance information of the top 20 applications retrieved using multiple retrieval approaches. On average, there are 81 judged applications per query. Each query-application pair was annotated by three users in a three-value scale: 0 for no satisfaction at all, 1 for partial satisfaction, and 2 for perfect satisfaction. The relevance score is then given by the average judgement of the annotators.

We decided to use this dataset since, to the best of our knowledge, it is the only publicly available one featuring relevance scores of query-application pairs. Furthermore, the results of previous studies on this dataset provide a baseline for comparison of our results.

4.2 Evaluation Approach

In order to compare our results with those reported in previous studies on the same dataset, we use the same evaluation metric as Park et al. [14], that is, the Normalized Discounted Cumulative Gain (NDCG) [11] at 3, 5, 10, and 20 top retrieved applications. NDCG is a widely used metric in the context of information retrieval to measure the effectiveness of search engine algorithms, by assessing whether the results are ordered by relevance. In the context of search for mobile applications, looking beyond the fifth result typically involves scrolling and, thus, the NDCG at 3 and 5 are the most important to consider.

Given a graded relevance scale of applications in a result set for a given query, to compute the corresponding NDCG, we start by computing the Discounted Cumulative Gain (DCG) of the result set:

$$\mathrm{DCG}_k = \sum_{i=1}^{k} \frac{\mathrm{rel}_i}{\log_2(i+1)} \tag{5}$$

where rel_i is the relevance of the i-th application in the result list for query q. This metric measures the gain of an application based on its position in the result list. The gain is then accumulated from the top of the list, with the gain of each result being discounted as the distance from the top increases. The NDCG is then obtained through normalization using the Ideal Discounted Cumulative Gain (IDCG), that is, the DCG of a perfectly sorted result list:

$$\mathrm{NDCG}_k = \frac{\mathrm{DCG}_k}{\mathrm{IDCG}_k} \tag{6}$$

As baselines, we use the results achieved using both BM25F [18] and Elasticsearch [2] when relying solely on matching with description texts, without topic information. That is, $w_d = 1, w_{st} = 0, w_{dt} = 0$ in Eqs. 2 and 3. This transforms BM25F into its single-field version, BM25. Additionally, we compare our results with the LBDM [22] and Google Play [9] results reported by Park et al. [14]. Since LBDM is able to take advantage of all the information captured by the topic model, its results provide an upper bound for performance when pairing topic information with keyword-matching with application descriptions. On the other hand, Google Play results serve as an indicator of the performance of current app store search engines, which rely on additional fields for matching and on popularity information for boosting.

4.3   Implementation Details

The application descriptions provided in the dataset contain HTML tags and escape characters. We used the html2text package [20] to convert them to plain text. Then, we used the spaCy parser [10] for sentence splitting, dependency parsing, POS tagging, and lemmatization. Since the set of English stopwords used by spaCy is too aggressive, we relied on the set defined in NLTK [12] while filtering the tokens. Additionally, for consistency with the experiments by Park et al. [14], we defined $\mathrm{df}_{min} = 5$ and $\mathrm{df}_{max} = 0.3$ for keyword-matching with the description. That is, we discarded tokens that appeared in less than 5 descriptions or in more than 30%.

To train the topic models, we used the parallelized LDA implementation provided by the gensim library [17]. Additionally, we performed a more aggressive low-frequency token filtering. While training the sentence-level topic model, we used $df_{min} = 10$, since, in this case, we considered the sentence frequency and lower values still included many terms that only occurred in a single description. While training the description-level topic model, we used $df_{min} = 0.01$, that is, we discarded tokens that appeared in less than 1% of the descriptions, in order to identify more generic topics. In terms of the number of topics, we defined $N_t = 300$ for the description-level model, for consistency with the experiments by Park et al. [14], and $N_t = 100$ for the sentence-level model, since for higher values there were topics that were not attributed to any application. While identifying the relevant terms for each topic, we defined a maximum number of terms $n = 20$ and a residual weight threshold $r = 0.01$. The same residual weight threshold was used to attribute topics to applications.

For keyword-matching with the description in BM25(F), we used the same parameters as Park et al. [14]. That is, $k1 = 4$, $k3 = 1000$, and $b = 0.4$. The remaining parameters – w_d , w_{st}, and w_{dt} for both BM25F and Elasticsearch, and tb for Elasticsearch – were tuned using grid search to maximize the mean of the four NDCG results, that is,

$$\bar{\text{NDCG}} = \frac{\sum_{k \in K} \text{NDCG}_k}{|K|}, K = \{3, 5, 10, 20\} \tag{7}$$

For that reason, the concrete values and their meaning are discussed in Sect. 5.

5   Results

Table 1 shows the NDCG results of our experiments, as well as the reference results achieved using LBDM and Google Play. In the context of search for mobile applications, the NDCG results lose relevance as the number of considered applications increases, since only a reduced set can be shown on screen at each time. Thus, we will focus this discussion on NDCG@3 results. However, since the parameters were tuned to maximize the mean results at the multiple values of k, we will also make some remarks regarding the results achieved when a higher number of applications is considered.

First of all, it is important to note that the baseline BM25 results are one percentage point lower than those reported by Park et al. [14] for $k \in \{3, 5\}$, in spite of using the same values for all the parameters. This is due to differences in preprocessing, especially regarding the filtering of tokens, which also considered frequency in reviews. This means that the results achieved using LBDM are also expected to be lower if using our preprocessing approach.

Overall, due to its focus on temporal performance, Elasticsearch performs worse than BM25(F). When considering textual descriptions, the decrease in performance is between two and three percentage points. However, when considering topic information only, the decrease is around 20 percentage points. This is due to the reduced vocabulary and the normalization factors applied by the Elasticsearch score function. On the other hand, since we consider the whole

Table 1. NDCG results of BM25(F) and Elasticsearch applied to textual descriptions (D), topic information (T) and their combination (D + T). The last block provides reference results reported by Park et al. [14].

Approach	NDCG@3	NDCG@5	NDCG@10	NDCG@20
BM25 (D)	0.569	0.540	0.523	0.537
Elasticsearch (D)	0.540	0.523	0.502	0.512
BM25F (T)	0.554	0.553	0.535	0.530
Elasticsearch (T)	0.341	0.342	0.356	0.370
BM25F (D + T)	0.574	0.542	0.527	0.544
Elasticsearch (D + T)	0.552	0.532	0.504	0.519
LBDM	0.584	0.563	0.543	0.565
Google Play	0.589	0.575	0.568	0.566

vocabulary and do not include normalization factors for topic information while computing the BM25F scores, its results are not penalized.

Comparing the results achieved using textual descriptions with those achieved using topic information, we can see that the performance of BM25(F) decreases 1.5 percentage points in terms of NDCG@3, but actually increases in terms of NDCG@5 and NDCG@10. This means that the set of topic keywords is an appropriate summary of the information provided by the description. However, these results were achieved when relying solely on the sentence-level topic keywords, that is, $w_{st} = 1$ and $w_{dt} = 0$. Including description-level topic keywords does not lead to improvement. On the other hand, the results using Elasticsearch were achieved using $w_{st} = 2w_{dt}$ and tb = 0.1, which means that the sentence-level topic keywords are still the most informative, but that the document-level topic keywords can provide complementary information.

As expected, the best results are achieved when combining the information provided by textual descriptions and topic information. However, in the case of BM25F, there is only improvement when topic information is given a reduced weight. More specifically, the results reported in Table 1 were achieved with $w_d = 0.96$, $w_{st} = 0$, and $w_{dt} = 0.04$. Several other configurations, including ones that also give weight to sentence-level topic information, lead to similar results. Still, the weights are always severely biased towards the textual descriptions. For instance, the parameters that maximized NDCG@3 in our experiments were $w_d = 0.98$, $w_{st} = 0.01$, and $w_{dt} = 0.01$. This means that the topic keywords are only used as complementary information that enable the retrieval of more relevant applications in specific cases. On the other hand, in the case of Elasticsearch, the mean NDCG was maximized with $w_{st} = 2w_{dt}$, $w_d = w_{dt}$, and tb = 0.5. This means that the relation between the weights of sentence- and description-level topic information is kept in relation to when the textual descriptions are not considered. Furthermore, although higher weight is given to topic information, the value of the tie breaker parameter shows that all fields have an important contribution to the score.

Overall, including topic information improves the performance of Elasticsearch by one percentage point in terms of both NDCG@3 and NDCG@5. However, by comparing the BM25F results with those of LBDM, even assuming that the performance of LBDM is expected to decrease with our preprocessing approach, we can see that the topic keywords are not able to capture all the information provided by the topic models. This happens because the representation of the topics in the form of their most relevant terms does not allow matching with similar keywords that are not as common. Finally, the performance of Google Play shows that additional fields, such as the application titles, and popularity information are relevant for delivering the best results for non-exact queries.

6 Conclusions

In this paper, we have explored how topic information can be represented in the form of keywords to be considered by mobile application retrieval approaches based on keyword-matching. This is important, since app store search engines have strict requirements in terms of temporal performance and availability, which, currently, are only fulfilled by highly distributed retrieval approaches based on multi-field keyword-matching and boosting.

We focused on application descriptions and trained two LDA models, one on whole descriptions and another on their sentences. While the first generates more generic topics, the second captures more fine-grained subjects. Then, we computed the topic mixtures of the application descriptions and represented the topic information of an application as the aggregate of the relevant terms for each topic in its mixture. The set of relevant terms for a topic is identified by approximating its term distribution by a negative exponential function and selecting the terms which appear before the inflection point.

The results of our experiments have shown that both sentence- and description-level topic information provides cues for application retrieval from non-exact queries, leading to improved performance. Furthermore, the topic keywords make a sufficiently informative summary of the information provided by the descriptions. However, they do not allow matching with similar keywords that are not as common. Thus, the performance is still lower than when performing retrieval based on topic similarity, which relies on all the information provided by the topic models. Thus, as future work, it would be interesting to assess whether including synonyms of the relevant terms that occur in the same context can enable matching with those less common keywords without introducing ambiguity.

Furthermore, it is important to assess whether the performance improvement observed by Park et al. [14] when leveraging review data can also be observed when the information captured by the topic models that merge description and review information is provided in the form of keywords.

Finally, it is important to assess how this approach behaves in combination with boosting factors based on popularity.

References

1. Apple: App Store (2008). https://www.apple.com/ios/app-store/
2. Banon, S.: Elasticsearch (2010). https://www.elastic.co/
3. Blei, D.M., Ng, A.Y., Jordan, M.I.: Latent Dirichlet allocation. J. Mach. Learn. Res. **3**, 993–1022 (2003). https://doi.org/10.5555/944919.944937
4. Burges, C.J.C.: From RankNet to LambdaRank to LambdaMART: an overview. Learning **11**(23–581), 81 (2010)
5. Cutting, D.: Apache Lucene (1999). https://lucene.apache.org/
6. Datta, A., Dutta, K., Kajanan, S., Pervin, N.: MobileWalla: a mobile application search engine. In: MobiCASE, pp. 172–187 (2011). https://doi.org/10.1007/978-3-642-32320-1_12
7. Datta, A., Kajanan, S., Pervin, N.: A mobile app search engine. Mob. Netw. Appl. **18**(1), 42–59 (2013). https://doi.org/10.1007/s11036-012-0413-z
8. Fang, H., Tao, T., Zhai, C.: A formal study of information retrieval heuristics. In: SIGIR, pp. 49–56 (2004). https://doi.org/10.1145/1008992.1009004
9. Google: Google Play (2008). https://play.google.com/
10. Honnibal, M., Montani, I.: spaCy 2: Natural Language Understanding with Bloom Embeddings, Convolutional Neural Networks and Incremental Parsing (2017). https://spacy.io/
11. Järvelin, K., Kekäläinen, J.: Cumulated gain-based evaluation of IR techniques. ACM Trans. Inf. Syst. **20**(4), 422–446 (2002). https://doi.org/10.1145/582415.582418
12. Loper, E., Bird, S.: NLTK: the natural language toolkit. In: ACL Workshop on Effective Tools and Methodologies for Teaching Natural Language Processing and Computational Linguistics. **1**, 63–70 (2002). https://doi.org/10.3115/1118108.1118117
13. Park, D.H., Fang, Y., Liu, M., Zhai, C.: Mobile app retrieval for social media users via inference of implicit intent in social media text. In: CIKM, pp. 959–968 (2016). https://doi.org/10.1145/2983323.2983843
14. Park, D.H., Liu, M., Zhai, C., Wang, H.: Leveraging user reviews to improve accuracy for mobile app retrieval. In: SIGIR, pp. 533–542 (2015). https://doi.org/10.1145/2766462.2767759
15. Pérez-Iglesias, J., Pérez-Agüera, J.R., Fresno, V., Feinstein, Y.Z.: Integrating the probabilistic models BM25/BM25F into lucene. Comput. Res. Repository arXiv:0911.5046 (2009)
16. Ponte, J.M., Croft, W.B.: A language modeling approach to information retrieval. In: SIGIR, pp. 275–281 (1998). https://doi.org/10.1145/290941.291008
17. Řehůřek, R., Sojka, P.: Software framework for topic modelling with large corpora. In: LREC Workshop on New Challenges for NLP Frameworks, pp. 45–50 (2010). https://doi.org/10.13140/2.1.2393.1847
18. Robertson, S., Zaragoza, H.: The probabilistic relevance framework: BM25 and beyond. Found. Trends® Inf. Retrieval **3**(4), 333–389 (2009). https://doi.org/10.1561/1500000019
19. Seeley, Y.: Apache Solr (2004). https://lucene.apache.org/solr/
20. Swartz, A.: html2text (2003). https://github.com/Alir3z4/html2text/
21. Trezentos, P.: Aptoide (2009). https://www.aptoide.com/
22. Wei, X., Croft, W.B.: LDA-based document models for Ad-Hoc retrieval. In: SIGIR, pp. 178–185 (2006). https://doi.org/10.1145/1148170.1148204

23. Yi, X., Allan, J.: A comparative study of utilizing topic models for information retrieval. In: Boughanem, M., Berrut, C., Mothe, J., Soule-Dupuy, C. (eds.) ECIR 2009. LNCS, vol. 5478, pp. 29–41. Springer, Heidelberg (2009). https://doi.org/10. 1007/978-3-642-00958-7_6
24. Zhuo, J., et al.: Semantic matching in app search. In: WSDM, pp. 209–210 (2015). https://doi.org/10.1145/2684822.2697046

A Graph Theory Approach to Fuzzy Rule Base Simplification

Caro Fuchs[1]([✉]) [iD], Simone Spolaor[2,3] [iD], Marco S. Nobile[1,3] [iD],
and Uzay Kaymak[1] [iD]

[1] School of Industrial Engineering, Eindhoven University of Technology,
Eindhoven, The Netherlands
{c.e.m.fuchs,m.s.nobile,u.kaymak}@tue.nl
[2] Department of Informatics, Systems and Communication,
University of Milano-Bicocca, Milan, Italy
simone.spolaor@disco.unimib.it
[3] SYSBIO.IT Centre for Systems Biology, Milan, Italy

Abstract. Fuzzy inference systems (FIS) gained popularity and found
application in several fields of science over the last years, because they
are more transparent and interpretable than other common (black-box)
machine learning approaches. However, transparency is not automatically achieved when FIS are estimated from data, thus researchers are
actively investigating methods to design interpretable FIS. Following this
line of research, we propose a new approach for FIS simplification which
leverages graph theory to identify and remove similar fuzzy sets from
rule bases. We test our methodology on two data sets to show how this
approach can be used to simplify the rule base without sacrificing accuracy.

Keywords: Fuzzy logic · Takagi–Sugeno fuzzy model · Data-driven
modeling · Open-source software · Python · Graph theory

1 Introduction

Fuzzy Inference Systems (FIS) are based on the fuzzy set theory introduced
by Zadeh [37]. FIS are universal approximators that can implement non-linear
mappings between inputs and output, designed to model linguistic concepts.
Owing to these characteristics, FIS have been successfully applied in a variety of
fields, including systems biology, automatic control, data classification, decision
analysis, expert systems, and computer vision [7,14,21,25,27,30]. One of the
main advantages provided by FIS over black-box methods, such as (deep) neural
networks, is that fuzzy models are (to a certain degree) transparent, and hence
open to interpretation and analysis.

The first FIS relied on the ability of fuzzy logic to model natural language and
were developed using expert knowledge [23]. The knowledge of human experts
was extracted and transformed into rules and membership functions. These FIS

© Springer Nature Switzerland AG 2020
M.-J. Lesot et al. (Eds.): IPMU 2020, CCIS 1237, pp. 387–401, 2020.
https://doi.org/10.1007/978-3-030-50146-4_29

are easy to interpret, but unfortunately, they cannot be easily used to model large and complex systems, since human knowledge is often incomplete and episodic.

In 1985, Takagi and Sugeno [33] proposed a method to construct self-learning FIS from data. The fuzzy rules underlying this kind of FIS are automatically generated from data, but follow the same if–else structure as the rules based on expert knowledge, thus making it possible to model large and complex systems. However, there is generally a loss of semantics when the FIS is constructed in this way, since the number of induced rules can be large, and the rules might become complex because of the number of considered variables. Therefore, many researchers are investigating the problem of designing interpretable fuzzy models [1,2,12].

When FIS are identified from data, it is common to obtain a system with a large number of highly overlapping fuzzy sets, that hardly allow for any interpretation. This hinders the user from labeling the fuzzy sets with linguistic terms and thus giving semantic interpretation to the model. This problem arises especially when Takagi and Sugeno (TS) [33] fuzzy models are determined based on input–output product space fuzzy clustering. Fuzzy rule base simplification has been proposed to reduce the complexity of such models in order to make them more amenable to interpretation [29].

In this paper we propose a new approach based on graph theory to simplify the fuzzy rule base by reducing the number of fuzzy sets in the model when a high overlap is detected between membership functions. Specifically, we combine Jaccard similarity and graph theory to determine which fuzzy sets can be simplified in the model. We name our approach Graph-Based Simplification (GRABS).

GRABS was implemented using the Python programming language [26], and it is part of pyFUME, a novel Python package developed to define FIS from data [9]. pyFUME provides a set of classes and methods to estimate the antecedent sets and the consequent parameters of TS fuzzy models. This information is then used to create an executable fuzzy model using the Simpful library [31], a Python library designed to handle fuzzy sets, fuzzy rules and perform fuzzy inference. pyFUME's source code and documentation can be downloaded from GITHUB at the following address: https://github.com/CaroFuchs/pyFUME.

In this study we investigate the pyFUME's GRABS functionality, testing the methodology on both synthetic and real data sets. Our results show that pyFUME produces interpretable models, written in a human-readable form, characterized by a tunable level of complexity in terms of separation of fuzzy sets.

The paper is structured as follows. We provide a theoretical background about FIS simplification in Sect. 2. The GRABS method is described in Sect. 3. Section 4 describes how to use GRABS in pyFUME. Some results of GRABS with pyFUME are shown in Sect. 5. We conclude the paper in Sect. 6.

2 Rule Base Simplification

When dealing with the interpretability of a fuzzy model, two main aspects have to be considered [1,2]: the readability and comprehensibility. The former depends on the complexity of the FIS structure, while the latter is tied to the semantics associated to it.

Rule base simplification is an approach to simplify the structure of a fuzzy system. Following the classification proposed in [15], methods for fuzzy rule base simplification can be divided into five categories:

1. **Feature reduction.** This category includes methods that rely on feature reduction by means of feature transformation [20] (also referred to as feature extraction) or feature selection [13]. Feature transformation consists of creating additional features from the given ones, or selecting a new set of features to replace the old one. Since feature transformation changes the underlying meaning of the features, this approach can make feature interpretation harder, ultimately resulting in a loss of semantics. Feature selection is not affected by this shortcoming, since it selects a subset of the most influential features, and discards features affected by noise or that do not contribute significantly to the accuracy of the FIS.

2. **Similarity-based simplification.** Methods belonging to this class perform a merging of similar rules and/or eliminate redundancy in the FIS. Similarity merging methods perform a merging of fuzzy sets representing comparable and analogous concepts, by exploiting some similarity measure (see e.g., [6, 17,29]). When the model shows high redundancy, this merging might result in some rules being equivalent and thus amenable to being merged, thereby reducing the number of rules as well. Compatible cluster merging algorithms try to combine similar clusters into a single one, in order to reduce the FIS rule base (e.g., [18]). Finally, methods for consistency checking [34] and inactivity checking [17] are employed to decrease the number of rules. In particular, consistency checking reduces the rule base by eliminating conflicting rules, while inactivity checking removes rules with a low firing strength, according to a predetermined threshold.

3. **Orthogonal transformation.** These methods reduce fuzzy rule bases by means of matrix computations. They achieve such reduction in two ways: either by taking into account the firing strength matrix and employing some metrics to estimate the impact of a rule on the FIS performance [36]; or by considering matrix decompositions (e.g., singular value decomposition) and removing the rules that correspond to the less important, smaller components and updating the membership functions accordingly [35].

4. **Interpolative reasoning.** Traditional fuzzy reasoning methods require the universe of discourse of input variables to be entirely covered by the fuzzy rule base. These methods do not perform well when input data fall in a non-covered region of the universe of discourse, as this does not trigger the firing of any rule and no consequences are drawn from the rule base. The first fuzzy interpolative reasoning method was proposed in [19] to overcome the above

mentioned limitation. This method consists in generating the conclusions of FIS with sparse rule bases by means of approximation. In the context of FIS simplification, fuzzy interpolation can be employed to reduce fuzzy rule bases. This is achieved by eliminating the rules that can be approximated through interpolation of neighboring rules.

5. **Hierarchical reasoning.** Methods adopting a hierarchical reasoning approach reorganize the fuzzy rule base structure in order to obtain a hierarchical fuzzy system [28]. Hierarchical fuzzy systems consist of several low-dimensional FISs, connected together according to some defined hierarchy. This approach was applied for example in [32], where the authors propose a hierarchical fuzzy system for the automatic control of an unmanned helicopter.

The GRABS approach proposed in this paper aims at simplifying the fuzzy rule base by eliminating redundant information from the overlapping fuzzy sets. Thus, our approach falls in the category of similarity-based rule base simplification. Methods belonging to this category need to assess the similarity between the fuzzy sets in the antecedents with a given measure, in order to remove similar sets. In [29], the authors suggest to adopt the Jaccard similarity index [16] to quantify such similarity between two fuzzy sets. Given two fuzzy sets A and B, the Jaccard index S is computed as follows:

$$S(A, B) = \frac{|A \cap B|}{|A \cup B|}, \tag{1}$$

where $|\cdot|$ denotes the cardinality of a fuzzy set, and the $\cap$ and $\cup$ operators represent the intersection and the union of fuzzy sets, respectively. The Jaccard similarity index takes values between 0 and 1, with 1 representing total similarity (i.e, perfectly overlapping fuzzy sets) and 0 disjoint fuzzy sets.

3   Graph-Based Rule Base Simplification

A graph is an abstract mathematical structure used to model pairwise relations between objects. A undirected graph is defined by a pair $G = (V, E)$, where V is a set of vertices (or nodes) connected by a set of edges E.

We represent the similarities between fuzzy sets of a same variable by using a graph. Specifically, each vertex $v \in V$ represents a fuzzy set. See for example Fig. 1a, where four fuzzy sets are defined on the universe of discourse. If the Jaccard similarity of two fuzzy sets exceeds a certain, user-specified, threshold σ, their two nodes are connected by an edge. This process is schematized in Fig. 1b, where the fuzzy sets 1 and 2 show high similarity and therefore are connected by an edge. Please observe that the graph now contains multiple connected components (three in this example).

Assume now that the fuzzy set 3 is also similar to the fuzzy sets 1 and 2. Then, by adding the corresponding additional edges $(3, 1)$ and $(3, 2)$, the graph changes as schematized in Fig. 1c. In particular, the largest component of the graph is

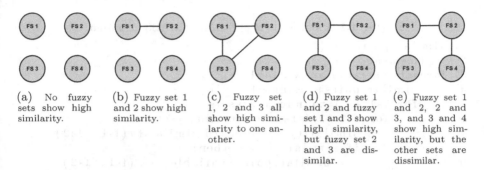

(a) No fuzzy sets show high similarity.

(b) Fuzzy set 1 and 2 show high similarity.

(c) Fuzzy set 1, 2 and 3 all show high similarity to one another.

(d) Fuzzy set 1 and 2 and fuzzy set 1 and 3 show high similarity, but fuzzy set 2 and 3 are dissimilar.

(e) Fuzzy set 1 and 2, 2 and 3, and 3 and 4 show high similarity, but the other sets are dissimilar.

Fig. 1. Graphs representing variables that each have four fuzzy sets.

complete, meaning that clusters 1, 2 and 3 are all similar to one another with respect to the specified threshold σ. Please note that the transitive closure is in general not valid in this context. For example, Fig. 1d shows a case where fuzzy set 1 is similar to both fuzzy set 2 and 3, but fuzzy set 2 and 3 are not similar to each other, according to Jaccard similarity. In this respect, the GRABS method deviates from the compatible cluster merging in [18], where transitive closure is imposed before merging, or from the similarity-based rule base simplification method in [29], where merging takes place iteratively, by combining only the most similar pair of fuzzy sets at each step.

According to our merging algorithm, the fuzzy sets of Fig. 1a should all be retained, since they are all dissimilar. However, in Fig. 1b both fuzzy set 1 and and 2 give the same information. Therefore, one of them can be dropped without losing (much) information and accuracy of the model. The same applies to Fig. 1c: only one of the three similar fuzzy sets can be retained to preserve all the information and the accuracy of the model. In Fig. 1d, dropping fuzzy set 1, 2 or 3 would lead to a loss of information, since fuzzy set 2 and 3 are dissimilar. Therefore, all three fuzzy sets should be retained. We also considered the possibility of multiple partially overlapping fuzzy sets (see Fig. 1e): in this circumstance, we do not allow the full removal of inner nodes characterized by a higher degree (i.e., FS1 and FS2), since that could lead to a fuzzy partitioning that does not span the full universe of discourse. These concepts represents the foundations of our graph-based rule base simplification algorithm. The aforementioned heuristic–that represents a trade-off between simplicity, computational costs, and accuracy in the simplification–seems to be effective for practical scenarios.

The pseudo-code of our GRABS methodology is shown in Listing 1.1. The algorithm begins by creating two empty dictionaries that will store the information about pairs of similar fuzzy sets (line 1) and the information about fuzzy sets replacements (line 2). Then, for each variable, we calculate the pair-wise Jaccard similarities of the associated fuzzy sets (lines 3–13). If the Jaccard similarity of a pair of fuzzy sets is above the user-defined threshold σ then that pair is added to the dictionary (lines 7–10). After the Jaccard similarities are assessed, the algorithm proceeds to build and analyze the graphs. Specifically, the algorithm

Listing 1.1. Pseudocode of pyFUME's GRABS algorithm.

```
1   similar_pairs ← {}
2   replacement ← {}
3   foreach variable in variables:
4       similar_pairs[variable] ← []
5       for fs1 ← 1 to num_fuzzysets:
6           for fs2 ← fs1+1 to num_fuzzysets:
7               similarity ← Jaccard_similarity(fs1, fs2)
8               if similarity > σ then:
9                   similar_pairs[variable] ← (fs1, fs2)
10              end if
11          end for
12      end for
13  end foreach
14  foreach variable, similar_clusters in similar_pairs:
15      G ← create_graph(similar_pairs)
16      SC ← G.get_components()
17      for component in SC:
18          if component.is_complete() then:
19              retained ← component.pick_one_node()
20              component.remove_node(retained)
21              foreach node in component
22                  replacement[(variable, node)] ← retained
23              end foreach
24          end if
25      end for
26  end foreach
```

iterates on variables (lines 14–26). For each variable, a graph is created by using the fuzzy sets stored in the similar_pairs dictionary (line 15). Then, all the components of the the graph are extracted (line 16) and, for each sub-component, the algorithm performs a completeness check (line 18). If the sub-component is complete, then all fuzzy sets are similar and can be simplified: one node is picked to be retained (line 19) and removed from the component (line 20). The remaining nodes (i.e., similar fuzzy sets) can be removed from the model. We store the information about the removed fuzzy sets in the replacement dictionary. To simplify the lookup in pyFUME, the keys of the dictionary are pairs (variable, removed fuzzy set) and the values are the retained fuzzy sets (line 22).

4   GRABS in pyFUME

pyFUME was designed to have an easy to use interface both for practitioners and researchers. Currently, pyFUME supports the following features.

1. Loading the input data.
2. Division of the input data into a training and test data set.

3. Clustering of the data in the input-output space by means of Fuzzy C-Means (FCM) clustering [4] or an approach based on Fuzzy Self-Tuning Particle Swarm Optimization (FST-PSO [8,24]).
4. Estimating the antecedent sets of the fuzzy model, using the method described in [10]. Currently, Gaussian (default option), double Gaussian and sigmoidal membership functions are supported
5. Estimating the consequent parameters of the first-order TS fuzzy model, implementing the functionalities described in [3].
6. The generation, using the estimated antecedents and consequents, of an executable fuzzy model based on Simpful, possibly exporting the source code as a separate, executable file.
7. Testing of the estimated fuzzy model, by measuring the Root Mean Squared Error (RMSE), Mean Squared Error (MSE) or Mean Absolute Error (MAE).

To use pyFUME to estimate a fuzzy model from data, the user simply has to call the `pyfume()` function and specify the path to the data and the number of clusters as input. Optionally, the user can diverge from default settings (for example to use a clustering approach based on FST-PSO [24] or normalizing the data) by choosing additional key-value pairs. More information on pyFUME's functionalities can be found in [9].

If a user wants to use GRABS in pyFUME to simplify the produced rule bases, an optional input argument `similarity_threshold` must be specified. Thanks to this parameter, the user can set any arbitrary threshold for fuzzy sets similarity, implicitly controlling the error tolerance. This allows our method to be applied, in principle, to any system. By default, this threshold is set to 1.0, which means that one of the fuzzy sets is only dropped if the Jaccard similarity (which is assessed by using (1)) is 1.0, i.e., the fuzzy sets are identical. Since membership functions estimated from data will hardly ever be identical, this means that by default the functionality is switched off. The user can activate the functionality by setting `similarity_threshold` to a lower number. For example, when `similarity_threshold` = 0.9, pairs of fuzzy sets that have a Jaccard similarity higher than 0.9 will be dropped.

Internally, pyFUME represents the relation between the fuzzy sets as graphs such as the ones shown in Fig. 1. After identifying all complete components, one vertex for each of them is randomly selected to be retained, while the others are dropped.

In practice, this means that from all overlapping fuzzy sets in the FIS, one set is randomly selected and retained in the model. The other overlapping sets are discarded and remapped to the fuzzy set that was retained. This means that multiple rules in a FIS can have the same fuzzy set for a certain variable in their antecedent.

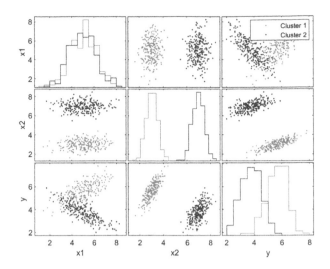

Fig. 2. Plots for the synthetic data set. On the diagonal, the distributions of the variables per cluster are visualized. Each off-diagonal plot is a scatter plot of a column of the data against another column of data.

5   Results

We use two data sets to show the effects of different threshold levels on the estimated fuzzy models. The first data set is synthetic and follows the same distributions as the data set described in [11], the second data set was downloaded from the UCI repository [22].

5.1   Example Case: Synthetic Data Set

For the first tests, we created a data set which contains 500 points and two variables (x_1 and x_2). In the data set there are two clusters, each containing 250 data points. For both clusters, variable x_1 follows a normal distribution $N(\mu, \sigma^2)$ with $\mu = 5$ and $\sigma = 1.2$. For cluster 1, variable x_2 follows the distribution $N(3, 0.5^2)$ and for cluster 2 the values for variable x_2 are drawn from $N(7, 0.5^2)$. The values for the output variable were calculated as $0.4 * x_1 + 1.2 * x_2 + \varepsilon$ for the first cluster and $-0.2 * x_1 + 0.9 * x_2 + \varepsilon$ for the second one. ε is random noise drawn from $N(0, 0.1^2)$. In Fig. 2 a matrix of scatter plots of the input and output variables, and (on the diagonal) frequency histograms of the data is shown.

Using pyFUME, we train first-order Takagi-Sugeno fuzzy models with two rules (and therefore two clusters) for this data set. Therefore, each input variable has two fuzzy sets. pyFUME's similarity threshold is varied from 0.0 to 1.0 in steps of 0.05, and for each level 100 models are built using 75% of the data as training data. All models are then evaluated in terms of Root Mean Square Error (RMSE) with the remaining 25% of the data.

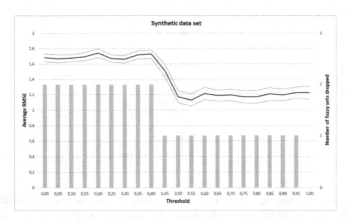

Fig. 3. The average and 95% confidence intervals of the RMSE for fuzzy models for the synthetic data set, build in pyFUME using different similarity thresholds (100 runs each).

In Fig. 3 the average RMSE and the 95% confidence interval for each threshold level are plotted. It can be observed that using very low values for the threshold results in worse performing models. When a similarity threshold of <0.45 is chosen, the two fuzzy sets for each of the variables are deemed similar, and therefore, one of them is dropped. Because of this, only one fuzzy set per variable remains, making both rules identical. As a result, the model does not separate the clusters anymore and behaves like a multiple regression model. This leads to a loss in accuracy.

Using any threshold level $\geqslant 0.45$ but <1.0 results in the merging of the two fuzzy sets for variable $x1$, since these fuzzy sets have a Jaccard similarity index of 0.97. Dropping one of these fuzzy sets does not result in loss of information, since variable $x1$ follows the same distribution in both cluster 1 and 2. Because of this, the RMSE does not decrease when one of the two fuzzy sets is dropped. This can be observed in Fig. 3.

In Fig. 4 the membership function of the fuzzy model that still contains all fuzzy sets is depicted. For variable x_1 indeed a large overlap can be observed for the fuzzy sets. Figure 5 shows the new membership functions for the fuzzy model when the similarity threshold is set to 0.75. Note that for variable x_1 only one set is left. In pyFUME the rules are now simplified to (bold highlights the change):

– RULE1 = IF (x1 IS cluster1) AND (x2 IS cluster1) THEN (OUTPUT IS fun1)
– RULE2 = IF (x1 IS cluster1) AND (x2 IS cluster2) THEN (OUTPUT IS fun2)

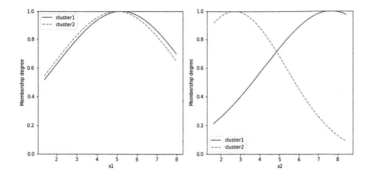

Fig. 4. The membership functions of a fuzzy model based on the synthetic data set. The similarity threshold was set to 1.0 and therefore, no fuzzy sets were dropped.

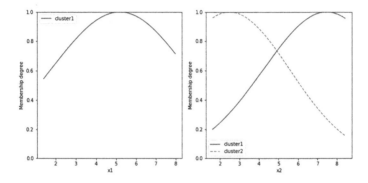

Fig. 5. The membership functions of a fuzzy model based on the synthetic data set. The similarity threshold was set to 0.75 and therefore, one of the fuzzy sets of the variable x_1 was dropped.

5.2 Example Case: NASA Data Set

The NASA data set [5] (downloaded from and described in the UCI repository [22]) consists of 1503 cases of different size NACA 0012 airfoils, which are airfoil shapes for aircraft wings developed by the National Advisory Committee for Aeronautics (NACA). Measurement were taken at various wind tunnel speeds and angles of attack. The span of the airfoil and the position of the observer were kept the same during data gathering in all of the experiments. During the experiments the frequency, angle of attack, chord length, free-stream velocity, and suction side displacement thickness of the NACA 0012 airfoils were recorded. These input variables should be mapped to the output variable, which is the scaled sound pressure level in decibels.

Again, the fuzzy models are build in pyFUME, and a 25% hold-out set is used for testing. To determine the similarity threshold value, different thresholds are tested. The results of this are plotted in Fig. 6 When exploring the effect of these different similarity thresholds on the accuracy of the fuzzy model, it can be observed that dropping fuzzy sets that show a similarity of more than 0.55 to another set does not result in significantly higher error rates. At this threshold, ten fuzzy sets are dropped.

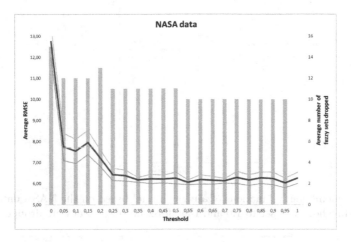

Fig. 6. The average and 95% confidence intervals of the RMSE for fuzzy models for the NASA data set, build in pyFUME using different similarity thresholds (100 runs each).

Figure 7 visualizes the membership functions of the NASA model before fuzzy sets were dropped. In this figure it can be observed that, except for the first variable, all variables have fuzzy sets that show high similarity. These fuzzy sets are dropped when the similarity threshold is set to 0.55, as can be seen in Fig. 8. This figure also shows that the variable 'chord_length' and (to a lesser extent) 'freestream_velocity' have membership functions that are similar to the universal set. This might indicate that these variables can be removed from the model, but that goes beyond the scope of this study. Then, the rule base is as follows (bold highlights the changes):

- RULE1 = IF (frequency IS cluster1) AND (angle_of_attack IS cluster1) AND (chord_length IS cluster1) AND (freestream_velocity IS cluster1) AND (suction_side_displacement_thickness IS cluster1) THEN (OUTPUT IS fun1)
- RULE2 = IF (frequency IS cluster2) AND (angle_of_attack IS cluster2) AND (chord_length IS cluster1) AND (freestream_velocity IS cluster1) AND (suction_side_displacement_thickness IS cluster2) THEN (OUTPUT IS fun2)
- RULE3 = IF (frequency IS cluster3) AND (angle_of_attack IS cluster2) AND (chord_length IS cluster1) AND (freestream_velocity IS cluster1) AND (suction_side_displacement_thickness IS cluster2) THEN (OUTPUT IS fun3)
- RULE4 = IF (frequency IS cluster4) AND (angle_of_attack IS cluster2) AND (chord_length IS cluster1) AND (freestream_velocity IS cluster1) AND (suction_side_displacement_thickness IS cluster2) THEN (OUTPUT IS fun4)

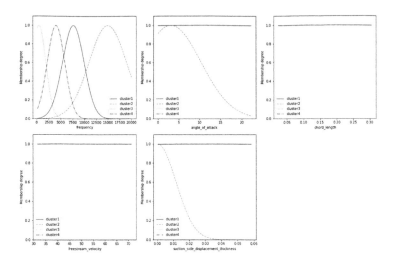

Fig. 7. The membership functions of a fuzzy model based on the NASA data set. The similarity threshold was set to 1.0 and therefore, no fuzzy sets were dropped.

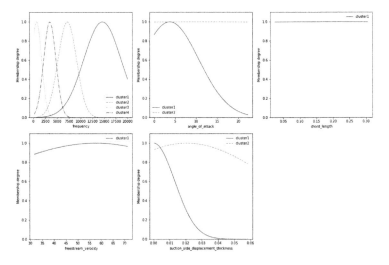

Fig. 8. The membership functions of a fuzzy model based on the NASA data set. The similarity threshold was set to 0.55 and therefore, ten fuzzy sets were dropped.

6   Conclusions

In this paper we introduced a novel graph theory based approach to simplify fuzzy rule bases called GRABS. By combining the Jaccard similarity and graph theory, we determine which fuzzy sets can be simplified in the model. The examples in this paper show that simplifying the model using this approach does not result in significant information and accuracy loss. Future studies will show how these result generalise to other data sets, and how this method compare to other simplification methods. The GRABS approach is implemented in the pyFUME package, whose source code and documentation is available at: https://github. com/CaroFuchs/pyFUME.

In [11] it is shown that when only one fuzzy set remains for a variable, the corresponding antecedent clause can be removed from all the rules in the fuzzy rule base. This improves the readability of the rule base even further. In future releases of pyFUME, we wish to implement the automatic detection and removal of these antecedent clauses. Moreover, we plan to implement a semi-automatic procedure to assign meaningful labels to fuzzy sets (e.g., low, medium, high), after performing simplification, in order to further improve the interpretation of the model. Finally, we will investigate the possibility of exploiting graph measures (e.g., degree centrality) as an alternative to detect fuzzy sets to be removed.

References

1. Alonso, J.M., Castiello, C., Mencar, C.: Interpretability of fuzzy systems: current research trends and prospects. In: Kacprzyk, J., Pedrycz, W. (eds.) Springer Handbook of Computational Intelligence, pp. 219–237. Springer, Heidelberg (2015). https://doi.org/10.1007/978-3-662-43505-2_14
2. Alonso, J.M., Magdalena, L.: Special issue on interpretable fuzzy systems (2011)
3. Babuška, R.: Fuzzy modelling and identification toolbox. Control Engineering Laboratory, Faculty of Information Technology and Systems, Delft University of Technology, Delft, The Netherlands, version 3 (2000)
4. Bezdek, J.C.: Models for pattern recognition. In: Bezdek, J.C., et al. (eds.) Pattern Recognition with Fuzzy Objective Function Algorithms. AAPR, pp. 1–13. Springer, Boston (1981). https://doi.org/10.1007/978-1-4757-0450-1_1
5. Brooks, T.F., Pope, D.S., Marcolini, M.A.: Airfoil self-noise and prediction (1989)
6. Cross, V., Setnes, M.: A study of set-theoretic measures for use with the generalized compatibility-based ranking method. In: 1998 Conference of the North American Fuzzy Information Processing Society - NAFIPS (Cat. No. 98TH8353), pp. 124–129, August 1998. https://doi.org/10.1109/NAFIPS.1998.715549
7. Fan, C.Y., Chang, P.C., Lin, J.J., Hsieh, J.: A hybrid model combining case-based reasoning and fuzzy decision tree for medical data classification. Appl. Soft Comput. 11(1), 632–644 (2011)
8. Fuchs, C., Spolaor, S., Nobile, M.S., Kaymak, U.: A swarm intelligence approach to avoid local optima in fuzzy c-means clustering. In: 2019 IEEE International Conference on Fuzzy Systems (FUZZ-IEEE), pp. 1–6. IEEE (2019)
9. Fuchs, C., Spolaor, S., Nobile, M.S., Kaymak, U.: pyFUME: a Python package for fuzzy model estimation. In: 2020 IEEE International Conference on Fuzzy Systems (FUZZ-IEEE) (2020, accepted)
10. Fuchs, C., Wilbik, A., Kaymak, U.: Towards more specific estimation of membership functions for data-driven fuzzy inference systems. In: 2018 IEEE International Conference on Fuzzy Systems (FUZZ-IEEE), pp. 1–8. IEEE (2018)
11. Fuchs, C., Wilbik, A., van Loon, S., Boer, A.-K., Kaymak, U.: An enhanced approach to rule base simplification of first-order Takagi-Sugeno fuzzy inference systems. In: Kacprzyk, J., Szmidt, E., Zadrożny, S., Atanassov, K.T., Krawczak, M. (eds.) IWIFSGN/EUSFLAT -2017. AISC, vol. 642, pp. 92–103. Springer, Cham (2018). https://doi.org/10.1007/978-3-319-66824-6_9

12. Guillaume, S.: Designing fuzzy inference systems from data: an interpretability-oriented review. IEEE Trans. Fuzzy Syst. **9**(3), 426–443 (2001)
13. Guyon, I., Elisseeff, A.: An introduction to variable and feature selection. J. Mach. Learn. Res. **3**(Mar), 1157–1182 (2003)
14. Ho, S.Y., Lee, K.C., Chen, S.S., Ho, S.J.: Accurate modeling and prediction of surface roughness by computer vision in turning operations using an adaptive neuro-fuzzy inference system. Int. J. Mach. Tools Manuf **42**(13), 1441–1446 (2002)
15. Huang, Z.: Rule model simplification. Ph.D. thesis, University of Edinburgh. College of Science and Engineering. School of Informatics (2006)
16. Jaccard, P.: Etude comparative de la distribution florale dans une portion des Alpes et du Jura. Impr, Corbaz (1901)
17. Jin, Y.: Fuzzy modeling of high-dimensional systems: complexity reduction and interpretability improvement. IEEE Trans. Fuzzy Syst. **8**(2), 212–221 (2000)
18. Kaymak, U., Babuska, R.: Compatible cluster merging for fuzzy modelling. In: Proceedings of the Fourth IEEE International Conference on Fuzzy Systems and The Second International Fuzzy Engineering Symposium, vol. 2, pp. 897–904. IEEE (1995)
19. Kóczy, L., Hirota, K.: Interpolative reasoning with insufficient evidence in sparse fuzzy rule bases. Inf. Sci. **71**(1–2), 169–201 (1993)
20. Kusiak, A.: Feature transformation methods in data mining. IEEE Trans. Electron. Packag. Manuf. **24**(3), 214–221 (2001)
21. Levrat, E., Voisin, A., Bombardier, S., Brémont, J.: Subjective evaluation of car seat comfort with fuzzy set techniques. Int. J. Intell. Syst. **12**(11–12), 891–913 (1997)
22. Lichman, M.: UCI machine learning repository (2013). http://archive.ics.uci.edu/ml
23. Mamdani, E.H., Assilian, S.: An experiment in linguistic synthesis with a fuzzy logic controller. Int. J. Man Mach. Stud. **7**(1), 1–13 (1975)
24. Nobile, M.S., Cazzaniga, P., Besozzi, D., Colombo, R., Mauri, G., Pasi, G.: Fuzzy self-tuning PSO: a settings-free algorithm for global optimization. Swarm Evol. Comput. **39**, 70–85 (2018)
25. Nobile, M.S., et al.: Fuzzy modeling and global optimization to predict novel therapeutic targets in cancer cells. Bioinformatics, btz868 (2019)
26. Oliphant, T.E.: Python for scientific computing. Comput. Sci. Eng. **9**(3), 10–20 (2007)
27. Polat, K., Güneş, S.: An expert system approach based on principal component analysis and adaptive neuro-fuzzy inference system to diagnosis of diabetes disease. Digital Sig. Proc. **17**(4), 702–710 (2007)
28. Raju, G., Zhou, J., Kisner, R.A.: Hierarchical fuzzy control. Int. J. Control **54**(5), 1201–1216 (1991)
29. Setnes, M., Babuska, R., Kaymak, U., van Nauta Lemke, H.R.: Similarity measures in fuzzy rule base simplification. IEEE Trans. Syst. Man Cybern. Part B (Cybern.) **28**(3), 376–386 (1998)
30. Shahin, M., Tollner, E., McClendon, R.: Ae-automation and emerging technologies: artificial intelligence classifiers for sorting apples based on watercore. J. Agric. Eng. Res. **79**(3), 265–274 (2001)
31. Spolaor, S., Fuchs, C., Cazzaniga, P., Kaymak, U., Besozzi, D., Nobile, M.S.: Simpful: a user-friendly python library for fuzzy logic (2020, submitted)
32. Sugeno, M., Griffin, M., Bastian, A.: Fuzzy hierarchical control of an unmanned helicopter. In: 17th IFSA World Congress, pp. 179–182 (1993)

33. Takagi, T., Sugeno, M.: Fuzzy identification of systems and its applications to modeling and control. IEEE Trans. Syst. Man Cybern. **1**, 116–132 (1985)
34. Xiong, N., Litz, L.: Reduction of fuzzy control rules by means of premise learning - method and case study. Fuzzy Sets Syst. **132**(2), 217–231 (2002)
35. Yam, Y., Baranyi, P., Yang, C.T.: Reduction of fuzzy rule base via singular value decomposition. IEEE Trans. Fuzzy Syst. **7**(2), 120–132 (1999)
36. Yen, J., Wang, L.: Simplifying fuzzy rule-based models using orthogonal transformation methods. IEEE Trans. Syst. Man Cybern. Part B (Cybern.) **29**(1), 13–24 (1999)
37. Zadeh, L.A.: Fuzzy sets. Inf. Control **8**(3), 338–353 (1965)

MaTED: Metadata-Assisted Twitter Event Detection System

Abhinay Pandya[1]($\boxtimes$) (iD), Mourad Oussalah[2] (iD), Panos Kostakos[1] (iD),
and Ummul Fatima[1]

[1] Center for Ubiquitous Computing, Faculty of ITEE, University of Oulu,
Oulu, Finland
{abhinay.pandya,panos.kostakos,ummul.fatima}@oulu.fi
[2] Center for Machine Vision and Signal Analysis, Faculty of ITEE,
University of Oulu, Oulu, Finland
mourad.oussalah@oulu.fi
http://ubicomp.oulu.fi, http://www.oulu.fi/cmvs

Abstract. Due to its asynchronous message-sharing and real-time capabilities, Twitter offers a valuable opportunity to detect events in a timely manner. Existing approaches for event detection have mainly focused on building a temporal profile of named entities and detecting unusually large bursts in their usage to signify an event. We extend this line of research by incorporating external knowledge bases such as DBPedia, WordNet; and exploiting specific features of Twitter for efficient event detection. We show that our system utilizing temporal, social, and Twitter-specific features yields improvement in the precision, recall, and DERate on the benchmarked Events2012 corpus compared to the state-of-the-art approaches.

Keywords: Twitter · Event detection · DBPedia · Microblogging ·
Social media

1 Introduction

Social media serve as important *social sensor* to capture the zeitgeist of the society. With real-time, online, asynchronous message-sharing supporting text, audio, video and images, these platforms offer valuable opportunities to detect events such as natural disasters and terrorist activity in a timely manner. Twitter is certainly a leader among microblogging platforms. With over 1.3 billion users and about 500 million messages (called *tweets*) posted per day and over 15 billion API calls per day, Twitter provides a massive source to detect event occurrences as they happen. Tweets are restricted to be maximum 280 characters in length and it is this design choice that enables information sharing extremely fast and in real-time. Owing to this scale and speed of information updates, Twitter is at the focus of attention to be monitored for new event detection. Also, Twitter's social network features allow user interactions which helps further in gaining an

© Springer Nature Switzerland AG 2020
M.-J. Lesot et al. (Eds.): IPMU 2020, CCIS 1237, pp. 402–414, 2020.
https://doi.org/10.1007/978-3-030-50146-4_30

insight on an event's reception by people (by analysing their sentiments/opinions expressed in their tweets).

While Twitter analytics in general may entail a comprehensive multi-modal approach, for example, to harness relevant information from media (images, videos, audio), we argue that since the textual part of the message is authored by the creator of the message, it is more reliable, authentic, and credible personal source to gain insights on the information shared. However, while analyzing language is as such challenging because of the inherent lack of structure in its expression, Twitter exacerbates this by supporting short message text, slang and emojis, user-created meta-information tags (hashtags), and URLs. Traditional methods in natural language processing (NLP) were designed to work with large discourses and perform poorly on social media text. This is owing to the following:

- Tweets are short mainly because of the restrictions imposed by the underlying platform (i.e. Twitter), but sometimes also under the influence of the cultural norms of the cyberspace[1];
- Tweets are often ungrammatical data type. Indeed, under the restriction of typing short messages from a mobile phone, grammar often takes a back seat. Also, because of the growth of social media encompassing the global scale, significant number of users are non-native speakers of English;
- Tweets are highly contextual, i.e., the message cannot be understood without the context – geopolitical, cultural, topical (current affairs), and even conversational context (e.g. replies to previous messages);
- Tweets use a lot of slang and emojis; exacerbating this, new slang words and emojis are invented and popularized every day and their meanings are volatile.

Indeed, tweets contain polluted content [18], and rumors [6], which negatively affect the performance of the event detection algorithms. Besides, only very few tweets actually carry a message about a newsworthy event [13]. These limitations motivate the current work, which aims to develop an enhanced Twitter event detection system that integrates external sources of information such as DBPedia [4] and WordNet [7]; exploits social network features of Twitter; and integrates knowledge obtained from the annotated resources (such as URLs cited in the tweets pointing to external web pages).

Topic Detection and Tracking (TDT) project [2] defines an event as "some unique thing that happens at some point in time". [5] defines an event as "a real-world occurrence e with an associated time period T_e and a time-ordered stream of Twitter messages M_e of substantial volume, discussing the occurrence and published during time T_e". We use these definitions as a guideline in developing our system of event detection from Twitter. We exploit specific features of Twitter that allow users to share others' tweets by re-tweeting them, quoting them, liking them; allows users to embed hashtags, URLs.

---

[1] https://blog.twitter.com/official/en_us/topics/product/2017/tweetingmadeeasier.html.

Inspired by the existing event detection methods for detecting and tracking event-related segments [10, 19, 31], our system is based on identifying important phrases from individual tweets and creating a temporal profile of these phrases to identify if they are bursty enough to signify an occurrence of an event. However, our system differs from the state-of-the-art in the following:

- We utilize DBPedia to efficiently identify named entities of import.
- We use WordNet to assist in identifying event-specific words and phrases.
- We expand the URLs embedded in tweets to find important phrases from the titles of the web pages that these URLs point to.
- We harness the meta-data from Twitter such as 'quoted status', 'retweet', 'liked', 'in reply to', and also, the social network of the Twitter user.

The rest of the paper is organized as follows. In Sect. 2, we discuss the work closely related to ours. Section 3 describes our system for Twitter event detection. Section 4 explains our experimental setup and results followed by discussion. Finally, in Sect. 5, we present our conclusions.

2   Related Work

Event detection from Twitter has been extensively studied in the past as evidenced by the rich body of work. [3, 14, 25, 30], among others. Especially, techniques for event detection from Twitter can be classified according to the event type (specified or unspecified events), detection method (supervised or unsupervised learning), and detection task (new event detection or retrospective event detection). However, most of the techniques described in the aforementioned surveys suffer from rigorous evaluation. On the other hand, a major acknowledged obstacle in measuring the accuracy and performance of an event detection methods is the lack of large-scale, benchmarked corpora. Some authors have created their own manually annotated datasets and made them available publicly [23, 26].

Our work focuses on unsupervised, unspecified event detection retrospectively from a large body of tweets. Among many approaches to event detection from Twitter such as keyword volume approach, topic modeling, and sentiment analysis based methods, our work is based on keyphrase/segment detection and tracking which aim to identify keyphrases/segments whose occurrences grow unusually within the corpus [8, 9, 11, 12, 16, 17, 29]. Some of the most related works to ours are by [1, 19, 27, 31].

EDCoW [31] proposed a three-step approach. First, a wavelet transform and auto-correlation are applied to measure the bursty energy of each word and words associated with high energies are retained as event features. Then, they measure the similarity between each pair of event features by using cross correlation. At last, modularity-based graph partitioning is used to detect the events, each of which contains a set of words with high cross correlation.

[19] presented a system called Twevent that analyzes the tweets by breaking them into non-overlapping segments and subsequently identifying bursty

segments. These bursty segments are then clustered to obtain event-related segments.

[23] contributed a large manually labeled corpus of 120 million tweets containing 506 events in 8 categories. They used Locality Sensitive Hashing (LSH) technique followed by cluster summarization and employed Wikipedia as an external knowledge source.

[1] employed a statistical analysis of historical usage of words for finding bursty words – those with burstiness degree above two standard deviation from the mean are selected clustered. However, their method was used to find localized events only.

[10] proposed mention-anomaly based approach incorporating social aspect of tweets by leveraging the creation frequency of mentions that users insert in tweets to engage discussion. [22] advocated the importance of named entities in Twitter event detection. They used a clustering technique which partitions tweets based upon the entities they contain, burst detection and cluster selection techniques to extract clusters related to ongoing real-world events.

Recently, [28] employed extracting a structured representation from the tweets' text using NLP, which is then integrated with DBpedia and WordNet in an RDF knowledge graph. Their system enabled security analysts to describe the events of interest precisely and declaratively using SPARQL queries over the graph.

3 Our System

Our system, Metadata-assisted Twitter Event Detection (MaTED) is an extension of a previous work Twevent [19]; however, our system makes use of several other features of Twitter ignored in previous research.

Figure 1 shows the architecture of MaTED which consists of four components: i) detection of important phrases from tweets; ii) creating temporal profiles of these phrases to identify bursty phrases; iii) clustering bursty phrases with an aim to group related phrases about an event, and iv) characterizing an event from the clusters obtained above. We parse the tweet JSON object after receiving it from a stream, and the first component of our system identifies important segments/phrases not just from the tweet text but also from the titles of the webpages that URL links in the tweet points to. Since event-related phrases are mostly named entities, we harness the DBPedia to extract such phrases. The resultant phrases along with tweet timestamps are then fed to the next component of our system which estimates their burstiness behavior using statistical modeling of their occurrence frequency. Subsequently, we group the event-related phrases using a graph-based clustering algorithm. In the rest of this section, we describe each component in detail following the order of their usage in our framework.

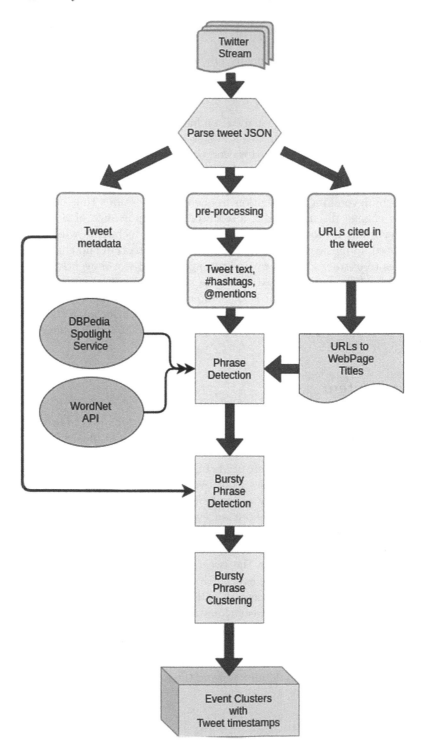

Fig. 1. MaTED system architecture

3.1 Identifying Important Phrases from Tweets

In this component, we parse the tweet JSON object to obtain tweet text, hashtags, URLs, user mentions and other available metadata. We then create a set of phrases/keywords to be monitored consisting of the following items:

- List of named entities obtained after inputting the tweet text (after preprocessing and cleaning) to DBPedia Spotlight [24] web service.
- List of wordings related to action or activity present in the original tweet message. These are identified as words that are either a direct or indirect hyponym of 'event.n.01' synset of WordNet.
- List of hashtags included in the tweet.
- For each URL that is cited in the tweet, we obtain the title text of the web page that URL is pointing to. We submit this title text to our locally running DBPedia Spotlight web service to obtain named entities and include these into the list of items to be monitored.
- For each user mention, we include the 'name' the user mention handle is associated to.

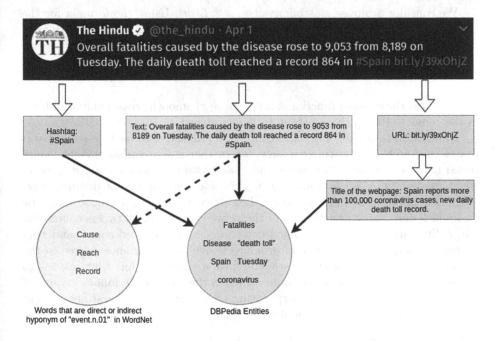

Fig. 2. An example of phrase extraction from tweets.

Figure 2 illustrates the overall process. The tweet text *"Overall fatalities caused by the disease rose to 9053 from 8189 on Tuesday. The daily death toll reached a record 864 in #Spain."* does not mention *coronavirus* which is important entity for event detection, tracking and monitoring. Our system fetches and processes the title of the webpage linked to the URL cited in the tweet (http://bit.ly/39xOhjZ). Concurrently, finding words from the tweet text that are direct or indirect hyponyms of the *event.n.01* synset of the WordNet finds important words to track/monitor ('cause', 'reach', and 'record').

3.2   Extracting Bursty Phrases

After creating a set of phrases from the dataset as indicated above, using a model proposed by Twevent [19], we find bursty phrases potentially indicative of an event. However, our model includes several other factors in finding burstiness score of a phrase than considered in [19]. Below we outline our method.

Let N_t denote the number of tweets published within the current time window t and $n_{i,t}$ be the number of tweets containing phrase i in t. The probability of observing i with a frequency $n_{i,t}$ can be modeled by a Binomial distribution $B(N_t, p_i)$ where p_i is the expected probability of observing phrase i in a random time window. Since N_t is very large in case of Twitter stream, this probability distribution can be approximated by a Normal distribution p with parameters $E[i|t] = N_t \times p_i$ and $\sigma(i|t) = \sqrt{N_t \times p_i \times (1 - p_i)}$.

We consider a phrase i as bursty if $n_{i,t} \geq E[i|t]$. Using the formula for the burstiness probability $P_b(i,t)$ for phrase i in time window t defined by [19]:

$$P_b(i,t) = S(10 \times \frac{n_{i,t} - (E[i|t] + \sigma[i|t]}{\sigma[i|t]}) \tag{1}$$

where $S()$ is the sigmoid function, and since $S(x)$ smooths reasonably well for x in the range $[-10,10]$, the constant 10 is introduced.

In addition to finding the importance of a phrase based on *how many times* it was used in the given time window, we further assign various weight values based on *who* authored the phrase, *how many times it was retweeted, quoted, liked, replied to*. More formally, let $u_{i,t}$ denote the number of distinct users authoring phrase i in time window t. Let retweet count of a phrase i in t be $rt_{i,t}$ which corresponds to the sum of the retweet counts of all tweets containing i in t. Similarly, let $l_{i,t}$ be the liked count, $q_{i,t}$ be the quoted count, and $rp_{i,t}$ be the 'replied to' count. Also, in order to assign an importance degree to the phrases used by those Twitter users who have a significant following, we assign the weight $fc_{i,t}$ as the follower count which is the sum of the follower count of all users using phrase i in t. Incorporating all the above, the burstiness weight $w_b(i,t)$ for a phrase i in t can be defined as:

$$w_b(i,t) = P_b(i,t) \cdot log(u_{i,t}) \cdot log(rt_{i,t}) \cdot log(l_{i,t}) \cdot log(q_{i,t}) \cdot log(rp_{i,t}) \cdot log(log(fc_{i,t})) \tag{2}$$

After finding the burstiness weight for all phrases, the top K are selected in decreasing order of weights. Empirically, we find that decreasing K results in low recall, while increasing K brings in a significant noise. [19] suggest an optimal value of K is set to $\sqrt{N_t}$.

3.3 Clustering Bursty Phrases

We adopt the approach by [19] without any modification to group bursty phrases to derive event-related clusters. Each time window is evenly split into M sub-windows $t = <t_1, t_2, \ldots, t_M>$. Let $n_t(i, m)$ be the tweet frequency of phrase i in the subwindow t_m and $T_t(i, m)$ be the concatenation of all the tweets in the subwindow t_m that contain phrase i. The similarity $sim_t(i_a, i_b)$ between phrases i_a and i_b in time window t is calculated as follows:

$$sim_t(i_a, i_b) = \sum w_t(i_a, m) w_t(i_b, m) \times sim(T_t(i_a, m), T_t(i_b, m)) \qquad (3)$$

where $sim(T_t(i_a, m), T_t(i_b, m))$ is TF-IDF similarity between tweets $T_t(i_a, m)$ and $T_t(i_b, m)$; and $w_t(i_a, m)$ is the fraction of frequency of segment i_a in the time subwindow t_m as calculated as follows:

$$w_t(i, m) = \frac{n_t(i, m)}{n_{(i,t)}} \qquad (4)$$

Using the above similarity measures, all the bursty phrases are clustered using a graph-based clustering algorithm [15]. In this method, all bursty phrases are considered as nodes and initially, all nodes are disconnected. An edge is added between phrases i_a and i_b if k-Nearest neighbors of i_a contain i_b and vice versa. All connected components of the resultant graph are considered as candidate event clusters. Each connected component is essentially a set of phrases which are related to a single event. Disconnected nodes (phrases) are discarded as non-significant.

3.4 Event Characterization

We characterize an event as a group of phrases associated to it. To visualize this, we adopt the approach by MABED [10]. An interface is designed to allow us to visualize the list of relevant tweets defining the event by 'clicking' on the event name.

4 Experiments and Results

We use the corpus collected by [23] Events2012 which contains 120 million tweets and 506 labeled events. These tweets were collected from Oct 10 till Nov 7, 2012 and were filtered to remove tweets containing more than 3 hashtags, 3 user mentions, or 2 URLs discarding them as spam [6]. However, not all tweets were available due to some users' data being not available because of account inactivation, privacy mode setting changes, etc. Our final dataset contain ~38 million tweets of which 127,356 are related to events. It should be noted that these tweets are limited to maximum 140 characters in length since the increased length (up to 280 characters) was introduced in late 2017. In Table 1 we show some of the important events in the Events2012 [23] dataset which have more than 1,500 tweets associated to them.

Table 1. Top events in Events2012 dataset. Each with more than 1500 tweets associated to it.

Event ID	Manually-annotated event description	#tweets
8	During US presidential debate, President Barack Obama tells candidate Mitt Romney he is the last person to get tough on China	12805
1	12 Oct 2012 – Paul Ryan spoke for 40 of the 90 min during Thursday night's vice presidential debate and managed to tell at least 24 myths during that time	6894
11	Barack Obama and Mitt Romney went head-to-head in the final Presidential Debate. Romney said no government that makes businesses successful!	5243
22	The Redbull Stratos jump was a space diving event. Felix Baumgartner flew many miles into the air above the south- western U.S. and then jumped, breaking several world records	2967
52	People react to incoming election results, threatening to leave the country if their favored candidate does not win	2289
14	In Major League Baseball, the San Francisco Giants defeat the Detroit Tigers in game four	1796

4.1 Pre-processing

We perform the following steps sequentially as part of preprocessing the tweet text and the titles of the webpages linked by the cited URLs in the tweet message:

1. We use the Stanford tokenizer[2] to tokenize the tweets.
2. Use of words like *cooooolll, awesommmme*, are sometimes used in tweets to emphasize emotion. We use a simple trick to normalize such occurrences. Namely, let n denote the number of such letters that have three or more consecutive occurrences in a given word. We first replace three or more consecutive occurrences of the same character with two occurrences. Then we generate $\binom{n}{2}$ prototypes that are at edit distance 1 (only delete operation, deleting only repeated character) and look for this prototype in the dictionary to find the word. For example, *coooooolllll → cooll → cool*.
3. We use an acronym dictionary from an online resource[3] to find expansions of the tokens such as *gr8, lol, rotfl*, etc.

After the pre-processing task, to obtain a list of named entities, we submit the text to the DBPedia Spotlight [24] web service. We chose DBPedia over other named-entity recognition software (such as Standford NER [21], OpenNLP [32], NLTK [20], etc.) because employing such tools yields phrases that induce noise in the resulting system.

---

[2] https://nlp.stanford.edu/software/tokenizer.htm.
[3] http://www.noslang.com.

4.2   Evaluation Metrics

As evaluation metrics, we used precision, recall, and DERate (duplicate event rate, proposed by [19]). *Precision* conveys the fraction of the detected events that are related to a realistic event. *Recall* indicates the fraction of events detected from the manually labeled ground truth set of events. However, if two detected events are related to the same realistic event within the same time window, then both are considered correct in terms of precision, but only one realistic event is considered in counting recall. Therefore, [19] defined a metric DERate to denote the fraction of events (in percentage values) that are duplicately detected among all events detected.

4.3   Baseline Methods

In order to evaluate our proposal, we compare our approach with closely related works: EDCoW [31], Twevent [19], NEED [22], and MABED [10]. Table 2 shows the comparative performance of our system to selected state-of-art approaches. For MABED, we modified their online available code to include hashtags, instead of user mentions to measure anomaly (ref as MABED+ht). Also, for our system, in order to observe the effect of WordNet words related to events, we conducted two sets of experiments where MaTED-WN is the system without using WordNet words. We share our source code and dataset used online.

4.4   Results and Discussion

Table 2 shows the results we obtained compared to the baseline method.

Table 2. Results on Events2012 dataset

Approach	EDCoW	Twevent	NEED	MABED	MABED+ht	MaTED-WN	MaTED
Precision	0.42	**0.81**	0.61	0.68	0.74	0.83	0.79
Recall	0.38	0.73	0.41	0.43	0.55	0.79	**0.82**
DERate(%)	–	15.6	19.4	17.4	16.7	14.7	**14.3**

Table 3 shows some of the events detected by MaTED that were not detected by any of the above systems.

Several parameters impact the performance of the resulting system and the results shown in Table 2 are obtained by an optimal combination of them. It is evident from Table 2 that the performance of existing bursty segment detection based systems is enhanced by including social and Twitter-specific features incorporated in our system. Especially, we notice a significant improvement in recall by including title texts of the web pages pointed to by the URLs in the tweets. A tweet is often a comment on the web page that is shared and therefore, by including the title text, the system incorporates a better context for the

Table 3. A sample of events that only our system could extract.

Date	Event	Phrases
17.10.2012	Alpha Centauri BB planet discovered	[solar system], [centauri], [earth], [exoplanet]
16.10.2012	Ford recalls Fiesta because of airbag issue	[Ford], [Fiesta], [airbag], [fault]
15.10.2012	Amanda Todd suicide because of cyber-bullying	[Amanda Todd], [bullying], [suicide], [RIP]

tweet. Further, because of misspellings, DBPedia Spotlight sometimes fail to find the named entity and in such cases, tracking event specific words from WordNet (total 7878 words) helps identify an event. For example, in Table 3, the event on 16.10.2012 about Ford would be missed if the word 'fault' was not included in the list of important key-phrases to be considered. Better results are observed for MABED+ht as opposed to the original MABED owing to the fact that hashtags are better indicators of events than user mentions. We attribute our system's less precision value than [19]'s system to including several more event-specific phrases from the web page titles, hashtags, and event-specific words from the WordNet resulting in a higher recall but at a slight loss of precision (0.79 as opposed to 0.81 of [19]). Finally, we also noticed that many events were not reported in the crowd-sourced ground-truth Events 2012 corpus. Event on 15.10.2012 about *Amanda Todd suicide* is one example of many events we found which were not included in the corpus.

5   Conclusion

A phenomenal growth in online social network services generate massive amounts of data posing a lot of challenges especially owing to the volume, variety, velocity, and veracity of the data. Concurrently, methods to detect events from social streams in an efficient, accurate, and timely manner are also evolving. In this paper, we build on an existing system Twevent [19] by incorporating external knowledge bases of DBPedia and WordNet together with exploiting user's mentions and hashtags contained in Twitter messages for efficient event detection. In addition, harnessing the fact that a tweet is often just a remark/comment on the news/information shared in the URL cited in it, we improve event detection performance by detecting and tracking important event-related phrases from the titles of the web pages linked to the URLs. We examined the effect of adding our novel features incrementally and concluded that our model outperforms the state-of-the-art on the benchmarked Events2012 [23] corpus. Future research includes investigating usage of distributed semantics (e.g., word embeddings) incorporated in a larger framework of a deep learning inspired model towards achieving higher accuracy on event detection from a massive-scale collection of social media messages.

Acknowledgment. This work is partly supported by EU YoungRes project (#823701) on polarization detection.

References

1. Abdelhaq, H., Sengstock, C., Gertz, M.: Eventweet: online localized event detection from Twitter. Proc. VLDB Endow. **6**(12), 1326–1329 (2013)
2. Allan, J.: Topic Detection and Tracking: Event-based Information Organization, vol. 12. Springer, Heidelberg (2012). https://doi.org/10.1007/978-1-4615-0933-2
3. Atefeh, F., Khreich, W.: A survey of techniques for event detection in Twitter. Comput. Intell. **31**(1), 132–164 (2015)
4. Auer, S., Bizer, C., Kobilarov, G., Lehmann, J., Cyganiak, R., Ives, Z.: DBpedia: a nucleus for a web of open data. In: Aberer, K., et al. (eds.) ASWC/ISWC -2007. LNCS, vol. 4825, pp. 722–735. Springer, Heidelberg (2007). https://doi.org/10.1007/978-3-540-76298-0_52
5. Becker, H., Naaman, M., Gravano, L.: Beyond trending topics: real-world event identification on Twitter. In: Fifth International AAAI Conference on Weblogs and Social Media (2011)
6. Castillo, C., Mendoza, M., Poblete, B.: Information credibility on Twitter. In: Proceedings of the 20th International Conference on World Wide Web, pp. 675–684. ACM (2011)
7. Fellbaum, C.: WordNet. The encyclopedia of applied linguistics (2012)
8. Fung, G.P.C., Yu, J.X., Yu, P.S., Lu, H.: Parameter free bursty events detection in text streams. In: Proceedings of the 31st International Conference on Very Large Data Bases, pp. 181–192. VLDB Endowment (2005)
9. Goorha, S., Ungar, L.: Discovery of significant emerging trends. In: Proceedings of the 16th ACM SIGKDD International Conference on Knowledge Discovery and Data Mining, pp. 57–64. ACM (2010)
10. Guille, A., Favre, C.: Mention-anomaly-based event detection and tracking in Twitter. In: 2014 IEEE/ACM International Conference on Advances in Social Networks Analysis and Mining (ASONAM 2014), pp. 375–382. IEEE (2014)
11. He, Q., Chang, K., Lim, E.P.: Analyzing feature trajectories for event detection. In: Proceedings of the 30th Annual International ACM SIGIR Conference on Research and Development in Information Retrieval, pp. 207–214. ACM (2007)
12. He, Q., Chang, K., Lim, E.P., Zhang, J.: Bursty feature representation for clustering text streams. In: Proceedings of the 2007 SIAM International Conference on Data Mining, pp. 491–496. SIAM (2007)
13. Hurlock, J., Wilson, M.L.: Searching Twitter: separating the tweet from the Chaff. In: Fifth International AAAI Conference on Weblogs and Social Media (2011)
14. Imran, M., Castillo, C., Diaz, F., Vieweg, S.: Processing social media messages in mass emergency: a survey. ACM Comput. Surv. (CSUR) **47**(4), 1–38 (2015)
15. Jarvis, R.A., Patrick, E.A.: Clustering using a similarity measure based on shared near neighbors. IEEE Trans. Comput. **100**(11), 1025–1034 (1973)
16. Kleinberg, J.: Bursty and hierarchical structure in streams. Data Min. Knowl. Disc. **7**(4), 373–397 (2003)
17. Kontostathis, A., Galitsky, L.M., Pottenger, W.M., Roy, S., Phelps, D.J.: A survey of emerging trend detection in textual data mining. In: Berry, M.W. (ed.) Survey of Text Mining, pp. 185–224. Springer, NewYork (2004). https://doi.org/10.1007/978-1-4757-4305-0_9

18. Lee, K., Eoff, B.D., Caverlee, J.: Seven months with the devils: a long-term study of content polluters on Twitter. In: Fifth International AAAI Conference on Weblogs and Social Media (2011)
19. Li, C., Sun, A., Datta, A.: Twevent: segment-based event detection from tweets. In: Proceedings of the 21st ACM International Conference on Information and Knowledge Management, pp. 155–164. ACM (2012)
20. Loper, E., Bird, S.: NLTK: the natural language toolkit. arXiv preprint cs/0205028 (2002)
21. Manning, C.D., Surdeanu, M., Bauer, J., Finkel, J.R., Bethard, S., McClosky, D.: The stanford CoreNLP natural language processing toolkit. In: Proceedings of 52nd Annual Meeting of the Association for Computational Linguistics: System Demonstrations, pp. 55–60 (2014)
22. McMinn, A.J., Jose, J.M.: Real-time entity-based event detection for Twitter. In: Mothe, J., et al. (eds.) CLEF 2015. LNCS, vol. 9283, pp. 65–77. Springer, Cham (2015). https://doi.org/10.1007/978-3-319-24027-5_6
23. McMinn, A.J., Moshfeghi, Y., Jose, J.M.: Building a large-scale corpus for evaluating event detection on Twitter. In: Proceedings of the 22nd ACM International Conference on Information & Knowledge Management, pp. 409–418. ACM (2013)
24. Mendes, P.N., Jakob, M., García-Silva, A., Bizer, C.: DBpedia spotlight: shedding light on the web of documents. In: Proceedings of the 7th International Conference on Semantic Systems, pp. 1–8. ACM (2011)
25. Panagiotou, N., Katakis, I., Gunopulos, D.: Detecting events in online social networks: definitions, trends and challenges. In: Michaelis, S., Piatkowski, N., Stolpe, M. (eds.) Solving Large Scale Learning Tasks. Challenges and Algorithms. LNCS (LNAI), vol. 9580, pp. 42–84. Springer, Cham (2016). https://doi.org/10.1007/978-3-319-41706-6_2
26. Petrovic, S.: Real-time event detection in massive streams (2013)
27. Phuvipadawat, S., Murata, T.: Breaking news detection and tracking in Twitter. In: 2010 IEEE/WIC/ACM International Conference on Web Intelligence and Intelligent Agent Technology, vol. 3, pp. 120–123. IEEE (2010)
28. Tonon, A., Cudré-Mauroux, P., Blarer, A., Lenders, V., Motik, B.: ArmaTweet: detecting events by semantic tweet analysis. In: Blomqvist, E., Maynard, D., Gangemi, A., Hoekstra, R., Hitzler, P., Hartig, O. (eds.) ESWC 2017. LNCS, vol. 10250, pp. 138–153. Springer, Cham (2017). https://doi.org/10.1007/978-3-319-58451-5_10
29. Wang, X., Zhai, C., Hu, X., Sproat, R.: Mining correlated bursty topic patterns from coordinated text streams. In: Proceedings of the 13th ACM SIGKDD International Conference on Knowledge Discovery and Data Mining, pp. 784–793. ACM (2007)
30. Weiler, A., Grossniklaus, M., Scholl, M.H.: Survey and experimental analysis of event detection techniques for Twitter. Comput. J. 60(3), 329–346 (2017)
31. Weng, J., Lee, B.S.: Event detection in Twitter. In: Fifth International AAAI Conference on Weblogs and Social Media (2011)
32. Wilcock, G.: Text annotation with OpenNLP and UIMA (2009)

Image-Based World-perceiving Knowledge Graph (WpKG) with Imprecision

Navid Rezaei[1], Marek Z. Reformat[1,2]([ID]), and Ronald R. Yager[3]([ID])

[1] University of Alberta, Edmonton, T6G 1H9, Canada
{nrezaeis,marek.reformat}@ualberta.ca
[2] University of Social Sciences, 90-113 Łódź, Poland
[3] Iona College, New Rochelle, NY 10801, USA
yager@panix.com

Abstract. Knowledge graphs are a data format that enables the representation of semantics. Most of the available graphs focus on the representation of facts, their features, and relations between them. However, from the point of view of possible applications of semantically rich data formats in intelligent, real-world scenarios, there is a need for knowledge graphs that describe contextual information regarding realistic and casual relations between items in the real world.

In this paper, we present a methodology of generating knowledge graphs addressing such a need. We call them *World-perceiving Knowledge Graphs – WpKG*. The process of their construction is based on analyzing images. We apply deep learning image processing methods to extract scene graphs. We combine these graphs, and process the obtained graph to determine importance of relations between items detected on the images. The generated WpKG is used as a basis for constructing possibility graphs. We illustrate the process and show some snippets of the generated knowledge and possibility graphs.

Keywords: Knowledge graph · Deep learning · Common sense · Possibility theory

1 Introduction

Knowledge graphs are composed of a set of triple relations, i.e. <*subject – predicate – object*>, where subjects and objects are items connected via predicates representing relations between them. The graphs are useful in representing data semantics and are employed in different applications, such as common-sense and causal reasoning [1,2], question-answering [3], natural language processing [4], and recommender systems [5]. Some examples of existing knowledge graphs are DBpedia [6], Wikidata [7], Yago [8], the now-retired Freebase [9], and WordNet [10]. The aforementioned knowledge graphs contain information about facts,

© Springer Nature Switzerland AG 2020
M.-J. Lesot et al. (Eds.): IPMU 2020, CCIS 1237, pp. 415–428, 2020.
https://doi.org/10.1007/978-3-030-50146-4_31

their features, and basic relations between them. They focus on people, geographical locations, movies, music, and organizations and institutions. They are missing a piece of information about everyday real-world items, their contexts, and arrangements.

From the human perspective, we can state that the visual information plays a significant role in human learning processes [11]. At the same time, the eye's information transfer rate is quite high [12] that makes a visual stimulus to be of significant importance in processes of gaining understanding about different items and how they are related to each other. Given the importance of visual data, it is appealing to develop systems that could observe, learn and create knowledge based on such data. Additionally, traditional knowledge graphs do not provide any degree of confidence associated with relations. It is assumed that all of them are equally important.

In this paper, we look at the task of creating knowledge graphs based on visual data. The idea is to process images, generate scene graphs from them, and aggregate these graphs. Graphs constructed in such a way contain knowledge about everyday objects, their contexts and their situational information, as well as information related to the importance of common-sense relations between multiple objects in their natural scenarios.

We call such a graph *World-perceiving Knowledge Graph*, *WpKG* in short. The quality and suitability of knowledge we retrieve from images depend on the capability of tools and methods we use for image processing. Processing an image means generating a scene graph representing relations between objects/entities present on this image. Once numerous images are processed, all scene graphs are aggregated. This alone allows us to treat the process of constructing graphs via aggregation as the human-like process of learning via processing of observed images.

We also look at a process of using knowledge graphs – *WpKGs* – to construct possibility graphs reflecting conditional dependencies between sets of entities as observed in their usual environments. The information about the importance of relations allows us to build possibilistic conditional distributions. They are used for processing and reasoning about entities and relations between them in their own relevant contexts. The included case study shows an application of the presented procedure to Visual Genome (VG) dataset [13].

2   Related Work

Extracting information from different media to create a knowledge graph has been examined in the literature. Yet, the area of focus of these works has been different: some of them focus on images, some on text, and some on a combination of both. Also, the methods used for information retrieval can be different – automatic or manual. A brief overview is presented in Subsect. 2.1.

Possibilistic knowledge bases and graphs are important forms representing uncertainty of data and information [14], and [15]. A set of basic definitions is included in the following subsections.

2.1   Knowledge Graph Construction

There is a number of different knowledge graph generation methods that focus on text as the source of information, such as NELL [16], ConceptNet [17], ReVerb [18], and Quasimodo [19]. Some other published approaches, such as WebChild KB [20,21] or LEVAN [22], extract knowledge from text and image captions or only from image objects without in-image relations. Probably, the most relevant work to our work is NEIL [23], which create a knowledge graph directly from images.

Compared to NEIL, our proposed automatic approach is capable of extracting much more types of object-to-object relations. Compared to ConceptNet, which represents an example of a semi-automatic method of retrieving knowledge from text, our proposed approach can extract common-sense relations based on only observing visual data.

2.2   Possibilistic Knowledge Base

A possibilistic base is a set of pairs (p, α) where p is a proposition, and α is a degree to which p is true and is in the interval $(0, 1)$ [14]. Let Ω be a set of interpretations of the real world, and possibilistic distribution π a mapping from Ω to the interval $(0, 1)$. An interpretation ω that satisfies p has $\pi(\omega) = 1$, and $1 - \alpha$ when ω fails to satisfy p. In summary:

$$\forall \omega \in \Omega, \pi_{\{p\ \alpha\}}(w) = 1 \qquad if\ \omega \models p$$
$$= 1 - \alpha\ otherwise$$

From now on, we identify the base as $\sum = \{(p_i, \alpha_i), i = 1, \dots, n\}$. Then all interpretations satisfying propositions in $\sum$ have the possibility degree of 1, while other interpretations are ranked based on the highest values of α associated with proposition they do not satisfy, i.e., $\forall \omega \in \Omega$:

$$\pi_{\sum}(w) = 1 \qquad\qquad\qquad\qquad if\ \omega \models \sum$$
$$= 1 - max\{\alpha_i : (p_i, \alpha_i) \in \sum\ and\ \omega \models \neg p_i\}\ otherwise$$

In other words, $\pi_{\sum}$ induces a necessity 'grading' of p_i that evaluates to what extent p_i is a consequence of the available knowledge. The necessity measure *Nec* is:

$$Nec_{\pi_{\sum}}(p_i) = 1 - max\{\pi_{\sum}(\omega) \mid \omega \models \neg p_i\}$$

Based on that, we can say that (p_i, α_i) is a plausible conclusion of $\pi_{\sum}$ if

$$Nec_{\pi_{\sum}}(p_i) > Nec_{\pi_{\sum}}(\neg p_i)$$

and $Nec_{\pi_{\sum}}(p_i) \geq \alpha_i$ [24].

A possibility distribution $\pi_{\sum}$ is *normal* if there is an interpretation ω that it totally possible, i.e., $\pi_{\sum}(\omega) = 1$.

2.3 Possibilistic Graph

A possibility graph ΠG is an acyclic directed graph [14]. The nodes of such a graph are associated with variables A_i, each with its domain D_i; while its edges represent dependencies between elements of nodes. For the case of binary variables, i.e., when $D_i = \{a_i, \neg a_i\}$, the assignment of value to the variable is called an interpretation ω. Let us denote a set of nodes that have edges connecting them to a node A_i as its parents: $Par(A_i)$. Possibility degrees Π associated with nodes are:

for each node A_i without a parent $Par(A_i) = 0$ prior possibility degrees associated with a single node are $\Pi(a)$ for every value $a \in D_i$ of the variable A_i; possibilities must satisfy the normalization condition: $max_{a \in D_i}$: $\Pi(a) = 1$.

for each node A_j with parent(s) $Par(A_j) \neq 0$ possibility degrees are conditional ones $\Pi(a|\omega_{Par(A_j)})$ where $a \in D_j$, and $\omega_{Par(A_j)}$ is an element of the Cartesian product of domains D_k of variables $A_k \in Par(A_j)$; as above, conditional possibilities must satisfy the normalization condition: $max_{a \in D_j} : \Pi(a|\omega_{Par(A_j)}) = 1$.

In our case, a conditional probability measure is defined using min:

$$\Pi(p|q) = 1 \qquad if \ \Pi(q \wedge p) = \Pi(q)$$
$$= \Pi(q \wedge p) \ otherwise$$

and obeys [14]:

$$\Pi(q \wedge p) = min\{\Pi(p|q), \Pi(q)\}$$

3 Generation of Image-Based *WpKG*

We introduce a systemic approach to generate knowledge graphs given visual data. Such graphs provide us with contextual information about objects present in the world with very limited input from humans. There are unique challenges associated with the generation of this type of graph. First, we need methods able to detect objects in images, and second, we require tools to extract relations between the detected objects.

Once we have the object recognition and relation extraction processes, we execute them on a set of images. The obtained triples – <*entity – relation – entity*> are aggregated into a single knowledge graph. The strength of relations is determined by the number of co-occurrences of objects with specific relations. The overall process is shown in Fig. 1.

Having a trained model, the process is liberated from specific visual data and its annotations. Additionally, more visual data can be processed using the proposed methodology and comprehensive context-specific knowledge graphs could be created.

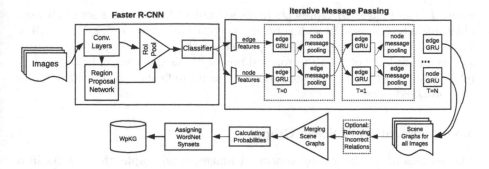

Fig. 1. Overall procedure for generation of a knowledge graph from images

3.1 Detection of Objects

To detect objects and their corresponding bounding boxes, we use the Faster R-CNN model [25]. In this model, the full image is passed through a convolutional neural network (CNN) to generate image features. To detect image features, usually a pre-trained CNN, such as VGG network [26], trained on ImageNet [27] is used. Given the image features as input, another neural network, called Region Proposal Network (RPN), predicts regions that may contain an object and their corresponding bounding boxes. This learning network is the principal contribution of the Faster R-CNN model compared to the Fast R-CNN model [28]. This results in an improvement of performance in both training and inference. The regions of interest (RoIs) are then mapped into the image feature tensor, and via application of a process called RoI Pooling the regions are downsampled to be fed to the next neural network. This allows for the prediction of image classes and their correct bounding boxes. Given the error losses from the classification and bounding box predictions, the entire network is trained end-to-end using backpropagation and stochastic gradient descent (SGD) [29]. An illustration of the process can be found in Fig. 1.

3.2 Identification of Relations Between Objects

Determining relations between objects is required to generate scene graphs and it can be done in several ways. There has been several publications that propose such methods as Iterative Message Passing [30], Neural Motifs [31], Graphical Contrastive Losses [32], and Factorizable Net [33]. In our work, we use the Iterative Message Passing model.

The Iterative Message Passing model predicts relations between objects detected by the Faster R-CNN model. Mathematically, a scene graph generation process means finding the optimal $\mathbf{x}^* = \arg\max_{\mathbf{x}} \Pr(\mathbf{x}|I, B_I)$ that maximizes the following probability function:

$$\Pr(\mathbf{x}|I, B_I) = \prod_{i \in V} \prod_{j \neq i} \Pr(x_i^{cls}, x_i^{bbox}, x_{i \to j}|I, B_I). \tag{1}$$

where I is an image, B_I represents proposed object boxes, $\mathbf{x}$ is a set of all variables, including classes, bounding boxes and relations ($\mathbf{x} = \{x_i^{cls}, x_i^{bbox}, x_{i \to j} | i = 1 \ldots n, j = 1 \ldots n, i \neq j\}$), with n representing the number of proposed boxes, x_i^{cls} as a class label of the i-th proposed box, x_i^{bbox} as the offset of bounding box relative to the i-th proposed box, and $x_{i \to j}$ as a predicate between the i-th and j-th proposed boxes.

3.3 Aggregation of Scene Graphs

The process of amalgamating generated image scene graphs that results in a single knowledge graph has a number of challenges: 1) establishing a unique identifier for each entity; 2) identifying the importance of connections; 3) dealing with missing values and incorrect data; and 4) keeping the knowledge graph updated in presence of new data.

In the specific case of the Visual Genome dataset, we use *synsets* from Word-Net to identify nodes and relations, as well as different meanings of a specific word. There are various methods to identify the uniqueness of words, such as using words occurring in natural language, grouping similar words with the same meaning, or trying to assign words to their specific synsets. Yet, another way is to keep words and phrases as they are and let their occurrence numbers show the importance of connections and nodes. Such a simple approach provides a good indication which relations are more likely to occur.

Another challenge is to mitigate missing or incorrect information. For example, the used methods/models could incorrectly label objects/relations and the processes could fail to find unique words or synsets. Even the hand-annotated data in the Visual Genome (VG) dataset [13], which is used for training, has missing and incorrect data [34]. The unknowns are reduced by relying on the information already present, such as recovering a missing synset based on an already-known name to synset relation or WordNet.

4   Image-Based *WpKG*: Experimental Studies

The Iterative Message Passing model [30] is trained on the VG dataset. It contains 108,077 images that capture everyday scenarios. For evaluation, only the most common 150 object categories and 50 predicates are used.

The Faster R-CNN model that is applied to detect objects and their bounding boxes is pre-trained on MS-COCO dataset. This dataset has 80 object categories. The training set is of size 80k images. Validation and test sets are 40k and 20k images, respectively. Around thirty percent of the VG dataset (test set) is used to detect objects and predict predicates. The subset has around 30,000 images. Running the process described in Sect. 3, a *WpKG* with 138 nodes and 7,287 relations is generated.

Neo4j [35] software is used to store and analyze the generated graph. It allows us to store object and relationship names and synsets, as well as occurrence numbers. Also, it visualizes a structure composed of triples *subject-predicate-object*.

One advantage of the generated *WpKG* is the existence of common sense relations occurring in the actual world extracted during the processing of visual data. The most important entities related to the entity of interest can be found by inspecting the strength of connections between them. One way to accomplish this is to measure how often these objects are associated with each other.

As an example, the entity *plate* together with the related entities is shown in Fig. 2 (a). As we can see, removing non-frequent relations leads to identification of tightly related objects relevant to the *plate*, Fig. 2 (b).

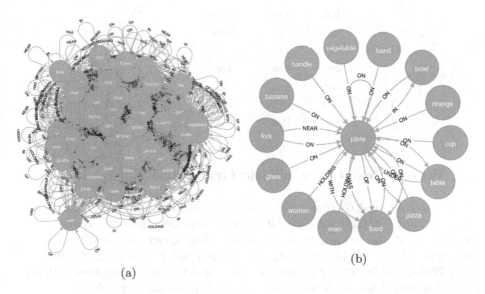

(a)

(b)

Fig. 2. Relationships to/from *plate* entity: with at least one instance and interrelationships between associated items (a); and at least 10 instances for each relationship and without the interrelationships (b).

A sample of relation occurrence statistics is shown in Table 1. Based on the analysis of visual data, we can find out about some common-sense knowledge, such as places where a *vase* can be placed, and what can be put into it. Most of the relations, such as flower-in-vase, make sense and agree with the crowd-sourced VG dataset. However, some relations, such as vase-in-vase, may not make sense. This could be a shortcoming of the method/model used for prediction of relations. Besides a better model, processing more images and detecting more types of relations and objects may improve the results.

The comparison of our method, which is based on image processing, with other relevant automatic and semi-automatic methods is demonstrated in Table 2.

Table 1. Three most common relations in *WpKG* generated using Faster R-CNN and iterative message passing models to recognize objects and predict relations, respectively.

	Subject	Predicate	Object	Occurrence number
	Woman	Wearing	Shirt	192
Woman	Woman	Holding	Umbrella	168
	Woman	Has	Hair	141
	Plate	On	Table	388
Plate	Plate	On	Plate	193
	Plate	On	Pizza	19
	Flower	In	Vase	173
Flower	Flower	On	Table	41
	Flower	On	Tree	15
	Vase	On	Table	116
Vase	Vase	In	Vase	44
	Vase	Has	Flower	31

5   *WpKG*-Based Possibilistic Graph and Base

The generated *WpKG*s consist of an enormous amount of nodes and relations. The relations – as built via aggregation of scene graph relations – contain information about the frequency of occurrence. This means that each relation is equipped with a weight indicating its strength and importance. For practical use, *WpKG* can be further processed and a subset of nodes together with relations between them can be used to construct a possibilistic graph.

Table 2. Comparison of relevant generated knowledge graphs from literature. Our method and NEIL are the ones that focus on in-image relations.

Method	In-image relations	Input source(s)	Relation types	Triples	Automation
NEIL [23]	**Yes**	Image	<10	$<10K$	**Automated**
ConceptNet [17]	No	Text	<100	**34M**	Semi-automated
LEVAN [22]	No	Text and Objects in Images	<10	$<100K$	**Automated**
WebChild KB 2.0 [21]	No	Text and Image/Video Captions	>1000	$>18M$	**Automated**
Quasimodo [19]	No	Text (logs and QA forums)	**Dynamic**	2.3M	**Automated**
Our work	**Yes**	Image	<50 **(Dynamic)**	$>7K$	**Automated**

5.1 Extracting Possibilistic Graph from *WpKG*

A *WpKG* is constructed with no constrains. It contains cycles, very strong and weak relations, as well as erroneous information due to the imperfection of used image processing tools. In that context, a possibilistic graph is more organized and 'clean'. Therefore, extracting nodes and edges from *WpKG* and building a graph that satisfies rules of the possibilistic graph (Sect. 2.3) seems important steps in utilizing generated *WpKG*s.

First, a proto-possibilistic graph is constructed. It is free of cycles and contains outwards relations linked to the entity of interest. The procedure used to extract relevant entities and connections is presented as Algorithms 1 and 2. The important aspects of this process are:

Algorithm 1, line 4 the value of *Depth* identifies the allowed length of a 'relation chain' at the process of building a graph;

Algorithm 2, line 6 the procedure *randomize_createGroups()* is crucial in the construction process: 1) randomization of a sequence of entities allows to generate graphs with different paths; once this is combined with a process presented in line 8 (explained below) it prevents the existence of cycles in the generated graph; 2) grouping of relations/predicates connected to the same object, i.e., prepositions/adjectives playing the role of relations; as illustration, see entities *flower, window, table, plant*, Fig. 3;

Algorithm 2, line 8 this allows to solve an issue of cycles, i.e., relations between pairs of entities *flower-vase, plant-vase* and *table-vase*, Fig. 3, would lead to cyclic directed graph; however, if a connection between both entities already exist, a new one – in the opposite direction – is not created.

Algorithm 1: Construction of Proto-Possibilistic Graph

Data:	Image-based **WpKG**; Seed_Entity; Depth
Result:	Proto-Possibilistic Graph

1 **begin**
2    $root \leftarrow Seed\_Entity$;
3    $d \leftarrow 1$;
4   call $CreateConn(root, d, Depth)$;
5   **return**;

The application of the presented procedure leads to a graph that is acyclic and direct. It also contains occurrences associated with each connection. The last step of constructing a possibilistic graph is to determine possibility degrees. To do so, all input connections to a given node are analyzed. The maximum value is identified and is used for normalization of all other occurrence values associated with inward connections to the node. This ensures satisfaction of the requirement of maximum possibility equal to 1.0 (Sect. 2.3).

Algorithm 2: Generation of Connections

```
1  CreateConn(r, d, Depth)
2  begin
3      if d <= Depth then
4          listChild ← create_list_children(r);
5          if listChild ≠ ∅ then
6              randomize_createGroups(listChild);
7              for e ∈ listChild do
8                  if e not_connected_to r then
9                      setupConnection(r, e);
10                 call CreateConn(e, d + 1, Depth);
11         else
12             return;
13     else
14         return;
15     return;
```

5.2 Construction of Possibilistic Base

The extracted possibilistic graph allows us to build a possibilistic knowledge base. Here, we follow the process presented in [14]. For that purpose, we consider the graph as a set of triples:

$$\Pi G = \{(a, P_a, \alpha) : \Pi(a|P_a) = \alpha\}$$

where a is an instance of A_i and P_a is the Cartesian product of domains D_k of variables $A_k \in Par(A_i)$. Each such triple can be represented as a formula:

$$(\neg a \vee \neg P_a, 1 - \alpha)$$

so, following [14], we have that the possibilistic knowledge base associated with ΠG defined as:

$$\sum = \{(\neg a_i \vee \neg P_{a_i}, 1 - \alpha_i) \ / \ (a_i, P_{a_i}, \alpha_i) \in \Pi G\}$$

6 Possibilistic Graph and Base: Experimental Studies

Let us illustrate the process of building a simple possibilistic graph and a possibilistic knowledge base. We apply the procedure to build a graph of facts related to the entity *vase*, and relations between this entity and other entities from the vase's environment.

Application of Algorithm 1 to the generated *WpKG* allows us to extract entities related to the entity of interest, *vase*. The Neo4j snapshot of *WpKG* with
vase and relations to 'relevant' entities is shown in Fig. 3(a). The version processed by the algorithm is shown in Fig. 3(b). It contains – marked as dashed
lines – the pairs *flower-vase*, *plant-vase*, and *table-vase* that could result in different graphs depending on the element of randomness embedded in the procedure
randomize_createGroups(), Algorithm 2.

The *WpKG* with occurrences assigned to connections allows us to determine
conditional degrees. We have simplified our graph, i.e, combined all inward connections to a node into a single one, as shown in Fig. 3(c). This graph is further
processed – the occurrence numbers are used to determine possibility values.
Based on the graph in Fig. 3(c), we build conditional possibility degrees. All of
them are presented in Tables 3, 4, and 5.

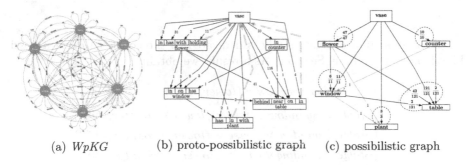

(a) *WpKG*   (b) proto-possibilistic graph   (c) possibilistic graph

Fig. 3. A fragment of *WpKG* and a possibilistic graph constructed based on it.

Table 3. Possibility degrees for *Vase, Flower, Counter,* and *Plant*.

(a) Π(Vase)

vase	1.
¬ vase	1.

(b) Π(Counter|Vase)

Counter|Vase	vase	¬ vase
counter	1.	1.
¬ counter	1.	1.

(c) Π(Flower|Vase)

Flower|Vase	vase	¬ vase
flower	1.	1.
¬ flower	1.	1.

(d) Π(Plant|Vase)

Plant|Vase	vase	¬ vase
plant	1.	1.
¬ plant	1.	1.

Table 4. Possibility degrees for *Window* – Π(Window|Vase, Flower)

Window|Vase, Flower	**vase**, ¬flower	¬vase, **flower**	Elsewhere
window	1	.545	1.
¬window	1.	1.	1.

Table 5. Possibility degrees for *Table* –Π(Table|Window, Vase, Flower, Counter)

Shelf|Window, Vase, Flower, Counter	**window,** ¬vase, ¬ flower, ¬counter	¬window, **vase,** ¬flower, ¬counter	¬window, ¬vase, **flower,** ¬counter	¬window, ¬vase, ¬ flower, **counter**	Elsewhere
table	.017	1.	.347	.017	1.
¬table	1.	1.	1.	1.	1.

The last step of our case study is dedicated to the construction of a possibilistic knowledge base, Sect. 5.2. As a result, we obtain:

$$\sum = \{ \ (\neg window \lor vase \lor \neg flower, \ .455),$$

$$(\neg table \lor \neg window \lor vase \lor flower \lor counter, \ .983),$$

$$(\neg table \lor window \lor \neg vase \lor flower \lor counter, \ .653),$$

$$(\neg table \lor window \lor vase \lor flower \lor \neg counter, \ .983) \ \}.$$

7   Conclusion

The paper focuses on the automatic construction of a knowledge graph – called *World-perceiving Knowledge Graph (WpKG)* – that contains results of the analysis of multiple images. Further, the generated *WpKG* is processed and multiple possibilistic graphs can be constructed based on it.

It is shown that using deep learning models, we can extract common-sense situational information about objects present in visual data. The trained neural networks may already know these relations implicitly, but extracting this knowledge in the form of a knowledge graph provides the ability to have this information explainable and explicit. The strength of the overall procedure depends on the capabilities of the applied learning model as well as the data it has been trained on. By improving the models themselves, the overall procedure can be improved.

Constructed *WpKG*s are contextualized by images used as an input to the presented process. A different graph will be obtained when images representing a specific geographical location are used, while a different graph will be built based on images illustrated a specific historical event. Also, multiple different possibilistic graphs can be created to reason about the correctness of contextual utilization of specific items and relations between them.

Given the adaptability of *WpKG* to new scenarios, context-aware and even time-variant knowledge graphs can be constructed. For example, processing car images from a specific country will lead to the construction of *WpKG* representing a very specific information related to cars' details and their contextual settings. Another important aspect that can be considered is time. It can affect both occurrences of relations and meanings of words linked to the nodes.

As future work, better models can be used to improve the overall construction process, biases can be reduced by implementing procedures to diversify the input images, and prediction of unknown objects can be added.

References

1. Sap, M., et al.: ATOMIC: an atlas of machine commonsense for if-then reasoning. In: AAAI, pp. 3027–3035 (2019)
2. Bosselut, A., Rashkin, H., Sap, M., Malaviya, C., Celikyilmaz, A., Choi, Y.: COMET: commonsense transformers for automatic knowledge graph construction. In: 57th ACL Meeting, pp. 4762–4779 (2019)
3. Hao, Y., et al.: An end-to-end model for question answering over knowledge base with cross-attention combining global knowledge. In: 55th ACL Meeting, pp. 221–231 (2017)
4. Kapanipathi, P., et al.: Infusing knowledge into the textual entailment task using graph convolutional networks. In: AAAI (2020)
5. Wang, J., Huang, P., Zhao, H., Zhang, Z., Zhao, B., Lee, D.L.: Billion-scale commodity embedding for e-commerce recommendation in Alibaba. In: KDD 2018 (2018)
6. Auer, S., Bizer, C., Kobilarov, G., Lehmann, J., Cyganiak, R., Ives, Z.: DBpedia: a nucleus for a web of open data. In: The Semantic Web, pp. 722–735 (2007)
7. Vrandečić, D., Krötzsch, M.: Wikidata: a free collaborative knowledgebase. Commun. ACM **57**(10), 78–85 (2014)
8. Rebele, T., Suchanek, F., Hoffart, J., Biega, J., Kuzey, E., Weikum, G.: YAGO: a multilingual knowledge base from Wikipedia, Wordnet, and Geonames. In: Groth, P., et al. (eds.) International Semantic Web Conference, pp. 177–185 (2016)
9. Bollacker, K., Evans, C., Paritosh, P., Sturge, T., Taylor, J.: Freebase: a collaboratively created graph database for structuring human knowledge. In: ACM SIGMOD, pp. 1247–1250 (2008)
10. Miller, G.A.: WordNet: a lexical database for English. Commun. ACM **38**, 39–41 (1995)
11. The importance of vision. Ophthalmology **94**, 9–13 (1987). https://doi.org/10.1016/S0161-6420(87)33553-5
12. Jacobson, H.: The informational capacity of the human eye. Science **113**(2933), 292–293 (1951)
13. Krishna, R., et al.: Visual genome: connecting language and vision using crowd-sourced dense image annotations. Int. J. Comput. Vis. **123**, 32–73 (2016)
14. Benferhat, S., Dubois, D., Garcia, L., Prade, H.: Possibilistic logic bases and possibilistic graphs. In: 15th Conference on Uncertainty in AI, pp. 57–64 (1999)
15. Gebhardt, J., Kruse, R.: Automated construction of possibilistic networks from data. J. Appl. Math. Comput. Sci. **6**(3), 101–136 (1996)

16. Mitchell, T., et al.: Never-ending learning. Commun. ACM **61**, 103–115 (2018)
17. Speer, R., Chin, J., Havasi, C.: ConceptNet 5.5: an open multilingual graph of general knowledge. In: AAAI, pp. 4444–4451 (2017)
18. Fader, A., Soderland, S., Etzioni, O.: Identifying relations for open information extraction. In: Conference on Empirical Methods in NLP, pp. 1535–1545 (2011)
19. Romero, J., Razniewski, S., Pal, K., Pan, J.Z., Sakhadeo, A., Weikum, G.: Commonsense properties from query logs and QA forums. In: CIKM 2019, pp. 1411–1420 (2019)
20. Tandon, N., de Melo, G., Suchanek, F.M., Weikum, G.: WebChild: harvesting and organizing commonsense knowledge from the web. In: WSDM 2014, pp. 523–532 (2014)
21. Tandon, N., de Melo, G., Weikum, G.: WebChild 2.0 : fine-grained commonsense knowledge distillation. In: 59th ACL Meeting, pp. 115–120 (2017)
22. Divvala, S.K., Farhadi, A., Guestrin, C.: Learning everything about anything: Webly-supervised visual concept learning. In: 2014 IEEE Conference on Computer Vision and Pattern Recognition, pp. 3270–3277 (2014)
23. Chen, X., Shrivastava, A., Gupta, A.: NEIL: extracting visual knowledge from web data. In: 2013 IEEE International Conference on Computer Vision, pp. 1409–1416 (2013)
24. Benferhat, S., Sossai, C.: Merging uncertain knowledge bases in a possibilistic logic framework. In: 14th Conference on Uncertainty in AI, pp. 8–15 (1998)
25. Ren, S., He, K., Girshick, R.B., Sun, J.: Faster R-CNN: towards real-time object detection with region proposal networks. IEEE Trans. Pattern Anal. Mach. Intell. **39**, 1137–1149 (2015)
26. Simonyan, K., Zisserman, A.: Very deep convolutional networks for large-scale image recognition. In: ICLR (2015)
27. Deng, J., Dong, W., Socher, R., Li, L.J., Li, K., Fei-Fei, L.: ImageNet: a large-scale hierarchical image database. In: IEEE Conference on Computer Vision and Pattern Recognition, pp. 248–255 (2009)
28. Girshick, R.: Fast R-CNN. In: IEEE Conference on Computer Vision, pp. 1440–1448 (2015)
29. LeCun, Y., et al.: Backpropagation applied to handwritten zip code recognition. Neural Comput. **1**(4), 541–551 (1989)
30. Xu, D., Zhu, Y., Choy, C.B., Fei-Fei, L.: Scene graph generation by iterative message passing. In: IEEE Conference on Computer Vision and Pattern Recognition, pp. 3097–3106 (2017)
31. Zellers, R., Yatskar, M., Thomson, S., Choi, Y.: Neural motifs: scene graph parsing with global context. In: IEEE/CVF Conference on Computer Vision and Pattern Recognition, pp. 5831–5840 (2018)
32. Zhang, J., Shih, K.J., Elgammal, A., Tao, A., Catanzaro, B.: Graphical contrastive losses for scene graph parsing, pp. 11535–11543 (2019)
33. Li, Y., Ouyang, W., Zhou, B., Shi, J., Zhang, C., Wang, X.: Factorizable net: an efficient subgraph-based framework for scene graph gen. In: IEEE/CVF Conference on Computer Vision and Pattern Recognition, pp. 335–351 (2018)
34. Gu, J., Zhao, H., Lin, Z., Li, S., Cai, J., Ling, M.: Scene graph generation with external knowledge and image reconstruction. In: IEEE/CVF Conference on Computer Vision and Pattern Recognition, pp. 1969–1978 (2019)
35. Neo4j Inc: Neo4j. https://neo4j.com

Machine Learning I

Possibilistic Estimation of Distributions to Leverage Sparse Data in Machine Learning

Andrea G. B. Tettamanzi[1]([✉]) [iD], David Emsellem[2], Célia da Costa Pereira[3] [iD], Alessandro Venerandi[4] [iD], and Giovanni Fusco[4] [iD]

[1] Université Côte d'Azur, CNRS, Inria, I3S, Sophia Antipolis, France
andrea.tettamanzi@univ-cotedazur.fr
[2] Kinaxia SA, Sophia Antipolis, France
david.emsellem@kcitylabs.fr
[3] Université Côte d'Azur, CNRS, I3S, Sophia Antipolis, France
celia.da-costa-pereira@univ-cotedazur.fr
[4] Université Côte d'Azur, CNRS, ESPACE, Nice, France
{alessandro.venerandi,giovanni.fusco}@univ-cotedazur.fr

Abstract. Prompted by an application in the area of human geography using machine learning to study housing market valuation based on the urban form, we propose a method based on possibility theory to deal with sparse data, which can be combined with any machine learning method to approach weakly supervised learning problems. More specifically, the solution we propose constructs a possibilistic loss function to account for an uncertain supervisory signal. Although the proposal is illustrated on a specific application, its basic principles are general. The proposed method is then empirically validated on real-world data.

Keywords: Possibility theory · Machine learning · Weakly supervised learning

1 Introduction

Supervised learning is the machine learning task of learning a function that maps an input to an output based on example input-output pairs [18]. Each example consists of an input record, which collects the values of a number of input variables, and the associated value of the output variable (also called the supervisory signal). The learnt function can then be used to "predict" the value of the output variable for new unlabeled input records, whose output value is not known.

In many real-world problems, obtaining a fully labeled dataset is expensive, difficult, or outright impossible. An entire subfield of machine learning, called weakly (or semi-) supervised learning has thus emerged, which studies how datasets where the supervisory signal is not available or completely known

M.-J. Lesot et al. (Eds.): IPMU 2020, CCIS 1237, pp. 431–444, 2020.
https://doi.org/10.1007/978-3-030-50146-4_32

for all the records can be used for learning. According to a useful taxonomy of classification problems that can arise in that field [10], four broad classes of problems can be identified:

- single-instance, single-label (SISL), which corresponds to the standard setting where all the examples consist of a single instance, to which a single (i.e., certain) class label is assigned;
- single-instance, multiple-label (SIML), where some examples, consisting of single instances, are assigned a (disjunctive) set of possible class labels, including, to an extreme, the set of all the class labels, which corresponds to the case of a missing supervisory signal;
- multiple instance, single-label (MISL), when examples may consist of sets of instances, being assigned a single label as a whole;
- multiple instance, multiple-label (MIML), when, in addition, some examples are assigned a (disjunctive) set of possible values.

This taxonomy of course assumes that the output variable of the underlying objects, which one seeks to predict, can only have a single true value. It should be mentioned that other problems exist, called *multi-label* classification [19], where, for a given underlying object, described by an input record, multiple values can be active at the same time, which would then be described by a *conjunctive* set of output values. In that case, the learnt function is set-valued.

The above taxonomy can be extended to regression or "predictive modeling" problems, where the "label" is a number, ranging on a discrete or continuous interval.

In the framework of an interdisciplinary research project applying machine learning and urban morphology theory to the investigation of the influence of the urban environment on the value of residential real estate [21], we faced the problem of incorporating into a predictive model uncertain information associated with prices, addressing issues of data sparsity, a problem that falls within the SIML category of the above taxonomy. This prompted us to propose an original method to deal with output variable uncertainty in predictive modeling, based on possibility theory, which we present below.

As it has been argued for example by Bouveyron *et al.* [1], while the problem of noise in data has been widely studied in the literature on supervised learning, the problem of label noise remains an important and unsolved problem in supervised classification. Nigam *et al.* [15] proposed Robust Mixture Discriminant Analysis (RMDA), a supervised classification whose aim is to detect inconsistent labels by comparing the labels given for labeled data set with the ones obtained through an unsupervised modeling based on the Gaussian mixture model. To solve the problem of automatic building extraction from aerial or satellite images with noise labels, Zhang *et al.* [23] propose to capture the relationship between the true label and the noisy label, a general label noise-adaptive (NA) neural network framework consisting of a combination of a base network with an additional probability transition module (PTM) introduced to capture the relationship between the true label and the noisy label. Other researchers prefer to focus on constructing a loss function that is robust to noise [9].

However, the uncertainty brought about by data sparsity in our problem is hard to characterize as a probability distribution, making the application of one of the probabilistic approaches to weak supervised learning proposed in the literature unattractive.

Vannoorenberghe *et al.*, instead, proposed a different approach [20] in which the induction of decision trees is based on the theory of belief functions. In their framework, it is supposed that the training examples have uncertain or imprecise labels. In the same spirit, Quost *et al.* [17] also proposed a belief-function-based framework to be used for supervised learning, in which the training data are associated with uncertain labels. They supposed that each example in the training data set is associated to a belief assignment that represents the actual knowledge of the actual class of the example and used a boosting method to solve the classification problem. Denœux *et al.* [4] introduce a category of learning problems in which the labels associated to the examples in the training data set are assessed by an expert and encoded in the form of a possibility distribution. Although this work is very relevant to what we are proposing here, the authors obtain their possibility distributions from human experts, which can be expensive and difficult, whereas the method we propose automatically computes those distributions from data.

Traditionally, housing market valuations are modeled, in a linear fashion, through a combination of intrinsic and extrinsic features evaluated for each dwelling. Although such linear models provide easy-to-read results, they are severely limited as they assume linearity and independence among variables. However, this might not be the case. For example, a specific variable might change its behaviour for different subsets of observations. More recently, researchers have applied Machine Learning (ML) techniques to study the same phenomenon [3,8,16]. However, their aim was mainly predictive. Thus, although linearity and independence among variables were tackled through the use of such algorithms, their results lacked interpretability. Finally, the intrinsic/extrinsic dichotomy does not hold when the goal of the analysis is the valuation of urban subspaces (like neighborhoods or street segments) instead of individual dwellings. To tackle these issues, we have devised an approach rooted in Urban Morphology to explain housing values at a fine level of spatial granularity, that of street segments and we designed a sequence of appropriate ML techniques that output interpretable results. To be more specific, the proposed approach, firstly, computes street-based measures of housing values, urban form, functions, and landscape and then models the relationship between them through an ensemble method comprising of Gradient Boosting (GB) [7], topological Moran's test [14], and SHAP [13], a recently developed technique to interpret outputs of ML algorithms. The approach has been used to explain the median valuation of street segments in the French Riviera, using housing transactions from the period 2008–2017, through more than 100 metrics of urban form, functions, and landscape.

One difficulty this approach runs into is that transaction data, which are the only source of observations of the output variables (the measures of housing values), are rather sparse at the scale of the street segment. One possible way to overcome this difficulty would be to limit our study to those street segments for which enough observations are available, e.g., at least ten transactions within the observation period; however, this drastically reduces the number of street segments that can be studied and introduces a bias toward neighborhoods having a relatively high turnover. An alternative and more attractive solution, which is the subject of this paper, is to use all available observations, while taking into account the uncertainty brought about by data sparsity.

Essentially, the solution we propose is to model the uncertainty relevant to the output variable within the framework of possibility theory and modify the loss function of the ML technique, used to model the phenomenon, so that it can weight the error based on the uncertainty of the output variable. This approach is very much in the same spirit as the fuzzy loss function proposed by Hüllermeier [11,12]. An important advantage of such solution is that it is readily transferable to any supervised or semi-supervised ML technique using a loss function (which is the case for the vast majority of such techniques).

The rest of the paper is organized as follows: Sect. 2 provides some background on possibility theory, which is required in order to understand the proposed approach; Sect. 3 states the main question we address, as it emerges from the real-estate price study that motivated our proposal. The proposed solution itself is presented in Sect. 4, while Sect. 5 discusses its empirical validation. Section 6 draws some conclusions and proposes some ideas for further research.

2 Background on Possibility Theory

Fuzzy sets [22] are sets whose elements have degrees of membership in $[0, 1]$. Possibility theory [6] is a mathematical theory of uncertainty that relies upon fuzzy set theory, in that the (fuzzy) set of possible values for a variable of interest is used to describe the uncertainty as to its precise value. At the semantic level, the membership function of such set, π, is called a *possibility distribution* and its range is $[0, 1]$. A possibility distribution can represent the available knowledge of an agent. $\pi(\mathcal{I})$ represents the degree of compatibility of the interpretation $\mathcal{I}$ with the available knowledge about the real world if we are representing uncertain pieces of knowledge. By convention, $\pi(\mathcal{I}) = 1$ means that it is totally possible for $\mathcal{I}$ to be the real world, $1 > \pi(\mathcal{I}) > 0$ means that $\mathcal{I}$ is only somehow possible, while $\pi(\mathcal{I}) = 0$ means that $\mathcal{I}$ is certainly not the real world.

A possibility distribution π is said to be normalized if there exists at least one interpretation $\mathcal{I}_0$ s.t. $\pi(\mathcal{I}_0) = 1$, i.e., there exists at least one possible situation which is consistent with the available knowledge.

Definition 1 *(Possibility and Necessity Measures). A possibility distribution π induces a* possibility measure *and its dual* necessity measure, *denoted by Π and N respectively. Both measures apply to a classical set $S \subseteq \Omega$ and are defined as follows:*

$$\Pi(S) = \max_{\mathcal{I} \in S} \pi(\mathcal{I}); \tag{1}$$

$$N(S) = 1 - \Pi(\bar{S}) = \min_{\mathcal{I} \in S}\{1 - \pi(\mathcal{I})\}. \tag{2}$$

In words, $\Pi(S)$ expresses to what extent S is consistent with the available knowledge. Conversely, $N(S)$ expresses to what extent S is entailed by the available knowledge. It is equivalent to the impossibility of its complement $\bar{S}$—the more $\bar{S}$ is impossible, the more S is certain. A few properties of Π and N induced by a normalized possibility distribution on a finite universe of discourse Ω are the following. For all subsets $A, B \subseteq \Omega$:

1. $\Pi(A \cup B) = \max\{\Pi(A), \Pi(B)\}$;
2. $\Pi(A \cap B) \leq \min\{\Pi(A), \Pi(B)\}$;
3. $\Pi(\emptyset) = N(\emptyset) = 0$; $\Pi(\Omega) = N(\Omega) = 1$;
4. $N(A \cap B) = \min\{N(A), N(B)\}$;
5. $N(A \cup B) \geq \max\{N(A), N(B)\}$;
6. $\Pi(A) = 1 - N(\bar{A})$ (duality);
7. $N(A) > 0 \Rightarrow \Pi(A) = 1$; $\Pi(A) < 1 \Rightarrow N(A) = 0$;

A consequence of these properties is that $\max\{\Pi(A), \Pi(\bar{A})\} = 1$. In case of complete ignorance on A, $\Pi(A) = \Pi(\bar{A}) = 1$.

3 Problem Statement

To make this paper self-contained, we briefly recall here some elements, relevant to the solution we are going to describe in Sect. 4, of the problem that motivated our proposal. The interested reader can refer to [21] for a more detailed explanation.

3.1 Pre-processing

Housing transactions of different years and housing typologies cannot be directly compared due to yearly inflation, housing market cycles (e.g., economic recession, upturn), and specific market behaviours affecting different housing types. For example, bigger properties tend to be sold less frequently as they are more expensive and subject to long term investments, while smaller properties, due to their relative lower valuations, tend to be exchanged more easily and tend to be subject of shorter-term investments. The average price per square meter tends also to be structurally higher for small flats for technical reasons (even the smallest flat needs sanitary and cooking equipment, which proportionally weigh more on the average price per surface unit compared to a larger property). The very notion of average price per square meter can thus be challenged when applied to such diverse housing markets. To address these issues, instead of the conventional price per square meter, our method requires to separate the transactions by year and housing type and compute ventiles of prices for each subset year of transaction - housing type. We consider such statistics as appropriate normalized values, which account for different market segments and years, thus making transactions comparable among them.

3.2   Computation of the Median of Values

Having classified each transaction in a ventile of value, the next step requires to first assign to each data point the street segment to which they belong and, second, aggregate the information on value at the street level through the computation of a measure of central tendency (i.e., median). Such measure provides information on the central point of the distribution of the value of the housing market, for each street. We perform such computation for the ensemble of each street segment and its immediate neighbouring streets. This for two reasons: firstly, transactions located in streets directly connected to one another tend to have similar valuations (due to the influence of the same location factors, presence of properties at the intersection of several streets segments, etc.); secondly, data on house prices tend to be quite sparse, even for several years, and thus a local interpolation allows us to increase the data coverage. Nevertheless, most street segments end up having less than ten transactions per housing type, which introduces uncertainty into the computation of the median statistics.

3.3   Street-Based Metrics of the Urban Environment and Landscape

To characterize the context of each street segment in the most comprehensive way possible, we compute a set of descriptors that quantify aspects of the urban fabric, street-network configuration, functions, housing stock, and landscape. Their definition is out of the scope of this paper.

4   Proposed Solution

In abstract terms, we can describe the problem as follows. We are given a sample of observed variates $x_1, x_2, \ldots, x_n$ and a number of probability distributions (the hypothesis space). We want to assess the possibility degree of these probability distributions based on the given sample, i.e., the degree to which they are compatible with the observations.

4.1   Possibility of a Price Distribution

We can limit ourselves to parametric families of distributions. In this case, this problem can be described as a sort of possibilistic parameter estimation. In the Incertimmo project, we consider distributions described by three deciles: $(d_1, d_5 = \text{median}, d_9)$. We recall that the ith decile d_i is the smallest number x satisfying $\Pr[X \leq x] \geq \frac{i}{10}$. By definition, $d_1 \leq d_5 \leq d_9$. Furthermore, for a probability distribution defined in the $[a, b]$ interval, we know that 10% of the probability mass is in the $[a, d_1]$ interval, 40% in the $(d_1, d_5]$ interval, 40% in the $(d_5, d_9]$ interval, and 10% in the $(d_9, b]$ interval. We make the additional simplifying assumption that the probability mass is uniformly distributed within each of the above intervals. While this might look like a very restrictive assumption, on the one hand it is motivated by the type of qualitative descriptions of price distribution that are of interest to the human geographers (i.e., by the application at hand) and, on the other hand, could easily be relaxed by selecting other parametric families of distributions without serious consequences on the proposed approach.

This yields a parametric family of probability distributions on the $[a, b]$ interval whose density (in the continuous case) or probability (in the discrete case) function is

$$f(x; d_1, d_5, d_9) = \begin{cases} \frac{1}{10(d_1-a)}, & a \leq x \leq d_1; \\ \frac{2}{5(d_5-d_1)}, & d_1 < x \leq d_5; \\ \frac{2}{5(d_9-d_5)}, & d_5 < x \leq d_9; \\ \frac{1}{10(b-d_9)}, & d_9 < x \leq b. \end{cases} \tag{3}$$

The degenerate cases that arise when $d_1 = d_5$, $d_5 = d_9$, or $d_9 = b$ are treated by adding to the result the contributions of the lines whose condition becomes empty.

In the specific application described in Sect. 3, as we have seen, the observed variates are ventiles of the general distribution of housing prices, taking up values in the discrete set $\{1, 2, \ldots, 20\}$.

Given the sample $x_1, x_2, \ldots, x_n$, it is easy to compute the probability that it is produced by a given distribution of the parametric family, having parameters (d_1, d_5, d_9). This is

$$\Pr[x_1, x_2, \ldots, x_n \mid (d_1, d_5, d_9)] = \prod_{i=1}^{n} f(x_i; d_1, d_5, d_9). \tag{4}$$

This probability is in fact a likelihood function over the distribution of price ventiles, which we will denote by $\mathcal{L}(d_1, d_5, d_9)$.

The link between likelihoods and possibility theory has been explored in [5]. The main result of that study was that possibility measures can be interpreted as the supremum of a family of likelihood functions. It should be stressed that this is an *exact* interpretation, not just an approximation. Based on this result, we transform the likelihood function $\mathcal{L}$ into a possibility distribution over the set of parametric probability distributions of the form (d_1, d_5, d_9).

Notice that the parameters d_1, d_5, d_9 of the probability distributions are the elementary events of this possibility space. A possibility distribution over that space is obtained by letting

$$\pi(d_1, d_5, d_9) = \mathcal{L}(d_1, d_5, d_9)/\mathcal{L}_{\max}, \tag{5}$$

where

$$\mathcal{L}_{\max} = \max_{1 \leq x \leq y \leq z \leq 20} \mathcal{L}(x, y, z),$$

with $x, y, z \in \{0, 1, \ldots, 20\}$, so that all the maximum-likelihood probability distributions have a possibility degree of 1, thus yielding a normalized possibility distribution. Alternatively, the logarithm of the likelihood could be used instead. We restrict the values of the three parameters to the set $\{0, 1, \ldots, 20\}$, because prices are relative and expressed in ventiles in the application at hand. Of course, for other applications, different ranges of values should be considered.

Now, specific values or intervals of one parameter will constitute complex events, i.e., sets of elementary events, and their possibility and necessity measures can be computed as usual, as the maximum of the possibilities of the distributions that fit the specification and $1 - \max$ of all the others, respectively. For instance, the possibility measure over the median price ventile (d_5) of a given street segment will be given, for all $d_5 \in \{1, \ldots, 20\}$, by

$$\Pi(d_5) = \max_{1 \leq x \leq d_5 \leq y \leq 20} \pi(x, d_5, y), \tag{6}$$

where $\pi(\cdot, \cdot, \cdot)$ is defined as in Eq. 5.

4.2   Loss Function Under Possibilistic Uncertainty

The error made by the model predicting that the median price for a street segment is in ventile $\hat{y}$ when all we know is the possibility distribution π over the probability distributions of the prices for that segment, can be defined as

$$L(\hat{y}, \pi) = \int_0^1 \min_{\Pi(y) \geq \alpha} (\hat{y} - y)^2 d\alpha, \tag{7}$$

where $\Pi(y)$ is the possibility measure of the distributions having y as their median. Equation 7 is based on an underlying square error function $e(y) = (\hat{y} - y)^2$, but it could be easily generalized to use an arbitrary error function.

In practice, if $\Lambda = (0 = \lambda_1, \lambda_2, \ldots, 1)$ is the list of possibility levels of π, such that $\forall i > 1, \exists (z, y, z) : \pi(z, y, z) = \lambda_i$, Eq. 7 can be rewritten as

$$L(\hat{y}, \pi) = \sum_{i=2}^{\|\Lambda\|} (\lambda_i - \lambda_{i-1}) \min_{\Pi(y) \geq \lambda_i} (\hat{y} - y)^2. \tag{8}$$

This loss function has been coded in Python in such a way that it could be provided as a custom evaluation function to an arbitrary machine learning method offering this possibility. Most of the loss computation requires iterating through all price distributions of the parametric family (with the three parameters $0 \leq d_1 \leq d_5 \leq d_9 \leq 20$, there are 1,540 of them). To optimize performance, we pre-computed the loss function into a lookup table. Since we are using Gradient Boosting Gradient descent (Newton version in XGBoost), we have provided also functions that return its gradient and Hessian.

5   Experiments and Results

In this section we report the experiments we carried out to validate our method. We use real-world data consisting of all the housing transactions made on the

French Riviera over the period 2008–2017.[1] Each record contains detailed intrinsic features of the dwelling that was sold/bought, including its address and the price paid. From these data, we compiled a dataset whose records correspond to street segments, described by more than 100 metrics of urban form, functions, and landscape, and a distribution of price ventiles for each type of dwelling.

Our goal is to compare the performance of a predictive model trained on this dataset, where the labels are uncertain, due to sparseness of transaction data, to the performance of a predictive model trained on a dataset where the labels are certain (in our case, estimated based on a sufficient number of transactions). If the model trained under uncertainty is able to obtain results similar to those of the model trained without uncertainty, this can be taken as evidence that our method is successful at compensating for the loss of transaction data.

5.1   Experimental Protocol

We proceeded as follows. From our dataset, we extracted the set of street segments having at least 10 recorded transactions. These are the street segments for which we consider that the distribution of prices can be estimated in a reliable way. Let us call this dataset D.

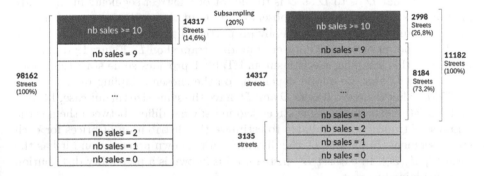

Fig. 1. A graphical illustration of the sampling mechanism used to create subsamples of the original dataset.

We then constructed a second dataset D' by randomly subsampling the transactions from the street segments of D, in such a way as to obtain a similar distribution of the number of transactions per street segment as in the full dataset (i.e., the dataset including also segments having fewer than 10 recorded transactions). For example, if in the full dataset 15% of the street segments has more than 10 recorded transactions (by the way, this subset of the full dataset is what we have called D), then also 15% of the street segments of D' has more than 10 transactions. In general, if the percentage of street segments in the full dataset that have n transactions is x, then D is samples so that the percentage of street

---

[1] Extracted from the PERVAL database, https://www.perval.fr/.

Table 1. Distribution of available transactions per street segment for different sub-sampling rates. Column "≥ 10" gives the number of street segments with at least 10 transactions, "$[3, 9]$" the number of street segments with 3 to 9 transactions, and so on.

Rate	≥ 10	$[3, 9]$	$[1, 2]$	$= 0$
100%	14,317	–	–	–
90%	13,404	913	–	–
80%	12,366	1,951	–	–
70%	11,125	3,190	2	–
60%	9,939	4,364	14	–
50%	8,527	5,695	95	–
40%	6,917	6,989	406	6
30%	5,055	8,050	1,149	63
20%	2,998	8,184	2,791	344
10%	814	6,013	5,610	1,880

segments in D' that have n transactions be as close as possible to x. This way, we can say that D' is to D as D is to the set of all street segments in the study area and, as a consequence, any observation about the predictive power on D of models trained on D' can provide, by proportional analogy, an indication of the predictive power on the full dataset of models trained on D. Figure 1 graphically illustrates this sampling mechanism and Table 1 provides some statistics about the dataset D' that we obtain depending on the chosen sampling rate.

Notice that the two datasets D and D' have the same size (in our case, 14,317) and consist of exactly the same street segments; what differs between them is the number of transactions available to estimate the distribution of prices for each street segment. In dataset D, the distribution is known precisely, and it has the form (d_1, d_5, d_9). In dataset D', instead, all is known is a possibility distribution π, as explained in Sect. 4.

5.2   Validation of the Possibility Distribution

To show that the possibility distribution π defined as per Eq. 5, as well as its associated possibility and necessity measures, does indeed qualitatively describe the actual price distribution, we studied the possibility (computed according to Eq. 6) of the observed median price ventile m of street segments. Ideally, $\Pi(m)$ should be 1 for every street segment, if π perfectly described how the prices are distributed.

Figure 2 shows the probability distribution of $\Pi(m)$ for different sampling rates of the set of transactions. We can observe that when the possibility distributions π of transaction prices for each street segment is constructed using all the recorded transactions (which are at least 10 for any one of the 14,317 street segments considered for this study), the median is assigned a possibility of 1 for most street segments, with some exceptions, which, upon inspection, turned out to be street segments whose price distribution is not unimodal. Since the parametric family of distribution used to fit the actual price distributions is unimodal, the most likely values for their parameters are those that make d_5 correspond to one of the modes. This is an intrinsic limitation introduced by the particular choice of a unimodal family of distribution, which was made to simplify the geographical interpretation of the result; however, despite this limitation, the results of the study seem to confirm the validity of the method used to construct the possibility distributions.

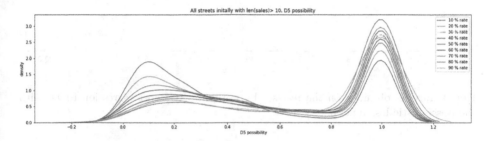

Fig. 2. Distribution of $\Pi(m)$ for different subsampling rates.

Unsurprisingly, as the sampling rate decreases, the number of street segments for which the median is assigned a high possibility decreases.

5.3   Empirical Test of the Method

To conduct our tests, we selected, among all possible predictive modeling methods, XGBoost [2], which is the one that gave the best results when applied to dataset D based on a critical comparison and benchmarking of the most popular methods available. The rationale of this choice is that we wanted our solution to "prove its mettle" on a very challenging task, namely to prevent the degradation of the accuracy of the strongest available method when the supervisory signal becomes uncertain.

We trained a model to predict the median of ventile prices (i.e., d_5) on dataset D by using XGBoost regression with the standard loss function and we trained another model on dataset D' by using XGBoost regression instrumented with the proposed possibilistic loss function. We compare the results given by these two models when applied to a test set consisting of street segments not used to train the models. We treat the model trained on dataset D as the ground truth and we measure the deviation of the model trained on D' from such a target.

As a measure of prediction error, we compute the RMSD of the median (d_5) predicted by the model trained on D' for each segment, with respect to the median of that segment in D. We used a sampling rate of 20% to generate D' from D, i.e., D' contains only 20% of the transactions available in D. We split D' into

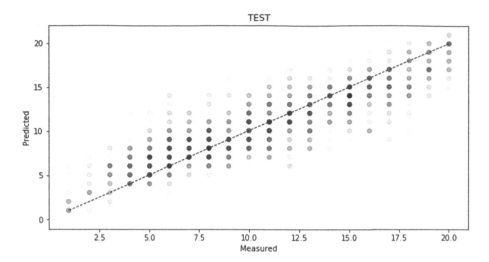

Fig. 3. Results obtained on the test set by applying XGBoost regression to D' with the possibilistic loss function.

a training set containing 80% of the street segments and a test set containing the remaining 20% of the street segments. After training the model on the training set, we obtain an RMSD of 2.778556 on the test set. Figure 3 shows a plot of predicted vs. actual price ventiles for the test set. For comparison, XGBoost using the standard loss function based on MSE trained on 80% of D (therefore, with full information), gives an RMSD of 2.283798 when tested on the remaining 20%. In other words, the possibilistic loss function allows the prediction error to increase by less than 22%, even though 80% of the transactions were removed from the dataset!

6   Conclusion

We have proposed a method based on possibility theory to leverage sparse data, which can be combined with any machine learning method to approach weakly supervised learning problems. The solution we propose constructs a possibilistic loss function, which can then be plugged into a machine learning method of choice, to account for an uncertain supervisory signal.

Our solution is much in the same spirit as the fuzzification of learning algorithms based on the generalization of loss function proposed in [11], in that it pursues model identification at the same time as data disambiguation, except that in our case ambiguity (i.e., uncertainty) affects the output variable only, which is only partially observable in the available data, while all the input variables are perfectly known and, thus, non-ambiguous. Furthermore, as in [11], the predictive model is evaluated by looking at how well its prediction fits the *most favorable* instantiation of the uncertain labels of the training data (this is the sense of the minimum operator in the possibilistic loss function definition).

The development of the method we presented has been motivated and driven by a very specific application, namely by the need to leverage sparse data in a human geography setting. However, its working principle is quite general and could be extended to suit other scenarios. Indeed, distilling a completely general method is the main direction for future work.

Acknowledgments. This research was funded by the IDEX "UCA JEDI", within the AAP Partenariat 2016 (action 6.5).

Andrea Tettamanzi has been supported by the French government, through the 3IA Côte d'Azur "Investments in the Future" project managed by the National Research Agency (ANR) with the reference number ANR-19-P3IA-0002.

The authors would like to thank Denis Overal, director of the R&D at Kinaxia, for his support and insightful suggestions.

References

1. Bouveyron, C., Girard, S.: Robust supervised classification with mixture models: learning from data with uncertain labels. Pattern Recogn. **42**(11), 2649–2658 (2009)
2. Chen, T., Guestrin, C.: XGBoost: a scalable tree boosting system. In: Proceedings of the 22nd ACM SIGKDD International Conference on Knowledge Discovery and Data Mining, KDD 2016, pp. 785–794. ACM, New York (2016)
3. Chiarazzo, V., Caggiani, L., Marinelli, M., Ottomanelli, M.: A neural network based model for real estate price estimation considering environmental quality of property location. Transp. Res. Procedia **3**, 810–817 (2014)
4. Denœux, T., Zouhal, L.M.: Handling possibilistic labels in pattern classification using evidential reasoning. Fuzzy Sets Syst. **122**(3), 409–424 (2001)
5. Dubois, D., Moral, S., Prade, H.: A semantics for possibility theory based on likelihoods. J. Math. Anal. Appl. **205**, 359–380 (1997)
6. Dubois, D., Prade, H.: Possibility Theory–An Approach to Computerized Processing of Uncertainty. Plenum Press, New York (1988)
7. Friedman, J.H.: Greedy function approximation: a gradient boosting machine. Ann. Stat. **29**(5), 1189–1232 (2001)
8. Gerek, I.H.: House selling price assessment using two different adaptive neuro-fuzzy techniques. Autom. Constr. **41**, 33–39 (2014)
9. Ghosh, A., Manwani, N., Sastry, P.S.: Making risk minimization tolerant to label noise. Neurocomputing **160**, 93–107 (2015)
10. Hernández-González, J., Inza, I., Lozano, J.A.: Weak supervision and other non-standard classification problems: a taxonomy. Pattern Recogn. Lett. **69**, 49–55 (2016)
11. Hüllermeier, E.: Learning from imprecise and fuzzy observations: data disambiguation through generalized loss minimization. Int. J. Approx. Reason. **55**(7), 1519–1534 (2014)
12. Hüllermeier, E., Destercke, S., Couso, I.: Learning from imprecise data: adjustments of optimistic and pessimistic variants. In: Ben Amor, N., Quost, B., Theobald, M. (eds.) SUM 2019. LNCS (LNAI), vol. 11940, pp. 266–279. Springer, Cham (2019). https://doi.org/10.1007/978-3-030-35514-2_20
13. Lundberg, S.M., Lee, S.I.: A unified approach to interpreting model predictions. In: Advances in Neural Information Processing Systems, pp. 4765–4774 (2017)

14. Moran, P.A.: Notes on continuous stochastic phenomena. Biometrika **37**(1/2), 17–23 (1950)
15. Nigam, N., Dutta, T., Gupta, H.P.: Impact of noisy labels in learning techniques: a survey. In: Kolhe, M.L., Tiwari, S., Trivedi, M.C., Mishra, K.K. (eds.) Advances in Data and Information Sciences. LNNS, vol. 94, pp. 403–411. Springer, Singapore (2020). https://doi.org/10.1007/978-981-15-0694-9_38
16. Park, B., Bae, J.K.: Using machine learning algorithms for housing price prediction: the case of Fairfax County, Virginia housing data. Expert Syst. Appl. **42**(6), 2928–2934 (2015)
17. Quost, B., Denœux, T.: Learning from data with uncertain labels by boosting credal classifiers. In: KDD Workshop on Knowledge Discovery from Uncertain Data, pp. 38–47. ACM (2009)
18. Russell, S.J., Norvig, P.: Artificial Intelligence: A Modern Approach. Prentice Hall, Upper Saddle River (2010)
19. Tsoumakas, G., Katakis, I.: Multi-label classification: an overview. Int. J. Data Warehouse. Min. **3**(3), 1–13 (2007)
20. Vannoorenberghe, P., Denœux, T.: Handling uncertain labels in multiclass problems using belief decision trees. In: Proceedings of the 9th International Conference on Information Processing and Management, pp. 1919–1926 (2002)
21. Venerandi, A., Fusco, G., Tettamanzi, A., Emsellem, D.: A machine learning approach to study the relationship between features of the urban environment and street value. Urban Sci. **3**(3), 25 (2019)
22. Zadeh, L.A.: Fuzzy sets. Inf. Control **8**, 338–353 (1965)
23. Zhang, Z., Guo, W., Li, M., Yu, W.: GIS-supervised building extraction with label noise-adaptive fully convolutional neural network. IEEE Geosci. Remote Sens. Lett. 1–5 (2020). https://doi.org/10.1109/LGRS.2019.2963065

Maximal Clique Based Influence Maximization in Networks

Nizar Mhadhbi[1($\boxtimes$)] and Badran Raddaoui[2($\boxtimes$)]

[1] CRIL - CNRS UMR 8188, University of Artois, Lens, France
mhadhbi.nizar@gmail.com
[2] SAMOVAR, Télécom SudParis, Institut Polytechnique de Paris, Évry, France
badran.raddaoui@telecom-sudparis.eu

Abstract. Influence maximization is a fundamental problem in several real life applications such as viral marketing, recommendation system, collaboration and social networks. Maximizing influence spreading in a given network aims to find the initially active vertex set of size k called seed nodes (or initial spreaders (In this paper, we use seed set and initial spreaders interchangeably.)) which maximizes the expected number of the infected vertices. The state-of-the-art local-based techniques developed to solve this problem are based on local structure information such as degree centrality, nodes clustering coefficient, and others utilize the whole network structure, such as k-core decomposition, and node betweenness. In this paper, we aim at solving the problem of influence maximization using maximal clique problem. Our intuition is based on the fact that the presence of a dense neighborhood around a node is fundamental to the maximization of influence. Our approach follows the following three steps: (1) discovering all the maximal cliques from the complex network; (2) filtering the set of maximal cliques; we then denote the vertices belonging to the rest of maximal cliques as superordinate vertices, and (3) ranking the superordinate nodes according to some indicators. We evaluate the proposed framework empirically against several high-performing methods on a number of real-life datasets. The experimental results show that our algorithms outperform existing state-of-the-art methods in finding the best initial spreaders in networks.

Keywords: Influence maximization · Maximal clique · Independent cascade model

1 Introduction

Influence maximization problem in social networks has become a hot topic in recent years due to the great deal of real-life applications concerned, such as viral marketing and disease spreading. One of its application in the field of viral marketing is to select a set of highly influential users to adopt a particular product and the goal here is to attract as much as possible of users for purchasing this product [4,6,10]. In order to model the process of the spread of an idea or

© Springer Nature Switzerland AG 2020
M.-J. Lesot et al. (Eds.): IPMU 2020, CCIS 1237, pp. 445–456, 2020.
https://doi.org/10.1007/978-3-030-50146-4_33

an information through a given network, Kempe et al. [15] proposed two models named Independent cascade model (ICM) and Linear threshold model (LTM). The independent cascade model is the most common model for information diffusion. In this model (Algorithm 1), the input are a graph in which every edge (u, v) is associated with a propagation probability p_{uv} (represents the probability that node v can be influenced by node u) and a set of initially activate nodes (seed set) and the diffusion probability called also activation threshold θ. In this model, nodes can have two states, either active or inactive. Nodes are allowed to switch from inactive to active but not in the other. As shown in Algorithm 1, the diffusion model starts with an initial set of active nodes $(t = 0)$. In time t, an active node u will get chance to activate its inactive neighbor v. v will become active if $p_{uv} \geq \theta$, otherwise u will not get any further chance to activate v. The process of diffusion stops when no further activation is possible. This method is called independent because the activation of a node does not depend on the history of active nodes. In the linear threshold model, the idea is that a node becomes active if a large part of its neighbours is active. More formally, each node u has a threshold t_u. The threshold represents the fraction of neighbours of u that must become active in order to active the node u. Influence maximization problem is the problem of assigning a subset of k users as seed nodes in a graph that could maximize the spread of influence by maximizing the expected number of influenced users.

2   Related Works

Choosing the best k initially active nodes in order to maximize the number of activated nodes at the end of the diffusion process had made a prominent place in several works. Numerous techniques have been developed for both efficient and effective influence maximization. Related works can be classified in to four categories: *local-based approaches, global-based approaches, community-based approaches, and approximation-based approaches*. Algorithms of the local-based category use the local information of the network in order to select the best k influential nodes. The first solution proposed in this category is to select nodes with higher number of neighbors. That is, select the nodes based on their degree scores. Domingos and Richardson [10,17] were the first to study this as an algorithmic problem. Algorithms in the global-based approaches exploit the information of the whole network. In this category, a plethora of centrality measures such as betweeness centrality [11,12], M-centrality [13] and coreness centrality [16] are proposed in order to rank nodes according to their topological importance in the network. M-Centrality measure combines the information on the position of the node in the network with the local information on its nearest neighborhood. The position is measured by the K-shell decomposition, and the degree variation in the neighborhood of the node quantifies the influence of the local context. Coreness is a well-established centrality index that focuses on the structure of networks. Authors in [16] found that the most efficient spreaders are those located within the core of the network as identified by the k-shell decomposition analysis. Betweenness centrality identifies key nodes in a network called bridges. A bridge is a node that has short paths to other nodes in the network. Despite the efficiency of the local-based and the global-based approaches in terms of time which is very fast, these approaches may result in less influence over the network. Studies in [3] showed that the degree-based and centrality-based approaches may result in less influence over the network. The reason behind it might be, these measures do not consider the effect of neighborhood. Indeed, a given group of connected nodes may have a high degree or a high centrality score, but if their adjacent nodes are overlapped then the information may not propagate through the rest of the network. Several approaches are proposed to deal with the problem of neighborhood overlapping. In [7], a faster method that considers the neighbors of each node is required in order to avoid overlapping. Algorithms of the category of community-based use the communities in the network as an intermediate step to select the most influencial nodes [8]. Authors of [8] improve the efficiency of influence maximization by incorporating information on the community structure of the network into the optimization process. They detect the community structure of the input network using the concept of (maximal) cliques problem. Algorithms in the category of approximation-based give the worst case bound for influence spread [15]. However, most of them suffer from the scalability issues, which means, with the increase of the network size, running time grows heavily.

Algorithm 1: Independent Cascade Model

Input: A network $G = (V, E)$ and $A_0 = \{v_0, ..., v_k\}$ denote the set of
  initially activated nodes; θ denote the activation threshold
Output: A set of final activated nodes
1 **while** $A_i \neq \emptyset$ **do**
2     $A_i \leftarrow$ *Newly activated nodes*;
3   **for** $u \in A_i$ **do**
4   **for** $v \in N_u$ **do**
5   **if** $\theta <= p_{uv}$ **then** $A_{i+1} \leftarrow A_{i+1} \cup \{v\}$;
6 **return** $A_0 \cup A_1 \ldots \cup A_i$

In this paper, by utilizing maximal clique problem, we propose *IMSN* (*Influence Maximization using Superordinate Nodes*), which is a novel algorithm for influence maximization in large networks. *IMSN* is based on superordinate nodes to look for the initial vertex set which maximizes the expected number of the infected vertices in the independent cascade model. *IMSN* starts by discovering all the maximal cliques from the complex network represented as a graph. We then denote the vertices belonging to the set of maximal cliques with size greater than or equal to α as superordinate vertices. As a next step, we propose two indicators to rank influential individuals in the networks. We then simulate the information spread using the complete random simulation used in [15]. We also compare the simulation results of our *IMSN* algorithm against two popular algorithms for influence maximization problem.

3   Background

Formally speaking, a complex network is generally abstracted as a graph with entities as the vertex set and the relationships (co-authorship, friendship, etc.) between them as the edge set. Graphs discussed in this paper are simple and undirected. Formally, an undirected graph is defined as a pair $G = (V, E)$ where V is a set of nodes and $E \subseteq V \times V$ is a set of edges. We denote by n (respectively m) the number of nodes (respectively edges) in G. For a node $u \in V$, we denote by N_u the set of neighbors of u, i.e., $N_u = \{v \in V : (u, v) \in E\}$. The *degree* of a node $u \in V$, denoted by d_u, is equal to $|N_u|$. The concept of maximal cliques is defined as follows:

Definition 1 (Clique, Maximal Clique). *Let $G = (V, E)$ be an undirected graph. Then, a clique of G is a subset of nodes $C \subseteq V$ such that whatever v_1 and v_2 belong to C, then the edge (v_1, v_2) belongs to E. A clique C of G is said maximal if for any $x \in V \backslash C$, $C \cup \{x\}$ is not a clique. The set of all maximal cliques of G will be denoted by C.*

A clique represents a densely connected structure in the graph, as such it can be used to recover the locally most related elements, useful for several data mining tasks such as clustering, frequent patterns and community mining [14].

Given a network G, an integer k and an activation threshold θ, in the problem of influence maximization we are looking for the initial active vertex set of size k which maximizes the expected number of the infected vertices. As shown by Kempe et al. in [15], the optimization problem is NP-complete. A plethora of work has emerged in order to solve the problem of influence maximization, for which we refer the reader to existing surveys [3].

4   Influence Maximization Based on Maximal Cliques

Maximal cliques are widely used in several real life applications. For instance, in anomaly detection, signals of rare events are defined as a set of large maximal cliques [5]. In data visualisation, maximal cliques are used to visualize a large graph where the cliques are grouped together in the display. In community detection problem, also known as graph clustering, a rigorous way to model communities is to consider maximal cliques, that is, maximal (with respect to set inclusion) subgraphs in which any pair of nodes is connected by an edge. The objective of this paper is to develop a solution for influence maximization in real networks based on maximal clique problem. To illustrate our method for influence maximization problem, we build our intuition from the following simple but very relevant principle: *a node can be a good infector in a network if multiple maximal cliques contain it*. This intuition is based on the fact that the presence of a dense neighborhood around a node is fundamental to the maximization of influence, because in this way the node can spread the information between dense regions of the network. The proposed *SNIM (Superordinate Nodes for Influence Maximization)* follows three phases (Fig. 1). The first phase detects the maximal cliques of the input network, where a maximal clique is a clique that cannot be extended by including one more adjacent vertex. Such maximal cliques are really the ideal communities structures, that one would like to find. In complex networks, a community structure is a subset of individuals who interact with each other more frequently than other individuals outside the community. In the second phase, maximal cliques with size smaller than a fixed threshold α are removed. This simple tactic may also find the most largest maximal cliques. We then denote the vertices belonging to the rest of maximal cliques with size greater than or equal to α as superordinate vertices. The main idea here is to select from each set of intersecting maximal cliques the most k influential nodes. To do this, we use some indicators that can find a group of nodes of size k that by acting all together maximize the expected number of influenced nodes at the end of the spreading process, formally called *Influence Maximization*.

Fig. 1. Overview of SNIM algorithm

Let us now introduce two simple but very relevant indicators which scores the superordinate nodes based on their connectivity to other nodes. The first indicator called superordinate vertex frequency is defined as follows:

$$\mathcal{F}(u) = \sum_{c \in C} \sigma_{u,c} \tag{1}$$

where the sum is over maximal cliques C obtained after the procedure of filtering, $\sigma_{u,c} = 1$ if $u \in c$ and 0 otherwise. This first indicator quantifies the ability of a node to connect different maximal cliques. Then, a high score of frequency of a given node is obtained if the node belongs to many maximal cliques. Let us now denote the set of intersecting maximal cliques on a given node u by I_u, the second indicator quantify the set of vertices that could be influenced directly by a given node, i.e, the number of nodes in the union of maximal cliques containing the node u. More formally, this indicator is defined as follows:

$$\mathcal{W}(u) = \left| \bigcup_{c \in I_u} c \right| \tag{2}$$

where the union is over maximal cliques containing the node u. It is clear in this second indicator that the more nodes in the union of the set of intersecting maximal cliques on node u, the higher value of $\mathcal{W}(u)$ is obtained.

Next, we show how to select the seed nodes using the two indicators (1) and (2). Let f be a function which assigns a number to each node in the graph. Such number is computed using the indicators (1) and (2) as follows:

$$f: \left\{ v \mapsto \mathcal{F}(v) * \mathcal{W}(v) \right.$$

The function f scores the superordinate nodes based on the principle that a node can be a good infector if it is a member of many maximal cliques and has a strong connection with the other nodes. The nodes with large value f can work as an infection bridge between different maximal cliques, since in real life a person or a company can be a good infector if it appears in many different areas of life [1].

Example 1. Let us consider the undirected network depicted in Fig. 2. Using the two indicators $\mathcal{F}$ and $\mathcal{W}$, for each node we have: $\{\mathcal{F}(1) = \mathcal{F}(2) = \mathcal{F}(7) = \mathcal{F}(8) = \mathcal{F}(11) = 2; \mathcal{F}(3) = \mathcal{F}(4) = \mathcal{F}(5) = \mathcal{F}(6) = 3; \mathcal{F}(10) = \mathcal{F}(12) = 4\}$ and $\{\mathcal{W}(1) = \mathcal{W}(2) = \mathcal{W}(7) = \mathcal{W}(8) = 4; \mathcal{W}(3) = \mathcal{W}(4) = \mathcal{W}(5) = \mathcal{W}(6) = \mathcal{W}(10) = \mathcal{W}(11) = \mathcal{W}(12) = 5\}$.

As a result, the two nodes 10 and 12 have the best value of f.

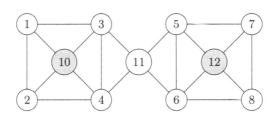

Fig. 2. Example of undirected graph

4.1   Algorithm

Algorithm 2 describes the general feature of our superordinate nodes based procedure to determine the initial spreaders in the graph. It proceeds as follows: first the set of maximal cliques are identified. Then, a procedure of filtering is done using a parameter α. As a next step, we rank all the superordinates nodes following the principle using the function f.

After that only k nodes are selected to be initial spreaders. The question now is how to select the k initial spreaders among all the superordinates nodes? The easiest solution would probably be a selection of the top k nodes with the

---

Algorithm 2: Superordinate Nodes for Influence Maximization

Input: A network $\mathcal{N} = (V, E)$, α, $k > 1$
Output: A set of initial spreaders

1 $C = maximalCliques(\mathcal{N})$;
2 $\mathcal{NS} = superordinatesNodes(C, \alpha)$;
3 **for** $u \in \mathcal{NS}$ **do**
4     $Compute(\mathcal{F}(u))$;
5     $Compute(\mathcal{W}(u))$;
6     $f(u) = \mathcal{F}(u) * \mathcal{W}(u)$;
7 $\mathcal{S} \leftarrow \emptyset$;
  // Top k not connected nodes
8 **while** $|\mathcal{S}| <= k - 1$ **do**
9   Let v the node with
10   the maximum score of f;
11   **if** $\forall u \in \mathcal{S}$ $(u, v) \notin E$ **then**
12     $\mathcal{S} \leftarrow \mathcal{S} \cup \{v\}$;
13     $\mathcal{NS} \leftarrow \mathcal{NS} \setminus \{v\}$;
  // Link-discount
14 **while** $|S| <= k - 1$ **do**
15   Let v the node with
16   the maximum score of f;
17    $\mathcal{S} \leftarrow \mathcal{S} \cup \{v\}$;
18   **for** $u \in N_v$ **do**
19   **if** $u \in \mathcal{NS}$ **then**
20      $\mathcal{W}(u) \leftarrow \mathcal{W}(u) - 1$;
21      $Recompute(f(u))$;
22    $\mathcal{NS} \leftarrow \mathcal{NS} \setminus \{v\}$;
23 **return** S

---

highest value of f. However, if two nodes having several friends in common are selected as initial spreaders then their influence will overlap and will cause negative effect for influence maximization. For example, suppose that the nodes 2 and 3 are selected as initial spreader in the graph presented in Fig. 2, since most of the neighbors of node 1 are also neighbors of node 2, then their influence will overlap and will cause negative effect for influence maximization. To overcome this drawback, we propose to select the initial spreader nodes following two methods. (1) The first method called $IMSN_{nc}$ (line 8–13 in Algorithm 2) in which we select the top k not connected nodes with the highest value of f. (2) Our second method called $IMSN_{ld}$ (line 14–22 in Algorithm 2) in which the main idea is that if one superordinate node is considered as seed, then the links connecting this node with the other superordinate nodes not yet chosen will be discounted, i.e., when considering the next node, the links connecting this node with the other superordinate nodes already in the seed set will be discounted.

Example 2. Let us consider the undirected depicted in Fig. 2 and let $k = 2$. Using the principle of 'top k not connected superordinate nodes' procedure, nodes 10 and 12 are considered as initial spreaders.

5   Experimental Evaluation

The proposed algorithm, referred to as $SNIM_{nc/ld}$ was written in Python. Given an input network as a set of edges, our algorithm starts by generating the set of maximal cliques. To detect maximal cliques, we consider the state-of-the-art algorithm proposed in [9]. We compare our algorithm with two popular algorithms in influence maximization problem, namely degree-based algorithm and degree-discount algorithm. The degree-based heuristic is commonly used in the sociology literature as estimates of a node's influence [18]. The degree-based heuristic chooses nodes v in order of decreasing degrees d_v in a given graph G. Authors of [2] and [18] used high-degree nodes as influential nodes. The General idea of the degree discount algorithm proposed in [7] is that if one node is considered as seed then the links connecting this node with the other node will not be counted as a degree, i.e., when considering the next node, the links connecting with the nodes already in the seed set will be discounted.

The comprehensive performance study conducts on two real world datasets, Amazon network and Dblp network. In each experiment, we vary parameters, of the diffusion model, to compare the influence spread (number of activated nodes) of four algorithms. For our experimental study, all algorithms have been run on a PC with an Intel Core i7 processor and 16 GB memory. We imposed 1 h time limit for all the methods.

Complete Simulation: To compute the expected number of infected vertices, we use the random simulation used in [15]. More specifically, for a given activation threshold θ and for each seed set identified by an algorithm, we simulate the independent cascade model 10000 times (number of iteration). At each time, we choose randomly the propagation probability p_{uv} between each two connected nodes u and v in the graph. The expected number of infected nodes is the total of activated nodes throughout the simulation process divided by the number of iteration.

Results on Amazon Instance: We discuss the influence maximization on a large real-life dataset, Amazon network [19]. It is a product network, where nodes denote the products. If a product i is frequently co-purchased with product j, the graph contains an undirected edge from i to j. Amazon network contains 334 863 nodes and 925 872 edges. We analyze the efficiency and influence spread of our algorithms ($IMSN_{nc}$ and $IMSN_{ld}$) with respect to different numbers of seeds and values of parameters. For the filtering procedure, we set the minimum size of the maximal cliques α to the value 3. Figure 3(c) shows the influence spread of different algorithms with different number of seeds on Amazon. The x-axis indicates the number of seeds and y-axis indicates influence spread. In most cases, $IMSN_{nc}$'s influence spread $> IMSN_{ld} >=$'s influence spread > Degree discount's influence spread > degree's influence spread. With the increasing number of seeds, $IMSN_{nc}$ get better influence spread than the baselines.

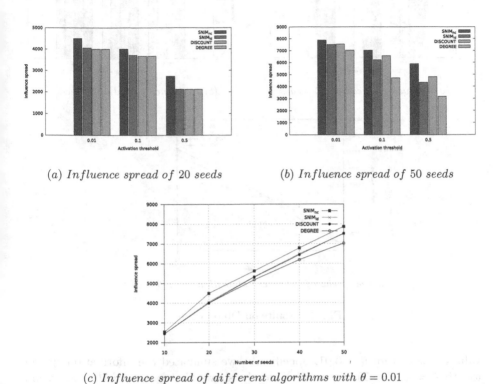

(a) *Influence spread of* 20 *seeds*   (b) *Influence spread of* 50 *seeds*

(c) *Influence spread of different algorithms with* $\theta = 0.01$

Fig. 3. Results on Amazon network

Figures 3(a) and 3(b) perform the influence spread of 20 seeds and 50 seeds with different θ values (0.01, 0.1 and 0.5). The x-axis indicates activation threshold and y-axis indicates influence spread. The results reflected in the figures show that although total influence spread of the four algorithms will decrease as θ increases. Notice that $IMSN_{nc}$ improves its influence spread with the increasing of θ.

Results on Dblp Instance: Now, we discuss the influence maximization on a second dataset, named Dblp [19]. It is a large real-life academic collaboration dataset in Computer Science. Each node in the undirected network represents an author. If an author i co-authored a paper with author j (they publish at least one paper together), the graph contains an undirected edge between i and j. Dblp network contains 317 080 nodes and 1 049 866 edges. For the filtering procedure, we set the minimum size of the maximal cliques α to the value 3. In Fig. 4(c), we report the influence spread of the four algorithms for different

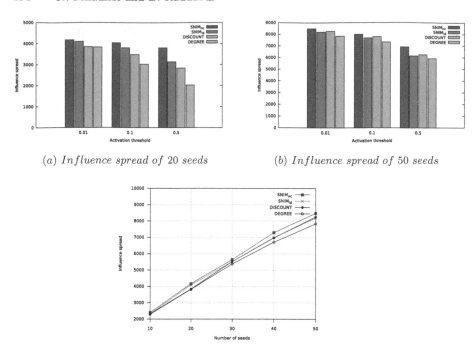

(a) *Influence spread of* 20 *seeds*   (b) *Influence spread of* 50 *seeds*

(c) *Influence spread of different algorithms with* $\theta = 0.01$

Fig. 4. Results on Dblp instance

values of k and for $\theta = 0.01$. Specifically, we simulated the information spread for $10 <= k <= 50$. Figure 4(c) clearly shows that our $SNIM_{nc}$ algorithm outperforms $SNIM_{ld}$, degree-based algorithm and degree-discount algorithm in most cases of k. Figures 4(a) and 4(b) perform the influence spread of 20 seeds and 50 seeds, respectively with different θ values $(0.01, 0.1$ and $0.5)$. The x-axis indicates activation threshold and y-axis indicates influence spread. The results reflected in the figures show that although total influence spread of the four algorithms will decrease as θ increase. It is clear in theses figures that $SNIM_{nc}$ algorithm outperforms the other algorithms in all cases of k and θ.

Overall, in terms of influence spread, $SNIM_{nc} > SNIM_{ld} >= degree - discount > degree - based$. In terms of scalability, our algorithm is able to maintain the same efficiency when the number of nodes and edges increase. Indeed, several efficient parallel algorithms to solve the problem of maximal cliques are proposed in the last decade.

6   Conclusion

In this paper, we proposed an algorithm for influence maximization problem in networks based on maximal clique problem. In particular, we make an original use of a particular concept of nodes called *superordinate nodes*. Then, we introduced two indicators in order to select the most influential nodes in the graph. There are many possible directions for future works. Possible improvements can be obtained by designing better indicator functions for superordinate nodes selection. Another direction is to extend our proposed framework in order to use other cohesive structures such as k-plex, k-truss, etc. We also plan to extend our method in order to deal with the problem of influence maximization in dynamic networks.

References

1. Community based influence maximization in the Independent Cascade Model. Zenodo, September 2018
2. Albert, R., Joong, H., Barabasi, A.: Error and attack tolerance of complex networks. Nature **406**(6794), 378–382 (2000)
3. Banerjee, S., Jenamani, M., Pratihar, D.K.: A survey on influence maximization in a social network. CoRR, abs/1808.05502 (2018)
4. Bass, F.M.: A new product growth for model consumer durables. Manag. Sci. **50**(12 Supplement), 1825–1832 (2004)
5. Berry, N., Ko, T., Moy, T., Smrcka, J., Turnley, J., Wu, B.: Emergent clique formation in terrorist recruitment, January 2016
6. Brown, J.J., Reingen, P.H.: Social ties and word-of-mouth referral behavior. J. Consum. Res. **14**(3), 350–362 (1987)
7. Chen, W., Wang, Y., Yang, S.: Efficient influence maximization in social networks. In: Proceedings of the 15th ACM SIGKDD International Conference on Knowledge Discovery and Data Mining, KDD 2009, pp. 199–208. ACM (2009)
8. Chen, Y., Zhu, W., Peng, W., Lee, W., Lee, S.: CIM: community-based influence maximization in social networks. ACM TIST **5**(2), 25:1–25:31 (2014)
9. Cheng, J., Zhu, L., Ke, Y., Chu, S.: Fast algorithms for maximal clique enumeration with limited memory. In: Proceedings of the 18th ACM SIGKDD International Conference on Knowledge Discovery and Data Mining, KDD 2012, pp. 1240–1248. ACM (2012)
10. Domingos, P., Richardson, M.: Mining the network value of customers. In: Proceedings of the Seventh ACM SIGKDD International Conference on Knowledge Discovery and Data Mining, KDD 2001, pp. 57–66. ACM, New York (2001)
11. Freeman, L.C.: A set of measures of centrality based on betweenness. Sociometry **40**(1), 35–41 (1977)
12. Freeman, L.C.: Centrality in social networks conceptual clarification. Soc. Netw. **1**, 215 (1978)
13. Ibnoulouafi, A., Haziti, M.E., Cherifi, H.: M-centrality: identifying key nodes based on global position and local degree variation. J. Stat. Mech: Theory Exp. **2018**(7), 073407 (2018)
14. Jabbour, S., Mhadhbi, N., Raddaoui, B., Sais, L.: Detecting highly overlapping community structure by model-based maximal clique expansion. In: IEEE International Conference on Big Data, Big Data 2018, Seattle, WA, USA, 10–13 December 2018, pp. 1031–1036 (2018)
15. Kempe, D., Kleinberg, J., Tardos, E.: Maximizing the spread of influence through a social network. In: Proceedings of the Ninth ACM SIGKDD International Conference on Knowledge Discovery and Data Mining, KDD 2003. ACM (2003)

16. Kitsak, M., et al.: Identification of influential spreaders in complex networks. Nat. Phys. **6**(11), 888–893 (2010)
17. Richardson, M., Domingos, P.: Mining knowledge-sharing sites for viral marketing. In: Proceedings of the Eighth ACM SIGKDD International Conference on Knowledge Discovery and Data Mining, KDD 2002, pp. 61–70 (2002)
18. Wasserman, S., Faust, K.: Social Network Analysis: Methods and Applications, vol. 8. Cambridge University Press, Cambridge (1994)
19. Yang, J., Leskovec, J.: Defining and evaluating network communities based on ground-truth. Knowl. Inf. Syst. **42**(1), 181–213 (2013). https://doi.org/10.1007/s10115-013-0693-z

A Probabilistic Approach for Discovering Daily Human Mobility Patterns with Mobile Data

Weizhu Qian[1]([✉]) [ID], Fabrice Lauri[1], and Franck Gechter[1,2]

[1] CIAD UMR 7533, Univ. Bourgogne Franche-Comté, UTBM, 90010 Belfort, France
weizhu.qian@utbm.fr
[2] Mosel LORIA UMR CNRS 7503, Université de Lorraine,
54506 Vandœuvre-lès-Nancy, France

Abstract. Analyzing human mobility with geo-location data collected from smartphones has been a hot research topic in recent years. In this paper, we attempt to discover daily mobile patterns using the GPS data. In particular, we view this problem from a probabilistic perspective. A non-parametric Bayesian modeling method, the Infinite Gaussian Mixture Model (IGMM) is used to estimate the probability density of the daily mobility. We also utilize the Kullback-Leibler (KL) divergence as the metrics to measure the similarity of different probability distributions. Combining the IGMM and the KL divergence, we propose an automatic clustering algorithm to discover mobility patterns for each individual user. Finally, the effectiveness of our method is validated on the real user data collected from different real users.

Keywords: Probabilistic model · Infinite Gaussian Mixture Model · Kullback-Leibler divergence · Human mobility

1 Introduction

Smartphone devices are equipped with multiple sensors that can record user behavior on the handsets. With the help of large-scale smartphone usage data, researchers are able to study human behavior in the real world. Since location information is one of the crucial aspects of human behaviors, investigating human mobility from mining mobile data has drawn the attentions of many researchers.

Previous research in this filed mainly focused on discovering the significant places or predicting the transition among the significant places [2,6,11]. However, these research neglected the data sampled at the places where one stays for a relatively short time period, for instance, amid the transitions. As opposed to this point of view, we suggest that these type of data is important for revealing human mobility patterns as well. In our work, the human mobility is recorded by the GPS modules embedded on the smartphone devices. It should be emphasized that GPS data (longitudes and latitudes) are not evenly distributed spatially

© Springer Nature Switzerland AG 2020
M.-J. Lesot et al. (Eds.): IPMU 2020, CCIS 1237, pp. 457–470, 2020.
https://doi.org/10.1007/978-3-030-50146-4_34

because one may stay longer at the significant places (i.e., a home or school) than at the less significant places (i.e., a restaurants or road). Thus, an appropriate description for human mobility is to treat the location of an individual as a set of data points randomly distributed in the space with respect to different probabilities. Moreover, in practice, the data collecting procedure may not be continuous all the time because the GPS module is turned off or does not function well sometimes. As a consequence, it arises the issue of data sparsity. These unique data characteristics prevent researchers adopting some conventional methods. Therefore, in our work, we adopt a probabilistic approach to describe the daily human mobility. As compared to conventional methods, we believe our approach can explore more information from the original GPS data and mitigate the impact of data sparsity.

The first step of the method is to estimate the probability density for each day's trajectories. For such a task, Gaussian Mixture Model [14] is a possible solution. However, the standard Gaussian Mixture Model needs to set the number of components in advance, which is tricky to implement because the trajectory data can be statistically heterogeneous and a fixed component number for all the daily trajectories is not appropriate either. To handle this problem, we adopt the Infinite Gaussian Mixture Model (IGMM) [13], in which the Dirichlet process prior is used to modify the mixed weights of components. Further, to measure the difference between different mobility probability densities, the Kullback-Leibler (KL) divergence [9] is used. The KL divergence is an asymmetric metric, which means the distance from distribution p to distribution q is not the same as the distance from distribution q to distribution p unless they are identical. We exploit the inequality property of the KL divergence to reveal the subordinate relationship of one trajectory to another. Finally, we devise a clustering algorithm using the IGMM with the KL divergence to discover the mobility patterns existing in human mobility data. More importantly, as compared to traditional methods, our clustering algorithm is automatic because it does not require a preset of the pattern number.

The reminder of the paper is organized as follows. Section 2 surveys the related work. Section 3 addresses the problem we are tackling in this paper. In Sect. 4, the proposed method is depicted. In Sect. 5 presents the conducted experiment and its results to evaluate our method with real user data. Finally, we conclude our paper and discuss about the future work in Sect. 6.

2   Related Work

A widespread topic is to predict human mobility with the smartphone usage contextual information, e.g., temporal information, application usage, call logs, WiFi status, Cell ID, etc. In [2] for instance, the researchers applied various machine learning techniques to accomplish prediction tasks such as the next-time slot location prediction and the next-place prediction. In particular, they exploited how different combinations of contextual features are related to smartphone usage can affect the prediction accuracy. Moreover, they also compared the predicting performances of the individual models and the generic models.

Another frequently-used method for such tasks is to use probabilistic models. By calculating the conditional probabilities between contextual features, [5] developed the contextual conditional models for the next-place prediction and visit duration prediction. In [4], the researchers presented the probabilistic prediction frameworks based on Kernel Density Estimation (KDE). [4] utilized conditional kernels density estimation to predict the mobility events while [12] devised different kernels for different context information types. In [11], the authors developed a location Hierarchical Dirichlet Process (HDP) based approach to model heterogeneous location habits under data sparsity.

Among the other possible approaches, [18] proposed a Hypertext Induced Topic Search based inference model for mining interesting locations and travel sequences using a large GPS dataset in a certain region. In [6], the authors employed the random forests classifiers to label different places without any geolocation information. [15] made use of nonlinear time series analysis of the arrival time and residence time for location prediction.

In particular, for clustering user trajectories, there exists several different methods. However, these conventional algorithms are not applicable to our objectives. For example, some researchers used K-means [1] in their work, whereas K-means can not handle the trajectories with complex shapes or noisy data because it is based on Euclidean distance. Besides, it also needs the pre-knowledge of the cluster numbers, which is not acquirable in many real-world cases. DBSCAN [17], a density-based clustering techniques, can deal with data with arbitrary shapes and does not require the number of cluster in advance. However, it still needs to set the minimum points number and neighbourhood radius to recognize the core areas and it treats the non-core data points as noise. As for the grid searching algorithm [5], it focus on detecting the stay points within a set of square regions, whereas it fails to reveal the mobility at a larger scale.

3   Problem Formulation

In this paper, our purpose is to discover the mobility patterns for each individual from the GPS location data. As shown in Fig. 1a, the mobility for one individual consists of many different trajectories (the data is from the MDC dataset, the detailed data description will be in following experiments). A trajectory here means that a set of GPS data points collected from the user's smartphone, however, we do not treat it as a sequence. We believe that one's daily mobility is rather regular and there are common mobility patterns shared among different daily trajectories. Generally, one may follow the regular daily itineraries, for instance, home-work place/school-home. Yet, on different days the daily itineraries may not be the same, for instance, on the way to home, one may take a detour to do shopping in a supermarket sometimes. Hence, our objective is to discover all the potential daily mobility from the data with location information.

We extract each day's trajectory from the all GPS trajectories from a user. Figure 1b reveals that daily trajectories recorded by GPS data are not distributed

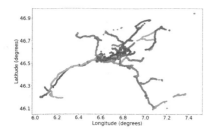 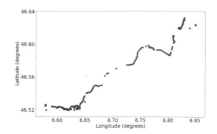

(a) All GPS trajectories from a randomly   (b) A randomly selected daily trajectory
selected user.   from the user.

Fig. 1. GPS data samples.

evenly in space. It may be caused by the data collecting procedure: some data collecting time period is actually relatively short (less than 24 h, in fact, only few hours sometimes), which leads to the data sparsity problem. In order to overcome this problem and exploit as much information as possible from the GPS data, we argue that a reasonable way to describe the daily trajectories is to estimate the probability density of the location data. The relationship among the trajectories can be represented by their probability densities. As a result, we can discover all the mobility patterns for each user. The tasks in this paper will be as follows:

- Task 1: to estimate the probability density for mobility for each day.
- Task 2: to measure the closeness between different trajectories.
- Task 3: to discover the similar mobility patterns.
- Task 4: to compare the proposed algorithm with other methods.

4   Proposed Method

4.1   Estimating Daily Trajectories Probability Density

We assume that the GPS location data points are distributed randomly spatially. The distribution of each day also consists of unknown number of heterogeneous sub-distributions. Therefore, one feasible method is to use mixed Gaussian models for estimating the probability density of daily mobility.

Gaussian Mixture Model. A Gaussian Mixture Model (GMM) is composed of a fixed number K of sub-components. The probability distribution of a GMM can be described as follows:

$$P(x) = \sum_{k=1}^{K} \pi_k P(x|\theta_k) \tag{1}$$

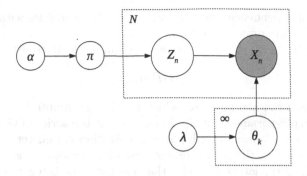

Fig. 2. The plate representation of Infinite Gaussian Mixture Model.

where, x is the observable variable, π_k is the assignment probability for each model, with $\sum_{k=1}^{K} \pi_k = 1, (0 < \pi_k < 1)$, and θ_k is the internal parameters of the base distribution.

Let z_n be the latent variables for indicating categories.

$$z_n \sim Categorical(z_n|\pi) \quad \sum_{k=1}^{K} z_{nk} = 1 \tag{2}$$

where, $z_n = [z_{n1}, z_{n2}, ..., z_{nk}, ..., z_{nK}]$, in which only one element $z_{nk} = 1$. It means x_n is correspondent to θ_k.

If the base distribution is a Gaussian, then:

$$P(x|\theta_k) = N(x|\mu_k, \Lambda_k^{-1}) \tag{3}$$

where, μ_k is the mean vector and Λ_k is the precision matrix.

Therefore, an observable sample x_n is drawn from GMM according to:

$$x_n \sim \prod_{k=1}^{K} N(x_n|\nu_k, \Lambda_k)^{z_{nk}} \tag{4}$$

As it is illustrated above, one crucial issue of GMM is to pre-define the number of components K. It is tricky because the probability distribution for each day's mobility is not identical. Thus, to define a fixed K for all mobility GMM models is not suitable in our case.

Infinite Gaussian Mixture Model. Alternatively, we resort to the Infinite Gaussian Mixture Model (IGMM) [13]. As compared to the finite Gaussian Mixture Model, by using a Dirichlet process (DP) prior, IGMM does not need to specify the number of components in advance. Figure 2 presents the graphical structure of the Infinite Gaussian Mixture Model.

In Fig. 2, the nodes represent the random variables and especially, the shaded node is observable and the unshaded nodes are unobservable. The edges represent the conditional dependencies between variables. The variables within the plates means that they are drawn repeatedly.

According to Fig. 2, the Dirichlet process can be depicted as:

$$G \sim DP(\alpha, G_0) \tag{5}$$

where, G is a random measure, which consists of infinite base measures G_0 and λ is the hyper-parameter of G_0. In our case, it is a series of Gaussian distributions. And $\alpha \sim Gamma(1,1)$ is the concentration parameter. N is the total samples number. θ_k is the parameters of base distribution. X_k is the observable data for θ_k. Z_k is the latent variables that indicates the category of X_k.

Alternatively, G can be explicitly depicted as follow:

$$G(\theta) = \sum_{k=1}^{\infty} \pi_k \delta_{\theta_k} \tag{6}$$

where, $\theta_k \sim G_0(\lambda)$, and δ is Dirac function. π_k determines the proportion weights of the clusters and the δ_{θ_k} is the prior of the θ_k to determine the location of clusters in space.

We choose the Stick-breaking process (SBP) [16] to implement the Dirichlet process as the prior of π_k. The Stick-breaking process can be described as follow:

$$\pi_k = \nu_k \prod_{j=1}^{k-1} (1 - \nu_j) \quad k \geq 2 \tag{7}$$

where, $\nu_k \sim Beta(1, \alpha)$.

Since $P(x|\theta)$ is Gaussian, $\theta = \{\mu, \Lambda\}$. Further, let G_0 be a Gaussian-Wishart distribution, then, $\mu_k, \Lambda_k \sim G_0(\mu, \Lambda)$. Therefore, similarly, drawing an observable sample x_n from IGMM can be described as follow:

$$x_n \sim \prod_{k=1}^{\infty} N(x_n|\nu_k, \Lambda_k^{-1})^{z_{nk}} \tag{8}$$

Variational Inference (VI) is used to solve the IGMM models. In contrast with Gibbs sampling, a Markov chain Monte Carlo (MCMC) method, VI is relatively faster which makes it salable to large datasets [3]. The results will be demonstrated in the later experiments.

4.2 Measure Daily Trajectories Similarities

The Kullback-Leibler (KL) divergence is a metric to evaluate the closeness between two distributions. For continuous variables, the KL divergence $D_{KL}(p||q)$ is the expectation of the logarithmic difference between the p and q with respect to probability p and vice versa. From (9) and (10), it can be seen that the KL divergence is non-negative and asymmetric. In many occasions, the inequality of the KL divergence is notorious. However, in our method, we take advantage of this characteristic of inequality to reveal the similarities among different trajectories instead of other symmetric metrics.

$$D_{KL}(p||q) = \int_{-\infty}^{\infty} p(x) \log(\frac{p(x)}{q(x)}) dx \tag{9}$$

$$D_{KL}(q||p) = \int_{-\infty}^{\infty} q(y) \log(\frac{q(y)}{p(y)}) dy \tag{10}$$

There is no closed form to implement the KL divergence by the definition of (9) and (10) for Gaussian Mixture Models. Therefore, we resort to the Monte Carlo simulation method proposed in [7]. Then, the KL divergence can be calculated via:

$$D_{KL_{MC}}(p||q) = \frac{1}{n} \sum_{i=1}^{n} \log(\frac{p(x_i)}{q(x_i)}) \tag{11}$$

$$D_{KL_{MC}}(q||p) = \frac{1}{n} \sum_{i=1}^{n} \log(\frac{q(y_i)}{p(y_i)}) \tag{12}$$

This method is to draw a large amount of i.i.d samples x_i from distribution p to calculate $D_{KL_{MC}}(p||q)$ according to (11) and $D_{KL_{MC}}(p||q) \rightarrow D_{KL}(p||q)$ as $n \rightarrow \infty$. It is the same for implementing (10) by using (12). The results will be demonstrated in the later experiments. Furthermore, if we define a representative trajectory for a mobility pattern then we can distinguish whether a new trajectory belong to this cluster by comparing it to the most representative trajectory. To this end, we need to set a threshold with a lower bound and an upper bound for the KL divergences, afterwards it can be used as the metrics to cluster the trajectories.

4.3   Discovering Mobility Patterns

The proposed algorithm is shown in Algorithm 1 and its variables are described in Table 1. The first step of the clustering algorithm is to calculate the probability densities by using the Infinite Gaussian Mixture Models. At this step, we create a list, in which the members are the probability densities of each trajectories. Then the first cluster is created with one trajectory as its first member and it also will be the first baseline trajectories used to compare with other trajectories. It may be replaced by other trajectories later. Afterwards, we select another daily trajectory in the list and calculate the KL divergences, both $D_{KL(p||q)}$ and $D_{KL(q||p)}$. The new trajectory is added to the current cluster if the minimum and maximum of the KL-divergences are smaller than the lower threshold and the upper threshold of the thresholds respectively at the same time. If the $D_{KL(p||q)}$ is smaller than $D_{KL(q||p)}$, the new trajectory become the new baseline for the current cluster. This step will be repeated until all the trajectories belonging to the current cluster are discovered at the end of this iteration. Then all the members of the current cluster are removed from the iteration because we assume that each trajectory can only be the member of one mobility pattern. At the start of new iteration, a new cluster is created, the above steps will be repeated until the list is empty.

Table 1. Variables description

Variable	Domain	Description
D	*Integer*	Total number of data collecting day
d	$\{1, 2, \ldots, D\}$	Index of data collecting day
X	$\{X_1, X_2, \ldots, X_d, \ldots, X_D\}$	Total GPS data (longitudes, latitudes
P	$\{P_1, P_2, \ldots, P_d, \ldots, P_D\}$	Probability density for X
M	$\{M_1, M_2, \ldots, M_k, \ldots M_K\}$	Total mobility patterns list
K	*Integer*	Total number of Discovered mobility patterns
M_k	$\{X_{k1}, X_{k2}, \ldots, X_{kn}\}$	Discovered mobility pattern sub-members list
Th	$\{$Lower bound, Upper bound$\}$	Thresholds for distinguishing patterns
D_{KL}	$\{D_{KL(p\|\|q)}, D_{KL(q\|\|p)}\}$	KL divergences

As it can be seen that our algorithm is designed to discover the latent mobility patterns automatically without the pre-knowledge of the number of existing patterns.

5 Experiments and Results

5.1 Data Description

We use the Mobile Data Challenge (MDC) dataset [8,10] to validate our method. This dataset records a comprehensive smartphone usage with fine granularity of time. The participants of the MDC dataset are up to nearly 200 and the data collection campaign lasted more than 18 months. This abundant information can be used to investigate individual mobility patterns in our research. We attempt to find the trajectories that belong to the same mobility patterns, therefor we focus on the spatial information of the GPS records, namely, the latitudes and longitudes, while the time-stamps of the data are not considered. In addition, the data we use is unlabeled and without any semantic information.

5.2 Experimental Setup

For the experiments, we randomly select 20 users with sufficient data. Each user's is segmented by the time range of one day. Generally, the data length of each day varies from less than 4 h to 24 h and most of them is less than 8 h.

Algorithm 1. Algorithm for Discovering Mobility Pattern

Input: X
Output: M

1: $P \leftarrow \text{IGMM}(X)$ ▷ probability density estimation

2: Initialize:$M = \{M_k\}$ ▷ create the mobility patterns set
3: **while** $P \neq \emptyset$ **do**
4:     $X_s = X_1$ ▷ set a baseline mobility for M_k
5:     $M_k = \{X_s\}$ ▷ create current pattern M_k
6:     **for** $d = 2, \ldots, D$ **do**
7:         $D_{KL} \leftarrow (P_s, P_d)$ ▷ measure similarity
8:         **if** $(\min(D_{KL}) < Th[0])$ & $(\max(D_{KL}) < Th[1])$ **then**   ▷ two patterns are similar
9:             add P_d to M_k ▷ add new member
10:            **if** $D_{KL}[0] > D_{KL}[1]$ **then**
11:                $P_s \leftarrow P_d$ ▷ change the baseline mobility
12:            **end if**
13:        **end if**
14:    **end for**
15:    remove $P_d \in M_k$ from P ▷ current pattern is finished

16:    create M_{k+1} ▷ find new mobility pattern
17:    add M_{k+1} to M
18: **end while**
19:
    return M

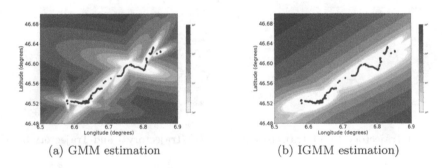

(a) GMM estimation   (b) IGMM estimation)

Fig. 3. Distribution estimation by GMM and IGMM (negative log-likelihood)

5.3   Experimental Results

Probability Density Estimation. Fig. 3a and Fig. 3b show the density estimation results obtained by the GMM and the IGMM, respectively. It can be seen
that, compared to the GMM, the result of the IGMM is less overfitting than the
GMM. It suggests that the IGMM is not affected by the number of components
and it infers more information from the original data and it is less influenced by

data sparsity. That is to say, on the same dataset, the computational results of the IGMM have higher fidelity. Hence, in our approach, we chose the IGMM to estimate probability density of daily mobility.

Measuring Daily Trajectories Similarities. As shown in Fig. 4, we select 5 daily trajectories from the data of one random user to demonstrate the KL divergences between different trajectories. The baseline trajectory is Trajectory 1 and the rest of trajectories are chosen to make comparisons.

Table 2. KL-divergences for different trajectories.

p	q	$D_{KL(p\|\|q)}$	$D_{KL(q\|\|p)}$
Trajectory 1	Trajectory 2	7.21	2.82
Trajectory 1	Trajectory 3	1.28	1.83
Trajectory 1	Trajectory 4	19.07	1269.47
Trajectory 1	Trajectory 5	3.08	996.17

The combinations are shown in Fig. 4 and the results are illustrated in Table 2. Trajectory 2 is nearly a sub-part of Trajectory 1, the KL divergence values are both small, thus Trajectory 2 and Trajectory 1 can be regarded to belong to the same mobility pattern. Trajectory 3 is very similar to Trajectory 1 and $D_{KL}(p\|\|q)$ almost equals to $D_{KL}(q\|\|p)$. Hence, they also are the members

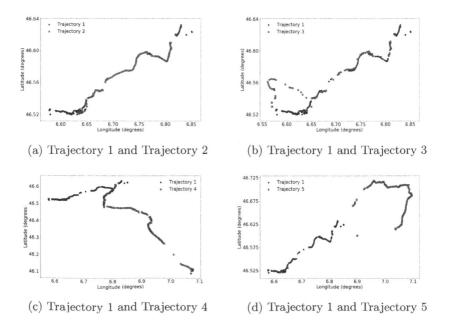

(a) Trajectory 1 and Trajectory 2   (b) Trajectory 1 and Trajectory 3

(c) Trajectory 1 and Trajectory 4   (d) Trajectory 1 and Trajectory 5

Fig. 4. Comparison between different trajectories.

of the same mobility pattern. Trajectory 4 shares a small part with Trajectory 1 whereas generally they are very different. $D_{KL}(p||q)$ and $D_{KL}(q||p)$ are both very large. Therefore, Trajectory 4 and Trajectory 1 are different patterns. For Trajectory 5 and Trajectory 1. $D_{KL}(p||q)$ is small but $D_{KL}(p||q)$ are very large. So they naturally are not in the same pattern. Finally, we can say that the Kl divergence can be used as the distance metrics to distinguish different trajectory patterns.

Discovering Daily Mobility Patterns. We run our algorithm on the data of the 20 users to discover their daily mobility patterns. The partial clustering results are demonstrated in Fig. 5. It proves our method is able to find different mobility patterns even under the condition of noise and discontinuity. Figure 6 shows that our approach is able not only to identify the different patterns in the daily trajectories data but also to find the most representative trajectories for each mobility pattern. Figure 7a shows the number of discovered mobility pattern for all the user in our experiments. Figure 7b depicts the number of members for each discovered mobility patterns for all users.

Comparing with Other Methods. In comparison with the IGMM-based model, we utilize Kernel Density Estimation (KDE) and a set of Gaussian Mixture Models with different numbers of components (GMM-n), to estimate the daily mobility probability densities in our proposed clustering algorithm. Since the GPS data are not labeled, which means that the ground truth is not available. In this case, we run our algorithm on all the trajectories collected from the 20 users and choose the mean log-likelihood, which indicts the reliability of the models, as a reasonable evaluation metrics. The results in Table 3 show that our method outperforms other conventional methods.

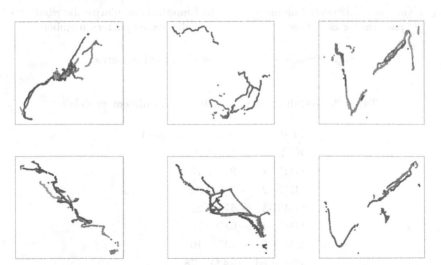

Fig. 5. Discovered mobility patterns from 3 random selected users. Different colors represents different days. (Color figure online)

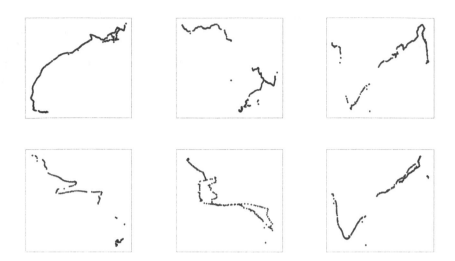

Fig. 6. Representative trajectories for each discovered mobility patterns.

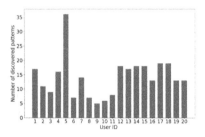
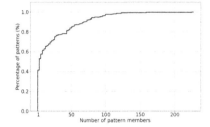

(a) Number of discovered mobility patterns for each user.

(b) Empirical cumulative distribution of discovered pattern members.

Fig. 7. Statistical analysis of discovered patterns.

Table 3. Overall mean log-likelihood for different models

Model	Mean log-likelihood
KDE	−51991.03
GMM-1	−26078.15
GMM-2	−38514.32
GMM-3	−52431.62
GMM-4	−63794.70
GMM-5	−73508.10
Proposed	**−24871.78**

6   Conclusion

In this work, we present a probabilistic approach to discover human daily mobility patterns based on GPS data collected by smartphones. In our approach, the human daily mobility is considered as sets of probability distributions. We argue that Infinite Gaussian Mixture Model is more appropriate than the standard Gaussian Mixture Model on this issue. Further, in order to find the similar trajectories, we use the Kullback-Leibler divergences as the distance metrics. Finally, we devise a novel automatic clustering algorithm combining the advantages of IGMM and the KL divergence so as to discover human daily mobility patterns. Our algorithm do not need the knowledge of the cluster number in advance. For validation, we conducted a set of experiments to prove the effectiveness of our method. For further study, we plan to use WiFi fingerprint data and other machine learning methods to study human mobility.

Acknowledgment. The research in this paper used the MDC Database made by Idiap Research Institute, Switzerland and owned by Nokia. The authors would like to thank the MDC team for providing the access to the database. The authors also would like to the financial support from the China Scholarship Council.

References

1. Ashbrook, D., Starner, T.: Using gps to learn significant locations and predict movement across multiple users. Pers. Ubiquit. Comput. **7**(5), 275–286 (2003)
2. Baumann, P., Koehler, C., Dey, A.K., Santini, S.: Selecting individual and population models for predicting human mobility. IEEE Trans. Mob. Comput. **17**(10), 2408–2422 (2018)
3. Blei, D.M., Jordan, M.I., et al.: Variational inference for dirichlet process mixtures. Bayesian Anal. **1**(1), 121–143 (2006)
4. Do, T.M.T., Dousse, O., Miettinen, M., Gatica-Perez, D.: A probabilistic kernel method for human mobility prediction with smartphones. Pervasive Mob. Comput. **20**, 13–28 (2015)
5. Do, T.M.T., Gatica-Perez, D.: Contextual conditional models for smartphone-based human mobility prediction. In: Proceedings of the 2012 ACM Conference on Ubiquitous Computing, pp. 163–172. ACM (2012)
6. Do, T.M.T., Gatica-Perez, D.: The places of our lives: visiting patterns and automatic labeling from longitudinal smartphone data. IEEE Trans. Mob. Comput. **13**(3), 638–648 (2014)
7. Hershey, J.R., Olsen, P.A.: Approximating the Kullback Leibler divergence between Gaussian mixture models. In: 2007 IEEE International Conference on Acoustics, Speech and Signal Processing-ICASSP 2007, vol. 4, pp. IV-317. IEEE (2007)
8. Kiukkonen, N., Blom, J., Dousse, O., Gatica-Perez, D., Laurila, J.: Towards rich mobile phone datasets: Lausanne data collection campaign. In: Proceedings of ICPS, vol. 68, Berlin (2010)
9. Kullback, S.: Information Theory and Statistics. Courier Corporation (1997)
10. Laurila, J.K., et al.: The mobile data challenge: big data for mobile computing research. Technical report (2012)

11. McInerney, J., Zheng, J., Rogers, A., Jennings, N.R.: Modelling heterogeneous location habits in human populations for location prediction under data sparsity. In: Proceedings of the 2013 ACM International Joint Conference on Pervasive and Ubiquitous Computing, pp. 469–478. ACM (2013)

12. Peddemors, A., Eertink, H., Niemegeers, I.: Predicting mobility events on personal devices. Pervasive Mob. Comput. **6**(4), 401–423 (2010)

13. Rasmussen, C.E.: The infinite Gaussian mixture model. In: Advances in Neural Information Processing Systems, pp. 554–560 (2000)

14. Reynolds, D.: Gaussian Mixture Models. Encyclopedia of Biometrics, pp. 827–832 (2015)

15. Scellato, S., Musolesi, M., Mascolo, C., Latora, V., Campbell, A.T.: NextPlace: a spatio-temporal prediction framework for pervasive systems. In: Lyons, K., Hightower, J., Huang, E.M. (eds.) Pervasive 2011. LNCS, vol. 6696, pp. 152–169. Springer, Heidelberg (2011). https://doi.org/10.1007/978-3-642-21726-5_10

16. Sethuraman, J.: A constructive definition of Dirichlet priors. Stat. Sinica **4**(2), 639–650 (1994)

17. Yu, C., et al.: Modeling user activity patterns for next-place prediction. IEEE Syst. J. **11**(2), 1060–1071 (2017)

18. Zheng, Y., Zhang, L., Xie, X., Ma, W.Y.: Mining interesting locations and travel sequences from GPS trajectories. In: Proceedings of the 18th International Conference on World Wide Web, pp. 791–800. ACM (2009)

Feature Reduction in Superset Learning Using Rough Sets and Evidence Theory

Andrea Campagner[1], Davide Ciucci[1(✉)] ⓘ, and Eyke Hüllermeier[2]

[1] Dipartimento di Informatica, Sistemistica e Comunicazione,
University of Milano–Bicocca, viale Sarca 336, 20126 Milan, Italy
davide.ciucci@unimib.it
[2] Department of Computer Science, Paderborn University, Paderborn, Germany

Abstract. Supervised learning is an important branch of machine learning (ML), which requires a complete annotation (labeling) of the involved training data. This assumption, which may constitute a severe bottleneck in the practical use of ML, is relaxed in weakly supervised learning. In this ML paradigm, training instances are not necessarily precisely labeled. Instead, annotations are allowed to be imprecise or partial. In the setting of superset learning, instances are assumed to be labeled with a set of *possible* annotations, which is assumed to contain the correct one. In this article, we study the application of *rough set theory* in the setting of superset learning. In particular, we consider the problem of feature reduction as a mean for *data disambiguation*, i.e., for the purpose of figuring out the most plausible precise instantiation of the imprecise training data. To this end, we define appropriate generalizations of decision tables and reducts, using information-theoretic techniques based on evidence theory. Moreover, we analyze the complexity of the associated computational problems.

Keywords: Feature selection · Superset learning · Rough sets · Evidence theory

1 Introduction

In recent years, the increased availability of data has fostered the interest in machine learning (ML) and knowledge discovery, in particular in *supervised learning* methodologies. These require each training instance to be annotated with a target value (a discrete label in classification, or a real number in regression). The annotation task is a fundamental component of the ML pipeline, and often a bottleneck in terms of cost. Indeed, the high costs caused by the standard annotation process, which may require the involvement of domain experts, have triggered the development of alternative annotation protocols, such as those based on *crowdsourcing* [4] or (semi-)*automated annotation* [12].

A different approach, which has attracted increasing attention in the recent years, is the combination of supervised and unsupervised learning techniques,

© Springer Nature Switzerland AG 2020
M.-J. Lesot et al. (Eds.): IPMU 2020, CCIS 1237, pp. 471–484, 2020.
https://doi.org/10.1007/978-3-030-50146-4_35

472   A. Campagner et al.

sometimes referred to as *weakly supervised learning* [30]. In this setting, training instances are not necessarily labeled precisely. Instead, annotations are allowed to be imprecise or partial.

A specific variant of weakly supervised learning is the setting of *superset learning* [9,16,18], where an instance x is annotated with a set S of (precise) candidate labels that are deemed *possible*. In other words, the label of x cannot be determined precisely, but is known to be an element of S. For example, an image could be tagged with {horse, pony, zebra}, suggesting that the animal shown on the picture is one of these three, though it is not exactly known which of them. Superset learning has been widely investigated under the classification perspective [10,15], that is, with the goal of training a predictive model that is able to correctly classify new instances, despite the weak training information. Nevertheless, the task of *feature selection* [6], which is of critical importance for machine learning in general, has not received much attention so far.

In this article, we study the application of *rough set theory* in the setting of superset learning. In particular, we consider the problem of feature reduction as a mean for *data disambiguation*, i.e., for the purpose of figuring out the most plausible precise instantiation of the imprecise training data. Broadly speaking, the idea is as follows: An instantiation that can be explained with a simple model, i.e., a model that uses only a small subset of features, is more plausible than an instantiation that requires a complex model. To this end, we will define appropriate generalizations of decision tables and reducts, using information-theoretic techniques based on evidence theory. Moreover, we analyze the complexity of the associated computational problems.

2 Background

In this section, we recall basic notions of rough set theory (RST) and evidence theory, which will be used in the main part of the article.

2.1 Rough Set Theory

Rough set theory has been proposed by Pawlak [19] as a framework for representing and managing uncertain data, and has since been widely applied for various problems in the ML domain (see [2] for a recent overview and survey). We briefly recall the main notions of RST, especially regarding its applications to feature reduction.

A decision table (DT) is a triple $DT = \langle U, Att, t \rangle$ such that U is a universe of objects and Att is a set of *attributes* employed to represent objects in U. Formally, each attribute $a \in Att$ is a function $a : U \to V_a$, where V_a is the domain of values of a. Moreover, $t \notin Att$ is a distinguished *decision* attribute, which represents the target decision (also labeling or annotation) associated with each object in the universe. We say that DT is *inconsistent* if the following holds: $\exists x_1, x_2 \in U, \forall a \in Att, a(x_1) = a(x_2)$ and $t(x_1) \neq t(x_2)$.

Given $B \subseteq Att$ we can define the *indiscernibility partition* with respect to B as $\pi_B = \{[x]_B \subset U \mid \forall x' \in [x]_B, \forall a \in B, a(x') = a(x)\}$. We say that $B \subseteq Att$ is a *decision reduct* for DT if $\pi_B \leq \pi_t$ (where the order $\leq$ is the refinement order for partitions, that is, π_t is a coarsening of π_B) and there is no $C \subsetneq B$ such that $\pi_C \leq \pi_t$. Then, evidently, a reduct of a decision table DT represents a set of non-redundant and necessary features to represent the information in DT. We say that a reduct R is *minimal* if it is among the smallest (with respect to cardinality) reducts.

Given $B \subseteq Att$ and a set $S \subseteq U$, a *rough approximation* of S (with respect to B) is defined as the pair $B(S) = \langle l_B(S), u_B(S) \rangle$, where $l_B(S) = \bigcup \{[x]_B \mid [x]_B \subseteq S\}$ is the *lower approximation* of S, and $u_B(s) = \bigcup \{[x]_B \mid [x]_B \cap S \neq \emptyset\}$ is the corresponding *upper approximation*.

Finally, given $B \subseteq Att$, the *generalized decision* with respect to B for an object $x \in U$ is defined as $\delta_B(x) = \{t(x') \mid x' \in [x]_B\}$. Notably, if DT is not inconsistent and B is a reduct, then $\delta_B(x) = t(x)$ for all $x \in U$.

We notice that in the RST literature, there exist several definitions of reduct. We refer the reader to [25] for an overview of such a list and a study of their dependencies. We further notice that, given a decision table, the problem of finding the minimal reduct is in general Σ_2^P-complete (by reduction to the *Shortest Implicant* problem [28]), while the problem of finding a reduct is in general NP-complete [23]. We recall that Σ_2^P is the complexity class defined by problems that can be verified in polynomial time given access to an oracle for an NP-complete problem [1].

2.2   Evidence Theory

Evidence theory (ET), also known as Dempster-Shafer theory or belief function theory, has originally been introduced by Dempster in [5] and subsequently formalized by Shafer in [21] as a generalization of probability theory (although this interpretation has been disputed [20]). The starting point is a *frame of discernment* X, which represents all possible states of a system under study, together with a *basic belief assignment* (bba) $m : 2^X \rightarrow [0,1]$, such that $m(\emptyset) = 0$ and $\sum_{A \in 2^X} m(A) = 1$. From this bba, a pair of functions, called respectively *belief* and *plausibility*, can be defined as follows:

$$Bel_m(A) = \sum_{B:B \subseteq A} m(B) \tag{1}$$

$$Pl_m(A) = \sum_{B:B \cap A \neq \emptyset} m(B) \tag{2}$$

As can be seen from these definitions, there is a clear correspondence between belief functions (resp., plausibility functions) and lower approximations (resp., upper approximations) in RST; we refer the reader to [29] for further connections between the two theories.

Starting from a bba, a probability distribution, called *pignistic probability*, can be obtained [26]:

$$P_{Bet}^m(x) = \sum_{A:x\in A} \frac{m(A)}{|A|} \tag{3}$$

Finally, we recall that appropriate generalizations of information-theoretic concepts [22], specifically the concept of *entropy* (which was also proposed to generalize the definition of reducts in RST [24]), have been defined for evidence theory. Most relevantly, we recall the definition of *aggregate uncertainty* [7]

$$AU(m) = \max_{p\in\mathcal{P}(m)} H(p), \tag{4}$$

where $H(p) = -\sum_{x\in X} p(x)log_2 p(x)$ is the Shannon entropy of p and $\mathcal{P}(m)$ the set of probability distributions p such that $Bel_m \leq p \leq Pl_m$; and the definition of *normalized pignistic entropy* (see [13] for the un-normalized definition)

$$H_{Bet}(m) = \frac{H(P_{Bet}^m)}{H(\hat{p}_m)}, \tag{5}$$

where $\hat{p}_m$ is the probability distribution that is uniform on the support of $P_{Bet}^m(x)$, i.e., on the set of elements $\{x \,|\, P_{Bet}^m(x) > 0\}$.

3   Superset Decision Tables and Reducts

In this section, we extend some key concepts of rough set theory to the setting of superset learning.

3.1   Superset Decision Tables

In superset learning, each object $x \in U$ is not associated with a single annotation $t(x) \in V_t$, but with a set S of candidate annotations, one of which is assumed to be the true annotation associated with x. In order to model this idea in terms of RST, we generalize the definition of a decision table.

Definition 1. *A superset decision table (SDT) is a tuple $SDT = \langle U, Att, t, d\rangle$, where $\langle U, Att, t\rangle$ is a decision table, i.e.:*

- *U is a universe of objects of interest;*
- *Att is a set of attributes (or features);*
- *t is the decision attribute (whose value, in general, is not known);*

and d, with $\{d\} \cap Att = \emptyset$, is a set-valued decision attribute, that is, $d : U \to \mathcal{P}(V_t)$ such that the superset property *holds: For all $x \in U$, the real decision $t(x)$ associated with x is in $d(x)$.*

The intuitive meaning of the set-valued information d is that, if $|d(x)| > 1$ for some $x \in U$, then the real decision associated with x (i.e. $t(x)$) is not known precisely, but is known to be in $d(x)$. Notice that Definition 1 is a proper generalization of decision tables: if $|d(x)| = 1$ for all $x \in U$, then we have a standard decision table.

Remark 1. In Definition 1, a set-valued decision attribute is modelled as a function $d : U \to \mathcal{P}(V_t)$. While this mapping is formally well-defined for a concrete decision table, let us mention that, strictly speaking, there is no functional dependency between x and $d(x)$. In fact, $d(x)$ is not considered as a property of x, but rather represents *information* about a property of x, namely the underlying decision attribute $t(x)$. As such, it reflects the epistemic state of the decision maker.

Definition 2. *An* instantiation *of an SDT $\langle U, Att, t, d \rangle$ is a standard DT $\langle U, Att, t' \rangle$ such that $t'(x) \in d(x)$ for all $x \in U$. The set of instantiations of SDT is denoted $\mathcal{I}(SDT)$.*

Based on the notion of SDT, we can generalize the notion of inconsistency.

Definition 3. *Let $B \subset Att$, then SDT is B-inconsistent if*

$$\exists x_1, x_2 \in U, \forall a \in B, a(x_1) = a(x_2) \text{ and } d(x_1) \cap d(x_2) = \emptyset. \tag{6}$$

We call such a pair x_1, x_2 inconsistent, otherwise it is consistent.

Thus, inconsistency implies the existence of (at least) two indiscernible objects with non-overlapping superset decisions. We say that an instantiation I is *consistent with a SDT S* (short, is consistent) if the following holds for all x_1, x_2: if x_1, x_2 are consistent in S, then they are also consistent in I.

3.2   Superset Reducts

Learning from superset data is closely connected to the idea of *data disambiguation* in the sense of figuring out the most plausible instantiation of the set-valued training data [8,11]. But what makes one instantiation more plausible than another one? One approach originally proposed in [9] refers to the principle of simplicity in the spirit of *Occam's razor* (which can be given a theoretical justification in terms of *Kolmogorov complexity* [14]): An instantiation that can be explained by a simple model is more plausible than an instantiation that requires a complex model. In the context of RST-based data analysis, a natural measure of model complexity is the size of the reduct. This leads us to the following definition.

Definition 4. *A set of attributes $R \subseteq Att$ is a superset reduct if there exists a consistent instantiation $\mathcal{I} = \langle U, Att, t \rangle$ such that R is a reduct for $\mathcal{I}$. We denote with RED_{super} the set of superset reducts. The minimum description length (MDL) instantiation is one of the consistent instantiations of SDT that admit a reduct of minimum size compared to all the reducts of all possible consistent instantiations. We will call the corresponding reduct MDL reduct.*

First of all, we briefly comment on the fact that the definition of MDL reduct generalizes the standard definition of (minimal) reduct. Indeed, in a classical decision table, there is only one possible instantiation, hence the MDL reduct is

---

Algorithm 1. The brute force algorithm for finding MDL reducts of a superset decision table S.

---

   procedure Brute-Force-MDL-Reduct(S: superset decision table)
      $reds \leftarrow \emptyset$
      $l \leftarrow \infty$
      $ists \leftarrow enumerate\text{-}instantiations(S)$
      for all $i \in ists$ **do**
         $tmp\text{-}reds \leftarrow find\text{-}shortest\text{-}reducts(i)$
         $len \leftarrow |red|$ where $red \in tmp\text{-}reds$
         if $len < l$ **then**
            $reds \leftarrow tmp\text{-}reds$
            $l \leftarrow len$
         else if $len = l$ **then**
            $reds \leftarrow reds \cup tmp\text{-}reds$
         end if
      end for
      return $reds$ ▷ The MDL reducts for S
   end procedure

---

exactly (one of) the minimal reducts of the decision table. Further, if we denote by RED_{MDL} the set of MDL reducts, then evidently $RED_{MDL} \subsetneq RED_{super}$.

An algorithmic solution to the problem of finding the MDL reduct for an SDT can be given as a brute force algorithm, which computes the reducts of all the possible instantiations, see Algorithm 1. It is easy to see that the worst case runtime complexity of this algorithm is exponential in the size of the input. Unfortunately, it is unlikely that an asymptotically more efficient algorithm exists. Indeed, if we consider the problem of finding *any* MDL reduct, then the number of instantiations of S is, in the general case, exponential in the number of objects, and for each such instantiation one should find the shortest reduct for the corresponding decision table, which is known to be in Σ_2^P. Interestingly, we can prove that the decisional problem MDL-reduct related to finding MDL-Reducts is also in Σ_2^P. That is, finding an MDL-Reduct is no more complex than finding a minimal reduct in standard decision tables.

Theorem 1. *MDL-Reduct is Σ_2^P-complete.*

Proof. We need to show that there is an algorithm for verifying instances of MDL-Reduct whose runtime is polynomial given access to an oracle for an NP-complete problem. Indeed, a certificate can be given by an instantiation I (whose size is clearly polynomial in the size of the input SDT) together with a reduct R for I, which is an MDL-reduct. Verifying whether R is a minimal reduct for I can then be done in polynomial time with an oracle for NP, hence the result. Further, as finding the minimal reduct for classical decision tables is Σ_2^P-complete (by reduction to the Shortest Implicant problem), MDL-Reduct is also complete.

While heuristics could be applied to speed up the computation of reducts [27] (specifically, to reduce the complexity of the *find-shortest-reducts* step in

Algorithm 1) the approach described in Algorithm 1 still requires enumerating all the possible instantiations. Thus, in the following section we propose two alternative definitions of reduct in order to reduce the computational costs.

4 Methods

In this section, we present the main results concerning the application of rough set and evidence theory towards feature reduction in the superset learning setting.

4.1 Entropy Reducts

We begin with an alternative definition of reduct, based on the notion of entropy [24], which simplifies the complexity of finding a reduct in SDT. Given a decision d, we can associate with it a pair of belief and plausibility functions. Let $v \in V_t$ and $[x]_B$ for $B \subseteq Att$ an equivalence class, then:

$$Bel_S(v|[x]_B) = \frac{|\{x' \in [x]_B : d(x') = \{v\}\}|}{|[x]_B|}$$

$$Pl_S(v|[x]_B) = \frac{|\{x' \in [x]_B : v \in d(x')\}|}{|[x]_B|}$$

For each $W \subseteq V_t$, the corresponding basic belief assignment is defined as

$$m(W|[x]_B) = \frac{|\{x' \in [x]_B : d(x') = W\}|}{|[x]_B|}. \tag{7}$$

Given this setting, we now consider two different entropies. The first one is the pignistic entropy $H_{Bet}(m)$ as defined in (5). As regards the second definition, we will not directly employ the AU measure (see Eq. (4)). This measure, in fact, corresponds to a quantification of the degree of conflict in the bba m, which is not appropriate in our context, as it would imply finding an instantiation which is maximally inconsistent. We thus define a modification of the AU measure that we call *Optimistic Aggregate Uncertainty* (OAU). This measure, which has already been studied in the context of superset decision tree learning [9] and soft clustering [3], is defined as follows:

$$OAU(SDT) = \min_{I \in \mathcal{I}(SDT)} H(p(I)), \tag{8}$$

where $p(I)$ is the probability distribution over the decision attribute induced by the instantiation $I \in \mathcal{I}$.

Let $B \subseteq Att$ be a set of attributes and denote by $IND_B = \{[x]_B\}$ the equivalence classes (granules) with respect to B. Let $d_{[x]_B}$ be the restriction of d on the equivalence class $[x]_B$. The entropy of d, conditional on B, is defined as

$$H_{Bet}(d|B) = \sum_{[x]_B \in IND_B} \frac{|[x]_B|}{|U|} H_{Bet}(d_{[x]_B}) = \sum_{[x]_B \in IND_B} \frac{|[x]_B|}{|U|} \frac{H(P_{Bet}^m(d_{[x]_B}))}{H(\hat{p}_m(d_{[x]_B}))} \tag{9}$$

$$OAU(d|B) = \sum_{[x]_B \in IND_B} \frac{|[x]_B|}{|U|} OAU(d_{[x]_B}) \tag{10}$$

Definition 5. *We say that $B \subseteq Att$ is*

- *an OAU super-reduct (resp., H_{Bet} super-reduct) if $OAU(d\,|\,B) \leq OAU(d\,|\,Att)$ (resp., $H_{Bet}(d\,|\,B) \leq H_{Bet}(d\,|\,Att)$);*
- *an OAU reduct (resp., H_{Bet} reduct) if no proper subset of B is also a super-reduct.*

Definition 6. *We say that $B \subseteq Att$ is*

- *an OAU ϵ-approximate super-reduct (resp., H_{Bet} ϵ-approximate super-reduct), with $\epsilon \in [0,1)$, if $OAU(d|B) \leq OAU(d\,|\,Att) - log_2(1-\epsilon)$ (resp., $H_{Bet}(d\,|\,B) \leq H_{Bet}(d\,|\,Att) - log_2(1-\epsilon)$);*
- *an OAU ϵ-approximate reduct (resp., H_{Bet} ϵ-approximate reduct) if no proper subset of B is also an ϵ-approximate super-reduct.*

Let $[x]_B$ be one of the granules with respect to an OAU-reduct. Then, the *OAU instantiation* with respect to $[x]_B$ is given by

$$dec_{O\!AU(B)}([x]_B) = \arg\max_{v \in V_t} \left\{ p(v) \,|\, p = \arg\min_{p \in P_{Bel}} H(p) \right\}, \tag{11}$$

that is, the most probable among the classes under the probability distribution which corresponds to the minimum value of entropy. Similarly, the H_{Bet} *instantiation* with respect to $[x]_B$ is given by

$$dec_{H_{Bet}(B)}([x]_B) = \arg\max_{v \in V_t} Bet_{Bel}(v) \tag{12}$$

The following example shows, for a simple SDT, the OAU reducts, MDL reducts, and H_{Bet} reducts and their relationships.

Example 1. Consider the superset decision table $SDT = \langle U = \{x_1, ..., x_6\}, A = \{w, x, v, z\}, d\rangle$ given in Table 1. We have $OAU(d\,|\,A) = OAU(d\,|\,B) = 0$ for $B = \{x, v\}$. Thus, B is an OAU reduct of SDT, as $OAU(d\,|\,x) = OAU(d\,|\,v) > 0$. Notice that $\{z\}$ is also an OAU reduct. The OAU instantiation given by $\{x, v\}$ is $dec_{x,v}(\{x_1, x_2\}) = dec_{x,v}(\{x_3, x_4\}) = 0$, $dec_{x,v}(\{x_5, x_6\}) = 1$, while the one given by $\{z\}$ is $dec_z(\{x_1, x_3, x_6\}) = 0$, $dec_z(\{x_2, x_4, x_5\}) = 1$.

On the other hand, $H_{Bet}(d\,|\,A) = \frac{1}{2}$, while $H_{Bet}(d\,|\,\{x, v\}) = 0.81$. Therefore, $\{x, v\}$ is not an H_{Bet} reduct. Notice that, in this case, there are no H_{Bet} reducts (excluding A). However, it can easily be seen that $\{x, v\}$ is an H_{Bet} approximate reduct when $\epsilon \geq 0.20$.

The MDL instantiation is $dec_{MDL}(\{x_1, x_3, x_6\}) = 0$, $dec_{MDL}(\{x_2, x_4, x_5\}) = 1$, which corresponds to the MDL reduct $\{z\}$. Thus, in this case, the MDL reduct is equivalent to one of the OAU reducts.

Table 1. An example of superset decision table

	w	x	v	z	d
x_1	0	0	0	0	0
x_2	0	0	0	1	$\{0,1\}$
x_3	0	1	1	0	0
x_4	0	1	1	1	$\{0,1\}$
x_5	0	1	0	1	1
x_6	0	1	0	0	$\{0,1\}$

In Example 1, it is shown that the MDL reduct is one of the OAU reducts. Indeed, we can prove that this holds in general.

Theorem 2. *Let R be an MDL reduct whose MDL instantiation is consistent. Then R is also an OAU reduct.*

Proof. As the instantiation corresponding to R is consistent, $OAU(d\,|\,R) = 0$. Thus R is an OAU reduct.

Concerning the computational complexity of finding the minimal OAU or one OAU, we have the following results.

Proposition 1. Finding the minimal OAU reduct for a consistent SDT is Σ_2^P-complete.

Proof. As any MDL reduct of a consistent SDT is also an OAU reduct and MDL reducts are by definition minimal, the complexity of finding a minimal OAU reduct is equivalent to that of finding MDL reducts, hence is Σ_2^P-complete.

On the other hand, as both OAU [3,9] and H_{Bet} can be computed in polynomial time, the following result holds for finding OAU (resp. H_{Bet}) reducts.

Theorem 3. *Finding an OAU (resp. H_{Bet}) reduct is NP-complete.*

On the other hand, as shown in Example 1, the relationship between MDL reducts (or OAU reducts) and H_{Bet} reducts is more complex as, in general, an OAU reduct is not necessarily a H_{Bet} reduct. In particular, one could be interested in whether an H_{Bet} exists and whether there exists an H_{Bet} reduct which is able to disambiguate objects that are not disambiguated when taking in consideration the full set of attributes *Att*. The following two results provide a characterization in the binary (i.e., $V_t = \{0,1\}$), consistent case.

Theorem 4. *Let $B \subseteq Att$ be a set of attributes, $[x_1]_{Att}, [x_2]_{Att}$ be two distinct equivalence classes (i.e., $[x_1]_{Att} \cap [x_2]_{Att} = \emptyset$) that are merged by B (i.e., $[x_1]_B = [x_1]_{Att} \cup [x_2]_{Att}$), that are not inconsistent and such that $|[x_1]_{Att}| = n_1 + m_1$, $|[x_2]_{Att}| = n_2 + m_2$, where the n_1 (resp., n_2) objects are such that $|d(x)| = 1$ and the m_1 (resp., m_2) objects are such that $|d(x)| = 2$. Then $H_{Bet}(d\,|\,B) \geq H_{Bet}(d\,|\,Att)$, with equality holding iff one of the following two holds:*

1. $m_1 = m_2 = 0$ and $n_1, n_2 > 0$;
2. $m_1, m_2 > 0$ and $n_1 \geq 0$, $n_2 = \frac{m_2 n_1}{m_1}$ (and, symmetrically when changing n_1, n_2).

Proof. A sufficient and necessary condition for $H_{Bet}(d \,|\, B) \geq H_{Bet}(d \,|\, Att)$ is:

$$\frac{n_1 + \frac{m_1 + m_2}{2} + n_2}{n_1 + m_1 + n_2 + m_2} \geq \max \left\{ \frac{n_1 + \frac{m_1}{2}}{n_1 + m_1}, \frac{\frac{m_2}{2} + n_2}{n_2 + m_2} \right\} \tag{13}$$

under the constraints $n_1, n_2, m_1, m_2 \geq 0$, as the satisfaction of this inequality implies that the probability is more peaked on a single alternative. The integer solutions for this inequality provide the statement of the Theorem. Further, one can see that the strict inequality is not achievable.

Corollary 1. *A subset $B \subseteq Att$ is an H_{Bet} reduct iff, whenever it merges a pair of equivalence classes, the conditions expressed in Theorem 4 are satisfied.*

Notably, these two results also provide an answer to the second question, that is, whether an H_{Bet} reduct can disambiguate instances that are not disambiguated when considering the whole attribute set Att. Indeed, Theorem 4 provides sufficient conditions for this property and shows that, in the binary case, disambiguation is possible only when at least one of the equivalence classes (w.r.t. Att) that are merged w.r.t. the reduct is already disambiguated. On the contrary, in the general n-ary case, disambiguation could happen also in more general situations. This is shown by the following example.

Example 2. Let $SDT = \langle U = \{x_1, ..., x_{10}\}, Att = \{a, b\}, d \rangle$ such that $\forall i \leq 5$, $d(x_i) = \{0, 1\}$ and $\forall i > 5, d(x_i) = \{1, 2\}$. Then, assuming the equivalence classes are $\{x_1, ..., x_5\}, \{x_6, ..., x_{10}\}$, it holds that $H_{Bet}(d \,|\, Att) = 1$.

Suppose further that $\pi_a = \{U\}$. Then $H_{Bet}(d \,|\, a) < 0.95 < H_{Bet}(d \,|\, Att)$ and hence a is a H_{Bet} reduct. Notice that Att is not able to disambiguate since

$$dec_{H_{Bet}(Att)}([x_1]_{Att}) = \{0, 1\}$$

$$dec_{H_{Bet}(Att)}([x_6]_{Att}) = \{1, 2\}.$$

On the other hand, $dec_{H_{Bet}(a)}(x_i) = 1$ for all $x_i \in U$. Notice that, in this case, $\{a\}$ would also be an OAU reduct (and hence a MDL reduct, as it is minimal).

A characterization of H_{Bet} reducts in the n-ary case is left as future work.

Finally, we notice that, while the complexity of finding OAU (resp. H_{Bet}) reducts is still NP-complete, even in the approximate case, these definitions are more amenable to optimization through heuristics, as they employ a quantitative measure of quality for each attribute. Indeed, a simple greedy procedure can be implemented, as shown in Algorithm 2, which obviously has polynomial time complexity.

Algorithm 2. An heuristic greedy algorithm for finding approximate entropy reducts of a superset decision table S.

---

procedure HEURISTIC-ENTROPY-REDUCT(S: superset decision table, ϵ: approximation level, $E \in \{OAU, H_{Bet}\}$)
    $red \leftarrow Att$
    $Ent \leftarrow E(d \,|\, red)$
    $check \leftarrow True$
    while check **do**
        Find $a \in red$ s.t. $\begin{cases} E(d \,|\, red \setminus \{a\}) \leq E(d \,|\, Att) - log_2(1 - \epsilon) \\ E(d \,|\, red \setminus \{a\}) \text{ is minimal} \end{cases}$
        if a exists **then**
            $red \leftarrow red \setminus \{a\}$
        else
            $check \leftarrow False$
        end if
    end while
    return red
end procedure

---

5 Conclusion

In this article we investigated strategies for the simultaneous solution of the feature reduction and disambiguation problems in the superset learning setting through the application of rough set theory and evidence theory. We first defined a generalization of decision tables to this setting and then studied a purely combinatorial definition of reducts inspired by the Minimum Description Length principle, which we called MDL reducts. After studying the computational complexity of finding this type of reducts, which was shown to be NP-hard, harnessing the natural relationship between superset learning and evidence theory, we proposed two alternative definitions of reducts, based on the notion of entropy. We then provided a characterization for both these notions in terms of their relationship with MDL reducts, their existence conditions and their disambiguation power. Finally, after having illustrated the proposed notions by means of examples, we suggested a simple heuristic algorithm for computing approximate entropy reducts under the two proposed definitions.

    While this paper provides a first investigation towards the application of RST for feature reduction in the superset learning setting, it leaves several interesting open problems to be investigated in future work:

- In Theorem 2, we proved that (in the consistent case) $RED_{MDL} \subset RED_{OAU}$, that is, every MDL reduct is also an OAU reduct. In particular, the MDL reducts are the minimal OAU reducts. As $RED_{MDL} \subsetneq RED_{super}$, the relationship between the OAU reducts and the superset reducts should be investigated in more depth. Specifically we conjecture the following:

Conjecture 1. For each SDT, $RED_{super} = R_{OAU}$.

While the inclusion $RED_{super} \subseteq RED_{OAU}$ is easy to prove in the consistent case, the general case should also be considered.

- In Theorem 4, we provided a characterization of H_{Bet} reducts in the binary consistent case, however, the behavior of this type of reducts should also be investigated in the more general setting, specifically with respect to the relationship between RED_{OAU} and $RED_{H_{Bet}}$.
- Given the practical importance of the superset learning setting, an implementation of the presented ideas and algorithms should be developed, in order to provide a computational framework for the application of the rough set methodology also to these tasks, in particular with respect to the implementation of algorithms (both exact or heuristic) for finding MDL or entropy reducts.

In closing, we would like to highlight an alternative motivation for the superset extension of decision tables in general and the search for reducts of such tables in particular. In this paper, the superset extension was motivated by the assumption of imprecise labeling: The value of the decision attribute is not known precisely but only characterized in terms of a set of possible candidates. Finding a reduct is then supposed to help disambiguate the data, i.e., figuring out the most plausible among the candidates. Instead of this "don't know" interpretation, a superset S can also be given a "don't care" interpretation: In a certain context characterized by x, all decisions in S are sufficiently good, or "satisficing" in the sense of March and Simon [17]. A reduct can then be considered as a maximally simple (least cognitively demanding) yet satisficing decision rule. Thus, in spite of very different interpretations, the theoretical problems that arise are essentially the same as those studied in this paper. Nevertheless, elaborating on the idea of reduction as a means for specifically finding satisficing decision rules from a more practical point of view is another interesting direction for future work.

References

1. Arora, S., Barak, B.: Computational Complexity: A Modern Approach. Cambridge University Press, Cambridge (2009)
2. Bello, R., Falcon, R.: Rough sets in machine learning: a review. In: Wang, G., Skowron, A., Yao, Y., Ślęzak, D., Polkowski, L. (eds.) Thriving Rough Sets. SCI, vol. 708, pp. 87–118. Springer, Cham (2017). https://doi.org/10.1007/978-3-319-54966-8_5
3. Campagner, A., Ciucci, D.: Orthopartitions and soft clustering: soft mutual information measures for clustering validation. Knowl.-Based Syst. **180**, 51–61 (2019)
4. Chang, J.C., Amershi, S., Kamar, E.: Revolt: collaborative crowdsourcing for labeling machine learning datasets. In: Proceedings of CHI 2017, pp. 2334–2346 (2017)
5. Dempster, A.P.: Upper and lower probabilities induced by a multivalued mapping. In: Yager, R.R., Liu, L. (eds.) Classic Works of the Dempster-Shafer Theory of Belief Functions, vol. 219, pp. 57–72. Springer, Heidelberg (2008). https://doi.org/10.1007/978-3-540-44792-4_3

6. Guyon, I., Elisseeff, A.: An introduction to variable and feature selection. J. Mach. Learn. Res. **3**(Mar), 1157–1182 (2003)
7. Harmanec, D., Klir, G.J.: Measuring total uncertainty in Dempster-Shafer theory: a novel approach. Int. J. Gen. Syst. **22**(4), 405–419 (1994)
8. Hüllermeier, E.: Learning from imprecise and fuzzy observations: data disambiguation through generalized loss minimization. Int. J. Approximate Reason. **55**(7), 1519–1534 (2014)
9. Hüllermeier, E., Beringer, J.: Learning from ambiguously labeled examples. Intell. Data Anal. **10**(5), 419–439 (2006)
10. Hüllermeier, E., Cheng, W.: Superset learning based on generalized loss minimization. In: Appice, A., Rodrigues, P.P., Santos Costa, V., Gama, J., Jorge, A., Soares, C. (eds.) ECML PKDD 2015. LNCS (LNAI), vol. 9285, pp. 260–275. Springer, Cham (2015). https://doi.org/10.1007/978-3-319-23525-7_16
11. Hüllermeier, E., Destercke, S., Couso, I.: Learning from imprecise data: adjustments of optimistic and pessimistic variants. In: Ben Amor, N., Quost, B., Theobald, M. (eds.) SUM 2019. LNCS (LNAI), vol. 11940, pp. 266–279. Springer, Cham (2019). https://doi.org/10.1007/978-3-030-35514-2_20
12. Johnson, D., Levesque, S., Zhang, T.: Interactive machine learning system for automated annotation of information in text, 3 February 2005. US Patent App. 10/630,854
13. Jousselme, A.-L., Liu, C., Grenier, D., Bossé, É.: Measuring ambiguity in the evidence theory. IEEE Trans. Syst. Man Cybern.-Part A: Syst. Hum. **36**(5), 890–903 (2006)
14. Li, M., Vitányi, P., et al.: An Introduction to Kolmogorov Complexity and Its Applications, 3rd edn. Springer, Heidelberg (2008). https://doi.org/10.1007/978-0-387-49820-1
15. Liu, L., Dietterich, T.: Learnability of the superset label learning problem. In: Proceedings of ICML 2014, pp. 1629–1637 (2014)
16. Liu, L., Dietterich, T.G.: A conditional multinomial mixture model for superset label learning. In: Advances in Neural Information Processing Systems, pp. 548–556 (2012)
17. March, J.G., Simon, H.A.: Organizations. Wiley, New York (1958)
18. Nguyen, N., Caruana, R.: Classification with partial labels. In: Proceedings of the 14th ACM SIGKDD, pp. 551–559 (2008)
19. Pawlak, Z.: Rough sets. Int. J. Comput. Inf. Sci. **11**(5), 341–356 (1982)
20. Pearl, J.: Reasoning with belief functions: an analysis of compatibility. Int. J. Approximate Reason. **4**(5–6), 363–389 (1990)
21. Shafer, G.: A Mathematical Theory of Evidence. Princeton University Press, Princeton (1976)
22. Shannon, C.E.: A mathematical theory of communication. Bell Syst. Tech. J. **27**(3), 379–423 (1948)
23. Skowron, A., Rauszer, C.: The discernibility matrices and functions in information systems. In: Słowiński, R. (ed.) Intelligent Decision Support, vol. 11, pp. 331–362. Springer, Heidelberg (1992). https://doi.org/10.1007/978-94-015-7975-9_21
24. Slezak, D.: Approximate entropy reducts. Fundam. Inform. **53**(3–4), 365–390 (2002)
25. Ślęzak, D., Dutta, S.: Dynamic and discernibility characteristics of different attribute reduction criteria. In: Nguyen, H.S., Ha, Q.-T., Li, T., Przybyła-Kasperek, M. (eds.) IJCRS 2018. LNCS (LNAI), vol. 11103, pp. 628–643. Springer, Cham (2018). https://doi.org/10.1007/978-3-319-99368-3_49

26. Smets, P., Kennes, R.: The transferable belief model. Artif. Intell. **66**(2), 191–234 (1994)
27. Thangavel, K., Pethalakshmi, A.: Dimensionality reduction based on rough set theory: a review. Appl. Soft Comput. **9**(1), 1–12 (2009)
28. Umans, C.: On the complexity and inapproximability of shortest implicant problems. In: Wiedermann, J., van Emde Boas, P., Nielsen, M. (eds.) ICALP 1999. LNCS, vol. 1644, pp. 687–696. Springer, Heidelberg (1999). https://doi.org/10.1007/3-540-48523-6_65
29. Yao, Y.Y., Lingras, P.J.: Interpretations of belief functions in the theory of rough sets. Inf. Sci. **104**(1–2), 81–106 (1998)
30. Zhou, Z.-H.: A brief introduction to weakly supervised learning. Natl. Sci. Rev. **5**(1), 44–53 (2018)

Graphical Causal Models and Imputing Missing Data: A Preliminary Study

Rui Jorge Almeida[1]([✉]), Greetje Adriaans[2], and Yuliya Shapovalova[3]

[1] Department of Quantitative Economics,
Department of Data Analytics and Digitization, School of Business and Economics,
Maastricht University, Maastricht, The Netherlands
rj.almeida@maastrichtuniversity.nl
[2] Department of Hepatology and Gastroenterology,
Maastricht Universitary Medical Centrum+, Maastricht University,
Maastricht, The Netherlands
greetje.adriaans@mumc.nl
[3] Institute for Computing and Information Sciences,
Radboud University, Nijmegen, The Netherlands
Yuliya.Shapovalova@ru.nl

Abstract. Real-world datasets often contain many missing values due to several reasons. This is usually an issue since many learning algorithms require complete datasets. In certain cases, there are constraints in the real world problem that create difficulties in continuously observing all data. In this paper, we investigate if graphical causal models can be used to impute missing values and derive additional information on the uncertainty of the imputed values. Our goal is to use the information from a complete dataset in the form of graphical causal models to impute missing values in an incomplete dataset. This assumes that the datasets have the same data generating process. Furthermore, we calculate the probability of each missing data value belonging to a specified percentile. We present a preliminary study on the proposed method using synthetic data, where we can control the causal relations and missing values.

Keywords: Missing data · Graphical causal models · Uncertainty in missing values

1 Introduction

Datasets of real-world problems often contain missing values. A dataset has partial missing data if some values of a variable are not observed. Incomplete datasets pose problems in obtaining reliable results when analyzing the data. Many algorithms require a complete dataset to estimate models. On the other hand, in certain real-world problems obtaining reliable and complete data can be a tedious and costly task and can hamper the desired goal of the problem. An example is e-health. E-health tools often contain standardized forms (*i.e.* questionnaires) to capture data. Yet the questionnaires at times are lengthy and

© Springer Nature Switzerland AG 2020
M.-J. Lesot et al. (Eds.): IPMU 2020, CCIS 1237, pp. 485–496, 2020.
https://doi.org/10.1007/978-3-030-50146-4_36

this imposes a burden on the patients' time, which leads to reduced amount of patients completing questionnaires [1] causing incomplete datasets regarding e-health.

Since the introduction of the electronic health record (EHR) in Dutch clinical health care, large amounts of digital data are created on a daily basis. Furthermore, due to the emerging implementation of e-health applications in Dutch health care, large amounts of health-related data are created not only inside but also outside clinical institutions (*e.g.* hospitals). For instance, *MyIBDcoach*, is an e-health tool developed for home monitoring of disease activity for inflammatory bowel disease, a chronic disease with a relapsing-remitting disease course [9]. Results analyzing data captured in this e-health tool have shown the potential to predict disease activity. These results could potentially aid timely intervention and better health care resource allocation as the frequency of outpatient clinic visits could be scaled according to the risk of increased disease activity within a patient [10,24]. Exploring the further potential of combined data, data captured in the EHR and e-health tools, could lead to new insights by analyzing these data in a meaningful way.

In clinical studies, that use observational data, the data are often obtained by extracting information from the EHR. In addition, observational data documented in longitudinal prospective cohort studies often make use of standardized forms to register admission data of the cohort participants and to register data of certain variables during follow-up. Therefore datasets of prospective cohort studies can be considered complete. Since incomplete e-health datasets could lead to unreliable prediction results, incomplete data could, therefore, be problematic when e-health tools are used as an integral part in the care pathway [7].

In this paper we investigate if graphical causal models can be used to impute missing values. Causal discovery aims to learn the causal relations between variables of a system of interest from data. Thus it is possible to make predictions of the effects of interventions, which is important for decision making. Graphical models can represent a multivariate distribution in the form of a graph. Causal models can be represented as graphical models and represent not only the distribution of the observed data but also the distributions under interventions.

Causal inference has been applied to combine information from multiple datasets [15,16], including observational and experimental data [13,18]. Causal discovery algorithms have been adapted to deal with missing data [6]. For example, [4] presents a modification of PC algorithm [20] to be able to handle missing data, [19] and [5] present different approach to deal with mixed discrete and continuous data. We take a different perspective.

Our goal is to use the information from a complete dataset (*e.g.* cohort studies) in the form of graphical causal models to impute missing values in an incomplete dataset (*e.g.* from e-health monitoring). This assumes that these datasets represent the same population and have the same data generating process, which is implicit in setting up cohort studies. The use of causal models allows preserving causal relationships present in data, without strict assumptions of a pre-specified data generating process. Furthermore, we explore the stochastic uncertainty in imputing missing values with the proposed method. We calculate the probability of each missing data value belonging to a specified percentile. Low or high percentiles can indicate risk situations, *e.g.* existence of an active disease in e-health monitoring. In this paper we present a preliminary study using synthetic data, where we can control the causal relations and for which there is ground truth for the missing values.

2   Preliminaries

2.1   Graphical Models and Causal Discovery

A causal structure is often represented by a graphical model. A graph G is an ordered pair $<V, E>$ where V is a set of vertices, and E is a set of edges [20]. The pairs of vertices in E are unordered in an undirected graph and ordered in a directed graph. A **directed graph** G contains only directed edges as illustrated in Fig. 1(b). A **directed acyclic graph** (DAG) often represents underlying causal structures in causal discovery algorithms [17]. On the other hand, a **mixed graph** can contain more than one type of an edge between to vertices. A DAG contains only directed edges and has no directed cycles. We call the **skeleton** of a DAG an undirected graph obtained by ignoring direction of the edges in the DAG itself. See Figs. 2(a) and 2(b) for illustration. Further, if there is a directed edge from X_1 to X_2 then X_1 is called to be **parent** of X_2, and X_2 is called to be **child** of X_1. If two vertices are joined by an edge they are called to be **adjacent**. A set of parents of a vertex X_2 is denoted by $\mathrm{pa}(X_2)$, in Fig. 2(a) $\mathrm{pa}(X_2) = \{X_1\}$ while $\mathrm{pa}(X_4) = \{X_2, X_3\}$. The joint distribution implied by Fig. 2(a) implies the following conditional probability relation:

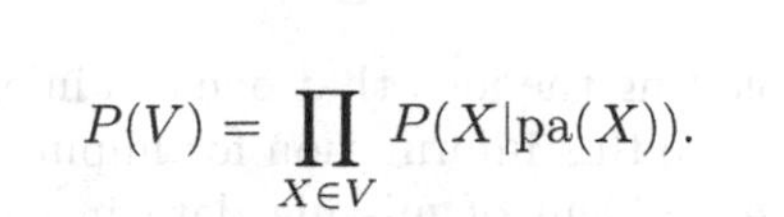

$$
\begin{array}{cc}
X_1 & X_2 \quad X_1 & X_2 \\
\circ\!\!-\!\!\!-\!\!\!-\!\!\circ & \circ\!\!-\!\!\!-\!\!\!-\!\!\rightarrow\circ \\
\text{(a) Undirected} & \text{(b) Directed}
\end{array}
$$

Fig. 1. Undirected and directed relationship between two variables

$$P(V) = \prod_{X \in V} P(X|\mathrm{pa}(X)). \tag{1}$$

Causal discovery connects the graphical theoretic approach and statistical theory. The DAG in Fig. 2(a) implies the following conditional distributions:

$$P(X|\mathrm{pa}(X)) = P\left(X \mid \bigcup_{X_j \in \mathrm{pa}(X)} \mathrm{pa}(X_j) \right), \tag{2}$$

e.g. $P(X_4|\mathrm{pa}(X_4)) = P(X_4|X_2, X_3) = P(X_4|X_1)$.

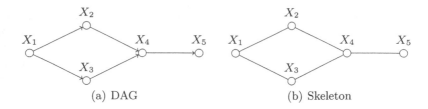

Fig. 2. Directed acyclic graph and its skeleton

A DAG encodes conditional independence relationships, which help us to reason about causality. A criteria known as **d-separation** is central in this type of inference, see for more details [20]. In particular, in any distribution P factorizing according to G, if X and Y are d-separated given Z then $X \perp\!\!\!\perp Y|Z$ in P. There are multiple algorithms that use d-separation rules to learn the graph structure; many of them are computationally intensive.

In this paper we use the PC algorithm[1] for causal discovery [20]. The idea of this algorithm is based on first forming the complete undirected graph, then removing the edges with zero-order conditional independence, then removing first-order conditional independence relations, etc. Thus, the PC algorithm heavily relies on testing conditional independence. Pearson's correlation is frequently used to test for conditional independence in the Gaussian case; other popular choices are, Spearman's rank correlation, or Kendall's tau. In addition, next to the correlation matrix, the PC algorithm requires a sample size as input. The estimate of the correlation matrix is more reliable with larger sample size, and thus we easier can reject the null hypothesis of conditional independence [5].

PC algorithm is widely applied in causal discovery algorithms and thus has been extended in various directions, including missing data cases. [3] consider causal discovery in DAGs with arbitrarily many latent and selection variables with the available R software package *pcalg* [11]. [8] use rank-based correlation and extend PC algorithm to Gaussian copula models. [4] extend this approach to mixed discrete and continuous data, while [5] further include missing data in this approach.

2.2   Graphical Models with Missing Data

In this paper we are exploiting the idea that one can infer causal structure from a cohort study and then use this information for imputing missing values in an incomplete dataset. The problem of missing data in causal inference is being studied in the literature quite extensively. [14] derive graphical conditions for recovering joint and conditional distributions and sufficient conditions for recovering causal queries. [22] consider different missingness mechanisms and present graphical representations of those. Usually, three missing mechanisms are considered in the literature [12]: missing completely at random (MCAR), missing at

---

[1] Named after its two inventors, Peter and Clark.

random (MAR), and not missing at random (NMAR). MCAR missingness mechanism imposes the least problems for statistical inference, while NMAR imposes most problems for statistical inference. It is important to note, that in our case there is no problem of identifying the type of missingness mechanism, however, it is useful to know and understand the distinction of missing mechanisms from the literature.

Similarly to [22] let us denote by D_{obs} observed part of the data and D_{mis} missing part of the data, and R the indicator matrix of missingness. The MCAR mechanism states that

$$P(R|D) = P(R|D_{obs}, D_{mis}) = P(R). \tag{3}$$

Equation (3) can be expressed in conditional independence statement as

$$R \perp\!\!\!\perp (D_{obs}, D_{mis}). \tag{4}$$

Thus, the missingness in this case is independent of both D_{obs} and D_{mis}. Further, MAR, a less restrictive mechanism, states that

$$P(R|D) = P(R|D_{obs}, D_{mis}) = P(R|D_{obs}), \tag{5}$$

where Eq. (5) can also be expressed in terms of a conditional independence statement

$$R \perp\!\!\!\perp D_{mis}|D_{obs}. \tag{6}$$

Thus, while the dependence between the observed data and missingness is allowed, the missingness R is independent of missing part of the data D_{mis} given information about the observed part of the data D_{obs}. Finally, for NMAR mechanism we have

$$P(R|D_{obs}, D_{mis}) \neq P(R|D_{obs}). \tag{7}$$

[22] propose a way to create m-graphs (graphs with missing data for all three mechanisms) and discuss graphical criteria for identification of means and regression coefficients. For us it is useful in a sense that while deciding on which parts of the data can be missing, we can impose requirement of identifiability.

3   Causal Models for Imputing Missing Data

In this paper, we propose using the causal information from a DAG, built from a complete sample, to impute missing values in another sample. The proposed method uses the causal discovery defined within a DAG and estimated relations between variables using the PC algorithm. The DAG and PC estimation provide the causal relations between the missing and observed variables. Once this relation is defined, the exact specification of causality between observed and missing values, together with the predictions of the missing values are obtained using nonparametric regressions. Nonparametric regressions are used to avoid assumptions on the specific functional relationship between variables.

As an illustration. Suppose that X_1, X_2, X_3, X_5 are observed in Fig. 2(a) while X_4 is missing. The DAG implies the following conditional probability relation:

$$P(X_4|\mathrm{pa}(X_4)) = P(X_4|X_2, X_3). \tag{8}$$

In case both X_4 and a parent, *e.g.* X_2, are missing, we use the following DAG-implied conditional probability relations to estimate the causal relationship between X_2 and X_1 in the training data, and obtain an estimate for the missing value of X_2. Impute the missing values of X_2 and X_4:

$$P(X_2|\mathrm{pa}(X_2)) = P(X_2|X_1) \tag{9}$$
$$P(X_4|\mathrm{pa}(X_4)) = P(X_4|X_2, X_3) = P(X_4|\mathrm{pa}(X_2), X_3) = P(X_4|X_1, X_3) \tag{10}$$

The iteration over parents of DAG implied conditionals continues until all conditioning variables are observed. When X_2, X_3, X_4 are all unobserved, we use the following DAG-implied conditional probability relation:

$$P(X_2|\mathrm{pa}(X_2)) = P(X_2|X_1) \tag{11}$$
$$P(X_3|\mathrm{pa}(X_3)) = P(X_3|X_1) \tag{12}$$
$$P(X_4|\mathrm{pa}(X_4)) = P(X_4|X_2, X_3) = P(X_4|\mathrm{pa}(X_2), \mathrm{pa}(X_3)) = P(X_4|X_1). \tag{13}$$

When the graph structure is more complicated than Fig. 3, the above procedure to obtain 'observed parents' of a missing value is more involved since backward iterations of $\mathrm{pa}(\cdot)$ are needed until none of the conditioned variables have missing values. To avoid this computational cost, we define the iterated parents of a missing observation. Let $X_{\mathrm{mis}} \subset \mathrm{pa}(X)$ denote the set of parents of X with missing values. The iterated parents of X, $\hat{\mathrm{pa}}(X)$ are defined as:

$$\hat{\mathrm{pa}}(X) = \begin{cases} \mathrm{pa}(X) & \text{if } X_{\mathrm{mis}} = \emptyset \\ (\mathrm{pa}(X) \setminus X_{\mathrm{mis}}) \cup X_1 & \text{otherwise,} \end{cases} \tag{14}$$

where the conditioning on variable X_1 is due to the graph structure in Fig. 3.

Given the conditional probability definitions in Eqs. (8)–(13), and the parent set definition in (14), we propose to obtain the predicted values of missing values using nonparametric regressions. For N observed data samples $X_{i,j}$ with $i = 1, \ldots, p$ and $j = 1, \ldots, N$, local linear regressions are estimated for each variable X_i in a training set. Each of these local linear regressions minimize the following:

$$\min_{\alpha,\beta} \sum_{n=1}^{N} (X_{i,j} - \alpha - \beta\,(\hat{\mathrm{pa}}(X_i) - \hat{\mathrm{pa}}(X)_{i,j}))^2\, K_h\,(\hat{\mathrm{pa}}(X_i) - \hat{\mathrm{pa}}(X)_{i,j}) \tag{15}$$

where $X_i = (X_{i,1}, \ldots, X_{i,N})'$ is the vector of observations from variable X_i, $\mathrm{pa}(X_i) = (\mathrm{pa}(X)_{i,1}, \ldots, \mathrm{pa}(X)_{i,N})$ and $\mathrm{pa}(X)_{i,j}$ denotes the jth observation from parents of X_i. In addition, $K_h\,(\mathrm{pa}(X_i) - \mathrm{pa}(X)_{i,j})$ is defined as a Gaussian kernel with $h = 1$, but the proposed methodology is applicable to other kernel specifications or similar nonparametric regression methods.

The imputation method we propose is based on estimating (15) for a DAG based on complete data, and predicting the missing values in an incomplete dataset. This imputation, denoted by $\hat{X}_{i,j}$ for variable i in observation j is calculated using the local linear regression results:

$$\hat{X}_{i,j} = \hat{\alpha} + \hat{\beta}\left(\hat{\text{pa}}(X_i) + \hat{\text{pa}}(X)_{i,j}\right), \tag{16}$$

where $\hat{\alpha}$ and $\hat{\beta}$ are obtained according to the minimization in (15). In addition, the Gaussian kernel defined for (15) implies local normality for all predicted values. We use this property to quantify the uncertainty of the imputed value in (16). Given a normal distribution $X_{i,j} \sim N(\hat{X}_{i,j}, \hat{\sigma}_{i,j}^2)$, we calculate the probability of $X_{i,j}$ belonging to a pre-specified percentile range $[p_1, p_2]$ as:

$$pr(p_1 < X_{i,j} \leq p_2) = \int_{p_1}^{p_2} \phi\left(X_{i,j}; \hat{X}_{i,j}, \hat{\sigma}_{i,j}^2\right) dX_{i,j} \tag{17}$$

where $\phi\left(x; \mu, \sigma^2\right)$ denotes the probability density function with mean μ, variance σ^2 evaluated at point x and $\hat{\sigma}_{i,j}^2$ is estimated as the variance of the regression errors. Please note that in this preliminary study, we ignore uncertainty when estimating the model parameters α and β.

4   Simulation Results for Imputing Missing Data

We illustrate the performance of the proposed method using a DAG with eight variables. The random graph is defined for 8 variables with conditional Gaussian distributions and the probability of connecting a node to another node with higher topological ordering is set as 0.3, following [3]. The true DAG and the estimated DAG are presented in Fig. 3. The structure of this DAG implies that variables 2, 3, 5, 6, 7 and 8 can be explained by parent variables or variable 1. Variable 4, on the other hand, is completely exogenous in this graph. Hence our methodology cannot be used to impute missing values of variable 4.

We simulate 5000 training observations and estimate the DAG using these training data. The estimated DAG is presented in the right panel of Fig. 3. Given the test data with 2000 observations, we create 9 incomplete test datasets with randomly missing values (MCAR) for 6 variables that have parents in the map, *i.e.* variables 2, 3, 5, 6, 7 and 8. These 9 incomplete test datasets differ in the probability of missing observations $q = 10\%, 20\%, \ldots, 90\%$. Each observation can have none, one or more missing variables, hence the total number of missing observations in each incomplete test dataset is 2000 or less, while the expected number of missing variables is $q \times 2000 \times 6$.

For each incomplete training dataset, we use the methodology in Sect. 3 to impute the missing values. We compare our method to other baseline models, namely replacing missing values by the sample average of the variable in the test data, excluding missing values; the MissForest method, a non-parametric missing value imputation based on random forests [21]; and MICE a multivariate imputation method based on fully conditional specification [23], as implemented

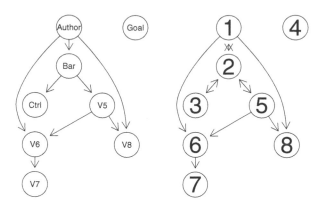

Fig. 3. True (left) and estimated (right) DAG for simulated data

in [2]. The mean squared errors of the proposed model and the baseline models are reported in Table 1. For missing values above $q = 40\%$, the proposed method performs better than all other models. For values below $q = 40\%$, the best performing model is MissForest, although the results appear to be comparable. The proposed model performance, measured by the MSE in Table 1, decreases with increasing q. This result is expected as the number of missing values for each observation increase with q. Since this increase implies that within an observation, it is more likely that the parents of a missing variable are also unobserved, hence there is an additional loss of information in the causal relations.

Table 1. MSE results from the proposed method and baseline models

	10%	20%	30%	40%	50%	60%	70%	80%	90%
Mean	1.47	1.49	1.56	1.51	1.48	1.52	1.50	1.51	1.52
DAG	0.92	0.99	1.03	**1.06**	**1.10**	**1.15**	**1.19**	**1.26**	**1.32**
MissForest	**0.91**	**0.96**	**1.02**	1.09	1.26	1.38	1.49	2.05	1.50
MICE	1.72	1.78	1.86	1.90	2.02	2.11	2.29	2.43	2.83

In addition to the overall results in Table 1, we present the errors for each variable for $q = 10\%$ and $q = 90\%$ in Fig. 4. For a small percentage of missing values, $q = 10\%$, the ranges of errors are clearly smaller in the proposed method compared to the mean baseline model. The MissForest model has some observations with a larger absolute error compared to the proposed method. Note that the variable-specific errors present the cases where the causal relations, hence the imputations are relatively less accurate. For variables 2 and 3, which have a single parent and a short link to variable 1, the obtained errors are relatively small in absolute values. Other variables, such as 6 and 8, have multiple parents, thus a higher probability of missing values in parents. When the missing parent information is replaced with the value of variable 1, some information is lost and the estimates will be less accurate. Figure 4 shows that this inaccuracy occurs especially for $q = 90\%$ where the probability of missing observations, hence the probability of missing parent information is high.

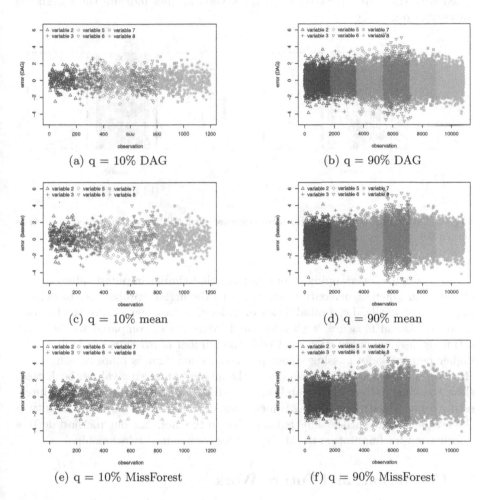

(a) q = 10% DAG   (b) q = 90% DAG

(c) q = 10% mean   (d) q = 90% mean

(e) q = 10% MissForest   (f) q = 90% MissForest

Fig. 4. Errors per variable from imputed missing values using the proposed method (DAG), the mean and MissForest baseline models.

Finally, we illustrate the uncertainty in the missing values, quantified using the imputed values. For each variable, we set four pre-defined percentiles of 0–10%, 10–50%, 50–90% and 90–100%, corresponding to the empirical percentiles of the training data. We then calculate percentile probabilities for missing value by applying Eq. 17 for the four pre-defined percentiles. Based on these percentile probabilities, the percentile with the highest probability is selected as the estimated percentile. In Fig. 5, we present the imputed data values and percentile estimates for variable 2 for two missing value probabilities, $q = 10\%$ and $q = 90\%$. For readability, we only present observations for which the estimated and true percentiles are different. In addition, estimated percentiles are indicated with the respective colors and thick vertical lines indicate the thresholds for correct percentiles.

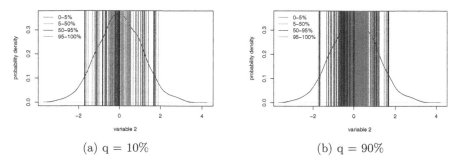

(a) q = 10%   (b) q = 90%

Fig. 5. Estimated percentiles for observations with different estimated and true percentiles

In this figure, overlapping colors indicate that similar imputed values can be classified in different percentiles according to the highest probability of belonging to a percentile. I.e. probabilities of belonging to a percentile can be used as an additional measure, with additional information, compared to the point estimates used as imputation. In Fig. 5, the number of overlaps are higher for a higher percentage of missing values $p = 90\%$, since there is more missing data. However, it appears that irrespective of the amount of missing values, both cases show the same pattern of overlap between estimated percentiles. This is an interesting result, since more missing values mean do not indicate more uncertainty in the estimated percentiles. This is likely due to the fact that our method derives information for imputation of missing values from causal relationships.

5   Conclusions and Future Work

In this paper we investigate if graphical causal models derived from complete datasets can be used to impute missing values in an incomplete dataset, assuming the same data generating process. We calculate the probability of each missing data value belonging to a specified percentile, to provide information on the uncertainty of the imputed values. We apply this methodology using synthetic data, where we can control the causal relations and missing values. We show that the proposed method performs better than a baseline model of imputing missing values by the mean in different simulation settings with different percentages of missing data. Furthermore, our model can still provide adequate information on missing values for very high percentages of missing values. Our results show that this methodology can be used in inputting missing values while providing information about the probability distribution of percentiles the missing value belongs to.

This is a preliminary study which opens many questions. In the future we want to investigate how to incorporate information on bidirectional causal relationships, different non-parametric models for imputing missing values and the relationship of this method with fully conditional specification.

References

1. Blankers, M., Koeter, M.W., Schippers, G.M.: Missing data approaches in eHealth research: simulation study and a tutorial for nonmathematically inclined researchers. J. Med. Internet Res. **12**(5), e54 (2010)
2. Buuren, S.V., Groothuis-Oudshoorn, K.: MICE: multivariate imputation by chained equations in R. J. Stat. Softw. **45**(3), 1–67 (2011)
3. Colombo, D., Maathuis, M.H., Kalisch, M., Richardson, T.S.: Learning high-dimensional directed acyclic graphs with latent and selection variables. Ann. Stat. **40**, 294–321 (2012)
4. Cui, R., Groot, P., Heskes, T.: Copula PC algorithm for causal discovery from mixed data. In: Frasconi, P., Landwehr, N., Manco, G., Vreeken, J. (eds.) ECML PKDD 2016. LNCS (LNAI), vol. 9852, pp. 377–392. Springer, Cham (2016). https://doi.org/10.1007/978-3-319-46227-1_24
5. Cui, R., Groot, P., Heskes, T.: Learning causal structure from mixed data with missing values using Gaussian copula models. Stat. Comput. **29**(2), 311–333 (2018). https://doi.org/10.1007/s11222-018-9810-x
6. Ding, P., Li, F., et al.: Causal inference: a missing data perspective. Stat. Sci. **33**(2), 214–237 (2018)
7. Gorelick, M.H.: Bias arising from missing data in predictive models. J. Clin. Epidemiol. **59**(10), 1115–1123 (2006)
8. Harris, N., Drton, M.: PC algorithm for nonparanormal graphical models. J. Mach. Learn. Res. **14**(1), 3365–3383 (2013)
9. de Jong, M., et al.: Development and feasibility study of a telemedicine tool for all patients with IBD: MyIBDcoach. Inflamm. Bowel Dis. **23**(4), 485–493 (2017)
10. de Jong, M.J., et al.: Telemedicine for management of inflammatory bowel disease (myIBDcoach): a pragmatic, multicentre, randomised controlled trial. Lancet **390**(10098), 959–968 (2017)
11. Kalisch, M., Mächler, M., Colombo, D., Maathuis, M.H., Bühlmann, P.: Causal inference using graphical models with the R package pcalg. J. Stat. Softw. **47**(11), 1–26 (2012). http://www.jstatsoft.org/v47/i11/
12. Little, R.J., Rubin, D.B.: Statistical Analysis with Missing Data, vol. 793. Wiley, Hoboken (2019)
13. Magliacane, S., Claassen, T., Mooij, J.M.: Joint causal inference on observational and experimental datasets. arXiv preprint arXiv:1611.10351 (2016)
14. Mohan, K., Pearl, J.: Graphical models for recovering probabilistic and causal queries from missing data. In: Advances in Neural Information Processing Systems, pp. 1520–1528 (2014)
15. Mooij, J., Heskes, T.: Cyclic causal discovery from continuous equilibrium data (2013). arXiv preprint arXiv:1309.6849

16. Mooij, J.M., Magliacane, S., Claassen, T.: Joint causal inference from multiple contexts. arXiv preprint arXiv:1611.10351 (2016)
17. Pearl, J., Verma, T.S.: A statistical semantics for causation. Stat. Comput. **2**(2), 91–95 (1992). https://doi.org/10.1007/BF01889587
18. Rau, A., Jaffrézic, F., Nuel, G.: Joint estimation of causal effects from observational and intervention gene expression data. BMC Syst. Biol. **7**(1), 111 (2013)
19. Sokolova, E., Groot, P., Claassen, T., von Rhein, D., Buitelaar, J., Heskes, T.: Causal discovery from medical data: dealing with missing values and a mixture of discrete and continuous data. In: Holmes, J.H., Bellazzi, R., Sacchi, L., Peek, N. (eds.) AIME 2015. LNCS (LNAI), vol. 9105, pp. 177–181. Springer, Cham (2015). https://doi.org/10.1007/978-3-319-19551-3_23
20. Spirtes, P., Glymour, C.N., Scheines, R., Heckerman, D.: Causation, Prediction, and Search. MIT Press, Cambridge (2000)
21. Stekhoven, D.J., Bühlmann, P.: Missforest-non-parametric missing value imputation for mixed-type data. Bioinformatics **28**(1), 112–118 (2012)
22. Thoemmes, F., Mohan, K.: Graphical representation of missing data problems. Struct. Eq. Model.: Multidiscip. J. **22**(4), 631–642 (2015)
23. Van Buuren, S.: Multiple imputation of discrete and continuous data by fully conditional specification. Stat. Methods Med. Res. **16**(3), 219–242 (2007)
24. Wintjens, D.S., et al.: Novel perceived stress and life events precede flares of inflammatory bowel disease: a prospective 12-month follow-up study. J. Crohn's Colitis **13**(4), 410–416 (2019)

Machine Learning II

Competitive Online Quantile Regression

Raisa Dzhamtyrova[1]([✉]) and Yuri Kalnishkan[1,2]

[1] Computer Science Department, Royal Holloway, University of London, Egham, UK
raisa.dzhamtyrova.2015@live.rhul.ac.uk
[2] Laboratory of Advanced Combinatorics and Network Applications,
Moscow Institute of Physics and Technology, Moscow, Russia

Abstract. Interval prediction often provides more useful information compared to a simple point forecast. For example, in renewable energy forecasting, while the initial focus has been on deterministic predictions, the uncertainty observed in energy generation raises an interest in producing probabilistic forecasts. One aims to provide prediction intervals so that outcomes lie in the interval with a given probability. Therefore, the problem of estimating the quantiles of a variable arises. The contribution of our paper is two-fold. First, we propose to apply the framework of prediction with expert advice for the prediction of quantiles. Second, we propose a new competitive online algorithm Weak Aggregating Algorithm for Quantile Regression (WAAQR) and prove a theoretical bound on the cumulative loss of the proposed strategy. The theoretical bound ensures that WAAQR is asymptotically as good as any quantile regression. In addition, we provide an empirical survey where we apply both methods to the problem of probability forecasting of wind and solar powers and show that they provide good results compared to other predictive models.

Keywords: Prediction with expert advice · Online learning · Sequential prediction · Weak Aggregating Algorithm · Quantile regression · Probabilistic forecasting

1 Introduction

Probabilistic forecasting attracts an increasing attention in sports, finance, weather and energy fields. While an initial focus has been on deterministic forecasting, probabilistic prediction provides a more useful information which is essential for optimal planning and management in these fields. Probabilistic forecasts serve to quantify the uncertainty in a prediction, and they are an essential ingredient of optimal decision making [4]. An overview of the state of the art methods and scoring rules in probabilistic forecasting can be found in [4]. Quantile regression is one of the methods which models a quantile of the response variable conditional on the explanatory variables [6].

Due to its ability to provide interval predictions, quantile regression found its niche in the renewable energy forecasting area. Wind power is one of the fastest growing renewable energy sources [3]. As there is no efficient way to store wind power, producing accurate wind power forecasts are essential for reliable operation of wind turbines.

© Springer Nature Switzerland AG 2020
M.-J. Lesot et al. (Eds.): IPMU 2020, CCIS 1237, pp. 499–512, 2020.
https://doi.org/10.1007/978-3-030-50146-4_37

Due to the uncertainty in wind power generation, there have been studies for improving the reliability of power forecasts to ensure the balance between supply and demand at electricity market. Quantile regression has been extensively used to produce wind power quantile forecasts, using a variety of explanatory variables such as wind speed, temperature and atmospheric pressure [7].

The Global Energy Forecasting Competition 2014 showed that combining predictions of several regressors can produce better results compared to a single model. It is shown in [9] that a voted ensemble of several quantile predictors could produce good results in probabilistic solar and wind power forecasting. In [1] the analogue ensemble technique is applied for prediction of solar power which slightly outperforms the quantile regression model.

In this paper we apply a different approach to combine predictions of several models based on the method of online prediction with expert advice. Contrary to batch mode, where the algorithm is trained on training set and gives predictions on test set, in online setting we learn as soon as new observations become available. One may wonder why not to use predictions of only one best expert from the beginning and ignore predictions of others. First, sometimes we cannot have enough data to identify the best expert from the start. Second, good performance in the past does not necessary lead to a good performance in the future. In addition, previous research shows that combining predictions of multiple regressors often produce better results compared to a single model [11].

We consider the adversarial setting, where no stochastic assumptions are made about the data generating process. Our approach is based on Weak Aggregating Algorithm (WAA) which was first introduced in [5]. The WAA works as follows: we assign initial weights to experts and at each step the weights of experts are updated according to their performance. The approach is similar to the Bayesian method, where the prediction is the average over all models based on the likelihood of the available data. The WAA gives a guarantee ensuring that the learner's loss is as small as best expert's loss up to an additive term of the form $C\sqrt{T}$, where T is the number of steps and C is some constant. It is possible to apply WAA to combine predictions of an infinite pool of experts. In [8] WAA was applied to the multi-period, distribution-free perishable inventory problem, and it was shown that the asymptotic average performance of the proposed method was as good as any time-dependent stocking rule up to an additive term of the form $C\sqrt{T}\ln T$.

The WAA was proposed as an alternative to the Aggregating Algorithm (AA), which was first introduced in [12]. The AA gives a guarantee ensuring that the learner's loss is as small as best expert's loss up to a constant in case of finitely many experts. The AA provides better theoretical guarantees, however it works with mixable loss functions, and it is not applicable in our task. An interesting application of the method of prediction with expert advice for the Brier loss function in forecasting of football outcomes can be found in [14]; it was shown that the proposed strategy that follows AA is as good as any bookmaker. Aggregating Algorithm for Regression (AAR) which competes with any expert from an infinite pool of linear regressions under the square loss was proposed in [13].

The contribution of our paper is two-fold. First, as a proof of concept, we apply WAA to a finite pool of experts to show that this method is applicable for this problem.

As our experts we pick several models that provide quantile forecasts and then combine their predictions using WAA. To the best of our knowledge prediction with expert advice was not applied before for the prediction of quantiles. Second, we propose a new competitive online algorithm Weak Aggregating Algorithm for Quantile Regression (WAAQR), which is as good as any quantile regression up to an additive term of the form $C\sqrt{T}\ln T$. For this purpose, we apply WAA to an infinite pool of quantile regressions. While the bound for the finite case can be straightforwardly applied to finite or countable sets of experts, every case of a continuous pool needs to be dealt with separately. We listed above a few results for different specific pools of experts, however there is no generic procedure for deriving a theoretical bound for the cumulative loss of the algorithm. WAAQR can be implemented by using Markov chain Monte Carlo (MCMC) method in a way which is similar to the algorithm introduced in [15], where AAR was applied to generalised linear regression class of function for making a prediction in a fixed interval. We derive a theoretical bound on the cumulative loss of our algorithm which is approximate (in the number of MCMC steps). MCMC is only a method for evaluating the integral and it can be replaced by a different numerical method. Theoretical convergence of the Metropolis-Hastings method in this case follows from Theorems 1 and 3 in [10]. Estimating the convergence speed is more difficult. With the experiments provided we show that by tuning parameters online, our algorithm moves fast to the area of high values of the probability function and gives a good approximation of the prediction.

We apply both methods to the problem of probabilistic forecasting of wind and solar power. Experimental results show a good performance of both methods. WAA applied to a finite set of models performs close or better than the retrospectively best model, whereas WAAQR outperforms the best quantile regression model that was trained on the historical data.

2   Framework

In the framework of prediction with expert advice we need to specify a *game* which contains three components: a space of outcomes Ω, a decision space Γ, and a loss function $\lambda : \Omega \times \Gamma \to \mathbb{R}$. We consider a game with the space of outcomes $\Omega = [A, B]$ and decision space $\Gamma = \mathbb{R}$, and as a loss function we take the pinball loss for $q \in (0, 1)$

$$\lambda(y, \gamma) = \begin{cases} q(y - \gamma), & \text{if } y \geq \gamma \\ (1 - q)(\gamma - y), & \text{if } y < \gamma \end{cases}. \tag{1}$$

This loss function is appropriate for quantile regression because on average it is minimized by the q-th quantile. Namely, if Y is a real-valued random variable with a cumulative distribution function $F_Y(x) = \Pr(Y \leq x)$, then the expectation $\mathbb{E}\lambda(Y, \gamma)$ is minimized by $\gamma = \inf\{x : F_Y(x) \geq q\}$ (see Sect. 1.3 in [6] for a discussion).

In many tasks predicted outcomes are bounded. For example, wind and solar power cannot reach infinity. Therefore, it is possible to have a sensible estimate for the outcome space Ω based on the historical information.

Learner works according to the following protocol:

Protocol 1

for t = 1, 2, ...
    nature announces signal $x_t \subseteq \mathbb{R}^n$
    learner outputs prediction $\gamma_t \in \Gamma$
    nature announces outcome $y_t \in \Omega$
    learner suffers loss $\lambda(y_t, \gamma_t)$
end for

The cumulative loss of the learner at the step T is:

$$L_T := \sum_{\substack{t=1,...,T: \\ y_t < \gamma_t}} (1-q)|y_t - \gamma_t| + \sum_{\substack{t=1,...,T: \\ y_t > \gamma_t}} q|y_t - \gamma_t|. \tag{2}$$

We want to find a strategy which is capable of competing in terms of cumulative loss with all prediction strategies $\mathcal{E}_\theta$, $\theta \in \mathbb{R}^n$ (called *experts*) from a given pool, which output $\xi_t(\theta)$ at step t. In a finite case we denote experts $\mathcal{E}_i$, $i = 1, \ldots, N$.

Let us denote L_T^θ the cumulative loss of expert $\mathcal{E}_\theta$ at the step T:

$$L_T^\theta := \sum_{\substack{t=1,...,T: \\ y_t < \xi_t(\theta)}} (1-q)|y_t - \xi_t(\theta)| + \sum_{\substack{t=1,...,T: \\ y_t > \xi_t(\theta)}} q|y_t - \xi_t(\theta)|. \tag{3}$$

3   Weak Aggregating Algorithm

In the framework of prediction with expert advice we have access to experts' predictions at each time step and the learner has to make a prediction based on experts' past performance. We use an approach based on the WAA since a pinball loss function $\lambda(y, \gamma)$ is convex in γ. The WAA maintains experts' weights $P_t(d\theta)$, $t = 1, \ldots, T$. After each step t the WAA updates the weights of the experts according to their losses:

$$P_t(d\theta) = \exp\left(-\frac{cL_{t-1}^\theta}{\sqrt{t}}\right) P_0(d\theta), \tag{4}$$

where $P_0(d\theta)$ is the initial weights of experts and c is a positive parameter.

Experts that suffer large losses will have smaller weights and less influence on futher predictions.

The prediction of WAA is a weighted average of the experts' predictions:

$$\gamma_t = \int_\Theta \xi_t(\theta) P_{t-1}^*(d\theta), \tag{5}$$

where $P_{t-1}^*(d\theta)$ are normalized weights:

$$P_{t-1}^*(d\theta) = \frac{P_{t-1}(d\theta)}{P_{t-1}(\Theta)},$$

where Θ is a *parameter space*, i.e. $\theta \in \Theta$.

In a finite case, an integral in (5) is replaced by a weighted sum of experts' predictions $\xi_t(i)$, $i = 1, \dots, N$.

In particular, when there are finitely many experts $\mathcal{E}_i$, $i = 1, \dots, N$ for bounded games the following lemma holds.

Lemma 1 (Lemma 11 in [5]). *For every $L > 0$, every game $\langle \Omega, \Gamma, \lambda \rangle$ such that $|\Omega| < +\infty$ with $\lambda(y, \gamma) \leq L$ for all $y \in \Omega$ and $\gamma \in \Gamma$ and every $N = 1, 2, \dots$ for every merging strategy for N experts that follows the WAA with initial weights $p_1, p_2, \dots, p_N \in [0, 1]$ such that $\sum_{i=1}^{N} p_i = 1$ and $c > 0$ the bound*

$$L_T \leq L_T^i + \sqrt{T} \left(\frac{1}{c} \ln \frac{1}{p_i} + cL^2 \right),$$

is guaranteed for every $T = 1, 2, \dots$ and every $i = 1, 2, \dots, N$.

After taking equal initial weights $p_1 = p_2 = \dots = p_N = 1/N$ in the WAA, the additive term reduces to $(cL^2 + (\ln N)/c)\sqrt{T}$. When $c = \sqrt{\ln N}/L$, this expression reaches its minimum. The following corollary shows that the WAA allows us to obtain additive terms of the form $C\sqrt{T}$.

Corollary 1 (Corollary 14 in [5]). *Under the conditions of Lemma 1, there is a merging strategy such that the bound*

$$L_T \leq L_T^i + 2L\sqrt{T \ln N}$$

is guaranteed.

Applying Lemma 1 for an infinite number of experts and taking a positive constant $c = 1$, we get the following Lemma.

Lemma 2 (Lemma 2 in [8]). *Let $\lambda(y, \gamma) \leq L$ for all $y \in \Omega$ and $\gamma \in \Gamma$. The WAA guarantees that, for all T*

$$L_T \leq \sqrt{T} \left(-\ln \int_\Theta \exp \left(-\frac{L_T^\theta}{\sqrt{T}} \right) P_0(d\theta) + L^2 \right).$$

4   Theoretical Bounds for WAAQR

In this section we formulate the theoretical bounds of our algorithm.

We want to find a strategy which is capable of competing in terms of cumulative loss with all prediction strategies $\mathcal{E}_\theta$, $\theta \in \Theta = \mathbb{R}^n$, which at step t output:

$$\xi_t(\theta) = x_t'\theta, \tag{6}$$

where x_t is a signal at time t. The cumulative loss of expert $\mathcal{E}_\theta$ is defined in (3).

Theorem 1. *Let $a > 0$, $y \in \Omega = [A, B]$ and $\gamma \in \Gamma$. There exists a prediction strategy for Learner such that for every positive integer T, every sequence of outcomes of length T, and every $\theta \in \mathbb{R}^n$ with initial distribution of parameters*

$$P_0(d\theta) = \left(\frac{a}{2}\right)^n e^{-a\|\theta\|_1} d\theta, \tag{7}$$

the cumulative loss L_T of Learner satisfies

$$L_T \leq L_T^\theta + \sqrt{T}a\|\theta\|_1 + \sqrt{T}\left(n \ln\left(1 + \frac{\sqrt{T}}{a} \max_{t=1,\dots,T}\|x_t\|_\infty\right) + (B - A)^2\right).$$

The theorem states that the algorithm predicts as well as the best quantile regression, defined in (6), up to an additive regret of the order $\sqrt{T} \ln T$. The choice of the regularisation parameter a is important as it affects the behaviour of the theoretical bound of our algorithm. Large parameters of regularisation increase the bound by an additive term $\sqrt{T}a\|\theta\|_1$, however the regret term has a smaller growth rate as time increases. As the maximum time T is usually not known in advance, the regularisation parameter a cannot be optimised, and its choice depends on the particular task. We discuss the choice of the parameter a in Sect. 6.2.

Proof. We consider that outcomes come from the interval $[A, B]$, and it is known in advance. Let us define the truncated expert $\tilde{\mathcal{E}}_\theta$ which at step t outputs:

$$\tilde{\xi}_t(\theta) = \begin{cases} A, & \text{if } x_t'\theta < A \\ x_t'\theta, & \text{if } A \leq x_t'\theta \leq B \\ B, & \text{if } x_t'\theta > B \end{cases}. \tag{8}$$

Let us denote $\tilde{L}_T^\theta$ the cumulative loss of expert $\tilde{\mathcal{E}}_\theta$ at the step T:

$$\tilde{L}_T^\theta := \sum_{t=1}^T \lambda(y_t, \tilde{\xi}_t(\theta)). \tag{9}$$

We apply WAA for truncated experts $\tilde{\mathcal{E}}_\theta$. As experts $\tilde{\mathcal{E}}_\theta$ output predictions inside the interval $[A, B]$, and predictions of WAA is a weighted average of experts' predictions (5), then each γ_t lies in the interval $[A, B]$.

We can bound the maximum loss at each time step:

$$L := \max_{y \in [A,B], \, \gamma \in [A,B]} \lambda(y, \gamma) \leq (B - A)\max(q, 1 - q) \leq B - A. \tag{10}$$

Applying Lemma 2 for initial distribution (7) and putting the bound on the loss in (10) we obtain:

$$L_T \leq \sqrt{T}\left(-\ln\left(\left(\frac{a}{2}\right)^n \int_{\mathbb{R}^n} e^{-\tilde{J}(\theta)} d\theta\right) + (B - A)^2\right), \tag{11}$$

where

$$\tilde{J}(\theta) := \frac{\tilde{L}_T^\theta}{\sqrt{T}} + a\|\theta\|_1. \tag{12}$$

For all $\theta, \theta_0 \in \mathbb{R}^n$ we have:

$$\sum_{\substack{t=1,\ldots,T: \\ y_t < x_t'\theta}} |x_t'\theta - y_t| \leq \sum_{\substack{t=1,\ldots,T: \\ y_t < x_t'\theta}} |x_t'\theta_0 - y_t| + \sum_{\substack{t=1,\ldots,T: \\ y_t < x_t'\theta}} |x_t'\theta - x_t'\theta_0| \tag{13}$$

$$\leq \sum_{\substack{t=1,\ldots,T: \\ y_t < x_t'\theta}} |x_t'\theta_0 - y_t| + \sum_{\substack{t=1,\ldots,T: \\ y_t < x_t'\theta}} \max_{t=1,\ldots,T} \|x_t\|_\infty \|\theta - \theta_0\|_1$$

$$\leq \sum_{\substack{t=1,\ldots,T: \\ y_t < x_t'\theta}} |x_t'\theta_0 - y_t| + T \max_{t=1,\ldots,T} \|x_t\|_\infty \|\theta - \theta_0\|_1.$$

Analogously, we have:

$$\sum_{\substack{t=1,\ldots,T: \\ y_t > x_t'\theta}} |x_t'\theta - y_t| \leq \sum_{\substack{t=1,\ldots,T: \\ y_t > x_t'\theta}} |x_t'\theta_0 - y_t| + T \max_{t=1,\ldots,T} \|x_t\|_\infty \|\theta - \theta_0\|_1. \tag{14}$$

By multiplying inequality (13) by $(1-q)$, inequality (14) by q and summing them, we have:

$$L_T^\theta \leq L_T^{\theta_0} + T \max_{t=1,\ldots,T} \|x_t\|_\infty \|\theta - \theta_0\|_1. \tag{15}$$

The cumulative loss of truncated expert $\tilde{\mathcal{E}}_\theta$ cannot exceed the cumulative loss of non-truncated expert $\mathcal{E}_\theta$ for all $\theta \in \mathbb{R}^n$:

$$\tilde{L}_T^\theta \leq L_T^\theta.$$

By dividing (15) by $\sqrt{T}$ and adding $a\|\theta\|_1$ to both parts, we have:

$$\tilde{J}(\theta) \leq J(\theta) \leq J(\theta_0) + \sqrt{T} \max_{t=1,\ldots,T} \|x_t\|_\infty \|\theta - \theta_0\|_1 + a(\|\theta\|_1 - \|\theta_0\|_1)$$

$$\leq J(\theta_0) + (\sqrt{T} \max_{t=1,\ldots,T} \|x_t\|_\infty + a)\|\theta - \theta_0\|_1,$$

where

$$J(\theta) := \frac{L_T^\theta}{\sqrt{T}} + a\|\theta\|_1.$$

Let us denote $b_T = \sqrt{T} \max_{t=1,\ldots,T} \|x_t\|_\infty + a$. We evaluate the integral:

$$\int_{\mathbb{R}^n} e^{-\tilde{J}(\theta)} d\theta \geq \int_{\mathbb{R}^n} e^{-(J(\theta_0)+b_T\|\theta-\theta_0\|_1)} d\theta$$

$$= e^{-J(\theta_0)} \int_{\mathbb{R}} \cdots \int_{\mathbb{R}} e^{-b_T \sum_{i=1}^n |\theta_i - \theta_{i,0}|} d\theta_i$$

$$= e^{-J(\theta_0)} \int_{\mathbb{R}} \cdots \int_{\mathbb{R}} \prod_{i=1}^n e^{-b_T|\theta_i - \theta_{i,0}|} d\theta_i$$

$$= e^{-J(\theta_0)} \prod_{i=1}^n \int_{\mathbb{R}} e^{-b_T|\theta_i - \theta_{i,0}|} d\theta_i = e^{-J(\theta_0)} \left(\frac{2}{b_T}\right)^n.$$

By putting this expression in (11) we obtain the theoretical bound.

Note that even though we apply WAA for truncated experts (8), we achieve the theoretical bound for prediction strategy that competes with a class of experts (6).

5   Prediction Strategy

A prediction of WAA (5) can be re-written as follows:

$$\gamma_T = \int_\Theta \tilde{\xi}_T(\theta) w^*_{T-1}(\theta) d\theta, \tag{16}$$

where

$$w^*_T(\theta) = Z w_T(\theta) = Z \exp\left(-\frac{1}{\sqrt{T}} \left(\sum_{\substack{t=1,\dots,T: \\ y_t < \tilde{\xi}_t(\theta)}} (1-q)|y_t - \tilde{\xi}_t(\theta)| \right.\right. \tag{17}$$

$$\left.\left. + \sum_{\substack{t=1,\dots,T: \\ y_t > \tilde{\xi}_t(\theta)}} q|y_t - \tilde{\xi}_t(\theta)| \right) - a\|\theta\|_1 \right).$$

and Z is the normalising constant ensuring that $\int_\Theta w^*_T(\theta) d\theta = 1$.

Integral (16) is a Bayesian mixture, where function $\xi_T(\theta)$ needs to be integrated with respect to the normalized distribution $w^*_T(\theta)$. It is possible to avoid the calculation of normalising constant Z as it is a computationally inefficient operation, and integrate function $\xi_T(\theta)$ from the unnormalized distribution $w_T(\theta)$. In order to calculate the integral (16), it is possible to use MCMC algorithms. A good introduction of MCMC for Machine Learning is in [2].

We will use Metropolis-Hastings algorithm for sampling parameters θ from the posterior distribution $\mathcal{P}$. As a proposal distribution we choose Gaussian distribution $\mathcal{N}(0, \sigma^2)$ with some chosen parameter σ. We start with some initial parameter θ^0 and at each step m we update:

$$\theta^m = \theta^{m-1} + \mathcal{N}(0, \sigma^2), \ m = 1, \dots, M,$$

where M is a maximum number of iterations in MCMC method.

The update parameter θ^m at step m is accepted with probability $\min\left(1, \frac{f_\mathcal{P}(\theta^m)}{f_\mathcal{P}(\theta^{m-1})}\right)$, where $f_\mathcal{P}(\theta)$ is the density function for the distribution $\mathcal{P}$ at point θ. At each step by accepting and rejecting the updates of parameters θ we move closer to the maximum of the density function. At the beginning it is common to use a 'burn-in' stage when the integral is not calculated till we will reach the area of high values of the density function $f_\mathcal{P}$. Thus, we perform integration only from the area with high density of $\mathcal{P}$. Some values of θ are accepted even when the calculated probability is less than 1, it allows the algorithm to move away from local minimum of the density function. Because we are interested only in the ratio of density functions of generated parameters, we can generate new parameters θ from the unnormalized posterior distribution $w_T(\theta)$ and avoid the weights normalization at each step which is more computationally efficient.

At time $t = 0$ the algorithm starts with the initial estimate of the parameters $\theta_0 = 0$. At each iteration $t > 0$ we start with parameter θ_{t-1}^M calculated at the previous step $t - 1$. It allows the algorithm to converge faster to the correct location of the main mass of the distribution.

WAAQR

Parameters: number $M > 0$ of MCMC iterations,

      standard deviation $\sigma > 0$,

      regularization coefficient $a > 0$

initialize $\theta_0^M := 0 \in \mathbb{R}^n$

define $w_0(\theta) := \exp(-a\|\theta\|_1)$

for $t = 1, 2, \dots$ do

 $\gamma_t := 0$

 define $w_t(\theta)$ by (17)

 read $x_t \in \mathbb{R}^n$

 initialize $\theta_t^0 = \theta_{t-1}^M$

 for $m = 1, 2, \dots, M$ do

  $\theta^* := \theta_t^{m-1} + \mathcal{N}(0, \sigma^2 I)$

  flip a coin with success probability

   $\min\left(1, w_{t-1}(\theta^*)/w_{t-1}(\theta_t^{m-1})\right)$

  if success then

   $\theta_t^m := \theta^*$

  else

   $\theta_t^m := \theta_{t-1}^m$

  end if

  $\gamma_t := \gamma_t + \tilde{\xi}_t(\theta_t^m)$

 end for

 output predictions $\gamma_t = \gamma_t/M$

end for

---

6 Experiments

In this section we apply WAA and WAAQR for prediction of wind and solar power and compare their performance with other predictive models. The data set is downloaded from Open Power System Data which provides free and open data platform for power system modelling. The platform contains hourly measurements of geographically aggregated weather data across Europe and time-series of wind and solar power. Our training data are measurements in Austria from January to December 2015, test set contains data from January to July 2016.[1]

---

[1] The code written in R is available at https://github.com/RaisaDZ/Quantile-Regression.

6.1   WAA

We apply WAA for three models: Quantile Regression (QR), Quantile Random Forests (QRF), Gradient Boosting Decision Trees (GBDT). These models were used in GEF-Com 2014 energy forecasting competition on the final leaderboard [9]. In this paper the authors argue that using multiple regressors is often better than using only one, and therefore combine multiple model outputs. They noted that voting was found to be particularly useful for averaging the quantile forecasts of different models.

We propose an alternative approach to combine different models' predictions by using WAA. We work according to Protocol 1: at each step t before seeing outcome y_t, we output our prediction γ_t according to (5). After observing outcome y_t, we update experts' weights according to (4).

To build models for wind power forecasting we use wind speed and temperature as explanatory variables. These variables have been extensively used to produce wind power quantile forecasts [7]. We train three models QR, QRF and GBDT on training data set, and then apply WAA using forecasts of these models on test data set. We start with equal initial weights of each model and then update their weights according to their current performance. We estimate the constant of WAA $c = 0.01$ using information about maximum losses on training set.

Figure 1 shows weights of each model for different quantiles depending on the current time step. We can see from the graph that for most of quantiles GBDT obtains the largest weights which indicates that it suffers smaller losses compared to other models. However, it changes for $q = 0.95$, where the largest weights are acquired by QR. It shows that sometimes we can not use the past information to evaluate the best model. The retrospectively best model can perform worse in the future as an underlying nature of data generating can change. In addition, different models can perform better on different quantiles.

Table 1 illustrates total losses of QR, QRF, GBDT, WAA and Average methods, where Average is a simple average of QR, QRF and GBDT. For the prediction of wind power, for $q = 0.25$ and $q = 0.50$ the total loss of WAA is slightly higher than the total loss of GBDT, whereas for $q = 0.75$ and $q = 0.95$ WAA has the smallest loss. In most cases, WAA outperforms Average method.

We perform similar experiments for prediction of solar power. We choose measurements of direct and diffuse radiations to be our explanatory variables. In a similar way, QR, QRF and GBDT are trained on training set, and WAA is applied on test data. Figure 2 illustrates weights of models depending on the current step. Opposite to the previous experiments, GBDT has smaller weights compared to other models for $q = 0.25$ and $q = 0.50$. However, for $q = 0.75$ and $q = 0.95$ weights of experts become very close to each other. Therefore, predictions of WAA should become close to Average method. Table 1 shows total losses of the methods. For $q = 0.25$ and $q = 0.5$ both QR and QRF have small losses compared to GBDT, and WAA follows their predictions. However, for $q = 0.75$ and $q = 0.95$ it is not clear which model performs better, and predictions of WAA almost coincide with Average method. It again illustrates that the retrospectively best model could change with time, and one should be cautious about choosing the single retrospectively best model for future forecasts.

Table 1. Total losses ($\times 10^3$)

	wind						solar				
q	QRF	GBDT	QR	Average	WAA	q	QRF	GBDT	QR	Average	WAA
0.25	538.5	**491.2**	516.6	500.3	493.0	0.25	**48.6**	98.3	53.1	63.8	50.1
0.5	757.0	**707.5**	730.7	714.0	709.0	0.5	70.5	110.7	**68.8**	79.1	69.2
0.75	668.3	610.7	633.9	616.6	**610.1**	0.75	63.5	67.6	59.3	58.7	**58.0**
0.95	270.5	222.1	217.5	216.0	**211.0**	0.95	29.2	26.1	23.2	21.0	**20.8**

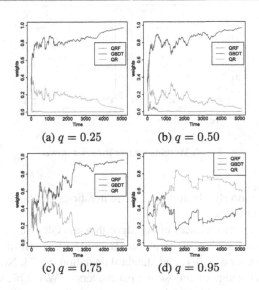

(a) $q = 0.25$ (b) $q = 0.50$

(c) $q = 0.75$ (d) $q = 0.95$

Fig. 1. Weights update for wind power

6.2 WAAQR

In this section we demonstrate the performance of our algorithm for prediction of wind power and compare it with quantile regression model. We train QR on training data set, and apply WAAQR on test set. First, we use training set to choose the parameters of our algorithm. Table 2 illustrates the acceptance ratio of new sampling parameters of our algorithm for $q = 0.5$. Increasing values of σ results in decreasing acceptance ratios of new sampling parameters θ. With large values of σ we move faster to the area of high values of density function while smaller values of σ can lead to more expensive computations as our algorithm would require more iterations to find the optimal parameters. Figure 3 illustrates logarithm of parameters likelihood $w(\theta)$ defined in (17) for $a = 0.1$ and $\sigma = 0.5$ and 3.0. We can see from the graphs that for $\sigma = 3.0$ the algorithm reaches maximum value of log-likelihood after around 800 iterations while for $\sigma = 0.5$ it still tries to find maximum value after 1500 iterations. Table 2 shows the total losses of WAAQR for different parameters a and σ. We can see that choosing the right parameters is very important as it notably affects the performance of WAAQR. It is important to keep track of acceptance ratio of the algorithm, as high acceptance ratio

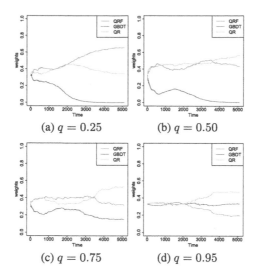

(a) $q = 0.25$ (b) $q = 0.50$

(c) $q = 0.75$ (d) $q = 0.95$

Fig. 2. Weights update for solar power

means that we move too slowly and need more iterations and larger 'burn-in' period to find the optimal parameters.

Now we compare performances of our algorithm and QR. We choose the parameters of WAAQR to be the number of iterations $M = 1500$, 'burn-in' stage $M_0 = 300$, regularization parameter $a = 0.1$, and standard deviation $\sigma = 3$. Note that even though we use the prior knowledge to choose the parameters of WAAQR, we start with initial $\theta_0 = 0$ and train our algorithm only on the test set. Figure 4 illustrates a difference between cumulative losses of QR and WAAQR. If the difference is greater than zero, our algorithm shows better results compared to QR. For $q = 0.25$ WAAQR shows better performance at the beginning, but after around 1000 iterations its performance becomes worse, and by the end of the period cumulative losses of QR and WAAQR are almost the same. We observe a different picture for $q = 0.5$ and $q = 0.75$: most of the time a difference between cumulative losses is positive, which indicates that WAAQR performs better than QR.

Figure 5 shows predictions of WAAQR and QR with $[25\%, 75\%]$ confidence interval for the first and last 100 steps. We can see from the graph, that initially predictions of WAAQR are very different from predictions of QR. However, by the end of period, predictions of both methods become very close to each other.

One of the disadvantages of WAAQR is that it might perform much worse with non-optimal input parameters of regularization a and standard deviation σ. If no prior knowledge is available, one can start with some reasonable values of input parameters and keep track of the acceptance ratio of new generated θ. If the acceptance ratio is too high it might indicate that the algorithm moves too slowly to the area of high values of the probability function of θ, and standard deviation σ should be increased. Another option is to take very large number of steps and larger 'burn-in' period.

Table 2. Acceptance ratio (AR) and total losses of WAAQR on training set

	AR					Loss			
$a \setminus \sigma$	0.5	1.0	2.0	3.0	$a \setminus \sigma$	0.5	1.0	2.0	3.0
0.1	0.533	0.550	0.482	0.375	0.1	1821.8	823.5	216.3	28.8
0.3	0.554	0.545	0.516	0.371	0.3	1806.2	844.9	265.3	62.7
0.5	0.549	0.542	0.510	0.352	0.5	1815.7	878.5	272.7	92.1
1.0	0.548	0.538	0.502	0.343	1.0	1810.4	877.5	379.3	116.9

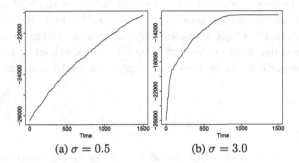

(a) $\sigma = 0.5$ (b) $\sigma = 3.0$

Fig. 3. Log-likelihood of parameters for $a = 0.1$.

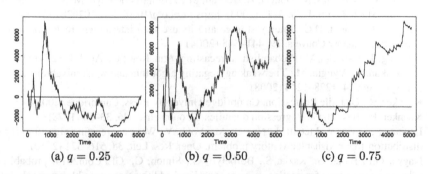

(a) $q = 0.25$ (b) $q = 0.50$ (c) $q = 0.75$

Fig. 4. Cumulative loss difference between QR and WAAQR

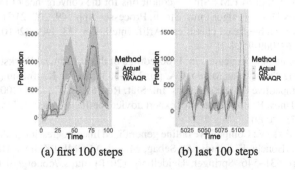

(a) first 100 steps   (b) last 100 steps

Fig. 5. Predictions with $[25\%, 75\%]$ confidence interval for WAAQR and QR

7 Conclusions

We proposed two ways of applying the framework of prediction with expert advice to the problem of probabilistic forecasting of renewable energy. The first approach is to apply WAA with a finite number of models and combine their predictions by updating weights of each model online based on their performance. Experimental results show that WAA performs close or better than the best model in terms of cumulative pinball loss function. It also outperforms the simple average of predictions of models. With this approach we show that it is reasonable to apply WAA for the prediction of quantiles.

Second, we propose a new competitive online algorithm WAAQR which combines predictions of an infinite pool of quantile regressions. We derive the theoretical bound which guarantees that WAAQR asymptotically performs as well as any quantile regression up to an additive term of the form $C\sqrt{T}\ln T$. Experimental results show that WAAQR can outperform the best quantile regression model that was trained on the historical data.

References

1. Alessandrini, S., Delle Monache, L., Sperati, S., Cervone, G.: An analog ensemble for short-term probabilistic solar power forecast. Appl. Energy **157**, 95–110 (2015)
2. Andrieu, C., de Freitas, N., Doucet, A., Jordan, M.I.: An introduction to MCMC for machine learning. Mach. Learn. J. **50**, 5–43 (2003). https://doi.org/10.1023/A:1020281327116
3. Barton, J.P., Infield, D.G.: Energy storage and its use with intermittent renewable energy. IEEE Trans. Energy Convers. **19**, 441–448 (2004)
4. Gneiting, T., Katzfuss, M.: Probabilistic forecasting. Ann. Rev. Stat. Appl. **1**, 125–151 (2014)
5. Kalnishkan, Y., Vyugin, M.: The weak aggregating algorithm and weak mixability. J. Comput. Syst. Sci. **74**, 1228–1244 (2008)
6. Koenker, R.: Quantile Regression. Cambridge University Press, Cambridge (2005)
7. Koenker, R., Bassett, G.: Regression quantiles. Econometrica **46**, 33–50 (1978)
8. Levina, T., Levin, Y., McGill, J., Nediak, M., Vovk, V.: Weak aggregating algorithm for the distribution-free perishableinventory problem. Oper. Res. Lett. **38**, 516–521 (2010)
9. Nagya, G.I., Barta, G., Kazia, S., Borbelyb, G., Simon, G.: GEFCom2014: probabilistic solar and wind power forecasting using ageneralized additive tree ensemble approach. Int. J. Forecast. **32**, 1087–1093 (2016)
10. Roberts, G.O., Smith, A.F.M.: Simple conditions for the convergence of the Gibbs sampler and Metropolis-Hastings algorithms. Stoch. Processes Appl. **49**, 207–216 (1994)
11. Rokach, L.: Ensemble-based classifiers. Artif. Intell. Rev. **33**, 1–39 (2010). https://doi.org/10.1007/s10462-009-9124-7
12. Vovk, V.: Aggregating strategies. In: Proceedings of the 3rd Annual Workshop on Computational Learning Theory, San Mateo, CA, pp. 371–383. Morgan Kaufmann (1990)
13. Vovk, V.: Competitive on-line statistics. Int. Stat. Rev. **69**(2), 213–248 (2001)
14. Vovk, V., Zhdanov, F.: Prediction with expert advice for the Brier game. J. Mach. Learn. Res. **10**, 2445–2471 (2009)
15. Zhdanov, F., Vovk, V.: Competitive online generalized linear regression under square loss. In: Balcázar, J.L., Bonchi, F., Gionis, A., Sebag, M. (eds.) ECML PKDD 2010. LNCS (LNAI), vol. 6323, pp. 531–546. Springer, Heidelberg (2010). https://doi.org/10.1007/978-3-642-15939-8_34

On the Analysis of Illicit Supply Networks Using Variable State Resolution-Markov Chains

Jorge Ángel González Ordiano[1](✉) (iD), Lisa Finn[2], Anthony Winterlich[2],
Gary Moloney[2], and Steven Simske[1] (iD)

[1] Colorado State University, Fort Collins, CO, USA
{jorge.gonzalez_ordiano,steve.simske}@colostate.edu
[2] Micro Focus International, Galway, Ireland
{finn,winterlich,gary.moloney}@microfocus.com

Abstract. The trade in illicit items, such as counterfeits, not only leads to the loss of large sums of private and public revenue, but also poses a danger to individuals, undermines governments, and—in the most extreme cases—finances criminal organizations. It is estimated that in 2013 trade in illicit items accounted for 2.5% of the global commerce. To combat illicit trade, it is necessary to understand its illicit supply networks. Therefore, we present in this article an approach that is able to find an optimal description of an illicit supply network using a series of Variable State Resolution-Markov Chains. The new method is applied to a real-world dataset stemming from the Global Product Authentication Service of Micro Focus International. The results show how an illicit supply network might be analyzed with the help of this method.

Keywords: Data mining · Markov Chain · Illicit trade

1 Introduction

Illicit trade is defined as the trade in illegal goods and services that have a negative impact on our economies, societies, and environments [12]. Two of the most prevalent forms of illicit trade are counterfeiting and piracy, whose negative effects have been studied by both the OECD and the ICC. The former estimates that in 2013 counterfeiting and piracy accounted for 2.5% of all world imports [10], while the latter assesses that by 2022 counterfeiting and piracy will drain 4.2 trillion dollars from the world economy and put 5.4 million jobs at risk. [1] The consequences of illicit trade go beyond the loss of public and private revenue. Counterfeit medicines, for instance, have caused a large number of malaria and tuberculosis related deaths [6], while counterfeit cigarettes, cd's, etc. have been linked to terrorist organizations [1]. These examples show the danger that

---

[1] iccwbo.org/global-issues-trends/bascap-counterfeiting-piracy/, Accessed:07-17-2019.

© Springer Nature Switzerland AG 2020
M.-J. Lesot et al. (Eds.): IPMU 2020, CCIS 1237, pp. 513–527, 2020.
https://doi.org/10.1007/978-3-030-50146-4_38

illicit trade poses to our communities. Therefore, finding ways to combat this type of trade is of paramount importance.

A possibility for battling illicit trade is through the disruption of its illicit supply networks (ISNs). Different methods on how to achieve this disruption are found in literature. Many articles deal with technologies for distinguishing between licit and illicit goods, such as the works of Dégardin et al. [3], Simske et al. [15], and Meruga et al. [9]. More closely related to the present article are those in which the ISNs are investigated directly. Some examples of this type of articles are shown by Giommoni et al., [5], Magliocca et al. [7], and Triepels et al. [16]. In the first, network analysis of the heroin trafficking networks in Europe is conducted. In the second, a simulation of the response of drug traffickers to interdiction is presented. In the third, international shipping records are used to create Bayesian networks able to detect smuggling and miscoding.

The goal of this article is to identify the locations in which illicit activity is more prevalent. To achieve this goal, we make use of Markov Chains, as they are a type of model that is useful at determining the amount of time that a system (i.e. a supply network) spends on a given state (i.e. a location) [13]. The first step for creating a Markov Chain is to define what the states, or nodes, of the model will be. These states can be defined at different resolution levels, as shown for instance in [8]. In this article, the states represent possible geographic locations within a supply network; which in turn can be defined in terms of countries, regions, continents, etc. Unfortunately, the state description (i.e. the Markov Chain design) that is best at modeling a given system is not immediately clear. To address this issue, we present in this work a new method that optimizes—in terms of a user-defined cost function—the design of a Markov Chain. Notice that the models created via this new method are referred to as Variable State Resolution-Markov Chains (VSR-MCs) to denote the fact that their states are a combination of the various possible descriptions. Furthermore, a real world dataset containing spatio-temporal information of serial code authentications is used to show how the new approach can be used to analyze ISNs. This dataset stems from the Global Product Authentication Service of Micro Focus International. The VSR-MC obtained with this data is then used to compare a licit supply network to its illicit counterpart. The results of this comparison offer insight on the locations in which illicit activity is more prevalent.

The remainder of this article is organized as follows: Sect. 2 offers preliminary information on Markov Chains. Section 3 shows the new method. Section 4 describes this article's experiment. Section 5 shows and discusses the obtained results and Sect. 6 contains the conclusion and outlook of this work.

2 Preliminaries

A Markov Chain can be defined as a discrete time random process $\{X_n : n \in \mathbb{N}_0\}$ whose random variables[2] only take values within a given state space, i.e. $x_n \in$

---

[2] Note that the common notation for random variables is used herein, i.e. random variables are written in uppercase and their realizations in lowercase.

S [2]. In this section, a state space—consisting of $K \in \mathbb{N}_{>1}$ different states—is defined as $S = \{s_k : k \in [1, K]\}$. In general, a Markov Chain can be viewed as a Markov Process with discrete time and state space.

The most important property of a Markov Chain is its lack of "memory", i.e. the probability of an outcome at time $n + 1$ depends only on what happens at time n [11]. This is better described by the following equation:

$$P(X_{n+1}|X_n = x_n, \ldots, X_1 = x_1) = P(X_{n+1}|X_n = x_n). \tag{1}$$

A Markov Chain is further characterized by its transition probability matrix $\mathbf{P}_n$; a matrix defined as:

$$\mathbf{P}_n = \begin{bmatrix} p_{11} & \cdots & p_{1K} \\ \vdots & \ddots & \vdots \\ p_{K1} & \cdots & p_{KK} \end{bmatrix}_n, \tag{2}$$

with the entries $p_{ij,n}$ representing the probability of transitioning from state s_i to state s_j at time n, i.e. $P(X_{n+1} = s_j|X_n = s_i)$. If the transition probabilities are independent of n (i.e. $\mathbf{P}_n = \mathbf{P}$), the Markov chain is called time homogeneous [11].

Additionally, the probabilities of X_0 being equal to each one of the states can be written in vector form as follows:

$$\boldsymbol{\pi}_0 = [P(X_0 = s_1), \cdots, P(X_0 = s_K)]^T = [\pi_{01}, \cdots, \pi_{0K}]^T, \tag{3}$$

where $\boldsymbol{\pi}_0$ is the start probability vector and π_{0k} is the probability of X_0 being equal to s_k.

Based on Eq. (1), (2), and (3), the probability of a sequence of events in a time homogeneous Markov Chain can be calculated as a multiplication of a start probability and the corresponding p_{ij} values [14]. For instance, the probability of the sequence $\{X_0 = s_1, X_1 = s_3, X_2 = s_2\}$ is given as:

$$P(X_0 = s_1, X_1 = s_3, X_2 = s_2) = \pi_{01} \cdot p_{13} \cdot p_{32}. \tag{4}$$

Interested readers are referred to [2] and [11] for more information on Markov Chains.

3   Variable State Resolution-Markov Chain

The method presented herein offers a novel alternative on how to optimize the design of a Variable State Resolution-Markov Chain (VSR-MC). The main difference between a traditional Markov Chain and a VSR-MC is the way in which the state space is defined. This difference stems from the fact that a state can be defined at different resolution scales, which are referred in this article as scales of connectivity. For example, a geographic location within a supply network can be described at a country or at a continent scale. Based on this idea, we define the state space of a VSR-MC as:

$$\begin{aligned} S &= \{\Phi_G(s_k) : k \in [1, K]\} \\ &= \{s_{G,k'} : k' \in [1, K_G]\} \text{ , with} \\ G &= \{g_{lr} : l \in [1, L], r \in [1, R_l]\}, \end{aligned} \tag{5}$$

where s_k represents the states, G is a set containing the groups (i.e. g_{lr}) in which the states can be clustered, $L \in \mathbb{N}_{>0}$ is the number of scales of connectivity, and $R_l \in \mathbb{N}_{>1}$ is the number of groups within the lth scale. Furthermore, $\Phi_G(s_k)$ is a function that defines $K_G \in \mathbb{N}_{>1}$ new states, which are referred to as $s_{G,k'}$. In other words, $\Phi_G(s_k)$ is defined as follows:

$$\Phi_G(s_k) = \begin{cases} s_k & \text{, if } s_k \notin G \\ g_{lr} : s_k \in g_{lr} & \text{, else .} \end{cases} \tag{6}$$

When defining G, it is important to consider that each s_k can only be **contained in either one or none of the groups within the set**. For the sake of illustration, Fig. 1 shows an example of possible high resolution states and their corresponding groups at different scales of connectivity.

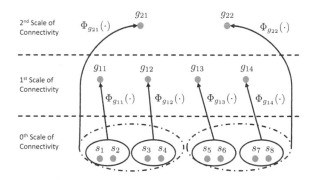

Fig. 1. Example of various states and their corresponding groups

Based on all previous aspects, it is clear that we can create different Markov Chains based on the combination of different states and groups. For instance, consider a case in which four states (i.e. s_1, s_2, s_3, and s_4) can be aggregated in two groups (i.e. $g_{11} = \{s_1, s_2\}$ and $g_{12} = \{s_3, s_4\}$). As shown in Table 1, the possible combinations result in four VSR-MCs with different scales of connectivity.

In general, the number of combinations (i.e. group sets G) that can be obtained with L scales of connectivity is given by the next equation:

$$N_c = 1 + \sum_{l=1}^{L} (l+1)^{R_l} - l^{R_l}, \tag{7}$$

Table 1. Possible state spaces of the Variable State Resolution-Markov Chains with four states s_1, s_2, s_3, and s_4, one scale of connectivity, and two groups $g_{11} = \{s_1, s_2\}$ and $g_{12} = \{s_3, s_4\}$. As given by Eq. (7), the number of possible group sets equals four.

G	$\{\}$	$\{g_{11}\}$	$\{g_{12}\}$	$\{g_{11}, g_{12}\}$
S	$\{s_1, s_2, s_3, s_4\}$	$\{g_{11}, s_3, s_4\}$	$\{s_1, s_2, g_{12}\}$	$\{g_{11}, g_{12}\}$

where N_c is the number of all possible combinations and R_l is again the number of groups within each scale.

After defining the group sets G, the probabilities of the group set-dependent transition matrices $(\mathbf{P}_G)$ and start probability vectors $(\boldsymbol{\pi}_{G,0})$ are calculated. These probabilities are obtained using a dataset containing N sequences of events of the system we want to model. In this article the sequences are described as:

$$\mathbf{x}_m = \{x_{mn} : n \in [0, N_m]\}, \tag{8}$$

in which $\mathbf{x}_m$ is the mth sequence within the dataset, $N_m \in \mathbb{N}_{>0}$ is a value that defines the sequence length, and x_{mn} is one of the realizations forming the sequence.

As mentioned at the beginning, the main goal is to find the scale of connectivity that will optimize the Markov chain architecture. In other words, we are interested in finding the VSR-MC that minimizes a problem-specific cost function $c(\cdot)$. This optimization problem can be described in general as:

$$G_{\text{opt}} = \underset{G}{\text{argmin}}\ c(G, S, \mathbf{P}_G, \boldsymbol{\pi}_{G,0}, \cdots), \tag{9}$$

where G_{opt} represents the optimal group set.

4 Experimental Study

4.1 Data

The dataset used comes from the Global Product Authentication Service (GPAS) of Micro Focus International. GPAS protects products in the marketplace by embedding a URL and unique serial number into a QR code placed on each product. The consumer is encouraged to scan the QR code which can authenticate their purchase in real-time. This dataset contains therefore spatio-temporal information of licit and illicit activity. To be more specific, it contains the authentication results (i.e. "True" or "False") of 1,725,075 unique serial codes.[3] In addition to the authentication, the dataset contains the geographic position (i.e. latitude and longitude) and the time at which each serial code was authenticated. Since many codes have been authenticated several times at different times and places, we assume that a reconstruction of the supply network is possible.

---

[3] The serial codes correspond to five different products. In this article, however, they are not separated by their product type, but are rather investigated as a single group.

In the present article, we are interested in analyzing licit and illicit serial codes that are authenticated a similar number of times at different geographic locations. Henceforth, the data is preprocessed as follows. First, all entries with missing geographic information, as well as all serial codes that do not change their position are removed from the dataset (i.e. 1,659,726). Afterwards, codes whose authentication result is sometimes "True" and sometimes "False" are also eliminated (i.e. 5,453). Note that the serial codes that have been removed are still of interest, as they can be used in the future for other type of analysis. For instance, serial codes that do not change position could be used to identify hot spots of serial code harvesting, while serial codes that change their authentication can be used to analyze locations in which the original licit codes might have been copied. As mentioned earlier, the serial codes we are considering here are the ones authenticated at different locations. We do this because we are interested in discovering the network architecture, and by inference the distribution channels, of the illicit actors. Finally, serial codes authenticated first and last at the exact same position, as well as those authenticated in more geographic positions than 99% of all serial codes are deleted (i.e. 3,897). The goals of this final step are the removal of serial codes that are suspect of being demos and the elimination of copied serial codes authenticated a huge number of times (i.e. with a clearly different behavior than licit serial codes).

The resulting dataset contains 55,999 unique serial codes, of which 31,989 are authenticated as "True", while 24,010 are authenticated as "False".

4.2   Description

The goal of this experiment is to find a VSR-MC able to accurately describe a licit and an illicit supply network. To do so, we create a series of Markov Chains with computed probabilities of state-state transitions for both licit and illicit serial codes. Then, we select the one that is best at classifying illicit activity as the one with the optimal scale of connectivity. To solve this classification problem and to obtain representative results, we create three different training/test set pairs, by randomly selecting—three separate times—50% of the unique serial codes as training set and the rest as test set.

We begin the experiment by defining three different ways in which the location of a serial code can be described, i.e. country, region, or continent. These descriptions are the scales of connectivity of the VSR-MC we are looking to create. Using the given geographic positions, we can easily determine the countries and continents where the serial codes were authenticated. The regions, in contrast, are calculated using a clustering algorithm, i.e. the affinity propagation algorithm [4]. This algorithm clusters the countries of a specific continent based on a similarity measure. The similarity measure we use here is the geographic proximity between the countries' centroids. For the sake of illustration, Table 2 shows the three scales of connectivity used in this article.

As Eq. (7) shows, the regional and continental descriptions can be used to create a staggering number of possible combinations; whose individual testing would

Table 2. Scales of connectivity

Continents	Regions	Countries
Africa	Africa 1	Djibouti, Egypt, Ethiopia, Kenya, Sudan
	Africa 2	Angola, Botswana, Mozambique, Namibia, South Africa, Zambia, Zimbabwe
	Africa 3	Algeria, Libya, Morocco, Tunisia
	Africa 4	Burundi, Congo - Kinshasa, Malawi, Rwanda, South Sudan, Tanzania, Uganda
	Africa 5	Benin, Burkina Faso, Côe d'Ivoire, Ghana, Mali, Niger, Nigeria, Togo
	Africa 6	Cape Verde, Gambia, Guinea, Guinea-Bissau, Liberia, Mauritania, Senegal, Sierra Leone
	Africa 7	Cameroon, Congo - Brazzaville, Gabon
	Africa 8	Comoros, Madagascar, Mauritius, Réunion, Seychelles
Asia	Asia 1	Bangladesh, Bhutan, India, Maldives, Myanmar (Burma), Nepal, Sri Lanka
	Asia 2	Japan, South Korea
	Asia 3	Bahrain, Iran, Kuwait, Oman, Qatar, Saudi Arabia, United Arab Emirates, Yemen
	Asia 4	Cambodia, Hong Kong SAR China, Laos, Macau SAR China, Taiwan, Thailand, Vietnam
	Asia 5	Brunei, Indonesia, Malaysia, Philippines, Singapore
	Asia 6	Armenia, Azerbaijan, Georgia, Iraq, Israel, Jordan, Lebanon, Palestinian Territories, Syria, Turkey
	Asia 7	China, Mongolia, Russia
	Asia 8	Afghanistan, Kazakhstan, Kyrgyzstan, Pakistan, Tajikistan, Turkmenistan, Uzbekistan
Europe	Europe 1	Albania, Bulgaria, Cyprus, Greece, Macedonia, Malta, Montenegro, Serbia
	Europe 2	Iceland
	Europe 3	Moldova, Romania, Ukraine
	Europe 4	France, Portugal, Spain
	Europe 5	Belgium, Denmark, Germany, Ireland, Luxembourg, Netherlands, United Kingdom
	Europe 6	Åland Islands, Finland, Norway, Sweden
	Europe 7	Austria, Bosnia & Herzegovina, Croatia, Czechia, Hungary, Italy, Monaco, Slovakia, Slovenia, Switzerland
	Europe 8	Belarus, Estonia, Latvia, Lithuania, Poland
North America	North America 1	United States
	North America 2	Canada
	North America 3	Mexico
	North America 4	Bahamas, Curaçao, Dominican Republic, Haiti, Jamaica
	North America 5	Barbados, British Virgin Islands, Guadeloupe, Martinique, Puerto Rico, Sint Maarten, Trinidad & Tobago
	North America 6	Costa Rica, El Salvador, Guatemala, Honduras, Nicaragua, Panama
Oceania	Oceania 1	Australia, Fiji, New Zealand, Samoa
	Oceania 2	Northern Mariana Islands, Palau, Papua New Guinea
South America	South America 1	Bolivia, Brazil, Paraguay
	South America 2	Colombia, Ecuador, Peru
	South America 3	Argentina, Chile, Uruguay
	South America 4	French Guiana, Guyana, Venezuela

be computationally infeasible. For this reason, the next paragraphs describe a two step alternative to deal with this issue.

Step 1: We begin by finding an optimal VSR-MC using only the regions and continents. In other words, we define the regions as the Markov states s_k and the continents as the groups at the first—and only—scale of connectivity, i.e. $g_{1r} \in \{$Africa, Asia, North America, South America, Oceania$\}$. Using these groups and Eq. (5), we define a total of 64 different state spaces S.

Afterwards, we calculate for each state space and each available training set the probabilities that are necessary to construct a Markov Chain. To be more specific, for each state space two start probability vectors and two transition probability matrices are calculated, for licit and illicit serial codes, respectively. These probabilities are based on the trajectories that are described herein as a sequence of geographic positions in which a serial code has been authenticated. Based on Eq. (8), the trajectories can be described as:

$$\mathbf{x}_m^\alpha = \{x_{mn}^\alpha : n \in [0, N_m]\} : \alpha = \{\text{licit}, \text{illicit}\}, \tag{10}$$

where α indicates if the trajectory corresponds to a licit or an illicit serial code.

Notice that we assume the Markov Chains to be homogeneous. Therefore, all sequences within the training sets that have an $N_m > 1$ are divided in N_m sequences of two realizations each. With these new set of sequences as well as Eq. (5), the start probabilities can be calculated:

$$\pi_{G,0i}^\alpha = \frac{1}{M^\alpha} \sum_{m=1}^{M^\alpha} I(x_{m0}^\alpha = s_{G,i}) : \alpha = \{\text{licit}, \text{illicit}\}, \ i \in [1, K_G], \tag{11}$$

where $\pi_{G,0i}^\alpha$ is the probability of a sequence starting at state $s_{G,i}$, M^α represents the number of available sequences (licit or illicit), and $I(\cdot)$ is a function that equals one if its condition is fulfilled and equals zero otherwise. Thereafter, the elements of the transition matrices can be obtained using Bayes's rule:

$$p_{G,ij}^\alpha = \frac{1}{M^\alpha \pi_{G,0i}^\alpha} \sum_{m=1}^{M^\alpha} I(x_{m0}^\alpha = s_{G,i} \cap x_{m1}^\alpha = s_{G,j})$$
$$: \alpha = \{\text{licit}, \text{illicit}\}, \ i, j \in [1, K_G], \tag{12}$$

with $p_{G,ij}^\alpha$ being the probability of transitioning from state $s_{G,i}$ to $s_{G,j}$.

So, using the previous values we can determine—in every state space—the probability of a serial code sequence if we assume it to be licit or illicit, i.e. $P_G(\mathbf{x}_m|\alpha) : \alpha = \{\text{licit}, \text{illicit}\}$. Thereafter, we can use the resulting probabilities to classify a serial code as illicit if $P_G(\mathbf{x}_m|\text{illicit}) \geq P_G(\mathbf{x}_m|\text{licit})$.

The method described previously is used to classify all codes within the test sets. Afterwards, the classification results are evaluated using the weighted F-Score, i.e.:

$$F_{\beta,G} = (1 + \beta^2) \frac{Q_{\text{p},G} \ Q_{\text{r},G}}{\beta^2 \ Q_{\text{p},G} + Q_{\text{r},G}}, \tag{13}$$

where $Q_{p,G}$ represents the precision, $Q_{r,G}$ is the recall, and β is a parameter that defines which of the former values is weighted more strongly. The value of β is set equal to 0.5 to give precision two times more importance than recall. This is done, since we are more interested in correctly identifying illicit serial codes (precision) than we are in flagging every possible one (recall).

After finishing the evaluation on each test set, the mean value and variance of the weighted F-Score are calculated, i.e. $\overline{F}_{\beta,G}$ and $\sigma^2_{\beta,G}$, respectively. With these values, the optimization problem described in Eq. (9) can be redefined as:

$$G_{\text{opt}} = \underset{G}{\text{argmin}} \; \gamma \, (1 - \overline{F}_{\beta,G}) + (1 - \gamma) \; \sigma^2_{\beta,G} : \gamma \in [0, 1]. \tag{14}$$

Notice that in this article the parameter γ is set equal to 0.5 to give both terms of the cost function an equal weight and to make the cost function less sensitive to noise. Solving Eq. (14) results in an optimal VSR-MC whose state space S (defined by G_{opt}) might be a combination of regions and continents.

Step 2: If the number of regions within the state space S is greater than zero, we can conduct an additional experiment to test if a country level description of the regions improves our modeling of the supply network. Note that our experiment uses a forward selection to reduce the number of combinations that need to be tested. We first create new state spaces by individually separating the regions within S into their corresponding countries. For instance, if S contains 6 regions we obtain 6 new state spaces. Afterwards, we test if some of these new state spaces result in a Markov Chain with a cost (cf. Eq (14)) that is lower than the one currently consider optimal. If so, we define the VSR-MC with the lowest cost as the new optimal one and its state space as the new optimal state space S. Afterwards, we repeat the previous steps again until none of the new Markov Chains result in a better cost or until there are no more regions within the state space. The result of this process is a VSR-MC with states that could stem from all of our available scales of connectivity (i.e. countries, regions, and continents). For the sake of simplicity, we will refer to the group set that maps the individual countries to the state space of this new optimal VSR-MC also as G_{opt}. It is worth noting, that since we are not testing all possible state spaces, the solution of our method may not be the global optimum. Nevertheless, we still consider our approach of dividing one region at a time to be acceptable. There are two main reasons for this: (i) we are able to improve the overall cost function testing only a small subset of all combinations; and (ii) we are able to increase the resolution of our network description, something that may improve our understanding of how the network operates.

After finding the best VSR-MC, we use all available data to recalculate the probabilities of the transition matrices to analyze the differences between the licit and the illicit supply networks with more detail. The analysis consists in calculating the limiting distributions of the licit and illicit transition matrices. These describe the probabilities of authenticating the serial codes at the different locations if we observe our system (i.e. the supply network) over a long period of time. In other words, these values can be interpreted as estimates of the amount of time that licit or illicit serial codes will spend on the different locations.

Therefore, a comparison of licit and illicit limiting distributions will allow us to estimate the locations where we expect illicit serial codes to spend more time. The comparison is based on a relative difference that is defined in this article as:

$$\Delta\pi'_{s_{G_{\text{opt}},i}} = \frac{\pi^{\text{licit}}_{s_{G_{\text{opt}},i}} - \pi^{\text{illicit}}_{s_{G_{\text{opt}},i}}}{\pi^{\text{licit}}_{s_{G_{\text{opt}},i}}}, \tag{15}$$

where $\pi^{\text{licit}}_{s_{G_{\text{opt}},i}}$ is the licit limiting distribution value of $s_{G_{\text{opt}},i}$, $\pi^{\text{illicit}}_{s_{G_{\text{opt}},i}}$ represents the illicit limiting distribution value of $s_{G_{\text{opt}},i}$, and $\Delta\pi'_{s_{G_{\text{opt}},i}}$ is the relative difference for state $s_{G_{\text{opt}},i}$.

After estimating the relative differences, we can test the difference between our approach and a simple descriptive analysis. This test consists in comparing the $\Delta\pi'_{s_{G_{\text{opt}},i}}$ values to benchmark relative differences (BRDs) calculated using descriptive statistics. To be more specific, the BRDs are also obtained with Eq. (15), but instead of using the limiting distribution values, we use the actual percentage of true and false authentications on the given states.

5 Results and Discussion

The results obtained on the three separate test sets by the VSR-MCs with only regions and continents as scales of connectivity (cf. Sect. 4.2; Step 1) are depicted in Fig. 2.

The first thing we notice when looking at Fig. 2 is that the standard deviations are relatively small. This not only means that the results on all test sets are similar, but also that our mean estimates are quite accurate, as the standard error of the mean is directly proportional to the standard deviation. In addition, Fig. 2 also shows that the precision does not appear to change when modifying the state space; as it is consistently around 90%. In contrast, the use of different group sets divides the recall in two distinct groups with different recall values; the first between 60 and 70% and the second between 80 and 90%. The decrease in recall is caused by considering Asia as a continent instead of looking at its individual regions. In other words, individual networks between Asian regions seem to play an important role in the accurate modeling of licit and illicit supply networks. Due to the recall, the weighted F-Score is also dependent on Asia being modeled as a single state or as individual regions. This result, i.e. that the scale of connectivity affects the quality of the supply network models, supports the use of this article's method (cf. Sect. 3). Therefore, we use Eq. (14) and the obtained weighted F-Scores to determine the scale of connectivity that will best describe the licit and illicit supply networks.

According to Eq. (14), the optimal VSR-MC is the one with states representing the continents of Africa, Europe, North America, South America, and Oceania, as well as the individual regions of Asia, i.e. $G_{\text{opt}} = \{$Africa, Europe, North America, South America, Oceania$\}$. This result shows again the importance that Asia appears to play in the accurate modeling of the supply networks.

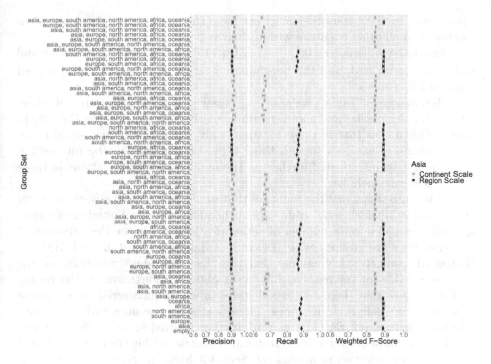

Fig. 2. Group set-dependent mean values and standard deviations of the precision, recall, and weighted F-score (cf. Eq. (13)) obtained on the three separate test sets

After finding the best VSR-MC, we can identify the regions that, when divided, improve our model (cf. Sect. 4.2; Step 2). Our method concludes that our description of the licit and illicit supply networks improve if we consider five of the eight Asian regions (i.e. Asia 1, Asia 5, Asia 6, Asia 7, and Asia 8) as individual countries. Therefore, we redefine the optimal group set as $G_{\mathrm{opt}} = \{$Asia 2, Asia 3, Asia 4, Africa, Europe, North America, South America, Oceania$\}$. This group set defines a new VSR-MC with a state space that combines the three scales of connectivity we considered in this article. Lastly, it is important to mention, that the cost of this new optimal VSR-MC (i.e. 0.047) is not only lower than the one obtained when Asia is divided purely into regions (i.e. 0.054), but also than the one obtained when Asia is divided purely into countries (i.e. 0.048).

Once the optimal scale of connectivity, given any limitations of our process, has been found, we recalculate the Markov Chain probabilities with all available data and use Eq. (15) to identify the states in which illicit serial codes are more prevalent. It is important to mention that having some of the countries as states results in the transition matrices having absorbing states; a type of state that complicates the calculation of the limiting distributions. Therefore to calculate the limiting distributions, we first group those countries with the "less absorbing" countries within their region. In this context, "less absorbing" refers to countries

524   J. Á. González Ordiano et al.

whose rows in their transition matrices have the least number of zeros when compared to all other countries within their region.

The relative differences $\Delta \pi'_{s_{G_{opt},i}}$ (cf. Eq. (15)), the number of times a serial code is authenticated, and the benchmark relative differences (BRDs) are all contained in Table 3.

When looking at Table 3, we notice a state with a relative difference of minus infinity (i.e. Mongolia) and another with a relative difference of one (i.e. Tajikistan). This means that in those locations only illicit or only licit serial codes were authenticated. Even though these types of results might be interesting, they will not be investigated further, as the number of authentications in those locations is extremely low.

Table 3 also shows that the countries that would have formed absorbing states are locations in which serial codes are authenticated a small number of times, specially compared to the number of authentications within the "less absorbing" countries they are grouped with (i.e. India, Malaysia, Pakistan, and Turkey). Henceforth, we can safely assume that the "less absorbing" countries are the ones responsible for the relative differences obtained. Furthermore, the results in Table 3 show that there are several states in which illicit serial codes appear to spend more time than licit ones. These states are the ones with a negative relative difference (cf. Eq. (15)) and are further referred to as "critical" states. The fact that most of these critical states are countries within the regions selected by our forward selection algorithm (cf. Sect. 4.2; Step 2), speaks in favor of our approach.

As Table 3 shows, Turkey is the most "critical" state, as its relative difference estimates that illicit serial codes will spend close to 1200% more time there than their licit counterparts. This is an extreme result that needs to be investigated further, for instance by identifying the reasons behind this outcome and/or by finding out if Turkey is again a critical state when looking at illicit activities, such as serial code harvesting. The critical states with the next three lowest relative differences are Georgia, Singapore, and Syria. There the limiting distribution values of an illicit serial code are between 200 and 300% higher than those of a licit one. However, we can also observe that the number of authentications occurring on those locations is quite low in comparison to other places. Henceforth, a further investigation of those locations may not be of extreme importance. In addition to the results mentioned above, there are several "critical" states with relative difference that can still be considered high, i.e. between 20 and 50%. Within these states, Europe and China are the ones with a considerably larger amount of authentications. Therefore, a more in depth study of these two locations could be interesting for future related works.

It is also important to mention that a state having a relative difference close to zero does not mean that it is free of illicit activity. For instance, Sri Lanka, North America, and South America have relative differences of just −0.07, −0.01, and 0, respectively, meaning that their limiting distribution values for licit and illicit serial codes are almost the same. In other words, states whose relative

Table 3. Number of authentications, relative limiting distribution difference $\Delta\pi'_{s_{G_{opt},i}}$, and benchmark relative difference (BRDs) of the states forming the optimal Variable State Resolution-Markov Chain; the absorbing states and their realization are shown in parentheses next to the name and realizations of their corresponding "less absorbing" countries

$s_{G_{opt},i}$	# of Authentications	$\Delta\pi'_{s_{G_{opt},i}}$	BRD
Mongolia	1	$-$Inf	$-$Inf
Turkey (Jordan, Armenia)	42064 (53, 5)	-11.69	-12.07
Georgia	16	-2.91	-1.72
Singapore	189	-2.51	-0.69
Syria	22	-2.41	-0.59
Europe	20621	-0.49	-1.87
Kazakhstan	114	-0.32	0.47
Pakistan (Kyrgyzstan, Turkmenistan, Uzbekistan)	1944 (1, 1, 5)	-0.32	-0.02
China	19923	-0.20	0.80
Sri Lanka	298	-0.07	0.42
North America	10991	-0.01	0.48
South America	24590	0.00	0.50
Asia 3	3606	0.25	0.51
India (Nepal, Maldives)	13669 (38, 29)	0.37	0.64
Russia	4475	0.42	0.59
Philippines	283	0.52	0.78
Bangladesh	282	0.53	0.51
Bhutan	279	0.53	-0.53
Africa	11756	0.56	0.58
Afghanistan	445	0.58	0.50
Asia 4	1815	0.65	0.73
Malaysia (Brunei)	1898 (16)	0.69	0.75
Indonesia	318	0.70	0.77
Palestinian Territories	113	0.71	0.28
Oceania	644	0.76	0.76
Israel	363	0.87	0.23
Asia 2	654	0.89	0.49
Lebanon	56	0.92	0.80
Azerbaijan	457	0.92	0.95
Iraq	32	0.98	0.91
Myanmar	92	0.99	0.98
Tajikistan	3	1.00	1.00

differences are close to zero may have a similar rate of licit and illicit activity and thus should be investigated further.

Lastly, we observe that some relative difference values vary significantly or do not agree to those calculated using simple descriptive statistics (i.e. the BRDs). For example, though China has the relative difference of a so-called "critical" state, its BRD is clearly above zero. Similarly, North and South America have BRDs that indicate more licit than illicit authentications, while their relative differences estimate instead similar rates. Note that though we are aware about the interesting results obtained for Bhutan (its BRD and its relative difference are complete opposites), no further investigation and analysis were conducted given its relatively small number of authentications. In general, the results obtained in this work demonstrate that modeling the spatio-temporal information of a supply network (as we do with our approach) leads to conclusions that are different from those obtained through a simple descriptive analysis. Furthermore, since our method models the behavior of the illicit supply network as a whole, we can argue that it is better suited at combating illicit trade than a descriptive analysis.

6   Conclusion and Outlook

This article presents a new approach for describing illicit supply networks based on Variable State Resolution-Markov Chain (VSR-MC) models. These type of models stem from the idea that a location within a supply network can be described at different scales of connectivity (e.g., countries, regions, continents).

The new method described herein is divided in two main steps. The first step creates a series of VSR-MCs that describe the same network using different state spaces, while the second uses a user-defined cost function to select the VSR-MC that best describes the network. The new method is applied to a dataset containing spatio-temporal information of licit and illicit activity. This dataset comes from the Global Product Authentication Service of Micro Focus International and contains information of the time and place in which licit and illicit serial codes have been authenticated. Applying our new method to this dataset results in Markov Chain models of the licit and illicit supply networks. The comparison of both networks enables us to ascertain the geographic locations in which illicit serial codes are expected to spend more time than their licit counterparts.

Even though this article shows a promising approach for analyzing illicit supply networks, there are still a number of aspects that have to be studied in future related works. For instance, in this article all scales of connectivity stem from grouping the countries based on their geographic proximity. Therefore, future works should investigate if better descriptions of the illicit supply networks can be obtained by clustering the countries based on other measures of similarity; such as, their number of free trade agreements, their culture, or their language. Such a study will allow us to better identify the aspects that drive illicit supply networks. Additionally, we should also use the method described herein to compare networks stemming from different forms of illicit trade, such as counterfeiting, serial code harvesting, and human trafficking. A comparison like this will enable us to identify both similarities and differences between different types of illicit trade. Moreover, future works should also investigate the use of nth order and non-homogeneous Markov Chains. Finally, we must compare our method to other approaches to better understand its advantages and limitations.

Acknowledgments. Jorge Ángel González Ordiano and Steven Simske acknowledge the support given by the NSF EAGER grant with the abstract number 1842577, "Advanced Analytics, Intelligence and Processes for Disrupting Operations of Illicit Supply Networks".

References

1. Bindner, L.: Illicit trade and terrorism financing. Centre d'Analyse du Terrorisme (CAT) (2016)
2. Collet, J.F.: Discrete Stochastic Processes and Applications. Universitext. Springer, Cham (2018). https://doi.org/10.1007/978-3-319-74018-8
3. Dégardin, K., Guillemain, A., Klespe, P., Hindelang, F., Zurbach, R., Roggo, Y.: Packaging analysis of counterfeit medicines. Forensic Sci. Int. **291**, 144–157 (2018)
4. Frey, B.J., Dueck, D.: Clustering by passing messages between data points. Science **315**(5814), 972–976 (2007)
5. Giommoni, L., Aziani, A., Berlusconi, G.: How do illicit drugs move across countries? A network analysis of the heroin supply to Europe. J. Drug Issues **47**(2), 217–240 (2017)
6. Mackey, T.K., Liang, B.A.: The global counterfeit drug trade: patient safety and public health risks. J. Pharm. Sci. **100**(11), 4571–4579 (2011)
7. Magliocca, N.R., et al.: Modeling cocaine traffickers and counterdrug interdiction forces as a complex adaptive system. Proc. Natl. Acad. Sci. **116**(16), 7784–7792 (2019)
8. Meidani, H., Ghanem, R.: Multiscale Markov models with random transitions for energy demand management. Energy Build. **61**, 267–274 (2013)
9. Meruga, J.M., Cross, W.M., May, P.S., Luu, Q., Crawford, G.A., Kellar, J.J.: Security printing of covert quick response codes using upconverting nanoparticle inks. Nanotechnology **23**(39), 395201 (2012)
10. OECD: Governance Frameworks to Counter Illicit Trade (2018)
11. Privault, N.: Understanding Markov Chains: Examples and Applications. Springer Undergraduate Mathematics Series. Springer, Singapore (2013). https://doi.org/10.1007/978-981-13-0659-4
12. Shelley, L.I.: Dark Commerce: How a New Illicit Economy is Threatening our Future. Princeton University Press, Princeton (2018). YBP Print DDA
13. Simske, S.: Meta-Analytics: Consensus Approaches and System Patterns for Data Analysis. Morgan Kaufmann, Burlington (2019)
14. Simske, S.J.: Meta-Algorithmics: Patterns for Robust, Low Cost, High Quality Sysstem Ebook Central (EBC). IEEE Press, Wiley, Chichester (2013)
15. Simske, S.J., Sturgill, M., Aronoff, J.S.: Comparison of image-based functional monitoring through resampling and compression. In: Proceedings of IEEE International Geoscience & Remote Sensing Symposium (2009)
16. Triepels, R., Daniels, H., Feelders, A.: Data-driven fraud detection in international shipping. Expert Syst. Appl. **99**, 193–202 (2018)

Deep Conformal Prediction
for Robust Models

Soundouss Messoudi$^{(\boxtimes)}$, Sylvain Rousseau⑩, and Sébastien Destercke⑩

HEUDIASYC - UMR CNRS 7253, Université de Technologie de Compiègne,
57 avenue de Landshut, 60203 Compiegne Cedex, France
{soundouss.messoudi,sylvain.rousseau,sebastien.destercke}@hds.utc.fr,
https://www.hds.utc.fr/

Abstract. Deep networks, like some other learning models, can associate high trust to unreliable predictions. Making these models robust and reliable is therefore essential, especially for critical decisions. This experimental paper shows that the conformal prediction approach brings a convincing solution to this challenge. Conformal prediction consists in predicting a set of classes covering the real class with a user-defined frequency. In the case of atypical examples, the conformal prediction will predict the empty set. Experiments show the good behavior of the conformal approach, especially when the data is noisy.

Keywords: Deep learning · Conformal prediction · Robust and reliable models

1 Introduction

Machine learning and deep models are everywhere today. It has been shown, however, that these models can sometimes provide scores with a high confidence in a clearly erroneous prediction. Thus, a dog image can almost certainly be recognized as a panda, due to an adversarial noise invisible to the naked eye [4]. In addition, since deep networks have little explanation and interpretability by their very nature, it becomes all the more important to make their decisions robust and reliable.

There are two popular approaches that estimate the confidence to be placed in the predictions of machine learning algorithms: Bayesian learning and Probably Approximately Correct (PAC) learning. However, both these methods provide major limitations. Indeed, the first one needs correct prior distributions to produce accurate confidence values, which is often not the case in real-world applications. Experiments conducted by [10] show that when assumptions are incorrect, Bayesian frameworks give misleading and invalid confidence values (i.e. the probability of error is higher than what is expected by the confidence level).

© Springer Nature Switzerland AG 2020
M.-J. Lesot et al. (Eds.): IPMU 2020, CCIS 1237, pp. 528–540, 2020.
https://doi.org/10.1007/978-3-030-50146-4_39

The second method, i.e. PAC learning, does not rely on a strong underlying prior but generates error bounds that are not helpful in practice, as demonstrated in [13]. Another approach that offers hedged predictions and does not have these drawbacks is conformal prediction [14].

Conformal prediction is a framework that can be implemented on any machine learning algorithm in order to add a useful confidence measure to its predictions. It provides predictions that can come in the form of a set of classes whose statistical reliability (the average percentage of the true class recovery by the predicted set) is guaranteed under the traditional identically and independently distributed (i.i.d.) assumption. This general assumption can be relaxed into a slightly weaker one that is exchangeability, meaning that the joint probability distribution of a sequence of examples does not change if the order of the examples in this sequence is altered. The principle of conformal prediction and its extensions will be recalled in Sect. 2.

Our work uses an extension of this principle proposed by [6]. They propose to use the density $p(x|y)$ instead of $p(y|x)$ to produce the prediction. This makes it possible to differentiate two cases of different uncertainties: the first predicts more than one label compatible with x in case of ambiguity and the second predicts the empty set $\emptyset$ when the model does not know or did not see a similar example during training. This approach is recalled in Sect. 2.3. However, the tests in [6] only concern images and Convolutional Neural Networks.

Therefore, the validity and interest of this approach still largely remains to be empirically confirmed. This is what we do in Sect. 3, where we show experimentally that this approach is very generic, in the sense that it works for different neural network architectures (Convolutional Neural Networks, Gated Recurrent Unit and Multi Layer Perceptron) and various types of data (image, textual, cross sectional).

2 Conformal Prediction Methods

Conformal prediction was initially introduced in [14] as a transductive online learning method that directly uses the previous examples to provide an individual prediction for each new example. An inductive variant of conformal prediction is described in [11] that starts by deriving a general rule from which the predictions are based. This section presents both approaches as well as the density-based approach, which we used in this paper.

2.1 Transductive Conformal Prediction

Let $z_1 = (x_1, y_1), z_2 = (x_2, y_2), \ldots, z_n = (x_n, y_n)$ be successive pairs constituting the examples, with $x_i \in X$ an object and $y_i \in Y$ its label. For any sequence $z_1, z_2, \ldots, z_n \in Z^*$ and any new object $x_{n+1} \in X$, we can define a *simple predictor* D such as:

$$D : Z^* \times X \to Y. \tag{1}$$

This simple predictor D produces a point prediction $D(z_1, \ldots, z_n, x_{n+1}) \in Y$, which is the prediction for y_{n+1}, the true label of x_{n+1}.

By adding another parameter $\epsilon \in (0,1)$ which is the probability of error called the *significance level*, this simple predictor becomes a *confidence predictor* Γ that can predict a subset of Y with a *confidence level* $1 - \epsilon$, which corresponds to a statistical guarantee of coverage of the true label y_{n+1}. Γ is defined as follows:

$$\Gamma : Z^* \times X \times (0,1) \to 2^Y, \tag{2}$$

where 2^Y denotes the power set of Y. This confidence predictor Γ^ϵ must be decreasing for the inclusion with respect to ϵ, i.e. we must have:

$$\forall n > 0, \quad \forall \epsilon_1 \geq \epsilon_2, \quad \Gamma^{\epsilon_1}(z_1, \ldots, z_n, x_{n+1}) \subseteq \Gamma^{\epsilon_2}(z_1, \ldots, z_n, x_{n+1}). \tag{3}$$

The two main properties desired in confidence predictors are (a) *validity*, meaning the error rate does not exceed ϵ for each chosen confidence level ϵ, and (b) *efficiency*, i.e. prediction sets are as small as possible. Therefore, a prediction set with fewer labels will be much more informative and useful than a bigger prediction set.

To build such a predictor, conformal prediction relies on a *non-conformity measure* A_n. This measure calculates a score that estimates how strange an example z_i is from a bag of other examples $\{z_1, \ldots, z_{i-1}, z_{i+1}, \ldots, z_n\}$. We then note α_i the non-conformity score of z_i compared to the other examples, such as:

$$\alpha_i := A_n(\{z_1, \ldots, z_{i-1}, z_{i+1}, \ldots, z_n\}, z_i). \tag{4}$$

Comparing α_i with other non-conformity scores α_j with $j \neq i$, we calculate a *p-value* of z_i expressing the proportion of less conforming examples than z_i, with:

$$\frac{|\{j = 1, \ldots, n : \alpha_j \geq \alpha_i\}|}{n}. \tag{5}$$

If the p-value approaches the lower bound $1/n$ then z_i is non-compliant to most other examples (an outlier). If, on the contrary, it approaches the upper bound 1 then z_i is very consistent.

We can then compute the p-value for the new example x_{n+1} being classified as each possible label $y \in Y$ by using (5). More precisely, we can consider for each $y \in Y$ the sequence $(\{z_1, \ldots, z_n, z_{n+1} = (x_{n+1}, y)\})$ and derive from that scores $\alpha_1^y, \ldots, \alpha_{n+1}^y$. We thus get a conformal predictor by predicting the set:

$$\Gamma^\epsilon(x_{n+1}) = \left\{ y \in Y : \frac{|\{i = 1, \ldots, n, n+1 : \alpha_i^y \geq \alpha_{n+1}^y\}|}{n+1} > \epsilon \right\}. \tag{6}$$

Constructing a conformal predictor therefore amounts to defining a non-conformity measure that can be built based on any machine learning algorithm called the *underlying algorithm* of the conformal prediction. Popular underlying algorithms for conformal prediction include Support Vector Machines (SVMs) and k-Nearest Neighbours (k-NN).

2.2 Inductive Conformal Prediction

One important drawback of Transductive Conformal Prediction (TCP) is the fact that it is not computationally efficient. When dealing with a large amount of data, it is inadequate to use all previous examples to predict an outcome for each new example. Hence, this approach is not suitable for any time consuming training tasks such as deep learning models. Inductive Conformal prediction (ICP) is a method that was outlined in [11] to solve the computational inefficiency problem by replacing the transductive inference with an inductive one. The paper shows that ICP preserves the validity of conformal prediction. However, it has a slight loss in efficiency.

ICP requires the same assumption as TCP (the i.i.d. assumption or the weaker assumption exchangeability), and can also be applied on any underlying machine learning algorithm. The difference between ICP and TCP consists of splitting the original training data set $\{z_1, \ldots, z_n\}$ into two parts in the inductive approach. The first part $D^{tr} = \{z_1, \ldots, z_l\}$ is called the *proper training set*, and the second smaller one $D^{cal} = \{z_{l+1}, \ldots, z_n\}$ is called the *calibration set*. In this case, the non-conformity measure A_l based on the chosen underlying algorithm is trained only on the proper training set. For each example of the calibration set $i = l + 1, \ldots, n$, a non-conformity score α_i is calculated by applying (4) to get the sequence $\alpha_{l+1}, \ldots, \alpha_n$. For a new example x_{n+1}, a non-conformity score α_{n+1}^y is computed for each possible $y \in Y$, so that the p-values are obtained and compared to the significance level ϵ to get the predictions such as:

$$\Gamma^\epsilon(x_{n+1}) = \{y \in Y : \frac{|\{i = l + 1, \ldots, n, n + 1 : \alpha_i \geq \alpha_{n+1}^y\}|}{n - l + 1} > \epsilon\}. \qquad (7)$$

In other words, this inductive conformal predictor will output the set of all possible labels for each new example of the classification problem without the need of recomputing the non-conformity scores in each time by including the previous examples, i.e., only α_{n+1} is recomputed for each y in Eq. (7).

2.3 Density-Based Conformal Prediction

The paper [6] uses a density-based conformal prediction approach inspired from the inductive approach and considers a density estimate $\hat{p}(x|y)$ of $p(x|y)$ for the label $y \in Y$. Therefore, this method divides labeled data into two parts: the first one is the *proper training* data $D^{tr} = \{X^{tr}, Y^{tr}\}$ used to build $\hat{p}(x|y)$, the second is the *calibration* data $D^{cal} = \{X^{cal}, Y^{cal}\}$ to evaluate $\{\hat{p}(x_i|y)\}$ and set $\hat{t}_y$ to be the empirical quantile of order ϵ of the values $\{\hat{p}(x_i|y)\}$:

$$\hat{t}_y = \sup\left\{t : \frac{1}{n_y} \sum_{\{z_i \in D_y^{cal}\}} I(\hat{p}(x_i|y) \geq t) \geq 1 - \epsilon\right\}, \qquad (8)$$

where n_y is the number of elements belonging to the class y in D^{cal}, and $D_y^{cal} = \{z_i \in D^{cal} : y_i = y\}$ is the subset of calibration examples of class y. For a new observation x_{n+1}, we set the conformal predictor Γ_d^ϵ such that:

$$\Gamma_d^\epsilon(x_{n+1}) = \{y \in Y : \hat{p}(x_{n+1}|y) \geq \hat{t}_y\}. \tag{9}$$

This ensures that the observations with low probability—that is, the poorly populated regions of the input space—are classified as $\emptyset$. This divisional procedure avoids the high cost of deep learning calculations in the case where the online approach is used. The paper [6] also shows that $|P(y \in \Gamma_d^\epsilon(x_{n+1})) - (1 - \epsilon)| \to 0$ with $\min_y n_y \to \infty$, which ensures the validity of the model. The training and prediction algorithms are defined in the Algorithms 1 and 2.

Algorithm 1. Training algorithm

Input: Training data $Z = (x_i, y_i)$, $i = 1 \dots n$, Class list $\mathcal{Y}$, Confidence level ϵ, Ratio p.
Initialize: $\hat{p}_{list} = list, \hat{t}_{list} = list$
for $y \in \mathcal{Y}$ **do**
$\quad X_y^{tr}, X_y^{cal} \leftarrow SubsetData(Z, \mathcal{Y}, p)$
$\quad \hat{p}_y \leftarrow LearnDensityEstimator(X_y^{tr})$
$\quad \hat{t}_y \leftarrow Quantile(\hat{p}_y(X_y^{cal}), \epsilon)$
$\quad \hat{p}_{list}.append(\hat{p}_y); \hat{t}_{list}.append(\hat{t}_y)$
end for
return $\hat{p}_{list}, \hat{t}_{list}$

Algorithm 2. Prediction algorithm

Input: Input to be predicted x, Trained $\hat{p}_{list}, \hat{t}_{list}$, Class list $\mathcal{Y}$.
Initialize: $C = list$
for $y \in \mathcal{Y}$ **do**
$\quad$ **if** $\hat{p}_y(x) \geq \hat{t}_y$ **then**
$\quad\quad C.append(y)$
$\quad$ **end if**
end for
return C

We can rewrite (9) so that it approaches (7) with a few differences, mainly the fact that Γ_d^ϵ uses a conformity measure based on density estimation (calculating how much an example is compliant with the others) instead of a non-conformity measure as in Γ^ϵ, with $\alpha_i^y = -\hat{p}(x_i|y)$ [14], and that the number of examples used to build the prediction set depends on y. Thus, Γ_d^ϵ can also be written as:

$$\Gamma^\epsilon(x_{n+1}) = \left\{ y \in Y : \frac{|\{z_i \in D_y^{cal} : \alpha_i^y \geq \alpha_{n+1}^y\}|}{n_y} > \epsilon \right\}. \tag{10}$$

The proof can be found in Appendix A.

The final quality of the predictor (its efficiency, robustness) depends in part on the density estimator. The paper [7] suggests that the use of kernel estimators gives good results under weak conditions.

The results of the paper show that the training and prediction of each label are independent of the other classes. This makes conformal prediction an adaptive method, which means that adding or removing a class does not require retraining the model from scratch. However, it does not provide any information on the relationship between the classes. In addition, the results depend on ϵ: when ϵ is small, the model has high precision and a large number of classes predicted for each observation. On the contrary, when ϵ is large, there are no more cases classified as $\emptyset$ and fewer cases predicted by label.

3   Experiments

In order to examine the effectiveness of the conformal method on different types of data, three data sets for binary classification were used. They are:

1. **CelebA** [8]: face attributes dataset with over 200,000 celebrity images used to determine if a person is a man (1) or a woman (0).
2. **IMDb** [9]: contains more than 50,000 different texts describing film reviews for sentiment analysis (with 1 representing a positive opinion and 0 indicating a negative opinion).
3. **EGSS** [1]: contains 10000 examples for the study of the electrical networks' stability (1 representing a stable network), with 12 numerical characteristics.

3.1   Approach

The overall approach followed the same steps as in density-based conformal prediction [6] and meets the conditions listed above (the i.i.d. or exchangeability assumptions). Each data set is divided into proper training, calibration and test sets. A deep learning model dedicated to each type of data is trained on the proper training and calibration sets. The before last dense layer serves as a feature extractor which produces a fixed size vector for each dataset and representing the object (image, text or vector). These feature vectors are then used for the conformal part to estimate the density. Here we used a gaussian kernel density estimator of bandwidth 1 available in Python's scikit-learn [12]. The architecture of deep learning models is shown in Fig. 1. It is built following the steps below:

1. Use a basic deep learning model depending on the type of data. In the case of CelebA, it is a CNN with a ResNet50 [5] pre-trained on ImageNet [2] and adjusted to CelebA. For IMDb, this model is a bidirectional GRU that takes processed data with a tokenizer and padding. For EGSS, this model is a multilayer perceptron (MLP).

2. Apply an intermediate dense layer and use it as a feature extractor with a vector of size 50 representing the object, and which will be used later for conformal prediction.
3. Add a dense layer to obtain the class predicted by the model (0 or 1).

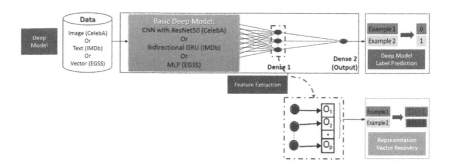

Fig. 1. Architecture of deep learning models.

Based on the recovered vectors, a Gaussian kernel density estimate is made on the proper training set of each class to obtain the values $P(x|y)$. Then, the calibration set is used to compute the density scores and sort them to determine the given ϵ threshold of all the values, thus delimiting the density region of each class. Finally, the test set is used to calculate the performance of the model. The code used for this article is available in Github[1].

The visualization of the density regions (Fig. 2) is done via the first two dimensions of a Principal Component Analysis. The results show the distinct regions of the classes 0 (in red) and 1 (in blue) with a non-empty intersection (in green) representing a region of random uncertainty. The points outside these three regions belong to the region of epistemic uncertainty, meaning that the classifier "does not know".

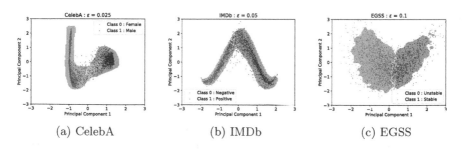

(a) CelebA   (b) IMDb   (c) EGSS

Fig. 2. Conformal prediction density regions for all datasets. (Color figure online)

---

[1] https://github.com/M-Soundouss/density_based_conformal_prediction.

3.2 Results on the Test Examples

To obtain more information on the results of this experiment, the accuracy of the models was calculated with different values ϵ between 0.01 and 0.5 when determining the threshold of conformal prediction density as follows:

- DL accuracy: the accuracy of the basic deep model (CNN for CelebA, GRU for IMDb or MLP for EGSS) on all the test examples.
- Valid conformal accuracy: the accuracy of the conformal model when one considers only the singleton predictions 0 or 1 (without taking into account the $\{0, 1\}$ and the empty sets).
- Valid DL accuracy: The accuracy of the basic deep model on the test examples that have been predicted as 0 or 1 by the conformal model.

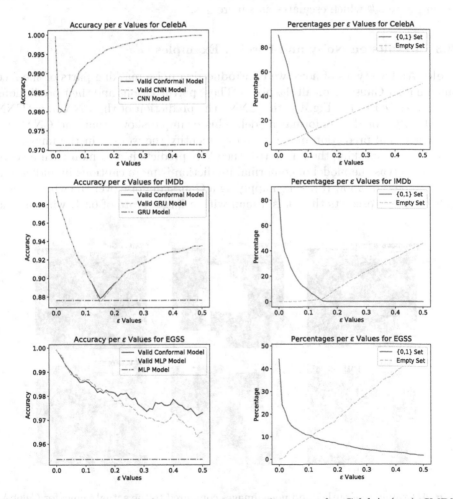

Fig. 3. The accuracy and the percentages according to ϵ for CelebA (top), IMDb (middle) and EGSS (bottom).

The percentage of empty sets ∅ and {0, 1} sets was also calculated from all the predictions of the test examples made by the conformal prediction model. The results are shown in the Fig. 3.

The results show that the accuracy of the valid conformal model and the accuracy of the valid basic deep learning model are almost equal and are better than the accuracy of the base model for all ϵ values. In our tests, the addition of conformal prediction to a deep model does not degrade its performance, and sometimes even improves it (EGSS). This is due to the fact that the conformal prediction model allows to abstain from predicting (empty set ∅) or to predict both classes for ambiguous examples, thus making it possible to have a more reliable prediction of the label. It is also noticed that as ϵ grows, the percentage of predicted {0, 1} sets decreases until it is no longer predicted (at $\epsilon = 0.15$ for CelebA for example). Conversely, the opposite is observed with the percentage of empty sets ∅ which escalates as ϵ increases.

3.3 Results on Noisy and Foreign Examples

CelebA: Two types of noise were introduced: a noise masking parts of the face and another Gaussian on all the pixels. These perturbations and their predictions are illustrated in the Fig. 4 with "CNN" the prediction of the CNN and "CNN + CP" that of the conformal model. This example shows that the CNN and the conformal prediction model correctly identify the woman in the image (a). However, by masking the image (b), the CNN predicts it as a man with a score of 0.6 whereas the model of conformal prediction is more cautious by indicating that it does not know (∅). When applying a Gaussian noise over the whole image (c), the CNN predicts that it is a man with a larger score of 0.91, whereas the

Fig. 4. Examples of outlier and noisy images compared to the actual image for CelebA.

conformal model predicts both classes. For outliers, examples (d), (e), and (f) illustrate the ability of the conformal model to identify different outliers as such ($\emptyset$) in contrast to the deep model that predicts them as men with a high score.

IMDb: The Fig. 5 displays a comparison of two texts before and after the random change of a few words (in bold) by other words in the model's vocabulary. The actual text predicted as negative opinion by both models becomes positive for the GRU after disturbance. Nevertheless, the conformal model is more cautious by indicating that it can be both cases ($\{0,1\}$). For the outlier example formed completely of vocabulary words, the GRU model predicts positive with a score of 0.99, while the conformal model says that it does not know ($\emptyset$).

Original Example	« **Every great** romantic comedy **needs** conflict between the romantic leads [...] **This story** falls completely flat in this **area** [...] suspense is flat, **there is** no anticipation, **and there really is** no allure [...] I was quite surprised. During the movie, I **expected them** more **to** play a game of checkers and chat **about** the weather than see **any** moving passion [...] While I'm a fan **of** both actors [...] The writing **was** very **weak, which** also **might** have **impacted the performances** [...] »	Label . Negative sentiment GRU : Negative sentiment (0.09) GRU + CP : Positive sentiment
Noisy Example	« **ambidexterous trentini** romantic comedy **dispassionately** conflict between the romantic leads [...] **fanout phoolan** falls completely flat in this **centerpiece** [...] suspense is flat, **binding** is no anticipation, **scroller** there **wiedzmin laudable thunderball** allure [...] I was quite surprised. During the movie, I **lives' 'are** more **subtextual** play **commemorations** game of checkers and chat **stratovarius** the weather than see **linking** moving passion [...] While I'm a fan **unti** both actors [...] The writing **ripoff's** very **nare releases** also **sharikov** have **maes** the **'sketching'** [...] »	Label : Negative sentiment , GRU : Positive sentiment (0.74) GRU + CP : {Positive sentiment , Negative sentiment }
Outlier Example	« wolverines 'sandwich' controversial posit homme subfunctions snowmobile symbiotic malamud challenge needle's personl witch's nonce wills' swooshes cobbled brash mcq wanky 'bought regenerated southstreet amazed ravenna 'mainly belyt hijixn shrugs deodorant mesquida anodynesprech romishness malice seldomely settling dispicable vocation [...] reduce macfarlane's disclosing officers' wiretapping balbao seagals ml3 dibnah romulan controls dolled maguire' [...] »	GRU : Positive sentiment (0.99) GRU + CP : $\emptyset$

Fig. 5. Examples of outlier and noisy texts compared to the original one for IMDb.

EGSS: The Fig. 6 displays a comparison of the positions of the test examples on the density regions before (a) and after (b) the addition of a Gaussian noise. This shows that several examples are positioned outside the density regions after the introduction of the disturbances. The outlier examples (c) created by modifying some characteristics of these test examples with extreme values (to simulate a sensor failure, for example) are even further away from the density regions, and recognized as such by the conformal model ($\emptyset$).

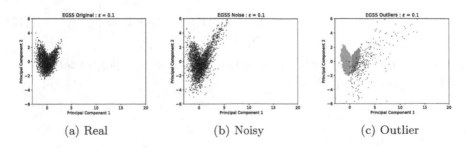

(a) Real   (b) Noisy   (c) Outlier

Fig. 6. Density visualization of real, noisy and outlier examples for EGSS.

4 Conclusions and Perspectives

We used the conformal prediction and the technique presented in [6] to have a more reliable and cautious deep learning model. The results show the interest of this method on different data types (image, text, tabular) used with different deep learning architectures (CNN, GRU and MLP). Indeed, in these three cases, the conformal model not only adds reliability and robustness to the deep model by detecting ambiguous examples but also keeps or even improves the performance of the basic deep model when it predicts only one class. We also illustrated the ability of conformal prediction to handle noisy and outlier examples for all three types of data. These experiments show that the conformal method can give more robustness and reliability to predictions on several types of data and basic deep architectures.

To improve the experiments and results, the perspectives include the optimization of density estimation based on neural networks. For instance, at a fixed ϵ the problem of finding the most efficient model arises that could be done by modifying the density estimation technique, but also by proposing an end-to-end, integrated estimation method. Also, it would be useful to compare the conformal prediction with calibration methods, for example, evidential ones that are also adopted for cautious predictions [3].

A Appendix

This appendix is to prove that Eqs. (9) and (10) in Sect. 2.3 are equivalent. We recall that Eq. (10) is

$$\Gamma^\epsilon(x_{n+1}) = \left\{ y \in Y : \frac{|\{z_i \in D_y^{cal} : \alpha_i^y \geq \alpha_{n+1}^y\}|}{n_y} > \epsilon \right\}. \tag{11}$$

We recall that Eq. (9) uses the "greater or equal" sign. Here we need to use the "greater" signs in Eqs. (12) and (13) to have an equivalence, which is

$$\Gamma_d^\epsilon(x_{n+1}) = \{ y \in Y : \hat{p}(x_{n+1}|y) > \hat{t}_y \}, \tag{12}$$

such that

$$\hat{t}_y = \sup \left\{ t : \frac{1}{n_y} \sum_{\{z_i \in D_y^{cal}\}} I(\hat{p}(x_i|y) > t) \geq 1 - \epsilon \right\}. \tag{13}$$

Let $f(t)$ be the decreasing function $f(t) = \frac{1}{n_y} \sum_{\{z_i \in D_y^{cal}\}} I(\hat{p}(x_i|y) > t)$.
Let us prove that (12) $\implies$ (11).

Since $\hat{t}_y$ is the upper bound such that $f(\hat{t}_y) \geq 1 - \epsilon$, then $\hat{p}(x_{n+1}|y)$ does not satisfy this inequality, thus

$$f(\hat{p}(x_{n+1}|y)) = \frac{1}{n_y} \sum_{\{z_i \in D_y^{cal}\}} I(\hat{p}(x_i|y) > \hat{p}(x_{n+1}|y)) < 1 - \epsilon$$

$$= \frac{1}{n_y} \sum_{\{z_i \in D_y^{cal}\}} 1 - I(\hat{p}(x_i|y) \leq \hat{p}(x_{n+1}|y)) < 1 - \epsilon$$

$$= 1 - \frac{1}{n_y} \sum_{\{z_i \in D_y^{cal}\}} I(\hat{p}(x_i|y) \leq \hat{p}(x_{n+1}|y)) < 1 - \epsilon$$

$$= \frac{1}{n_y} \sum_{\{z_i \in D_y^{cal}\}} I(\hat{p}(x_i|y) < \hat{p}(x_{n+1}|y)) > \epsilon \qquad (14)$$

Since $\hat{p}(x_{n+1}|y))$ is a conformity score, whereas α_i^y is a non-conformity score, we can write $\hat{p}(x_{n+1}|y)) = -\alpha_i^y$ [14]. So (14) becomes

$$\frac{1}{n_y} \sum_{\{z_i \in D_y^{cal}\}} I(\alpha_i^y \geq \alpha_{n+1}^y) > \epsilon \implies \frac{|\{z_i \in D_y^{cal} : \alpha_i^y \geq \alpha_{n+1}^y\}|}{n_y} > \epsilon$$

This shows that (12) $\implies$ (11).

Let us now prove that (11) $\implies$ (12). Using the indicator function of the complement, and changing the non-conformity score into a conformity score as shown before, we can simply find that

$$\frac{|\{z_i \in D_y^{cal} : \alpha_i^y \geq \alpha_{n+1}^y\}|}{n_y} > \epsilon \implies \frac{1}{n_y} \sum_{\{z_i \in D_y^{cal}\}} I(\hat{p}(x_i|y) > \hat{p}(x_{n+1}|y)) < 1 - \epsilon$$

Using the same function f, we then have

$$f(\hat{p}(x_{n+1}|y)) < 1 - \epsilon. \qquad (15)$$

Let us show by contradiction that $\hat{p}(x_{n+1}|y) > \hat{t}_y$. Suppose that $\hat{p}(x_{n+1}|y) \leq \hat{t}_y$. Since f is a decreasing function, we have $f(\hat{p}(x_{n+1}|y)) \geq f(\hat{t}_y)$. By the definition of $\hat{t}_y$, we have $f(\hat{t}_y) \geq 1 - \epsilon$. Thus $f(\hat{p}(x_{n+1}|y)) \geq f(\hat{t}_y) \geq 1 - \epsilon$. However, this contradicts (15). So we proved that (11) $\implies$ (12), which concludes the proof.

References

1. Arzamasov, V.: UCI electrical grid stability simulated data set (2018). https://archive.ics.uci.edu/ml/datasets/Electrical+Grid+Stability+Simulated+Data+
2. Deng, J., Dong, W., Socher, R., Li, L.J., Li, K., Fei-Fei, L.: ImageNet: a large-scale hierarchical image database. In: CVPR (2009)

3. Denoeux, T.: Logistic regression, neural networks and Dempster-Shafer theory: a new perspective. Knowl.-Based Syst. **176**, 54–67 (2019)
4. Goodfellow, I.J., Shlens, J., Szegedy, C.: Explaining and harnessing adversarial examples. In: 3rd International Conference on Learning Representations, ICLR 2015, San Diego, CA, USA, 7–9 May 2015, Conference Track Proceedings (2015). http://arxiv.org/abs/1412.6572
5. He, K., Zhang, X., Ren, S., Sun, J.: Deep residual learning for image recognition. In: CVPR, pp. 770–778 (2016)
6. Hechtlinger, Y., Póczos, B., Wasserman, L.: Cautious deep learning. arXiv preprint arXiv:1805.09460 (2018)
7. Lei, J., Robins, J., Wasserman, L.: Distribution-free prediction sets. J. Am. Stat. Assoc. **108**(501), 278–287 (2013)
8. Liu, Z., Luo, P., Wang, X., Tang, X.: Deep learning face attributes in the wild. In: Proceedings of International Conference on Computer Vision (ICCV), December 2015
9. Maas, A.L., Daly, R.E., Pham, P.T., Huang, D., Ng, A.Y., Potts, C.: Learning word vectors for sentiment analysis. In: Proceedings of the 49th Annual Meeting of the Association for Computational Linguistics: Human Language Technologies, vol. 1, pp. 142–150 (2011)
10. Melluish, T., Saunders, C., Nouretdinov, I., Vovk, V.: Comparing the Bayes and typicalness frameworks. In: De Raedt, L., Flach, P. (eds.) ECML 2001. LNCS (LNAI), vol. 2167, pp. 360–371. Springer, Heidelberg (2001). https://doi.org/10.1007/3-540-44795-4_31
11. Papadopoulos, H.: Inductive conformal prediction: theory and application to neural networks. In: Tools in Artificial Intelligence. IntechOpen (2008)
12. Pedregosa, F., et al.: Scikit-learn: machine learning in python. J. Mach. Learn. Res. **12**, 2825–2830 (2011)
13. Proedrou, K., Nouretdinov, I., Vovk, V., Gammerman, A.: Transductive confidence machines for pattern recognition. In: Elomaa, T., Mannila, H., Toivonen, H. (eds.) ECML 2002. LNCS (LNAI), vol. 2430, pp. 381–390. Springer, Heidelberg (2002). https://doi.org/10.1007/3-540-36755-1_32
14. Vovk, V., Gammerman, A., Shafer, G.: Algorithmic Learning in a Random World. Springer, Heidelberg (2005). https://doi.org/10.1007/b106715

Continuous Analogical Proportions-Based Classifier

Marouane Essid[1], Myriam Bounhas[1,2]([✉]), and Henri Prade[3]

[1] Larodec Lab, ISG de Tunis, Tunis, Tunisia
marouane.essid@gmail.com, myriam_bounhas@yahoo.fr
[2] Emirates College of Technology, Abu Dhabi, United Arab Emirates
[3] CNRS, IRIT, Université Paul Sabatier, Toulouse, France
prade@irit.fr

Abstract. Analogical proportions, often denoted $A : B :: C : D$, are statements of the form "A is to B as C is to D" that involve comparisons between items. They are at the basis of an inference mechanism that has been recognized as a suitable tool for classification and has led to a variety of analogical classifiers in the last decade. Given an object D to be classified, the basic idea of such classifiers is to look for triples of examples (A, B, C), in the learning set, that form an analogical proportion with D, on a maximum set of attributes. In the context of classification, objects A, B, C and D are assumed to be represented by vectors of feature values. Analogical inference relies on the fact that if a proportion $A : B :: C : D$ is valid, one of the four components of the proportion can be computed from the three others. Based on this principle, analogical classifiers have a cubic complexity due to the search for all possible triples in a learning set to make a single prediction. A special case of analogical proportions involving only three items A, B and C are called *continuous* analogical proportions and are of the form "A is to B as B is to C" (hence denoted $A : B :: B : C$). In this paper, we develop a new classification algorithm based on continuous analogical proportions and applied to numerical features. Focusing on pairs rather than triples, the proposed classifier enables us to compute an unknown midpoint item B given a pair of items (A, C). Experimental results of such classifier show an efficiency close to the previous analogy-based classifier while maintaining a reduced quadratic complexity.

Keywords: Classification · Analogical proportions · Continuous analogical proportions

1 Introduction

Reasoning by analogy establishes a parallel between two situations. More precisely, it enables us to relate two pairs of items (a, b) and (c, d) in such way that "a is to b as c is to d" on a comparison basis. This relationship, often noted $a : b :: c : d$, expresses a kind of equality between the two pairs, i.e., the

© Springer Nature Switzerland AG 2020
M.-J. Lesot et al. (Eds.): IPMU 2020, CCIS 1237, pp. 541–555, 2020.
https://doi.org/10.1007/978-3-030-50146-4_40

two items of the first pair are similar and differ in the same way as the two items of the second pair. The case of numerical (geometric) proportions where we have an equality between two ratios (i.e., $a/b = c/d$) is at the origin of the name "analogical proportions". Analogical proportions, when d is unknown, provides an extrapolation mechanism, which with numbers yields $d = (b \times c)/a$, and $d = b + c - a$ in case of arithmetic proportions (such that $a - b = c - d$). The analogical proportions-based extrapolation has been successfully applied to classification problems [4,8]. The main drawback of algorithms using analogical proportions is their cubic complexity.

A particular case of analogical proportions, named *continuous* analogical proportions, is obtained when the two central components are equal, namely they are statements of the form "a is to b as b is to c". In case of numerical proportions, if we assume that b is unknown, it can be expressed in terms of a and c as $b = \sqrt{a \times c}$ in the geometric case and $b = (a + c)/2$ in the arithmetic case. Note that similar inequalities hold in both cases: $min(a, c) \leq \sqrt{a \times c} \leq max(a, c)$ and $min(a, c) \leq (a + c)/2 \leq max(a, c)$. This means that the continuous analogical proportion induces a form of interpolation between a and c in the numerical case by involving an intermediary value that can be obtained from a and c. A continuous analogical proportions-based interpolation was recently proposed as a way of enlarging a training set (before applying some standard classification methods), and led to good results [2]. In contrast to extrapolation, interpolation with analogy-based classifiers has a quadratic complexity.

In this paper, we investigate the efficiency for classification of using such approach. The paper is organized as follows. Section 2 provides a short background on analogical proportions and more particularly on continuous ones. Then Sect. 3 surveys related work on analogical extrapolation. Section 4 presents the proposed interpolation approach for classification. Finally, Sect. 5 reports the results of our algorithm.

2   Background on Analogical Proportions

An analogical proportion is a relationship on X^4 between 4 items $A, B, C, D \in X$. This 4-tuple, when it forms an analogical proportion is denoted $A : B :: C : D$ and reads "A is to B as C is to D". Both relationships "is to" and "as" depend on the nature of X [9]. As it is the case for numerical proportions, the relation of analogy still holds when the pairs (A, B) and (C, D) are exchanged, or when central items B and C are permuted (see [11] for other properties). In the following subsections, we recall analogical proportions in the Boolean setting (i.e., $X \in \mathbb{B} = \{0, 1\}$)) and their extension for nominal and for real-valued settings (i.e., $X \in [0, 1]$), before considering the special case of *continuous* analogical proportions.

2.1   Analogical Proportions in the Boolean Setting

Let us consider four items A, B, C and D, respectively described by their binary values $a, b, c, d \in \mathbb{B} = \{0, 1\}$. Items A, B, C and D are in analogical proportion,

Table 1. Truth table for analogical proportion

a	b	c	d	$a:b::c:d$
0	0	0	0	1
0	0	1	1	1
0	1	0	1	1
1	0	1	0	1
1	1	0	0	1
1	1	1	1	1

which is denoted $A : B :: C : D$ if and only if $a : b :: c : d$ holds true (it can also be written $a : b :: c : d = 1$ or simply $a : b :: c : d$). The truth table (Table 1) shows the six possible assignments for a 4-tuple to be in analogical proportion, out of sixteen possible configurations.

Boolean analogical proportions can be expressed by the logical formula:

$$a : b :: c : d = (a \wedge \neg b \equiv c \wedge \neg d) \wedge (\neg a \wedge b \equiv \neg c \wedge d) \tag{1}$$

See [10, 12] for justification. This formula holds true for the 6 assignments shown in the truth table. It reads "a differs from b as c differs from d and b differs from a as d differs from c", which fits with the expected meaning of analogy. An equivalent formula is obtained by negating the two sides of the first and the second equivalence in formula (1):

$$a : b :: c : d = (a \rightarrow b \equiv c \rightarrow d) \wedge (b \rightarrow a \equiv d \rightarrow c) \tag{2}$$

Items are generally described by vectors of Boolean values rather than by a single value. A natural extension for vectors in $\{0,1\}^n$ of the form $\boldsymbol{x} = (x_1, \cdots, x_n)$ is obtained component-wise as follows:

$$\boldsymbol{a} : \boldsymbol{b} :: \boldsymbol{c} : \boldsymbol{d} \text{ iff } \forall i \in [1, n], a_i : b_i :: c_i : d_i \tag{3}$$

2.2 Nominal Extension

When a, b, c, d take their values in a finite set $\mathcal{D}$ (with more than 2 elements), we can derive three patterns of analogical proportions in the nominal case, from the six possible assignments for analogical proportions in the Boolean case. This generalization is thus defined by:

$$a : b :: c : d = 1 \text{ iff } (a, b, c, d) \in \{(s, s, s, s), (s, t, s, t), (s, s, t, t) | s, t \in \mathcal{D}\} \tag{4}$$

$$a : b :: c : d = 0 \text{ otherwise}$$

2.3 Multiple-Valued Extension

In case items are described by numerical attributes, it will be necessary to extend the logic modeling underlying analogical proportions in order to support a numerical setting. a, b, c, d are now real values normalized in the interval

[0, 1] and their analogical proportion $a : b :: c : d$ is extended from $\mathbb{B}^4$ to $[0,1]^4$. Analogical proportions are no longer valid or invalid but the extent to which they hold is now a matter of degree. For example, if a, b, c, d have 1, 0, 1 and 0.1 as values respectively, we expect that $a : b :: c : d$ has a high value (close to 1) since 0.1 is close to 0.

Table 2. Multi-valued extension

Operator	Extension
Negation: $\neg a$	$1 - a$
Implication: $a \rightarrow b$	$min(1, 1 - a + b)$
Conjunction: $a \wedge b$	$min(a, b)$
Equivalence: $a \equiv b$	$min(a \rightarrow b, b \rightarrow a) = 1 - |a - b|$

The extension of the logical expression of analogical proportions to the multiple-valued case requires the choice of appropriate connectives for preserving desirable properties [5]. To extend expression (2), conjunction, implication and equivalence operators are then replaced by the multiple valued connectives given in Table 2. This leads to the following expression P:

$$P(a,b,c,d) = a : b :: c : d = \begin{cases} 1 - |(a - b) - (c - d)|, \\ \quad \text{if } a \geq b \text{ and } c \geq d \text{ or } a \leq b \text{ and } c \leq d \\ 1 - max(|a - b|, |c - d|), \\ \quad \text{if } a \leq b \text{ and } c \geq d \text{ or } a \geq b \text{ and } c \leq d \end{cases} \quad (5)$$

When a, b, c, d are restricted to $\{0, 1\}$, the last expression coincide with the definition for the Boolean case (given by (1)), which highlights the agreement between the extension and the original idea of analogical proportion. For the interval $[0, 1]$, we have $P(a, b, c, d) = 1$ as soon as $a - b = c - d$ and as we expected, we get a high value for the 4-tuple $(1, 0, 1, 0.1)$, indeed $1 : 0 :: 1 : 0.1 = 0.9$.

Moreover, since we have $|(1-a)-(1-b)| = |b-a| = |a-b|$, $|(1-a-(1-b)) - (1-c-(1-d))| = |(b-a)-(d-c)| = |(c-d)-(a-b)| = |(a-b)-(c-d)|$, and $1-s \geq 1-t \Leftrightarrow s \leq t$, it is easy to check a remarkable *code independence* property: $a : b :: c : d = (1 - a) : (1 - b) :: (1 - c) : (1 - d)$. Code independence means that 0 and 1 play symmetric roles, and it is the same to encode an attribute positively or negatively.

As items are commonly described by vectors, we can extend the notion of analogical proportion to vectors in $[0, 1]^n$.

$$P(\mathbf{a}, \mathbf{b}, \mathbf{c}, \mathbf{d}) = \frac{\sum_{i=1}^{n} P(a_i, b_i, c_i, d_i)}{n} \quad (6)$$

where $P(a_i, b_i, c_i, d_i)$ refers to expression (5)).

Let us observe that $P(a, b, c, d) = 1$ (i.e. $a : b :: c : d$ holds) if and only if the analogical proportion holds perfectly on every component:

$$P(a, b, c, d) = 1 \text{ iff } \forall i \in [1, n], P(a_i, b_i, c_i, d_i) = 1 \qquad (7)$$

2.4 Inference with Analogical Proportions

Analogical proportion-based inference relies on a simple principle:if four Boolean vectors a, b, c and d make a valid analogical proportion component-wise between their attribute values, then it is expected that their class labels also make a valid proportion [4].

$$\frac{a : b :: c : d}{cl(a) : cl(b) :: cl(c) : cl(d)} \qquad (8)$$

where $cl(x)$ denotes to the class value of x.

It means that the classification of a Boolean vector d is only possible when the equation $cl(a) : cl(b) :: cl(c) : x$ is solvable[1] (the classes of a, b, c are known as they belong to the sample set), and the analogical proportion $a : b :: c : d$ holds true. If these two criteria are met, we assign x to $cl(d)$.

In the numerical case, where a, b, c, d are 4 real-valued vectors over $[0, 1]^n$ (the numerical values are previously normalized), the inference principle strictly clones the Boolean setting:

$$\frac{P(a, b, c, d) = 1}{cl(a) : cl(b) :: cl(c) : cl(d)} \qquad (9)$$

In practice, the resulting degree $P(a, b, c, d)$ is rarely equal to 1 but should be close to 1. Therefore Eq. (9) has to be adapted for a proper implementation.

2.5 Continuous Analogical Proportions

Continuous analogical proportions, denoted $a : b :: b : c$, are ternary relations which are a special case of analogical proportions. This enables us to calculate b using a pair (a, c) only, rather than a triple as in the general case. In $\mathbb{B}$ the unique solutions of equations $0 : x :: x : 0$ and $1 : x :: x : 1$ are respectively $x = 0$ and $x = 1$, while $0 : x :: x : 1$ or $1 : x :: x : 0$ have no solution.

Drawing the parallel with the Boolean case, we deduce that the only solvable equation for the nominal case is $s : x :: x : s$, having $x = s$ as solution, while $s : x :: x : t$ $(s \neq t)$ has no solution.

Contrary to these trivial cases, the multi-valued framework (Eq. (5)) is richer. We have

$$P(a, b, c) = a : b :: b : c = \begin{cases} 1 - |a + c - 2b|, \\ \quad \text{if } a \geq b \text{ and } b \geq c \text{ or } a \leq b \text{ and } b \leq c \\ 1 - max(|a - b|, |b - c|), \\ \quad \text{if } a \leq b \text{ and } b \geq c \text{ or } a \geq b \text{ and } b \leq c \end{cases} \qquad (10)$$

---

[1] Indeed the nominal equation $s : t :: t : x = 1$ has no solution if $s \neq t$.

We notice that for $b = (a+c)/2$, we have $a : b :: b : c = 1$ which fits the statement "A is to B as B is to C". As we expect, we get a higher value of analogy (closer to 1) as b tends to $(a + c)/2$. Computing continuous analogy for items described by vectors is exactly the same as for the general case (i.e., for real-valued setting $P(\boldsymbol{a}, \boldsymbol{b}, \boldsymbol{c}) = \frac{\sum_{i=1}^{n} P(a_i, b_i, c_i)}{n}$).

Applying analogy-based inference for numerical values with continuous analogical proportions, we obtain:

$$\frac{P(\boldsymbol{a}, \boldsymbol{b}, \boldsymbol{c}) = 1}{cl(\boldsymbol{a}) : cl(\boldsymbol{b}) :: cl(\boldsymbol{b}) : cl(\boldsymbol{c})} \tag{11}$$

One may wonder if continuous analogical proportions could be efficient enough compared to general analogical proportions. As already said, $a : b :: c : d$ holds at degree 1 if and only if $a - b = c - d$ (from which one can extrapolate $d = c + b - a$). Now consider two continuous proportions: $a - b = b - c$ (which corresponds to the interpolation $b = (a + c)/2$) and $b - c = c - d$ (which gives the interpolation $c = (b + d)/2$). Adding each side of the two proportions yields $a - c = b - d$, which is equivalent to $a - b = c - d$. In this view, two intertwined interpolations may play the role of an extrapolation. However the above remark applies only to numerical values, but not to Boolean ones.

3   Related Works on Analogical Proportions and Classification

Continuous analogical proportions have been recently applied to enlarge a training set for classification by creating artificial examples [2]. A somewhat related idea can be found in Lieber et al. [6] which extended the paradigm of classical Case-Based Reasoning by either performing a restricted form of interpolation to link the current case to pairs of known cases, or by extrapolation exploiting triples of known cases.

In the classification context, the authors in [3] introduce a measure of oddness with respect to a class that is computed on the basis of pairs made of two nearest neighbors in the same class; this amounts to replace the two neighbors by a fictitious representative of the class. Moreover, some other works have exploited analogical proportions to deal with classification problems. Most noteworthy are those based on using analogical dissimilarity [1] and applied to binary and nominal data and later the analogy-based classifier [4] applied to binary, nominal and numerical data. In the following subsections, we especially review these two latter works as they seem the closest to the approach that we are developing in this paper.

3.1   Classification by Analogical Dissimilarity

Analogical dissimilarity between binary objects is a measure that quantifies how far a 4 tuple (a, b, c, d) is from being in an analogical proportion. This is equivalent to the minimum number of bits to change in a 4-tuple to achieve a perfect analogy, thus when a 4-tuple is in analogical proportion, its analogical dissimilarity is zero. So for the next three examples of 4-tuples, we have $AD(1,1,1,1) = 0$, $AD(0,1,1,1) = 1$ and finally $AD(0,1,1,0) = 2$. In $\mathbb{B}$ the value of an analogical dissimilarity is in $[0,2]$. When dealing with vectors a, b, c and d in $\mathbb{B}^m$, analogical dissimilarity is defined as $\sum_{j=1}^{m} AD(a_j, b_j, c_j, d_j)$, in this case an analogical dissimilarity value belongs to the interval $[0, 2m]$.

A classifier based on analogical dissimilarity is proposed in [1]. Given a training set S, and a constant k specifying the number of the least dissimilar triples, the basic algorithm for classifying an instance $x \notin S$ in a naive way, using analogical dissimilarities is as follows:

1. For each triple (a, b, c) having a solution for the class equation $cl(a)$: $cl(b) :: cl(c) : x$, compute the analogical dissimilarity $AD(a, b, c, x)$.
2. Sort these triples by ascending order of their analogical dissimilarity $AD(a, b, c, x)$.
3. If the k-th triple of the list has the value p, then let the k'-th triple be the last triple of this list with the value p.
4. For the first k'-th triples, solve the class equation and apply a voting strategy on the obtained class labels.
5. Assign to x, the winner class.

This procedure may be said naive since it looks for every possible triple from the training set S in order to compute the analogical dissimilarity $AD(a, b, c, x)$, therefore it has a complexity of $O(n^3)$, n being the number of instances in the training set. To optimize this procedure, the authors propose the algorithm FADANA which performs an off line pre-processing on the training set in order to speed up on line computation.

3.2 Analogical Proportions-Based Classifier

In a classification problem, objects A, B, C, D are assumed to be represented by vectors of attribute values, denoted a, b, c, d. Based on the previously defined AP inference, analogical classification rely on the idea that, if vectors a, b, c and d form a valid analogical proportion componentwise for all or for a large number of attributes (i.e., $a : b :: c : d$), this still continue hold for their corresponding class labels. Thus the analogical proportion between classes $cl(a)$: $cl(b) :: cl(c) : x$ may serve for predicting the unknown class $x = cl(d)$ of the new instance d to be classified. This is done on the basis of triples (a, b, c) of examples in the sample set that form a valid analogical proportion with d.

In a brute force way, AP-classifier proposed in [4], looks for all triples (a, b, c) in the training set whose class equation $cl(a) : cl(b) :: cl(c) : x$ have a possible solution l . Then, for each of these triples, compute a truth value $P(a, b, c, d)$ as the average of the truth values obtained in a componentwise manner using Eq. (5) (P can also be computed using the conservative extension, introduced in [5]). Finally, assign to d the class label having the highest value of P.

An optimized algorithm of this brute force procedure has been developed in [4] in which the authors rather search for suitable triples (a, b, c) by constraining c to be one of the k nearest neighbours of d.

This algorithm processes as follows:

1. Look for each triple (a, b, c) in the training set s.t: $c \in N_k(d)$.
2. Solve $cl(a) : cl(b) :: cl(c) : x$.
3. If the previous analogical equation on classes has a solution l, increment the credit $credit(l)$ with $P(a, b, c, d)$ as $credit(l)+ = P(a, b, c, d)$.
4. Assign to d the class label having the highest credit as $cl(d) = argmax_l(credit))$.

4   Continuous Analogical Proportions-Based Classifier

Extrapolation and interpolation have been recognized as suitable tools for prediction and classification [6]. Continuous analogical proportions rely on the idea that if three items a, b and c form a valid analogical proportion $a : b :: b : c$, this may establish the basic for interpolating b in case a and c are known. As introduced in Sect. 2, in the numerical case b can be considered as the *midpoint* of (a, c) and may simply be computed from a and c.

In this section, we will show how *continuous analogical proportions* may help to develop an new classification algorithm dealing with numerical data and leading to a reduced complexity if compared to the previous Analogical Proportions-based classifiers.

4.1   Basic Procedure

Given a training set $S = \{(o_i, cl(o_i)\}$, s.t. the class label $cl(o_i)$ is known for each $o_i \in S$, the proposed algorithm aims to classify a new object $b \notin S$ whose label $cl(b)$ is unknown. Objects are assumed to be described by numerical attribute values. The main idea is to predict the label $cl(b)$ by interpolating labels of other objects in the training set S. Unlike algorithms previously mentioned in Sect. 3, continuous analogical proportions-based interpolation enables us to perform prediction using pairs of examples instead of triples. The basic idea is to find all pairs $(a, c) \in S^2$ with known labels s.t. the equation $cl(a) : x :: x : cl(c)$ has a solution l, l being a potential prediction for $cl(b)$. If this equation is solvable, we should also check that the continuous analogical proportion holds on each feature j. Indeed we have $a : b :: b : c$ if and only if $\forall j, \quad a_j : b_j :: b_j : c_j$ (i.e., for each feature j, b_j is being the exact midpoint of the pair (a_j, c_j), $b_j = (a_j + c_j)/2$).

As it is frequent to find multiple pairs (a, c) which may build a valid continuous analogical proportion with b with different solutions for the equation

$cl(a) : x :: x : cl(c)$, it is necessary to set up a *voting* procedure to aggregate the potential labels for b. This previous process can be described by the following procedure:

1. Find pairs (a, c) such that the equation $cl(a) : x :: x : cl(c)$ has a valid solution l.
2. If the continuous analogical proportion $a : b :: b : c$ is also valid, increment the score $ScoreP(l)$ for label l.
3. Assign to b the label l having the highest $ScoreP$.

4.2 Algorithm

As already said, the simplest way is to consider pairs (a, c) for which the analogical equation $cl(a) : x :: x : cl(c)$ is solvable and the analogical proportion $a : b :: b : c$ is valid.

However, unlike for Boolean features, where $a : b :: b : c$ may hold for many pairs (a, c), it is not really the case for numerical features. In fact, $P(a, b, c) = 1$ does not occur frequently. To deal with such situation in the numerical case, *AP*-classifiers [4] cumulate individual analogical credits $P(a, b, c, d)$ to the amount $CreditP(l)$ each time the label l is a solution for the equation $cl(a) : cl(b) :: cl(c) : x$. Even though learning from the entire sample space is often beneficial (in contrast to k-NN principle which is based on a local search during learning), considering all pairs for prediction may seem unreasonable as this could blur the results. Instead of blindly considering all pairs (a, c) for prediction, we suggest to adapt the analogical inference, defined by Eq. (9), in such way to consider only pairs (a, c) whose analogical score $P(a, b, c)$ exceeds a certain threshold θ.

$$\frac{P(a, b, c) > \theta}{cl(a) : cl(b) :: cl(b) : cl(c)} \tag{12}$$

This threshold is fixed on an empirical basis. Determining which threshold fits better with each type of dataset is still has to be investigated. The case of unclassified instances may be more likely to happen because of a conflict between multiple classes (i.e., $max(ScoreP)$ is not unique) rather than because of no pairs were found to made a proper classification. That's why we propose to record the best analogical score $bestP(l)$, and even the number of pairs having this best value $vote(l)$ in order to avoid this conflicting situation.

Algorithm 1. CAP-classifier for numerical data

Input:a training set S, object b $\notin S$, a threshold θ
for each label l **do** $ScoreP(l) = 0$, $bestP(l) = 0$, $vote(l) = 0$ **end for**
for each pair $(a, c) \in S^2$ **do**
  if $cl(a) : x :: x : cl(c)$ has solution l **then**
    $p = P(a, b, c)$
    if $p \geq \theta$ **then**
      $ScoreP(l) = ScoreP(l) + 1$
    else if $bestP(l) < p$ **then**
      $bestP(l) = p$
      $vote(l) = 1$
    else if $bestP(l) = p$ **then**
      $vote(l) = vote(l) + 1$
    end if
  end if
end for
$maxScore = max(ScoreP(l))$
if $unique(maxScore, ScoreP(l))$ **then**
  return $argmax_l(ScoreP(l))$
else
  $maxBest = max(bestP(l))$
  if $unique(maxBest, bestP(l))$ **then**
    return $argmax_l(bestP(l)), l \in argmax_l(ScoreP(l))$
  else
    return $argmax_l(vote(l)), l \in argmax_l(bestP(l)), l \in argmax_l(ScoreP(l))$
  end if
end if

---

5 Experimentations and Discussion

In this section, we aim to evaluate the efficiency of the proposed algorithm to classify numerical data. For this aim, we test the CAP-classifier on a variety of datasets from the U.C.I. machine learning repository [7], we provide its experimental results and compare them to the AP-classifier [4] as well as to the state of the art ML classifiers, especially, k-NN, C4.5, JRIP and SVM classifiers.

5.1 Datasets for Experiments

The experimentations are done on datasets from the U.C.I. machine learning repository [7]. Table 3 presents a brief description of the numerical datasets selected for this study. Datasets with numerical attributes must be normalized before testing to fit the multi-valued setting of analogical proportion. A numeric attribute value r is rescaled into the interval $[0, 1]$ as follows:

$$r_{rescaled} = \frac{r - r_{min}}{r_{max} - r_{min}}$$

r_{min} and r_{max} being the maximum and the minimum value of the attribute in the training set. We experiment over the following 9 datasets:

- "Diabetes", "W.B. Cancer", "Heart", "Ionosphere" are binary class datasets.
- "Iris", "Wine", "Sat.Image", "Ecoli" and "Segment" datasets are multiple class problems.

Table 3. Description of numeric datasets

Datasets	Instances	Numerical attrs.	Classes
Diabetes	768	8	2
W. B. Cancer	699	9	2
Heart	270	13	2
Ionosphere	351	34	2
Iris	150	4	3
Wine	178	13	3
Satellite Image	1090	36	6
Ecoli	336	7	8
Segment	1500	19	7

5.2 Testing Protocol

In terms of protocol, we apply a standard 10 fold cross-validation technique. As usual, the final accuracy is obtained by averaging the 10 different accuracies for each fold.

However, we have to tune the parameter θ of the CAP-classifier as well as parameter k for AP-classifier and the ones of the classical classifiers (with which we compare our approach) before performing this cross-validation.

For this end, in each fold we keep only the corresponding training set (i.e. which represents 90% of the full dataset). On this training set, we again perform an inner 10-fold cross-validation with diverse values of the parameter. We then select the parameter value providing the best accuracy. The tuned parameter is then used to perform the initial cross-validation. As expected, these tuned parameters change with the target dataset. To be sure that our results are stable enough, we run each algorithm (with the previous procedure) 5 times so we have 5 different parameter optimizations. The displayed parameter β is the average value over the 5 different values (one for each run). The results shown in Table 4 are the average values obtained from 5 rounds of this complete process.

5.3 Results for CAP-Classifiers

In order to evaluate the efficiency of our algorithm, we compare the average accuracy over five 10-fold CV to the following existing classification approaches:

Table 4. Results of *CAP*-classifier, *AP*-classifier and other ML classifiers obtained with the best parameter β

Datasets	CAP-classifier		AP-classifier		k-NN		C4.5		JRIP		SVM (RBF)		SVM (Poly)	
	acc.	β	acc.	β	acc.	β	acc.	β	acc.	β	acc.	β	acc.	β
Diabetes	72.81	0.906	73.28	11	73.42	11	74.73	0.2	74.63	5	77.37	(8192, 3.051E−5)	77.34	(0.5, 1)
Cancer	96.11	0.825	97.01	4	96.70	3	94.79	0.2	95.87	4	96.74	(2, 2)	96.92	(2, 1)
Heart	81.63	0.693	81.90	10	82.23	11	78.34	0.2	78.52	4	79.98	(32, 0.125)	83.77	(0.5, 1)
Ionosphere	86.44	0.887	90.55	1	88.80	1	89.56	0.1	89.01	5	94.70	(2, 2)	89.28	(0.03125, 2)
iris	95.73	0.913	94.89	5	94.88	3	94.25	0.2	93.65	6	94.13	(32768, 0.5)	96.13	(512, 1)
Wine	96.85	0.832	98.12	9	97.75	7	94.23	0,1	94.99	8	98.20	(32768, 2)	98.53	(2, 1)
Sat image	95.60	0.991	94.96	1	94.88	1	92.71	0.1	92.77	3	96.01	(8, 2)	95.11	(0.5, 4)
Ecoli	86.01	0.93	83.32	7	85.37	5	82.60	0.2	81.56	5	87.50	(2, 8)	87.50	(8, 1)
Segment	96.91	1	96.84	1	96.76	1	95.77	0.2	94.55	6	96.98	(2048, 0.125)	97.14	(8, 4)
Average	89.79		90.10		90.09		88.55		88.39		91.29		91.30	

- **IBk:** implements k-NN, using manhattan distance and the tuned parameter is the number of nearest neighbours during the inner cross-validation with the values $k = 1, 2, ..., 11$.
- **C4.5:** implements a generator of pruned or unpruned C4.5 decision tree. the tuned parameter is the confidence factor used for pruning with the values $C = 0.1, 0.2, ..., 0.5$.
- **JRip:** implements the rule learner RIPPER (Repeated Incremental Pruning to Produce Error Reduction) an optimized version of IREP. The number of optimization runs with the values $O = 2, 4, ..., 10$ is tuned during the inner cross-validation.
- **SVM:** an implementation of the Support Vector Machine classifier. We use SVM with both RBF and polynomial kernels and the tuned parameters are, successively *gamma* for the RBF Kernel, with $\gamma = 2^{15}, 2^{-13}, ..., 2^3$ and the degree for the polynomial kernel, $d = 1, 2, ..., 10$. The complexity parameter $C = 2^{-5}, 2^{-3}, ..., 2^{15}$ is also tuned.
- **AP-classifier:** implements the analogical proportions-based classifier with the tuned parameter k with k being the number of nearest neighbours $k = 1, 2, ..., 11$.
- **CAP-classifier:** We test the classifier and we tune the threshold θ with values $\theta = 0.5, 0.6, ..., 1$.

Results for AP-classifier as well as for classic ML classifiers are taken from [4], ML classifiers results are initially obtained by applying the free implementation of Weka software. Table 4 shows these experimental results.

Evaluation of CAP-Classifier and Comparison with Other ML Classifiers: If we analyse the results of CAP-classifier, we can conclude that:

- As expected, the threshold θ of the CAP-classifier change with the target dataset.
- The average θ is approximately equal to 0.89. This proves that CAP-classifier obtains its highest accuracy only if the selected pairs, useful for predicting the class label, are relatively in analogy with the item to be classified.

- For "Iris", "Ecoli", "Sat.Image" and "Segment" datasets, CAP-classifier performs better than AP-classifier, and even slightly better than SVM (polynomial kernel) on the "Sat.Image" dataset, which proves the ability of this classifier to deal with multi-class datasets (up to 8 class labels for these datasets).
- Moreover, we note that for most tested datasets, the optimized θ is close to 1. This fits our first intuition that CAP-classifier performs better when the selected pairs (a, c) form a valid continuous analogical proportion with b on all (case when $\theta = 1$) or maximum set of attributes (case when $\theta \approx 1$).
- CAP-classifier performs slightly less than AP-classifier for datasets "Diabetes", "Cancer" and "Ionosphere" which are binary classification problems. We may expect that extrapolation, involving triples of examples and thus larger set of the search space is more appropriate for prediction than interpolation using only pairs for such datasets. Identifying the type of data that fits better with each kind of approaches is subject to further instigation.
- For the rest of the datasets, CAP-classifier performs in the same way as the AP-classifier or k-NN. CAP-classifier achieves good results with a variety of datasets regardless the number of attributes (e.g., "Iris" with only 4 attributes, "Sat. image" with 36 attributes).
- As it may be expected, using triples of items for classification is more informative than pairs since more examples are compared against each other in this case. Even though, CAP-classifier performs approximately the same average accuracy as *AP-classifier* exploiting triples ($89, 79\% \approx 90, 10\%$) while keeping a lower complexity if compared to classic AP-classifiers. These results highlight the interest of continuous analogical proportions for classification.

Nearest Neighbors Pairs. In this sub-section, we would like to investigate better the characteristics of the pairs used for classification. For this reason, we check if voting pairs (a, c) are close or not to the item b to be classified. To do that, we compute the proportion of pairs that are close to b among all voting pairs. If this proportion is rather low, we can conclude that the proposed algorithm is able to correctly classify examples b using pairs (a, c) for which b is just the midpoint of a and c without being necessarily in their proximity.

From a practical point, we adopt this strategy:

- Given an item b to be classified.
- Search for the k nearest neighbors $NN = \{n_1, n_2, ...n_k\}$ of b. In practice, we consider to test with $k = 5, 10$.
- Compute the percentage of voting pairs (a, c) that are among the k nearest neighbors of b, i.e. $min(D(a, b), D(b, c)) \leq D(n_k, b)$, $D(x, y)$ being the distance between items x and y. If this percentage is low, it means that even if voting pairs (a, c) remain far to the item b, the proposed interpolation-based approach succeeds to guess the correct label for b.

The results are shown in Table 5. In this supplementary experiment, we only consider testing examples whose voting pairs (a, c) have a continuous analogical proportion $P(a, b, c)$ exceeding the threshold θ (see last column in Table 5).
From these results we can note:

- For $k = 5$ (first column), the proportion of pairs (a, c) (among those exceeding the threshold) that are in the neighborhood of b (those (a, c) that are closest to b than its neighbor n_5) is less than 10% for all tested datasets except for "Wine" which is little higher. This demonstrates that for these datasets, the CAP-classifier exploits the entire space of pairs for prediction, indeed most of examples are predicted thanks to pairs (a, c) that are located outside of the neighborhood of b.
- Even when the number of nearest neighbors k is extended to 10, this proportion remains low for most of the datasets. Especially for "Diabetes" and "Ecoli", the percentage of pairs in the neighborhood of b is close to 5%. For other datasets, this percentage is less than 20%.
- Note that the behavior of our algorithm is quite different from the k-NN classifier. While this latter computes the similarity between the example b to be classified and those in the training set, then classifies this example in the same way as its closest neighbors, our algorithm evaluates to what extent b is in continuous analogy with the pairs in the training set (these pairs are not necessarily in the proximity), then classifies it as the winning class having the highest number of voting pairs.
- These last results show that voters (a, c) remain far from to the item b to be classified.

Table 5. Proportion of pairs (a, c) that are nearest neighbors to b

Datasets	% of pairs (a, c) that are among the 5 neighbors of b	% of pairs (a, c) that are among the 10 neighbors of b	% of examples b for which $P(a, b, c) > \theta$
Diabetes	4.03%	5.98%	80.42%
Cancer	5.35%	8.29%	94.32%
Heart	6.85%	9.01%	95.04%
Ionosphere	5.53%	11.60%	63.17%
Iris	8.19%	14.67%	94.13%
Wine	14.65%	18.78%	87.85%
Ecoli	4.55%	6.88%	90.03%

6 Conclusion

This paper studies the ability of continuous analogical proportions, namely statements of the form a *is to* b *as* b *is to* c, to classify numerical data and presents a classification algorithm for this end. The basic idea of the proposed approach is to search for all pairs of items, in the training set, that build a continuous analogical proportion on *all* or most of the features with the item to be classified. An analogical value is computed for each of these pairs and only those pairs whose score exceeds a given threshold are kept and used for prediction. In case no such pairs could be found for each class label, the best pair having the highest analogical value is rather used. Finally, the class label with the best score is assigned to the example to be classified. Experimental results show the interest of the CAP-classifier for classifying numerical data. In particular the proposed algorithm may slightly outperform some state-of-the-art ML algorithms (such as: k-NN, C4.5 and JRIP), as well as the AP-classifier on some datasets. This leads to conclude that for classification, building analogical proportions with three objects (using continuous analogical proportions) instead of four enables to get an overall average accuracy close to that of previous AP-classifier while reducing the complexity to be quadratic instead of being cubic.

References

1. Bayoudh, S., Miclet, L., Delhay, A.: Learning by analogy: a classification rule for binary and nominal data. In: Proceedings of the International Joint Conference on Artificial Intelligence, IJCAI 2007, Hyderabad, India, 6–12 January, pp. 678–683 (2007)
2. Bounhas, M., Prade, H.: An analogical interpolation method for enlarging a training dataset. In: Ben Amor, N., Quost, B., Theobald, M. (eds.) SUM 2019. LNCS (LNAI), vol. 11940, pp. 136–152. Springer, Cham (2019). https://doi.org/10.1007/978-3-030-35514-2_11
3. Bounhas, M., Prade, H., Richard, G.: Oddness-based classification: a new way of exploiting neighbors. Int. J. Intell. Syst. **33**(12), 2379–2401 (2018)
4. Bounhas, M., Prade, H., Richard, G.: Analogy-based classifiers for nominal or numerical data. Int. J. Approx. Reason. **91**, 36–55 (2017)
5. Dubois, D., Prade, H., Richard, G.: Multiple-valued extensions of analogical proportions. Fuzzy Sets Syst. **292**, 193–202 (2016)
6. Lieber, J., Nauer, E., Prade, H., Richard, G.: Making the best of cases by approximation, interpolation and extrapolation. In: Cox, M.T., Funk, P., Begum, S. (eds.) ICCBR 2018. LNCS (LNAI), vol. 11156, pp. 580–596. Springer, Cham (2018). https://doi.org/10.1007/978-3-030-01081-2_38
7. Metz, J., Murphy, P.M.: UCI Repository (2000). ftp://ftp.ics.uci.edu/pub/machine-learning-databases
8. Miclet, L., Bayoudh, S., Delhay, A.: Analogical dissimilarity: definition, algorithms and two experiments in machine learning. J. Artif. Intell. Res. **32**, 793–824 (2008)
9. Miclet, L., Bayoudh, S., Delhay, A., Mouchére, H.: De l'utilisation de la proportion analogique en apprentissage artificiel. In: Actes des Journées Intelligence Artificielle Fondamentale, IAF 2007, Grenoble, 2–3 July 2007 (2007). http://www.cril.univ-artois.fr/konieczny/IAF07/
10. Miclet, L., Prade, H.: Handling analogical proportions in classical logic and fuzzy logics settings. In: Sossai, C., Chemello, G. (eds.) ECSQARU 2009. LNCS (LNAI), vol. 5590, pp. 638–650. Springer, Heidelberg (2009). https://doi.org/10.1007/978-3-642-02906-6_55
11. Prade, H., Richard, G.: Analogical proportions: another logical view. In: Bramer, M., Ellis, R., Petridis, M. (eds.) Research and Development in Intelligent Systems, pp. 121–134. Springer, London (2010). https://doi.org/10.1007/978-1-84882-983-1_9
12. Prade, H., Richard, G.: Analogical proportions: from equality to inequality. Int. J. Approx. Reason. **101**, 234–254 (2018)

Evaluation of Uncertainty Quantification in Deep Learning

Niclas Ståhl$^{(\boxtimes)}$ ⓘ, Göran Falkman ⓘ, Alexander Karlsson ⓘ,
and Gunnar Mathiason ⓘ

School of Informatics, University of Skövde, Högskolevägen 28, 54145 Skövde, Sweden
niclas.stahl@his.se

Abstract. Artificial intelligence (AI) is nowadays included into an increasing number of critical systems. Inclusion of AI in such systems may, however, pose a risk, since it is, still, infeasible to build AI systems that know how to function well in situations that differ greatly from what the AI has seen before. Therefore, it is crucial that future AI systems have the ability to not only function well in known domains, but also understand and show when they are uncertain when facing something unknown. In this paper, we evaluate four different methods that have been proposed to correctly quantifying uncertainty when the AI model is faced with new samples. We investigate the behaviour of these models when they are applied to samples far from what these models have seen before, and if they correctly attribute those samples with high uncertainty. We also examine if incorrectly classified samples are attributed with an higher uncertainty than correctly classified samples. The major finding from this simple experiment is, surprisingly, that the evaluated methods capture the uncertainty differently and the correlation between the quantified uncertainty of the models is low. This inconsistency is something that needs to be further understood and solved before AI can be used in critical applications in a trustworthy and safe manner.

1 Introduction

Much of the great progress of AI in the last years is due to the development of *deep learning (DL)* [14]. However, one big problem with DL methods is that they are considered to be "black box" methods, which are difficult to interpret and understand. This becomes problematic when DL algorithms are taking more and more critical decisions, impacting the daily life of people and no explanation is given for why a certain decision is taken. There are some researchers, for example Samek et al. [21] and Montavon et al. [18], that currently address this problem and try to make DL models interpretable. This problem is far from solved and it is an important research direction since it is likely that many critical decisions taken by AI based algorithms in the near future will be in consensus with a human [8]. Examples of such decisions would, for example, be those of an autonomous car with a human driver and those of a doctor using an

© Springer Nature Switzerland AG 2020
M.-J. Lesot et al. (Eds.): IPMU 2020, CCIS 1237, pp. 556–568, 2020.
https://doi.org/10.1007/978-3-030-50146-4_41

image to determine if a patient has skin cancer or not [2,4,22]. In these cases, the AI will support and enhance the human decision maker. Here, the human can act in contradiction to what is suggested by the AI, and, thus, prevent erroneous decisions taken by the AI. In both of the named cases, such erroneous decisions could potentially cause a lot of human suffering and even fatalities. However, there are problems where the AI cannot be supervised by a human and, consequently, the AI itself needs to be able to determine when there is a risk of an incorrect decision.

While wrongly taken decisions can be decreased by better models and more and better training data, it is infeasible to cover all possible situations for all but the most trivial problems. Consequently, a system built with an ML model will always encounter situations that differ from all the previous samples used for training. In order to be trustworthy, in this case, it is crucial that the model shows that it encounters an unknown situation where it is forced to extrapolate its knowledge and emphasises that its outcome, therefore, is uncertain [12]. However, as pointed out by Gal and Ghahramani [3], Richter and Roy [20] and Lakshminarayanan et al. [13], it is a challenging and still open problem to quantify the uncertainty of deep learning models. Hendrycks and Gimpel [5] do, for example, show that deep learning models that use the softmax activation function in the last layer are bad at estimating prediction uncertainty and often produce overconfident predictions. It is not difficult to imagine that such overconfident predictions can lead to catastrophic outcomes in some of the previously mentioned cases, such as in the medical domain. Therefore, it is an important research direction to find methods that allow for the quantification of uncertainty in the provided predictions. In this paper, we do not propose such an approach, but do instead evaluate existing models that have been proposed to solve this problem. This is done in order to further understand their limitations and to highlight the differences that arise when different models are selected.

When quantifying the uncertainty, it is essential that methods consider both the epistemic uncertainty (the uncertainty that arises due to lack of observed data) and the aleatory uncertainty (the uncertainty that arises due to underlying random variations within the collected data) [7]. But, it is also important to differentiate between these two causes of uncertainty. In the latter case, there is an observed variation among the samples and, hence, the uncertainty can be well quantified and all risks can be assessed. It is therefore possible to take a well-informed decision, knowing the uncertainty. This is not the case for epistemic uncertainty, where the uncertainty arises due to lack of data and, hence, the model is forced to extrapolate its knowledge. When the model extrapolates it takes an uniformed decision, which can be far from optimal.

To further highlight this problem, this paper examine how well current methods for the quantification of uncertainty manage to show the uncertainty that arises from out of the distribution samples. Two experiments are therefore conducted. In the first experiment, deep learning models that has been proposed to support quantification of predictive uncertainty and that can be used for classification of data are evaluated. These models are: a *deep neural network* with a *softmax* output layer, an *ensemble* of deep neural networks [13] and a *deep Bayesian neural network* [3], where two separate ways to quantify the uncertainty are used for the softmax model. The first treats the output as a probability while the second method considers the gradient information. The result from these models are compared to another deep learning approach for the detection of out of distribution samples, namely an *autoencoder*. Of these methods, there are two, the Bayesian neural network and the ensemble of neural networks, that are able to disentangle the epistemic and the aleatory uncertainties. A second experiment is therefore conducted, with these two methods, in order to see if the results can be further improved when the uncertainty is split into an aleatoric and an epistemic part.

With these experiments, we show that there is a clear difference in how the investigated methods quantify the uncertainty and what samples they considered to be uncertain. The correlations between the quantified uncertainty of the different models are also very low, showing that there is an inconsistency in the uncertainty quantification. Thus, there is a need for further study of uncertainty in deep learning methods before these can be applied in real world applications in an absolutely safe way.

2   Method

The different models and their setups are first described in this section. Since the targets of the different models differ, there is a need to quantify the uncertainty of these models differently. The second part of this section will therefore describe different ways to quantify the uncertainty. In this section, the motivation behind the selection of how to quantify the uncertainty in the experiments is also given. The last part of this section then describes the experimental setup for all experiments.

2.1   Models

Different deep learning models for classification and uncertainty quantification are used in the conducted experiments. They are all described below, together with the corresponding architecture and parameter settings that are used in the experiments. How the uncertainty is quantified is described in Sect. 2.2–2.3.

Softmax Deep Neural Network. The softmax function is often used in neural networks to fuzzily assign a sample to a given class [1]. Thus, the softmax will give the proportional belief of how much a sample belongs to a given class. This is often used as an approximation of the probability for how likely it is that a sample belongs to a given class. The softmax function is given by:

$$p(y = z | x, \omega) = \frac{f_z^\omega(x)}{\sum\limits_{z' \in Z} f_{z'}^\omega(x)}, \tag{1}$$

where z is the given output class, which belongs to the set of all possible outcomes, Z. X is the input sample and $f_z^\omega(x)$ is an arbitrary function, parameterised by ω, giving the support that x belongs to class z. Equation (1) allows us to find the probability distribution of all possible outcome classes. This distribution can be used to quantify the uncertainty, as described in Sect. 2.2.

Recent studies, for example by Oberdiek et al. [19] and Vasudevan et al. [26], suggest that the uncertainty of the model is reflected by the stability of the model, where the stability can be measured by the gradients of the parameters. Hence, the stability of the model is given by:

$$\nabla_\omega \mathcal{L} = \nabla_\omega l\left(\hat{y}_i, f^\omega(x_i)\right), \tag{2}$$

where $\mathcal{L}$ is the loss of the model given by an arbitrary loss function l, $\hat{y}_i$ is the predicted class for the i:th sample and $f^\omega(x_i)$ is the prediction from the model that is parameterised by ω. We follow the same experimental setup as Oberdiek et al. [19] and use the negative log-likelihood for the predicted class as the loss function. In this case, Eq. 2 can be written as

$$\nabla_\omega \mathcal{L} = \nabla_\omega - \log\left(p(y_i = \hat{y}_i | x_i, \omega)\right). \tag{3}$$

Furthermore, we use a deep neural network as the underlying model, that is, $p(y_i = \hat{y}_i | x_i, \omega)$ in Eq. 3 is given by Eq. 1.

Bayesian Neural Network. We consider Bayesian neural networks to be neural networks that have a posterior distribution of weights instead of a single point estimate. The same definition is, for example, used by Gal and Ghahramani [3]. Hence, the training of a Bayesian neural network consists of finding a good estimate to the probability distribution $p(\omega|X, Y)$, where ω is the network weights and X is the set of inputs and Y is the set of outputs. It is however, unfeasible to find the exact solution to $p(\omega|X, Y)$ and, hence, an estimate must be used instead. In this paper we approximate $p(\omega|X, Y)$ with a network that uses dropout [25] during both the training and the testing phase. This is the same approach as Gal and Ghahramani [3]. With an estimated posterior distribution, $p(\omega|X, Y)$, multiple network weights can be sampled. Hence, many likely network weights can be used for predictions, which would allow for a smaller risk of overfitting and a greater diversity in the output. The final classification of a sample x of a Bayesian neural network is given by:

$$p(y|x, \Omega) = \frac{1}{M} \sum_{i=0}^{M} f^{\omega_i}(x), \tag{4}$$

where M is the total number of samples, f is a neural network parameterized by ω_i that is the i:th sample from the posterior distribution, Ω, of network weights.

Ensemble of Neural Networks. Ensemble methods are learning algorithms that consist of a set of models. Each model makes its own prediction independently of the other models in the ensemble. The final prediction is then derived from the composition of all models in the ensemble. We use an ensemble of neural networks, such as the one presented by Maqsood et al. [17]. Such ensembles have been shown to be good at quantifying the predictive uncertainty, as shown by Lakshminarayanan et al. [13]. The classification of a sample x by the ensemble is the average prediction over all classifiers. Hence, the prediction, y, is given by:

$$p(y|x, w_0, \ldots, w_M) = \frac{1}{M} \sum_{i=0}^{M} f^{w_i}(x),$$ (5)

where M is the number of networks in the ensemble and w_i is the parametrisation of the i:th classifier, f, in the ensemble. Note the similarity with the deep Bayesian neural network as given in Eq. 4.

Autoencoder. An autoencoder is a neural network that has the same number of input neurons as output neurons. This network consists of two parts: an encoder that compresses the input to a compressed representation of the sample, with an as low loss of information as possible, and a decoding part that decompresses the compressed representation to the original representation [6]. These parts are jointly trained and, hence, the encoder is forced to learn an encoding scheme that the decoder can decompress. Therefore, these two models will learn how to collaborate, but only on data that is similar to the data they see during training. This means that the encoder would not be able to encode novel data in such a way that it can be reconstructed by the decoder. This can be exploited to detect how much a new sample diverges from an initial distribution. Thus, the uncertainty of a prediction in a predictive model may be quantified by the reconstruction error of a sample given to the autoencoder. This is, for example, done by Leibig et al. [16], and the same approach will be used in our experiments.

2.2 Uncertainty Quantification

There are multiple ways that uncertainty can be quantified. Kendall and Gal [9], for example, quantifies the uncertainty as the variance in the predictions. We follow the same approach as Lakshminarayanan et al. [13] and use the Shannon entropy [23]:

$$H(y, X) = - \sum_i p(y = i|X) \log p(y = i|X)$$ (6)

as a measure of the uncertainty in the predictions of the models specified in Sect. 2.1. This design choice is mainly selected to enable the comparison between the uncertainties of the softmax network and the other models, since the softmax network does not have any variation in its predictions.

It is, however, not possible to use this uncertainty metric for the experiments that consider the gradient information. In this case, we use the approach suggested by Oberdiek et al. [19], namely to use the Euclidean norm of the gradients. The quantified uncertainty of a prediction is then given by $||\nabla_w L||_2$, where $\nabla_w L$ is described in Eq. 3.

It is also impossible to measure the entropy of predictions in the autoencoder (described in Sect. 2.1) since the autoencoder does not provide any predictions. Instead, we quantify the uncertainty by measuring the Euclidean distance between the original sample and the encoded and decoded sample:

$$\|X, \hat{X}\| = \sqrt{\sum_{i=1}^{n} (x_i - \hat{x}_i)^2}, \qquad (7)$$

here x_i is the original value of the i:th feature of x and $\hat{x}_i$ is the reconstructed value for the same feature. Hence, Eq. 7 measures how well the autoencoder manages to encode the vector and then decode the sample vector X of length n.

2.3 Heteroscedastic Aleatoric Uncertainty

The aleatoric uncertainty can be divided into two sub-categories: *heteroscedastic* and *homoscedastic* uncertainty. Heteroscedastic uncertainty assumes that the aleatoric uncertainty is data dependent and, thus, that the uncertainty varies over different inputs. Hence, models that can capture the heteroscedastic uncertainty are useful when the uncertainty is greater in some areas of the input space. Such is the case in the MNIST dataset [15], where some of the digits are badly written and the output class is uncertain.

The heteroscedastic uncertainty in the models will be treated in the same way by Kendall and Gal [9] and furthermore described as in Kendall et al. [10]. Here, the expected variance of the noise in the output is modelled by the noise parameter σ. This parameter will be dependent on the input and the models will learn how to predict it, given some particular input. In the presented multiclass setting, the loss with included heteroscedastic uncertainty can be approximated with:

$$\mathcal{L}(\omega, x, y) = \frac{1}{\sigma^2} \mathcal{L}_{ce}(softmax(f^\omega(x)), y, \omega) + \log(\sigma), \qquad (8)$$

where $\mathcal{L}_{ce}$ is the categorical cross entropy loss and f^ω is the model, parameterized by ω and predicting logits to the softmax function. The predicted logits are assumed to be drawn from a Gaussian distribution with a variance of σ, where σ is dependent on the input x. This loss function is used in the second part of the experiments, where it is examined how well the Bayesian neural network and the ensemble of neural networks can capture and separate epistemic and aleatoric uncertainty.

2.4 Experiment Setup

All previously described models are trained on the MNIST dataset, which is a dataset that contains 70,000 samples of hand-written digits [15]. The predefined and commonly used training and test split, which uses 10,000 samples in the test set, is used in our experiments as well. A randomly selected validation set, consisting of 10% of the training set, is also used to prevent overfitting of the models when training. All models are then evaluated on the MNIST test set and the uncertainty of their predictions is quantified and split a set of correctly classified samples and a set of incorrectly classified samples. The main hypothesis is that the uncertainty would be much greater in the set of incorrectly classified samples. These models are then applied to the manually cleaned notMNIST[1] set that consist of 19,000 characters, set in different fonts. The objective of the models is to detect that these samples are very different from the original training data and attribute them with a high uncertainty. Since the autoencoder is not used to perform any classification, we decided to use a feed forward neural network that uses the latent encoding to predict the class of the output.

Parameter Settings. Both the Bayesian neural network and the softmax deep neural network have two layers with 800 neurons each. This is the same network architecture used by Simard et al. [24]. The inference in the Bayesian neural network is conducted in the same way as described by Gal and Ghahramani [3], with 60% dropout rate. The networks in the ensemble are each trained on bootstrap samples, which have 60% less samples than the original dataset. Since the amount of data is reduced, we also reduce the number of neurons in each layer to 40 % of the size of the Bayesian neural network. Hence, the ensemble will consist of 50 networks where each network has two layers of 320 neurons.

The autoencoder used in the experiment has 7 layers with the following number of neurons: 1000, 250, 50, 10, 50, 250, 1000. This is the same setup as used by Wang et al. [27]. All models are trained using the ADAM optimisation algorithm [11] with the commonly used learning rate of 0.001, to minimise the binary cross entropy error between the model predictions and the targeted classes.

3   Experimental Results

All presented deep learning methods are trained on the MNIST dataset and then evaluated on a smaller test set from MNIST as well as the notMNIST dataset. Some examples of samples from these datasets are shown in Fig. 1. The accuracy of all models are approximately the same and in line with what is expected from a two layered neural network model and the MNIST dataset [24]. The Bayesian neural network is, for example, the best performing model with an error rate of 1.3%, while softmax is the worst with an error rate of 1.6%.

Unlike the accuracy, there is a great difference in how the uncertainty of the models are quantified. This can be seen in Fig. 2, where the distributions of the quantified uncertainties are shown. The distributions over the uncertainty is split into three different distributions: the distribution over the uncertainty for correctly classified samples, the distribution over the uncertainty for incorrectly classified samples and the distribution over the uncertainty for samples from the notMNIST dataset. If a model acts as desirable, it should separate these three classes and thus, that the distributions in Fig. 2 are disjoint. This optimal case

---

[1] Available at: http://yaroslavvb.blogspot.co.uk/2011/09/notmnist-dataset.html.

corresponds to that the model correctly detects digits that are easy to classify and attribute them with a low uncertainty. Furthermore, the model detects odd looking digits and correctly attribute them with medium uncertainty, since the classification may be erroneous. Also, when a sample that is clearly not a number, the model should attribute it to an even higher uncertainty, since the prediction is extrapolated far from what the model knows. However, this implies that correct and possibly incorrect predictions can be identified by the quantified uncertainty and real digits can, thus, be filtered from non digits. This is not the case, the distributions do indeed overlap, as can be seen by studying the 95% quantiles for the distributions in Fig. 2.

The consensus of the uncertainties of the models are measured by their Pearson correlation (see Fig. 4). The measurements show a strong correlation between gradient information and the softmax predictions, but no strong correlation besides that. The quantified uncertainty of the softmax neural network is even negatively correlated to the autoencoder.

The uncertainty is furthermore divided into an epistemic and an aleatoric part, as shown in Fig. 3. The expected result would be that the notMNIST samples would have much greater epistemic uncertainty than all other samples, while the misclassified MNIST samples would have a greater aleatoric uncertainty. However, this can only partially be observed, since both the notMNIST and the misclassified MNIST samples show a high epistemic uncertainty.

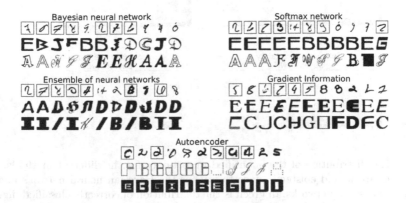

Fig. 1. Examples of the behaviour of the evaluated methods tested on the MNIST and the notMNIST datasets. The first row, for the given method, consists of the eleven most uncertain predictions from the MNIST dataset. Incorrectly classified examples are marked red. The second and the third row show the eleven most certain and uncertain examples from the notMNIST dataset, respectively. (Color figure online)

4   Discussion

The results show that all the evaluated methods quantify the uncertainty differently. The results, furthermore, support the previous observation by Hendrycks

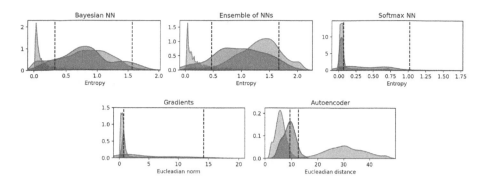

Fig. 2. The distribution of the quantified uncertainty for the different methods. In green is the distribution of correctly classified digits in the MNIST dataset. In blue is the distribution of incorrectly classified digits and in red is the distribution of the quantified uncertainty for samples from the notMNIST dataset. The 95% quantile of the quantified uncertainty of all samples from the MNIST dataset is marked with the dashed line to the left. The right dashed line is the 95% quantile of the quantified uncertainty when only the misclassified samples are considered. (Color figure online)

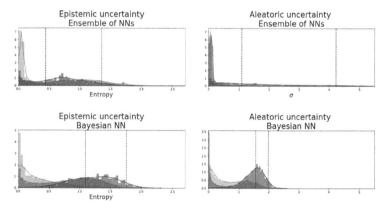

Fig. 3. The distribution of the quantified uncertainty for the different methods, split up into aleatoric and epistemic uncertainty for the Bayesian neural network and the ensemble of neural networks. In green is the distribution of correctly classified digits in the MNIST dataset. In blue, the incorrectly classified digits in MNIST and, in red is the distribution of the quantified uncertainty for samples from the notMNIST dataset. The 95% quantile of the quantified uncertainty of all samples from the MNIST dataset is marked with the dashed line to the left. The right dashed line is the 95% quantile of the quantified uncertainty when only the misclassified samples are considered. (Color figure online)

and Gimpel [5] that deep learning models that only use the softmax activation function to quantify the uncertainty are overconfident when faced with out of the distribution samples. The same holds true when the gradient information of the softmax neural network is used to quantify the uncertainty.

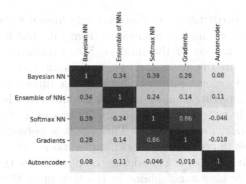

Fig. 4. The Pearson correlation between the quantified uncertainty of all tested methods. Many of the methods are very weakly correlated and the softmax neural network is even negatively correlated to the uncertainty of the autoencoder.

When the results in Fig. 1 were furthered studied, we identified several interesting behaviours of the models. The models that are used for classification extract knowledge about the shape of digits and apply it to the notMNIST data. Both the Bayesian neural network and the ensemble of neural networks do, for example, pick up the curvy shape of a "B" and interpret this as the digit "3" and, hence, the models are certain of the output in these cases. The round shape and the empty middle of the letter "D" being classified as the digit "0" is another example of the extrapolation of features into the new domain. The two methods that are based on a softmax neural network do an even cruder extrapolation and classify everything with a straight horizontal line at a certain height as a "7" (all the first eight samples shown in Fig. 1 for the softmax network and the gradient information is classified the digit "7").

No model achieved the optimal goal of quantifying the uncertainty in such a way that it separates the three different cases of input: digits that could be classified correctly, digits that could not be classified correctly and characters from the notMNIST set. However, the autoencoder correctly uses the uncertainty quantification to separate all notMNIST samples from the MNIST samples, while the Bayesian neural network and the ensemble of neural networks can correctly separate classified MNIST samples from the other two cases. It can, consequently, be efficient to use an autoencoder as a first filtering step to remove all out of the distribution samples. Another method, such as a Bayesian neural network, can then be used to perform safer classifications, where the uncertainty quantification can be used to identify possibly misclassified samples. There is no downside of such a combination of models, besides the slightly higher computational cost. It is, therefore, an interesting future research question how different models can be combined in order to handle and distinguish between the different cases that may cause these models to be uncertain.

A surprising observation is that the quantified uncertainties of most of the models are weakly correlated. All three models that are used for prediction are, for example, weakly correlated to the autoencoder, which is considered to capture the initial distribution well. Since we only measure the linear correlation it is difficult to draw any major conclusions from this, but it still gives us some insights into the behaviour of the models. We, therefore, suggest that these methods do not capture the uncertainty that arises due to the extrapolation, but instead finds fuzzy decision boundaries between the different classes and, hence, are able to spot odd looking samples between the different classes. However, this implies that there is no guarantee that predictions on out of the distribution samples will be considered uncertain. This poses a potential risk when using these kind of models in critical real world applications.

The use of an autoencoder is a good way to approximate the distance between a new sample and its closest neighbour in the training set. This is a promising result since the autoencoder is more efficient, when considering the computational complexity, compared to finding the closest neighbour in the training set and calculating the Euclidean distance. The computational complexity of finding the closest neighbour in the training set grows linearly in terms of the cardinality of the training dataset, while the computational complexity of the autoencoder is constant. Hence, it appears that the autoencoder correctly discovers when a model is faced with a sample that is far from what the model has seen before and, hence, forces the model to extrapolate. Thus, an autoencoder could potentially be used to detect when a sample would force a predictive model to extrapolate, if trained with the same data.

Splitting up the uncertainty into an epistemic and an aleatoric part and then use the epistemic uncertainty to detect outliers is not a successful approach in the performed experiments. While we expect the epistemic uncertainty to be much higher for such samples, it is not the case, since both the badly written MNIST digits and the notMNIST samples are attributed with a high epistemic uncertainty. However, the notMNIST samples distinguish themselves from the rest by having a very low aleatoric uncertainty. Hence, the outlier samples distinguish themselves from the rest by having a low aleatoric uncertainty, rather than having a high epistemic uncertainty. The combined epistemic and aleatoric uncertainty can therefore be used to detect the notMNIST samples. The reason why outlier samples are attributed with a low uncertainty can be seen in Eq. 8. Since the models are good at predicting the outcome, the expected cross entropy loss would be rather small. Hence, it is more beneficial for the model to minimise the $\log(\sigma)$ term for new unknown samples than to expect a large cross entropy error.

5   Conclusion and Summary

In this paper, several models for the quantification of uncertainty are evaluated. Even though the experimental setup is rather basic, it is shown that there is no consensus in the uncertainty of the models and that they capture different dimensions of the uncertainty. This problem is likely to persist, and may even be worse, when more advanced models are used or when more complicated problems are tackled. It is shown that the uncertainty quantification of some models (the Bayesian neural network and the ensemble of neural networks) can be used to distinguish between samples that are easy to classify and those that are difficult. Hence, these models quantify the uncertainties around the hyperplanes separating the different classes. The autoencoder, on the other hand, is good at quantifying the uncertainty that arises due to the extrapolation of points far from the training distribution. The performed experiments show that it can be beneficial to split up the uncertainty into an epistemic and an aleatoric part. However, the notMNIST samples did not differentiate themselves from the rest by having much higher epistemic uncertainty than the other samples, as was expected. Instead, the notMNIST samples stood out by having the combination of a high epistemic uncertainty and a low aleatoric uncertainty. However, none of the models managed to separate the three different cases of samples that were studied, namely correctly classified samples, incorrectly classified samples and samples that are far from the training distribution. On the others hand, as described above, some methods succeeded partially, and managed to separate one of the cases from the other. It can, therefore, be beneficial to use several models in real world applications to capture all uncertainties that may arise, in order to build safer AI systems.

References

1. Bridle, J.S.: Training stochastic model recognition algorithms as networks can lead to maximum mutual information estimation of parameters. In: Touretzky, D.S. (ed.) Advances in Neural Information Processing Systems 2, pp. 211–217. Morgan-Kaufmann (1990)
2. Esteva, A., et al.: Dermatologist-level classification of skin cancer with deep neural networks. Nature **542**(7639), 115 (2017)
3. Gal, Y., Ghahramani, Z.: Dropout as a Bayesian approximation: representing model uncertainty in deep learning. In: Balcan, M.F., Weinberger, K.Q. (eds.) Proceedings of The 33rd International Conference on Machine Learning. Proceedings of Machine Learning Research, vol. 48, pp. 1050–1059. PMLR, New York (2016)
4. Gerdes, J.C., Thornton, S.M.: Implementable ethics for autonomous vehicles. In: Maurer, M., Gerdes, J.C., Lenz, B., Winner, H. (eds.) Autonomes Fahren, pp. 87–102. Springer, Heidelberg (2015). https://doi.org/10.1007/978-3-662-45854-9_5
5. Hendrycks, D., Gimpel, K.: A baseline for detecting misclassified and out-of-distribution examples in neural networks. arXiv preprint arXiv:1610.02136 (2016)
6. Hinton, G.E., Zemel, R.S.: Autoencoders, minimum description length and Helmholtz free energy. In: Cowan, J.D., Tesauro, G., Alspector, J. (eds.) Advances in Neural Information Processing Systems 6, pp. 3–10. Morgan-Kaufmann (1994)
7. Hüllermeier, E., Waegeman, W.: Aleatoric and epistemic uncertainty in machine learning: a tutorial introduction. arXiv preprint arXiv:1910.09457 (2019)
8. Jarrahi, M.H.: Artificial intelligence and the future of work: human-AI symbiosis in organizational decision making. Bus. Horiz. **61**(4), 577–586 (2018)
9. Kendall, A., Gal, Y.: What uncertainties do we need in Bayesian deep learning for computer vision? In: Guyon, I., et al. (eds.) Advances in Neural Information Processing Systems 30, pp. 5574–5584. Curran Associates, Inc. (2017)

10. Kendall, A., Gal, Y., Cipolla, R.: Multi-task learning using uncertainty to weigh losses for scene geometry and semantics. In: Proceedings of the IEEE Conference on Computer Vision and Pattern Recognition, pp. 7482–7491 (2018)
11. Kingma, D., Ba, J.: Adam: a method for stochastic optimization. arXiv preprint arXiv:1412.6980 (2014)
12. Kuleshov, V., Fenner, N., Ermon, S.: Accurate uncertainties for deep learning using calibrated regression. arXiv preprint arXiv:1807.00263 (2018)
13. Lakshminarayanan, B., Pritzel, A., Blundell, C.: Simple and scalable predictive uncertainty estimation using deep ensembles. In: Guyon, I., et al. (eds.) Advances in Neural Information Processing Systems 30, pp. 6402–6413. Curran Associates, Inc. (2017)
14. LeCun, Y., Bengio, Y., Hinton, G.: Deep learning. Nature **521**(7553), 436–444 (2015)
15. LeCun, Y., Bottou, L., Bengio, Y., Haffner, P., et al.: Gradient-based learning applied to document recognition. Proc. IEEE **86**(11), 2278–2324 (1998)
16. Leibig, C., Allken, V., Ayhan, M.S., Berens, P., Wahl, S.: Leveraging uncertainty information from deep neural networks for disease detection. Sci. Rep.-UK **7**(1), 17816 (2017)
17. Maqsood, I., Khan, M.R., Abraham, A.: An ensemble of neural networks for weather forecasting. Neural Comput. Appl. **13**(2), 112–122 (2004). https://doi.org/10.1007/s00521-004-0413-4
18. Montavon, G., Samek, W., Müller, K.R.: Methods for interpreting and understanding deep neural networks. Digit. Sig. Process. **73**, 1–15 (2018)
19. Oberdiek, P., Rottmann, M., Gottschalk, H.: Classification uncertainty of deep neural networks based on gradient information. In: Pancioni, L., Schwenker, F., Trentin, E. (eds.) ANNPR 2018. LNCS (LNAI), vol. 11081, pp. 113–125. Springer, Cham (2018). https://doi.org/10.1007/978-3-319-99978-4_9
20. Richter, C., Roy, N.: Safe visual navigation via deep learning and novelty detection. In: Robotics: Science and Systems Conference. Robotics: Science and Systems Foundation, July 2017
21. Samek, W., Wiegand, T., Müller, K.R.: Explainable artificial intelligence: understanding, visualizing and interpreting deep learning models. arXiv preprint arXiv:1708.08296 (2017)
22. Shalev-Shwartz, S., Shammah, S., Shashua, A.: Safe, multi-agent, reinforcement learning for autonomous driving. arXiv preprint arXiv:1610.03295 (2016)
23. Shannon, C.E.: A mathematical theory of communication. Bell Syst. Tech. J. **27**(3), 379–423 (1948)
24. Simard, P.Y., Steinkraus, D., Platt, J.C.: Best practices for convolutional neural networks applied to visual document analysis. In: Proceedings of the Seventh International Conference on Document Analysis and Recognition, pp. 958–963 (2003)
25. Srivastava, N., Hinton, G.E., Krizhevsky, A., Sutskever, I., Salakhutdinov, R.: Dropout: a simple way to prevent neural networks from overfitting. J. Mach. Learn. Res. **15**(1), 1929–1958 (2014)
26. Vasudevan, V.T., Sethy, A., Ghias, A.R.: Towards better confidence estimation for neural models. In: ICASSP 2019–2019 IEEE International Conference on Acoustics, Speech and Signal Processing (ICASSP), pp. 7335–7339. IEEE (2019)
27. Wang, Y., Yao, H., Zhao, S.: Dropout: a simple way to prevent neural networks from overfitting. Neurocomputing **184**, 232–242 (2016)

XAI

Performance and Interpretability in Fuzzy Logic Systems – Can We Have Both?

Direnc Pekaslan$^{(\boxtimes)}$, Chao Chen, Christian Wagner, and Jonathan M. Garibaldi

University of Nottingham, Nottingham NG8 1BB, UK
{direnc.pekaslan,chao.chen,christian.wagner,
jon.garibaldi}@nottingham.ac.uk

Abstract. Fuzzy Logic Systems can provide a good level of interpretability and may provide a key building block as part of a growing interest in explainable AI. In practice, the level of interpretability of a given fuzzy logic system is dependent on how well its key components, namely, its rule base and its antecedent and consequent fuzzy sets are understood. The latter poses an interesting problem from an optimisation point of view – if we apply optimisation techniques to optimise the parameters of the fuzzy logic system, we may achieve better performance (e.g. prediction), however at the cost of poorer interpretability. In this paper, we build on recent work in non-singleton fuzzification which is designed to model noise and uncertainty 'where it arises', limiting any optimisation impact to the fuzzification stage. We explore the potential of such systems to deliver good performance in varying-noise environments by contrasting one example framework - ADONiS, with ANFIS, a traditional optimisation approach designed to tune all fuzzy sets. Within the context of time series prediction, we contrast the behaviour and performance of both approaches with a view to inform future research aimed at developing fuzzy logic systems designed to deliver both – high performance and high interpretability.

Keywords: Non-singleton fuzzy system · Interpretability · ADONiS · ANFIS · Parameter tuning

1 Introduction

A key aspect of the vision of interpretable artificial intelligence (AI) is to have decision-making models which can be understood by humans. Thus, while an AI may deliver good performance, providing an insight of the decision process is also an important asset for the given model. Even though the interpretability of AI is widely acknowledged to be a critical issue, it still remains as a challenging task [17].

Fuzzy set (FS) theory introduced by Zadeh [34], establishes the basis for Fuzzy Logic Systems (FLSs). Zadeh introduced them to capture aspects of human reasoning and in FLSs are frequently being referred to as 'interpretable'.

© Springer Nature Switzerland AG 2020
M.-J. Lesot et al. (Eds.): IPMU 2020, CCIS 1237, pp. 571–584, 2020.
https://doi.org/10.1007/978-3-030-50146-4_42

The main rationale for the latter is that FSs are generally designed in respect to linguistic labels and are interconnected by linguistic rules, which can provide insight into 'why/how results are produced' [28]. This capacity for interpretability is one of the main assets of fuzzy logic and is often one of the key motivations to use FLSs in decision-making [4].

While FLSs are considered to possess mechanisms which can provide a good degree of interpretability, research establishing the latter has been comparatively limited. Only in recent years an increasing number of studies have started to focus on fundamental questions such as what *interpretability* is, in general, and in particular in respect to FLSs? From a complexity point of view, how many rules or how many variables per rule is interpretable? Or from a semantic point of view, to which degree are properties of the partitioning of the variables (e.g. *completeness, distinguishability or complementarity*) key for interpretable FLSs? [1,12,15,19] These studies show that the interpretability of FLSs depends on their various components i.e. the number of rules, the structure of the rule set and the actual interpretability of each rule - which in turn depends on how meaningful the actual FSs are, i.e. how well they reflect the model which the interpreting stakeholder has in mind when considering the given linguistic label [12,13,28].

Traditionally, AI models use statistical optimisation techniques to tune parameters based on a data-driven approach. While these optimisation procedures provide performance benefits, they commonly do not consider whether the resulting model is interpretable or not. This poses an interesting question for the optimisation or tuning of FLS: can we use statistical optimisation to tune FLS parameters without negatively affecting the given FLSs interpretability? I.e., can we have both: interpretability and good performance?

There are several established approaches to tune FLSs using statistical optimisation. Here, ANFIS (adaptive-network-based fuzzy inference system), introduced by Jang [14], and later extended in [6] for interval type-2 fuzzy logic system has been one of the most popular. ANFIS uses statistical optimisation to update FLS parameters based on a given training dataset with the objective to deliver good performance, i.e. minimum error. However, during the optimisation, ANFIS does not consider aspects of interpretability [27], for example potentially changing antecedent and consequent sets drastically in ways which do not align with stakeholders' expectations.

This paper explore whether and how we can design FLSs which can preserve their interpretability while also providing the required degrees of freedom for statistical tuning to deliver good performance.

To achieve the latter, we focus on Non singleton FLSs (NSFLSs) [5,22], which are designed to model disturbance affecting a system through its inputs within the (self-contained) fuzzification stage. Recently, NSFLS approaches have received increasing attention [10,11,21,24–26,29,31,32], with a particular focus on the development of FLSs which 'model uncertainty where it arises', i.e. FLSs which model input uncertainty directly and only within the input fuzzification stage. The latter provides an elegant modelling approach which avoids changing otherwise unrelated parameters (e.g. antecedent or consequent FSs) in respect to disturbance affecting a systems' inputs. Most recently, the ADONiS framework [23] was proposed, where input noise is estimated and the fuzzification stage is adapted at run-time, delivering good performance in the face of varying noise conditions. As noted, ADONiS limits tuning to the fuzzification stage, leaving rules (which can be generated based on experts insights or in a data-driven way) 'untouched', thus providing a fundamental requirement for good interpretability.

In this paper, we compare and contrast the effects of employing both the ANFIS optimisation and the ADONiS adaptation frameworks in response to varying noise levels in a time-series prediction context. We do not aim to explore which approach delivers the best time series prediction (for that, many other machine learning methods are available), but rather, how the resulting FLSs compare after tuning, when both approaches deliver good or at least reasonable results. Specifically, we focus on the degree to which the key parameters – antecedents and consequents are preserved (we maintain an identical rule set to enable systematic comparison), and thus to which degree the original interpretability of a FLS can be preserved post-tuning using such approaches.

The structure of this paper is as follows. Section 2 gives a brief overview of singleton, non-singleton FSs, as well as the ADONiS and ANFIS models. Section 3 introduces methodology including details of the rule generation, training and testing. Section 4 provides detailed steps of the conducted experiments and a discussion of the findings. In Sect. 6, the conclusions of experiments with possible future work directions are given.

2 Background

2.1 Singleton Non-singleton Type-1 Fuzzy Sets

In the fuzzification step of fuzzy models, a given crisp input is characterised as membership function membership function (MF). Generally in singleton fuzzification, the given input x is represented by singleton MF.

When input data contain noise, it may not be appropriate to represent them as singleton MFs, as there is a possibility of the actual value being distorted by this noise. In this case, the given input x is mapped to non-singleton MFs with a support where membership degree achieves maximum value at x. Two samples of non-singleton MFs -under relatively low and high noise- can be seen in Fig. 1a.

Conceptually, the given input is assumed to be likely to be correct, but because of existing uncertainty, neighbouring values also have potential to be correct. As we go away from the input value, the possibility of being correct decreases. As shown in Fig. 1a the width of the non-singleton input is associated with the uncertainty levels of the given input.

2.2 ADONiS

The recently proposed ADONiS [23] framework provides two major advantages over non-singleton counterpart models: (i) in the fuzzification step, it captures

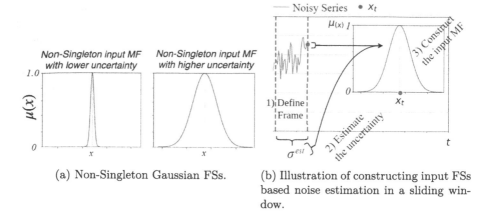

(a) Non-Singleton Gaussian FSs.

(b) Illustration of constructing input FSs based noise estimation in a sliding window.

Fig. 1. Different non-singleton FSs and ADONiS framework structure

input uncertainty through an online learning method–which utilises a sequence of observations to continuously update the input Fuzzy Sets (ii) in the inference engine step, it handles the captured uncertainty through the *sub*-NS [24] method to produce more reasonable firing strengths.

Therefore, the ADONiS framework enables us to model noise and uncertainty 'where it arises' and also to limit any optimisation impact to the fuzzification and inference steps. In doing so, ADONiS limits tuning to the fuzzification stage and remain rules (which can be generated based on experts insights or in a data-driven way) 'untouched', thus providing a fundamental requirement for good interpretability. –if rules and sets were understood well initially.

The general framework structure of the ADONiS framework can be summarised in the following four steps:

1. Defining a frame size to collect a sequence of observations. For example, when using sensors, such as in a robotics context, the size of the frame may be selected in respect to the sampling rate of the sensors or based on a fixed time frame.
2. In the defined frame, the uncertainty estimation of the collected observation is implemented. Different uncertainty estimation techniques can be implemented in the defined frame.
3. Non-singleton FS is formed by utilising the estimated uncertainty around the collected input. For example, in this paper, Bell shaped FSs are used and the detected uncertainty is utilised to define the width of these FSs.
4. In the inference engine step of NSFLSs, interaction between the input and antecedent FSs results in the rule firing strengths which in turn determines the degree of truth of the consequents of individual rules. In this step, in this paper, the *sub*-NS technique [24] is utilised to determine the interaction and thus firing strength between input and antecedent FSs.

The overall illustration of ADONiS can be seen in Fig. 1b and for details, please refer the [23, 24].

2.3 ANFIS

Neuro-fuzzy models are designed to combine the concept of artificial neural networks with fuzzy inference systems. As one common model, ANFIS is widely used in many applications to improve the performance of fuzzy inference systems [2,3,9,16]. With ANFIS, model parameters are 'fine-tuned' during optimisation procedures to obtain more accurate approximation than a predefined fuzzy system. An ANFIS illustration with seven antecedents can be seen in Fig. 2 [6].

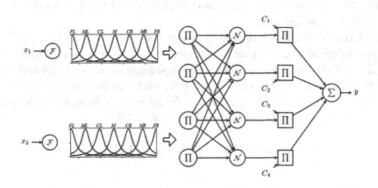

Fig. 2. ANFIS structure

3 Methodology

A Mackey-Glass (MG) time series is generated and 1009 noise-free values are obtained for t from 100 to 1108. One of the common models for noise is additive white Gaussian noise [20]. Three different signal-to-noise ratios (20 dB, 5 dB and 0 dB) are used to generate noisy time series with additive Gaussian white noise. These four (noise-free and noisy) datasets are split into 70% (training) and 30% (testing) samples to be used in different variants of the experiments. In the MG generation, τ value is set to be 17 to exhibit chaotic behaviour.

3.1 Rule Generation

In the literature there are many different rule generation techniques, either expert-driven or data-driven [8,18]. In this paper, one of the most commonly used techniques for FLS rule generation – the *one-pass* Wang-Mendel method is utilised. Even though in the case of interpretability assessment, Wang-Mendel

576   D. Pekaslan et al.

may not be the best approach to generate rules, in order to make a base rule set for both ADONiS and ANFIS and make a fair comparison, we choose to use *one-pass* Wang-Mendel method. In the future, different rule reduction algorithms or other rule generation techniques can be investigated. By following similar FLS architecture in [33], the rule generation is implemented as follows:

First, the domain of the training set $[x_{min}, x_{max}]$ is defined. In order to capture all inputs (including the ones which are outside of the input domain), the defined domain is expanded by 10% and the *cut-off* procedure is implemented for the inputs which are outside of this domain.

Then the input domain is evenly split into seven regions, and bell-shaped antecedents are generated. As shown in Fig. 3, these are named as *Further Left (FL), Medium Left (ML), Close Left (CL), Medium (M), Close Right (CR), Medium Right (MR), Further Right (FR)*.

As in [33], nine past values are used as inputs and the following (10th) value is predicted, i.e. the output.

After forming the input-output pairs as $((\mathbf{x}^1 : y^1), (\mathbf{x}^2 : y^2), ..., (\mathbf{x}^N : y^N),)$ each input value within the pair is assigned to the corresponding antecedent FS $(FL, .., M, .., FR)$. As practised in the Wang-Mendel *one-pass* method, the same seven FSs are used for the consequent FSs, and the outputs (y^i) are assigned to the corresponding FSs $(FL, .., M, .., FR)$ as well. A sample of the generated rules can be seen in (1). For details, please refer [33].

$$R^1 = \textbf{IF } x_1 \, is \, MR \, \textbf{AND}... \, x_9 \, is \, M \, \textbf{THEN} \, y_1 \, is \, CR \qquad (1)$$

3.2 Training

ADONiS. When implementing ADONiS, no formal optimisation procedure is used. Therefore, previously established antecedents $(FL, .., M, .. FR)$ and model rules remain untouched.

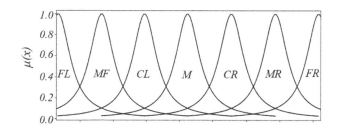

Fig. 3. An illustration of the seven antecedent MFs used.

ANFIS Optimisation. In ANFIS implementation, each of the seven antecedent MFs are assigned an input neuron (See Fig. 2) [6]. Then, the gradient descent optimisation technique is implemented to update the antecedent

MF parameters and the consequent linear functions. In the meantime, the least-squares estimation method [30] is used to update the parameters of consequent linear functions in each training epoch. During each epoch, the antecedent FS parameters are updated for each input. Therefore, while beginning with only seven antecedents, after optimisation, many different antecedent FSs may be generated–with associated increase in model complexity.

3.3 Performance Evaluation

In order to assess the noise handling capability of each model, we calculate the difference between model predictions and noise-free data values at each time-point. Both ADONiS and ANFIS performances are measured by using the common root-mean-squared error (RMSE) and in addition, the recently proposed Unscaled Mean Bounded Relative Absolute Error (UMBRAE) [7]. UMBRAE combines the best features of various alternative measures without suffering their common issues.

To use UMBRAE, a benchmark method needs to be selected. In this paper, the benchmark method simply uses the average of input values as predictions. With UMBRAE, the performance of a proposed method can be easily interpreted: when $UMBRAE$ is equal to 1, the proposed method performs approximately the same as the benchmark method; when $UMBRAE < 1$, the proposed method performs better than the benchmark method; when $UMBRAE > 1$, the proposed method performs worse than the benchmark method.

4   Experiment and Results

In total, $4 \times 4 = 16$ different experimental scenarios are implemented, using different noise levels in both rule generation/optimisation and testing phases. Specifically, four different training sets (noise-free, 20, 5 and 0 dB) and four different testing sets (noise-free, 20, 5 and 0 dB) are used–to represent a variety of potential real-world noise levels. In each experiment of ADONiS, the first 700 values are used to generate rules and the remaining 300 values are used for testing. Note that as ADONiS uses 9 inputs to construct input FSs, the first 9 values of the testing set are omitted, leaving only the final 291. In ANFIS, while using the exact same rules as ADONiS, the first 400 data pairs are used as the training set; the following 300 data pairs are used as a validation set; and the final 291 of the remaining 300 data pairs are used as testing set.

4.1   Experiment 1: Noise-Free Rule Generation

In the first experiment, the rule set is generated using the noise-free time series dataset. Four different testing datasets (noise-free, 20, 5 and 0 dB) are used.

Results of the ADONiS prediction experiment, with noise free testing, can be seen on the left hand side of Fig. 4a. Note that since there is no noise in the testing dataset, the generated input FSs tend to be a singleton FS. Thus, the traditional singleton prediction is implemented in this particular experiment.

After completing noise-free testing, and using the same rule set (from the noise-free training dataset), the 20 dB testing dataset is used in the prediction experiment of ADONiS. The RMSE result of this experiment is shown in Fig. 4a. Thereafter, the remaining 5 dB and 0 dB testing datasets are used with the same rule set–RMSE results are shown in Fig. 4a.

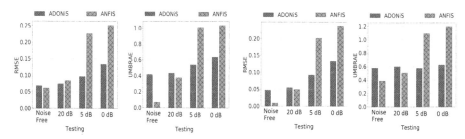

(a) Experiment 1: Noise-Free rule genera- (b) Experiment 2: 20dB Noisy rule gener-
tion   ation

Fig. 4. RMSE and UMBRAE results for both Experiment 1 and Experiment 2. (Color figure online)

Following the ADONiS prediction (with noise-free rule set and four different testing datasets), ANFIS optimisation is carried out on the previously generated rule parameters and the antecedent parameters are updated in the *'black-box'* manner. Then, these updated antecedents are used in the prediction of noise-free testing dataset. The results of this experiment are shown in Fig. 4a as orange bars. Overall, as can be seen, ANFIS outperform ADONiS significantly in this particular experiment.

Thereafter, the same updated rules from the noise-free training dataset, are used with the 20 dB testing dataset. The performance of ANFIS is reported in Fig. 4a. As can be seen, ADONiS and ANFIS have similar performances under the noise-free and 20 dB noisy testing variant.

Following this, 5 dB and 0 dB noisy datasets were used in testing–RMSE results are illustrated in Fig. 4a. As shown, in both of these noisy conditions, ADONiS outperform ANFIS substantially.

As the second error measure, UMBRAE is calculated between the prediction and noise-free input datasets. These sets of experiment results can be seen right hand side of Fig. 4a.

4.2   Experiment 2: 20 dB Noisy Rule Generation

In the four sets used in this experiment, rule generation is completed by using the 20 dB noisy time series dataset. The resulting rules are then used in ADONiS predictions on the noise-free, 20 dB, 5 dB and 0 dB noisy datasets. The RMSE experiment results are shown in Fig. 4b.

After ADONiS implementation, ANFIS optimisation is implemented on the antecedents' parameters, according to the 20 dB noisy training dataset. Then the ANFIS predictions are performed on the same four (noise-free, 20 dB, 5 dB and 0 dB) different datasets. These prediction results are illustrated in Fig. 4b. These findings show a clear trend that under noise-free or low-noise conditions, ADONiS and ANFIS provide similar performances. Under higher noise levels (5 and 0 dB), ADONiS has a clear performance advantage.

Equivalent results, as evaluated using the UMBRAE error measure, are illustrated on the right hand side of Fig. 4b.

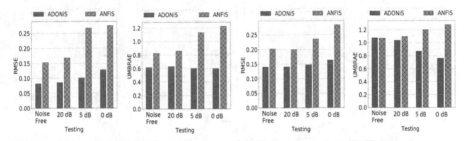

(a) Experiment 3: 5 dB Noisy rule genera-(b) Experiment 4: 0 dB Noisy rule generation   tion

Fig. 5. RMSE and UMBRAE results for both Experiment 3 and Experiment 4.

4.3 Experiment 3 and 4: 5 and 0 dB Noisy Rule Generation

The same procedures from the previous experiments are followed. First, rules are generated, based upon the 5 dB noisy time series datasets. Next, ADONiS performance is tested with the four (noise-free, 20 dB, 5 dB and 0 dB) datasets. Afterwards, ANFIS optimisation is used to update the antecedent parameters and ANFIS predictions are completed on the same four (noise-free, 20 dB, 5 dB and 0 dB) testing datasets. 5 dB rule generation results are shown in Fig. 5a for RMSE and UMBRAE.

Thereafter, 0 dB rule generation is completed and the four different testing results are illustrated in Figs. 5b.

5 Discussion

Overall, the interpretability of a fuzzy model builds upon several components i.e. rules, antecedents and/or consequent numbers, and the semantics at the fuzzy partitioning level. Traditionally, while optimisation techniques may provide a better performance, it leads to changing the parameters (i.e. antecedents FSs) based on a training dataset which results in a less interpretable model. However, since FLSs have mechanisms to provide interpretability, the changing of these parameters in a data-driven way can deteriorate the interpretability of models by causing for example a loss of complementarity, coverage or distinguishability of FSs across a universe of discourse and thus the meaningfulness of the used FSs. Conversely, tuning parameters in the fuzzification step can maintain the interpretability as well as provide a performance benefits.

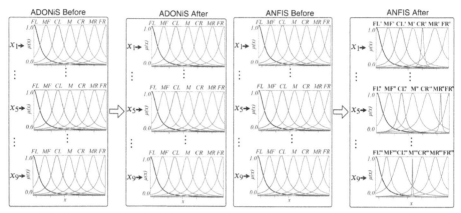

(a) A comparison of the used antecedent in ADONiS.

(b) A comparison of used antecedents in ANFIS.

Fig. 6. The used antecedent FSs in both model ADONiS and ANFIS.

Regarding the rule generation in the experiments, while different approaches have been introduced [18], in this paper we follow the well established Wang-Mendel [33] rule generation technique. We acknowledge that other approaches may be equally or more viable for example in the given domain of time series prediction, nevertheless, for this paper, our key objective was to generate one basic rulebase which is maintained identical across all FLSs, thus providing a basis for systematic comparison. Further, we note that the specific antecedent and consequent FSs used here are selected arbitrarily (to evenly partition the domain of the variables), and thus are not meaningful in a traditional linguistic sense. However, in this paper, we consider the preservation of the original shape of the FSs (post-tuning) as important (as it is that shape which will be meaningful in applications of FLSs such as in medical decision making, common control applications, etc.).

In the experiment, we first explore the ADONiS model which targets the fuzzification step by limiting the optimisation effect but handling noise 'where it arises'. Second, traditional ANFIS optimisation is used. In this section, after a brief performance comparison, the interpretability is discussed for both models.

Overall, when all the results are scrutinised all together (Figs. 4a, 4b, 5a, 5b), it can be seen that ADONiS and ANFIS provide comparable performance. While ANFIS shows better performance in the noise-free training and noise-free cases, especially under high levels of noise, ADONiS' performance is better than that of the ANFIS-tuned FLS.

In the experiments, by following the structure in [33], the input domain is divided into 7 antecedents, from *Further Left* to *Further Right* ($FL, .., M, .., FR$) (See Fig. 3) and each input is assigned with these antecedents as shown on the left hand side of Fig. 6a. The same rule set is generated once. In the ADONiS approach, no optimisation procedure is performed offline (all tuning is done online through adaptation) and all the rules, antecedents, consequents remain intact. As can be seen on the right hand side of Fig. 6a, the same antecedents and consequents are used in the testing stage for ADONiS. Here, the input uncertainty is captured and handled throughout the fuzzification and inference engine process rather than optimising antecedent or consequent parameters. We note that this is intuitive as changes affecting the inputs should not affect the linguistic models of antecedents and consequents - preserving interpretability. For example, when a rule is examined in (2), all the *Medium* (M) MFs are the same as in Fig. 3 and it can be observed that the given sample inputs x_1 and x_9 are processed using the same MFs.

$$IF\ x_1\ is\ M... x_5\ is\ MR... x_9\ is\ M\ THEN\ y_1\ is\ CR \qquad (2)$$

On the other hand, in the ANFIS implementation, although the same rules are used (see the left hand side of Fig. 6b) the optimisation procedure focuses on the antecedent parameters. Thus, the parameters are changed in respect to the training data, changing the antecedent and thus necessarily making it different to the original (considered interpretable) model (see the right hand side of Fig. 6b). This overall can affect both the semantics and the complexity at the fuzzy partitioning level. For example, the *Medium* MF is changed through the optimisation procedure. As can be seen in Fig. 6b and rule (3), the *Medium'* (M') and *Medium'''* (M''') are not the same for inputs x_1 and x_9 which inhibits the interpretability of the model.

$$IF\ x_1\ is\ M'... x_5\ is\ MR''... x_9\ is\ M'''\ THEN\ y_1 \qquad (3)$$

Therefore, overall, these initial results show that while both models can provide comparable prediction results under different levels of noise, tuning parameters in the fuzzification stage only can help to maintain the semantic meaningfulness (completeness, distinguishability and complementarity) of the used antecedent FSs which can overall provide a more 'interpretable' FLS model in contrast to a 'brute force' optimisation approach such as offered by traditional optimisation approaches for FLSs such as ANFIS.

6   Conclusions

One of the main motivations to use FLSs is their capacity for interpretability ability which is highly related to both complexity (number and structure of rules, variables) and semantic (completeness, distinguishability or complementarity at the level of the fuzzy set partitions) aspects. In regards to the performance of FLSs, while optimisation techniques can be applied to deliver improved performance, such optimisation has traditionally lead to changes of the same key parameters which are vital for interpretability, thus delivering improved performance at the cost of poorer interpretability. In this paper, we explore the possibility of automatically tuning an FLS to deliver good performance, while *also* preserving its valuable interpretable structure, namely the rules (kept constant), antecedents and consequents. Through a detailed set of time series prediction experiments, the potential of the ADONiS framework, which handles input noise where it arises, is explored in comparison to a traditional ANFIS optimisation approach. The behaviour and performance of both approaches is analysed with a view to inform future research aimed at developing FLSs with both high performance and high interpretability.

We believe that these initial results highlight a very interesting research direction for FLSs which can maintain interpretability by modelling complexity only in specific parts of their structure. Future work will concentrate on expanding the experimental evaluation with different rule generation techniques and datasets while broadening the capacity for optimisation beyond the specific design of ADONiS. Also, the use of interpretability indices will be explored to compare/contrast different model efficiently in regards to performance and interpretability ability.

References

1. Alonso, J.M., Magdalena, L., González-Rodríguez, G.: Looking for a good fuzzy system interpretability index: an experimental approach. Int. J. Approx. Reason. **51**(1), 115–134 (2009). https://doi.org/10.1016/j.ijar.2009.09.004. http://www.sciencedirect.com/science/article/pii/S0888613X09001418
2. Atsalakis, G.S., Valavanis, K.P.: Forecasting stock market short-term trends using a neuro-fuzzy based methodology. Expert Syst. Appl. **36**(7), 10696–10707 (2009). https://doi.org/10.1016/j.eswa.2009.02.043.   http://www.sciencedirect.com/science/article/pii/S0957417409001948
3. Azadeh, A., Saberi, M., Gitiforouz, A., Saberi, Z.: A hybrid simulation-adaptive network based fuzzy inference system for improvement of electricity consumption estimation. Expert Syst. Appl. **36**(8), 11108–11117 (2009). https://doi.org/10.1016/j.eswa.2009.02.081.   http://www.sciencedirect.com/science/article/pii/S0957417409002413
4. Bodenhofer U., Bauer P.: A formal model of interpretability of linguistic variables. In: Casillas, J., Cordón, O., Herrera, F., Magdalena, L. (eds.) Interpretability Issues in Fuzzy Modeling. STUDFUZZ, vol. 128, pp. 524–545. Springer, Heidelberg (2003). https://doi.org/10.1007/978-3-540-37057-4_22

5. Cara, A.B., Wagner, C., Hagras, H., Pomares, H., Rojas, I.: Multiobjective optimization and comparison of nonsingleton type-1 and singleton interval type-2 fuzzy logic systems. IEEE Trans. Fuzzy Syst. **21**(3), 459–476 (2013). https://doi.org/10. 1109/TFUZZ.2012.2236096
6. Chen, C., John, R., Twycross, J., Garibaldi, J.M.: An extended ANFIS architecture and its learning properties for type-1 and interval type-2 models. In: Proceedings IEEE International Conference on Fuzzy Systems, pp. 602–609 (2016)
7. Chen, C., Twycross, J., Garibaldi, J.M.: A new accuracy measure based on bounded relative error for time series forecasting. PLOS ONE **12**(3), 1–23 (2017)
8. Chi, Z., Wu, J., Yan, H.: Handwritten numeral recognition using self-organizing maps and fuzzy rules. Pattern Recogn. **28**(1), 59–66 (1995). https://doi.org/10. 1016/0031-3203(94)00085-Z. http://www.sciencedirect.com/science/article/pii/ 003132039400085Z
9. Esfahanipour, A., Aghamiri, W.: Adapted neuro-fuzzy inference system on indirect approach TSK fuzzy rule base for stock market analysis. Expert Syst. Appl. **37**(7), 4742–4748 (2010). https://doi.org/10.1016/j.eswa.2009.11.020. http://www.sciencedirect.com/science/article/pii/S0957417409009622
10. Fu, C., Sarabakha, A., Kayacan, E., Wagner, C., John, R., Garibaldi, J.M.: A comparative study on the control of quadcopter UAVs by using singleton and nonsingleton fuzzy logic controllers. In: 2016 IEEE International Conference on Fuzzy Systems (FUZZ-IEEE), pp. 1023–1030, July 2016. https://doi.org/10.1109/FUZZ-IEEE.2016.7737800
11. Fu, C., Sarabakha, A., Kayacan, E., Wagner, C., John, R., Garibaldi, J.M.: Input uncertainty sensitivity enhanced nonsingleton fuzzy logic controllers for long-term navigation of quadrotor uavs. IEEE/ASME Trans. Mechatron. **23**(2), 725–734 (2018). https://doi.org/10.1109/TMECH.2018.2810947
12. Gacto, M.J., Alcalá, R., Herrera, F.: Interpretability of linguistic fuzzy rule-based systems: an overview of interpretability measures. Inf. Sci. **181**(20), 4340–4360 (2011). https://doi.org/10.1016/j.ins.2011.02.021. http://www.sciencedirect.com/ science/article/pii/S0020025511001034
13. Garibaldi, J.M., Ozen, T.: Uncertain fuzzy reasoning: a case study in modelling expert decision making. IEEE Trans. Fuzzy Syst. **15**(1), 16–30 (2007)
14. Jang, J.S.: ANFIS: adaptive-network-based fuzzy inference system. IEEE Trans. Syst. Man Cybern. **23**(3), 665–685 (1993)
15. Jin, Y.: Fuzzy modeling of high-dimensional systems: complexity reduction and interpretability improvement. IEEE Trans. Fuzzy Syst. **8**(2), 212–221 (2000). https://doi.org/10.1109/91.842154. http://epubs.surrey.ac.uk/7644/
16. Khanesar, M.A., Kayacan, E., Teshnehlab, M., Kaynak, O.: Extended Kalman filter based learning algorithm for type-2 fuzzy logic systems and its experimental evaluation. IEEE Trans. Ind. Electron. **59**(11), 4443–4455 (2012)
17. Kim, B., et al.: Interpretability beyond feature attribution: quantitative testing with concept activation vectors (TCAV). arXiv preprint arXiv:1711.11279 (2017)
18. Wang, L.-X.: The WM method completed: a flexible fuzzy system approach to data mining. IEEE Trans. Fuzzy Syst. **11**(6), 768–782 (2003). https://doi.org/10. 1109/TFUZZ.2003.819839
19. Liu, H., Gegov, A., Cocea, M.: Complexity control in rule based models for classification in machine learning context. In: Angelov, P., Gegov, A., Jayne, C., Shen, Q. (eds.) Advances in Computational Intelligence Systems. AISC, vol. 513, pp. 125–143. Springer, Cham (2017). https://doi.org/10.1007/978-3-319-46562-3_9
20. Liu, X., Tanaka, M., Okutomi, M.: Single-image noise level estimation for blind denoising. IEEE Trans. Image Process. **22**(12), 5226–5237 (2013)

21. Mohammadzadeh, A., Kayacan, E.: A non-singleton type-2 fuzzy neural network with adaptive secondary membership for high dimensional applications. Neurocomputing **338**, 63–71 (2019). https://doi.org/10.1016/j.neucom.2019.01.095. http://www.sciencedirect.com/science/article/pii/S0925231219301882
22. Mouzouris, G.C., Mendel, J.M.: Nonlinear time-series analysis with non-singleton fuzzy logic systems. In: Computational Intelligence for Financial Engineering. Proceedings of the IEEE/IAFE 1995, pp. 47–56. IEEE (1995)
23. Pekaslan, D., Wagner, C., Garibaldi, J.M.: Adonis - adaptive online non-singleton fuzzy logic systems. IEEE Trans. Fuzzy Syst. 1 (2019). https://doi.org/10.1109/TFUZZ.2019.2933787
24. Pekaslan, D., Garibaldi, J.M., Wagner, C.: Exploring subsethood to determine firing strength in non-singleton fuzzy logic systems. In: IEEE World Congress on Computational Intelligence (2018)
25. Pourabdollah, A., Wagner, C., Aladi, J.H., Garibaldi, J.M.: Improved uncertainty capture for nonsingleton fuzzy systems. IEEE Trans. Fuzzy Syst. **24**(6), 1513–1524 (2016). https://doi.org/10.1109/TFUZZ.2016.2540065
26. Pourabdollah, A., John, R., Garibaldi, J.M.: A new dynamic approach for non-singleton fuzzification in noisy time-series prediction. In: 2017 IEEE International Conference on Fuzzy Systems (FUZZ-IEEE), pp. 1–6. IEEE (2017)
27. Rajab, S.: Handling interpretability issues in anfis using rule base simplification and constrained learning. Fuzzy Sets Syst. **368**, 36–58 (2019). https://doi.org/10.1016/j.fss.2018.11.010. Theme: Fuzzy Systems and Learning
28. Razak, T.R., Garibaldi, J.M., Wagner, C., Pourabdollah, A., Soria, D.: Interpretability and complexity of design in the creation of fuzzy logic systems - a user study. In: 2018 IEEE Symposium Series on Computational Intelligence (SSCI), pp. 420–426, November 2018. https://doi.org/10.1109/SSCI.2018.8628924
29. Ruiz-García, G., Hagras, H., Pomares, H., Ruiz, I.R.: Toward a fuzzy logic system based on general forms of interval type-2 fuzzy sets. IEEE Trans. Fuzzy Syst. **27**(12), 2381–2395 (2019)
30. Sorenson, H.W.: Least-squares estimation: from Gauss to Kalman. IEEE Spectr. **7**(7), 63–68 (1970)
31. Tellez-Velazquez, A., Cruz-Barbosa, R.: A CUDA-streams inference machine for non-singleton fuzzy systems. Concurr. Comput.: Pract. Exp. **30** (2017). https://doi.org/10.1002/cpe.4382
32. Wagner, C., Pourabdollah, A., McCulloch, J., John, R., Garibaldi, J.M.: A similarity-based inference engine for non-singleton fuzzy logic systems. In: 2016 IEEE International Conference on Fuzzy Systems (FUZZ-IEEE), pp. 316–323. IEEE (2016)
33. Wang, L.X., Mendel, J.M.: Generating fuzzy rules by learning from examples. IEEE Trans. Syst. Man Cybern. **22**(6), 1414–1427 (1992)
34. Zadeh, L.: Fuzzy sets. Inf. Control **8**(3), 338–353 (1965). https://doi.org/10.1016/S0019-9958(65)90241-X

Explaining the Neural Network: A Case Study to Model the Incidence of Cervical Cancer

Paulo J. G. Lisboa$^{(\boxtimes)}$, Sandra Ortega-Martorell, and Ivan Olier

Department of Applied Mathematics, Liverpool John Moores University,
Liverpool L3 3AF, UK
p.j.lisboa@ljmu.ac.uk

Abstract. Neural networks are frequently applied to medical data. We describe how complex and imbalanced data can be modelled with simple but accurate neural networks that are transparent to the user. In the case of a data set on cervical cancer with 753 observations excluding, missing values, and 32 covariates, with a prevalence of 73 cases (9.69%), we explain how model selection can be applied to the Multi-Layer Perceptron (MLP) by deriving a representation using a General Additive Neural Network.

The model achieves an AUROC of 0.621 CI [0.519,0.721] for predicting positive diagnosis with Schiller's test. This is comparable with the performance obtained by a deep learning network with an AUROC of 0.667 [1]. Instead of using all covariates, the Partial Response Network (PRN) involves just 2 variables, namely the number of years on Hormonal Contraceptives and the number of years using IUD, in a fully explained model. This is consistent with an additive non-linear statistical approach, the Sparse Additive Model [2] which estimates non-linear components in a logistic regression classifier using the backfitting algorithm applied to an ANOVA functional expansion.

This paper shows how the PRN, applied to a challenging classification task, can provide insights into the influential variables, in this case correlated with incidence of cervical cancer, so reducing the number of unnecessary variables to be collected for screening. It does so by exploiting the efficiency of sparse statistical models to select features from an ANOVA decomposition of the MLP, in the process deriving a fully interpretable model.

Keywords: Explainable machine learning · FATE · KDD · Medical decision support · Cervical cancer

1 Introduction

This paper is about explainable neural networks, illustrated by an application of a challenging data set on cervical cancer screening that is available in the UCI repository [3]. The purpose of the paper is to describe a case study of the interpretation of a neural network by exploiting the same ANOVA decomposition that has been used in statistics to infer sparse non-linear functions for probabilistic classifiers [2].

We will show how a shallow network, the Multi-Layer Perceptron (MLP) can be fully explained by formulating it as a General Additive Neural Network (GANN). This methodology has a long history [4]. However, to our knowledge there is no method to

© Springer Nature Switzerland AG 2020
M.-J. Lesot et al. (Eds.): IPMU 2020, CCIS 1237, pp. 585–598, 2020.
https://doi.org/10.1007/978-3-030-50146-4_43

derive the GANN from data, rather a model structure needs to be assumed or hypothesized from experimental data analysis. In this paper we use a mechanistic model to construct the GANN and show that, for tabular data i.e. high-level features that are typical of applications to medical decision support, a transparent and parsimonious model can be obtained, whose predictive performance comparable i.e. well within the confidence interval for the AUROC, with that obtained an alternative, opaque, deep learning neural network applied to the same data set [1].

Fairness, Accountability Transparency and Ethics (FATE) in AI [5] is emerging as a priority research area that relates to the importance of human-centered as a key enabler for practical application in risk-related domains such as clinical practice. Blind spots and bias in models e.g. due to artifacts and spurious correlations hidden in observational data, can undermine the generality of data driven models when they are used to predict for real-world data and this may have legal implications [6].

There different approaches that may be taken to interpret neural networks, in particular. These include derivation of rules to unravel the inner structure of deep learning neural networks [7] and saliency methods [8] to determine the image elements to which the network prediction is most sensitive.

An additional aspect of data modelling that is currently very much understudied is the assessment of the quality of the data. Generative Adversarial Networks have been used to quantify sample quality [9].

Arguably the most generic approach machine explanation is the attribution of feature influence with additive models. A unified framework for this class of models has been articulated [10]. This includes as a special case the approach of Local Interpretable Model Agnostic Explanations (LIME) [11].

However, it is acknowledged in [10] that General Additive Models (GAMs) are the most interpretable because the model is itself the interpretation, and this applies to data at a global level, not just locally.

Recently there has been a resurgence of interest in GAMs [11, 12] in particular through implementations as GANNs. These models sit firmly at the interface between computational intelligence and traditional statistics, since they permit rigorous computation of relevant statistical measures such as odds ratios for the influence of specific effects [12].

A previously proposed framework for the construction of GANNs from MLPs will be applied to carry out model selection and so derive the form of the GANN from a trained MLP. This takes the form of a Partial Response Network (PRN) whose classification performance on multiple benchmarking data sets matches that of deep learning but with much sparser and directly interpretable features [13].

This paper reports a specific case study of the application of PRN to demonstrate how it can interpret the MLP as a GAM, providing complete transparency about the use of the data by the model, without compromising model accuracy as represented by the confidence interval of the AUROC. Our results are compared with those from a state-of-the-art feature selection method for non-linear classification [2].

Moreover, the model selection process itself will generate insights about the structure of the data, illustrating the value of this approach for knowledge discovery in databases (KDD).

2   Data Description

2.1   Data Collection

Cervical cancer is a significant cause of mortality among women both in developed and developing countries world-wide [1]. It is unusual among cancers for being closely associated with contracting the Human Papillomavirus (HPV) [14] which is strongly influenced by sexual activity. This makes cervical cancer one of the most avoidable cancers, through lifestyle factors and by vaccination.

Screening for possible incidence of the cancer is a public health priority, with potential for low-cost screening to be effective. The data set used in this study was acquired for this purpose.

The data were collected from women who attended the Hospital Universitario de Caracas in Caracas [3]. Most of the patients belong to the lowest socioeconomic status, which comprises the population at highest risk. They are all sexually active. Clinical screening includes cytology, a colposcopic assessment with acetic acid and the Schiller test (Lugol's iodine solution). This is the most prevalent diagnostic index and is the choice for the present study.

2.2   Data Pre-processing

The data comprise records from a random sample of patients presenting between 2012 and 2013 (n = 858) [1, 3]. There is a wide age range and a broad set of indicators of sexual activity, several of which overlap in what they measure. Four target variables are reported, including the binary outcome of Schiller's test.

This data set is challenging, first because of severe class imbalance, which is typical in many medical diagnostic applications. The number of positive outcomes in the initial data sample is just 74 cases for Schiller's test, 44 for a standard cytology test and 35 for Hinselmann's test.

Secondly, the data include a range of self-reported behavioural characteristics, where noise levels may be significant. Third, some of the variables were problematic for data analysis. The report of STD: cervical condylomatosis comprises all zero values. STD: vaginal condylomatosis, pelvic inflammatory disease, genital herpes, molluscum contagiosum, AIDS, HIV, Hepatitis B, syphilis and HPV are all populated in <2.5% of all cases. For this reason, these variables were removed from the study as they are unlikely to provide statistical significance in predictive modelling and their low prevalence can cause numerical instabilities for model optimisation.

The number of pregnancies was deemed to be less informative about sexual behaviour than the number of sexual partners, so this was also excluded.

In total 105 rows of data had 20 or more of the 32 covariate values missing. While these values can be imputed, such a large proportion of covariates for individual observations can bias the study, since missingness can be informative. For this reason, these rows were removed from the data.

Among the selected variables, several pairs of covariates measure the same indicator in binary form and as an ordinal count. This applies to variables Smokes,

Hormonal Contraceptives, IUD and STDs. Consequently, the initial pool of covariates in this study comprises 9 variables. They are:

- Number of sexual partners;
- Age of first sexual intercourse;
- Years since first sexual intercourse, derived by subtracting the previous covariate from Age;
- Number of years smoking;
- Number of years taking Hormonal Contraceptives
- Number of years using IUDs;
- STD: condylomatosis;
- Number of STDs;
- Number of diagnosed STDs.

The dataset used in this study is a reduced cohort (n = 753) with marginal values summarized in Table 1. The prevalence of missing data in the study sample is now much reduced, especially as the number of pregnancies is not used. The maximum proportion of missing is 4.1% for IUD (years).

Table 1. Summary statistics of the sample population for Cervical Cancer screening. {} indicates a binary variable. [] shows the range of the variable.

Variable	Median [Min, Max]	Missing values
Age	26 [13, 84]	0
Number of sexual partners	2 [1, 28]	14
First sexual intercourse	17 [10, 32]	6
Number of pregnancies	2 [0, 11]	47
Smokes	0 {0, 1}	10
Smokes (years)	0 [0, 37]	10
Smokes (packs/year)	0 [0, 37]	10
Hormonal Contraceptives	1 {0, 1}	13
Hormonal Contraceptives (years)	0.5 [0, 30]	13
IUD	0 {0, 1}	16
IUD (years)	0 [0, 19]	31
STDs	0 {0, 1}	0
STDs (number)	0 [0, 4]	0
STDs: condylomatosis	0 {0, 1}	0
STDs: Number of diagnosis	0 [0, 3]	0

Missing values were imputed with the sample median. The reason for this is that the standardisation used in the following section maps the median value of every covariate to zero, which has the effect of discarding that instance from the gradient descent weight updates, so minimising the impact of unknown information in the training of the MLP.

3   Partial Response Network Methodology

In binary classification, GAMs model the statistical link function appropriate for a Bernoulli error distribution. This is the logit, hence the inverse of the familiar sigmoid function. An appropriate objective function is the equally familiar log-likelihood cost.

In order to control for overfitting of the original MLP, we apply regularisation using Automatic Relevance Determination [15]. This model evaluates the strength of weight decay using a Bayesian estimator, which enables a different weight decay parameter to be used for the fan-out weights linked to each input node. This results in soft model selection, that is to say a modulation of the weight values that compresses towards zero the weights linked to the less informative input variables.

Input variables are divided by the standard deviation and shifted by the median value, so that the median is represented by zero. This is important because in a Taylor expansion of the logit function about the median values, setting an individual variable to the median causes all of the terms involving that variable in the Taylor expansion to vanish. It is then possible to capture much of the most significant terms by systematically setting all bar one covariate to zero, then all but each pair of covariates to zero, and so on.

The MLP response when all but a few variables are zero is called the Partial Response and the GANN obtained by mapping the partial responses onto its weights, forms the Partial Response Network (PRN) [13].

The functional form of the PRN is given by the well-known statistical decomposition of multivariate effects into components with fewer variables, represented by the ANOVA functional model [2] shown in Eq. (1):

$$logit(P(C|x)) \approx \varphi(0) + \sum_i \varphi_i(x_i) + \sum_{i \neq j} \varphi_{ij}(x_i, x_j) + O(x_i, x_j, x_i) \qquad (1)$$

where the partial responses $\varphi_k(\bullet)$ are evaluated with all variables held fixed at zero except for one or two indexed as follows:

$$\varphi(0) = logit(P(C|0)) \qquad (2)$$

$$\varphi_i(x_i) = logit(P(C|(0, .., x_i, .., 0))) - \varphi(0) \qquad (3)$$

$$\varphi_{ij}(x_i, x_j) = logit(P(C|(0, .., x_i, .., x_j, ..0))) - \varphi_i(x_i) - \varphi_j(x_j) - \varphi(0) \qquad (4)$$

The derivation of the PRN proceeds as follows:

1. Train an MLP for binary classification;
2. Obtained the univariate and bivariate partial responses in Eqs. (2)–(4).
3. Apply the Lasso to the partial responses;
4. Construct a second MLP as a linear combination of the partial responses so as to replicate the functionality of the Lasso. Each partial response, whether univariate or bivariate, is represented by a modular structure comprising the same number of hidden nodes as the original MLP. The modules are assembled into a single multi-layer structure represented as a GANN, shown in Fig. 1.
5. Re-train the resulting multi-layer network.

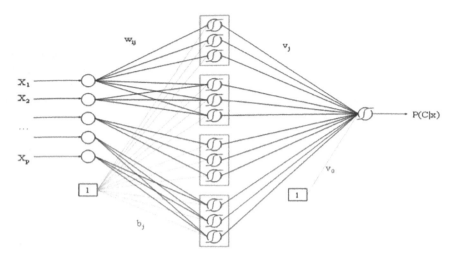

Fig. 1. Representation of the Partial Response Network as General Additive Neural Network (GANN). The weight values are derived from a trained MLP and re-calibrated by further training of the network as a GANN.

The mapping of the partial responses onto the GANN requires matching the weights and bias terms as follows:

1. Univariate partial responses

$$v_j \rightarrow v_j * (\beta_k - \beta_{kl}) \tag{5}$$

$$v_0 \rightarrow (v_0 - logit(P(C|0))) * (\beta_k - \beta_{kl}) \tag{6}$$

2. Bivariate partial response

$$v_0 \rightarrow (v_0 - logit(P(C|0))) * (\beta_k - \beta_{kl}) \tag{7}$$

$$v_0 \rightarrow (v_0 - logit(P(C|0))) * (\beta_k - \beta_{kl}) \tag{8}$$

The main limitation of the model as currently used is that it is restricted to univariate effects and bivariate interactions. However, in many medical applications, this is likely to suffice. The method can be extended to higher order interactions but it will generate a combinatorially large number of partial responses.

4 Experimental Results

This section explains how model selection took place and describes the models obtained with the PRN applied to the Cervical Cancer screening data set described in Sect. 2. The variables used in the model are the subset of Table 1 that is listed in 2.2 and the target variable is the outcome of Schiller's diagnostic test for cervical cancer.

Given the low prevalence of positive outcome, 73 out of the 753 cases retained (prevalence = 9.69%) the results presented are all for out-of-sample data using 2-fold cross validation. This choice of number of folds is motivated by the need to retain a meaningful number of events in each fold.

Model selection consisted of an iterative process of removing the least frequently occurring variable or set of variables at each stage in the process. Table 2 shows the frequency of occurrence of each covariate in the partial responses selected by the PRN. It also shows the average AUROC for 10 random starts.

Table 2. Model selection with the PRN applied to the Cervical Cancer screening dataset. φ_i; φ_{ij}: variable present in a univariate/bivariate partial response.

# var	AUC	#Sex partners	Age first sexual Inter	Smokes (Yrs)	Hormonal Contraceptives (Yrs)	IUD (Yr)	# STD	STD: condylomatosis
p = 9	**0.585**							
φ_i		1	1	1	3	1	1	3
φ_{ij}		1	6	13	12	7	7	
p = 5	**0.621**							
φ_i		–	–	–	4	4	3	2
φ_{ij}		–	–	16	10	8	5	9
p = 4	**0.593**							
φ_i		–	–	–	4	5	–	–
φ_{ij}		–	–	5	6	5	4	–
p = 3	**0.635**							
φ_i		–	–	–	9	10	5	–
φ_{ij}		–	–	–	9	9	2	–
p = 2	**0.621**							
φ_i		–	–	–	9	10	–	–
φ_{ij}		–	–	–	8	8	–	–

The results in Table 1 can be compared with those from a sparse non-linear statistical classifier, the Sparse Additive Model (SAM). This is an additive non-linear model that estimates component functions in an ANOVA decomposition using the backfitting algorithm that is standard for GAMs. It combines that with l_1 regularisation similar to the Lasso [2]. This provides the attractive property of convex optimisation, so that the model only needs to be estimated once.

In contrast, neural network models are not convex and so require multiple estimation. By interpreting the MLP in the form of a GAM with sparse features, the PRN model considerably reduces the variability in classification performance that is typical of the MLP, providing more consistent results.

However, correlations between variables can result in multiple models with very similar predictive power. This is the case for the present data set.

The SAM identified {#Years sexual intercourse; Smokes (years); STDs} for fold 1 and {Hormonal Contraceptives (years); IUD (years); STDs: condylomatosis; STDs} for fold 2 as univariate models; {STDs} for fold 1 and {IUD (years); STDs: condylomatosis; STDs; Number of sexual partners*IUD (years); #Years sexual intercourse*Hormonal Contraceptives (years); STDs: condylomatosis*STDs} when interaction terms were included.

The AUROCs for SAM in 2-fold cross validation are 0.599 and 0.565, respectively.

5 Discussion

The variable subsets extracted with model selection using the PRN model are all consistent with the previously cited work on this data set, and indeed with cervical screening literature.

The iterative process for feature selection applied in the previous section made use of the variability of the MLP under random starts to explore the space of predictive features in the presence of correlated variables. This enable the identification of stable features that could be applied for both folds to build a model with a consistent explanation. These two features are Hormonal Contraceptives (years) and IUD (years).

It cannot be claimed that these are the only predictive variables or indeed the best. However, they are a representative subset that achieves a high predictive model with parsimony, as can be seen from both the size of the derived feature set and high AUROC compared with the SAM.

Equally of interest is the shape of the partial responses and their stability under 2-fold cross validation, shown in Figs. 2, 3, 4 and 5.

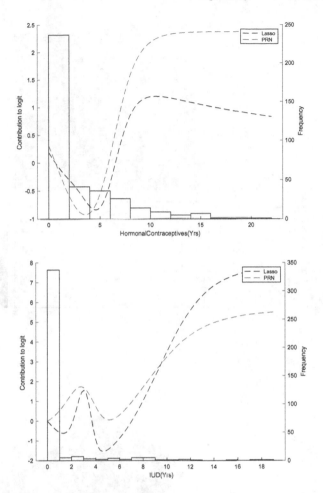

Fig. 2. Two univariate responses identified in the first fold. The abscissa measures the contribution of the individual covariate to the logit response. The histogram represents the empirical distribution of the covariate across the study population. The curves show the response derived from the initial MLP and after re-training with the PRN.

The partial responses are remarkably consistent given the challenges posed by the low prevalence and high noise in the data. Differences are apparent in areas of low data density, which is to be expected. Further work will involve quantifying the uncertainty about these estimates.

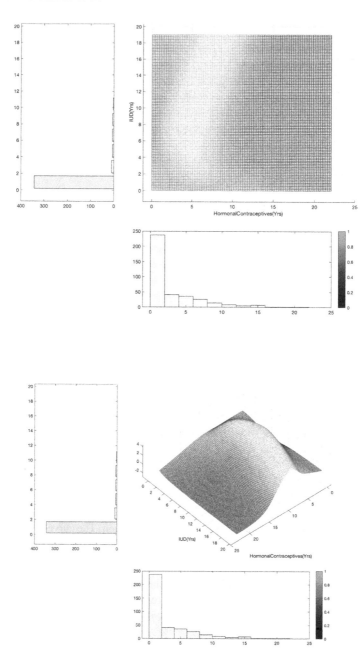

Fig. 3. Bivariate response found to be significant in the first fold of the data. The response is shown as a heat map and as a 3-d surface.

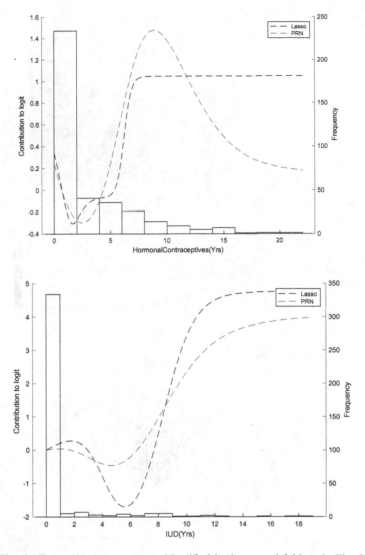

Fig. 4. Two univariate responses identified in the second fold, as in Fig. 2.

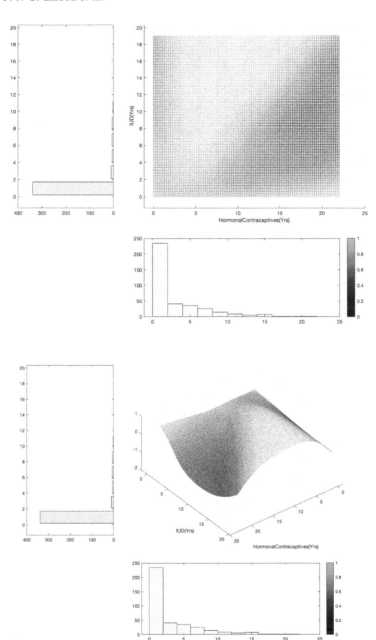

Fig. 5. Bivariate response found to be significant in the second fold of the data, as in Fig. 3.

6 Conclusion

The initial pool of 9 variables contains redundant information. This causes instability in neural network models, as several different models will capture information with similar predictive value. However, an iterative approach to feature selection can produce a stable sparse model.

It is perhaps remarkable how the same predictive information is contained in a small number of covariates compared with the size of the original pool. Bearing in mind that the typical standard deviation of the AUROC is 0.05, making the 95% confidence interval 0.10, the AUROC values for all models listed in Table 2 are comparable. Indeed, the average performance for ten random starts equals that of the best cross-validated model, 0.621 CI [0.519,0.721]. The overall performance figure is also consistent with the deep learning models in [1] and with a statistical approach to non-linear classification with an ANOVA decomposition, the SAM [2].

The main conclusion of this paper is that it is possible to break the black box that is the standard MLP, using it to derive a more interpretable structure as a GANN. Using partial responses is a common way to interpret non-linear statistical models. Here, it is shown that the responses can themselves be used directly in modelling, with little or no compromise in predictive performance.

The result is a small model that explains a large and complex data set in terms of variable dependencies that clinicians can understand and integrate into their reasoning models. Iterative modeling is necessary because of the inherent redundancy in the data set, but the sequence of models obtained is itself informative about the association with outcome for individual and pairs of covariates.

Ultimately, the PRN model shows that it is possible to be sure that the model is right for the right reasons. Moreover, the covariate dependencies provide the ability to diagnose flaws in the data, whether because of sampling bias or artifacts in observational cohorts.

It is concluded that the PRN approach can add significant insight and modelling value to the analysis of tabular data in general, and in particular medical data.

References

1. Fernandes, K., Chicco, D., Cardoso, J.S., Fernandes, J.: Supervised deep learning embeddings for the prediction of cervical cancer diagnosis. PeerJ Comput. Sci. **4**, e154 (2018). https://doi.org/10.7717/peerj-cs.154
2. Ravikumar, P., Lafferty, J., Liu, H., Wasserman, L.: Sparse additive models. J. Roy. Stat. Soc.: Ser. B (Stat. Methodol.) **71**(5), 1009–1030 (2009)
3. Fernandes, K., Cardoso, J.S., Fernandes, J.: Transfer learning with partial observability applied to cervical cancer screening. In: Alexandre, L.A., Salvador Sánchez, J., Rodrigues, J. M.F. (eds.) IbPRIA 2017. LNCS, vol. 10255, pp. 243–250. Springer, Cham (2017). https://doi.org/10.1007/978-3-319-58838-4_27. Accessed 8 Feb 2020
4. de Waal, D.A., du Toit, J.: Generalized additive models from a neural network perspective. In: Proceedings of the 7th IEEE International Conference on Data Mining, ICDM 2007, Omaha, Nebraska, pp. 265–270. IEEE (2007)

5. Holsting, K., Vaughan, J.W., Daumé III, H., Dudík, M., Wallach, H.: Improving fairness in machine learning systems: what do industry practitioners need? ACM CHI Conference on Human Factors in Computing Systems (2019)
6. Barocas, S., Selbst, A.: 10 Big data's disparate impact. Calif. Law Rev. (2016). http://dx.doi.org/10.15779/Z38BG31
7. Gomez, S., Despraz, J., Pena-Reyes, C.A.: Improving neural network interpretability via rule extraction. In: ICANN 2018. LNCS, vol. 11139, pp. 811–813. Springer, Heidelberg (2018)
8. Borji, A.: Saliency Prediction in the Deep Learning Era: Successes, Limitations, and Future Challenges. IEEE Trans PAMI (2019). arXiv:1810.03716v3
9. Zhou, Z., et al.: Activation Maximization Generative Adversarial Nets arXiv:1703.02000 (2017)
10. Lundberg, S., Lee, S.-I.: A unified approach to interpreting model predictions. In: Advances in Neural Information Processing Systems, vol. 30, pp. 4765–4774 (2017)
11. Ribeiro, M.T., Singh, S., Guestrin, C.: Why should i trust you? In: Proceedings of the 22nd ACM SIGKDD International Conference on Knowledge Discovery and Data Mining - KDD 2016, pp. 1135–1144 (2016)
12. Brás-Geraldes, C., Papoila, A., Xufre, P.: Odds ratio function estimation using a generalized additive neural network. Neural Comput. Appl. (8) (2020, in press). https://www.springerprofessional.de/en/neural-computing-and-applications-8-2020/17871790
13. Lisboa, P.J.G., Ortega-Martorell, S., Cashman, S., Olier, I.: The Partial Response Network. arXiv:1908.05978 (2019)
14. Burd, E.M.: Human papillomavirus and cervical cancer. Clin. Microbiol. Rev. 16(1), 1–17 (2003)
15. MacKay, D.J.C.: The evidence framework applied to classification networks. Neural Comput. 4, 720–736 (1992)

Image Processing

Thin Structures Segmentation Using Anisotropic Neighborhoods

Christophe Ribal[(✉)], Nicolas Lermé, and Sylvie Le Hégarat-Mascle

Université Paris-Saclay, ENS Paris-Saclay, CNRS, SATIE, 94235 Cachan, France
{christophe.ribal,nicolas.lerme,
sylvie.le-hegarat}@universite-paris-saclay.fr

Abstract. Bayesian and probabilistic models are widely used in image processing to handle noise due to various alteration phenomena. To benefit from the spatial information in a tractable way, Markov Random Fields (MRF) are often assumed with isotropic neighborhoods, that is however at the detriment of the preservation of thin structures. In this study, we aim at relaxing this assumption on stationarity and isotropy of the neighborhood shape in order to get a prior probability term that is relevant not only within the homogeneous areas but also close to object borders and within thin structures. To tackle the issue of neighborhood shape estimation, we propose to use tensor voting, that allows for the estimation of structure direction and saliency at various scales. We propose three main ways to derive anisotropic neighborhoods, namely shape-based, target-based and cardinal-based neighborhood. Then, having defined the neighborhood field, we introduce an energy that will be minimized using graph cuts, and illustrate the benefits of our approach against the use of isotropic neighborhoods in the applicative context of crack detection. First results on such a challenging problem are very encouraging.

Keywords: Thin structures · Segmentation · Anisotropic neighborhoods · Superpixels · Graph cuts

1 Introduction

Image segmentation is a challenging task in the computer vision field, which deals with the problem of partitioning an image (or video) into multiple regions with labels that may later be used in higher level tasks, like object classification, detection or tracking. This is an ill-posed problem since at the pixel level, such operation is prone to noise, corrupted data and all kind of optic phenomena altering the original image, resulting in multiple valid solutions. A common way to overcome these difficulties is to take into account spatial relationships between close pixels in order to favor some solutions exhibiting slow variations in the label field. Classically, one may model the 2-dimensional field of labels as a MRF [10] and compute the segmentation using the Maximum A Posteriori (MAP).

© Springer Nature Switzerland AG 2020
M.-J. Lesot et al. (Eds.): IPMU 2020, CCIS 1237, pp. 601–612, 2020.
https://doi.org/10.1007/978-3-030-50146-4_44

Variational approaches are widely used to provide solutions minimizing a functional which uses energy terms representing data fidelity and regularization terms. The numerous energy models for reducing the impact of image artifacts over the output segmentation nevertheless tend to share a common drawback: They behave poorly on thin structures[1], because of the small size and complex geometry of the latter with respect to neighborhood ones. The early removal of such structures is a well known effect of Total Variation (TV) regularization (e.g. in image reconstruction [20]) and Potts regularization (e.g. in image segmentation [14]). Thin structures are however ubiquitous in a number of applications (such as medical imaging or quality control) and detecting them as accurately as possible is therefore of great interest.

Alternatively, superpixel decomposition methods have been developed for grouping pixels sharing similar radiometric intensities into regions of controlled spatial extent. Superpixel partitions are generally seen as oversegmentations that preserve small structures but also noise. The benefits of superpixel decomposition is thus to drastically reduce the number of elements to process while keeping the geometrical information that is often lost with multi-resolution approaches and leaving noise removal for further processing steps.

Dealing with further processing, a major drawback of a superpixel segmentation is that the usual hypothesis of a regular lattice is lost (i.e. pixels are all of the same size and shape). As a result, image segmentation approaches taking advantage of superpixels must cope with these problems and introduce new spatial relationships. This induces a neighborhood construction step even for isotropic neighborhoods: For instance, a simple criterion is that superpixels are considered as neighbors when they share a common border [6,9,15,21]. The authors of [21] propose to minimize an energy using graph cuts on the adjacency graph obtained from the watershed of the input image. In this graph, edges connecting two adjacent regions are weighted upon their common border length, similarly to [6]. Those neighborhood fields based on adjacency do not favor specific orientations of neighborhoods with respect to superpixel context and/or location. The approach of [11] is to gather all superpixels whose centroid belongs to a disc centered on it and is therefore isotropic as in most of the other superpixel-based segmentation approaches.

At the pixel level however, anisotropic approaches have been introduced to minimize the alteration of thin structures by regularization processes [7,13]. This corresponds to the relaxation of the isotropic hypothesis often introduced when formulating the problem as a MRF. As an alternative to the weighted Total Variation (TV) [19], the authors of [17] introduce a directional TV approach, based on a "vesselness feature" which aims to detect thin structures. Finally, since we believe that structure orientation estimation is a key aspect of anisotropic regularization approaches, let us mention different ways to estimate it: tensor voting approaches [5,16], vesselness operators like RORPO [18], the Frangi vesselness [8], or structure aware regression filters [23] to perform structure-dependent

---

[1] In n-dimensional images, thin structures or tubular structures are characterized by a significantly smaller size in at least one of their n dimensions.

image smoothing. By analogy with typical probabilistic modeling, the uniform hypothesis widely used in the absence of prior knowledge corresponds to an isotropic neighborhood, and the specific prior distribution corresponds to an anisotropic neighborhood which can be derived from the observation of the local orientation in our case.

We thus propose a methodology that both allows for the relaxation of the isotropic neighborhood which is all the more relevant when we consider the superpixel level, and provides regularized results robust to noise. We consider in this context the construction of elliptic neighborhoods, that originate from [7] and [11], and of two path-based neighborhoods. Similarly to [17], we expect these anisotropic neighborhoods to take into account the orientations of image's structures. Thus, we introduce a new field embedding these orientations computed from tensor voting [16]. Finally, we formulate the segmentation problem in an energy minimization framework, and solve it using graph cuts.

The rest of the paper is organized as follows. The problem formulation is presented in Sect. 2 and the construction of isotropic and anisotropic neighborhoods is detailed in Sect. 3. Section 4 introduces the energy terms implemented and Sect. 5 compares our results against those obtained with isotropic neighborhoods on real and simulated images. Finally, Sect. 6 outlines the contributions of the paper and discusses future work.

2   Problem Definition

A superpixel is a group of pixels, defined by their coordinates in the n-dimensional space, $n \in \mathbb{N}_{>0}$ the set of positive integers ($n = 2$ in the experiments presented in Sect. 5). Since each pixel belongs to one and only one superpixel, the set $\mathcal{S}$ of $K \in \mathbb{N}_{>0}$ superpixels is a partition of the original image. Denoting by $\mathcal{P}$ the set of pixels, then each superpixel s is an element of $\mathcal{S} \subset 2^{\mathcal{P}}$, i.e. $s \in 2^{\mathcal{P}}$ where 2^X denotes the powerset of a set X. The partition constraint implies that $\forall p \in \mathcal{P}, \exists! s \in \mathcal{S}$ such that $p \in s$. Notice that the shape of any superpixel is also usually constrained to be composed of a single connected component. We denote by $\mathbb{F}$ the feature space holding the spectral information associated to any pixel or superpixel, for instance $\mathbb{R}$ (grayscale images), $\mathbb{R}^3$ (color images) or a higher dimensional space (hyperspectral images).

To stress that our approach can apply equally to an image of pixels or of superpixels, we define the position and the feature vector of a superpixel $s \in \mathcal{S}$. In our case (but other choices could have been done depending on the application), they are the barycenter of the coordinates (in n-dimensional space) of the pixels that compose s and the feature barycenter (in $\mathbb{F}$) of these same pixels. Given a finite set $\mathcal{C} = \{1, \ldots, C\}$ of $C \in \mathbb{N}_{>0}$ classes, segmenting the image is equivalent to finding a field of labels $\mathbf{u} \in \mathcal{C}^{\mathcal{S}}$.

We use the MAP criterion to assign, given the image of superpixels (with superpixels possibly reduced to a single pixel), denoted $\mathbf{I} \in \mathbb{F}^{\mathcal{S}}$, a label $u_s \in \mathcal{C}$ to $s, \forall s \in \mathcal{S}$. To this end, we set up a functional E, to be minimized over the field of labels $\mathbf{u} \in \mathcal{C}^{\mathcal{S}}$, that encompasses different priors on the labeling $\mathbf{u}$:

$$E(\mathbf{u}, V) = E_1(\mathbf{u}) + \alpha E_2(\mathbf{u}, V), \tag{1}$$

where $\alpha \in \mathbb{R}_{>0}$ is a parameter controlling the balance between the data fidelity term E_1 and the smoothness term E_2, and $V : \mathcal{S} \to 2^{\mathcal{S}}$ is the neighborhood field that is fixed. Note that E_1 only depends on the image data and on $\mathbf{u}$. Smoothness prior on the labeling $\mathbf{u}$ yields the smoothness term E_2 that is itself based on neighborhood field definition. In this study, we only consider the second order cliques, and we denote by $\mathcal{N} \subset \mathcal{S}^2$ the set of second order cliques of superpixels. Note that using such a definition, the superpixels of any pair $(s, t) \in \mathcal{N}$ are not required to have a common boundary. For any superpixel $s \in \mathcal{S}$, we define the neighborhood $V(s)$ of s as $V(s) = \{t \in \mathcal{S} \mid (s, t) \in \mathcal{N}\}$.

With the relaxation of isotropy and stationarity constraints on V comes the need to introduce additional priors. First, we formulate the hypothesis that the structure of the neighborhood of a superpixel depends on the structure of the objects in the image. We model such information of structure through a symmetric second order tensor field $\mathcal{T} \in (\mathbb{R}^{n \times n})^{\mathcal{S}}$ and assume that it can be built from the field of labels $\mathbf{u}$. Other considered priors yield different ways to construct the neighborhood field V, presented in the next section.

3   Proposed Neighborhoods Construction

In this section, we aim at defining the neighborhood field $V : \mathcal{S} \to 2^{\mathcal{S}}$, possibly anisotropic and non stationary. To each site $s \in \mathcal{S}$, we associate a set of sites $V(s) \in 2^{\mathcal{S}}$, where s is either a superpixel or a pixel, such definition being consistent with $\mathcal{S} = \mathcal{P}$. To underline the genericity of our formulation, we consider both cases in our experiments.

The construction of our anisotropic neighborhoods aims at encouraging strong relationships between sites aligned with respect to the directions of the thin structures of the image. As explained in Sect. 1, the characteristics thereby depicted for these structures, namely orientation and saliency, may be retrieved from vesselness operators [8,18,23] or Tensor Voting [16]. In this study, we consider this latter approach where a scale parameter $\sigma \in \mathbb{R}_{>0}$ sets the span of the voting field. Whatever the way they have been estimated, let us represent the thin structure features in a field of second order tensors $\mathcal{T} \in (\mathbb{R}^{n \times n})^{\mathcal{S}}$. For any site $s \in \mathcal{S}$, local orientation and saliency of structure are derived from the eigenvectors and the eigenvalues of the tensor $\mathcal{T}_s$. Eigenvectors are ranked by decreasing order of their corresponding eigenvalue. More precisely, for any site $s \in \mathcal{S}$, the construction of V is achieved using a set of vectors, $(\vec{v}_i(\mathcal{T}_s))_{i=0}^{n-1}$, where $\vec{v}_i(\mathcal{T}_s) \in \mathbb{R}^n$ is collinear with the ith eigenvector with its norm being equal to the ith eigenvalue, $\forall i \in \{0, \ldots, n-1\}$.

We distinguish two families of anisotropic neighborhoods, namely shape-based neighborhoods and path-based neighborhoods, both compared (see Sect. 5) against the following neighbourhoods: Stawiaski's boundary-based neighborhood [21] and Giraud's neighborhood [11]. Note that the latter can be seen as an isotropic restriction of our shape-based neighborhood. Path-based neighborhoods stem from the idea of adapting the neighborhood structure to 1-dimensional thin structures, represented by paths. Formally, for any $k \in \mathbb{N}_{>0}$, we define a path of cardinality k as a set of sites $(s_1, \ldots, s_k) \in \mathcal{S}^k$ such that, in our case, (s_i, s_{i+1}) have a common boundary (see [21]), $\forall i \in \{1, \ldots, k-1\}$. Moreover, we denote by $\mathscr{P}_K$ the set of paths of cardinality $K \in \mathbb{N}_{>0}$ and by $\mathscr{P}$ the set of paths of any cardinality $k \in \mathbb{N}_{>0}$. In what follows, we detail three different ways to construct the neighborhood field V using the tensor field $\mathcal{T}$.

3.1 Shape-Based Neighborhoods

We are inspired by [11], using superpixel centroid relative locations and n-dimensional shapes instead of discs to settle the shape-based neighborhoods (**shape**). Whenever the centroid of a superpixel belongs to the computed neighborhood shape of a second one, it is added to the neighborhood of the latter. For computational reasons, we discretize the orientations of the parametric shapes based on the one of $\vec{v_0}(\mathcal{T}_s)$ for any site $s \in \mathcal{S}$, which boils down to the use of a dictionary of neighborhood shapes. Notice that the neighborhood $V(s)$ of any site $s \in \mathcal{S}$ is not necessarily connected with such an approach.

3.2 Target-Based Neighborhoods

Target-based neighborhood (**target**) is a path-based neighborhood that aims at constructing the neighborhood $V(s)$ of a site $s \in \mathcal{S}$ by connecting it to two distant sites $t_0, t_1 \in \mathcal{S}$ (named "target") through paths of minimal energy. Hence, the connectedness along these paths (and so $V(s)$) is thus ensured by definition. We propose to find these paths in two stages. Firstly, for any $j \in \{0, 1\}$, targets connecting s are found with

$$t_j^* \in \underset{t \in \widetilde{V}(s)}{\operatorname{argmin}} \|I(s) - I(t)\|_2^2 - \beta \|\vec{st}\|^2 \operatorname{sign}\left((-1)^j \langle \vec{v_0}(\mathcal{T}_s), \vec{st} \rangle\right), \qquad (2)$$

where $\beta \in \mathbb{R}_{>0}$ is a free parameter, $\widetilde{V}(s)$ denotes a **shape**-based neighborhood (see Sect. 3.1), $\operatorname{sign}(.)$ denotes the sign of a real number, $\langle ., . \rangle$ denotes the scalar product, $\|.\|$ denotes the Euclidean norm and $\vec{st}$ denotes the vector connecting any pair of sites $(s, t) \in \mathcal{S}^2$. In Eq. (2), the first term favors the sites s and t to have similar image intensities while the second term favors far targets from s that are aligned with $\vec{v_0}(\mathcal{T}_s)$.

Secondly, paths of minimal energy connecting the site $s \in \mathcal{S}$ to either targets t_0^* or t_1^* (see Eq. (2)) are obtained with

$$p_j^* \in \underset{p=(s_1=s,\ldots,s_{\sharp p}=t_j^*) \in \mathscr{P}}{\operatorname{argmin}} \sum_{i=1}^{\sharp p-1} \|I(s_i) - I(s_{i+1})\|_2^2, \qquad (3)$$

where $\sharp$ stands for the cardinality of a set. The term to be minimized in Eq. (3) is large when image intensities of successive sites along a path are dissimilar and small otherwise. Finally, the neighborhood $V(s)$ of the site s can be now constructed as follows: $V(s) = (p_0^* \cup p_1^*) \setminus \{s\}$.

3.3 Cardinal-Based Neighborhoods

Cardinal-based neighborhood (**cardinal**) is a path-based neighborhood that aims at constructing the neighborhood $V(s)$ of a site $s \in \mathcal{S}$ by finding two paths of minimal energy starting from s, in opposite directions (according to $\overrightarrow{v_0}(\mathcal{T}(s))$) and of fixed length $K \in \mathbb{N}_{>0}$. For any $j \in \{0,1\}$, these paths are obtained with

$$p_j^* \in \underset{p=(s_1=s,\ldots,s_K) \in \mathscr{P}_K}{\operatorname{argmin}} l_C^j(p). \tag{4}$$

In the above expression, $l_C^j(p)$ provides a measure of the length of the path p starting from s. For any path $p = (s_1 = s, \ldots, s_K) \in \mathscr{P}_K$, $l_C^j(p)$ is defined by

$$l_C^j(p) = \sum_{i=2}^{K} \|I(s) - I(s_i)\|_2^2 + \beta' \phi_j(\overrightarrow{v_0}(\mathcal{T}_s), \overrightarrow{ss_i}), \tag{5}$$

where $\beta' \in \mathbb{R}_{>0}$ is a free parameter and

$$\phi_j(\overrightarrow{u}, \overrightarrow{v}) = \begin{cases} \arccos\left(\left|\frac{\langle \overrightarrow{u}, \overrightarrow{v} \rangle}{\|\overrightarrow{u}\| \|\overrightarrow{v}\|}\right|\right) & \text{if } (-1)^j \langle \overrightarrow{u}, \overrightarrow{v} \rangle > 0, \\ +\infty & \text{otherwise}, \end{cases}$$

measures the angle between the vectors $\overrightarrow{u}$ and $\overrightarrow{v}$ and discriminates whether the scalar product between them is positive or not. The first term of Eq. (5) encourages the image intensities of any site s_i to be similar to s while the second term aims at aligning the path with $\overrightarrow{v_0}(\mathcal{T}_s)$. This allows for ensuring that two paths in opposite directions are selected to establish the neighborhood $V(s)$ of s. Finally, the neighborhood $V(s)$ of the site s can be now constructed as follows: $V(s) = (p_0^* \cup p_1^*) \setminus \{s\}$. In the next section, the segmentation model using anisotropic and isotropic neighborhoods is detailed.

4   Proposed Model

The data fidelity term $E_1(\mathbf{u})$ in the functional $E(\mathbf{u}, V)$ (see Eq. (1)) is the energy term derived from the likelihood $\mathbb{P}(I \mid \mathbf{u})$. At the pixel level, popular models rely on statistical assumptions, especially by assuming site conditional independence. Relying on the same assumption but at the superpixel level, the probability $\mathbb{P}(I \mid \mathbf{u})$ is the product, over $\mathcal{S}$, of probabilities $\mathbb{P}(I(s) \mid u_s)$, $I(s) \in \mathbb{F}$ being the observation and $u_s \in \mathcal{C}$ the class of s.

In this study, we adopt a color model assuming that image intensities are Gaussian-distributed for each class $c \in \mathcal{C}$ with mean value $\mu_c \in \mathbb{F}$ and standard deviation $\sigma_c \in \mathbb{R}_{>0}$ [4]. Then, the data term E_1^s for any superpixel $s \in \mathcal{S}$ and any label $u_s \in \mathcal{C}$ is written

$$E_1^s(u_s) = \frac{\|I(s) - \mu_{u_s}\|_2^2}{2\sigma_{u_s}^2} + \log(\sigma_{u_s}),$$

and, for any $\mathbf{u} \in \mathcal{C}^{\mathcal{S}}$, E_1 in Eq. (1) is:

$$E_1(\mathbf{u}) = \sum_{s \in \mathcal{S}} E_1^s(u_s). \tag{6}$$

The energy term $E_2(\mathbf{u}, V)$ corresponds to the smoothness prior on the label field $\mathbf{u}$ and requires the definition of a neighborhood field V, as introduced in Sect. 2. Then, this neighborhood field being fixed, we assume $\mathbf{u}$ is an MRF so that a prior probability on $\mathbf{u}$ can be computed from 'elementary' energy terms.

In this study, we adopt the Potts model [25], weighted according to the strength of interaction between neighboring superpixels. The definition of $E_2(\mathbf{u}, V)$ is thus the following:

$$E_2(\mathbf{u}, V) = \sum_{s \in \mathcal{S}} \sum_{p \in V(s)} W(s, p) \mathbb{1}_{\{u_s \neq u_p\}},$$

where $\mathbb{1}_{\{a \neq b\}} = \begin{cases} 1 \text{ when } a \neq b, \\ 0 \text{ otherwise.} \end{cases}$ and $W : \mathcal{N} \to \mathbb{R}_{>0}$ is a weighting function.

For instance, in our implementation of the neighborhood of [21], the weighting function W is defined for any pair $(s, p) \in \mathcal{N}$ as $W(s, p) = \frac{\partial(s,p)}{\partial(s)} \in]0, 1]$, where $\partial(s, p)$ and $\partial(s)$ denote the common boundary between s and p and the perimeter of s, respectively. In the other neighborhood fields we compare, the cliques $\mathcal{N}$ can connect non adjacent superpixels. Thus, we propose to define for any pair $(s, p) \in \mathcal{N}$ the weighting function W as $W(s, p) = (\sharp V(s))^{-1} \in]0, 1]$.

The Potts model preserves its properties, in particular submodularity, and the data fidelity term E_1 is convex. Numerous works have proven the efficiency of graph cuts [3,12]. According to [12] and [3] respectively, the energy function defined in Eq. (1) can be exactly minimized when $\sharp \mathcal{C} = 2$ (this is the case in our experiments) and approximately minimized when $\sharp \mathcal{C} > 2$.

Finally, note that the estimation of the neighborhood field V itself requires a segmentation $\mathbf{u}$. In this study, we use the blind segmentation (i.e. $\alpha = 0$).

5   Numerical Experiments

5.1   Application Framework and Parameter Setting

Let us introduce the data and experiments carried out within our application context, namely crack detection. We aim at segmenting a crack, which is a thin structure over a highly textured and noisy background, e.g. some asphalt road or concrete wall as in the cracktree dataset [27]. In this study, in addition to images from this dataset, we consider a simulated image with arbitrary shapes and textured noise as shown in Fig. 1. Images intensities are normalized in $[0, 1]$.

A variety of algorithms for generating superpixels exist and exhibit different properties [1,22]. Besides, the requirement of providing a partition of the image into connected sets of pixels, the main desirable properties are the preservation of image boundaries, the control of the compactness of superpixels and their number, in addition to computational efficiency of the algorithms. In order to study the benefit of our approach also regardless the superpixel decomposition, we propose a "perfectly shaped" set of superpixels generated from the dilated ground truth. Then, for results derivation using actual superpixels, we require the following properties for superpixels: good compactness to be efficiently modeled by their centroid, regularity in size while at the same time allowing the grouping of crack pixels into thin superpixels. Given those prerequisites, we have considered Extended Topology Preserving Segmentation superpixels (ETPS) [26] after image smoothing with median filtering with a square window of size (7×7).

Concerning the construction of anisotropic neighborhoods, the parameters are fixed so that there are 6 neighbors per superpixel in average. For **shape**, this reduces to setting the ellipse's area to 7 times the mean area of a superpixel, while their flattening is set to 0.6. With **cardinal**, we set $K = 4$ and $\beta' = 5 \times 10^{-3}$. Finally for **target**, $\beta = 5 \times 10^{-3} \times \beta_R$ where $\beta_R \in \mathbb{R}_{>0}$ is the radius of the shape-based neighborhood ellipsis. Notice, these parameters are related to the scale of the thin structures to detect with respect to the superpixel size, so that some priors can help setting them.

Finally, we estimate the mean μ_c of each class $c \in \mathcal{C}$ (here, $\sharp\mathcal{C} = 2$) from the ground truth, and we set $\sigma_c = 1$ for simplicity. To compensate the effect of variation of texture size due to perspective, we set the mean value of classes as an affine function of the vertical position of the superpixel. Future works can handle the parameter estimation in an unsupervised context.

5.2   Quantitative and Qualitative Evaluation

Because our ground truth can be composed of 1 pixel width objects, in order to distinguish between slight mislocation errors and non-detection of some parts of the cracks, we compute the F-measure (FM) at scale $\epsilon = 2$, based on the number of true positives (TP), false positives (FP) and false negatives (FN), like in [2,24]. In addition, the crack region and the non-crack area being highly unbalanced (in favor to the non-crack area), we use a high value of $\gamma = 5$ in FM to increase sensitivity to FN with respect to FP:

$$\text{FM}(\gamma) = \left(1 + \frac{\gamma^2 FN}{(1+\gamma^2)TP} + \frac{FP}{(1+\gamma^2)TP}\right)^{-1} \in [0,1]. \tag{7}$$

The results are presented at pixel level in Fig. 1 and at superpixel level in Fig. 2, respectively. For each image, that corresponds to a set of parameters including the type of superpixels and the type of neighborhood, we select the best result, according to the FM criterion, among the results obtained varying the parameters σ (the scale parameter of tensor voting) and α (the regularization parameter in Eq. (1)). The automatic estimation of these parameters along with the analysis of their impact on performance (FM criterion for instance) will be carried in future work in order to make the proposed method more effective.

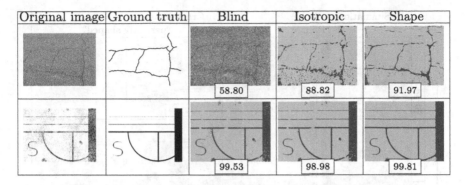

Original image	Ground truth	Blind	Isotropic	Shape
		58.80	88.82	91.97
		99.53	98.98	99.81

Fig. 1. Evaluation performance against ground truths at pixel level for a crack image (top row) and a simulated one (bottom row). The three last columns are segmentations without regularization ("blind") and with regularization (isotropic with 4-connectivity or **shape**-based anisotropic neighborhoods with ellipses). For each image, in both regularized cases, the results achieving the largest FM with respect to tensor voting scale σ and regularization parameter α, are depicted. FM measurements are also provided in percents for $\gamma = 5$.

At pixel level, Fig. 1 illustrates the clear improvement of the quality of the results with the use of anisotropic neighborhoods. In the first image of crack, anisotropic regularization allows for enhancing the continuity of the detected cracks, even if some small gaps still fragment it. In the simulated image, the improvement is significant with the correct segmentation of the six discontinuities in the cracks, without loss of precision on more complex shapes.

However, at superpixel level (Fig. 2), while exhibiting better blind results thanks to the averaging of information at pixel level, superpixel anisotropic neighborhoods seem to suffer in general from the fact that it is difficult to establish the right neighborhood V even with a correct estimation of its orientation (see last column). Our experiments reveal that even if we are far from "Optimal" neighborhood performances, path-based neighborhoods tend to outperform the shape-based ones. Unfortunately, the anisotropic approach benefits exhibited in Fig. 2 do not seem to improve the segmentation of the crack image in a so significant way: Path-based approach outperforms the other approaches when superpixels are perfectly shaped, but are still sensitive to the degradation of the quality of the superpixels.

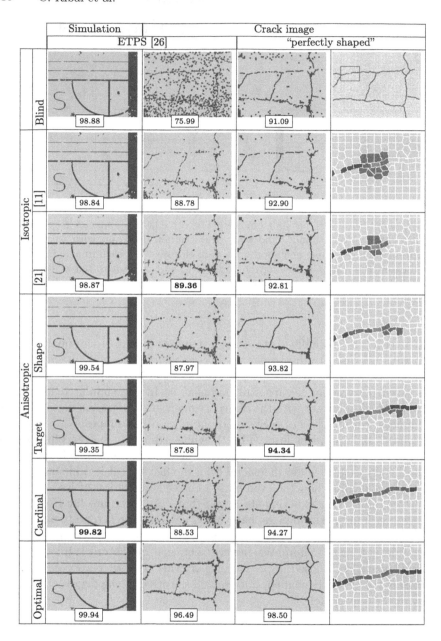

Fig. 2. Evaluation performance against ground truths at superpixel level for a crack image and a simulated one. The segmentations achieving the largest FM with respect to parameters σ and α for $\gamma = 5$, are depicted. The last two columns correspond to the use of the "perfectly shaped" superpixel. The last column represents, for one site highlighted in blue, the sites that are in its neighborhood (in red), depending on the method used for constructing the latter one: Each row shows a different type of neighborhood, specified in header lines. The last row is a ground truth-based neighborhood for comparison purpose. (Color figure online)

6   Conclusion

In this paper, we introduced three anisotropic neighborhoods, in order to make them able to fit the thin structures of the image and thus to improve segmentation results. They rely on the estimation of the orientations of such structures, based here on tensor voting that is efficient in estimating dense map of orientations from a sparse field of labeled sites in the blind segmentation. We then perform the minimization of our energy functional via graph cuts.

We tested our results with a simulated image and an actual difficult crack image, to validate the improvements brought by anisotropic regularization. While our results exhibit a high gain of performances at pixel level, super-pixel segmentation suffers from the challenging task to estimate neighborhood at superpixel level, that seems to weaken the benefits of anisotropic regularization.

Finally, we plan to investigate the possible refinement of the neighborhood field estimation after computing the regularized segmentation to introduce an alternative minimization procedure, and to explore extensions of our approach with thin structures in shape from focus in 3D-space [20] for future works.

References

1. Achanta, R., Susstrunk, S.: Superpixels and polygons using simple non-iterative clustering. In: Proceedings of IEEE Conference on CVPR, pp. 4895–4904 (2017)
2. Aldea, E., Le Hégarat-Mascle, S.: Robust crack detection for unmanned aerial vehicles inspection in an a-contrario decision framework. J. Electron. Imaging $24(6)$, 1–16 (2015)
3. Boykov, Y., Veksler, O., Zabih, R.: Fast approximate energy minimization via graph cuts. IEEE Trans. Pattern Anal. Mach. Intell. $23(11)$, 1222–1239 (2001)
4. Chan, T., Vese, L.: Active contours without edges. IEEE Trans. Image Process. $10(2)$, 266–277 (2001)
5. Tang, C.-K., Medioni, G., Lee, M.-S.: N-dimensional tensor voting and application to epipolar geometry estimation. IEEE Trans. Pattern Anal. Mach. Intell. $23(8)$, 829–844 (2001)
6. Cui, B., Xie, X., Ma, X., Ren, G., Ma, Y.: Superpixel-based extended random walker for hyperspectral image classification. IEEE Trans. Geosci. Remote Sens. $56(6)$, 1–11 (2018)
7. Favaro, P.: Recovering thin structures via nonlocal-means regularization with application to depth from defocus. In: 2010 IEEE Computer Society Conference on Computer Vision and Pattern Recognition, pp. 1133–1140 (2010)
8. Frangi, A.F., Niessen, W.J., Vincken, K.L., Viergever, M.A.: Multiscale vessel enhancement filtering. In: Wells, W.M., Colchester, A., Delp, S. (eds.) MICCAI 1998. LNCS, vol. 1496, pp. 130–137. Springer, Heidelberg (1998). https://doi.org/10.1007/BFb0056195
9. Fulkerson, B., Vedaldi, A., Soatto, S.: Class segmentation and object localization with superpixel neighborhoods. In: IEEE International Conference on Computer Vision, pp. 670–677 (2009)
10. Geman, S., Geman, D.: Stochastic relaxation, Gibbs distributions, and the Bayesian restoration of images. IEEE Trans. Pattern Anal. Mach. Intell. $6(6)$, 721–741 (1984)

11. Giraud, R., Ta, V.T., Bugeau, A., Coupe, P., Papadakis, N.: SuperPatchMatch: an algorithm for robust correspondences using superpixel patches. IEEE Trans. Image Process. **26**(8), 4068–4078 (2017)
12. Kolmogorov, V., Zabih, R.: What energy functions can be minimized via graph cuts? IEEE Trans. Pattern Anal. Mach. Intell. **26**(2), 147–159 (2004)
13. Le Hégarat-Mascle, S., Kallel, A., Descombes, X.: Ant colony optimization for image regularization based on a nonstationary Markov modeling. IEEE Trans. Image Process. **16**(3), 865–878 (2007)
14. Lermé, N., Le Hégarat-Mascle, S., Malgouyres, F., Lachaize, M.: Multilayer joint segmentation using MRF and graph cuts. J. Math. Imaging Vis. 1–21 (2020)
15. Liu, Y., Condessa, F., Bioucas-Dias, J.M., Li, J., Du, P., Plaza, A.: Convex formulation for multiband image classification with superpixel-based spatial regularization. IEEE Trans. Geosci. Remote Sens. **56**, 1–18 (2018)
16. Medioni, G., Tang, C.K., Lee, M.S.: Tensor voting: theory and applications. In: Proceedings of RFIA (2000)
17. Merveille, O., Miraucourt, O., Salmon, S., Passat, N., Talbot, H.: A variational model for thin structure segmentation based on a directional regularization. In: 2016 IEEE International Conference on Image Processing (ICIP), pp. 4324–4328 (2016)
18. Merveille, O., Talbot, H., Najman, L., Passat, N.: Curvilinear structure analysis by ranking the orientation responses of path operators. IEEE Trans. Pattern Anal. Mach. Intell. **40**(2), 304–317 (2018)
19. Miraucourt, O., Jezierska, A., Talbot, H., Salmon, S., Passat, N.: Variational method combined with Frangi vesselness for tubular object segmentation. In: Computational & Mathematical Biomedical Engineering (CMBE), pp. 485–488 (2015)
20. Ribal, C., Lermé, N., Le Hégarat-Mascle, S.: Efficient graph cut optimization for shape from focus. J. Vis. Commun. Image Representation **55**, 529–539 (2018)
21. Stawiaski, J., Decencière, E.: Region merging via graph-cuts. Image Anal. Stereol. **27**(1), 39 (2011)
22. Stutz, D., Hermans, A., Leibe, B.: Superpixels: an evaluation of the state-of-the-art. Comput. Vis. Image Understand. **166**, 1–27 (2018)
23. Su, Z., Zeng, B., Miao, J., Luo, X., Yin, B., Chen, Q.: Relative reductive structure-aware regression filter. J. Comput. Appl. Math. **329**, 244–255 (2018)
24. Vandoni, J., Le Hégarat-Mascle, S., Aldea, E.: Crack detection based on a Marked Point Process model. In: International Conference on Pattern Recognition (ICPR), pp. 3933–3938 (2016)
25. Wu, F.Y.: The Potts model. Rev. Modern Phys. **54**(1), 235–268 (1982)
26. Yao, J., Boben, M., Fidler, S., Urtasun, R.: Real-time coarse-to-fine topologically preserving segmentation. In: Proceedings of the IEEE Conference on Computer Vision and Pattern Recognition, pp. 2947–2955 (2015)
27. Zou, Q., Cao, Y., Li, Q., Mao, Q., Wang, S.: CrackTree: automatic crack detection from pavement images. Pattern Recogn. Lett. **33**(3), 227–238 (2012)

Dempster-Shafer Parzen-Rosenblatt Hidden Markov Fields for Multichannel Image Segmentation

Mohamed El Yazid Boudaren$^{(\boxtimes)}$, Ali Hamache, Islam Debicha,
and Hamza Tarik Sadouk

Ecole Militaire Polytechnique, PO Box 17, 16111 Bordj El Bahri, Algiers, Algeria
boudaren@gmail.com

Abstract. Theory of evidence has been successfully used in many areas covering pattern recognition and image processing due to its effectiveness in both information fusion and reasoning under uncertainty. Such notoriety led to extension of many existing Bayesian tools such as hidden Markov models, extensively used for image segmentation. This paper falls under this category of frameworks and aims to propose a new hidden Markov field that better handles nonGaussian forms of noise, designed for multichannel image segmentation. To this end, we use a recent kernel smoothing- based noise density estimation combined with a genuine approach of mass determination from data. The proposed model is validated on sampled and real remote sensing images and the results obtained outperform those produced by conventional hidden Markov fields.

Keywords: Data classification · Dempster-Shafer theory · Hidden Markov Field · Multichannel image segmentation

1 Introduction

Multichannel image analysis and processing have gained more interest among the image and signal processing community following the development of computing technologies [3,10,12,15]. The purpose of multichannel image classification, considered in this paper, is to produce a thematic map indicating the membership of each pixel in a specific class based on two sources: the spectral information and the spatial information. The first is represented by the different image channels. Each channel corresponds to an interval of the electromagnetic spectrum, where a dedicated sensor is used to measure the intensity of the spectrum received over this interval. The use of spectral information for image classification can be very effective especially in the supervised context because one has a knowledge base used at the learning stage. The interest of taking into consideration the second source of information, namely the contextual dependence, was quickly noticed. Image modeling through hidden Markov fields takes into account such dependencies which improves the classification performance [6,9,13,14].

© Springer Nature Switzerland AG 2020
M.-J. Lesot et al. (Eds.): IPMU 2020, CCIS 1237, pp. 613–624, 2020.
https://doi.org/10.1007/978-3-030-50146-4_45

A more elaborated classification model should perform at both levels to produce significantly best class maps. For this purpose, we propose a new hidden Markov model that better handles general forms of noise, typically nonGaussian. More explicitly, we propose to adopt an evidential approach for estimating the noise parameters, which thus allows a better use of spectral information in a Markovian context towards a more effective multichannel image classification.

The remainder of this paper is organized as follows. Section 2 briefly recalls Dempster- Shafer theory, Parzen- Rosenblatt density estimation and hidden Markov fields. Section 3 describes the proposed approaches and related estimation tasks. Experimental results are presented and discussed in Sect. 4. Concluding remarks and future directions are given in Sect. 5.

2   Preliminaries

In this section, we briefly recall some basic notions of Dempster-Shafer theory and Hidden Markov Fields.

2.1   Dempster-Shafer Theory

Data fusion particularly enhances the quality of decision when more than one source of information are available. This is mainly due to the possibility of increasing the amount of relevant information by exploiting redundancy and complementariness among sources. One powerful and flexible mathematical tool that has shown its usefulness in this area is Dempster-Shafer theory (DST) [16], [17] that generalizes the Bayesian frame by allowing on one hand to reap a consensus decision from all information sources; and on the other hand, to handle information uncertainty within each information source. Hence, DST has been applied in many fields [4,5,7,11]. In what follows, we give a quick overview about the DST concepts that will be needed for the sake of this paper.

Let $\Omega = \{\omega_1, ..., \omega_K\}$, and let $\mathcal{P}(\Omega) = \{A_1, ..., A_Q\}$ be its power set, with $Q = 2^K$. A function M defined from $\mathcal{P}(\Omega)$ to $[0, 1]$ is called a "basic belief assignment" (bba) if $M(\emptyset) = 0$ and $\sum_{A \in \mathcal{P}(\Omega)} M(A) = 1$. A bba M defines then a "plausibility" function Pl from $\mathcal{P}(\Omega)$ to $[0, 1]$ by $Pl(A) = \sum_{A \cap B \neq \emptyset} M(B)$, and a "credibility" function Cr from $\mathcal{P}(\Omega)$ to $[0, 1]$ by $Cr(A) = \sum_{B \subset A} M(B)$. Also, both aforementioned functions are linked by $Pl(A) + Cr(A^c) = 1$. Furthermore, a probability function p can be considered as a particular case for which $Pl = Cr = p$.

When two $bbas$ M_1 and M_2 describe two pieces of evidence, we can fuse them using the so called "Dempster-Shafer fusion" (DS fusion), which gives $M = M_1 \oplus M_2$ defined by:

$$M(A) = (M_1 \oplus M_2)(A) \propto \sum_{B_1 \cap B_2 = A} M_1(B_1) M_2(B_2) \tag{1}$$

Finally, an evidential bba M can be transformed into a probabilistic one using Smets method, according to which each mass of belief $M(A)$ is equally distributed among all elements of A, leading to the so called "pignistic probability", Bet, given by:

$$Bet(\omega_i) = \sum_{\omega_i \in A \subseteq \Omega} \frac{M(A)}{|A|} \tag{2}$$

where $|A|$ is the number of elements of Ω in A.

2.2 Parzen-Rosenblatt Dempster-Shafer Classifier

In this subsection, we briefly recall the Parzen-Rosenblatt Dempster-Shafer (PRDS) Classifier proposed in [8]. To this end, let us assume we have a sample of N prelabeled multiattribute data $(Z_1, ..., Z_N)$ where each datum $Z_n = (X_n, Y_n)$ with $X_n \in \Omega = \{\omega_1, ..., \omega_K\}$ being the label, and $Y_n = (Y_n^1, ..., Y_n^P) \in \mathbb{R}^P$ being the P-attribute observation. The problem is then to estimate the label of any new observation $Y_{n'}$ that is optimal with respect to some criterion.

In what follows, we recall the training and classification procedures. According to the PRDS scheme, training consists in estimating for each class $\omega_k \in \Omega$ and for each attribute p $(1 < p < P)$, the associated Parzen-Rosenblatt density $\hat{f}_k^p$. For further weighting sake, 5-fold cross-validation classification is achieved based on each attribute (taken alone) using the above Parzen-Rosenblatt PDFs according to maximum likelihood.

For a given new observation $Y_{n'}$, partial report about the identity of $X_{n'}$ can be made at each individual attribute level through a mass function M^p, on $P(\Omega)$, generated based on the Parzen-Rosenblatt PDF estimated at the training stage. Such reports are then combined, typically using DS- fusion to reap a consensus report M. Final decision is then be deduced through the Pignistic transform applied to M. In the following, we describe our approach step by step. For more details, the reader may refer to [8].

2.3 Hidden Markov Fields

Let S be a finite set, with $Card(S) = N$, and let $(Y_s)_{s \in S}$ and $(X_s)_{s \in S}$ be two collections of random variables, which will be called "random fields". We assume that Y is observable with each Y_s taking its values in $\mathbb{R}$ (or $\mathbb{R}^m$) whereas X is hidden with each X_s taking its values from a finite set of "classes" or "labels". Such situation occurs in image segmentation problem, which will be used in this paper as illustrative frame. Realizations of such random fields will be denoted using lowercase letters. We deal with the problem of the estimation of $X = x$ from $Y = y$. Such estimation subsumes the distribution of (X, Y) to be beforehand defined.

In hidden Markov fields (HMFs) context, the field X is assumed Markovian with respect to a neighborhood system $\mathfrak{N} = (\mathfrak{N}_s)_{s \in S}$. X is then called a Markov random field (MRF) defined by

$$p(X_s = x_s | (X_t)_{t \in S, t \neq s}) = p(X_s = x_s | (X_t)_{t \in \mathfrak{N}_s}) \tag{3}$$

Under some conditions usually assumed in digital imagery, the Hammersley-Clifford theorem [2] establishes the equivalence between an MRF, defined with respect to the neighbourhood system $\mathfrak{N}$, and a Gibbs field with potentials associated with $\mathfrak{N}$. Such potentials, describing the elementary relationships within the neighbourhood, are computed with respect to the system of cliques C, where a clique $c \in C$ is a subset of S which is either a singleton or a set of pixels mutually neighbors with respect to $\mathfrak{N}$. Setting $x_c = (X_s)_{s \in c}$, $\phi_c(x_c)$ denotes the potential associated to the clique c.

Finally, the distribution of X is given by

$$p(X = x) = \gamma \exp \left[- \sum_{c \in C} \phi_c(x_c) \right] \tag{4}$$

where γ is a normalizing constant which is impossible to compute in practice given the very high number of possible configurations K^N. The quantity $E(x) = \sum_{c \in C} \phi_c(x_c)$ is called "energy" and can also be expressed locally through $E_s(x_s) = \sum_{c \ni x_s} \phi_c(x_c)$. Hence, the local conditional probability of (3) becomes

$$p\left(X_s = x_s | (X_t)_{t \in S, t \neq s}\right) = \gamma_s \exp\left[-E_s(x_s) \right]$$

where γ_s is a computable normalizing constant.

To define the distribution of Y conditional on X, two assumptions are usually set:

(i) the random variables $(Y_s)_{s \in S}$ are independent conditional on X;
(ii) the distribution of each Y_s conditional on X is equal to its ditribution conditional on X_s.

When these two assumptions hold, the noise distribution is fully defined through K distributions $(f_i)_{1 \leq i \leq K}$ on $\mathbb{R}$ where f_i denotes the density, with respect to the Lebesgue measure on $\mathbb{R}$, of the distribution of Y_s conditional on $X_s = \omega_i$: $p(Y_s = y_s | x_s = \omega_i) = f_i(y_s)$. Then we have

$$p(Y = y | X = x) = \prod_{s \in S} f_{x_s}(y_s) \tag{5}$$

that can equivalently be written as

$$p(Y = y | X = x) = \exp \left[\sum_{s \in S} \log f_{x_s}(y_s) \right] \tag{6}$$

Since $p(x, y) = p(x)p(y|x)$, we obtain

$$p(X = x, Y = y) = \gamma \exp - \left[\sum_{c \in C} \phi_c(x_c) - \sum_{s \in S} \log f_{x_s}(y_s) \right] \tag{7}$$

Hence, according to (7), the couple (X, Y) is a Markov field and also is the distribution of X conditional on $Y = y$. This allows to sample a realization of

X according to its posterior distribution $p(x|y)$ and hence, to apply Bayesian techniques like maximum posterior marginal (MPM) and maximum a posteriori (MAP).

The feasibility of the different estimations of interest in HMFs stems from the possibility of sampling realizations of the hidden process X from $Y = y$ according to the posterior distribution $p(x|y)$, and which is possible when this latter distribution is of Markov form. On the other hand, the Markovianity of this latter distribution relies itself on the assumption that the random variables $(Y_s)_{s \in S}$ are independent conditionally on X.

3   Proposed Approach

The problem considered in this paper is to derive a thematic map from a multichannel (typically remote sensing) image. As described in the previous section, the hidden Markov field (HMF) model allows to find a hidden field X representing in this case the thematic map, from an observed field Y representing the observed multichannel image. The novelty in this paper is to adopt the Pignistic probabilities provided by PRDS classifier [8] (after combination of different channel reports) instead of $f_{x_s}(y_s)$ in Eq. (5). Thus, one need to achieve a training process on the prelabeled set of pixels (typically a prelabeled image of subimage) to derive noise densities associated to different channels which will later produce parameters of spectral information. Such prelabeled data are not available however, given that we deal with unsupervised classification. Then, the estimation of both spectral and spatial parameters is achieved in an unsupervised iterative way. More explicitly, one starts by coarsely perform an initial clustering which will service as a basis for initial parameters estimation. Indeed, when an initial realization of X is available, one can perform training according to PRDS to derive noise densities and, at the same time, estimation of spatial parameters as in conventional HMF context.

In what follows, we describe PRDS training, parameter estimation and labeling procedures.

3.1   Training

As specified before, let us consider a set of prelabeled multichannel pixels. We recall that such data may be available through an initial coarse clustering providing a realization of X and then iteratively through successive updates of X during parameter estimation, as we are going to see later. Hence, training will be concerned exclusively with spectral information. More explicitly, we use PR-DS classifier to estimating for each class $\omega_k \in \Omega$ and for each channel p $(1 < p < P)$, the Parzen-Rosenblatt density $\hat{f}_k^p$ as described in the previous section. Let us now show how the estimated Parzen-Rosenblatt densities will produce spectral parameters. In other words, we demonstrate how one can replace the noise densities $f_{x_s}(y_s)$ in Eq. (5) for a given $Y_n \in \mathbb{R}^P$.

Step 1: Generation of Mass Functions. To define the mass associated to channel p, let us consider the rank function δ_p defined from $\{1, .., K\}$ to Ω such as $\delta_p(k)$ is the k-ranked element of Ω in terms of $\hat{f}^p$, i.e. $\hat{f}^p_{\delta_p(1)}(Y_n^p) \le \hat{f}^p_{\delta_p(2)}(Y_n^p) \le \dots \le \hat{f}^p_{\delta_p(K)}(Y_n^p)$. Then, M^p is derived as follows:

$$\begin{cases} M^p(\Omega) \propto K\hat{f}^p_{\delta_p(1)}(Y_n^p) \\ \\ M^p(\{\omega_{\delta_p(k)}, ..., \omega_{\delta_p(K)}\}) \propto (K - k + 1)[\hat{f}^p_{\delta_p(k)}(Y_n^p) - \hat{f}^p_{\delta_p(k-1)}(Y_n^p)], \text{ for } k > 1 \end{cases} \tag{8}$$

Step 2: Combination of Mass Functions. Mass functions associated to different attributes are then combined into one collaborative mass $M = \bigoplus_{p=1}^{P} M^p$:

$$M(B) \propto \sum_{\cap_{p=1}^{P} B_p = B} \left[\prod_{p=1}^{P} M^p(B_p) \right], \text{ for } B, B_p \in \mathcal{P}(\Omega) \tag{9}$$

Step 3: Deriving Noise Density. Based on M, the noise density is then computed according to the Pignistic transform:

$$f_k(y_n) = \sum_{A \ni \omega_k} \frac{M(A)}{|A|} \tag{10}$$

3.2   Parameter Estimation

In this framework, we adopted ICE algorithm for parameter estimation. At each iteration i, a realization of X is simulated using Gibbs sampler. Then, we use the Derin and Elliott method for estimating spatial parameters ϕ_i; and the PRDS method for spectral parameters η_i (which are noise densities as described in Step 3 above). The algorithm stops when an end criterion is reached. The parameter estimation procedure is illustrated through Fig. 1.

3.3   Labeling

Once parameter set θ is estimated by the ICE method [1] while the MPM estimator is used to infer X. Using the Gibbs sampler, T realization $x^1, x^2, \dots, x^T$ of X are simulated according to $p(X|Y = y)$. Then, one estimates $\hat{p}(x_s = \omega \mid y)$ of each X_s from the realizations $x^1, x^2 \dots, x^N$. Finally, for each pixel x_s, one chooses the class whose number of appearances in the simulations is the highest.

4   Evaluation of the Proposed Approach

To validate our approach, we assess its performance in unsupervised segmentation of multichannel images against the conventional HMF model. To this end, we consider two series of experiments. The first series deal with synthetic images whereas the second series deals with a real multichannel remote sensing image.

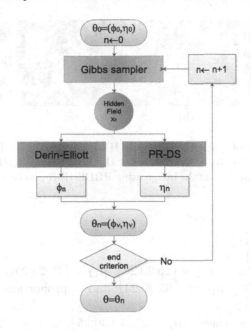

Fig. 1. Parameter estimation process.

4.1   Unsupervised Segmentation of Sampled Multichannel Images

To sample a synthetic multichannel image, we use Gibbs sampler with the following parameters: $\beta = [1, 1, 1]$, $\alpha_h = \alpha_{vb} = 2I$ and $\alpha_d = 4I$. Then, the obtained image is noised considering multidimensional Gaussian mixture densities in two different ways, considering two different sets of noise parameters η:

- Image 1: noisy version of the reference image, with a two-dimensional mixture noise of four Gaussians and a variance-covariance matrix $\Sigma = \begin{bmatrix} 28 & 3 \\ 3 & 28 \end{bmatrix}$.

  * For the first class $\mu_{11} = [10\ 20]$, $\mu_{12} = [25\ 20]$, $\mu_{13} = [40\ 20]$, $\mu_{14} = [50\ 20]$, $\Sigma_{11} = \Sigma_{12} = \Sigma_{13} = \Sigma_{14} = \Sigma$ and a proportion of mixture $p_1 = [0.25\ 0.25\ 0.25\ 0.25]$;

  * For the second class $\mu_{21} = [10\ 30]$, $\mu_{22} = [25\ 30]$, $\mu_{23} = [40\ 30]$, $\mu_{24} = [50\ 30]$, $\Sigma_{21} = \Sigma_{22} = \Sigma_{23} = \Sigma_{24} = \Sigma$ and a proportion of mixture $p_2 = [0.25\ 0.25\ 0.25\ 0.25]$;

  * For the third class $\mu_{31} = [10\ 40]$, $\mu_{32} = [25\ 40]$, $\mu_{33} = [40\ 40]$, $\mu_{34} = [50\ 40]$, $\Sigma_{31} = \Sigma_{32} = \Sigma_{33} = \Sigma_{34} = \Sigma$ and a proportion of mixture $p_3 = [0.25\ 0.25\ 0.25\ 0.25]$;

- Image 2: noisy image with a three-dimensional mixture noise of five Gaussians and a variance-covariance matrix for each element of the mixture for each class c: $\Sigma_{c1} = \Sigma_{c2} = \Sigma_{c3} = \Sigma_{c4} = \Sigma_{c5} = \Sigma$ such as $c \in \{1, 2, 3\}$ and

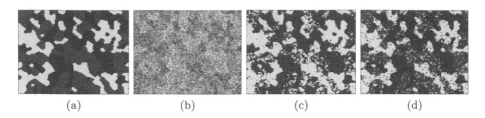

(a)   (b)   (c)   (d)

Fig. 2. Classification of synthetic data using HMF and PRDS-HMF. (a) Original class image. (b) Noisy image. (c) Classification of the image using HMF, success rate $\tau = 58.3740\%$. (d) Classification of the image using PRDS-HMF, success rate $\tau = 84.2285\%$.

$$\Sigma = \begin{bmatrix} 28 & 3 & 1 \\ 3 & 28 & 3 \\ 1 & 3 & 28 \end{bmatrix}.$$

* For the first class $\mu_{11} = \begin{bmatrix} 10 & 20 & 20 \end{bmatrix}$, $\mu_{12} = \begin{bmatrix} 25 & 20 & 20 \end{bmatrix}$, $\mu_{13} = \begin{bmatrix} 40 & 20 & 20 \end{bmatrix}$, $\mu_{14} = \begin{bmatrix} 50 & 20 & 20 \end{bmatrix}$, $\mu_{15} = \begin{bmatrix} 65 & 20 & 20 \end{bmatrix}$ and a proportion of mixture $p_1 = \begin{bmatrix} 0.2 & 0.2 & 0.2 & 0.2 & 0.2 \end{bmatrix}$;

* For the second class $\mu_{21} = \begin{bmatrix} 10 & 30 & 25 \end{bmatrix}$, $\mu_{22} = \begin{bmatrix} 25 & 30 & 25 \end{bmatrix}$, $\mu_{23} = \begin{bmatrix} 40 & 30 & 25 \end{bmatrix}$, $\mu_{24} = \begin{bmatrix} 50 & 30 & 25 \end{bmatrix}$, $\mu_{25} = \begin{bmatrix} 65 & 30 & 25 \end{bmatrix}$ and a proportion of mixture $p_2 = \begin{bmatrix} 0.2 & 0.2 & 0.2 & 0.2 & 0.2 \end{bmatrix}$;

* For the third class $\mu_{31} = \begin{bmatrix} 10 & 40 & 30 \end{bmatrix}$, $\mu_{32} = \begin{bmatrix} 25 & 40 & 30 \end{bmatrix}$, $\mu_{33} = \begin{bmatrix} 40 & 40 & 30 \end{bmatrix}$, $\mu_{34} = \begin{bmatrix} 50 & 40 & 30 \end{bmatrix}$, $\mu_{35} = \begin{bmatrix} 65 & 40 & 30 \end{bmatrix}$ and a proportion of mixture $p_3 = \begin{bmatrix} 0.2 & 0.2 & 0.2 & 0.2 & 0.2 \end{bmatrix}$;

Then, unsupervised segmentation is performed using conventional HMFs; and the proposed PRDS- HMF. The results obtained are illustrated in Fig. 2 and Fig. 3 where the noisy multichannel images are depicted in monochannel gray level by averaging the channels' intensities for illustrative purpose. The segmentation accuracy rates obtained confirm the interest of the proposed model with respect to the classic HMF. The supremacy of the proposed model is mainly due to the possibility of considering more general forms of noise by the PRDS-

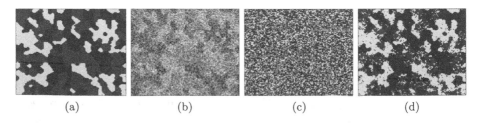

(a)   (b)   (c)   (d)

Fig. 3. Classification of synthetic data using HMF and PRDS-HM. (a) Original class image. (b) Noisy image. (c) Classification of the image using HMF, success rate $\tau = 32.7332\%$. (d) Classification of the image using PRDS-HMF, success rate $\tau = 80.8350\%$.

HMF thanks to the kernel smoothing technique that makes it possible to fit any form of noise instead of assuming it Gaussian as in HMF context.

4.2 Unsupervised Segmentation of Multichannel Remote Sensing Image

In this series of experiments, we consider a multichannel image of the Landsat-7 satellite ETM+ sensor acquired on June 03, 2001. The acquisition was made in an area around the city of Algiers. The image used has a resolution of 30 m × 30 m and a size of 256 × 256. The area of study includes 4 classes: (i) Urban Dense (UD); (ii) Urban Less Dense (ULD); (iii) Barren Land (BL); and (iv) Vegetation (V).

Figure 4 represents the 6 bands at the gray scale image.

Fig. 4. Different channel observations of the studied image.

To quantitatively assess the performance of the proposed approach against the conventional HMF model, we have a partial ground truth (see Fig. 4).

Qualitative assessment of the results obtained shows that the thematic map provided by the proposed approach contains less salt and pepper effect. This is confirmed by the quantitative assessment in terms of overall accuracy and kappa metrics. Indeed, the PRDS- HMF yields an accuracy rate of 79% (resp. a kappa of 0.70) against an accuracy of 68% (resp. a kappa of 0.5) by the conventional HMF model (Figs. 5 and 6).

Fig. 5. Partial ground truth of the studied area

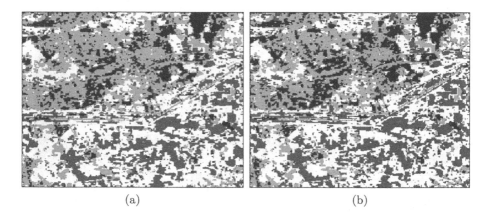

(a)   (b)

Fig. 6. Classification of a multichannel image using HMF and PRDS-HMF: Urban Dense (red), Urban Less Dense (orange), Barren Land (Yellow) and Vegetation (Green). (a) Thematic map obtained by PRDS-HMF: accuracy= 79.42%, Kappa= 0.70 (b) Thematic map obtained by HMF: accuracy= 68.23%, Kappa= 0.55. (Color figure online)

Table 1. Confusion matrix obtained by PRDS-HMF.

	(UD)	(ULD)	(BL)	(V)	Truth	Recall
(UD)	41	0	0	1	42	97.62%
(ULD)	19	84	7	1	111	75.68%
(BL)	0	2	192	31	225	85.33%
(V)	4	22	27	123	176	69.89%
Classification	64	108	226	156	554	
Precision	64.06%	77.78%	84.96%	78.85%		

Table 2. Confusion matrix obtained by HMF.

	(UD)	(ULD)	(BL)	(V)	Truth	Recall
(UD)	41	0	0	1	42	97.62%
(ULD)	43	60	7	1	111	54.05%
(BL)	1	1	164	59	225	72.89%
(V)	7	16	40	113	176	64.20%
Classification	92	77	211	174	554	
Precision	44.57%	77.92%	77.73%	64.94%		

Confusing matrices obtained by PRDS-HMF and HMF classifications are also given in Tables 1 and 2. We can confirm that the proposed HMF outperforms the plain one in terms of both precision and recall per each class. It is worth mentioning that a better modeling of noise allows also to a better estimation of spatial parameters. In fact, parameter estimation is an iterative process in which a good perception of noise leads to a better parameter estimation of spatial features.

5   Conclusion

In this paper, we proposed a new hidden Markov field model designed for unsupervised segmentation of multichannel images. The main novelty of the proposed model relies in the use of Dempster-Shafer theory and Parzen-Rosenblatt window for noise density estimation which makes it possible to model general forms of multidimensional noise. To assess the performance of the proposed PRDS-HMF, experiments were conducted on both synthetic and real multichannel images. The results obtained confirmed its interest with respect to the conventional HMF model. A possible future direction of this approach would be to consider more general Markov models with the same extension.

References

1. Benboudjema, D., Pieczynski, W.: Unsupervised image segmentation using triplet Markov fields. Comput. Vis. Image Underst. **99**(3), 476–498 (2005)
2. Besag, J.: Spatial interaction and the statistical analysis of lattice systems. J. Roy. Stat. Soc. Ser. B **6**, 192–236 (1974)
3. Borges, J.S., Bioucas-Dias, J.M., Marcal, A.R.: Bayesian hyperspectral image segmentation with discriminative class learning. IEEE Trans. Geosci. Remote Sens. **49**(6), 2151–2164 (2011)
4. Boudaren, M.E.Y., An, L., Pieczynski, W.: Dempster-Shafer fusion of evidential pairwise Markov fields. Int. J. Approximate Reason. **74**, 13–29 (2016)
5. Denœux, T.: 40 years of Dempster-Shafer theory. Int. J. Approximate Reason. **79**, 1–6 (2016)

6. Ghamisi, P., Benediktsson, J.A., Ulfarsson, M.O.: Spectral-spatial classification of hyperspectral images based on hidden Markov random fields. IEEE Trans. Geosci. Remote Sens. **52**(5), 2565–2574 (2014)
7. Guo, H., Shi, W., Deng, Y.: Evaluating sensor reliability in classification problems based on evidence theory. IEEE Trans. Syst. Man Cybern. Part B (Cybern.) **36**(5), 970–981 (2006)
8. Hamache, A., et al.: Uncertainty-aware Parzen-Rosenblatt classifier for multi-attribute data. In: Destercke, S., Denoeux, T., Cuzzolin, F., Martin, A. (eds.) BELIEF 2018. LNCS (LNAI), vol. 11069, pp. 103–111. Springer, Cham (2018). https://doi.org/10.1007/978-3-319-99383-6_14
9. Jackson, Q., Landgrebe, D.A.: Adaptive Bayesian contextual classification based on Markov random fields. IEEE Trans. Geosci. Remote Sens. **40**(11), 2454–2463 (2002)
10. Jia, X., Richards, J.A.: Segmented principal components transformation for efficient hyperspectral remote-sensing image display and classification. IEEE Trans. Geosci. Remote Sens. **37**(1), 538–542 (1999)
11. Jones, R.W., Lowe, A., Harrison, M.J.: A framework for intelligent medical diagnosis using the theory of evidence. Knowl.-Based Syst. **15**(1), 77–84 (2002)
12. Li, J., Bioucas-Dias, J.M., Plaza, A.: Hyperspectral image segmentation using a new Bayesian approach with active learning. IEEE Trans. Geosci. Remote Sens. **49**(10), 3947–3960 (2011)
13. Li, J., Bioucas-Dias, J.M., Plaza, A.: Spectral-spatial hyperspectral image segmentation using subspace multinomial logistic regression and Markov random fields. IEEE Trans. Geosci. Remote Sens. **50**(3), 809–823 (2012)
14. Mercier, G., Derrode, S., Lennon, M.: Hyperspectral image segmentation with Markov chain model. In: 2003 IEEE International Geoscience and Remote Sensing Symposium. Proceedings (IEEE Cat. No. 03CH37477), IGARSS 2003, vol. 6, pp. 3766–3768. IEEE (2003)
15. Plaza, A., et al.: Recent advances in techniques for hyperspectral image processing. Remote Sens. Environ. **113**, S110–S122 (2009)
16. Shafer, G.: A Mathematical Theory of Evidence, vol. 1. Princeton University Press, Princeton (1976)
17. Shafer, G.: A mathematical theory of evidence turns 40. Int. J. Approximate Reason. **79**, 7–25 (2016)

Automatic Detection of Symmetry in Dermoscopic Images Based on Shape and Texture

Vincent Toureau[1], Pedro Bibiloni[1,2], Lidia Talavera-Martínez[1,2],
and Manuel González-Hidalgo[1,2]

[1] Department of Mathematics and Computer Science, Soft Computing,
Image Processing and Aggregation (SCOPIA) Research Group,
University of the Balearic Islands, 07122 Palma de Mallorca, Spain
{p.bibiloni,l.talavera,manuel.gonzalez}@uib.es
[2] Balearic Islands Health Research Institute (IdISBa), 07010 Palma, Spain

Abstract. In this paper we present computational methods to detect the symmetry in dermoscopic images of skin lesions. Skin lesions are assessed by dermatologists based on a number of factors. In the literature, the asymmetry of lesions appears recurrently since it may indicate irregular growth. We aim at developing an automatic algorithm that can detect symmetry in skin lesions, as well as indicating the axes of symmetry. We tackle this task based on skin lesions' shape, based on their color and texture, and based on their combination. To do so, we consider symmetry axes through the center of mass, random forests classifiers to aggregate across different orientations, and a purposely-built dataset to compare textures that are specific of dermoscopic imagery. We obtain 84–88% accuracy in comparison with samples manually labeled as having either 1-axis symmetry, 2-axes symmetry or as being asymmetric. Besides its diagnostic value, the symmetry of a lesion also explains the reasons that might support such diagnosis. Our algorithm does so by indicating how many axes of symmetry were found, and by explicitly computing them.

Keywords: Dermoscopic images · Skin lesion · Computational methods · Symmetry detection · Shape · Texture · Color · Machine learning · Random forest

1 Introduction

Skin cancer is a disease caused by the abnormal and uncontrolled proliferation of melanocytes—cells that pigment the skin—that have undergone a genetic

This work was partially supported by the project TIN 2016-75404-P AEI/FEDER, UE. L. Talavera-Martínez also benefited from the fellowship BES-2017-081264 conceded by the *Ministry of Economy, Industry and Competitiveness* under a program co-financed by the *European Social Fund*.

© Springer Nature Switzerland AG 2020
M.-J. Lesot et al. (Eds.): IPMU 2020, CCIS 1237, pp. 625–636, 2020.
https://doi.org/10.1007/978-3-030-50146-4_46

mutation. This disease is one of the most widespread around the world as it represents 40% of all cancers [6]. There are several types of malignant skin cancers (basal cell carcinoma, squamous cell carcinoma, etc.) but the most aggressive and deadliest one is known as melanoma. In Europe, cutaneous melanoma represents 1–2% of all malignant tumors [3] but its estimated mortality in 2018 was 3.8 per 100.000 men and women per year [2].

This type of disorder is characterized by the development of a skin lesion which usually presents an irregular shape, asymmetry and a variety of colors, along with a history of changes in size, shape, color and/or texture. Based on this, experts designed protocols, the so-called diagnostic methods, to quantify the malignancy of the lesions. Some examples are pattern analysis, the ABCD rule, the 7-point checklist, and the Menzies method. In these, the asymmetry of the lesion plays an essential role towards the assessment of the lesion. However, each of them defines symmetry in a slightly different way. While according to the Menzies method, benign lesions are associated to symmetric patterns in all axes through the center of the lesion, disregarding shape symmetry [4]. Another example: regarding the ABCD rule, there might be symmetry in 0, 1 or 2 perpendicular axes when evaluating not only the contour, but also its colors and structures. Moreover, the assessment of symmetry might be altered by the individual judgment of the observers, which depends on their experience and subjectivity [4].

The increasing incidence of melanoma over the past decades along with the desire to overcome the variability in interpretation have promoted the development of computer-aided diagnosis systems. They provide reproducible diagnosis of the skin lesions as an aid to dermatologists and general practitioners in the early detection of melanoma.

There are several general-purpose techniques to calculate symmetry in the computer vision field, as presented in [13]. A few techniques have been applied to the detection of asymmetry in skin lesions in dermoscopic images. Seidenari et al. [14] quantify the asymmetry as the appearance of an irregular color distribution in patches within the lesion. Also, Clawson et al. [5], following the same line, further integrates Fourier descriptors into a shape asymmetry quantifier. Other authors, such as Kjoelen et al. [9] and Hoffman et al. [7], estimate the asymmetry by computing the nonoverlapping areas of the lesion after *folding* the image along the best axis of symmetry, taking into account grayscale texture and color.

However, as far as we know, the study of the presence of asymmetry in skin lesions has been used to classify lesions as malignant or benign in diagnostic aid systems. In most articles, the approaches that calculate the symmetry of the lesions do so in an integrated way in an automated system that extracts other features, simultaneously. This fact hinders both the symmetry evaluation and the interpretability of the results, such as reporting what is the impact of finding asymmetry towards the classification of a specific lesion as malignant.

Hence, our objective is two fold. First, to study the symmetry of lesions from three different points of view. The first focuses on the shape of symmetry of the lesion, while the second is based on the symmetry of the textures (including colors). Finally, these two approaches are combined into the third one. Second, to compare and analyze their impact, as well as to quantitatively assess their performance.

Dataset of Dermoscopic Images

In order to complete the clinical analysis and the diagnosis of skin lesions at its earliest stage, physicians commonly employ a technique called dermoscopy. It is an in-vivo, non-invasive imaging technique based on a specific optical system with light that amplifies the lesion, which has previously been covered with mineral oil, alcohol or water to avoid the reflection of light on it and increase the transparency of the outermost layer of the epidermis. Dermoscopy has been shown to improve the diagnostic accuracy up to 10–30% [10] compared to simple clinical observation. In some cases, dermoscopy can capture digital images of skin lesions, providing more detailed information of morphological structures and patterns compared to normal images of skin lesions.

From the available databases of dermoscopic images we decided to use the PH^2 database [11], which contains 200 dermoscopic images annotated by expert dermatologists. For each image, relevant information about the manual segmentation and the clinical diagnosis of the skin lesion, as well as some dermoscopic features, such as asymmetry, colors, and dermoscopic structures, are available. It is worth mentioning that the symmetry of the lesion is evaluated by clinicians according to the ABCD rule, and therefore, concerning its distribution of contour, color and structures simultaneously. There are three possible labels for this parameter: 0 for fully symmetric lesions, 1 for asymmetric lesions with respect to one axis, 2 for asymmetric lesions with respect to two axes.

In Sects. 2, 3 and 4 we detailed the three different approaches used to study the symmetry of the lesions, based respectively on shape descriptors, texture descriptors and a combination of both. In Sect. 5 we discuss on the results obtained, and conclude with the main strengths and limitations of our approach.

2   Shape-Based Method to Assess Symmetry of Skin Lesions

In this section we present the first of three computational approaches towards assessing the symmetry of skin lesions. In particular, we focus exclusively on the shape of the lesion.

We parameterize the candidates to be axes of symmetry as the lines through the center of mass. Such lines are the only ones that split any continuous two-dimensional figure in two parts of the same area. The assumption that symmetry axes contain the center of mass is convenient: they become characterized by their angle with respect to the horizontal axis, α. Also, the center of mass is easy to compute.

To assess whether an axis divides *symmetrically* the lesion we employ the Jaccard index [8]. We consider that a line is a perfect axis of symmetry if the second half is equal to the reflection of the first one with respect to the axis. Since we are dealing with the shape of the lesion, being *equal* refers to whether a pair of pixels are both tagged as lesion or both tagged as skin. Let ℓ be a line of the plane, M_+, M_- the two halves in which ℓ splits the lesion region. Let R_ℓ denote the reflection with respect to ℓ and let $|A|$ denote the area of a region. Then, we define the shape-based symmetry index of a line ℓ, $S_1(\ell)$ to be:

$$S_1(\ell) = \frac{|M_+ \cap R_\ell(M_-)|}{|M_+ \cup R_\ell(M_-)|}.$$

The final assessment of the symmetry within the skin lesion based on shape is based on a random forest classifier. We consider a pencil of N lines through the center of mass, $\ell_{180° \cdot k/N}$, for $k = 0, \ldots, N - 1$. Then, we obtain their shape-based symmetry index, $S(\ell_{180° \cdot k/N})$. A random forest classifier aggregates all the indices into a final answer, being either "no symmetry", "1-axis symmetry" or "2-axes symmetry". We remark that substituting the learning classifier with experimentally-set fixed thresholds achieves worse quantitative results, but provides the insight of which lines represent the main and perpendicular axes of symmetry, if there are any.

Qualitative results of this method are found in Fig. 1. In it, we show an accurately classified sample (left) and a wrongly classified one (right). The latter presents a symmetric shape, but it also shows some inner structures that are not symmetric with respect to one of the perpendicular axes.

In our implementation, we used $N = 20$, and 10 trees in the classifier. This is a fast algorithm, whose execution time is typically in the range 1–2 s.

3   Texture-Based Method to Assess Symmetry of Skin Lesions

To assess how two halves fold symmetrically with respect to their texture, we need to assess how similar two textures appear to be. Corresponding pixels are not required to be equal, but to have been drawn from a similar statistical distribution. Moreover, the distributions we found are specific: textures in dermoscopic images are not necessarily similar to textures found in other computer vision tasks. We consider a patch-based approach: we assess similarity of textures in two locations based on a local neighbourhood of them. This approach led to the creation of a dataset containing pairs of similar and different textures, introduced in the following.

Dataset to Discriminate Texture in Dermoscopic Skin Lesions

The texture of skin lesions play an important role in its symmetry. As previously mentioned, its symmetry is jointly based on the shape of the lesion, and the appearance of similar structures and patterns. To discriminate such patterns, we must be able to compare the local texture in different locations of the

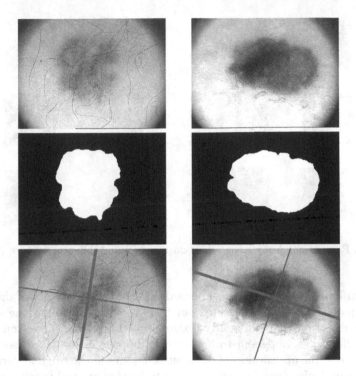

Fig. 1. Dermoscopic image (top), lesion mask (middle) and symmetry axes based on shape (bottom), of two samples.

image. However, textures found in dermoscopic images are specific, not having the same statistical distributions that textures found in textile, piles of similar objects or other settings. To discriminate such textures, we propose the extraction of a dataset from dermoscopic images, providing pairs of similar and different patches. Each patch, a $n \times n$-pixel region cropped from the original image, contains information about the local texture in a specific point.

We extract pairs of patches from a dermoscopic image in a fully automatic way. We require not only the dermoscopic image, but also a segmentation of the lesion. In order to obtain *pairs of patches with similar textures*, (p_A, p_B), we randomly select two locations that are very close, under the restriction that both patches are completely inside the lesion or completely outside. We remark that both patches are largely overlapping. To obtain *pairs of patches with different textures*, (p_A, p_B), we randomly select one patch completely inside the lesion, and the other one outside the lesion. In this case, there will be no overlap between them, and we will assume that they represent regions with a different underlying texture. All the samples of our dataset will be those patches, $x = (p_A, p_B)$, and the reference data will be whether they are of the first or the second, $y \in \{\text{'Similar', 'Different'}\}$ type. In Fig. 2 we show similar and different patches extracted using this strategy.

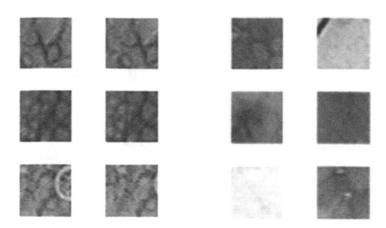

Fig. 2. Pairs of patches with similar texture (left) and different texture (right).

Several limitations must be acknowledged. First, the samples of differently-textured patches are biased towards our task: they do not follow the same statistical distribution that differently-textured patches *inside the lesion* do. This negative effect does not seem to have a huge impact in its usage (see Sect. 3), possibly due to the appearance of different skin tones in the original dermoscopic images. Second, close patches are assumed to be similar, and patches inside the lesion are assumed to be different to patches outside of it. While this is not necessarily the case, it has proven to hold the vast majority of times. Third, a learning classifier that compares them could cheat on solving the task, using the more basic approach of detecting as positive only those patches that present a large overlap.

To overcome the second and third limitation, we limit ourselves to manually select texture-relevant features. We use the Gray Level Co-occurrence Matrix (GLCM) with two-pixel distance and a horizontal orientation to extract five texture features. They are dissimilarity, correlation, energy, contrast, and homogeneity. Correlation is understood as a measure of the linear dependence of gray levels between pixels at the specified distance. The energy feature measures the brightness of the images as well as the repetition of subunits. Contrast refers to the local gray level variations, while the homogeneity is a measure of the smoothness of the gray level distribution. Finally, we extract the 25th, 50th and 75th percentiles of the marginal distribution of the RGB channels of the pixels.

The patches were randomly extracted from the PH^2 database of dermoscopic images, selecting 10 pairs with similar texture and 10 pairs with different texture for each of the 200 images in the database. It includes a manually segmented lesion, which we used to check whether patches were completely inside or outside the lesion. Given the original resolution of images, 764×576, we set the size of patches to be 32×32, as a good trade-off to obtain a region representative enough but whose texture is approximately uniform. Besides considering the features mentioned above, we decided to use shallow-learning algorithms only, specifically selecting random forest due to the amount and diversity of features.

Aggregation of Patch Similarity

Using the newly created patch dataset, we trained a random forest classifier T that, given two patches p_A, p_B, extract the features introduced in Sect. 3 and then estimates whether they represented the same texture or not. Such classifier is represented as:

$$T(p_A, p_B) \in \{0, 1\}.$$

Let us continue assuming that an axis of symmetry contains the center of mass. We define a line to be an axis of symmetry with respect to its texture if symmetric patches present the same texture. We thus define the texture-based symmetry index of a line ℓ, $S_2(\ell)$, as:

$$S_2(\ell) = \frac{1}{N} \sum_{i=1}^{N} T(p_+^{i)}, p_-^{i)}),$$

where N is the amount of patches that can be extracted from the intersection of the upper region and its reflected lower half, $M_+ \cup R_\ell(M_-)$; $p_+^{i)}$ is the i-th patch from the upper region M_+, and $p_-^{i)}$ is its corresponding patch from the lower region M_- with respect to the symmetry axis ℓ.

Similarly to the shape-based symmetry detector, we consider N equidistributed lines and aggregate their texture-based symmetry index with a different random forest classifier. Also, substituting such classifier with experimentally-set fixed thresholds provides the insight of which lines are the main and perpendicular axes of symmetry, if there are any.

Results of the above-mentioned procedure are found in Fig. 3. In it, we show a sample with symmetric textures (right) and a sample with differently textured matched regions (left). The result of the patch-based classifier is encoded as a semi-transparent circle, ranging from green (similar) to red (non-similar) patches. Also, we emphasize that we select partially overlapped patches, and we restrict the patch selection procedure to those regions within the lesion.

At an implementation level, several details must be mentioned. The patch-based classifier, $T(p_A, p_B)$, is a random forest classifier with 200 trees and two outputs—either similar or different. Its inputs are a list of $2 \cdot n$ features: n features extracted from the first patch p_A, and n features from the second one p_B. We used $n = 5 + 3 \cdot 3$, extracting 5 features from the grey-level co-occurrence matrix, and 3 quantiles of each of the channels R, G and B (see Sect. 3). Also, we used $N = 10$ lines, and 32×32-pixel patches. These parameters were experimentally selected. The second classifier, used to aggregate information across different orientations, is also a random forest with 100 trees and, as in the case of the shape-based classifier, three outputs. This algorithm requires, with non-optimized code, around 40–50 s.

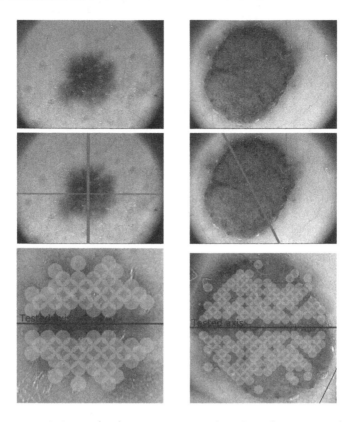

Fig. 3. Dermoscopic image (top), symmetry axes based on their texture (middle) and patch-based comparison of a specific axis (cropped and rotated, bottom), of two samples.

4   Combined Method to Assess Symmetry of Skin Lesions

Finally, to answer the hypothesis of whether both shape and texture are actively contributing towards the symmetry of the lesion—as identified by a human expert—, we combine both symmetry indicators. Following the same reasoning, given a pencil of lines $(\ell_\alpha)_\alpha$ we compute for each line ℓ_α its shape-based symmetry index, $S_1(\ell_\alpha)$ and its texture-based symmetry index, $S_2(\ell_\alpha)$. We aggregate these two lists with a 10-tree random forest classifier to output a final decision as either "no symmetry", "1-axis symmetry" or "2-axes-symmetry".

5   Analysis of Results and Conclusions

We presented a constructive approach towards symmetry detection, somehow similar to an ablation study. First, the symmetry detection has been addressed using only the shape of the lesion. However, as shown in Fig. 1 (right), taking into account only the shape can not provide satisfactory results. Experts take into account the shapes, but also the textures and colors of the inner parts of the lesion to define symmetry. Considering this, the information loss is too high, which implies that more information must be used to be more accurate. This is why the symmetry of textures and colors has to be included and studied.

In the second approach, only texture and color symmetry has been considered to determine the presence of symmetry in skin lesions. We remark that texture and color are hardly separable, since the former is defined in terms of changes of the latter. In this case two random forest classifiers have been used: to assess the symmetry of two 32×32 patches, and to aggregate information across different orientations. Qualitatively, the similarity map tends to be reliable and the axes of symmetry are never aberrant regarding textures.

In the following, we present the results obtained with the aforementioned methods and conclude with some final remarks.

5.1   Results

In this section, we present the experimental settings and results obtained. This is done quantitatively to add up to the qualitative results contained in Figs. 1 and 3. We do so by comparing with the manually labelled data of the PH^2 dataset. It contains, for each dermoscopic sample, a tag indicating either "no symmetry", "1 axis of symmetry" or "2 axes of symmetry". Therefore, we have a 3-class classification problem, where the three class are: no symmetry, 1-axis symmetry and 2-axis symmetry. In the PH2 database we have the following distribution labeled by experts: 26% for the first class, 15.5% for the second class, and 58.5% for the last class.

To assess the success of our method we consider its accuracy. That is, the ratio of correctly classified samples. We emphasize that we are considering a 3-class classification problem, so binary metrics (such as recall or F-measure) can not be computed. The results obtained by the different algorithms are presented in Table 1.

Table 1. Accuracy of the three-class classifiers.

Classifier	Accuracy
Based on shape	86%
Based on texture	84%
Based on both	88%

We train and validate the algorithms with disjoint datasets. We split the 200 images in the training set (first 50) and the validation set (last 150). No randomization was applied to the samples since their symmetry is not ordered. We remark that the validation set is larger since we aim at estimating the generalization capacity of the model with low variance. Metrics in Table 1 have been computed over the validation set. Any other learning stage, including the classifier employed to compare patches, have been trained only with the training set.

Table 1 summarizes the results obtained with each of the methods described in Sects. 2, 3 and 4. As can be observed, these classifiers provide satisfactory quantitative results. We emphasize the subjective nature of this task: in contrast to, for instance, assessing the malignancy of the lesion, the symmetry of a lesion is measured against the perceptive criteria of a human expert.

The superiority of the shape-based approach over the texture-based one is not contradictory: the latter purposely neglects information regarding the shape. That is, it exclusively uses pairs of patches such that both of them are located within the lesion, disregarding the fact that there may or may not exist additional pairs of patches such that first one represents healthy skin and the second one the lesion. Such pairs have not been provided to the texture-based method in order to avoid implicitly using information derived from the shape of the lesion.

The images leading to classification errors are different in the shape-based and texture-based methods. This implies that the two sources of information do not provide equivalent results.

As one would expect, classification based on both shape and on texture has superior results. The final accuracy reaches up to 88% which defines a reliable model that may be used in real applications. Other models may be used, but considering the features size and the quality of the results given by the random forest classifiers with a minimalist tuning, they seem to be appropriate to solve the problem raised in this work.

5.2   Strengths and Limitations

We have addressed the computational problem of symmetry evaluation in terms of (i) shape, and (ii) texture (including its color) as is considered by dermatologists [1,12]. This provides the clinician with a comprehensible and interpretable tool, that indicates the presence of symmetry axis and its location. Asymmetry of skin lesions is an important indicator of the presence of irregular growth in skin lesion. Thus, it contributes substantially to its diagnosis. In the ABCD rule of dermoscopy, for example, asymmetry is the parameter that contributes with a larger coefficient to the ABCD-based diagnosis [12]. We emphasize that the aim of this work is to detect the presence of symmetry (or not) in a dermoscopic image of a skin lesion, rather than the classification of the lesion as either malignant or benign.

The algorithms designed deal with the symmetry of skin lesions, which is an important indicator of uncontrolled growth of cells. They treat the symmetry as it is evaluated by the experts considering at the same time its shape, texture and colors. Therefore, the output provided by the algorithm is interpretable by experts. The algorithms in this paper can be freely accessed online at the website http://opendemo.uib.es/dermoscopy/, as well as using it as a the standalone python package dermoscopic_symmetry.

The shape-based algorithm is faster than the texture-based one: 1–2 s and 40–50 s to process a medium-sized dermoscopic image. Although the code could be optimized, the complexity of the latter is much higher. The shape-based method can be used for real-time applications, whereas both could be used off-line or for knowledge distillation into a faster classifier.

Both shape and texture information seem to be necessary towards assessing the skin lesion symmetry. The rationale lies on the fact that irregular growth—the malignancy cue looked after—may cause both types of effects. However, given the quantitative metrics in Table 1, both texture and shape provide a large amount of information.

A limitation of this work lies on the biases in the patch dataset introduced in Sect. 3. First, we assume that close regions present similar textures, which does not always hold. Second, we have a very limited amount of *interesting* different textures. Due to the automatic selection of patches, we can only assume that two patches represent a different texture if one of them is within the lesion and the other outside of it. This means that, in each pair of different patches, one of them was extracted from the skin, whereas we are later comparing two patches that are within the bounds of the lesion.

Finally, this study is biased towards light-skin patients: it has been quantitatively contrasted against the labels of the PH^2 dataset.

References

1. Dermnet New Zealand trust. https://www.dermnetnz.org/cme/dermoscopy-course/dermoscopic-features/. Accessed 12 Nov 2019
2. European Cancer Information System. https://ecis.jrc.ec.europa.eu/index.php. Accessed 12 Nov 2019
3. Melanoma Molecular Map Project. http://www.mmmp.org/MMMP/welcome.mmmp. Accessed 12 Nov 2019
4. Argenziano, G., et al.: Dermoscopy of pigmented skin lesions: results of a consensus meeting via the Internet. J. Am. Acad. Dermatol. **48**(5), 679–693 (2003)
5. Clawson, K.M., Morrow, P.J., Scotney, B.W., McKenna, D.J., Dolan, O.M.: Determination of optimal axes for skin lesion asymmetry quantification. In: 2007 IEEE International Conference on Image Processing, vol. 2, p. II-453. IEEE (2007)
6. Filho, M., Ma, Z., Tavares, J.M.: A review of the quantification and classification of pigmented skin lesions: from dedicated to hand-held devices. J. Med. Syst. **39**(11), 177 (2015)
7. Hoffmann, K., et al.: Diagnostic and neural analysis of skin cancer (DANAOS). A multicentre study for collection and computer-aided analysis of data from pigmented skin lesions using digital dermoscopy. Br. J. Dermatol. **149**(4), 801–809 (2003)
8. Jaccard, P.: The distribution of the flora in the alpine zone. New Phytol. **11**(2), 37–50 (1912)
9. Kjoelen, A., Thompson, M.J., Umbaugh, S.E., Moss, R.H., Stoecker, W.V.: Performance of AI methods in detecting melanoma. IEEE Eng. Med. Biol. Mag. **14**(4), 411–416 (1995)
10. Mayer, J.: Systematic review of the diagnostic accuracy of dermatoscopy in detecting malignant melanoma. Med. J. Austral. **167**(4), 206–210 (1997)

11. Mendonça, T., Ferreira, P.M., Marques, J.S., Marcal, A.R., Rozeira, J.: Ph 2-a dermoscopic image database for research and benchmarking. In: 2013 35th Annual International Conference of the IEEE Engineering in Medicine and Biology Society (EMBC), pp. 5437–5440. IEEE (2013)
12. Ruela, M., Barata, C., Marques, J.S.: What is the role of color symmetry in the detection of Melanomas? In: Bebis, G., et al. (eds.) ISVC 2013. LNCS, vol. 8033, pp. 1–10. Springer, Heidelberg (2013). https://doi.org/10.1007/978-3-642-41914-0_1
13. Schmid-Saugeon, P.: Symmetry axis computation for almost-symmetrical and asymmetrical objects: application to pigmented skin lesions. Med. Image Anal. 4(3), 269–282 (2000)
14. Seidenari, S., Pellacani, G., Grana, C.: Asymmetry in dermoscopic melanocytic lesion images: a computer description based on colour distribution. Acta dermato-venereologica 86(2), 123–128 (2006)

Temporal Data Processing

Modeling the Costs of Trade Finance During the Financial Crisis of 2008–2009: An Application of Dynamic Hierarchical Linear Model

Shantanu Mullick[1]([✉]) [iD], Ashwin Malshe[2], and Nicolas Glady[3]

[1] Eindhoven University of Technology, 5612AZ Eindhoven, The Netherlands
s.mullick@tue.nl
[2] University of Texas at San Antonio, San Antonio, TX 78249, USA
[3] Telecom Paris, 91120 Palaiseau, France

Abstract. The authors propose a dynamic hierarchical linear model (DHLM) to study the variations in the costs of trade finance over time and across countries in dynamic environments such as the global financial crisis of 2008–2009. The DHLM can cope with challenges that a dynamic environment entails: nonstationarity, parameters changing over time and cross-sectional heterogeneity. The authors employ a DHLM to examine how the effects of four macroeconomic indicators – GDP growth, inflation, trade intensity and stock market capitalization - on trade finance costs varied over a period of five years from 2006 to 2010 across 8 countries. We find that the effect of these macroeconomic indicators varies over time, and most of this variation is present in the year preceding and succeeding the financial crisis. In addition, the trajectory of time-varying effects of GDP growth and inflation support the "flight to quality" hypothesis: cost of trade finance reduces in countries with high GDP growth and low inflation, during the crisis. The authors also note presence of country-specific heterogeneity in some of these effects. The authors propose extensions to the model and discuss its alternative uses in different contexts.

Keywords: Trade finance · Financial crisis · Bayesian methods · Time series analysis

1 Introduction

Trade finance consists of borrowing using trade credit as collateral and/or the purchase of insurance against the possibility of trade credit defaults [2, 4]. According to some estimates more than 90% of trade transactions involve some form of credit, insurance, or guarantee [7], making trade finance extremely critical for smooth trades. After the global financial crisis of 2008–2009, the limited availability of international trade finance has emerged as a potential cause for the sharp decline in global trade [4, 13, 21].[1] As a result, understanding how trade finance costs varied over the period in and

---

[1] See [27] for counter evidence.

© Springer Nature Switzerland AG 2020
M.-J. Lesot et al. (Eds.): IPMU 2020, CCIS 1237, pp. 639–652, 2020.
https://doi.org/10.1007/978-3-030-50146-4_47

around the financial crisis has become critical for policymakers to ensure adequate availability of trade finance during crisis periods in order to mitigate the severity of the crisis.[2] In addition, as the drivers of trade finance may vary across countries, it is important to account for heterogeneity while studying the effect of these drivers on trade finance [20].

A systematic study of the drivers of trade finance costs can be challenging: modeling the effects of these drivers in dynamic environments (e.g., a financial crisis) requires one to have a method that can account for non-stationarity, changes in parameters over time as well as account for cross-sectional heterogeneity [42]. First, nonstationarity is an important issue in time-series analysis of observational data [36, 42].[3] The usual approach to address nonstationarity requires filtering the data in the hope of making the time-series mean and covariance stationary.[4] However, methods for filtering time series, such as first differences can lead to distortion in the spectrum, thereby impacting inferences about the dynamics of the system [22]. Further, filtering the data to make the time-series stationary can (i) hinder model interpretability, and (ii) emphasize noise at the expense of signal [43].

Second, the effect of the drivers of trade finance costs changes over time [10]. These shifts happen due to time-varying technological advances, regulatory changes, and evolution of the banking sector competitive environment, among others. As we are studying 2008–2009 global financial crisis, many drivers of the costs may have different effects during the crisis compared to the pre-crisis period. For example, during the crisis many lenders may prefer borrowers with the top most quality, thus exhibiting a "flight to quality" [12]. To capture changes in model parameters over time, studies typically use either (1) moving windows to provide parameter paths, or (2) perform a before-and-after analysis. However, both these methods suffer from certain deficiencies. Models that yield parameter paths [11, 32] by using moving windows to compute changes in parameters over time leads to inefficient estimates since, each time, only a subset of the data is analyzed. These methods also presents a dilemma in terms of selection of the length of the window as short windows yield unreliable estimates while long windows imply coarse estimates and may also induce artificial autocorrelation.

Using before-and-after analysis [9, 25, 38] to study parameter changes over time implies estimating different models before and after the event. The 'after' model is estimated using data from after the event under the assumption that this data represents the new and stabilized situation. A disadvantage of this approach is the loss in

---

[2] Such as the World Trade Organization (WTO), the World Bank (WB), and the International Monetary Fund (IMF).

[3] The studies that used surveys for understanding the impact of financial crisis on trade finance costs [30] are also susceptible to biases present in survey methods. First, survey responses have subjective components. If this subjectivity is common across the survey respondents, a strong bias will be present in their responses. For example, managers from the same country tend to exhibit common bias in their responses [8]. Second, survey responses are difficult to verify. Managers may over- or under-estimate their trade finance costs systematically, depending on the countries where their firms operate. Finally, survey research is often done in one cross-section of time, making it impossible to capture the variation over time.

[4] Methods like vector autoregression (VAR) often filter data to make it stationary [15, 17, 34].

statistical efficiency as a result of ignoring effects present in part of the data. Further, this approach assumes that the underlying adjustment (due to events, such as the financial crisis) occurs instantaneously. However, in practice, it may take time for financial markets to adjust before it reaches a new equilibrium. This also serves to highlight the drawback of the approach in assuming time-invariant parameters for the 'before' model, as well as for the 'after' model.

Third, the effects of the drivers of trade finance cost may vary across countries [28], and we need to account for this heterogeneity. A well accepted way to incorporate heterogeneity is by using hierarchical models that estimate country-specific effects of the drivers of trade finance cost [40]. However, as hierarchical models are difficult to embed in time-series analysis [24], studies tend to aggregate data across cross-sections which leads to aggregation biases in the parameter estimates [14].

Nonstationarity, time-varying parameters and cross-sectional heterogeneity render measurement and modeling of factors that impact the dependent variable of interest—in our case, cost of trade finance—challenging in dynamic environments (such as a financial crisis). Therefore, we propose a dynamic hierarchical linear model (DHLM) that addresses all these three concerns and permits us to explain the variations in trade finance costs *over* several years, while also allowing us to detect any variation *across* countries, if present.

Our DHLM consists of three levels of equations. At the higher level, *Observation Equation* specifies, for each country in each year, the relationship between trade finance costs and a set of macroeconomic variables (e.g., inflation in the country). The coefficients of the predictors in the *Observation Equation* are allowed to vary across cross-section (i.e., countries) and over time. Next, in the *Pooling Equation* we specify the relationship between the country-specific time-varying coefficients (i.e., parameters) from the *Observation Equation* to a new set of parameters that vary over time, but are common across countries. Thus, the *Pooling Equation* enables us to capture the "average" time-varying effect of macroeconomic variables on trade finance cost. Finally, this "average" effect can vary over time and is likely to depend on its level in the previous period. The *Evolution Equation,* which is the lowest level of the DHLM, captures these potential changes in the "average" effects of the macroeconomic variables in a flexible way through a random walk.

We employ our DHLM to study how the effects of four macroeconomic variables— GDP growth, trade intensity, inflation, and stock market capitalization—on trade finance costs varied across 8 nations over a period of five years from 2006 to 2010.[5] Although the objective of our paper is to introduce a model that can address the different challenges outlined earlier, our model estimates provide several interesting insights. We find that the effect of macroeconomic indicators on the cost of trade finance varies over time and that most of this variation is present in the years preceding and succeeding the financial crisis. This is of interest to policymakers in deciding how long to implement interventions designed to ease the cost of trade finance. In addition, the trajectory of time-varying effects of GDP growth and inflation are consistent with

---

[5] Stock market capitalization is scaled by GDP. Trade intensity is the ratio of a country's annual total trade and GDP. We use the terms "Trade Intensity" and "Trade/GDP" interchangeably.

the "flight to quality" story [12]: during the crisis, cost of trade finance reduces in countries that have high GDP growth and low inflation. The time-varying effects of trade intensity is also consistent with our expectations, but the time-varying effect of market capitalization is counter-intuitive. Finally, we also note heterogeneity in the trajectory of the country-specific time-varying effects, primarily for the effects of stock market capitalization and trade intensity.

This research makes two contributions. First, we introduce a new model to the finance literature to study the evolution in the drivers of trade finance costs over time in dynamic environments such as a financial crisis, while also allowing the impact due to these drivers to be heterogeneous across countries. Our modeling approach addresses concerns related to nonstationarity, time-varying model parameters and cross-sectional heterogeneity that are endemic to time-series analysis of dynamic environments. Our model can be adopted to study evolution of various other variables such as financial services costs and global trade. Our model can also be extended to a more granular level to incorporate firm-level heterogeneity by using a second pooling equation. Doing this can pave the way to identify the characteristics of companies which may need assistance during a financial crisis. Thus, our research can remove subjectivity in extending benefits to the affected exporters and importers. Even large scale surveys may not be able to provide such granular implications to policy makers. Second, our research has substantive implications. Using a combination of data from Loan Pricing Corporation's Dealscan database and the World Bank, we complement the finance literature by empirically studying the evolution of the drivers of trade finance cost. We find that the impact of these drivers varies over time, with a large part of the variation present in the years preceding and succeeding the financial crisis. To the best of our knowledge, we are the first to study the time-varying impact of these macro-economic drivers on trade finance and this is of use to policy makers in deciding how long to extend benefits to parties affected by the crisis.

The paper proceeds as follows. In the first section we describe the DHLM. We provide the theoretical underpinnings necessary to estimate the model. Next we describe the data and variables used in the empirical analysis. In the fourth section we provide detailed discussion of the results. We conclude the paper with the discussion of the findings.

2 Model Development

We specify trade finance cost of a country as a function of country-specific macroeconomic variables and country-specific time-varying parameters using a DHLM. The DHLM has been used by previous studies in marketing and statistics [19, 26, 33, 35, 39] to estimate time-varying parameters at the disaggregate level (e.g., at the level of a brand or store). A DHLM is a combination of Dynamic linear models (DLM) which estimates time-varying parameters at an aggregate level [5, 6], and a Hierarchical Bayesian (HB) model which estimates *time-invariant* parameters at the disaggregate level [31]. The DHLM and the HB model both have a hierarchical structure which permits us to pool information across different countries to arrive at overall aggregate-level inferences. Shrinking of the country-specific parameters to an "average" effect of the key variables across country has been used by other researchers to estimate country-specific tourism marketing elasticity [39] and to estimate store-level price elasticity [31].

We specify trade finance cost of a country as a function of country-level variables GDP growth, inflation, stock market capitalization and trade:

$$Trade\_finance\_cost_{it} = \alpha_{it} + \beta_{it}GDP\_growth_{it} + \gamma_{it}Inflation_{it} +$$
$$\delta_{it}Stock\_market\_capitalization_{it} + \zeta_{it}Trade\_intensity_{it} + u_1, \tag{1}$$

where $Trade\_finance\_cost_{it}$ is the cost of trade finance of country i at time t, GDP_growth_{it} is the GDP growth of country i at time t, $Inflation_{it}$ is the Inflation of country i at time t, $Stock\_market\_capitalization_{it}$ is the stock market capitalization of country i at time t, $Trade\_intensity_{it}$ is the intensity of trade of country i at time t, α_{it}, β_{it}, γ_{it}, δ_{it} and ζ_{it} are country-specific time-varying coefficients and u_1 is the error term.

In order to specify the equations in a compact manner, we cast Eq. 1 as the observation equation of the DHLM. A DHLM also consists of a pooling equation and an evolution equation, and we specify these three equations below.[6]

We specify the observation equation as:

$$y_t = F_{1t}\theta_{1t} + v_{1it} \; ; \; where \; v_{1it} \sim N\left(0, \sigma_{v_1,i}^2 I_1\right) \tag{2}$$

An observation y_t is defined as a vector that consists of country-specific trade finance cost at time t, whereas F_{1t} is a matrix that contains the country-specific macro-economic variables at time t. The vector of parameters θ_{1t} contains all the country-specific time-varying parameters defined in Eq. 1: α_{it}, β_{it}, γ_{it}, δ_{it} and ζ_{it}.

The error term v_{1it} is multivariate normal and is allowed to have a heteroskedastic variance $\sigma_{v_1,i}^2$, and I_1 an identity matrix of appropriate dimension. We specify y_t, F_{1t}, and θ_{1t} similar to [17, 35].

We specify the pooling equation as:

$$\theta_{1t} = F_{2t}\theta_{2t} + v_{2t} \; ; \; where \; v_{2t} \sim N\left(0, \sigma_{v_2}^2 I_2\right) \tag{3}$$

We specify the country-specific time-varying parameters θ_{1t} as a function of a new set of parameters θ_{2t} that vary only in time. This hierarchical structure pools information across countries at every point in time, and thus θ_{2t} represent the "average" time-varying effect. Hence, F_{2t} is the matrix of 0's and 1's which allows us to specify the relationship between the average time-varying parameters θ_{2t} and the country-specific time-varying parameters θ_{1t}. The error distribution v_{2t} is multivariate normal, and I_2 an identity matrix of appropriate dimension.

We specify how the average time-varying parameters, θ_{2t}, evolves over time. We follow the dynamic linear models (DLM) literature [43] and model the evolution of these parameters over time as a random walk.

---

[6] Note, in Eqs. (2) to (6) matrices are specified in **bold**.

We specify the evolution equation as:

$$\theta_{2t} = G\theta_{2,t-1} + w_t \ ; \ where \ w_t \sim N\left(0, \sigma_w^2 I_3\right) \tag{4}$$

The random walk specification requires G to be an identity matrix and w_t is a multivariate normal error, and I_3 an identity matrix of appropriate dimension.

3   Estimation

We compute the full joint posterior of the set of parameters (θ_{1t}, θ_{2t}, and the variance parameters $\sigma_{v_1,i}^2$, $\sigma_{v_2}^2$, and σ_w^2) conditional on observed data. To generate the posteriors of the parameters we used the Gibbs sampler [16]. In the interest of space, we refer the reader to [26] for more details. As a robustness check, we estimate our DHLM on simulated data to check if our sampler is able to recover the parameters. The model we use to simulate the data in similar to the one [26] used for their simulation study. We find that our sampler performs well and recovers the parameters used to simulate the data. Space constraints prevent us from including further details.

4   Data

For the empirical tests, the data are derived from two sources. The information on trade finance costs is obtained from Loan Pricing Corporation's Dealscan database. The information on macroeconomic variables for the countries is obtained from the World Bank. We briefly describe the data sources.

4.1   Dealscan

Dealscan provides detailed information on loan contract terms including the spread above LIBOR, maturity, and covenants since 1986. The primary sources of data for Dealscan are attachments on SEC filings, reports from loan originators, and the financial press [41]. As it is one of the most comprehensive sources of syndicated loan data, prior literature has relied on it to a large extent [1, 3, 16, 23, 41].

The Dealscan data records, for each year, the loan deals a borrowing firm makes. In some instances, a borrower firm may make several loan deals in a year. To focus on trade finance, we limit the sample to only those loans where the purpose was identified by Dealscan as one of the following: Trade Finance, CP Backup, Pre-Export, and Ship Finance. Our trade finance costs are measured as the loan price for each loan facility, which equals the loan facility's at-issue yield spread over LIBOR (in basis points). Due to the limited number of observations, we don't differentiate between different types of loans. Instead, the trade finance costs are averaged across different types of loans such as revolver loans, term loans, and fixed-rate bonds.

4.2 The World Bank Data

We use the World Bank data to get information on the economic and regulatory climate, and extent of development of the banking sector of the countries where the borrowing firms are headquartered. The economic and regulatory climate of a country is captured by GDP growth, inflation, stock market capitalization, and trade intensity.

Countries with high GDP growth are likely to face lower cost of trade finance, particularly during the financial crisis. As a high GDP growth is an indicator of the health of the economy, during the financial crisis lenders are likely to move their assets to these economies. Countries with higher inflation will likely have higher cost of trade finance as the rate of returns on the loans will incorporate the rate of inflation. We include stock market capitalization scaled by GDP as a proxy for the capital market development in the country. Countries with higher stock market capitalization are likely to have more developed financial markets. Therefore, the cost of trade finance in such markets is likely to be lower. Finally, we include total trade for the country scaled by the country's GDP as a measure of trade intensity. We expect that countries with a higher trade intensity will face a higher trade finance cost since a greater reliance on trade may make a country more risky during a crisis.

4.3 Merging the Two Datasets

As our objective is to study the phenomenon at the national level, we need to merge these two data sets. As our data from Dealscan contains trade finance costs at the firm level in a given year, we use the average of the trade finance costs at the level of a borrowing firm's home country to derive country-specific trade finance costs. This permits us to merge the data from Dealscan with macro-economic data from World Bank. Our interest is in modelling trade finance costs around the financial crisis of 2008–2009. Therefore, we use a 5-year time series starting in 2006 and ending in 2010. This gives us a reasonable window that contains pre-crisis, during the crisis, and post-crisis periods. While we would like to use a longer window, we are constrained by the number of years for which the data are available to us from Dealscan. After merging the two databases, our final sample consists of eight countries for which we have information on trade finance costs as well as macroeconomic indicators for all the five years. The eight countries are: Brazil, Ghana, Greece, Russia, Turkey, Ukraine, United Kingdom (UK), and the United States (USA).

4.4 Descriptive Analysis

We report the descriptive statistics for the sample in Table 1. Average trade finance costs are approximately 190 basis points above LIBOR. Mean GDP growth is just 2.57%, reflecting the lower growth during the financial crisis. Although average inflation is at 10.53%, we calculated the median inflation to be a moderate 6.55%. On average stock market capitalization/GDP ratio is around 63% while trade/GDP ratio is around 54%. More detailed summary statistics for the trade finance costs are depicted in Fig. 1.

Table 1. Descriptive statistics.

Variables	N	Mean	St. Deviation
Trade finance cost above LIBOR (basis points)	40	189.43	155.50
GDP growth %	40	2.57	5.16
Inflation %	40	10.53	13.32
Stock market cap/GDP	40	62.96	43.65
Trade/GDP	40	54.11	22.28

Figure 1 captures the variation in trade finance cost over time and across 8 countries. We find countries experience a large increase in trade finance costs from 2008 to 2009. Also, except for Greece, these costs came down in 2010 from their peak in 2009. This suggests that the crisis impacted trade finance costs uniformly in our sample. We also see heterogeneity across countries in the manner in which these costs evolve over time.

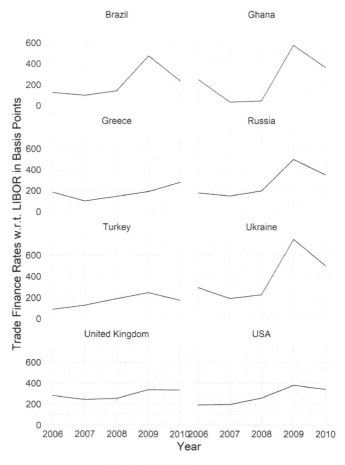

Fig. 1. Trade finance costs

We also tested for multicollinearity among the independent variables, GDP growth, Inflation, Stock market capitalization and Trade intensity. We specified a panel data regression model (i.e., without time-varying parameters) and calculated the Variance Inflation Factors (VIFs). The VIFs we get for GDP growth, Inflation, Stock market capitalization and Trade intensity are 1.13, 1.15, 1.42 and 1.36 respectively. As the VIFs are less than 10, we can conclude that multicollinearity is not a concern [44].

5 Results

In this section, we present the main results based on our DHLM, and subsequently compare our model to the benchmark HB model in which the parameters do not vary over time.

5.1 Main Findings Based on the DHLM

We estimate our model using the Gibbs sampler [18]. We use 200,000 iterations and use the last 20,000 iterations for computing the posterior, while keeping every tenth draw. We verified the convergence of our Gibbs sampler by using standard diagnostics: (1) We plotted the autocorrelation plot of the parameters and see that the autocorrelation goes to zero [40] and (2) we plot and inspect the posterior draws of our model parameters and find that they resemble a "fat, hairy caterpillar" that does not bend [29].

We first present the estimates for the Pooling Equation (θ_2) which are summarized in Fig. 2. These estimates represents the "average" effect across countries of the four macroeconomic variables, GDP growth, Inflation, Stock market capitalization and the Trade/GDP ratio. In Fig. 2, each of the four panels depict the "average" effect, over time, of the macro-economic variables on the cost of trade finance. The dotted lines depict the 95% confidence interval (CI). We discuss these "average" time-varying effects in the subsequent paragraphs.

We see that for all four macro-economic variables, the effects vary over time. In addition, a large part of the variations occur between 2007 to 2009, the 2 year span during which the financial crisis happened. Our estimates will interest policy makers as it implies that interventions to alleviate the impact of the crisis should start before its onset and should continue for some time after it has blown over.

We find that GDP Growth has a negative effect on Trade finance costs and this effect becomes more negative over time, especially during the years 2006 to 2009. Our result implies that countries with high GDP Growth faced monotonically decreasing cost of trade finance in the years before and during the financial crisis, and can be explained by the "flight to quality" hypothesis advanced in the finance literature [12].

Inflation has a positive effect on the cost of trade finance and this effect become more positive over time, especially during 2007 to 2009 which are the year preceding the crisis and the year of the crisis. Our result implies that countries with high inflation faced monotonically increasing costs of trade finance from 2007 to 2009 and is also consistent with the "flight to quality" theory.

Stock Market Capitalization has a positive effect on the cost of trade finance. This effect seems somewhat counterintuitive as we used Stock Market Capitalization as a

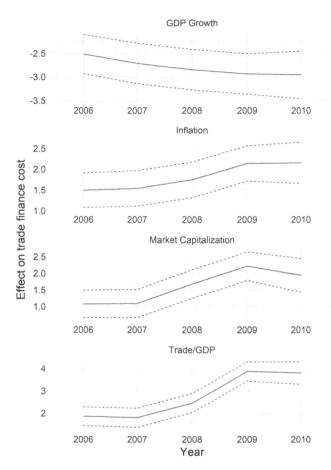

Fig. 2. Estimates of Pooling Equation (θ_2). Notes: Solid line depicts the estimate. Dotted lines indicate the 95% confidence interval.

proxy for development of financial markets and one would expect that during the financial crisis trade finance costs would decrease as financial markets became more developed.

We note that the Trade/GDP ratio has a positive effect on the cost of trade finance, and this effect becomes more positive between the years 2007 to 2009, similar to the pattern we noticed for the effects of inflation. Since this variable measures the intensity of trade of a country, our results indicate that, during the financial crisis, a greater reliance on trade leads to higher costs of trade finance. This is expected since higher reliance on trade may make a country more risky in a financial crisis. Countries with higher trade intensity are also exposed to higher counterparty risks.

Our model can also estimate country-specific time-varying parameters presented in the Observation Equation (θ_1). These estimates underscore the advantage of using a model such as ours, since with only 40 observations of our dependent variable, we are able to estimate 200 estimates which are country-specific and time-varying.[7] We note some heterogeneity in the country-specific trajectory of the effects of Stock Market Capitalization and Trade Intensity. For example, we see that for some countries such as Ghana, Russia and Greece, the effect of trade/GDP ratio on the cost of trade finance witnesses a steeper increase compared to other countries such as USA and Ukraine in 2008 to 2009, the year of the crisis; we are unable to present these results due to space constraints. However, these findings offer directions for future research.

5.2 Model Comparison

To assess model fit, we compare the forecasting accuracy of our proposed model to the benchmark Hierarchical Bayesian (HB) model which has time-invariant parameters. We specify the HB model as follows:

$$Y = X_1\mu_1 + \varepsilon_1 \; ; \; where \, \varepsilon_1 \sim N(0, \, V_{\varepsilon_1}) \, and \tag{5}$$

$$\mu_1 = X_2\mu_2 + \varepsilon_2 \; ; \; where \, \varepsilon_2 \sim N(0, V_{\varepsilon_2}) \tag{6}$$

The above specification is similar to the DHLM with the major difference being that the parameters now do not vary over time. The dependent variables (Y) and independent variables (X_1) are the same as those in the proposed model, while X_2 is a matrix that adjusts the size of μ_1 to that of μ_2.

We compare the model fit by computing the out-of-sample one-step-ahead forecast of our proposed model and the benchmark model. We calculate the mean absolute percentage error (MAPE), which is a standard fit statistic for model comparison [17, 35]. We find that the MAPE of our proposed model is 21.11, while that of the benchmark HB model is 42.68. Thus our proposed model forecasts more accurately than the benchmark HB model.

6 Discussion and Conclusion

In this research, we attempt to shed light on the following question: How can we develop a model that would permit us to examine variations in trade finance costs over time in dynamic environments (such as a financial crisis), while also accounting for possible variations across countries? We addressed this question by proposing a DHLM model that can cope with the three challenges present when modeling data from dynamic environments: nonstationarity, changes in parameters over time and

---

[7] We have 200 estimates since we have 8 countries and 5 time periods, and 5 independent variables (including the intercept). This large number of parameters can be estimated due to our model structure: (i) a first order Markov process relates the parameter at time t to the parameter at $t - 1$, and the parameters across countries at a time t are tied together using a Hierarchical Bayesian structure.

cross-sectional heterogeneity. Our model estimates detect variation over time of the macroeconomic drivers o trade finance, which are of interest to policy makers in deciding when and for how long to schedule interventions to alleviate the impact of a financial crisis. Further, the trajectory of the time-varying effects of the macroeconomic indicators are in line with our expectations. We also note some degree of country-specific heterogeneity in the manner in which these drivers evolve over time, and a detailed scrutiny of these findings may prove fertile ground for future research.

The DHLM can be easily scaled up thereby allowing us to extend our analysis. First, we can add another level in the model hierarchy by specifying a second pooling equation. This would permit us to study the problem at the firm level since evidence suggests that – during the crisis - firms from developing countries and financially vulnerable sectors faced higher trade finance costs [13, 30], and one can use recent NLP approaches [37] to gather firm information across different data sources. Second, more macroeconomic variables can be added in the observation equation. In addition, our model can be used to study other contexts that face dynamic environments such as financial services costs and global trade.

The suitability of our model for dynamic environments also implies that it can also be used to study the impact of the recent coronavirus (COVID-19) on financial activities, since reports from the European Central Bank have suggested that the virus can lead to economic uncertainty. In many ways, the way the virus impacts the economy is similar to that of the financial crisis: There is no fixed date on which the interventions starts and ends – unlike, for example, the imposition of a new state tax – and its impact may vary over time as the virus as well as people's reaction to it gains in strength and then wanes and it would be interesting to model these time-varying effects to see how they evolve over time.

References

1. Acharya, V.V., Almeida, H., Campello, M.: Aggregate risk and the choice between cash and lines of credit: aggregate risk and the choice between cash and lines of credit. J. Finan. **68**(5), 2059–2116 (2013)
2. Ahn, J.: A theory of domestic and international trade finance. In: IMF Working Papers, pp. 1–35 (2011)
3. Almeida, H., Campello, M., Hackbarth, D.: Liquidity mergers. J. Finan. Econ. **102**(3), 526–558 (2011)
4. Amiti, M., Weinstein, D.E.: Exports and financial shocks. Q. J. Econ. **126**(4), 1841–1877 (2011)
5. Ataman, M.B., Van Heerde, H.J., Mela, Carl F.: The long-term effect of marketing strategy on brand sales. J. Mark. Res. **47**(5), 866–882 (2010)
6. Ataman, M.B., Mela, C.F., van Heerde, Harald J.: Building brands. Mark. Sci. **27**(6), 1036–1054 (2008)
7. Auboin, M.: Boosting trade finance in developing countries: what link with the WTO? Working paper, SSRN eLibrary (2007)
8. Baumgartner, Hans, Steenkamp, J.-B.E.M.: Response styles in marketing research: a cross-national investigation. J. Mark. Res. **38**(2), 143–156 (2001)

9. Berger, A.N., Turk-Ariss, R.: Do depositors discipline banks and did government actions during the recent crisis reduce this discipline? An international perspective. J. Finan. Serv. Res. **48**(2), 103–126 (2015). https://doi.org/10.1007/s10693-014-0205-7
10. Beck, T., Demirgüç-Kunt, A., Levine, R.: Financial institutions and markets across countries and over time-data and analysis. In: World Bank Policy Research Working Paper Series (2009)
11. Bronnenberg, B.J., Mahajan, V., Vanhonacker, W.R.: The emergence of market structure in new repeat-purchase categories: the interplay of market share and retailer distribution. J. Mark. Res. **37**(1), 16–31 (2000)
12. Caballero, R.J., Krishnamurthy, A.: International and domestic collateral constraints in a model of emerging market crises. J. Monet. Econ. **48**(3), 513–548 (2001)
13. Chor, D., Manova, K.: Off the cliff and back? Credit conditions and international trade during the global financial crisis. J. Int. Econ. **87**(1), 117–133 (2012)
14. Christen, M., Gupta, S., Porter, J.C., Staelin, R., Wittink, D.R.: Using market-level data to understand promotion effects in a nonlinear model. J. Mark. Res. **34**, 322–334 (1997)
15. Colicev, A., Malshe, A., Pauwels, K., O'Connor, P.: Improving consumer mindset metrics and shareholder value through social media: the different roles of owned and earned media. J. Mark. **82**(1), 37–56 (2018)
16. Dahiya, S., John, K., Puri, M., Ramírez, G.: Debtor-in-possession financing and bankruptcy resolution: empirical evidence. J. Finan. Econ. **69**(1), 259–80 (2003)
17. Dekimpe, M.G., Hanssens, D.M.: The persistence of marketing effects on sales. Mark. Sci. **14**(1), 1–21 (1995)
18. Gelfand, A.E., Smith, A.F.M.: Sample-based approaches to calculating marginal densities. J. Am. Stat. Assoc. **85**, 398–409 (1990)
19. Gopinath, S., Thomas, J.S., Krishnamurthi, L.: Investigating the relationship between the content of online word of mouth, advertising, and brand performance. Mark. Sci. **33**(2), 241–258 (2014)
20. Greene, W.H.: Econometric Analysis 7th Edition International edition (2003)
21. Haddad, M., Ann H., Hausman, C.: Decomposing the great trade collapse: products, prices, and quantities in the 2008–2009 crisis. National Bureau of Economic Research (2010)
22. Hamilton, J.D.: Time Series Analysis. Princeton University Press, Princeton (1994)
23. Haselmann, R., Wachtel, P.: Foreign banks in syndicated loan markets. J. Bank. Finan. **35** (10), 2679–2689 (2011)
24. Horváth, C., Wieringa, J.E.: Combining time series and cross sectional data for the analysis of dynamic marketing systems. Working paper, University of Groningen (2003)
25. Kadiyali, V., Vilcassim, N., Chintagunta, P.: Product line extensions and competitive market interactions: an empirical analysis. J. Econ. **89**(1–2), 339–363 (1998)
26. Landim, F., Gamerman, D.: Dynamic hierarchical models: an extension to matrix-variate observations. Comput. Stat. Data Anal. **35**(1), 11–42 (2000)
27. Levchenko, A.A., Lewis, L.T., Tesar, L.L.: The collapse of international trade during the 2008–09 crisis. In search of the smoking gun. IMF Econ. Rev. **58**(2), 214–253 (2010). https://doi.org/10.1057/imfer.2010.11
28. Levine, R., Loayza, N., Beck, T.: Financial intermediation and growth: Causality and causes. J. Monet. Econ. **46**, 31–77 (2000)
29. Lunn, D., Jackson, C., Best, N., Spiegelhalter, D., Thomas, A.: The BUGS Book: A Practical Introduction to Bayesian Analysis. Chapman and Hall/CRC, Boco Raton (2012)
30. Malouche, M.: Trade and trade finance developments in 14 developing countries post September 2008-A World Bank Survey. In: World Bank Policy Research Working Paper Series (2009)

31. Montgomery, A.L.: Creating micro-marketing pricing strategies using supermarket scanner data. Mark. Sci. **16**(4), 315–337 (1997)
32. Mela, C.F., Gupta, S., Lehmann, D.R.: The long-term impact of promotion and advertising on consumer brand choice. J. Mark. Res. **34**, 248–261 (1997)
33. Mullick, S., Glady, N., Gelper, S.: Price elasticity variations across locations, time and customer segments: an application to the self-storage industry. Working paper, Available at SSRN 3285521(2020)
34. Mullick, S., Raassens, N., Hans, H., Nijssen, E.J.: Reducing food waste through digital platforms: a quantification of cross-side network effects. Working paper (2020). https://research.tue.nl/en/publications/defining-the-leader-and-the-follower-in-a-two-sided-market-proble
35. Neelamegham, R., Chintagunta, P.K.: Modeling and forecasting the sales of technology products. Quant. Mark. Econ. **2**(3), 195–232 (2004). https://doi.org/10.1023/B:QMEC.0000037077.02026.50
36. Nelson, J.P.: Consumer bankruptcies and the bankruptcy reform act: a time-series intervention analysis, 1960–1997. J. Finan. Serv. Res. **17**(2), 181–200 (2000). https://doi.org/10.1023/A:1008166614928
37. Paalman, J., Mullick, S., Zervanou, K., Zhang, Y.: Term based semantic clusters for very short text classification. In: Proceedings of the International Conference Recent Advances in Natural Language Processing, pp. 878–887 (2019). https://doi.org/10.26615/978-954-452-056-4_102
38. Pauwels, K., Srinivasan, S.: Who benefits from store brand entry? Mark. Sci. **23**(3), 364–390 (2004)
39. Peers, Y., van Heerde, H.J., Dekimpe, M.G.: Marketing budget allocation across countries: the role of international business cycles. Mark. Sci. **36**(5), 792–809 (2017)
40. Rossi, P.E., Allenby, G.M., McCulloch, R.: Bayesian Statistics and Marketing. Wiley, Hoboken (2012)
41. Sufi, A.: Information asymmetry and financing arrangements: evidence from syndicated loans. J. Finan. **62**(2), 629–668 (2007)
42. Heerde, V., Harald, J., Mela, C., Manchanda, P.: The dynamic effect of innovation on market structure. J. Mark. Res. **41**(2), 166–183 (2004)
43. West, M., Harrison, J.: Bayesian Forecasting and Dynamic Models. Springer, New York (1997). https://doi.org/10.1007/b98971
44. Yoder, B.J., Pettigrew-Crosby, R.E.: Predicting nitrogen and chlorophyll content and concentrations from reflectance spectra (400–2500 nm) at leaf and canopy scales. Remote Sens. Environ. **53**(3), 199–211 (1995)

Dynamic Pricing Using Thompson Sampling with Fuzzy Events

Jason Rhuggenaath[(✉)], Paulo Roberto de Oliveira da Costa, Yingqian Zhang, Alp Akcay, and Uzay Kaymak

Eindhoven University of Technology, 5612 AZ Eindhoven, The Netherlands
{j.s.rhuggenaath,p.r.d.oliveira.da.costa,yqzhang,
a.e.akcay,u.kaymak}@tue.nl

Abstract. In this paper we study a repeated posted-price auction between a single seller and a single buyer that interact for a finite number of periods or rounds. In each round, the seller offers the same item for sale to the buyer. The seller announces a price and the buyer can decide to buy the item at the announced price or the buyer can decide not to buy the item. In this paper we study the problem from the perspective of the buyer who only gets to observe a stochastic measurement of the valuation of the item after he buys the item. Furthermore, in our model the buyer uses fuzzy sets to describe his satisfaction with the observed valuations and he uses fuzzy sets to describe his dissatisfaction with the observed price. In our problem, the buyer makes decisions based on the probability of a fuzzy event. His decision to buy or not depends on whether the satisfaction from having a high enough valuation for the item out weights the dissatisfaction of the quoted price. We propose an algorithm based on Thompson Sampling and demonstrate that it performs well using numerical experiments.

Keywords: Dynamic pricing · Bayesian modeling · Exploration-exploitation trade-off · Probability of fuzzy events

1 Introduction

In this paper we study a repeated posted-price auction [17] between a single seller and a single buyer. In a repeated posted-price auction there is a single seller and a single buyer that interact for a finite number of periods or rounds. In each round, the seller offers the same item for sale to the buyer. The seller announces a price and the buyer can decide to buy the item at the announced price or the buyer can decide not to buy the item.

The main motivation for this work comes from the domain of online advertising. A large fraction of online advertisements (ads) are sold on online ad exchanges via auctions and this has led to a lot of research on revenue management in online advertising, see e.g. [1,2,5,6,19,21,23,25]. Most ads are sold via second-price auctions, where the winner pays the second highest bid or a

© Springer Nature Switzerland AG 2020
M.-J. Lesot et al. (Eds.): IPMU 2020, CCIS 1237, pp. 653–666, 2020.
https://doi.org/10.1007/978-3-030-50146-4_48

reserve price (whichever is larger), and no sale occurs if all of the bids are lower than the reserve price. However, a significant fraction of auctions only involve a single bidder [5, 6, 19] and this reduces to a posted-price auction when reserve prices announced: the seller sets a reserve price and the buyer decides whether to accept or reject it.

There are three main differences between this paper and previous work. First, unlike in previous work, we study repeated posted-price auctions from the perspective of the buyer that aims to maximize his expected utility, instead of from the perspective of the seller that aims to maximize his revenue. Second, previous papers assume that the buyer knows his valuation in each round. This valuation can either be a fixed value or an independently and identically distributed (i.i.d.) draw from a fixed distribution. In this paper, we do not make this assumption and study a version of the problem where the buyer does not know the distribution of his valuation and the buyer only observes a stochastic measurement of the valuation after he buys the item. Third, previous papers on dynamic pricing and auctions do not model imprecision and vagueness associated with the valuation of the item. In our setting, the buyer uses fuzzy sets to describe whether the valuation of the item is large or small. Furthermore, we use fuzzy sets to describe the dissatisfaction associated with the observed price. In our problem, the buyer makes decisions based on the probability of a fuzzy event [28]. His decision to buy or not depends on whether the satisfaction from having a high enough valuation for the item out-weights the dissatisfaction of the quoted price.

The goal of the buyer is to design a policy that has low *regret*, defined as the gap between the utility of a clairvoyant who has full information about all of the stochastic distributions and the utility achieved by a buyer facing an unknown distribution.

Our proposed algorithm uses Thompson Sampling [24] to balance the exploration-exploitation trade-off. Thompson Sampling is an idea that dates back to the work in [24] and has recently been analyzed theoretically in the context of multi-armed bandit problems [3, 4, 15]. To the best of our knowledge this technique has not been applied to exploration-exploitation trade-offs involving fuzzy sets and fuzzy events.

We summarize the main contributions of this paper as follows:

- To the best of our knowledge, we are the first to study exploration-exploitation trade-offs involving fuzzy sets and fuzzy events in the context of a dynamic pricing problem.
- We show how Thompson Sampling can be used to design tractable algorithms that can be used to dynamically learn the probability of fuzzy events over time.
- Experimental results show that our proposed method performs very well as the regret grows very slowly.

The remainder of this paper is organized as follows. In Sect. 2 we discuss the related literature. Section 3 provides a formal formulation of the problem. In Sect. 4 we present the our proposed algorithm. In Sect. 5 we perform experiments in order to assess the quality of our proposed algorithm. Section 6 concludes our work and provides some interesting directions for further research.

2   Related Literature

The problem considered in this paper is often referred to as a dynamic pricing problem in the operations research and management science community [9], and as a posted-price auction problem in the computer science community [17]. In the standard dynamic pricing problem there is a seller who wants to maximize revenue over some selling horizon by choosing prices in an optimal way. However, the precise relationship between price demand in unknown. This gives rise to the so-called exploration-exploitation trade-off. We refer the reader to [9] and [17] for a detailed overview of the dynamic pricing/posted-price auction problem. The main differences between this paper and existing works is as follows. First, most of the literature of dynamic pricing and learning focuses on perspective of the seller (see e.g. [8,13,16,17,22]). Similarly, most of the literature on posted-price auctions focuses on perspective of the seller (see e.g. [5,6,19]). However, in this paper, we focus on the buyer perspective. Second, most of the literature models the problem in a probabilistic way and does not consider combining fuzzy sets and online learning. While there are some papers that consider dynamic pricing and fuzzy logic such as [12,26] and fuzzy demand [14,20], these papers do not study exploration-exploitation trade-offs like we do in this paper.

From a methodological point of view, this paper is related to the literature on exploration-exploitation trade-offs, in particular, the literature related to multi-armed bandit problems [7,10,18]. In the traditional multi-armed bandit problem [10,18] there is a finite set of actions, called *arms*, and each arm yields a stochastic reward. Play proceeds for a number of rounds, and in each round, precisely one arm can be selected. The goal in a multi-armed bandit problem is to learn which sequence of arms to select in order to maximize the expected cumulative reward over a number of rounds. Two of the main design principles for solving multi-armed bandit problems are (i) optimism in the face of uncertainty [7] and (ii) Thompson Sampling [24] or *probability matching*. Thompson Sampling is an idea that dates back to the work in [24]. Thompson Sampling has recently been analyzed theoretically in the context of multi-armed bandit problems [3,4,15]. Furthermore, this idea has been fruitfully applied in other online decision making problems, see e.g. [11]. However, to the best of our knowledge this technique has not been applied to exploration-exploitation trade-offs involving fuzzy sets and fuzzy events.

3   Problem Formulation

We consider a single buyer and a single seller that interact for T rounds. An item is repeatedly offered for sale by the seller to the buyer over these T rounds. In each round $t = 1, \ldots, T$, a price p_t is offered by the seller and a decision

$a_t \in \{0, 1\}$ is made by the buyer: a_t takes value 1 when the buyer accepts to buy at that price, and a_t takes value 0 otherwise. In every round t there is a stochastic measurement $v_t \in [0, 1]$ for the item. The value of v_t is an i.i.d. draw from a distribution $\mathcal{D}$ and has expectation $\nu = \mathbb{E}\{v_t\}$. The measurement v_t is only revealed to the buyer if he buys the item in round t, i.e., the buyer only observes the value of the measurement after he buys the item. We assume that the buyer does not know the distribution $\mathcal{D}$.

From the buyer perspective we assume that there is imprecision and vagueness associated with the valuation of the item. We assume that the buyer uses $M_v \in \mathbb{N}$ fuzzy sets [27,29] to describe his valuation for the item. Denote these fuzzy sets by $V^1, \ldots, V^{M_v}$. The membership function is given by $\mu_v^m(x)$ for $m = 1, \ldots, M_v$ and maps a stochastic measurement to values in $[0, 1]$. For example, the buyer could use three fuzzy sets that describe valuations that can be low, medium or high.

We assume that the buyer also uses fuzzy sets to describe his dissatisfaction with a particular price. More specifically, the buyer uses $M_p \in \mathbb{N}$ fuzzy sets to describe his dissatisfaction for buying the item at a particular price. Denote these fuzzy sets by $D^1, \ldots, D^{M_p}$. The membership function is given by $\mu_p^k(x)$ for $k = 1, \ldots, M_p$ and maps the price to values in $[0, 1]$. For example, the buyer could use three fuzzy sets to describe that his dissatisfaction can be low, medium or high.

We make the following assumption on the membership functions.

Assumption 1. *The membership functions $\mu_v^m(x)$ for $m = 1, \ldots, M_v$ and $\mu_p^k(x)$ for $k = 1, \ldots, M_p$ are fixed for all rounds $t = 1, \ldots, T$.*

In our problem, the buyer makes decisions based on the probability of a fuzzy event. The definition for the probability of a fuzzy event is given by Definition 1 and is due to [28].

Definition 1 ([28]). *Let A be an arbitrary fuzzy set with membership function $\mu_A(x)$. The probability of the fuzzy event associated with the fuzzy set A with respect to a distribution $\mathcal{D}$ is defined as $\mathbb{P}\{A\} = \mathbb{E}_{\mathcal{D}}\{\mu_A\}$. Here $\mathbb{E}_{\mathcal{D}}$ denotes the expectation under distribution $\mathcal{D}$.*

If the buyer had complete information about the distribution $\mathcal{D}$, then he can calculate the probability of the fuzzy events $\mathbb{P}\{V^m\}$ for $m = 1, \ldots, M_v$ associated with the fuzzy sets $V^1, \ldots, V^{M_v}$. We assume that the buyer uses a function $F^V : [0, 1]^{M_v} \to [0, 1]$ that combines the probabilities $\mathbb{P}\{V^m\}$ for $m = 1, \ldots, M_v$ into an aggregated score $S_V \in [0, 1]$.

Similarly, we assume that the buyer uses a function $F^P : [0, 1]^{M_p} \to [0, 1]$ that combines membership values for the fuzzy sets $D^1, \ldots, D^{M_p}$ into an aggregated score $S_P \in [0, 1]$.

If the buyer had complete information about the distribution $\mathcal{D}$, we assume that the buyer makes decisions according to the rule described in Assumption 2.

Assumption 2. *Under complete information about the distribution $\mathcal{D}$, the buyer uses the following rule to make decisions:*

- *observe the price p_t in round t.*
- *calculate $S_P(p_t) = F^P(\mu_p^1(p_t), \ldots, \mu_p^{M_p}(p_t))$.*
- *compare the value of $S_P(p_t)$ with the value of $S_V = F^V(\mathbb{P}\{V^1\}, \ldots, \mathbb{P}\{V^{M_v}\})$.*
- *if $S_V \geq S_P(p_t)$, then buy the item.*
- *if $S_V < S_P(p_t)$, then do not buy the item.*

The intuition behind this rule is that the value of S_V represents the total aggregated degree of satisfaction with the item and that $S_P(p_t)$ represents the total aggregated dissatisfaction with the price p_t. The decision rule described above indicates that the buyer would only buy the item if his total satisfaction with the item outweighs the dissatisfaction at price p_t.

Note that the value of S_V depends on the function F^V and the probabilities $\mathbb{P}\{V^1\}, \ldots, \mathbb{P}\{V^{M_v}\}$: if the same function F^V is used in all the rounds, then S_V is fixed. Furthermore, note that the value of S_V is unknown to the buyer since the distribution $\mathcal{D}$ is unknown.

The expected utility of the buyer in round t is given by $u_t = a_t \cdot (S_V - S_P(p_t))$. In other words, if the buyer purchases the item ($a_t = 1$) the utility is the difference between the total aggregated degree of satisfaction and the total aggregated dissatisfaction with the price. Otherwise, the utility is zero. The objective of the buyer is to maximize his expected utility over the T rounds.

For a fixed sequence $p_1, \ldots, p_T$ of observed prices and a fixed sequence of decisions $a_1, \ldots, a_T$ by the buyer, the *regret* of the buyer over T rounds is defined as

$$R_T = \sum_{t=1}^{T} \max\{S_V - S_P(p_t), 0\} - \sum_{t=1}^{T} a_t \cdot (S_V - S_P(p_t)). \tag{1}$$

The expected regret over T rounds is defined as

$$\mathcal{R}_T = \mathbb{E}\{R_T\}, \tag{2}$$

where the expectation in Eq. (2) is taken with respect to possible randomization in the selection of the actions $a_1, \ldots, a_T$.

Note that (for a fixed sequence $p_1, \ldots, p_T$ of observed prices) the objective of maximizing expected utility is equivalent to minimizing the expected regret. The goal is to make decisions that are as close as possible to the decisions that are prescribed by the decision rule in Assumption 2. Since the distribution $\mathcal{D}$ is unknown to the buyer, he faces an exploration-exploitation trade-off. In order to gain information and estimate the value of S_V he needs to buy the item a number of times (exploration). However, he also wants to use all information collected so far and only buy the item if $S_V \geq S_P(p_t)$ holds (exploitation). In this paper, we seek to develop algorithms that ensure that the value of $\mathcal{R}_T$ grows slowly as the problem horizon T increases.

4   Proposed Algorithm

In this section we discuss our proposed algorithm. We refer to our algorithm as
TS-PFE (Thompson Sampling for Probabilities of Fuzzy Events). Our algorithm
uses Thompson Sampling [24] to balance the exploration-exploitation trade-off.
The pseudo-code for TS-PFE is given by Algorithm 1.

Algorithm 1: TS-PFE

Require: Parameters of prior distribution (a_0^m, b_0^m) for $m = 1, \ldots, M_v$,
function F^V, function F^P, membership functions $\mu_v^m(x)$ for $m = 1, \ldots, M_v$
and $\mu_p^k(x)$ for $k = 1, \ldots, M_p$.
1: **for** $m = 1$ **to** M_v **do**
2:    Set $a^m = a_0^m$. Set $b^m = b_0^m$.
3: **end for**
4: Set $t = 1$.
5: Purchase item in first round.
6: Observe v_1.
7: **for** $m = 1$ **to** M_v **do**
8:    Set $a^m = a^m + \mu_v^m(v_1)$. Set $b^m = b^m + (1 - \mu_v^m(v_1))$.
9: **end for**
10: **for** $t \in \{2, \ldots, T\}$ **do**
11:    Observe price p_t of item.
12:    **for** $m = 1$ **to** M_v **do**
13:       Sample $\hat{\theta}^m \sim \mathcal{B}(a^m, b^m)$.
14:    **end for**
15:    Set $\hat{S}_V = F^V(\hat{\theta}^1, \ldots, \hat{\theta}^{M_v})$.
16:    Set $S_P(p_t) = F^P(\mu_p^1(p_t), \ldots, \mu_p^{M_p}(p_t))$.
17:    **if** $\hat{S}_V \geq S_P(p_t)$ **then**
18:       Purchase item at price p_t.
19:       Observe v_t.
20:       **for** $m = 1$ **to** M_v **do**
21:          Set $a^m = a^m + \mu_v^m(v_t)$. Set $b^m = b^m + (1 - \mu_v^m(v_t))$.
22:       **end for**
23:    **end if**
24: **end for**

Our algorithm adopts a Bayesian framework for handling the unknown prob-
abilities of fuzzy events $\mathbb{P}\{V^1\}, \ldots, \mathbb{P}\{V^{M_v}\}$. Let $Z \sim \mathcal{B}(a, b)$ denote a random
variable Z that follows a Beta distribution with parameters a and b and with
expectation $\mathbb{E}\{Z\} = \frac{a}{a+b}$.

We associate a Beta distributed random variable θ^m with the probability
$\mathbb{P}\{V^m\}$ for $m = 1, \ldots, M_v$. At the start of our algorithm, $\theta^m \sim \mathcal{B}(a_0^m, b_0^m)$
indicates the prior distribution. Every time that the buyer buys the item, he
observes an i.i.d. draw v_t from the distribution $\mathcal{D}$. This draw is subsequently
used to update the distribution of θ^m. We can use the draw v_t in order to

learn the membership values corresponding to the fuzzy sets at the value of v_t. We subsequently use this information to update our estimate of $\mathbb{P}\{V^m\}$ for $m = 1, \ldots, M_v$.

In order to decide to buy the item or not, we sample values $\hat{\theta}^m$ from the posterior distributions $\mathcal{B}(a^m, b^m)$ and use these sampled values in order to form an estimate $\hat{S}_V$ of S_V. This is the Thompson Sampling step. Intuitively, the buyer buys the item with probability $\mathbb{P}_t\{S_V \geq S_P(p_t)\}$ and where the probability measure $\mathbb{P}_t$ is induced by the posterior distribution $\mathcal{B}(a^m, b^m)$ in round t. It is precisely this Thompson Sampling step that balances exploration and exploitation.

5   Experiments

In this section we conduct experiments in order to test the performance of our proposed algorithm. In total we have three experimental settings, which we describe in more detail below.

5.1   General Settings and Performance Metrics

In all our experiments the prior parameters are set to $(a_0^m = 1, b_0^m = 1)$ for $m = 1, \ldots, M_v$ which corresponds to a uniform distribution on $[0, 1]$. The problem horizon is set to $T = 5000$.

In order to measure the performance of the methods, we consider two performance metrics. Our main performance metric is the cumulative regret which is defined as $R_T = \sum_{t=1}^{T} \max\{S_V - S_P(p_t), 0\} - \sum_{t=1}^{T} a_t \cdot (S_V - S_P(p_t))$.

The second performance metric is the fraction of rounds in which the "best action" was selected. Here "best action" means the action that the buyer would have taken if he had complete information about the distribution $\mathcal{D}$. That is, if the buyer makes decisions according to the rule described in Assumption 2.

We run 500 independent simulations and all performance metrics are averaged over these 500 simulations.

5.2   Experimental Setting I

In this setting v_t is drawn from an uniform distribution on $[a - 0.3, a + 0.3]$, where $a = 0.5$. The price p_t is an i.i.d. draw from the distribution specified in Table 1 with $a = 0.5$.

We use three fuzzy sets V^1, V^2, V^3 for the valuation of the buyer and so $M_v = 3$. The three fuzzy sets describe valuations that can be low (V^1), medium (V^2) or high (V^3). We use a single fuzzy set to express the dissatisfaction with the price and so $M_p = 1$. The membership functions are displayed in Fig. 1 and Fig. 2. The interpretation is that for high prices the membership value will be 1 and the buyer is very dissatisfied. For intermediate values the membership value is between 0 and 1, and indicates that the buyer is partially dissatisfied. For low values of the price, the buyer is not dissatisfied.

For the function F^V we use $F^V = 0.25 \cdot \mathbb{P}\{V^2\} + 0.75 \cdot \mathbb{P}\{V^3\}$. For the function F^P we use $F^P = \mu_p^1(p_t)$.

Table 1. Distribution of prices.

Value	Probability
$a - 0.2$	$\frac{1}{11}$
$a - 0.1$	$\frac{1}{11}$
$a - 0.05$	$\frac{1}{11}$
$a - 0.02$	$\frac{1}{11}$
a	$\frac{2}{11}$
$a + 0.05$	$\frac{1}{11}$
$a - 0.2$	$\frac{3}{11}$
$a - 0.3$	$\frac{1}{11}$

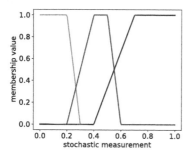

Fig. 1. Membership functions $\mu_v^m(x)$ for $m = 1, \ldots, M_v$.

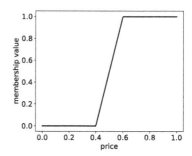

Fig. 2. Membership functions $\mu_p^k(x)$ for $k = 1, \ldots, M_p$.

The results for the cumulative regret and the fraction of the rounds in which the best action is selected are displayed in Fig. 3 and Fig. 4. The results indicate that TS-PFE performs relatively well as the cumulative regret grows very slowly as the number of rounds increases. In Fig. 4 we can see that TS-PFE learns relatively quickly as it selects the "best action" in at least 90% of the first 500 rounds. Furthermore, this percentage only increases as the number of rounds increases indicating that TS-PFE makes less and less mistakes.

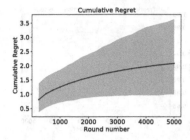
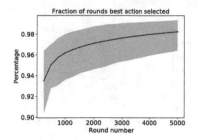

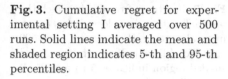

Fig. 3. Cumulative regret for experimental setting I averaged over 500 runs. Solid lines indicate the mean and shaded region indicates 5-th and 95-th percentiles.

Fig. 4. Fraction of rounds in which the optimal action is selected for experimental setting I averaged over 500 runs. Solid lines indicate the mean and shaded region indicates 5-th and 95-th percentiles.

5.3 Experimental Setting II

In this setting v_t is drawn from an uniform distribution on $[a-0.3, a+0.3]$, where $a = 0.3$. The price p_t is an i.i.d. draw from the distribution specified in Table 1 with $a = 0.3$. The membership function used to express the dissatisfaction with the price is different compared to experimental setting I and is displayed in Fig. 5. The membership functions for V^1, V^2, V^3 are the same as in experimental setting I and are displayed in Fig. 1. All other settings are the same as in experimental setting I.

The results for the cumulative regret and the fraction of the rounds in which the best action is selected are displayed in Fig. 6 and Fig. 7. The results are qualitatively similar as those reported in Fig. 3 and Fig. 4. Again we see that TS-PFE is able to learn relatively quickly to make the correct decisions.

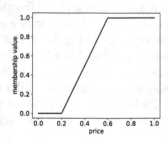

Fig. 5. Membership functions $\mu_p^k(x)$ for $k = 1, \ldots, M_p$.

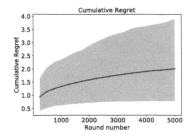
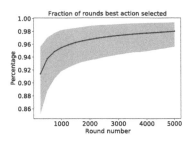

Fig. 6. Cumulative regret for experimental setting II averaged over 500 runs. Solid lines indicate the mean and shaded region indicates 5-th and 95-th percentiles.

Fig. 7. Fraction of rounds in which the optimal action is selected for experimental setting II averaged over 500 runs. Solid lines indicate the mean and shaded region indicates 5-th and 95-th percentiles.

5.4 Experimental Setting III

In this setting we test how the algorithm performs when the function F^V changes during the problem horizon. In this setting v_t is drawn from an uniform distribution on $[a - 0.3, a + 0.3]$, where $a = 0.3$. The price p_t is an i.i.d. draw from the distribution specified in Table 2 with $a = 0.3$. The membership function used to express the dissatisfaction with the price is different compared to experimental setting I and is displayed in Fig. 5. The membership functions for V^1, V^2, V^3 are the same as in experimental setting I and are displayed in Fig. 1.

For the function F^V we use $F^V = 0.25 \cdot \mathbb{P}\{V^2\} + 0.75 \cdot \mathbb{P}\{V^3\}$ in rounds $t < 2500$ and we take F^V to be the harmonic mean of $\mathbb{P}\{V^2\}$ and $\mathbb{P}\{V^3\}$ in rounds $t \geq 2500$. The problem horizon is set to $T = 10000$. All other settings are the same as in experimental setting I. With these settings the optimal action is not to buy in rounds $t < 2500$ and to buy at price 0.30 in rounds $t \geq 2500$.

The results for the cumulative regret and the fraction of the rounds in which the best action is selected are displayed in Fig. 8 and Fig. 9. The results are qualitatively similar as those reported in Fig. 3 and Fig. 4. The level of regret is in general higher in Fig. 8 compared to Fig. 3 and 6, which indicates the the problem is harder. Also, the percentiles in Figs. 8 and 9 indicate that the problem is harder. However, the overall pattern of regret is similar to the other experimental settings. Again we see that TS-PFE is able to learn relatively quickly to make the correct decisions, even if the function F^V changes during the problem horizon.

Table 2. Distribution of prices.

Value	Probability
0.3	$\frac{1}{2}$
0.5	$\frac{1}{2}$

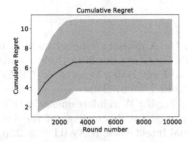

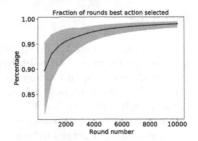

Fig. 8. Cumulative regret for experimental setting III averaged over 500 runs. Solid lines indicate the mean and shaded region indicates 5-th and 95-th percentiles.

Fig. 9. Fraction of rounds in which the optimal action is selected for experimental setting III averaged over 500 runs. Solid lines indicate the mean and shaded region indicates 5-th and 95-th percentiles.

6   Conclusion

In this paper we study a repeated posted-price auction between a single seller and a single buyer that interact for a finite number of periods or rounds. In this paper we study the problem from the perspective of the buyer who only gets to observe a stochastic measurement of the valuation of the item after he buys the item. In our model, the buyer uses fuzzy sets to describe his satisfaction with the observed valuations and he uses fuzzy sets to describe his dissatisfaction with the observed price. In our problem, the buyer makes decisions based on the probability of a fuzzy event. His decision to buy or not depends on whether the satisfaction from having a high enough valuation for the item out-weights the dissatisfaction of the quoted price. To the best of our knowledge, we are the first to study exploration-exploitation trade-offs involving fuzzy sets and fuzzy events in the context of a dynamic pricing problem. We show how Thompson Sampling can be used to design tractable algorithms that can be used to dynamically learn the probability of fuzzy events over time.

One direction for future work is to investigate whether the ideas used in this paper can be used in other exploration-exploitation problems involving fuzzy sets and probabilities based on fuzzy events. For example, it would be interesting to investigate whether the ideas can be extended or adapted to a setting of group decision making with stochastic feedback and preferences described by fuzzy sets. One could for example try to learn to make decisions online such that the preferences of the members of the group are/ remain close to each other.

Acknowledgements. We would like to thank Rick Augustinus Maria Gilsing, Jonnro Erasmus, Anasztázia Junger, Peipei Chen, Bambang Suratno, and Onat Ege Adalı for interesting discussions, which helped improve this paper considerably.

References

1. Afshar, R.R., Zhang, Y., Firat, M., Kaymak, U.: A decision support method to increase the revenue of ad publishers in waterfall strategy. In: 2019 IEEE Conference on Computational Intelligence for Financial Engineering Economics (CIFEr), pp. 1–8, May 2019. https://doi.org/10.1109/CIFEr.2019.8759106

2. Afshar., R.R., Zhang., Y., Firat., M., Kaymak., U.: A reinforcement learning method to select ad networks in waterfall strategy. In: Proceedings of the 11th International Conference on Agents and Artificial Intelligence, ICAART, vol. 2, pp. 256–265. INSTICC, SciTePress (2019). https://doi.org/10.5220/0007395502560265

3. Agrawal, S., Goyal, N.: Analysis of Thompson sampling for the multi-armed bandit problem. In: Mannor, S., Srebro, N., Williamson, R.C. (eds.) Proceedings of the 25th Annual Conference on Learning Theory. Proceedings of Machine Learning Research, vol. 23, pp. 39.1–39.26. PMLR, Edinburgh, 25–27 June 2012. http://proceedings.mlr.press/v23/agrawal12.html

4. Agrawal, S., Goyal, N.: Further optimal regret bounds for Thompson sampling. In: Carvalho, C.M., Ravikumar, P. (eds.) Proceedings of the Sixteenth International Conference on Artificial Intelligence and Statistics. Proceedings of Machine Learning Research, vol. 31, pp. 99–107. PMLR, Scottsdale, 29 Apr–01 May 2013. http://proceedings.mlr.press/v31/agrawal13a.html

5. Amin, K., Rostamizadeh, A., Syed, U.: Learning prices for repeated auctions with strategic buyers. In: Proceedings of the 26th International Conference on Neural Information Processing Systems, NIPS 2013, vol. 1, pp. 1169–1177. Curran Associates Inc., USA (2013). http://dl.acm.org/citation.cfm?id=2999611.2999742

6. Amin, K., Rostamizadeh, A., Syed, U.: Repeated contextual auctions with strategic buyers. In: Ghahramani, Z., Welling, M., Cortes, C., Lawrence, N.D., Weinberger, K.Q. (eds.) Advances in Neural Information Processing Systems vol. 27, pp. 622–630. Curran Associates, Inc. (2014). http://papers.nips.cc/paper/5589-repeated-contextual-auctions-with-strategic-buyers.pdf

7. Auer, P., Cesa-Bianchi, N., Fischer, P.: Finite-time analysis of the multiarmed bandit problem. Mach. Learn. **47**(2), 235–256 (2002). https://doi.org/10.1023/A:1013689704352

8. Besbes, O., Zeevi, A.: On the (surprising) sufficiency of linear models for dynamic pricing with demand learning. Manag. Sci. **61**(4), 723–739 (2015). https://doi.org/10.1287/mnsc.2014.2031

9. den Boer, A.V.: Dynamic pricing and learning: historical origins, current research, and new directions. Surv. Oper. Res. Manag. Sci. **20**(1), 1–18 (2015). https://doi.org/10.1016/j.sorms.2015.03.001. http://www.sciencedirect.com/science/article/pii/S1876735415000021

10. Bubeck, S., Cesa-Bianchi, N.: Regret analysis of stochastic and nonstochastic multi-armed bandit problems. Found. Trends® Mach. Learn. **5**(1), 1–122 (2012). https://doi.org/10.1561/2200000024

11. Chapelle, O., Li, L.: An empirical evaluation of Thompson sampling. In: Proceedings of the 24th International Conference on Neural Information Processing Systems, NIPS 2011, pp. 2249–2257. Curran Associates Inc., USA (2011). http://dl.acm.org/citation.cfm?id=2986459.2986710

12. Gen-dao, L., Wei, L.: Dynamic pricing of perishable products with random fuzzy demand. In: 2010 International Conference on Management Science Engineering 17th Annual Conference Proceedings, pp. 191–199, November 2010. https://doi.org/10.1109/ICMSE.2010.5719804

13. Harrison, J.M., Keskin, N.B., Zeevi, A.: Bayesian dynamic pricing policies: learning and earning under a binary prior distribution. Manag. Sci. **58**(3), 570–586 (2012). https://doi.org/10.1287/mnsc.1110.1426
14. Kao, C., Hsu, W.K.: A single-period inventory model with fuzzy demand. Comput. Math. Appl. **43**(6), 841–848 (2002). https://doi.org/10.1016/S0898-1221(01)00325-X. http://www.sciencedirect.com/science/article/pii/S089812210100325X
15. Kaufmann, E., Korda, N., Munos, R.: Thompson sampling: an asymptotically optimal finite-time analysis. In: Bshouty, N.H., Stoltz, G., Vayatis, N., Zeugmann, T. (eds.) ALT 2012. LNCS (LNAI), vol. 7568, pp. 199–213. Springer, Heidelberg (2012). https://doi.org/10.1007/978-3-642-34106-9_18
16. Keskin, N.B., Zeevi, A.: Chasing demand: learning and earning in a changing environment. Math. Oper. Res. **42**(2), 277–307 (2017). https://doi.org/10.1287/moor.2016.0807
17. Kleinberg, R., Leighton, T.: The value of knowing a demand curve: bounds on regret for online posted-price auctions. In: Proceedings of the 44th Annual IEEE Symposium on Foundations of Computer Science, FOCS 2003, p. 594. IEEE Computer Society, Washington, DC (2003). http://dl.acm.org/citation.cfm?id=946243.946352
18. Lai, T.L., Robbins, H.: Asymptotically efficient adaptive allocation rules. Adv. Appl. Math. **6**(1), 4–22 (1985)
19. Mohri, M., Medina, A.M.: Optimal regret minimization in posted-price auctions with strategic buyers. In: Proceedings of the 27th International Conference on Neural Information Processing Systems, NIPS 2014, vol. 2, pp. 1871–1879. MIT Press, Cambridge (2014). http://dl.acm.org/citation.cfm?id=2969033.2969036
20. Petrović, D., Petrović, R., Vujošević, M.: Fuzzy models for the newsboy problem. Int. J. Prod. Econ. **45**(1), 435–441 (1996). https://doi.org/10.1016/0925-5273(96)00014-X. http://www.sciencedirect.com/science/article/pii/092552739600014X. Proceedings of the Eighth International Symposium on Inventories
21. Rhuggenaath, J., Akcay, A., Zhang, Y., Kaymak, U.: Optimizing reserve prices for publishers in online ad auctions. In: 2019 IEEE Conference on Computational Intelligence for Financial Engineering Economics (CIFEr), pp. 1–8, May 2019. https://doi.org/10.1109/CIFEr.2019.8759123
22. Rhuggenaath, J., de Oliveira da Costa, P.R., Akcay, A., Zhang, Y., Kaymak, U.: A heuristic policy for dynamic pricing and demand learning with limited price changes and censored demand. In: 2019 IEEE International Conference on Systems, Man and Cybernetics (SMC), pp. 3693–3698, October 2019. https://doi.org/10.1109/SMC.2019.8914590
23. Rhuggenaath, J., Akcay, A., Zhang, Y., Kaymak, U.: Optimal display-ad allocation with guaranteed contracts and supply side platforms. Comput. Ind. Eng. **137**, 106071 (2019). https://doi.org/10.1016/j.cie.2019.106071. http://www.sciencedirect.com/science/article/pii/S0360835219305303
24. Thompson, W.R.: On the likelihood that one unknown probability exceeds another in view of the evidence of two samples. Biometrika **25**(3–4), 285–294 (1933). https://doi.org/10.1093/biomet/25.3-4.285
25. Wang, J., Zhang, W., Yuan, S.: Display advertising with real-time bidding (RTB) and behavioural targeting. Found. Trends® Inf. Retrieval **11**(4–5), 297–435 (2017). https://doi.org/10.1561/1500000049
26. Xiong, Y., Li, G., Fernandes, K.J.: Dynamic pricing model and algorithm for perishable products with fuzzy demand. Appl. Stoch. Models Bus. Ind. **26**(6), 758–774 (2010). https://doi.org/10.1002/asmb.816

27. Zadeh, L.: Fuzzy sets. Inf. Control **8**(3), 338–353 (1965). https://doi.org/10.1016/S0019-9958(65)90241-X. http://www.sciencedirect.com/science/article/pii/S0019995865902241X

28. Zadeh, L.: Probability measures of fuzzy events. J. Math. Anal. Appl. **23**(2), 421–427 (1968). https://doi.org/10.1016/0022-247X(68)90078-4. http://www.sciencedirect.com/science/article/pii/0022247X6890

29. Zadeh, L.: The concept of a linguistic variable and its application to approximate reasoning–i. Inf. Sci. **8**(3), 199–249 (1975). https://doi.org/10.1016/0020-0255(75)90036-5. http://www.sciencedirect.com/science/article/pii/0020025575900365

Electrical Power Grid Frequency Estimation with Fuzzy Boolean Nets

Nuno M. Rodrigues[1], Joao P. Carvalho[2]([✉]), Fernando M. Janeiro[3],
and Pedro M. Ramos[1]

[1] Instituto de Telecomunicações, Instituto Superior Técnico, Universidade de Lisboa,
Lisbon, Portugal
{nuno.medeiros.rodrigues,pedro.m.ramos}@tecnico.ulisboa.pt
[2] INESC-ID, Instituto Superior Técnico, Universidade de Lisboa, Lisbon, Portugal
joao.carvalho@inesc-id.pt
[3] Instituto de Telecomunicações, Universidade de Évora, Évora, Portugal
fmtj@uevora.pt

Abstract. Power quality analysis involves the measurement of quantities that characterize a power supply waveform such as its frequency. The measurement of those quantities are regulated by internationally accepted standards from IEEE or IEC. Monitoring the delivered power quality is even more important due to recent advances in power electronics and also due to the increasing penetration of renewable energies in the electrical power grid. The primary suggested method by IEC to measure the power grid frequency is to count the number of zero crossings in the voltage waveform that occur during 0.2 s. The standard zero crossing method is usually applied to a filtered signal that has a non deterministic and frequency dependent delay. For monitoring the power grid a range between 42.5 and 57.5 Hz should be considered which means that the filter must be designed in order to attenuate the delay compensation error. Fuzzy Boolean Nets can be considered a neural fuzzy model where the fuzziness is an inherent emerging property that can ignore some outliers acting as a filter. This property can be useful to apply zero crossing without false crossing detection and estimate the real timestamp without the non deterministic delay concern. This paper presents a comparison between the standard frequency estimation, a Goertzel interpolation method, and the standard method applied after a FBN network instead of a filtered signal.

Keywords: Power quality · Frequency estimation · Zero crossing · Neural networks · FBN

This work was supported by national funds through FCT, Fundação para a Ciência e a Tecnologia, under project UIDB/50021/2020, PhD program reference SFRH/BD/130327/2017 and is funded by FCT/MCTES through national funds and when applicable co-funded EU funds under the project UIDB/EEA/50008/2020.

M.-J. Lesot et al. (Eds.): IPMU 2020, CCIS 1237, pp. 667–679, 2020.
https://doi.org/10.1007/978-3-030-50146-4_49

1   Introduction

The quality of the electrical power grid is of utmost importance for normal operation of electrical equipment [1–3]. Due to its importance, power quality has been regulated in various internationally recognized standards. Recent advances in power electronics and the ever increasing penetration of renewable energies that require distributed power converters makes the monitoring of the delivered power quality even more important [4,5]. Monitored parameters include the frequency of the supplied power waveform, its RMS amplitude, the existence of harmonics, the effect of noise and the existence of distortions and transients.

Power quality measurements are typically performed only at predefined locations and not on a regular basis, usually to settle disputes between utility companies and consumers. The main reason for this is that commercial power quality analyzers are very accurate, but expensive and bulky. The alternative are smaller analyzers that are less expensive but have worse specifications. Therefore, there is an increasing demand for portable analyzers that can be easily and more universally deployed, with better accuracy than those that are presently available. The advances in the processing power of digital signal processors, available memory, and sensors' availability, have boosted the interest on the development of embedded power quality analyzers that are low-cost, yet very powerful [6].

Frequency estimation and tracking methods are an active research topic in many scientific areas as in power quality assessment [7]. The most basic spectral based methods performs a FFT but to have good spectral resolutions the computational cost increases. There are methods that can improve frequency estimation performing an interpolation with the calculated DFT bins [8],[9]. But for some applications, as in power quality, there is no need to compute a full FFT. Approaches as in [10] that uses Goertzel filters [11] or in [12] use a warped DFT in order to select only a defined spectral area to perform frequency estimation.

For some specific applications power grid frequency must be obtained in a 10 cycle time span, that in a 50 Hz electrical power grid system corresponds to 0.2 s period as specified in IEC standard 61000-4-30 [13]. The fundamental frequency is the number of integral cycles counted during the considered time interval, divided by the cumulative duration of those cycles. When using the zero crossing (ZC) method, harmonics, interharmonics and noise should be attenuated to avoid false zero crossings in frequency estimation. A particularly effective solution is to digitally low-pass filter the acquired waveform to attenuate the unwanted effects [14]. A disadvantage of this method is the delay that the filter introduces between the filtered and unfiltered signals. This delay, that is frequency dependent, must be compensated to estimate the real zero crossing timestamp. Another method to perform frequency estimation is based on the Goertzel algorithm with interpolation for frequency estimation. The Goertzel algorithm [11] is an efficient method to estimate individual components of the signal Discrete Fourier Transform (DFT). In order to estimate the power grid frequency, an interpolation algorithm based on the Interpolated Discrete Fourier Transform (IpDFT) of [8] is applied.

In this paper we study an alternative approach to estimate power grid frequency based on zero crossing detection that uses Fuzzy Boolean Nets (FBN) [15]. Natural or Biological neural systems have a certain number of features that leads to their learning capability when exposed to sets of experiments from the real outside world. They also have the capability to use the learnt knowledge to perform reasoning in an approximate way. FBN are a neural fuzzy model where the fuzziness is an inherent emerging property that were developed with the goal of exhibiting those systems' properties. FBN previous studies have shown promising results in learning and interpolating in applications such as pultrusion wastes [16] or dataflow management systems [17]. In [16] FBN method allows to find the best balance between the material parameters that maximizes the strength of the final composite. In [17] is used to augment performance, rationalization of resources, and task prioritization of dataflows based in probabilistic results of the networks output.

The contents of this paper is as follows: Sect. 2 describes the three different methods in order to perform a frequency estimation; Sect. 3 presents results for simulated data with some PQ events that can affect ZC detection; Sect. 4 compares the results between the three methods from the electrical power grid; In Sect. 5 the conclusions about the comparison of the methods applied.

2   Frequency Estimation Methods

2.1   Filtered Signal with Zero Crossing

According to IEC 61000-4-30 standard [13], a signal power frequency measure is the number of integral periods divided by the duration of those periods within the considered time interval. This measure can be accomplished by estimating the zero crossing timestamps trough interpolation between the acquired samples. To avoid false zero crossings caused by PQ events, the acquired data should be filtered. The digital filter introduces a delay which offsets the timestamps of the filtered signal zero crossings that is dependent on the signal frequency but can be corrected in order to have a real timestamp [14]. In Fig. 1 an example about the effect of filtering a signal is shown.

2.2   Goertzel and Interpolation

The Goertzel algorithm [11] is an efficient method to compute a single DFT tone without having to perform a full FFT. Since the power grid frequency bounds have a limited range, only a few selected spectral components are needed. After the Goertzel components computation, the Interpolated Discrete Fourier Transform (IpDFT) [8] is applied. IpDFT is a spectral based method that estimates the signal frequency based on the calculation of the signal FFT, selecting the highest amplitude spectral component, then its largest neighbour and interpolating them. With Goertzel and IpDFT is possible to achieve an accurate frequency estimation with a lower computational cost.

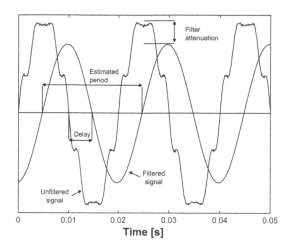

Fig. 1. Example of frequency estimation using digital filtering and zero crossing detection. The input signal is filtered what causes a frequency dependant delay. With a filtered signal is possible to estimate the real zero crossing compensating the delay.

In Fig. 2 5 DFT bins of a sine signal with spectral leakage are shown. Bold lines represents 5 Goertzel outputs G_f that are used to perform a frequency estimation. In this example the higher amplitudes are G_{50} and G_{55} which means that these are the DFT bins to be used in IpDFT algorithm.

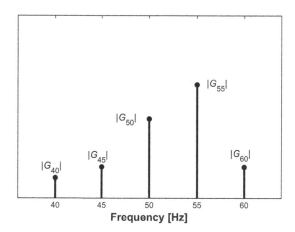

Fig. 2. Example of a signal with spectral leakage where the true frequency is not centered in none of the DFT bins. Interpolation is performed using G_{50} and G_{55} in order to obtain the signal frequency (53 Hz).

2.3   Using FBN to Estimate Power Grid Frequency

FBN [15] exhibit the natural or biological neural systems features that lead to a learning capability when exposed to sets of experiments from the real world. They also have the capability to use newly gained knowledge to perform approximate qualitative reasoning in the form of "if...then" rules. As in natural systems, FBN are robust and immune to individual neuron or connection errors and present good generalization capabilities that automatically minimize the importance of imbalances and sparseness in the training data. FBN use a Hebbian learning process and are capable of learning and implementing any possible multi-input single-output function of the type: $[0,1]^n \times [0,1]$.

FBN consist of neurons that are grouped into areas of concepts (variables). Meshes of weightless connections between antecedent neuron outputs and consequent neuron inputs are used to perform 'if-then' inference between areas. Neurons are binary and the meshes are formed by individual random connections (like in nature). Each neuron comprises m inputs for each antecedent area, and up to $(m+1)^N$ internal unitary memories, where N is the number of antecedents. $(m+1)^N$ corresponds to maximum granularity. When stimulated, the value of each concept is given by the activated/total neurons ratio. For rules with N antecedents and a single consequent, each neuron has $N \times m$ inputs.

Inference proceeds as follows: The single operation carried out by each neuron is the combinatorial count of activated inputs from every antecedent. For all counting combinations, neurons compare the sampled values with the ones in their unitary memory (FF). If the FF that corresponds to the sampled value of all antecedents contains the value "1", then the neuron output is also "1". Otherwise, the neuron output is "0". As a result of the inference process (which is parallel), each neuron assumes a boolean value, and the inference result will be given by the neural activation ratio in the consequent area.

Learning is performed by exposing the net to the data input and by modifying the internal binary memories of each consequent neuron according to the activation of the m inputs (per antecedent) and the state of that consequent neuron. Each experiment will set or reset one binary memory of each individual neuron. Due to its probabilistic nature, the FBN must be repeatedly exposed to the same training data for a minimum number of times (r). The optimization of r is not critical since FBN cannot be overtrained. Thus, it is only necessary to guarantee a minimum value that depends on the net parameters $(m, N,$ granularity) and sparsity of the training data set.

The idea is to use a 1-input/1-output FBN $(N = 1)$ as a filter without causing a delay. We start by feeding the network with a small set of training points (timestamp/amplitude) sampled during one period; The second step consists in letting the FBN interpolate/estimate the amplitude values for the whole period (based on the few training points), and letting it infer the timestamps of the zero crossing points; Finally we use a standard ZC procedure to estimate frequency.

3   Simulated Results

Different events, such as transients, harmonics or noise can occur in electrical power grids. In this section we present how FBNs behave reagarding ZC detection when simulating such effects. All tests were performed training and testing the FBN network with 100 repetitions.

3.1   Cosine

The first test consists in a fundamental cosine with 0.9 Vpp and 50 Hz. On the left side of Fig. 3 a simulated one period signal is presented. 25 evenly spaced points were automatically selected to train the FBN network (as shown on right side of Fig. 3).

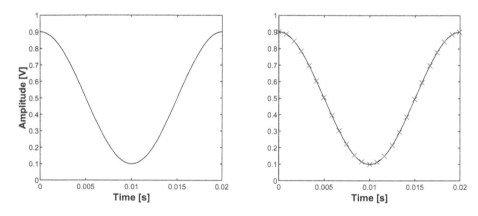

Fig. 3. On the left side a simulated cosine with 0.9 Vpp and 50 Hz is presented. 25 points were used to train the FBN network.

The system was tested with several parameter configurations, and the best results were obtained using areas with 250 neurons, a sample size $m = 60$, and maximum granularity $(m + 1)^N = 61$.

Since the main signal is a cosine, the zero crossings should be detected at 0.005 s and 0.015 s. The FBN indicated zero crossings at 0.0049 s and 0.015 s.

3.2   Sum of Cosines with Same Phase

The fundamental frequency in electrical power grids is 50 Hz, but harmonics can be present. In this simulation two harmonics (3rd and 5th) were added. The 3rd harmonic has an amplitude of 5% of the fundamental signal and the 5th 6% V that are the maximum amplitudes allowed by the IEEE standard. Figure 4 presents the effects of harmonics in the fundamental signal. Since this frequencies are in phase, zero crossing should not be affected.

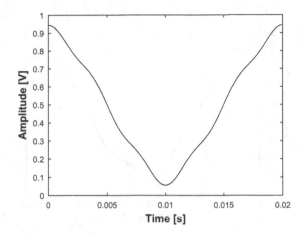

Fig. 4. Representation of a sum of cosines in multiples of fundamental frequency with standard maximum amplitudes.

The zero crossings should still be detected at 0.005 s and 0.015 s. Using the same configuration as in the previous section, the FBN obtained zero crossings at 0.0045 s and 0.0151 s.

3.3   Sum of Cosines with Different Phases

Harmonics out of phase with the fundamental signal can affect the ZC detection. A filter attenuates higher frequencies, so, the ZC detection is performed only on the fundamental one. In this case the FBN network will be trained with points that can not represent the fundamental zero cross. In Fig. 5 is presented a sum of signals with different phases and in Fig. 6 the obtained result.

In this test the first crossing was detected at 0.0047 s and the second one at 0.0150 s.

3.4   Cosine with Transients

Transients are another PQ event that could affect the frequency estimation. If a transient occurs without causing a false zero crossing transition, there should not be any problem with frequency estimation. But, as presented in Fig. 7, if a transient occurs near a zero cross transition this event can lead to a false transition getting a wrong frequency estimation.

As is shown in Fig. 8 the training points, represented as red crosses, were affected by the transients. But the FBN network output, represented as the black line, followed the fundamental frequency behavior. This result shows the capability of ignoring some outliers making possible the ZC detection. In this test the zero crossings were detected at 0.0049 s and 0.0153 s.

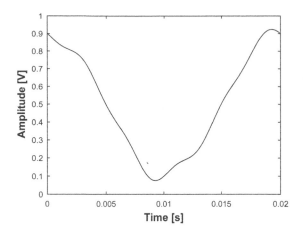

Fig. 5. Representation of a sum of signals with different phases in multiples of fundamental frequency.

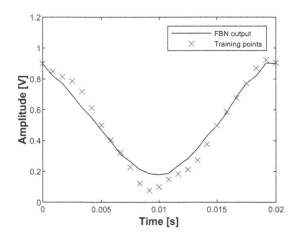

Fig. 6. FBN network output for the situation presented in Fig. 5. FBN output trained with 25 points. Areas of 250 neurons, each neuron performing $m = 60$ samples and maximum granularity. Red crosses represents the training points and the black line represents the FBN network output. (Color figure online)

3.5   Cosine with Noise

Another event that is common in electrical power grid is the presence of noise. As in the previous test, noise can affect zero crossing detection if occurs near a zero cross transition. This test was performed with a noise of 30 dB and the signal to estimate its zero crossing is represented in Fig. 9.

In Fig. 10 the output of FBN netowork is shown. Network training points, represented as red crosses, were not following the fundamental signal as it can be seen in the signal minimum peak. But the output of FBN, represented as the

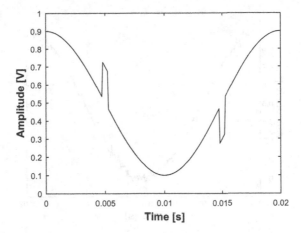

Fig. 7. Representation of transients in a cosine signal on the zero crossings.

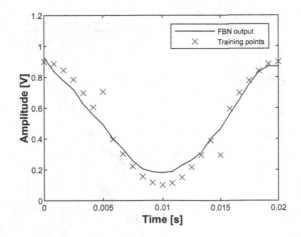

Fig. 8. FBN network output for the situation presented in Fig. 7. FBN output trained with 25 points. Areas of 250 neurons, each neuron performing $m = 60$ samples and maximum granularity. Red crosses represents the training points and the black line represents the FBN network output. (Color figure online)

black line, did not follow those transitions. This results shows that the FBN network can attenuate noisy effects. In this simulation zero crossings were detected at 0.0049 s and 0.0153 s.

4 Frequency Estimation with Power Grid Dataset

Once the best FBN parameters for ZC detection were selected (using the examples presented in the previous section), the system was tested on a real dataset. This dataset was acquired in Instituto Superior Técnico - Taguspark with the

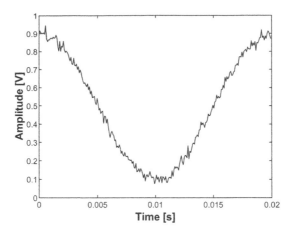

Fig. 9. Representation of a cosine with a measured noise of 30 dB.

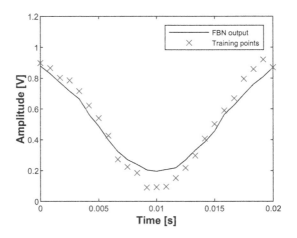

Fig. 10. FBN network output for the situation presented in Fig. 9. FBN output trained with 25 points. Areas of 250 neurons, each neuron performing $m = 60$ samples and maximum granularity. Red crosses represents the training points and the black line represents the FBN network output. (Color figure online)

analog input module NI 9215. Data was acquired with a sampling frequency of 12.5 kHz, which gives a dataset containing 250 points per period (considering a 50 Hz frequency). Figure 11 shows an example of a period of a real power grid signal (that corresponds to around 0.02 s). In this example it is possible to see some fluctuations in the signal peaks.

As stated in section I, ZC is performed every 200 ms which translates into 10 periods if a fundamental signal with 50 Hz is considered. In Fig. 12 the three methods are compared since the beginning of the process until t = 2 s, the equivalent of 10 frequency estimations. "Goertzel with interpolation" starts and ends

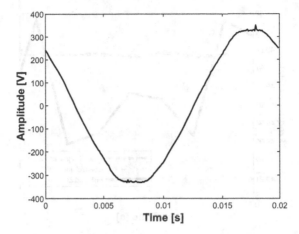

Fig. 11. A period of a real power grid signal acquired with a 12.5 kHz sampling frequency during 0.02 s. In this example some transients occurred near the peaks.

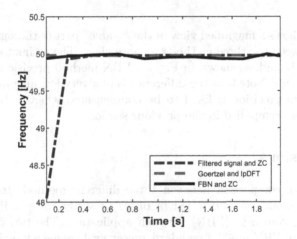

Fig. 12. Frequency estimation along 2 s in all methods. Digital Filter method starts with a poor estimation due to the time that digital filter takes to establish but then gets a closer estimation to the Goertzel method. FBN with a maximum error of around 0.05 Hz

without visible fluctuations in this scale. The delay in digital filtering method was already compensated in order to have a real timestamp but other effects are shown. Filters need some time before the output is stable, and that is the reason why the filtering method starts with a bad frequency estimation. The FBN method provides good results, but not as precise as any of the other methods. In this segment, the maximum estimation error, when compared with standard and Goertzel methods, was around 0.05 Hz.

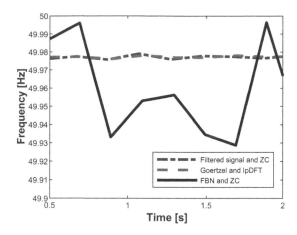

Fig. 13. Zoomed in view: comparison of the three methods during the stable part of the signal.

Figure 13 shows a magnified view of the "stable" part of the signal, where the difference between the three methods is more clear. The estimated frequency is around 49.98 Hz and, as shown in Fig. 12, FBN method precision is lower than the other methods. Note that the difference is of a very low magnitude, and that this decrease in precision is liked to be compensated when in the presence of events as those exemplified in the previous section.

5   Conclusions

In this paper a comparison between three different methods to estimate the power grid frequency is performed in order to study the possibility of using Fuzzy Boolean Networks (FBN) for such application. The two other methods used to validate FBN are the standard power grid frequency estimation which corresponds to counting the number of periods within a given time frame after filtering the input signal, and Goertzel with interpolation.

The main goal of this comparison was to understand the advantages and limitations of using FBN in electrical power grid frequency estimation.

A compromise between time and accuracy was done and better results were obtained training and testing the network 100 times, using 25 training points, an area size of 250 neurons, $m = 60$ samples per neuron, and maximum granularity. Comparing with other methods FBN results were marginally worse, but FBN can ignore outliers caused by noise or transients, as shown in Fig. 8 and Fig. 10. In addition, since the network is trained with only 25 points this method could be useful when a low number of measuring points is available. One unavoidable limitation of using FBN is the training and testing time. This makes them unsuitable for real-time applications unless using dedicated hardware.

References

1. Arrillaga, J., Watson, N.R., Chen, S.: Power System Quality Assessment, 1st edn. Wiley, Hoboken (2000)
2. Arrillaga, J., Smith, B.C., Watson, N.R., Wood, A.R.: Power System Harmonic Analysis. Wiley, Hoboken (1997). ISBN: 978-0-471-97548-9
3. Fuchs, E.F., Masoum, M.A.S.: Power Quality in Power Systems and Electrical Machines, 1st edn, February 2008. ISBN: 978-0-12-369536-9
4. Guerrero, J.M.: Guest editorial special issue on power quality in smart grids. IEEE Trans. Smart Grid 8(1), 379–381 (2017). https://doi.org/10.1109/TSG.2016.2634482
5. Montoya, F.G., García-Cruz, A., Montoya, M.G., Agugliaro, F.M.M.: Power quality techniques research worldwide: a review. Renew. Sustain. Energy Rev. 54, 846–856 (2016). https://doi.org/10.1016/j.rser.2015.10.091
6. Ramos, P.M., Radil, T., Janeiro, F.M., Serra, A.C.: DSP based power quality analyzer using new signal processing algorithms for detection and classification of disturbances in a single-phase power system. In: IMEKO TC4, Iasi, Romania, pp. 433–438 (2007)
7. Ramos, P.M., Serra, A.C.: Comparison of frequency estimation algorithms for power quality assessment. Measurement 42(9), 1312–1317 (2009). https://doi.org/10.1016/j.measurement.2008.04.013
8. Schoukens, J., Pintelon, R., Van Hamme, H.: The interpolated fast Fourier transform: a comparative study. IEEE Trans. Instr. Measur. 41, 226–232 (1992). https://doi.org/10.1109/19.137352
9. Gong, C., Guo, D., Zhang, B., Aijun, L.: Improved frequency estimation by interpolated DFT method, 29, 4112–4116 (2012). https://doi.org/10.1016/j.proeng.2012.01.629
10. Chen, W., Dehner, L.G.: A low-cost and high performance solution to frequency estimation for GSM/EDGE. In: Texas Wireless Symposium, pp. 6–10 (2005). http://citeseerx.ist.psu.edu/viewdoc/download?doi=10.1.1.128.7776&rep=rep1&type=pdf
11. Goertzel, G.: An algorithm for the evaluation of finite trigonometric series. Am. Math. Mon. 65, 34–35 (1958). https://doi.org/10.2307/2310304
12. Franz, S., Mitra, S., Doblinger, G.: Frequency estimation using warped discrete Fourier transform, 83, 1661–1671 (2003). https://doi.org/10.1016/S0165-1684(03)00079-3
13. IEC 61000-4-30, Electromagnetic compatibility (EMC) - Part 4–30: Testing and measurement techniques - Power quality measurement methods, 3.0 Edition, February 2015
14. Rodrigues, N.M., Janeiro, F.M., Ramos, P.M.: Digital filter performance for zero crossing detection in power quality embedded measurement systems. In: 2018 IEEE International Instrumentation and Measurement Technology Conference (I2MTC), pp. 1–6, May 2018. https://doi.org/10.1109/I2MTC.2018.8409701
15. Tomé, J., Carvalho, J.P.: Fuzzy Boolean nets - a nature inspired model for learning and reasoning. Fuzzy Sets Syst. 253, 1–27 (2014). https://doi.org/10.1016/j.fss.2014.04.020
16. Castro, A.M., et al.: An integrated recycling approach for GFRP pultrusion wastes: recycling and reuse assessment into new composite materials using Fuzzy Boolean nets. J. Clean. Prod. 66, 420–430 (2014). https://doi.org/10.1016/j.jclepro.2013.10.030
17. Esteves, S., Silva, J.N., Carvalho, J.P., Veiga, L.: Incremental dataflow execution, resource efficiency and probabilistic guarantees with Fuzzy Boolean nets. J. Parallel Distribu. Comput. 79–80, 52–66 (2015). https://doi.org/10.1016/j.jpdc.2015.03.001

Fuzzy Clustering Stability Evaluation of Time Series

Gerhard Klassen$^{(\boxtimes)}$ ⓘ, Martha Tatusch ⓘ, Ludmila Himmelspach ⓘ,
and Stefan Conrad ⓘ

Heinrich Heine University, Universitätsstr. 1, 40225 Düsseldorf, Germany
{gerhard.klassen,martha.tatusch,ludmila.himmelspach,stefan.conrad}@hhu.de

Abstract. The discovery of knowledge by analyzing time series is an important field of research. In this paper we investigate multiple multivariate time series, because we assume a higher information value than regarding only one time series at a time. There are several approaches which make use of the granger causality or the cross correlation in order to analyze the influence of time series on each other. In this paper we extend the idea of mutual influence and present FCSETS (**F**uzzy **C**lustering **S**tability **E**valuation of **T**ime **S**eries), a new approach which makes use of the membership degree produced by the fuzzy c-means (FCM) algorithm. We first cluster time series per timestamp and then compare the relative assignment agreement (introduced by Eyke Hüllermeier and Maria Rifqi) of all subsequences. This leads us to a stability score for every time series which itself can be used to evaluate single time series in the data set. It is then used to rate the stability of the entire clustering. The stability score of a time series is higher the more the time series sticks to its peers over time. This not only reveals a new idea of mutual time series impact but also enables the identification of an optimal amount of clusters per timestamp. We applied our model on different data, such as financial, country related economy and generated data, and present the results.

Keywords: Time series analysis · Fuzzy clustering · Evaluation

1 Introduction

The analysis of sequential data – so called time series (TS) – is an important field of data mining and already well researched. There are many different tasks, but the identification of similarities and outliers are probably among the most important ones. Clustering algorithms try to solve exactly these problems. There are various approaches for extracting information from time series data with the help of clustering. While some methods deal with parts of time series, so called subsequences [2], others consider the whole sequence at once [9,28], or transform them to feature sets first [17,34]. In some applications clusters may overlap, so that membership grades are needed, which enable data points to belong to more

© Springer Nature Switzerland AG 2020
M.-J. Lesot et al. (Eds.): IPMU 2020, CCIS 1237, pp. 680–692, 2020.
https://doi.org/10.1007/978-3-030-50146-4_50

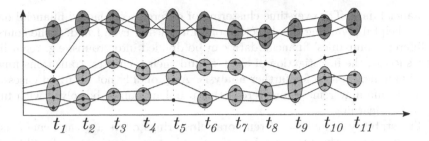

Fig. 1. Example for an over-time clustering of univariate time series [32]. The blue clusters are more stable over time than the red ones.

than one cluster to different degrees. These methods fall into the field of *fuzzy clustering* and they are used in time series analysis as well [24].

However, in some cases the exact course of time series is not relevant but rather the detection of groups of time series that follow the same trend. Additionally, time-dependent information can be meaningful for the identification of patterns or anomalies. For this purpose it is necessary to cluster the time series data per time point, as the comparison of whole (sub-)sequences at once leads to a loss of information. For example, in case of the euclidean distance the mean distance over all time points is considered. In case of Dynamic Time Warping (DTW) the smallest distance is relevant. The information at one timestamp has therefore barely an impact. The approach of clustering time series per time point enables an advanced analysis of their temporal correlation, since the behavior of sequences to their cluster peers can be examined. In the following this procedure will be called *over-time clustering*. An example is shown in Fig. 1. Note, that for simplicity reasons only univariate time series are illustrated. However, over-time clustering is especially valuable for multivariate time series analysis.

Unfortunately new problems like the right choice of parameters arise. Often the comparison of clusterings with different parameter settings is difficult since there is no evaluation function which distinguishes the quality of clusterings properly. In addition, some methods, such as outlier detection, require good clustering as a basis, whereby the quality can contextually be equated with the stability of the clusters.

In this paper, we focus on multiple multivariate time series with same length and equivalent time steps. We introduce an evaluation measure named FCSETS (**F**uzzy **C**lustering **S**tability **E**valuation of **T**ime **S**eries) for the over-time stability of a fuzzy clustering per time point. For this purpose our approach rates the over-time stability of all sequences considering their cluster memberships. To the best of our knowledge this is the first approach that enables the stability evaluation of clusterings and sequences regarding the temporal linkage of clusters.

Over-time clustering can be helpful in many applications. For example, the development of relationships between different terms can be examined when tracking topics in online forums. Another application example is the analysis of financial data. The over-time clustering of different companies' financial data can be helpful regarding the detection of anomalies or even fraud. If the courses of different companies' financial data can be divided into groups, e.g. regarding their success, the investigation of clusters and their members' transitions might be a fundamental step for further analysis. As probably not all fraud cases are known (some may remain uncovered) this problem cannot be solved with fully supervised learning.

The stability evaluation of temporal clusterings offers a great benefit as it not only enables the identification of suitable hyper-parameters for different algorithms but also ensures a reliable clustering as a basis for further analysis.

2   Related Work

In the field of time series analysis, different techniques for clustering time series data were proposed. However, to the best of our knowledge, there does not exist any approach similar to ours. The approaches described in [8,19,28] cluster entire sequences of multiple time series. This procedure is not well suited for our context because potential correlations between subsequences of different time series are not revealed. Additionally, the exact course of the time series is not relevant, but rather the trend they show. The problem of not recognizing interrelated subsequences also persists in a popular method where the entire sequences are first transformed to feature vectors and then clustered [17]. Methods for clustering streaming data like the ones proposed in [14] and [25] are not comparable to our method because they consider only one time series at a time and deal with other problems such as high memory requirements and time complexity. Another area related to our work is community detection in dynamic networks. While approaches presented in [12,13,26,36] aim to detect and track local communities in graphs over time, the goal of our method is finding a stable partitioning of time series over the entire period so that time series following the same trend are assigned to the same cluster.

In this section, first we briefly describe the fuzzy c-means clustering algorithm that we use for clustering time series objects at different time points. Then, we refer on the one hand to related work with regard to time-independent evaluation measures for clusterings. Finally, we describe a resampling approach for cluster validation and a fuzzy variant of the Rand index that we use in our method.

2.1   Fuzzy C-Means (FCM)

Fuzzy c-means (FCM) [4,7] is a partitioning clustering algorithm that is considered as a fuzzy generalization of the hard k-means algorithm [22,23]. FCM partitions an unlabeled data set $X = \{x_1, ..., x_n\}$ into c clusters represented by their prototypes $V = \{v_1, ..., v_c\}$. Unlike k-means that assigns each data point to exactly one cluster, FCM assigns data points to clusters with membership degrees $u_{ik} \in [0,1]$, $1 \leq i \leq c$, $1 \leq k \leq n$. FCM is a probabilistic clustering algorithm which means that its partition matrix $U = [u_{ik}]$ must satisfy two conditions given in (1).

$$\sum_{i=1}^{c} u_{ik} = 1 \quad \forall k \in \{1, ..., n\},$$

$$\sum_{k=1}^{n} u_{ik} > 0 \quad \forall i \in \{1, ..., c\}. \tag{1}$$

Since we focus on partition matrices produced by arbitrary fuzzy clustering algorithms, we skip further details of FCM and refer to the literature [4].

2.2 Internal Evaluation Measures

Many different *external* and *internal* evaluation measures for evaluating clusters and clusterings were proposed in the literature. In the case of the external evaluation, the clustering results are compared with a ground truth which is already known. In the internal evaluation, no information about the actual partitioning of the data set is known, so that the clusters are often evaluated primarily on the basis of characteristics such as compactness and separation.

One metric that evaluates the compactness of clusters is the *Sum of Squared Errors*. It calculates the overall distance between the data points and the cluster prototype. In the case of fuzzy clustering, these distances are additionally weighted by the membership degrees. The better the data objects are assigned to clusters, the smaller the error, the greater the compactness. However, this measure does not explicitly take the separation of different clusters into account.

There are dozens of fuzzy cluster validity indices that evaluate the compactness as well as the separation of different clusters in the partitioning. Some validity measures use only membership degrees [20,21], other include the distances between the data points and cluster prototypes [3,5,11,35]. All these measures cannot be directly compared to our method because they lack a temporal aspect. However, they can be applied in FCSETS for producing an initial partitioning of a data set for different time points.

2.3 Stability Evaluation

The idea of the resampling approach for cluster validation described in [30] is that the choice of parameters for a clustering algorithm is optimal when different partitionings produced for these parameter settings are most similar to each other. The *unsupervised cluster stability value* $s(c)$, $c_{min} \leq c \leq c_{max}$, that is used in this approach is calculated as average pairwise distance between m partitionings:

$$s(c) = \frac{\sum_{i=1}^{m-1} \sum_{j=i+1}^{m} d(U_{ci}, U_{cj})}{m \cdot (m-1)/2}, \tag{2}$$

where U_{ci} and U_{cj}, $1 \leq i < j \leq m$, are two partitionings produced for c clusters and $d(U_{ci}, U_{cj})$ is an appropriate similarity index of partitionings. Our stability measure is similar to the unsupervised cluster stability value but it includes the temporal dependencies of clusterings.

Since we deal with fuzzy partitionings, in our approach we use a modified version of the *Hüllermeier-Rifqi Index* [18]. There are other similarity indices for comparing fuzzy partitions like *Campello's Fuzzy Rand Index* [6] or *Frigui Fuzzy Rand Index* [10] but they are not reflexive.

The *Hüllermeier-Rifqi Index (HRI)* is based on the *Rand Index* [29] that measures the similarity between two hard partitions. The Rand index between two hard partitions $U_{c \times n}$ and $\tilde{U}_{\tilde{c} \times n}$ of a data set X is calculated as the ratio of all concordant pairs of data points to all pairs of data points in X. A data pair (x_k, x_j), $1 \leq k, j \leq n$ is concordant if either the data points x_k and x_j are assigned to the same cluster in both partitions U and $\tilde{U}$, or they are in different clusters in U and $\tilde{U}$. Since fuzzy partitions allow a partial assignment of data points to clusters, in [18], the authors proposed an equivalence relation $E_U(x_k, x_j)$ on X for the calculation of the assignment agreement of two data points to clusters in a partition:

$$E_U(x_k, x_j) = 1 - \frac{1}{2} \sum_{i=1}^{c} |u_{ik} - u_{ij}|. \tag{3}$$

Using the equivalence relation $E_U(x_k, x_j)$ given in Formula (3), the Hüllermeier-Rifqi index is defined as a normalized degree of concordance between two partitions U and $\tilde{U}$:

$$\mathrm{HRI}(U, \tilde{U}) = 1 - \frac{1}{n(n-1)} \sum_{k=1}^{n} \sum_{j=k+1}^{n} |E_U(x_k, x_j) - E_{\tilde{U}}(x_k, x_j)|. \tag{4}$$

In [31], Runkler has proposed the *Subset Similarity Index (SSI)* which is more efficient than the Hüllermeier-Rifqi Index. The efficiency gain of the Subset Similarity Index is achieved by calculating the similarity between cluster pairs instead of the assignment agreement of data point pairs. We do not use it in our approach because we evaluate the stability of a clustering over time regarding the *team spirit* of time series. Therefore, in our opinion, the degree of the assignment agreement between time series pairs to clusters at different time stamps contributes more to the stability score of a clustering than the similarity between cluster pairs.

3   Fundamentals

In this chapter we clarify our understanding of some basic concepts regarding our approach. For this purpose we supplement the definitions from [32]. Our method considers multivariate time series, so instead of a definition with real values we use the following definition.

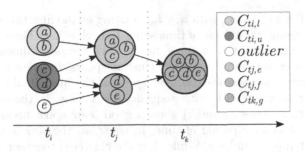

Fig. 2. Illustration of transitions of time series $T_a, .., T_e$ between clusters over time [32].

Definition 1 (Time Series). *A time series $T - o_{t_1}, ..., o_{t_n}$ is an ordered set of n real valued data points of arbitrary dimension. The data points are chronologically ordered by their time of recording, with t_1 and t_n indicating the first and last timestamp, respectively.*

Definition 2 (Data Set). *A data set $D = T_1, ..., T_m$ is a set of m time series of same length n and equal points in time.*

The vectors of all time series are denoted as the set $O = \{o_{t_1,1}, ..., o_{t_n,m}\}$. With the second index indicating the time series the data point originates from. We write O_{t_i} for all data points at a certain point in time.

Definition 3 (Cluster). *A cluster $C_{t_i,j} \subseteq O_{t_i}$ at time t_i, with $j \in \{1, ..., k_{t_i}\}$ with k_{t_i} being the number of clusters at time t_i, is a set of similar data points, identified by a cluster algorithm.*

Definition 4 (Fuzzy Cluster Membership). *The membership degree $u_{C_{t_i,j}}(o_{t_i,l}) \in [0,1]$ expresses the relative degree of belonging of the data object $o_{t_i,l}$ of time series T_l to cluster $C_{t_i,j}$ at time t_i.*

Definition 5 (Fuzzy Time Clustering). *A fuzzy time clustering is the result of a fuzzy clustering algorithm at one timestamp. In concrete it is the membership matrix $U_{t_i} = [u_{C_{t_i,j}}(o_{t_i,l})]$.*

Definition 6 (Fuzzy Clustering). *A fuzzy clustering of time series is the overall result of a fuzzy clustering algorithm for all timestamps. In concrete it is the ordered set $\zeta = U_{t_1}, ..., U_{t_n}$ of all membership matrices.*

4 Method

An obvious disadvantage of creating clusters for every timestamp is the missing temporal link. In our approach we assume that clusterings with different parameter settings show differences in the connectedness of clusters and that this connection can be measured. In order to do so, we make use of a stability function. Given a fuzzy clustering ζ, we first analyze the behavior of every subsequence of a time series $T = o_{t_1}, ..., o_{t_i}$, with $t_i \leq t_n$, starting at the first timestamp. In this way we rate a temporal linkage of time series to each other. Time series that are clustered together at all time stamps, have a high temporal linkage, while time series which often separate from their clusters' peers, indicate a low temporal linkage. One could say we rate the *team spirit* of the individual time series and therefore their cohesion with other sequences over time. In the example shown in Fig. 2, the time series T_a and T_b show a good team spirit because they move together over the entire period of time. In contrast, the time series T_c and T_d show a lower temporal linkage. While they are clustered together at time points t_i and t_k, they are assigned to different clusters in between at time point t_j. After the evaluation of the individual sequences, we assign a score to the fuzzy clustering ζ, depending on the over-time stability of every time series.

Let U_{t_i} be a fuzzy partitioning of the data objects O_{t_i} of all times series in k_{t_i} clusters at time t_i. Similar to the equivalence relation in Hüllermeier-Rifqi Index, we compute the relative assignment agreement of the data objects $o_{t_i,l}$ and $o_{t_i,s}$ of two time series T_l and T_s, $1 \leq l, s \leq m$ to all clusters in partitioning U_{t_i} at time t_i as follows

$$E_{U_{t_i}}(o_{t_i,l}, o_{t_i,s}) = 1 - \frac{1}{2} \sum_{j=1}^{k_{t_i}} |u_{C_{t_i,j}}(o_{t_i,l}) - u_{C_{t_i,j}}(o_{t_i,s})|. \tag{5}$$

Having the relative assignment agreement of time series at timestamps t_i and t_r, $t_1 \leq t_i < t_r \leq t_n$, we calculate the difference between the relative assignment agreements of time series T_l and T_s by subtracting the relative assignment agreement values:

$$D_{t_i,t_r}(T_l, T_s) = |E_{U_{t_i}}(o_{t_i,l}, o_{t_i,s}) - E_{U_{t_r}}(o_{t_r,l}, o_{t_r,s})|. \tag{6}$$

We calculate the stability of a time series T_l, $1 \leq l \leq m$, over all timestamps as an averaged weighted difference between the relative assignment agreements to all other time series as follows:

$$stability(T_l) = 1 - \frac{2}{n(n-1)} \sum_{i=1}^{n-1} \sum_{r=i+1}^{n} \frac{\sum_{s=1}^{m} E_{U_{t_i}}(o_{t_i,l}, o_{t_i,s})^m D_{t_i,t_r}(T_l, T_s)^2}{\sum_{s=1}^{m} E_{U_{t_i}}(o_{t_i,l}, o_{t_i,s})^m}. \tag{7}$$

In Formula (7) we weight the difference between the assignment agreements $D_{t_i,t_r}(T_l, T_s)$ by the assignment agreement between pairs of time series at the earlier time point because we want to damp the large differences for stable time series caused by supervention of new peers. On the other hand we aim to penalize the time series that leave their cluster peers while changing cluster membership at a later time point.

Finally, we rate the over-time stability of a clustering ζ as the averaged stability of all time series in the data set:

$$FCSETS(\zeta) = \frac{1}{m} \sum_{l=1}^{m} stability(T_l). \tag{8}$$

As we already stated, the over-time stability of the entire clustering depends on the stability of all time series regarding staying together in a cluster with times series, that follow the same trend.

5   Experiments

In the following, we present the results on an artificially generated data set, that demonstrates a meaningful usage of our measure and shows the impact of the stability evaluation. Additionally, we discuss experiments on two real world data sets. One consists of financial figures from balance sheets and the other one contains country related economy data. In all cases fuzzy c-means was used with different parameter combinations for the number of clusters per time point.

5.1   Artificially Generated Data Set

In order to show the effects of a rating based on our stability measure, we generated an artificial data set with time series that move between two separated groups. Therefore, at first, three random centroids with two features $\in [0,1]$ were placed for time point 1. These centroids were randomly shifted for the next timestamps whereby the maximal distance of a centroid at two consecutive time points could not exceed 0.05 per dimension. Afterwards 3, 4 and 5 time series were assigned to these centroids, respectively. This means that the data points of a time series for each time point were placed next to the assigned centroid with a maximal distance of 0.1 per feature. Subsequently, sequences with random transitions between two of the three clusters were inserted. Therefore 3 time series (namely 1, 2 and 3) were generated, that were randomly assigned to one of the two clusters at every time point. All together, a total of 4 time points and 15 time series were examined.

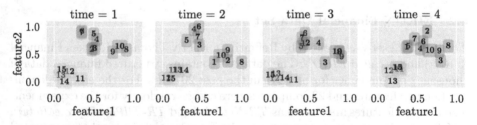

Fig. 3. Result of the most stable clustering on the artificially generated data set. (Color figure online)

To find the best stability score for the data set, FCM was used with various settings for the number of clusters per time point. All combinations with $k_{t_i} \in [2,5]$ were investigated. Figure 3 shows the resulting fuzzy clustering with the highest FCSETS score of 0.995. For illustration reasons the clustering was

Table 1. Stability scores for the generated data set depending on k_{t_i}.

k_{t_1}	k_{t_2}	k_{t_3}	k_{t_4}	FCSETS score
2	2	2	2	**0.995**
2	3	2	2	0.951
2	3	3	2	0.876
2	3	3	3	0.829
3	3	2	2	0.967
3	3	3	3	0.9
2	3	4	5	0.71
5	3	4	2	0.908
3	10	3	10	0.577

defuzzyfied. Although it might seem intuitive to use a partitioning with three clusters at time points 1 and 2, regarding the over-time stability it is beneficial to choose only two clusters. This can be explained by the fact that there are time series that move between the two apparent groups of the upper (blue) cluster. The stability is therefore higher when these two groups are clustered together.

In Table 1 a part of the corresponding scores for the different parameter settings of k_{t_i} are listed. As shown in Fig. 3, the best score is achieved with k_{t_i} being set to 2 for all time points. The worst score results with the setting $k_{t_1} = 2$, $k_{t_2} = 3$, $k_{t_3} = 4$ and $k_{t_4} = 5$. The score is not only decreased because the upper (blue) cluster is divided in this case, but also because the number of clusters varies and therefore sequences get separated from their peers. It is obvious that the stability score is negatively affected, if the number of clusters significantly changes over time. This influence is also expressed by the score of 0.577 for the extreme example in the last row.

5.2   EIKON Financial Data Set

The first data set was released by Refinitiv (formerly Thomson Reuters Financial & Risk) and is called *EIKON*. The database contains structured financial data of thousands of companies for more than the past 20 years. For the ease of demonstration two features and 23 companies were chosen randomly for the experiment. The selected features are named as *TR-NetSales* and *TR-TtlPlanExpectedReturn* by Thomson Reuters and correspond to the net sales and the total plan expected return, which are figures taken from the balance sheet of the companies. Since it is a common procedure in economics, we divided the features by the company's total assets and normalized them afterwards with a min-max-normalization.

We generated the clusterings for all combinations of k_{t_i} from two to five clusters per timestamp. Selected results can be seen in Table 2. The actual maximum retrieved from the iterations (in the third row) is printed bold. The worst score can be found in the last row and represents an unstable clustering. It can be seen

Table 2. Stability scores for the EIKON financial data set depending on k_{t_i}.

k_{t_1}	k_{t_2}	k_{t_3}	k_{t_4}	k_{t_5}	k_{t_6}	k_{t_7}	k_{t_8}	FCSETS score
2	2	2	2	2	2	2	2	0.929
3	3	3	3	3	3	3	3	0.9
3	2	2	2	2	2	2	2	**0.945**
5	4	3	2	2	2	2	2	0.924
2	2	4	3	2	4	5	5	0.72

that the underlying data is well separated into three clusters in the first point in time and into two clusters at the following timestamps. This is actually a rare case but can be explained with the selection of features and companies. Actually *TR-TtlPlanExpectedReturn* is rarely provided by Thomson Reuters and the fact that we only chose companies which got complete data for all regarded points in time. This may have diminished the number of companies which might have lower membership degrees.

5.3 GlobalEconomy Data Set

The next data set originates from www.theglobaleconomy.com [1], which is a website that provides economic data of the past years for different countries. Again, two features were selected randomly for this experiment and were normalized with a min-max-normalization. Namely the features are the "Unemployment Rate" and the "Public spending on education, percent of GDP". For illustration reasons, we considered only a part of the countries (28) for the years from 2010 to 2017.

Table 3. Stability scores for the GlobalEconomy data set depending on k_{t_i}.

k_{t_1}	k_{t_2}	k_{t_3}	k_{t_4}	k_{t_5}	k_{t_6}	k_{t_7}	k_{t_8}	FCSETS score
2	2	2	2	2	2	2	2	**0.978**
3	3	3	3	3	3	3	3	0.963
3	2	2	2	2	2	2	2	0.945
5	3	4	2	2	2	2	2	0.955
2	3	2	2	4	5	5	5	0.837

The results are shown in Table 3. It can be seen that the best score is achieved with two clusters at every point in time. Evidently the chosen countries can be well separated into two groups at every point in time. More clusters or different numbers of clusters for different timestamps performed worse. In this experiment we also iterated over all combinations of k_{t_i} for the given points in time. The bold printed maximum, and the minimum, which can be found in the last row of the table, represent the actual maximum and minimum within the range of the iterated combinations.

6 Conclusion and Future Work

In this paper we presented a new method for analyzing multiple multivariate time series with the help of fuzzy clustering per timestamp. Our approach defines a new target function for sequence-based clustering tasks, namely the stability of sequences. In our experiments we have shown that this enables the identification of optimal k_{t_i}s per timestamp and that our measure can not only rate time series and clusterings but also can be used to evaluate the stability of data sets. The latter is possible by examining the maximum achieved $FCSETS$ score. Our approach can be applied whenever similar behavior for groups of time series can be assumed. As it is based on membership degrees, clusterings with overlapping clusters and soft transitions can be handled. With the help of our evaluation measure a stable over-time clustering can be achieved, which can be used for further analysis such as outlier detection.

Future work could include the development of a fuzzy clustering algorithm which is based on our formulated target function. The temporal linkage could therefore already be taken into account when determining groups of time series. Another interesting field of research could be the examination of other fuzzy clustering algorithms like the Possibilistic Fuzzy c-Means algorithm [27]. This algorithm can also handle outliers which can be handy for certain data sets. In the experiment with the GlobalEconomy data set we faced the problem, that one outlier would form a cluster on its own in every point in time. This led to very high FCSETS scores. The handling of outliers could overcome such misbehavior. Future work should also include the application of our approach to incomplete data, since appropriate fuzzy clustering approaches already exist [15,16,33]. We have faced this problem when applying our algorithm to the EIKON financial data set. Also, the identification of time series that show a good team spirit for a specific time period could be useful in some applications and might therefore be investigated. Finally, the examination and optimization of FCSETS' computational complexity would be of great interest as it currently seems to be fairly high.

Acknowledgement. We thank the Jürgen Manchot Foundation, which supported this work by funding the AI research group *Decision-making with the help of Artificial Intelligence* at Heinrich Heine University Düsseldorf.

References

1. Global economy, world economy. https://www.theglobaleconomy.com/
2. Banerjee, A., Ghosh, J.: Clickstream clustering using weighted longest common subsequences. In: Proceedings of the Web Mining Workshop at the 1st SIAM Conference on Data Mining, pp. 33–40 (2001)

3. Beringer, J., Hüllermeier, E.: Adaptive optimization of the number of clusters in fuzzy clustering. In: Proceedings of the IEEE International Conference on Fuzzy Systems, pp. 1–6 (2007)
4. Bezdek, J.C.: Pattern Recognition with Fuzzy Objective Function Algorithms. Kluwer Academic Publishers, Norwell (1981)
5. Bouguessa, M., Wang, S., Sun, H.: An objective approach to cluster validation. Pattern Recogn. Lett. **27**, 1419–1430 (2006)
6. Campello, R.: A fuzzy extension of the rand index and other related indexes for clustering and classification assessment. Pattern Recogn. Lett. **28**(7), 833–841 (2007)
7. Dunn, J.C.: A fuzzy relative of the isodata process and its use in detecting compact well-separated clusters. J. Cybern. **3**(3), 32–57 (1973)
8. Ernst, J., Nau, G.J., Bar-Joseph, Z.: Clustering short time series gene expression data. Bioinformatics **21**(suppl-1), i159–i168 (2005)
9. Ferreira, L.N., Zhao, L.: Time series clustering via community detection in networks. Inf. Sci. **326**, 227–242 (2016)
10. Frigui, H., Hwang, C., Rhee, F.C.H.: Clustering and aggregation of relational data with applications to image database categorization. Pattern Recogn. **40**(11), 3053–3068 (2007)
11. Fukuyama, Y., Sugeno, M.: A new method of choosing the number of clusters for the fuzzy c-mean method. In: Proceedings of the 5th Fuzzy Systems Symposium, pp. 247–250 (1989)
12. Granell, C., Darst, R., Arenas, A., Fortunato, S., Gomez, S.: Benchmark model to assess community structure in evolving networks. Phys. Rev. E **92**, 012805 (2015)
13. Greene, D., Doyle, D., Cunningham, P.: Tracking the evolution of communities in dynamic social networks. In: Proceedings - 2010 International Conference on Advances in Social Network Analysis and Mining, ASONAM 2010, vol. 2010, pp. 176–183 (2010)
14. Guha, S., Meyerson, A., Mishra, N., Motwani, R., O'Callaghan, L.: Clustering data streams: theory and practice. IEEE Trans. Knowl. Data Eng. **15**(3), 515–528 (2003)
15. Hathaway, R., Bezdek, J.: Fuzzy c-means clustering of incomplete data. IEEE Trans. Syst. Man Cybern. Part B (Cybern.) **31**, 735–44 (2001)
16. Himmelspach, L., Conrad, S.: Fuzzy c-means clustering of incomplete data using dimension-wise fuzzy variances of clusters. In: Carvalho, J.P., Lesot, M.-J., Kaymak, U., Vieira, S., Bouchon-Meunier, B., Yager, R.R. (eds.) IPMU 2016. CCIS, vol. 610, pp. 699–710. Springer, Cham (2016). https://doi.org/10.1007/978-3-319-40596-4_58
17. Huang, X., Ye, Y., Xiong, L., Lau, R.Y., Jiang, N., Wang, S.: Time series k-means: a new k-means type smooth subspace clustering for time series data. Inf. Sci. **367–368**, 1–13 (2016)
18. Hüllermeier, E., Rifqi, M.: A fuzzy variant of the rand index for comparing clustering structures. In: Proceedings of the Joint 2009 International Fuzzy Systems Association World Congress and 2009 European Society of Fuzzy Logic and Technology Conference, pp. 1294–1298 (2009)
19. Izakian, H., Pedrycz, W., Jamal, I.: Fuzzy clustering of time series data using dynamic time warping distance. Eng. Appl. Artif. Intell. **39**, 235–244 (2015)
20. Kim, Y.I., Kim, D.W., Lee, D., Lee, K.: A cluster validation index for GK cluster analysis based on relative degree of sharing. Inf. Sci. **168**, 225–242 (2004)

21. Le Capitaine, H., Frelicot, C.: A cluster-validity index combining an overlap measure and a separation measure based on fuzzy-aggregation operators. IEEE Trans. Fuzzy Syst. **19**, 580–588 (2011)
22. Lloyd, S.: Least squares quantization in PCM. IEEE Trans. Inf. Theory **28**(2), 129–137 (1982)
23. MacQueen, J.: Some methods for classification and analysis of multivariate observations. In: Proceedings of the Fifth Berkeley Symposium on Mathematical Statistics and Probability, vol. 1, pp. 281–297. University of California Press (1967)
24. Möller-Levet, C.S., Klawonn, F., Cho, K.-H., Wolkenhauer, O.: Fuzzy clustering of short time-series and unevenly distributed sampling points. In: R. Berthold, M., Lenz, H.-J., Bradley, E., Kruse, R., Borgelt, C. (eds.) IDA 2003. LNCS, vol. 2810, pp. 330–340. Springer, Heidelberg (2003). https://doi.org/10.1007/978-3-540-45231-7_31
25. O'Callaghan, L., Mishra, N., Meyerson, A., Guha, S., Motwani, R.: Streaming-data algorithms for high-quality clustering. In: Proceedings of IEEE International Conference on Data Engineering, p. 685 (2001)
26. Orlinski, M., Filer, N.: The rise and fall of spatio-temporal clusters in mobile ad hoc networks. Ad Hoc Netw. **11**(5), 1641–1654 (2013)
27. Pal, N., Pal, K., Keller, J., Bezdek, J.: A possibilistic fuzzy c-means clustering algorithm. IEEE Trans. Fuzzy Syst. **13**, 517–530 (2005)
28. Paparrizos, J., Gravano, L.: k-shape: efficient and accurate clustering of time series. In: Proceedings of the 2015 ACM SIGMOD International Conference on Management of Data, SIGMOD 2015, pp. 1855–1870. ACM, New York (2015)
29. Rand, W.M.: Objective criteria for the evaluation of clustering methods. J. Am. Stat. Assoc. **66**(336), 846–850 (1971)
30. Roth, V., Lange, T., Braun, M., Buhmann, J.: A resampling approach to cluster validation. In: Härdle, W., Rönz, B. (eds.) COMPSTAT, pp. 123–128. Springer, Heidelberg (2002). https://doi.org/10.1007/978-3-642-57489-4_13
31. Runkler, T.A.: Comparing partitions by subset similarities. In: Hüllermeier, E., Kruse, R., Hoffmann, F. (eds.) IPMU 2010. LNCS (LNAI), vol. 6178, pp. 29–38. Springer, Heidelberg (2010). https://doi.org/10.1007/978-3-642-14049-5_4
32. Tatusch, M., Klassen, G., Bravidor, M., Conrad, S.: Show me your friends and i'll tell you who you are. finding anomalous time series by conspicuous cluster transitions. In: Le, T.D., et al. (eds.) AusDM 2019. CCIS, vol. 1127, pp. 91–103. Springer, Singapore (2019). https://doi.org/10.1007/978-981-15-1699-3_8
33. Timm, H., Döring, C., Kruse, R.: Different approaches to fuzzy clustering of incomplete datasets. Int. J. Approx. Reason. **35**, 239–249 (2004)
34. Truong, C.D., Anh, D.T.: A novel clustering-based method for time series motif discovery under time warping measure. Int. J. Data Sci. Anal. **4**(2), 113–126 (2017). https://doi.org/10.1007/s41060-017-0060-3
35. Xie, X.L., Beni, G.: A validity measure for fuzzy clustering. IEEE Trans. Pattern Anal. Mach. Intell. **13**(8), 841–847 (1991)
36. Zakrzewska, A., Bader, D.: A dynamic algorithm for local community detection in graphs. In: 2015 IEEE/ACM International Conference on Advances in Social Networks Analysis and Mining (ASONAM), pp. 559–564 (2015)

Text Analysis and Processing

Creating Classification Models from Textual Descriptions of Companies Using Crunchbase

Marco Felgueiras[1], Fernando Batista[1,2]([✉]) [iD], and Joao Paulo Carvalho[2,3] [iD]

[1] ISCTE - Instituto Universitário de Lisboa, Lisbon, Portugal
mfmfs@iscte-iul.pt
[2] INESC-ID Lisboa, Lisbon, Portugal
{fmmb,joao.carvalho}@inesc-id.pt
[3] Universidade de Lisboa, Lisbon, Portugal

Abstract. This paper compares different models for multilabel text classification, using information collected from Crunchbase, a large database that holds information about more than 600000 companies. Each company is labeled with one or more categories, from a subset of 46 possible categories, and the proposed models predict the categories based solely on the company textual description. A number of natural language processing strategies have been tested for feature extraction, including stemming, lemmatization, and part-of-speech tags. This is a highly unbalanced dataset, where the frequency of each category ranges from 0.7% to 28%. Our findings reveal that the description text of each company contain features that allow to predict its area of activity, expressed by its corresponding categories, with about 70% precision, and 42% recall. In a second set of experiments, a multiclass problem that attempts to find the most probable category, we obtained about 67% accuracy using SVM and Fuzzy Fingerprints. The resulting models may constitute an important asset for automatic classification of texts, not only consisting of company descriptions, but also other texts, such as web pages, text blogs, news pages, etc.

Keywords: Text mining · Multilabel classification · Text classification · Document classification · Machine learning · Crunchbase

1 Introduction

We live in a digital society where data grows day by day, most of it consisting of unstructured textual data. This creates the need of processing all this data in order to be able to collect useful information from it. Text classification may be considered a relatively simple task, but it plays a fundamental role in a variety of systems that process textual data. E-mail spam detection is one of

This work was supported by national funds through FCT, Fundação para a Ciência e a Tecnologia, under project UIDB/50021/2020.

the most well-known applications of text classification, where the main goal consists of automatically assigning one of two possible labels (spam or ham) to each message. Other well-known text classification tasks, nowadays receiving increasingly importance, include sentiment analysis and emotion detection, that consist of assign a positive/negative sentiment or an emotion to a text (e.g. happiness, anger, sadness, ...).

Crunchbase is the largest companies' database in the world, containing a large variety of up-to-date information about each company. Founded in 2007 by Michael Arrington, originally, it was the data storage for its mother company TechCrunch. Until 2015, TechCrunch was the owner of the Crunchbase data, but by that time Crunchbase decoupled itself from TechCrunch to focus on its own products. Crunchbase database contains up-to-date details about over 600000 companies, including a short description, a detailed description, number of employees, headquarters regions, contacts, market share, and the current areas of activity.

This paper compares different approaches for multilabel text classification, using recent information collected from Crunchbase. Each company is labeled with one or more categories, from a subset of 46 possible categories, and the proposed models predict the set of associated categories based solely on the company textual description. In order to address the multilabel problem, two classification strategies have been tested using different classification methods: a) we have created 46 binary models, one for each one of the categories, where the set of categories for a given description is achieved by combining the result of the 46 models; b) we have created a single model that gives the most probable categories for a given description. The resulting models may constitute an important asset for automatic classification of texts that can be applied, not only company descriptions, but to other texts, such as web pages, text blogs, news pages, etc. The work here described extends the work described in [2] to multilabel classification, and constitutes a more challenging task, since each record is associated with one or more categories.

This document is structured as follows: Sect. 2 overviews the related literature, focusing on the most commonly used methods and features to solve similar text classification problems. Section 3 describes the data extraction procedure, the resulting dataset, and the corresponding data pre-processing. Section 5 describes our experiments and the corresponding achieved results. Finally, Sect. 6 presents our final conclusions and pinpoints future research directions.

2   Related Work

Text based classification has become a major researching area, specially because it can be used for a large number of applications. The existing literature in text classification is vast, but most of the studies consider only a small number of possible categories.

Most text classification approaches are based on Supervised Learning. [14] applied machine learning to classify movie reviews from Internet Movie Database

(IMDb) by sentiment. Also in [9] an experiment to spam detection in customer reviews took place to check if false opinions were given to a product. Text classification has also been extensively applied to social media. The work described in [11] applies several algorithms to tweets trying to find "trending topics", and [22] used Twitter information to develop an automated detection model to find rumors and misinformation in social media, achieving an accuracy of about 91%. These are examples of binary classification problems, but a bigger challenge arises when it comes to multiple categories, also known as multi-class. The work described in [3] presents a strategy based on binary maximum entropy classifiers for automatic sentiment analysis and topic classification over Spanish Twitter data. Both tasks involve multiple classes, and each tweet may be associated with multiple topics. Different configurations have been explored for both tasks, leading to the use of cascades of binary classifiers for sentiment analysis and a one-vs-all strategy for topic classification, where the most probable topics for each tweet were selected.

The performance and overall simplicity of Naive Bayes makes it a very attractive alternative for several classification tasks [13]. [8] used a Naive Bayes Classifier for author attribution applied to a dataset called AAAT dataset (i.e Authorship attribution of Ancient Arabic Texts) obtaining results up to 96% classification accuracy. Naive Bayes results are mainly obtained from an unreal assumption of independence. For this, there has been a major focus on investigating the algorithm itself. Recently, [23] used a Naive Bayes on 20 newsgroups, and compared the Multinomial, Bernoulli and Gaussian variants of Naive Bayes approaches.

The work described by [7] reports the use of Fuzzy Fingerprints to find an author of a text document using a large Dataset of newspaper articles from more than 80 distinct authors, achieving about 60% accuracy. Also [18] and [5] make use of the same technique to solve a multi-class classification problem when trying to find events and twitter topics using textual data.

The work reported by [19] demonstrates that the use of Support Vector Machines (SVM) outperform many other methods when applied to text classification problems. The work described in [17] compares Naive Bayes and SVMs, using two well-known datasets with different sample sizes in multiple experiments, and concludes that SVMs outperform Naive Bayes by a large margin, giving a much lower error rate, at that time the lowest for the given sets of data. Also in [1] a text classification problem with a large number of categories is used to compare SVMs and Artificial Neural Networks (ANNs). The results are very clear for both recall and precision, both indicating the differences in performance of the SVM and ANN. The SVM once again outperforms ANN, suggesting that SVMs as more suitable for this type of problems, not only because they achieve better performance, but also because they are less computationally complex. Additionally, [1] also compares two sets of features, a large and a reduced feature set, concluding that, using SVMs, the small feature set achieves much better performance.

In what concerns features, one of the most common ways to represent textual data is the bag-of-words approach [6], a very simple and efficient way to quickly feed an algorithm and check its potential behavior. This method consists in a simple breakdown of a sentence into a set of words that are part of it. It usually achieves a decent performance, and in some cases, if the dataset is already very rich in terms of features it can be a good implementation. This type of approach assumes that words are independent, and do not consider the context where the word was used, losing the syntactic structure and semantic meaning of the sentence. When the data is sparse this technique may not be adequate, but it is possible to use a similar technique, based on n-grams, that preserves some of the local context of a word. The work reported by [10] describes experiments using n-grams (bag-of-n-grams), consisting in a n-size moving window (usually 1,2 or 3) along each sentence and collect the unique combination of words along with its count. Bag-of-words are compared to n-grams approaches and show a large improvement over the entire set of experiments. Another well-known and successful weighting scheme commonly used for Text Classification is Term Frequency - Inverse Document Frequency (TF-IDF). An alternative to bag of words, n-grams and TF-IDF is the use of word embeddings. Word embeddings are a much more complex technique that attempts to encode the semantics of the words in the text. Common implementations, such as word2vec, allow the use of embeddings in text classification often with good results. For example, [12] combines TF-IDF and word2vec and achieves more than 90% accuracy while processing a news dataset.

A set of Natural Language Processing (NLP) techniques are commonly used for extracting relevant features from a sentence. One of them is *lemmatization*, a way to prepare the text for further usage, and a widely used technique when working with text classifiers [15,21,24]. Unlike *stemming*, it is not just the process of removing the suffix of a word, it also considers the morphological structure of the word. *Stemming* is a similar approach that is much lighter to run, it does not look into the morphosyntactic form of a word [20]. *Part-of-Speech tagging* is another NLP technique commonly applied to text classification tasks, that consists of assigning a part-of-speech tag (e.g., noun, adjective, verb, adverb, preposition, conjunction, pronoun, etc.) to each word. For example, [16] use part-of-speech to approach a multi-class classification problem for Amazon product reviews.

An attempt to extract to automatically extract information from an older version Crunchbase has been done in [2]. At the time, Crunchbase contained around 120K companies, each classified to one out of 42 possible categories. The dataset also contained category *"Other"*, that grouped a vast number of other categories. The paper performs experiments using SVMs, Naive Bayes, TF-IDF, and Fuzzy fingerprints. To our knowledge, no other works have reported text classification tasks over a Crunchbase dataset.

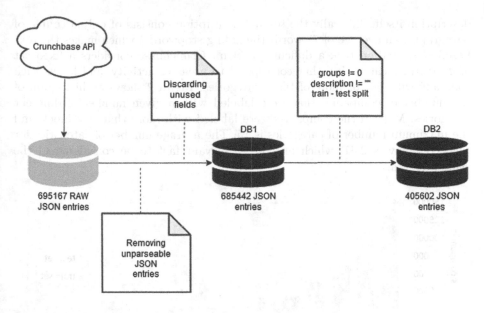

Fig. 1. Filtering and data transformation diagram.

3   Corpus

In this work we use a subset of a dataset extracted from Crunchbase containing up-to-date information of about 685000 companies. Crunchbase exposes a REST API that can be used to integrate all the information that is provided at Crunchbase for external applications. This API is accessible for researching purposes and Crunchbase Team, upon request. Crunchbase kindly provided full access for their data for 6 months, while developing this work. The data was extracted using the Crunchbase API and stored into SQLite3 databases. Figure 1 presents the data retrieval and filtering procedures. During the data retrieval stage we verified that some of the original responses were not retrieved properly, so they were removed. The extracted JSON entries contain a lot of information, but only a small portion of that information is relevant for our task: URL of the company, company name, description, short description, categories, and fine categories. DB1 contains only parseable records, containing only the relevant fields for our task. Finally, DB2 contains the data used in our experiments, were entries that did not belong to any category or that did not contain any description were filtered out. The final database contains a total of 405602 records, that have been randomly shuffled and stored into two different tables: *train*, containing 380602 records, will be used from training our models; and *test*, containing 25000 records, that will be used for evaluating our models.

Each record contains a textual description of the company, that explains the company for whoever wants to have a brief notion of what it does and the areas that it belongs, and a short description, which is a summary of the description itself. Typically the textual description consists of only a couple of paragraphs, an average of 77 words (including stopwords) which makes the text based classification task a difficult problem. Crunchbase considers a fixed set of 46 distinct categories, that correspond to areas of activity, and labels each company with one or more of those categories. Figure 2 shows an histogram of the number of companies that were labeled with a given number of different categories. Most of the companies were labeled with more than one label, and the maximum number of categories is 15. The average number of categories for each company is 2.41, which may be a relevant fact to be considered in the evaluation stage.

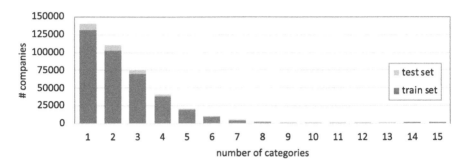

Fig. 2. Histogram of the number of companies labeled with a given number of categories.

Each category can also be decomposed into a number of fixed *fine categories*. The category is wider (e.g *Software*) while the *fine categories* are more specific (e.g., *Augmented Reality, Internet, Software, Video Games, Virtual Reality*). Each *fine category* can be present in more than one category, for instance *"Alumni"* appears as a *fine category* for *"Internet Services"*, *"Community and lifestyle"*, *"Software"*, and many other categories. Also, *"Consumer"* appears in *"Administrative Services"*, *"Hardware"* and *"Real Estate"*, among others. The analysis performed in this paper considers only the 46 wider categories.

Figure 3 presents the number of companies that have been labeled with a given group, revealing a highly unbalanced dataset, where the frequency of each category ranges from 28% (*Software*) to 0.7% (*Navigation and Mapping*). The *"Software"* category is assigned to over 100K records, while 17 categories occur in less than 10K companies. It is also important to note that even the second most represented category, *Internet Services*, corresponds to only 56% of the most represented category.

4   Experimental Setup

This section presents the details about the experimental setup. It starts by describing the corpora pre-processing steps, then it presents the adopted classification methods, and finally presents the used evaluation metrics.

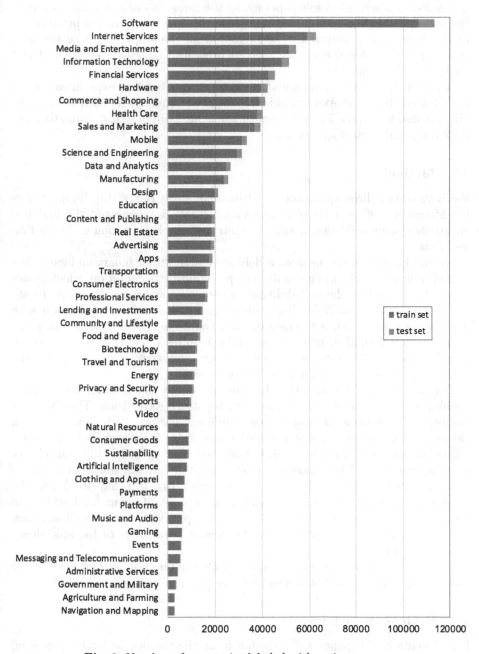

Fig. 3. Number of companies labeled with a given group.

4.1   Corpora Pre-processing and Feature Extraction

Experiments described in this paper model the categories of a company based on the text available in the company description. We have removed the punctuation marks, lower-cased every token, and removed all the stopwords from the text, by using the list of stopwords included with Natural Language Toolkit (NLTK) for the English language.

Concerning the text representation features, our baseline experiments use a bag-of-words representation, considering the word frequency for each description. We have also tested the TF-IDF weighting scheme, word bigrams, lemmatization, stemming, and part-of-speech tags.

4.2   Methods

We have tested three approaches: multinomial Naive Bayes [13], Support Vector Machines [19], and Fuzzy Fingerprints [2]. We implemented the first two approaches using scikit-learn, and used our own implementation of Fuzzy Fingerprints.

Naive Bayes is one of the most widely used methods for binary and multiclass text classification. The method outputs a probabilistic distribution, which makes it possible to analyse the probability of the outcome and to easily define thresholds. The multinomial Naive Bayes classifier is suitable for classification with discrete features, which is the case of word counts for text classification, given that the multinomial distribution normally requires integer feature counts. In practice, fractional counts such as TF-IDF may also work, but our experiments using TF-IDF achieved much worse performances.

Support Vector Machines (SVM) were introduced as a solution for a binary problem with two categories associated with pattern recognition. The SVM calculates the best decision limit between different vectors, each belonging to a category. Based on the limit minimization principle [4] for a given vector space where the goal is to find the decision boundary that splits the different classes or categories. SVM based models are often used in text classification problems since they behave quite well when used in supervised learning problems. The good results are due to the high generalization capacity of the method, which can be particularly interesting when trying to solve problems in bigger dimensions. Every experiment here described use the default parameters of the scikit-learn implementations.

Fuzzy fingerprints experiments use the *Pareto* function with $K = 4000$. For more information about the method refer to [2].

4.3   Evaluation Metrics

Experiments in this paper are evaluated using the metrics: accuracy, precision, recall and F1-score, defined as:

$$Accuracy = \frac{true\ positives + true\ negatives}{total\ predictions}$$

$$Precision = \frac{true\,positives}{true\,positives + false\,positives}$$

$$Recall = \frac{true\,positives}{true\,positives + false\,negatives}$$

$$F1 - score = \frac{2 * precision * recall}{precision + recall}$$

In order to calculate the overall metrics, we must consider the micro-average and macro-average versions of the above performance metrics. A macro-average computes the metric independently for each category and then takes the average (hence treating all categories equally), whereas a micro-average aggregates the contributions of all categories to compute the average metric. Micro-average metrics are usually preferable metric for unbalanced multi-class classification tasks.

5 Experiments and Results

This section presents two sets of experiments. Section 5.1 describes experiments with binary classification models, one for each category, where each model predicts a category. Section 5.2 presents a number of experiments, considering only one model in a multi-class scenario.

5.1 Binary Classification Models

Our first set of experiments consists of creating a model for each one of the categories. In order to train each model in a binary fashion, we have selected every companies labeled with the corresponding label as positive samples, and all the other companies as negative samples. In this scenario, the performance of each model can be evaluated individually, but micro-average or macro-average metrics must be used in order to assess the global performance.

Table 1 presents the most relevant micro-average results. Concerning the multinomial Naive Bayes, the best results were achieved using the word frequency, as expected (see Sect. 4.2). The performance of the SVM-based models improved when moving from the word frequency to the TF-IDF weighting scheme. However, the performance did not improve after introducing other NLP-based features, such as lemmatization, stemming, part-of-speech tags, or bigrams. The Fuzzy fingerprints did not produce interesting results, but this was an expected result due to the small size of the descriptions and the fact that they were developed for multi-class problems, and usually only are advantageous when dealing with large number of classes [7, 18]. Overall, the best result, considering the F1-score, was achieved by combining TF-IDF with SVMs.

Table 2 shows the performance of each individual classifier, revealing that the performance does not necessarily degrades for the most highly unbalanced categories.

Table 1. Classification results for each method, using different features.

Experiments	Accuracy	Precision	Recall	F1-score
Multinomial Naive Bayes (word frequency)	0.951	0.547	**0.439**	**0.487**
SVM				
Word frequency	0.950	0.537	0.412	0.467
TF-IDF weights	0.959	0.696	**0.420**	**0.524**
TF-IDF + Lemmas	0.960	0.702	0.416	0.522
TF-IDF + Stemming	0.959	**0.704**	0.410	0.518
TF-IDF + POS tags	0.959	0.701	0.405	0.513
TF-IDF + word bigrams	0.960	0.703	0.417	0.523
TF-IDF + word bigrams + POS tags	0.959	0.701	0.405	0.513
Fuzzy fingerprints (Word frequency)	–	0.204	**0.786**	0.324

The achieved results can not be directly compared with the results described in [2], not only because the evaluation sets differ, but also because different metrics and different modeling approaches are being used. However, it is interesting to note that SVMs, which did not perform well in that work, are now the best performing method (by far).

5.2 Multi-class Classification

Our second set of experiments consists of creating a single model that is able to provide the most probable category. In order to train such a model, we have duplicated each entry as many times as the number of corresponding category labels. So, each company labeled with n categories was duplicated once for each one of the categories, and used for training each individual category. Such a model may be useful for automatically guessing the best category for a given text and also provide the top best categories. The performance of the model cannot be easily evaluated, once the number of possible categories varies for each company. So, we have evaluated the performance of correctly predicting one of the categories, which may correspond to the best category only. In this scenario, Accuracy becomes the most adequate metric, and Precision and Recall do not apply. For this experiment we used TF-ICF (Term Frequency - Inverse Class Frequency) [18], a process similar to TF-IDF, in the Fuzzy Fingerprints approach.

Table 3 shows the classification performance when predicting the top category, for each one of the three methods. Our baseline corresponds to always guessing the most frequent category. In this multi-class experiment SVM and Fuzzy Fingerprints perform very similarly, but the SVM is around 16x slower. The multinomial Naive Bayes runs very fast, but the performance is more than 3% (absolute) lower than the other two approaches.

Table 2. Classification performance by group for SVM + TF-IDF

	Samples	Precision	Recall	F1-score
Software	6929	0.683	0.560	0.616
Internet Services	3956	0.584	0.297	0.393
Media and Entertainment	3338	0.710	0.485	0.576
Information Technology	3108	0.565	0.263	0.359
Financial Services	2767	0.826	0.670	0.740
Hardware	2630	0.658	0.349	0.456
Commerce and Shopping	2527	0.676	0.394	0.498
Health Care	2521	0.841	0.704	0.767
Sales and Marketing	2387	0.732	0.457	0.563
Mobile	2017	0.582	0.319	0.412
Science and Engineering	1949	0.731	0.428	0.540
Data and Analytics	1595	0.629	0.279	0.386
Manufacturing	1576	0.656	0.473	0.550
Design	1305	0.630	0.274	0.381
Education	1226	0.804	0.591	0.681
Content and Publishing	1233	0.643	0.354	0.456
Real Estate	1231	0.770	0.532	0.629
Advertising	1156	0.678	0.396	0.500
Apps	1190	0.449	0.100	0.164
Transportation	1155	0.749	0.435	0.551
Consumer Electronics	1084	0.534	0.122	0.198
Professional Services	1018	0.679	0.302	0.418
Lending and Investments	933	0.639	0.424	0.510
Community and Lifestyle	888	0.549	0.114	0.188
Food and Beverage	844	0.772	0.610	0.682
Biotechnology	766	0.747	0.560	0.640
Travel and Tourism	723	0.791	0.488	0.604
Energy	754	0.775	0.580	0.663
Privacy and Security	666	0.726	0.362	0.483
Sports	607	0.735	0.448	0.557
Video	563	0.648	0.393	0.489
Natural Resources	579	0.717	0.522	0.604
Consumer Goods	571	0.656	0.294	0.406
Sustainability	574	0.680	0.389	0.494
Artificial Intelligence	509	0.742	0.316	0.444
Clothing and Apparel	470	0.769	0.496	0.603
Payments	409	0.640	0.347	0.450
Platforms	375	0.429	0.048	0.086
Music and Audio	403	0.784	0.496	0.608
Gaming	358	0.651	0.453	0.534
Events	367	0.671	0.294	0.409
Messaging and Telecommunications	313	0.539	0.220	0.313
Administrative Services	272	0.574	0.129	0.210
Government and Military	220	0.511	0.109	0.180
Agriculture and Farming	222	0.725	0.392	0.509
Navigation and Mapping	173	0.476	0.116	0.186

Table 3. Multi-class classification results.

	Accuracy	Execution time (s)
Most frequent category (Baseline)	0.280	
Multinomial Naive Bayes	0.646	11.37
SVM	0.678	1010.35
Fuzzy Fingerprints	0.672	62.99

6   Conclusions and Future Work

This paper describes multi-label text classification experiments over a dataset containing more than 400000 records about companies extracted from Crunchbase. We have performed experiments using three classification approaches, multinomial Naive Bayes, SVM, and Fuzzy fingerprints, and considering different combinations of text representation features. Our dataset is highly unbalanced since the frequency of each category ranges from 28% to 0.7%. Nevertheless, our findings reveal that the description text of each company contains features that allow to predict its area of activity, expressed by its corresponding categories, with about an overall performance of 70% precision, and 42% recall. When using a multi-class approach, the accuracy for predicting the most probable category is above 65%.

We are planning to improve this work by considering additional evaluation metrics for ranking problems, such as precision@k, recall@k and f1@k, that may be suitable for measuring the multi-label performance. Additionally, we are also planning to introduce features based on embeddings and to compare the reported methods with other neural network classification approaches.

References

1. Basu, A., Walters, C., Shepherd, M.: Support vector machines for text categorization. In: Proceedings of the 36th Annual Hawaii International Conference on System Sciences, HICSS 2003, pp. 1–7 (2003). https://doi.org/10.1109/HICSS.2003.1174243
2. Batista, F., Carvalho, J.P.: Text based classification of companies in CrunchBase. In: 2015 IEEE International Conference on Fuzzy Systems (FUZZ-IEEE), pp. 1–7 (2013). https://doi.org/10.1109/FUZZ-IEEE.2015.7337892
3. Batista, F., Ribeiro, R.: Sentiment analysis and topic classification based on binary maximum entropy classifiers. Procesamiento de Lenguaje Nat. **50**, 77–84 (2013). http://journal.sepln.org/sepln/ojs/ojs/index.php/pln/article/view/4662
4. Cortes, C., Vapnik, V.: Support-vector networks. Mach. Learn. **20**(3), 273–297 (1995). https://doi.org/10.1007/BF00994018
5. Czarnowski, I., Jędrzejowicz, P.: An approach to rbf initialization with feature selection. In: Angelov, P., et al. (eds.) Intelligent Systems 2014. AISC, vol. 322, pp. 671–682. Springer, Cham (2015). https://doi.org/10.1007/978-3-319-11313-5_59
6. Harris, Z.S.: Distributional structure. Word **10**(2–3), 146–162 (1954)

7. Homem, N., Carvalho, J.P.: Authorship identification and author fuzzy "fingerprints". In: Annual Conference of the North American Fuzzy Information Processing Society - NAFIPS, pp. 180–185 (2011). https://doi.org/10.1109/NAFIPS.2011.5751998

8. Howedi, F., Mohd, M.: Text classification for authorship attribution using naive bayes classifier with limited training data. Comput. Eng. Intell. Syst. **5**(4), 48–56 (2014). http://iiste.org/Journals/index.php/CEIS/article/view/12132

9. Jindal, N., Liu, B.: Review spam detection. In: Proceedings of the 16th International Conference on World Wide Web, pp. 1189–1190 (2007)

10. Joulin, A., Grave, E., Bojanowski, P., Mikolov, T.: Bag of tricks for efficient text classification. arXiv preprint arXiv:1607.01759 (2016)

11. Lee, K., Palsetia, D., Narayanan, R., Patwary, M.M.A., Agrawal, A., Choudhary, A.: Twitter trending topic classification. In: 2011 IEEE 11th International Conference on Data Mining Workshops, pp. 251–258. IEEE (2011)

12. Lilleberg, J., Zhu, Y., Zhang, Y.: Support vector machines and Word2vec for text classification with semantic features. In: Proceedings of 2015 IEEE 14th International Conference on Cognitive Informatics and Cognitive Computing, ICCI*CC 2015, pp. 136–140 (2015). https://doi.org/10.1109/ICCI-CC.2015.7259377

13. Murphy, K.P., et al.: Naive bayes classifiers. Univ. Br. Columbia **18**, 60 (2006)

14. Pang, B., Lee, L., Vaithyanathan, S.: Thumbs up?: sentiment classification using machine learning techniques. In: Proceedings of the ACL-02 Conference on Empirical Methods in Natural Language Processing, vol. 10, pp. 79–86. ACL (2002)

15. Plisson, J., Lavrac, N., Mladenic, D., et al.: A rule based approach to word lemmatization. In: Proceedings of IS, vol. 3, pp. 83–86 (2004)

16. Pranckevicius, T., Marcinkevicius, V.: Application of logistic regression with part-of-the-speech tagging for multi-class text classification. In: 2016 IEEE 4th Workshop on Advances in Information, Electronic and Electrical Engineering, AIEEE 2016 - Proceedings, pp. 1–5 (2017). https://doi.org/10.1109/AIEEE.2016.7821805

17. Rennie, J.D.M., Rifkin, R.: Improving multiclass text classification with the support vector machine. Technical report, October 2001, Massachusetts Institute of Technology AI Memo 2001–026 (2001). http://dspace.mit.edu/handle/1721.1/7241

18. Rosa, H., Batista, F., Carvalho, J.P.: Twitter topic fuzzy fingerprints. In: WCCI2014, FUZZ-IEEE, 2014 IEEE World Congress on Computational Intelligence,International Conference on Fuzzy Systems, pp. 776–783. IEEE Xplorer, Beijing, July 2014

19. Sain, S.R., Vapnik, V.N.: The Nature of Statistical Learning Theory, vol. 38. Springer, Heidelberg (2006). https://doi.org/10.2307/1271324

20. Sharma, D., Cse, M.: Stemming algorithms: a comparative study and their analysis. Int. J. Appl. Inf. Syst. **4**(3), 7–12 (2012)

21. Toman, M., Tesar, R., Jezek, K.: Influence of word normalization on text classification. In: Proceedings of InSciT, pp. 354–358 (2006). http://www.kiv.zcu.cz/research/groups/text/publications/inscit20060710.pdf

22. Vosoughi, S.: Automatic detection and verification of rumors on Twitter. Ph.D. thesis, Massachusetts Institute of Technology (2015)

23. Xu, S.: Bayesian naive bayes classifiers to text classification. J. Inf. Sci. **44**(1), 48–59 (2018)

24. Zhang, D., Chen, X., Lee, W.S.: Text classification with kernels on the multinomial manifold. In: SIGIR 2005–28th Conference on Research and Development in Information Retrieval, pp. 266–273 (2005). https://doi.org/10.1145/1076034.1076081

Automatic Truecasing of Video Subtitles Using BERT: A Multilingual Adaptable Approach

Ricardo Rei[1], Nuno Miguel Guerreiro[1], and Fernando Batista[2(✉)]

[1] Unbabel, Lisbon, Portugal
{ricardo.rei,nuno.guerreiro}@unbabel.com
[2] INESC-ID Lisboa & ISCTE - Instituto Universitário de Lisboa, Lisbon, Portugal
fernando.batista@iscte-iul.pt

Abstract. This paper describes an approach for automatic capitalization of text without case information, such as spoken transcripts of video subtitles, produced by automatic speech recognition systems. Our approach is based on pre-trained contextualized word embeddings, requires only a small portion of data for training when compared with traditional approaches, and is able to achieve state-of-the-art results. The paper reports experiments both on general written data from the European Parliament, and on video subtitles, revealing that the proposed approach is suitable for performing capitalization, not only in each one of the domains, but also in a cross-domain scenario. We have also created a versatile multilingual model, and the conducted experiments show that good results can be achieved both for monolingual and multilingual data. Finally, we applied domain adaptation by finetuning models, initially trained on general written data, on video subtitles, revealing gains over other approaches not only in performance but also in terms of computational cost.

Keywords: Automatic capitalization · Automatic truecasing · BERT · Contextualized embeddings · Domain adaptation

1 Introduction

Automatic Speech Recognition (ASR) systems are now being massively used to produce video subtitles, not only suitable for human readability, but also for automatic indexing, cataloging, and searching. Nonetheless, a standard ASR system usually produces text without punctuation and case information, which makes this representation format hard to read [12], and poses problems to further

This work was supported by national funds through FCT, Fundação para a Ciência e a Tecnologia, under project UIDB/50021/2020 and by PT2020 funds, under the project "Unbabel Scribe: AI-Powered Video Transcription and Subtitle" with the contract number: 038510. The authors have contributed equally to this work.

M.-J. Lesot et al. (Eds.): IPMU 2020, CCIS 1237, pp. 708–721, 2020.
https://doi.org/10.1007/978-3-030-50146-4_52

automatic processing. The capitalization task, also known as truecasing [13,18], consists of rewriting each word of an input text with its proper case information given its context. Many languages distinguish between uppercase and lowercase letters, and proper capitalization can be found in many information sources, such as newspaper articles, books, and most of the web pages. Besides improving the readability of texts, capitalization provides important semantic clues for further text processing tasks. Different practical applications benefit from automatic capitalization as a preprocessing step, and in what concerns speech recognition output, automatic capitalization may provide relevant information for automatic content extraction, and Machine Translation (MT).

Unbabel combines the speed and scale of automatic machine translation with the authenticity that comes from humans, and is now dealing with an increasing demand for producing video subtitles in multiple languages. The video processing pipeline consists of a) processing each video with an ASR system adapted to the source language, b) manual post-edition of the ASR output by human editors, and c) perform the translation for other languages, first by using a customized MT system, and then by using humans to improve the resulting translations. Recovering the correct capitalization of the words coming from the speech transcripts constitutes an important step in our pipeline due to its impact on the post-edition time, performed by human editors, and on the MT task output. Automatic Video subtitles may contain speech recognition errors and other specific phenomena, including disfluencies originated by the spontaneous nature of the speech and other metadata events, that represent interesting practical challenges to the capitalization task.

This paper describes our approach for automatically recovering capitalization from video subtitles, produced by speech recognition systems, using the BERT model [8]. Experiments are performed using both general written data and video subtitles, allowing for assessment of the impact of the specific inner structural style of video subtitles in the capitalization task.

The paper is organized as follows: Sect. 2 presents the literature review. Section 3 describes the corpora and pre-processing steps used for our experiments. Section 4 presents our approach and the corresponding architecture, as well as the evaluation metrics. Section 5 presents the results achieved, both on a generic domain (monolingual and multilingual) and in the specific domain of video subtitles. Finally, Sect. 6 presents the most relevant conclusions and pinpoints a number of future directions.

2 Related Work

Capitalization can be viewed as a lexical ambiguity resolution problem, where each word has different graphical forms [10,30], by considering different capitalization forms as spelling variations. Capitalization can also be viewed as a sequence tagging problem, where each lowercase word is associated with a tag that describes its capitalization form [6,14,15,18].

A common approach for capitalization relies on n-gram language models estimated from a corpus with case information [10,15,18]. Common classification approaches include Conditional Random Fields (CRFs) [27] and Maximum Entropy Markov Models (MEMM) [6]. A study comparing generative and discriminative approaches can be found in [2]. The impact of using increasing amounts of training data as well as a small amount of adaptation is studied in [6]. Experiments on huge corpora sets, from 58 million to 55 billion tokens, using different n-gram orders are performed in [10], concluding that using larger training data sets leads to increasing improvements in performance, but the same tendency is not achieved by using higher n-gram order language models. Other related work, in the context of MT systems, exploit case information both from source and target sentences of the MT system [1,23,27].

Recent work on capitalization has been reported by [21,25]. [25] proposes a method for recovering capitalization for long-speech ASR transcriptions using Transformer models and chunk merging, and [21] extends the previous model to deal with both punctuation and capitalization. Other recent advances are reported by [29] for Named Entity Recognition (NER), a problem that can be tackled with similar approaches.

Pre-trained transformer models such as BERT [8] have outperformed previous state-of-the-art solutions in a wide variety of NLP tasks [7,8,19]. For most of these models, the primary task is to reconstruct masked tokens by uncovering the relation between those tokens and the tokens surrounding. This pre-train objective proved to be highly effective for token-level tasks such as NER. Bearing this in mind, in this paper, we will follow the approach proposed in [3–5] and address the capitalization task as a sequence tagging problem similar to NER and show that, as in that task, BERT can also achieve state-of-the-art results for capitalization.

3   Corpora

Constructing an automatic translation solution focused on video content is a complex project that can be subdivided into several tasks. In this work, we are focusing on enriching the transcription that comes from the ASR system, by training a model prepared to solve the truecasing problem. This process is of paramount importance for satisfactory machine translation task output and would ultimately alleviate the post-edition time performed by human editors.

3.1   Datasets

Experiments performed in the scope of this paper use internal data (hereinafter referred as *domain* dataset) produced by the ASR system and subsequently post-edited by humans in order to correct bad word transcripts, introduce capitalization and punctuation, and properly segment the transcripts to be used for video subtitling.

Table 1. Source sentence and target tags construction for a given sentence. Note that apart from lowercasing all tokens from the target, punctuation was also stripped to create the source sentence.

Target	Automatic Truecasing of Video Subtitles using BERT: A multilingual adaptable approach
Source	Automatic truecasing of video subtitles using bert a multilingual adaptable approach
Target tags	T T L T T L U U L L L

In order to establish a comparison with a structured out-of-domain training corpus, we use the Europarl V8 corpus. This corpus is composed of parallel sentences which allows for coherent studies in terms of complexity across different languages. As one of the main objectives is that of building a single model that can be used for several languages, we also constructed a dataset composed by sentences in four different languages (English, Spanish, French and Portuguese) in such a way that there are no parallel sentences across different languages.

The dataset composed by English-only sentences will be hereinafter referred as *monolingual* dataset whereas the one composed by sentences in different languages will be referred as *multilingual* dataset.

3.2 Pre-processing

Considering that we want to build a model prepared to receive the outputs from the ASR system and automatically solve the truecasing problem, we removed all punctuation but apostrophes and hyphens which are extensively used in the languages considered for this study. This is an important step towards building a reliable model, since the ASR outputs' punctuation is not consistently trustworthy. For fair comparisons with the generic dataset, punctuation was also removed from its data. Moreover, metadata events such as sound representations (e.g: "laughing") are removed from the domain dataset.

The problem of truecasing is approached as a sequence tagging problem [6, 14,15,18]. Thus, the source sentences for both datasets are solely composed by lowercased tokens, whereas the target sequences for both datasets are composed by the ground truth tags. A tag "U" is attributed to an uppercase token, a tag "T" is attributed to a *title* token (only the first letter is uppercase) and a tag "L" to all the remaining tokens. An example of this procedure can be seen in Table 1. We observed that for the monolingual and multilingual datasets, as the first token tag corresponds to "T" in the vast majority of their sentences, the model would capitalize the first token just for its position. As we do not want to rely on positional information to solve the truecasing problem, if a sentence starts with a *title* token, we do not consider that token during training/testing. Statistics on the size of the train and test set for each dataset (domain, monolingual and multilingual), absolute frequency of each tag and the ratio of not-lowercased tags

Table 2. Size of the train and test set for each dataset.

Dataset		Number of sentences	"L" tags	"U" tags	"T" tags	Not–"L" ratio (%)
Domain	Train	127658	904288	38917	94767	14.78
	Test	10046	76200	3420	8179	15.22
Generic	Train	1908970	42936838	861199	4953879	13.54
	Test	99992	2246120	43236	267972	13.86
Multilingual	Train	1917932	46420467	624757	4532864	11.11
	Test	99985	2399968	29564	240390	11.25

for each dataset is displayed in Table 2. The not–"L" ratio is relevant since the datasets are unbalanced as "L" tags are much more frequent.

4   Approach Description and Evaluation Metrics

As pre-trained text encoders have been consistently improving the state of the art on many NLP tasks, and since we are approaching the problem as a sequence tagging problem, we decided to use the BERT [8] model. The BERT base model is a 12-layer encoder-only bidirectional model based on the Transformer [26] architecture with 768 hidden units that was trained for masked word prediction and on next sentence prediction on a large corpus of plain unlabelled text. We refer to [8] for further details of the model.

4.1   Architecture

Given an input sequence $x = [x_0, x_1, \ldots, x_n]$, the BERT encoder will produce an embedding $e_{x_j}^{(\ell)}$ for each token x_j and each layer ℓ.

In [24], it is revealed that the BERT model captures, within the network, linguistic information that is relevant for downstream tasks. Thus, it is beneficial to combine information from several layers instead of solely using the output of the last layer. To do so, we used the approach in [17,22] to encapsulate information in the BERT layers into a single embedding for each token, e_{x_j}, whose size is the same as the hidden size of the model. This embedding will be computed as a weighted sum of all layer representations:

$$e_{x_j} = \gamma \sum_{\ell=0}^{12} e_{x_j}^{(\ell)} \cdot \text{softmax}\,(\boldsymbol{\alpha})^{(\ell)} \tag{1}$$

where γ is a trainable scaling factor and $\boldsymbol{\alpha} = \left[\alpha^{(1)}, \alpha^{(2)}, \ldots, \alpha^{(12)}\right]$ are the layer scalar trainable weights which are kept constant for every token. Note that this computation can be interpreted as **layer-wise attention mechanism**. So, intuitively, higher $\alpha^{(\ell)}$ values are assigned to layers that hold more relevant

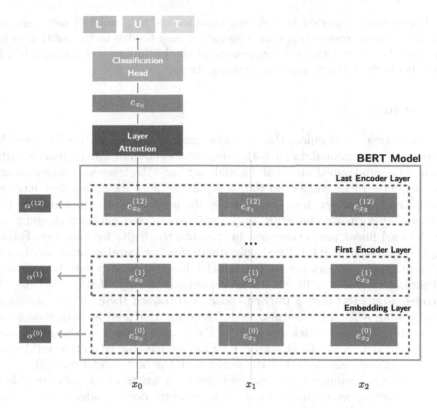

Fig. 1. The architecture of our solution. The output of the BERT model is computed using the Layer Attention block respective to (1). The scalar weights respective to each layer are trained simultaneously with the rest of the model.

information to solve the task. In order to redistribute the importance through all the model layers, we used **layer dropout**, devised in [17], in which each weight $\alpha^{(\ell)}$ is set to $-\infty$ with probability 0.1. This will also prevent overfitting of the model to the information captured in any single layer.

Finally, the embeddings are fed to a classification head composed by a feed-forward neural network which will down-project the size-768 token embedding e_{x_j} to a size-3 logits vector. This vector will then be fed to a softmax layer to produce a probability distribution over the possible tags and the position index of the maximum value of the vector will be considered as the predicted tag (Fig. 1).

4.2 Evaluation Metrics

All the evaluation presented in this paper uses the performance metrics: F1-score and Slot Error Rate (SER) [20]. Only capitalized words (not lowercase) are considered as slots and used by these metrics. Hence, the capitalization SER is computed by dividing the number of capitalization errors (misses and false alarms) by the number of capitalized words in the reference.

Experiments reported here do not consider the first word of each sentence whenever the corresponding case information may be due to its position in the sentence. So, every titled word appearing at the beginning of a sentence will be excluded both at the training and testing stages.

5 Results

In this section, we compare the results of experiments ran on the Europarl V8 corpus and on domain data for both monolingual and multilingual models. After loading the pre-trained model and initializing both the layer-wise attention and the feed-forward projection on top, we split the network parameters into two groups; encoder parameters, composed by the layer-wise attention and the pre-trained transformer architecture, and classification-head parameters, composed by the final linear projection used to compute the logits for each tag. Following the approach in [11,17] we apply discriminative learning rates for the two different groups of parameters. For the classification-head parameters we used a learning rate of 3×10^{-5} with a dropout probability of 0.1. We froze the encoder parameters during the first epoch, and trained them on the subsequent epochs using a 1×10^{-5} learning rate. The optimizer used in both groups was Adam [16]. We use a batch size of 8 for the models trained on the generic and domain datasets, and a batch size of 16 for the models trained on the multilingual dataset. At test time, we select the model with the best validation SER.

In order to evaluate if the models trained on written text data are able to transfer capabilities to in-domain data, we perform domain adaptation by fine-tuning the monolingual models on in-domain data.

We implemented all the models using either the `bert-base-uncased` (for the models trained on monolingual English data) or `bert-base-multilingual` `-uncased` (for the models trained on multilingual data) text encoders from the Huggingface library [28] as the pre-trained text models and we ran all experiments making use of the Pytorch Lightning wrapper [9].

5.1 Experiments on Europarl Data

For both generic and multilingual datasets, we train models under four settings: *+1.9M* (correspondent to the datasets in Table 2), *200K* (200,000 training sentences), *100K* (100,000 training sentences) and *50K* (50,000 training sentences). We will be referring to the models trained on monolingual data as **monoligual models**, and the models trained on multilingual data as **multilingual models**. Moreover, we trained a Bidirectional Long Short-Term Memory (BiLSTM) with a CRFs model on the entire monolingual dataset (+1.9M setting), which will be referred to as **baseline model** since we used its evaluation results as the baseline.

Monolingual Setting. Results are shown in Table 3. We observe that the monolingual model performs better than the baseline model for all training settings. This is evidence that our approach using pre-trained contextual embed-

Table 3. Results for the monolingual models evaluated on the **generic** test set.

Model architecture	Training setting	SER	F1-score
Baseline (BiLSTM + CRF)	+1.9M	0.1480	0.9200
Monolingual	+1.9M	**0.0716**	**0.9753**
	200K	0.0775	0.9717
	100K	0.0800	0.9701
	50K	0.0850	0.9682

Table 4. Evaluation on the **multilingual** test set.

Model	Training setting	SER	F1-score
Multilingual	+1.9M	**0.1040**	**0.9579**
	200K	0.1206	0.9472
	100K	0.1240	0.9447
	50K	0.1312	0.9379

Table 5. Evaluation on the **monolingual** test set.

Model	Training setting	SER	F1-score
Monolingual	+1.9M	**0.0716**	**0.9753**
Multilingual		0.0761	0.9690
Baseline		0.1480	0.9200

dings is not only able to achieve better results, but it also manages to do so using only a small portion of the data when compared to the baseline model.

Multilingual Setting. Results are shown in Tables 4 and 5. As expected, results for the monolingual model are better than the ones obtained by the multilingual model. Nevertheless, the multilingual model trained on its +1.9M setting outperforms all the models trained on monolingual data under all settings but the +1.9M setting, although this could be happening because the multilingual train dataset has more English individual sentences than the monolingual 200 K setting dataset. The results are evidence that a multilingual model which holds information on several languages is able to achieve similar results to a monolingual model and outperforms previous state-of-the-art solutions trained and tested in an equivalent monolingual setting.

Comparison with the Baseline Model. Results show that both the monolingual and multilingual models outperform the results obtained using the baseline model even when training on a small portion of the available data. Thus, further experiments will be solely evaluated on the models based on our architecture.

5.2 Experiments on Domain Data

All the experiments reported in this section make use of the domain datasets described in Table 2. First, we trained a model using the pre-trained contextual embeddings from `bert-base-uncased` and another using those from `bert-base-multilingual-uncased` on the domain training dataset. We will be referring to these models as **in-domain models**. Then, we perform domain

Table 6. Evaluation on the **domain** test set.

Model	Training setting	SER	F1-score
Domain		**0.2478**	**0.8489**
Monolingual	+1.9M	0.3128	0.7855
	200K	0.3136	0.7827
	100K	0.3071	0.7927
	50K	0.3193	0.7853
Multilingual	+1.9M	0.3285	0.7715
	200K	0.3355	0.7825
	100K	0.3372	0.7794
	50K	0.3530	0.7716

adaptation by loading the obtained models from experiments on the Europarl data and training them with in domain data.

In-domain Models. Results are shown in Table 6. Recalling the dataset statistics from Table 2, the domain dataset is comparable, in terms of number of sentences, with the monolingual dataset for the 50K training setting. Comparing the in-domain model and generic model for this setting, when tested on data from the same distribution that they trained on, we observe that there is a significant offset between the evaluation metrics. This is evidence that there are structural differences between the generic and the domain data. This notion is supported by the evaluation results of the generic and multilingual models initially trained on Europarl data on the domain test set and will be furtherly explored next.

Structural Differences Between Domain and Europarl Data. By observing both datasets, we noticed some clear structural differences between them. For one, since the original samples were segmented (the average number of tokens per sample is 6.74 for the domain training data and 24.19 for the generic training data), there is much less context preservation in the domain data. Moreover, the segmentation for subtitles is, in some cases, made in such a way that multiple sentences fit into a single subtitle, i.e, a single training sample (see Table 7). Since, as we previously remarked, we did not want to use the ASR outputs' punctuation, the truecasing task is hardened as it is difficult for the model to capture when a subtitle ends and a new one start for there can be a non-singular number of ground-truth capitalized tags assigned to words that are not typically capitalized. Note that recovering the initial sentences from the subtitles, i.e, the pre-segmentation transcripts, would be a difficult and cumbersome task. Moreover, different chunks of the ASR outputs' have been post-edited by different annotators which creates some variance in the way the segmentation and capitalization are done for some words (e.g: the word "Portuguese" is written as "portuguese" and attributed the tag "L" two times out of the eleven times it appears in the training data set). Last, when compared with the Europarl data, the in-domain data is significantly more disfluent. This is mainly due to the spontaneous nature of the source speech, since the ASR outputs are respective to content from video-sharing platforms that is considerably more unstructured (e.g: "Oh my God. Two more. You did it man!").

Table 7. In the example below, extracted from the domain data, we observe that the segmentation caused the capitalized tag respective to "The" to appear in the middle of the subtitle. Since we are not using any punctuation information, this significantly hardens the task of capitalization for this word. It is also noticeable that the length of each subtitle is small, hampering the use of context to solve the task.

ASR output	"After that it's all good, you get on the plane, and you're away. The airport is key to the start of a good beginning to the holiday."
Segmented subtitles	After that it's all good, you get on the plane, and you're away. The airport is key to the start of a good beginning to the holiday
Target tags	T L L L L L L L L L L L L L L L T L L L L L L L L L L L L

In-domain Adaptation. Given our interest in evaluating the ability to transfer capabilities from the models trained on generic data, we fine-tuned the monolingual models on in-domain data. These models will be referred to as **adapted models**. All four models trained on Europarl data, one for each training setting, are adapted to the domain data. Results shown in Table 8 reveal that all adapted models but the one initially trained in the total setting on Europarl data outperform the domain model. Moreover, the results shown in Fig. 2 indicate that by reducing the original training dataset size, we obtain models that are not only

Table 8. Evaluation results on the **domain** test set.

Model	Training setting	SER	F1-Score
In-domain		**0.2478**	**0.8489**
Monolingual	+1.9M	0.3128	0.7855
	200K	0.3136	0.7827
	100K	**0.3071**	**0.7927**
	50K	0.3193	0.7853
Adapted	+1.9M	0.2482	0.8467
	200K	0.2445	**0.8599**
	100K	0.2469	0.8503
	50K	**0.2404**	0.8540

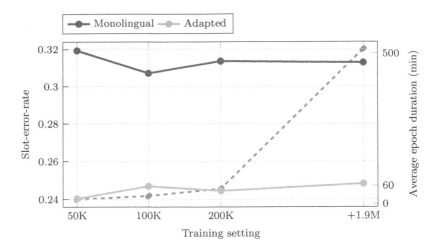

Fig. 2. In dashed, we represent the average duration of an epoch for the initial training of the monolingual model and, in full lines, we represent the SER for the monolingual and adapted models as a function of the original training dataset size. Domain adaptation is the most successful for the model that initially trained faster.

faster to train but also more adaptable to in-domain data, since they are not as prone to overfitting to the training data inner structural style as models that are trained on bigger training datasets.

5.3 Layer-Wise Attention Mechanism

All our models contain a layer-wise dot-product attention mechanism to compute the encoder output as a combination of the output of several encoder layers. This attention mechanism is devised in such a way that layer scalar weights are trained jointly with the encoder layers. By observing Fig. 3, it is clear that some layers contain more significant information than others for solving the truecasing task. Moreover, the effect of fine-tuning the monolingual model on domain data is also felt on the trained weights, in such a way that, generally, its original weight distribution approaches the in-domain model weight distribution.

Center of Gravity.   To better interpret the weight distributions in Fig. 3, we computed the **center of gravity** metric as in [24] for each of the models. Intuitively, higher values indicate that the relevant information for the truecasing task is captured in higher layers. Results are shown in Table 9, and, as expected, they are similar across all the trained models. Moreover, comparing with the results obtained for this metric in [24], we observe that the truecasing task center of gravity is very similar to that of the NER task (6.86). This result further supports the previously mentioned notion of similarity between the task at hand and the NER task.

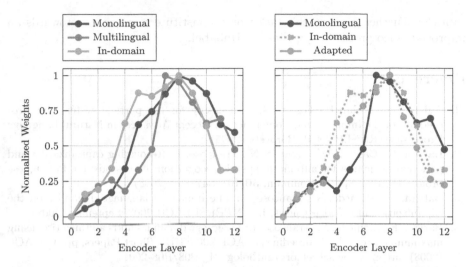

Fig. 3. Normalized weights distribution for the in-domain model, the monolingual and multilingual models trained with the +1.9M setting, and the adapted model initially trained with that same setting. Weight distributions are similar across different models. Moreover, by observing the adapted model weight distribution, we notice that, as expected, the adaptation process brings the weight distribution of the monolingual closer to that of the in-domain model.

Table 9. Center of gravity for the monolingual, adapted model and multilingual trained with the 50K setting.

Model	Training setting	Center of gravity
Monolingual	+1.9M	7.48
Multilingual		7.40
Adapted		6.93
In-domain		7.05

6 Conclusions and Future Work

We made use of pre-trained contextualized word embeddings to train monolingual and multilingual models to solve the truecasing task on transcripts of video subtitles produced by ASR systems. Our architecture, which makes use of a layer attention mechanism to combine information in several encoder layers, yielded consistent and very satisfactory results on the task at hand, outperforming previous state-of-the-art solutions while requiring less data. By performing domain adaptation, we furtherly improved these results, underscoring the notion that models initially trained on less data can adapt better and faster to in-domain data. In the future, we expect improvements on the task by addressing capitalization and punctuation simultaneously in a multitask setting and by making use of additional context by recovering the initial transcripts from the segmented subtitles. Further gains on this task would constitute a major step towards an improved video processing pipeline for Unbabel.

References

1. Agbago, A., Kuhn, R., Foster, G.: Truecasing for the portage system. In: Proceedings of the International Conference on Recent Advances in Natural Language Processing (RANLP 2005), Borovets (2005)
2. Batista, F., Caseiro, D., Mamede, N., Trancoso, I.: Recovering capitalization and punctuation marks for automatic speech recognition: case study for Portuguese broadcast news. Speech Commun. **50**(10), 847–862 (2008)
3. Batista, F., Mamede, N., Trancoso, I.: The impact of language dynamics on the capitalization of broadcast news. In: INTERSPEECH 2008, September 2008
4. Batista, F., Mamede, N., Trancoso, I.: Language dynamics and capitalization using maximum entropy. In: Proceedings of ACL 2008: HLT, Short Papers, pp. 1–4. ACL (2008). http://www.aclweb.org/anthology/P/P08/P08-2001
5. Batista, F., Moniz, H., Trancoso, I., Mamede, N.J.: Bilingual experiments on automatic recovery of capitalization and punctuation of automatic speech transcripts. IEEE Trans. Audio Speech Lang. Process. Spec. Issue New Front. Rich Transcr. **20**(2), 474–485 (2012). https://doi.org/10.1109/TASL.2011.2159594
6. Chelba, C., Acero, A.: Adaptation of maximum entropy capitalizer: little data can help a lot. In: Proceedings of the Conference on Empirical Methods in Natural Language Processing (EMNLP 2004) (2004)
7. Conneau, A., et al.: Unsupervised cross-lingual representation learning at scale. arXiv preprint arXiv:1911.02116 (2019)
8. Devlin, J., Chang, M.W., Lee, K., Toutanova, K.: BERT: pre-training of deep bidirectional transformers for language understanding. In: NAACL-HLT (2019)
9. Falcon, W.E.A.: Pytorch lightning. https://github.com/PytorchLightning/pytorch-lightning (2019)
10. Gravano, A., Jansche, M., Bacchiani, M.: Restoring punctuation and capitalization in transcribed speech. In: Proceedings of the IEEE International Conference on Acoustics, Speech, and Signal Processing (ICASSP 2009), Taipei (2009)
11. Howard, J., Ruder, S.: Universal language model fine-tuning for text classification. In: Proceedings of the 56th Annual Meeting of the Association for Computational Linguistics (Volume 1: Long Papers), pp. 328–339. Association for Computational Linguistics, Melbourne, July 2018. https://doi.org/10.18653/v1/P18-1031. https://www.aclweb.org/anthology/P18-1031
12. Jones, D., et al.: Measuring the readability of automatic speech-to-text transcripts. In: Proceedings of EUROSPEECH, pp. 1585–1588 (2003)
13. Jurafsky, D., Martin, J.H.: Speech and Language Processing: An Introduction to Natural Language Processing, Computational Linguistics, and Speech Recognition, 2nd edn., Prentice Hall PTR (2009)
14. Khare, A.: Joint learning for named entity recognition and capitalization generation. Master's thesis, University of Edinburgh (2006)
15. Kim, J.H., Woodland, P.C.: Automatic capitalisation generation for speech input. Comput. Speech Lang. **18**(1), 67–90 (2004)
16. Kingma, D.P., Ba, J.: Adam: a method for stochastic optimization (2014)
17. Kondratyuk, D., Straka, M.: 75 languages, 1 model: parsing universal dependencies universally (2019). https://www.aclweb.org/anthology/D19-1279

18. Lita, L.V., Ittycheriah, A., Roukos, S., Kambhatla, N.: tRuEcasIng. In: Proceedings of the 41st Annual Meeting on ACL, pp. 152–159. ACL (2003)
19. Liu, Y., et al.: Roberta: a robustly optimized BERT pretraining approach. ArXiv abs/1907.11692 (2019)
20. Makhoul, J., Kubala, F., Schwartz, R., Weischedel, R.: Performance measures for information extraction. In: Broadcast News Workshop (1999)
21. Nguyen, B., et al.: Fast and accurate capitalization and punctuation for automatic speech recognition using transformer and chunk merging (2019)
22. Peters, M.E., et al.: Deep contextualized word representations (2018)
23. Stüker, S., et al.: The ISL TC-STAR spring 2006 ASR evaluation systems. In: Proceedings of the TC-STAR Workshop on Speech-to-Speech Translation, Barcelona, June 2006
24. Tenney, I., Das, D., Pavlick, E.: BERT rediscovers the classical NLP pipeline (2019)
25. Thu, H.N.T., Thai, B.N., Nguyen, V.B.H., Do, Q.T., Mai, L.C., Minh, H.N.T.: Recovering capitalization for automatic speech recognition of Vietnamese using transformer and chunk merging. In: 2019 11th International Conference on Knowledge and Systems Engineering (KSE), pp. 1–5 (2019)
26. Vaswani, A., et al.: Attention is all you need (2017)
27. Wang, W., Knight, K., Marcu, D.: Capitalizing machine translation. In: HLT-NAACL, pp. 1–8. ACL (2006)
28. Wolf, T., et al.: HuggingFace's transformers: state-of-the-art natural language processing. ArXiv abs/1910.03771 (2019)
29. Yadav, V., Bethard, S.: A survey on recent advances in named entity recognition from deep learning models. ArXiv abs/1910.11470 (2018)
30. Yarowsky, D.: Decision lists for lexical ambiguity resolution: application to accent restoration in Spanish and French. In: Proceedings of the 2nd Annual Meeting of the Association for Computational Linguistics (ACL 1994), pp. 88–95 (1994)

Feature Extraction with TF-IDF and Game-Theoretic Shadowed Sets

Yan Zhang[1]([✉]), Yue Zhou[1], and JingTao Yao[2]

[1] School of Computer Science and Engineering, California State University,
San Bernardino, CA, USA
{Yan.Zhang,Yue.Zhou}@csusb.edu
[2] Department of Computer Science, University of Regina, Regina, SK, Canada
jtyao@cs.uregina.ca

Abstract. TF-IDF is one of the most commonly used weighting metrics for measuring the relationship of words to documents. It is widely used for word feature extraction. In many research and applications, the thresholds of TF-IDF for selecting relevant words are only based on trial or experiences. Some cut-off strategies have been proposed in which the thresholds are selected based on Zipf's law or feedbacks from model performances. However, the existing approaches are restricted in specific domains or tasks, and they ignore the imbalance of the number of representative words in different categories of documents. To address these issues, we apply game-theoretic shadowed set model to select the word features given TF-IDF information. Game-theoretic shadowed sets determine the thresholds of TF-IDF using game theory and repetition learning mechanism. Experimental results on real world news category dataset show that our model not only outperforms all baseline cut-off approaches, but also speeds up the classification algorithms.

Keywords: Feature extraction · TF-IDF · Text classification · Game-theoretic shadowed sets

1 Introduction

Term Frequency-Inverse Document Frequency, or TF-IDF, is one of the most commonly used weighting metrics for measuring the relationship of words and documents. It has been applied to word feature extraction for text categorization or other NLP tasks. The words with higher TF-IDF weights are regarded as more representative and are kept while the ones with lower weights are less representative and are discarded. An appropriate selection of word features is able to speed up the information retrieval process while preserving the model performance. However, for many works, the cutoff values or the thresholds of TF-IDF for selecting relevant words is only based on guess or experience [6,11,12]. Zipf's law is used to select words whose IDF exceeds a certain value [10]; Lopes et al. [9] proposed a cut-off policy by balancing the precision and recall from the model performance.

© Springer Nature Switzerland AG 2020
M.-J. Lesot et al. (Eds.): IPMU 2020, CCIS 1237, pp. 722–733, 2020.
https://doi.org/10.1007/978-3-030-50146-4_53

Despite their success, those cut-off policies have certain issues. The cut-off policy that the number of words to keep is determined by looking at the precision and recall score of the model can be restricted in specific domains or tasks. In addition, the number of relevant words may vary in different categories of documents in certain domains. For instance, there exists an imbalance between the number of representative positive words and negative words in many sentiment classification tasks. Thus, a cut-off policy that is able to capture such imbalance is needed.

To address these issues, we employ game-theoretic shadowed sets (GTSS) to determine the thresholds for feature extraction. GTSS, proposed by Yao and Zhang, is a recent promising model for decision making in the shadowed set context [22]. We calculate the difference of TF-IDF for each word between documents as the measurement of relevance, and then use GTSS to derive the asymmetric thresholds for word extraction. GTSS model aims to determine and explain the thresholds from a tradeoff perspective. The words with the difference of TF-IDF less than β or greater than α are selected. We regard the words whose difference of TF-IDF are between α and β as neutral. These words can be safely removed since they can not contribute much in text classification.

The results of our experiments on a real world news category dataset show that our model achieves significant improvement as compared with different TF-IDF based cut-off policies. In addition, we show our model can achieve comparable performance as compared to the model using all words' TF-IDF as features, while greatly speed up the classification algorithms.

2 Related Work

TF-IDF is the most commonly used weighting metrics for measuring the relationship of words and documents. By considering the word or term frequency (TF) in the document as well as how unique or infrequent (IDF) a word in the whole corpus, TF-IDF assigns higher values to topic representative words while devalues the common words. There are many variations of TF-IDF [19,20]. In our experiments, we use the basic form of TF-IDF and follow the notation given in [7]. The TF-IDF weighted value $w_{t,d}$ for the word t in the document d is thus defined as:

$$w_{t,d} = tf_{t,d} \times \log_{10}(\frac{N}{df_t})$$

(1)

where $tf_{t,d}$ is the frequency of word t in the document d, N is the total number of documents in the collection, and df_t is the number of documents where word t occurs in.

TF-IDF measures how relevant a word to a certain category of documents, and it is widely used to extract the most representative words as features for text classification or other NLP tasks. The extraction is often done by selecting top n words with the largest TF-IDF scores or setting a threshold below which the words are regarded as irrelevant and discarded. But an issue arises about how to choose such cut-off point or threshold so as to preserve the most relevant words. Many works choose such threshold only based on trial or experience [6,11,12,23]. On the other hand, some approaches address the issue. Zipf's law is used to select the words whose IDF exceeds a certain value in order to speed up information retrieval algorithms [4,10]. Lopes et al. [9] proposed a cut-off policy which determines the number of words to keep by balancing precision and recall in downstream tasks. However, such cut-off points should not be backward induced by the performance of downstream task; rather, the thresholds should be derived before feeding the extracted words to the classifier to speed up the model without reducing the performance. In addition, for certain domains, the number of relevant words may vary in different categories of documents. For instance, the number of words relevant to positive articles and the number of words relevant to negative articles are often imbalanced in many sentiment analysis tasks. Therefore, the cut-off points or thresholds may also vary in different categories. By observing these drawbacks, we attempt to find asymmetric thresholds of TF-IDF for feature extraction by using game-theoretic shadowed sets.

3   Methodology

In this section, we will introduce our model in details. Our model aims to find an approach of extracting relevant words and discarding less relevant words based on TF-IDF information so as to speed up learning algorithms while preserving the model performance. We first calculate the difference of TF-IDF for each word between documents as one single score to measure the degree of relevance, and then use game-theoretic shadowed sets to derive the asymmetric thresholds for words extraction.

3.1   TF-IDF Difference as Relevance Measurement

Consider a binary text classification task with a set of two document classes $C = \{c_1, c_2\}$. For each word t, we calculate the difference of TF-IDF weighted value between document c_1 and c_2 as:

$$DW_t = w_{t,c_1} - w_{t,c_2} = (tf_{t,c_1} - tf_{t,c_2}) \times \log_{10}(\frac{N}{df_t}). \tag{2}$$

We use DW_t to measure how relevant or representative the word t is to the document classes. The greater the magnitude of DW_t, the more representative the word t is to distinguish the document categories. A large positive value of DW_t indicates that a word t is not common word and more relevant to the document c_1, while a significant negative value shows the word t is representative to document c_2. If DW_t is closed to zero, then we regard the word t as neutral. In the next section, we will choose the cut-off thresholds for selecting the most representative word features by using the Game-theoretic Shadowed Sets method. For convenience, we here normalize the DW_t with min-max linear transformation.

3.2   Shadowed Sets

A shadowed set S in the universe U maps the membership grades of the objects in U to a set $\{0, [0,1], 1\}$, i.e., $S : U \rightarrow \{0, [0,1], 1\}$ a pair of thresholds (α, β) while $0 \leq \beta < \alpha \leq 1$ [13]. Shadowed sets are viewed as three-valued constructs induced by fuzzy sets, in which three values are interpreted as full membership, full exclusion, and uncertain membership [15]. Shadowed sets can capture the essence of fuzzy sets at the same time reducing the uncertainty from the unit interval to a shadowed region [15]. The shadowed set based three-value approximations are defined as a mapping from the universe U to a three-value set $\{0, \sigma, 1\}$, if a single value σ $(0 \leq \sigma \leq 1)$ is chosen to replace the unit interval $[0, 1]$ in the shadowed sets, that is [3],

$$T_{(\alpha,\beta)}(\mu_A(x)) = \begin{cases} 1, & \mu_A(x) \geq \alpha, \\ 0, & \mu_A(x) \leq \beta, \\ \sigma, & \beta < \mu_A(x) < \alpha. \end{cases} \tag{3}$$

The membership grade $\mu_A(x)$ of an object x indicates the degree of the object x belonging to the concept A or the degree of the concept A applicable to x [21].

Given a concept A and an element x in the universe U, if the membership grade of this element $\mu_A(x)$ is greater than or equal to α, the element x would be considered to belong to the concept A. An elevation operation elevates the membership grade $\mu_A(x)$ to 1 which represents a full membership grade [14]. If the membership grade $\mu_A(x)$ is less than or equal to β, the element x would not be considered to belong to the concept A. An reduction operation reduces the membership grade $\mu_A(x)$ to 0 which represents a null membership grade [14]. If the membership grade $\mu_A(x)$ is between α and β, the element x would be put in a shadowed area, which means it is hard to determine if x belongs to concept A. $\mu_A(x)$ is mapped to σ which represents the highest uncertainty, that is we are far more confident about including an element or excluding an element in the concept A. The membership grades between α and σ are reduced to σ; The membership grades between σ and β are elevated to σ. We get two elevated areas, $E_1(\mu_A)$ and $E_\sigma(\mu_A)$, and two reduced areas, $R_0(\mu_A)$ and $R_\sigma(\mu_A)$ shown as the dotted areas and lined areas in Fig. 1 (a). Figure 1 (b) shows the shadowed set based three-value approximation after applying the elevation and reduction operations on all membership grades.

The vagueness is localized in the shadowed area as opposed to fuzzy sets where the vagueness is spread across the entire universe [5,16].

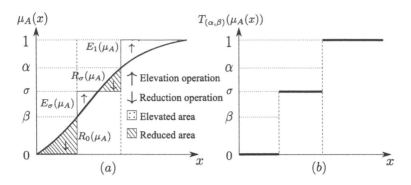

Fig. 1. A shadowed set based three-value approximation

3.3   Error Analysis

Shadowed set based three-value approximations use two operations, the elevation and reduction operations, to approximate the membership grades $\mu_A(x)$ to a three-value set $\{0, \sigma, 1\}$. Given an element x with the membership grade $\mu_A(x)$, the elevation operation changes the membership grade $\mu_A(x)$ to 1 or σ. The reduction operation changes the membership grade $\mu_A(x)$ to 0 or σ. These two operations change the original membership grades and produce the elevated and reduced areas which show the difference between the original membership grades and the mapped values 1, σ, and 0, as shown in Fig. 1(a). These areas can be viewed as the elevation and the reduction errors, respectively. The elevation operation produces two elevation errors $E_1(\mu_A)$ and $E_\sigma(\mu_A)$; the reduction operation produces two reduction errors $R_0(\mu_A)$ and $R_\sigma(\mu_A)$, that is

- The elevation error E_1 is produced when the membership grade $\mu_A(x)$ is greater than or equal to α (i.e., $\mu_A(x) \geq \alpha$), and the elevation operation elevates $\mu_A(x)$ to 1. We have $E_1(\mu_A(x)) = 1 - \mu_A(x)$.
- The elevation error E_σ is produced when $\beta < \mu_A(x) < \sigma$, and the elevation operation elevates $\mu_A(x)$ to σ. We have $E_\sigma(\mu_A(x)) = \sigma - \mu_A(x)$.
- The reduction error R_0 is produced when $\mu_A(x) \leq \beta$, and the reduction operation reduces $\mu_A(x)$ to 0. We have $R_0(\mu_A(x)) = \mu_A(x)$.
- The reduction error R_σ is produced when $\sigma < \mu_A(x) < \alpha$, and the reduction operation reduces $\mu_A(x)$ to σ. We have $R_\sigma(\mu_A(x)) = \mu_A(x) - \sigma$.

The elevation errors $E_{(\alpha,\beta)}(\mu_A)$ is the sum of two elevation errors produced by elevation operation. The total reduction errors $R_{(\alpha,\beta)}(\mu_A)$ is the sum of two reduction errors produced by reduction operation. For discrete universe of discourse, we have a collection of membership values. The total elevation and reduction errors are calculated as [22],

$$E_{(\alpha,\beta)}(\mu_A) = E_1(\mu_A) + E_\sigma(\mu_A)$$

$$= \sum_{\mu_A(x) \geq \alpha} E_1(\mu_A(x)) + \sum_{\beta < \mu_A(x) < \sigma} E_\sigma(\mu_A(x))$$

$$= \sum_{\mu_A(x) \geq \alpha} (1 - \mu_A(x)) + \sum_{\beta < \mu_A(x) < \sigma} (\sigma - \mu_A(x)), \qquad (4)$$

$$R_{(\alpha,\beta)}(\mu_A) = R_0(\mu_A) + R_\sigma(\mu_A)$$

$$= \sum_{\mu_A(x) \leq \beta} R_0(\mu_A(x)) + \sum_{\sigma < \mu_A(x) < \alpha} R_\sigma(\mu_A(x))$$

$$= \sum_{\mu_A(x) \leq \beta} \mu_A(x) + \sum_{\sigma < \mu_A(x) < \alpha} (\mu_A(x) - \sigma). \qquad (5)$$

Given a fixed σ, the elevation and reduction errors change when the thresholds (α, β) change. No matter which threshold changes and how they change, the elevation and reduction errors always change in opposite directions [22]. The decrease of one type of errors inevitably brings the increase of the other type of errors. The balanced shadowed set based three-value approximations are expected to represent a tradeoff between the elevation and reduction errors.

3.4   Game-Theoretic Shadowed Sets

Game-theoretic shadowed sets (GTSS) use game theory to determine the thresholds in the shadowed set context. The obtained thresholds represent a tradeoff between two different types of errors [22]. GTSS use a game mechanism to formulate games between the elevation and reduction errors. The strategies performed by two players are the changes of thresholds. Two game players compete with each other to maximize their own payoffs. A repetition learning mechanism is adopted to approach a compromise between two players by modifying game formulations repeatedly. The resulting thresholds are determined based on the game equilibria analysis and selected stopping criteria.

Game Formulation. Three elements should be considered when formulating a game G, i.e., game player set O, strategy profile set S, and utility functions u, $G = (O, S, u)$ [8,17]. The game players are the total elevation and reduction errors which are denoted by E and R, i.e., $O = \{E, R\}$.

The strategy profile set is $S = S_E \times S_R$, where $S_E = \{s_1, s_2, ..., s_{k1}\}$ is a set of possible strategies for player E, and $S_R = \{t_1, t_2, ..., t_{k2}\}$ is a set of possible strategies for player R. We select (σ, σ) as the initial threshold values, which represent that we do not have any uncertainty on all membership grades and we have the smallest shadowed area. Starting from (σ, σ), we gradually make α and β further to each other and increase the shadowed area. c_E and c_R are two constant change steps, denoting the quantities that two players E and R use to change the thresholds, respectively. For example, we set the initial threshold values $(\alpha, \beta) = (0.5, 0.5)$. The player E performs increasing α and the player R performs decreasing β. When we set $c_E = 0.01$ and

$c_R = 0.02$, we have $S_E = \{\alpha$ no change, α increases $0.01, \alpha$ increases $0.02\}$, and $S_R = \{\beta$ no change, β decreases $0.02, \beta$ decreases $0.02\}$.

The payoffs of players are $u = (u_E, u_R)$, and u_E and u_R denote the payoff functions of players E and R, respectively. The payoff functions $u_E(\alpha, \beta)$ and $u_R(\alpha, \beta)$ are defined by the elevation and reduction errors, respectively, that is,

$$u_E(\alpha, \beta) = E_{(\alpha,\beta)}(\mu_A),$$
$$u_R(\alpha, \beta) = R_{(\alpha,\beta)}(\mu_A), \tag{6}$$

where $E_{(\alpha,\beta)}(\mu_A)$ and $R_{(\alpha,\beta)}(\mu_A)$ are defined in Eqs. (4) and (5). We try to minimize the elevation and reduction errors, so both players try to minimize their payoff values.

We use payoff tables to represent two-player games. Table 1 shows a payoff table example in which both players have 3 strategies.

Table 1. An example of a payoff table

		R		
		β	$\beta \searrow c_R$	$\beta \searrow 2c_R$
E	α	$\langle u_E(\alpha, \beta), u_R(\alpha, \beta) \rangle$	$\langle u_E(\alpha, \beta - c_R),$ $u_R(\alpha, \beta - c_R) \rangle$	$\langle u_E(\alpha, \beta - 2c_R),$ $u_R(\alpha, \beta - 2c_R) \rangle$
	$\alpha \nearrow c_E$	$\langle u_E(\alpha + c_E, \beta),$ $u_R(\alpha + c_E, \beta) \rangle$	$\langle u_E(\alpha + c_E, \beta - c_R),$ $u_R(\alpha + c_E, \beta - c_R) \rangle$	$\langle u_E(\alpha + c_E, \beta - 2c_R),$ $u_R(\alpha + c_E, \beta - 2c_R) \rangle$
	$\alpha \nearrow 2c_E$	$\langle u_E(\alpha + 2c_E, \beta),$ $u_R(\alpha + 2c_E, \beta) \rangle$	$\langle u_E(\alpha + 2c_E, \beta - c_R),$ $u_R(\alpha + 2c_E, \beta - c_R) \rangle$	$\langle u_E(\alpha + 2c_E, \beta - 2c_R),$ $u_R(\alpha + 2c_E, \beta - 2c_R) \rangle$

Repetition Learning Mechanism. The involved players are trying to maximize their own payoffs in the formulated games. But one player's payoff is effected by the strategies performed by the other player. The balanced solution or game equilibrium is a strategy profile from which both players benefit. This game equilibrium represents both players reach a compromise or tradeoff on the conflict. The strategy profile (s_i, t_j) is a pure strategy Nash equilibrium, if for players E and R, s_i and t_j are the best responses to each other [17], this is,

$$\forall s_k \in S_E, \quad u_E(s_i, t_j) \geqslant u_E(s_k, t_j), \quad \text{where } s_i, s_k \in S_E \text{ and } k \neq i, t_j \in S_R,$$
$$\forall t_l \in S_R, \quad u_R(s_i, t_j) \geqslant u_R(s_i, t_l), \quad \text{where } t_j, t_l \in S_R \text{ and } l \neq j, s_i \in S_E. \tag{7}$$

The above equations can be interpreted as a strategy profile such that no player would like to change his/her strategy or they would loss benefit if deriving from this strategy profile, provided this player has the knowledge of other player's strategies.

The equilibrium of the current formulated game means the threshold pair corresponding this equilibrium are the best choices within the current strategy sets. We have to check if there are some threshold pairs near the current equilibrium that are better than the current ones. Thus we repeat the games with the updated initial thresholds. We may be able to find more suitable thresholds with repetition of thresholds modification.

We define the stopping criteria so that the iterations of games can stop at a proper time. There are many possible stopping criteria. For example, the payoff of each player is beyond a specific value; the thresholds (α, β) violate the constraint $0 \leq \beta \leq \sigma \leq \alpha \leq 1$; the current game equilibrium does not improve the payoffs gained by both players under the initial thresholds; no equilibrium exists. In this research, we compare the payoffs of both players under the initial thresholds and the thresholds corresponding to the current equilibrium. We set the stopping criteria as one of the players increases its payoff values, or there does not exist a pure strategy Nash equilibrium in the current game.

3.5 Feature Extraction

We now select the most representative words by applying the thresholds (α, β) derived in previous sections. The words with normalized DW_t being greater than the upper threshold α and less than the lower threshold β will be kept as our word features for text classification while the rest words are discarded.

4 Experiments

4.1 Dataset and Evaluation Metrics

We evaluate our approach on the HuffPost news category dataset [18]. This dataset consists of 200,853 news headlines with short descriptions from HuffPost website during the year 2012 to 2018. It contains 31 categories of news such as politics, entertainment, business, healthy living, art, and so forth. We use the largest two categories, the 32,739 politics news and 14,257 entertainment news, as the binary text classification data in our experiments. The news text is obtained by concatenating the news headline and the corresponding short description. We extract 381449 words from these selected news. We use 80% data for training and 20% for testing, and adopt accuracy and F1 scores as metrics for model evaluation.

4.2 Deriving Thresholds with GTSS

We first normalize DW_t using min-max normalization linear transformation. The distribution of normalized DW_t is shown in Fig. 2 Almost 80% of words have the normalized DW_t 0.548054 so we set $\sigma = 0.5481$ aiming to minimize the errors produced by mapping all DW_t values to three values $\{0, \sigma, 1\}$ via game-theoretic shadowed set model. If we set σ as other value instead of 0.548054, mapping the large amount of DW_t 0.548054 to σ definitely will produce more errors.

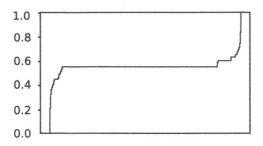

Fig. 2. Normalized DW_t information

We formulate a competitive game between the elevation and reduction errors to obtain the thresholds. The game players are elevation and reduction errors, i.e., $O = \{E, R\}$. The strategy profile set is $S = \{S_E \times S_R\}$. The game is being played with the initial thresholds $(\alpha, \beta) = (0.55, 0.54)$. Here, $\alpha = 0.55$ is the smallest value that greater than $\sigma = 0.5481$, and $\beta = 0.54$ is the largest value that less than $\sigma = 0.5481$. The player E and R try to increase α and decrease β with the change steps as 0.01 and 0.02, respectively. The strategy set of E is $S_E = \{\alpha$ no change, α increases 0.01, α increases 0.02$\}$. The corresponding α values are 0.55, 0.56, and 0.57. The strategy set of R is $S_R = \{\beta$ no change, β decreases 0.02, β decreases 0.4$\}$. The corresponding α values are 0.54, 0.52, and 0.5. The payoff functions are defined in Eqs. (4) and (5). Table 2 is the payoff table. The cell at the right bottom corner is the game equilibrium whose strategy profile is (α increases 0.02, β decreases 0.04). The payoffs of the players are (17689, 8742). We set the stopping criterion as one of players' payoff increases. When the thresholds change from (0.55, 0.54) to (0.57, 0.5), the elevation error is decreased from 17933 to 17689, and the reduction error is decreased from 10472 to 8742. We repeat the game by setting $(0.57, 0.5)$ as the initial thresholds.

Table 2. The payoff table

		R		
		β	$\beta \downarrow 0.02$	$\beta \downarrow 0.04$
E	α	<17933, 10472>	<17965, 9624>	<18034, 8733>
	$\alpha \uparrow 0.01$	<17690, 10476>	<17722, 9628>	<17792, 8738>
	$\alpha \uparrow 0.02$	<17588, 10480>	<17620, 9632>	**<17689, 8742>**

The competitive games are repeated four times. The result is shown in Table 3. In the fourth iteration, we find out that the payoff value of player E increases. The repetition of game is stopped and the final result is the initial thresholds of the fourth game $(\alpha, \beta) = (0.61, 0.42)$.

Table 3. The repetition of game

	Initial(α, β)	Result(α, β)	Payoffs	both decrease?
1	(0.55, 0.54)	(0.57, 0.5)	<17689, 8742>	√
2	(0.57, 0.5)	(0.59, 0.46)	<17686, 6516>	√
3	(0.59, 0.46)	(0.61, 0.42)	<7832, 3934>	√
4	(0.61, 0.42)	(0.63, 0.38)	<8221, 2693>	×

We got $(\alpha, \beta) = (0.61, 0.42)$, which means we keep the words with DW_t greater than 0.61 and less then 0.42, and discard the words with DW_t between 0.61 and 0.42.

4.3 Baselines

We calculate the DW_t value for each single word and bi-gram, and then use Support Vector Machine (SVM) [1,2] as our unique classifier to compare our approach with: (1) **ALL**, in which we keep all words with no feature extraction; (2) **Sym-Cutoff**, in which the symmetrical cut-off values are drawn purely based on a simple observation of the statistical distribution of DW_t; (3) **Sym-N-Words**, where we select $2n$ words given n smallest DW_t and n largest DW_t such that $2n$ is approximately equal to the total number of words extracted with our approach.

4.4 Results

We show the model performance of different extraction approaches on the new category dataset in Table 4. Our model is named as "Asym-GTSS-TH". From the results, we can observe that: (1) Our approach achieves superior performance compared with Sym-Cutoff which is purely based on a guess given TF-IDF distribution. It verifies our claim that the GTSS can better capture the pattern of TF-IDF and provide a more robust range for selecting relevant words given TF-IDF for text classification; (2) The Sym-N-Words approach achieves close performance as ours, because it takes the advantage of the information of the number of words to keep derived with our thresholds. However, our approach still outperforms the Sym-N-words approach since it evenly selects relevant words for document c_1 and c_2. It indicates that there exists imbalance of representative words between different categories of documents which is better captured by our model; (3) Compared with using all words' TF-IDF score as input, our model discards more than 52% words and speed up the process of classification while preserving the performance.

Table 4. Summary of the performance of different feature extraction approaches

	Accuracy	F1-score
ALL	94.7	92.3
Sym-Cutoff	91.9	88.2
Sym-N-Words	94.5	91.7
Asym-GTSS-TH	94.6	91.9

5   Conclusion

In this paper, we propose a feature extraction approach based on TF-IDF and game-theoretic shadowed sets in which the asymmetric thresholds for selecting relevant words are derived by repetitive learning on the difference of TF-IDF for each word between documents. Our model can explore the pattern of TF-IDF distribution as well as capture the imbalance of the number of representative words in different categories. The experimental results on the news category dataset show that our model can achieve improvement as compared to other cut-off policies and speed up the information retrieval process. In the future, we will explore the consistency of our model performance on more real world datasets and test the generalization ability of our GTSS model on different metrics that measures the relevance of words, such as BNS and Chi-square.

References

1. Ben-Hur, A., Horn, D., Siegelmann, H.T., Vapnik, V.: Support vector clustering. J. Mach. Learn. Res. **2**(12), 125–137 (2001)
2. Cortes, C., Vapnik, V.: Support-vector networks. Mach. Learn. **20**(3), 273–297 (1995)
3. Deng, X.F., Yao, Y.Y.: Decision-theoretic three-way approximations of fuzzy sets. Inf. Sci. **279**, 702–715 (2014)
4. Grossman, D.A., Frieder, O.: Information Retrieval: Algorithms and Heuristics, vol. 15. Springer, Netherlands (2012). https://doi.org/10.1007/978-1-4020-3005-5
5. Grzegorzewski, P.: Fuzzy number approximation via shadowed sets. Inf. Sci. **225**, 35–46 (2013)
6. Jing, L.P., Huang, H.K., Shi, H.B.: Improved feature selection approach TFIDF in text mining. In: Proceedings of the International Conference on Machine Learning and Cybernetics, vol. 2, pp. 944–946. IEEE (2002)
7. Jurafsky, D., Martin, J.H.: Speech and Language Processing: An Introduction to Natural Language Processing, Computational Linguistics, and Speech Recognition. Pearson Education, Inc., New Jersey (2008)
8. Leyton-Brown, K., Shoham, Y.: Essentials of game theory: a concise multidisciplinary introduction. Synth. Lect. Artif. Intell. Mach. Learn. **2**(1), 1–88 (2008)
9. Lopes, L., Vieira, R.: Evaluation of cutoff policies for term extraction. J. Braz. Comput. Soc. **21**(1), 1–13 (2015). https://doi.org/10.1186/s13173-015-0025-0
10. Manning, C.D., Raghavan, P., Schütze, H.: Introduction to Information Retrieval. Cambridge University Press, Cambridge (2008)

11. Milios, E., Zhang, Y., He, B., Dong, L.: Automatic term extraction and document similarity in special text corpora. In: Proceedings of the 6th Conference of the Pacific Association for Computational Linguistics, pp. 275–284. Citeseer (2003)
12. Özgür, A., Özgür, L., Güngör, T.: Text categorization with class-based and corpus-based keyword selection. In: Yolum, I., Güngör, T., Gürgen, F., Özturan, C. (eds.) ISCIS 2005. LNCS, vol. 3733, pp. 606–615. Springer, Heidelberg (2005). https://doi.org/10.1007/11569596_63
13. Pedrycz, W.: Shadowed sets: representing and processing fuzzy sets. IEEE Trans. Syst. Man Cybern. **28**(1), 103–109 (1998)
14. Pedrycz, W.: Shadowed sets: bridging fuzzy and rough sets. In: Rough Fuzzy Hybridization: A New Trend in Decision-Making, pp. 179–199 (1999)
15. Pedrycz, W.: From fuzzy sets to shadowed sets: interpretation and computing. Int. J. Intell. Syst. **24**(1), 48–61 (2009)
16. Pedrycz, W., Vukovich, G.: Granular computing with shadowed sets. Int. J. Intell. Syst. **17**(2), 173–197 (2002)
17. Rasmusen, E.: Games and Information: An Introduction to Game Theory. Blackwell, Oxford (1989)
18. Misra, R.: News category dataset from HuffPost website (2018). https://doi.org/10.13140/RG.2.2.20331.18729
19. Salton, G., Buckley, C.: Term-weighting approaches in automatic text retrieval. Inf. Process. Manag. **24**(5), 513–523 (1988)
20. Salton, G., McGill, M.J.: Introduction to Modern Information Retrieval. McGraw-Hill Inc. (1986)
21. Zadeh, L.A.: Toward extended fuzzy logic - a first step. Fuzzy Sets Syst. **160**(21), 3175–3181 (2009)
22. Zhang, Y., Yao, J.T.: Game theoretic approach to shadowed sets: a three-way tradeoff perspective. Inf. Sci. **507**, 540–552 (2020)
23. Zuo, Z., Li, J., Anderson, P., Yang, L., Naik, N.: Grooming detection using fuzzy-rough feature selection and text classification. In: 2018 IEEE International Conference on Fuzzy Systems (FUZZ-IEEE), pp. 1–8. IEEE (2018)

To BERT or Not to BERT Dealing with Possible BERT Failures in an Entailment Task

Pedro Fialho[1,3](✉) , Luísa Coheur[1,2] , and Paulo Quaresma[1,3]

[1] INESC-ID Lisboa, Lisbon, Portugal
peter.fialho@gmail.com
[2] Instituto Superior Tecnico, Universidade de Lisboa, Lisbon, Portugal
[3] Universidade de Évora, Évora, Portugal

Abstract. In this paper we focus on an Natural Language Inference task. Being given two sentences, we classify their relation as NEUTRAL, ENTAILMENT or CONTRADICTION. Considering the achievements of BERT (Bidirectional Encoder Representations from Transformers) in many Natural Language Processing tasks, we use BERT features to create our base model for this task. However, several questions arise: can other features improve the performance obtained with BERT? If we are able to predict the situations in which BERT will fail, can we improve the performance by providing alternative models for these situations? We test several strategies and models, as alternatives to the standalone BERT model in the possible failure situations, and we take advantage of semantic features extracted from Discourse Representation Structures.

Keywords: Natural Language Inference · Feature engineering · Failure prediction model

1 Introduction

Natural Language Inference (NLI) is a known task in Natural Language Processing (NLP)[1]. It can be implemented as a classification task in which the model needs to decide about the relation between a pair of sentences. Usual categories are ENTAILMENT, NEUTRAL and CONTRADICTION.

BERT (Bidirectional Encoder Representations from Transformers) [7] is a state-of-the-art language model that has shown impressive performance on many NLP tasks. Here, we take advantage of BERT to perform NLI. However, we also implement other NLI classifiers, based on lexical and semantic features that we extract from the Discourse Representation Structures obtained for each pair of sentences we want to classify. Then, we implement two strategies to detect possible failures. The first is based on the fact that BERT has lower results in ENTAILMENT and CONTRADICTION situations. Therefore, we run BERT and directly accept the NEUTRAL labels, while other classifiers are employed

© Springer Nature Switzerland AG 2020
M.-J. Lesot et al. (Eds.): IPMU 2020, CCIS 1237, pp. 734–747, 2020.
https://doi.org/10.1007/978-3-030-50146-4_54

in the other cases. In addition, we also implement several models that try to predict when BERT will fail. In the latter cases, other models are employed. Results show that we can improve results with the models based on lexical and semantic features.

This paper is organized as follows: Sect. 2 presents related work and Sect. 3 our models. Section 4 describes the experimental setup and Sect. 5 the results. Finally, Sect. 6 presents the main conclusions and future work.

2   Related Work

A benchmark for systems aimed at Recognizing Textual Entailment (RTE) was initially developed in the PASCAL challenge series [2]. The RTE task is to detect entailment between a premise and an hypothesis, while a related task is to detect NLI, where target labels are ENTAILMENT, CONTRADICTION and NEU-TRAL (no semantic relation).

NLI is represented in the SICK corpus [13], composed by 10000 pairs of sentences, seeded from corpora of image and video captions, and expanded by rule based transformations to introduce particular linguistic phenomena, such as negations. SICK is annotated by crowd-sourcing, and was the target of a shared task on the Semeval evaluation series [12].

Following SICK, the much larger SNLI [5] corpus was released, containing 570000 examples also seeded from a corpus of captions and annotated by crowd-sourcing, but instead expanded by crowd-sourcing. SNLI inspired the creation of other corpora on NLI, for instance the e-SNLI corpus [6] that augments SNLI with natural language explanations for the annotations, or the MultiNLI corpus [21], that follows the same design procedure and size of SNLI, but instead of captions includes sentence pairs from other text genres and sources, such as fiction books or transcripts of conversations. MultiNLI is one of the targets of the GLUE benchmark [20], that evaluates systems for their joint performance on multiple Natural Language Understanding (NLU) tasks.

Various forms of assessing NLI are presented in the mentioned shared tasks and benchmarks. However, as modern machine learning architectures partic-ularly leverage large data collections, recent approaches suitable for NLI are mostly applied to corpora such as SNLI or MultiNLI, both for their greater size and complexity. One of such approaches is the BERT model [7].

BERT generates a dynamic embedding according to the context in which a word is employed, and may even generate the embedding of a sentence pair, if the aim is to verify entailment on the pair [7]. Training a BERT model is expensive on time and resources, but models based on Wikipedia were made available in its original release.

The BERT model achieves competitive results on various NLU tasks, as shown from its performance on the GLUE benchmark [7], but also specifically in NLI, such as when applied only to MultiNLI [7], to SNLI [22], or to the recent CommitmentBank corpus [10] which is part of the SuperGLUE benchmark [19], that supersedes GLUE.

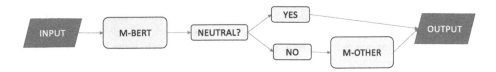

Fig. 1. Simple model – M-BERT directly used to detect NEUTRAL relations.

Recent studies on the generalization of various models, including BERT, suggest that performance is only consistent when assessed within the same benchmark [18], from combining train and test sets of different corpora. Other works focus specifically on BERT failures in NLI, such as in [14] to hypothesize that the success of BERT relies on the occurrence of certain linguistic patterns in the data, or in [10] to suggest that BERT does not implicitly learn linguistic priors and is mostly driven by statistical regularities. To the best of our knowledge, the performance of BERT in the SICK corpus was not yet evaluated.

3   Entailment and Failure Models

In this section we describe the models we use to perform the NLI task and the strategies we have implemented to predict when BERT will fail.

The BERT model, trained to perform NLI, uses BERT embeddings as features. From now on we will call M-BERT to this model. A set of lexical and semantic features, alone or associated with BERT embeddings, are also used to train several classifiers that perform NLI. We call M-OTHER to these models. Our semantic features are based on Discourse Representation Structures (DRS), that is, a formal representation of meaning that follows the Discourse Representation Theory [11].

Our first strategy (from now on STRATEGY 1) takes advantage of M-BERT results to decide which are the possible failure conditions. We have observed that BERT has lower results in both ENTAILMENT or CONTRADICTION situations. Thus, we run M-BERT and accept all the NEUTRAL labels, according to it. For the remaining labels we run the M-OTHER models, trained in the NLI task, but in a corpus that only has ENTAILMENT or CONTRADICTION labels. Figure 1 depicts this strategy.

We also implement a second strategy (from now on STRATEGY 2) in which we train several models that try to predict when BERT will fail. The previous mentioned lexical and semantic features, along with BERT, are used by these models. We call M-FAIL to these models. Here, the idea is the following: if a model of type M-FAIL predicts that BERT will fail, then the previous models, trained in the NLI task, are used instead of M-BERT. Figure 2 illustrates this strategy.

Finally, instead of using a single M-FAIL model to predict M-BERT failure, we consider the predictions of the different M-FAIL models. Three options are considered:

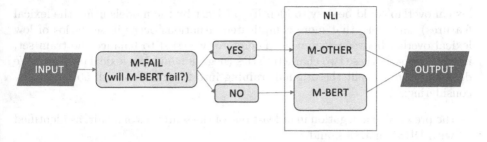

Fig. 2. Pipeline with M-FAIL models.

- *at least one*: if one model from the M-FAIL family returns an M-BERT failure, we will consider that M-BERT will fail;
- *majority voting*: if the majority of the models from the M-FAIL family returns an M-BERT failure, we will consider that M-BERT will fail;
- *all*: if all the models from the M-FAIL family returns a M-BERT failure, we will consider that M-BERT will fail

Considering the previous scheme, the different M-OTHER models will be used if M-BERT is expected to fail.

4 Experimental Setup

4.1 Corpora

Our experiments rely on the SICK corpus [13] for English. As previously said, sentences in SICK are image captions obtained by crowd-sourcing. Each instance in SICK, that is, each pair of sentences, is labelled as NEUTRAL, ENTAIL-MENT or CONTRADICTION regarding the semantic relation between the two sentences. For instance, the pair composed by the sentences "Three kids are jumping in the leaves" and "Three boys are jumping in the leaves", is labeled as ENTAILMENT, while the former sentence paired with "Three kids are sitting in the leaves" is labeled as NEUTRAL. An example of a pair labeled as CONTRADICTION in SICK is the pair composed by the sentences "Nobody is riding the bicycle on one wheel" and "A person is riding the bicycle on one wheel".

We follow the partitions suggested in [13], but 5 SICK instances were discarded as the DRS parser, Boxer [4], was unable to process them. Therefore, our train, development and test set have 4436, 495 and 4904 pairs of sentences, respectively. Notice that the train set is unbalanced, as 2522 pairs are labelled as NEUTRAL, 1274 as ENTAILMENT and 640 as CONTRADICTION.

Balancing the Training Data. In preliminary experiments, we have observed that when a negation was involved in a sentence, the classifiers found more difficult to return the appropriate label . In addition, we consider that a strong lexical overlap could be easy to identify (at least by the models using the lexical features), and thus, that more complicated situations occur in scenarios of low lexical overlap between sentences. Therefore, we tried to balance the train set, in what respects these two characteristics (negation and low lexical overlap). We decided, then, to split the original training into 2 partitions with 50% each, by considering:

- the presence of a negation in at least one of the sentences of a pair, as identified with DRS semantics, and
- low lexical overlap, as identified by a Jaccard score lower than 0.6 or a BLEU score lower than 0.5.

In the 2522 NEUTRAL instances in the original train set, 416 have a negation and 2188 have low lexical overlap. In the 1274 ENTAILMENT instances, 10 have a negation and 634 have low lexical overlap. Finally, from the 640 CON-TRADICTION instances, 575 have a negation and 318 have low lexical overlap. Hence, training instances that contain a negation are almost equally distributed among NEUTRAL and CONTRADICTION classes, and most of the examples from these classes have low lexical overlap. Negations are almost not employed in examples of the ENTAILMENT class, and there are as much examples with low lexical overlap as those with high lexical overlap. We split the original train set in two, each containing 50% of the examples from each class, and 50% of the examples that comply with the above features. For instance, the first set contains 319 examples of the CONTRADICTION class, of which 287 employ a negation and 166 have low lexical overlap.

Building Corpora for Strategy 1 and 2. In order to implement STRATEGY 1, the one that takes advantage of M-BERT results, we removed from the train corpus the NEUTRAL relation and train the M-OTHER models in order to distinguish ENTAILMENT from CONTRADICTION situations.

Concerning STRATEGY 2, and in order to create a reference to train the M-FAIL models (the FAIL-CORPUS), we split the training set in two (as previously described). In the first half we trained M-BERT. Then, we run it on the second half, to build the corpus to train the M-FAIL model: every time M-BERT successfully labelled an NLI relation, the associated sentence pair was labeled as 1; it was labeled as 0 otherwise. As usually, the development set was used for tuning (and first tests) and the test set for the final evaluation. Figure 3 details these partitions.

As we will see, since M-BERT model is successful in most examples, the dataset to train M-FAIL models is unbalanced. Therefore, to train M-FAIL models we discard examples where BERT succeeded until reaching the same number of examples where BERT failed to identify the entailment class, hence obtaining a balanced FAIL-CORPUS.

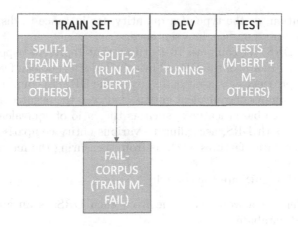

TRAIN SET		DEV	TEST
SPLIT-1 (TRAIN M-BERT+M-OTHERS)	SPLIT-2 (RUN M-BERT)	TUNING	TESTS (M-BERT + M-OTHERS)

FAIL-CORPUS (TRAIN M-FAIL)

Fig. 3. Corpus partition.

4.2 Evaluation Metrics

The performance of our system on entailment detection is measured with Accuracy, Precision, Recall and F-measure (F1, as we consider precision and recall to have the same weight/importance). All metrics produce values between 0 and 1, where greater values are better, hence we report results in percentages.

As the entailment task on SICK configures a multi class classification setup, and the Precision, Recall and F1 metrics are based on the assumption that a positive label exists (as in binary classification), we calculate such metrics using an average of scores from binary classifications, one for each class such that the positive label represents belonging to the class. We chose to average by a weighted mean that considers the number of instances of each class, since class distribution is imbalanced in SICK.

Our definition of accuracy also considers class imbalance. In a multi class setting, the accuracy is defined per class, and obtained by dividing each element in the diagonal of the confusion matrix (true positives per class) by the sum of elements in the corresponding row (the total number of examples of a class). The balanced accuracy is the arithmetic mean of the per class accuracy values.

4.3 Features

Lexical Features. We employ the INESC-ID@ASSIN [9] system that generates almost 100 features for a pair of sentences, based on the lexical aspects of their words or by using some similarity measure. Examples of such features are the length of the longest sentence, or the BLEU [16] metric.

Semantic Features. We obtain DRSs from the Boxer framework [4], containing semantic aspects for each sentence, such as the implicit entities resulting from pronoun resolution, or the type of a quantity, for instance to distinguish parts of a date from other numbers in a sentence.

Given two DRS, we compute 16 features that represent aspects shared by both or occurring in any of the DRS. These include: a) boolean features, such as to indicate the presence of a negation in any of the DRS; b) count based features, such as for the number of equivalent entities between the negated subsets of each DRS; c) percentage based features, such as the ratio of equivalent entities and total entities in both DRS, according to various entity comparison techniques, and; d) distance based features, such as from measuring the mean gap between dates from each DRS.

Entities within DRS are considered:

- not equivalent, if a word pair, one from each DRS, is an antonym in the WordNet [8] database;
- equivalent, if it is a synonym in WordNet;
- equivalent if the cosine of their FastText [3] embeddings is greater than 0.4. This threshold was chosen by observation, and as a compromise between the cosines for synonyms and antonyms sampled from WordNet.

Any technique for entity comparison results in 2 features, one for the count of entities matched and the other for the percentage of entities matched in the total count of entities of both DRS.

Other than entities, a DRS is also composed of conditions, defined as relations between a source and a target entity. We consider the target entities from a pair of conditions of the same type, one from each DRS, as equivalent if the source entities are also equivalent according to matched entities from the previously mentioned entity comparison techniques. Thus, relative to conditions, we consider two entities as equivalent if employed in the same type of condition, with the same role and paired with equivalent entities.

BERT Embeddings. We employ the base and uncased version of BERT pre-trained models for English only, as provided with the original BERT release[1], which produces embeddings with 768 dimensions. For such model, we lowercased text and removed accents from sentence pairs before input to BERT.

4.4   Tools and Model Configuration

Machine learning and data processing is mostly provided by scikit-learn [17]. All models are trained using Support Vector Machines (SVM) with a linear kernel, from the LIBLINEAR implementation. To obtain the final model for a certain combination of features, 7 different models are trained, corresponding to different values for the C parameter, sampled from a logarithmic scale between 0.001 and 1000. The model with optimal C parameter is further calibrated to maximize the performance of the SVM [15]. All model tuning is evaluated on the SICK development set.

---

[1]   https://github.com/google-research/bert.

Lexical and semantic features are linearly scaled with various approaches, according to the type of feature or feature vector. For instance, for all feature vectors, values greater than 1 are scaled to the 0 to 1 range, while for feature vectors that include BERT we do not employ feature centering around zero, since BERT features are sparse.

5   Results and Discussion

As previously said, M-BERT and M-OTHER models were trained in the first partition of the training set and evaluated in the test set. M-FAIL models were trained in the FAIL-CORPUS. In this section, we will identify each model according to the features that they use; we will use "b" for BERT features, "l" for the lexical features and "d" for the DRS ones.

5.1   M-BERT and M-OTHER Results

Results obtained by M-BERT and M-OTHER models can be seen in Table 1.

Table 1. Performance in the entailment task of the different models.

Features	Accuracy	Precision	Recall	F1
b (M-BERT)	78.62%	80.47%	80.53%	80.46%
b+d	79.57%	81.16%	81.18%	81.13%
b+l	**79.98%**	81.73%	81.77%	**81.71%**
b+l+d	78.56%	79.96%	79.87%	79.89%
l	67.78%	74.96%	75.18%	74.58%
d	74.16%	75.99%	76.06%	75.93%
l+d	76.72%	78.92%	78.92%	78.79%

The two best results differ from the others in at least 1% of accuracy, and almost the same for F1, and correspond to M-OTHER models trained on combinations of BERT embeddings with lexical or semantic features (b+l and b+d, respectively). M-BERT is the third best result.

Other than BERT features, the most informative features of the M-OTHER model based on semantic features include the previously described features for the count of matched entities according to DRS conditions and the percentage of matched entities from lexical semantics heuristics.

The most informative lexical features in the b+l model include various count based features, after scaled to the 0 to 1 range. The only non scaled feature in such set is the cosine distance between vector representations of trigram sequences for each sentence.

Table 2. STRATEGY 1.

Features	Accuracy	Precision	Recall	F1
b	78.80%	80.45%	80.55%	80.46%
b+l	78.83%	80.47%	80.57%	80.48%
b+d	78.85%	80.53%	80.61%	80.53%
b+l+d	78.88%	80.56%	80.63%	80.56%
l	70.10%	76.21%	76.20%	76.02%
d	**78.91%**	80.85%	80.79%	**80.76%**
l+d	**78.91%**	80.85%	80.79%	**80.76%**

5.2 Strategy 1 Results

Table 2 shows the results obtained by following STRATEGY 1.

Of the 4904 instances in the test set, 58% were predicted as neutral by M-BERT, and the remaining were classified by models trained only on ENTAILMENT and CONTRADICTION instances.

The best result was obtained from the model based on semantic features, or lexical and semantic features combined, while the worst result, with less 4% of F1 performance, is from the model based only on lexical features. In the l+d model, the only semantic feature of its most informative set is the count of matched entities according to heuristics, while lexical features in this set are once again mostly count based features.

5.3 M-FAIL Results

Table 3 shows the results obtained by the different M-FAIL models.

Table 3. M-FAIL results

Features	Accuracy	Precision	Recall	F1
b	58.20%	84.78%	60.86%	70.86%
b+l	59.12%	84.94%	65.61%	74.03%
b+d	59.21%	85.00%	65.48%	73.97%
b+l+d	59.28%	85.00%	65.91%	74.25%
l	59.48%	84.87%	69.11%	76.19%
d	58.47%	84.11%	73.41%	**78.39%**
l+d	**59.88%**	85.01%	70.03%	76.80%

M-BERT predicts the correct entailment class on 80% of the test set instances, hence the accuracy of M-FAIL models mostly represent their ability to predict that M-BERT will correctly identify the entailment class of a given example, which is low. However, F1 is more robust to such imbalanced situations, since it considers recall, and better represents the ability of M-FAIL to identify either of the classes.

Considering F1, the best model to identify that a given example has the properties to be correctly classified by M-BERT, is based on semantic features.

The second best model, by a distance of more than 1%, also involves semantic features, but combined with lexical features. However, the only semantic feature in the most informative features for the second best model is once again the count of matched entities according to heuristics, while lexical features in such set include less count based features than in previous experiments, although still in greater number among the top 10.

5.4   Strategy 2 Results

Table 4 shows the top-10 results considering the best combination between M-FAIL and M-OTHER models, considering STRATEGY 2, that is, a M-FAIL model predicts that BERT will fail and an M-OTHER model is activated in those situations. We will represent these combinations by m1/m2 in which m1 is an M-FAIL model or ensemble and m2 is an M-OTHER.

Table 4. STRATEGY 2 results

M-FAIL / M-OTHER features	Accuracy	Precision	Recall	F1
d / b+l	79.43%	81.28%	81.32%	81.26%
b+l / b+l	79.77%	81.39%	81.44%	81.39%
b+l+d / b+l	79.79%	81.41%	81.46%	81.41%
l / b+l	79.69%	81.44%	81.48%	81.42%
l+d / b+l	79.77%	81.48%	81.53%	81.47%
b+d / b+l	79.80%	81.47%	81.53%	81.47%
b / b+l	79.80%	81.57%	81.63%	81.56%
All / b+d	79.56%	81.11%	81.14%	81.09%
Majority voting / b+l	79.77%	81.44%	81.48%	81.43%
All / b+l	**79.85%**	81.61%	81.67%	**81.60%**

Results of classifying an instance with M-BERT according to at least one M-FAIL model are not shown in Table 4, since in such setting 88.87% of the test examples are classified with M-BERT, which results in performance similar to using the standalone M-BERT on the full test set (i.e., without M-FAIL models), hence lower than shown.

For the remaining settings, both from using a single M-FAIL model or an ensemble of M-FAIL models, M-BERT is employed to classify at least 32.99% of the test examples, in any of the "all" ensemble setting, and at most 70.07%, in any setting using only the M-FAIL model based on semantic features.

5.5   Results According to the Labels

Just to give an idea of how the best results relate with the different labels, Table 5 shows the results of the best model (d or l+d; see Table 2) according to STRATEGY 1, and Table 6 shows the results of the best model (all / b+l; see Table 4) according to STRATEGY 2.

Table 5. Performance per entailment label, of the best result with STRATEGY 1.

Label	Accuracy	Precision	Recall	F1
NEUTRAL	85.80%	82.77%	85.80%	84.26%
ENTAILMENT	71.56%	72.75%	71.56%	72.15%
CONTRADICTION	79.35%	89.26%	79.35%	84.01%

Table 6. Performance per entailment label, of the best result with STRATEGY 2.

Label	Accuracy	Precision	Recall	F1
NEUTRAL	86.66%	83.58%	86.66%	85.09%
ENTAILMENT	72.27%	75.06%	72.27%	73.64%
CONTRADICTION	80.62%	86.84%	80.62%	83.61%

In both cases, entailment relation is the most difficult to identify.

6   Conclusion and Future Work

We have presented several classifiers that perform NLI. Along with state-of-the-art BERT, other features were considered. We also implemented a model that tries to predict when BERT will fail. Various experiments here presented suggest that our semantic features are able to improve results, for instance in distinguishing ENTAILMENT from CONTRADICTIONS, as seen in results for STRATEGY 1. Moreover, we presented data analysis and manipulation techniques to better leverage a corpus for supervision of our models, and a novel approach to assess NLI by training a classifier to predict when a typically successful model might fail.

Machine learning in our experiments was based on linear SVM, to achieve the best performance for the least computation time and resources. However, as future work, we plan to experiment with non linear kernels, and other machine learning algorithms, such as decision trees or an ensemble of different models.

Our setup is adaptable to other corpora or features, but human supervision is required on balancing the training data and building the FAIL-CORPUS, to prevent extreme cases on particular corpora, for instance an empty FAIL-CORPUS due to sucess of M-BERT. As such, future work also includes assessing the performance of our strategies in other corpora, and inspection of models with low performance, such as the M-FAIL models, by example analysis.

Acknowledgements. This work was supported by national funds through FCT, Fundação para a Ciência e Tecnologia, under project UIDB/50021/2020 and by FCT's INCoDe 2030 initiative, in the scope of the demonstration project AIA, "Apoio Inteligente a empreendedores (chatbots)", which also supports the scholarship of Pedro Fialho.

References

1. Androutsopoulos, I., Malakasiotis, P.: A survey of paraphrasing and textual entailment methods. J. Artif. Int. Res. **38**(1), 135–187 (2010)
2. Bar-Haim, R., Dagan, I., Szpektor, I.: Benchmarking applied semantic inference: the PASCAL recognising textual entailment challenges. In: Dershowitz, N., Nissan, E. (eds.) Language, Culture, Computation. Computing - Theory and Technology. LNCS, vol. 8001, pp. 409–424. Springer, Heidelberg (2014). https://doi.org/10.1007/978-3-642-45321-2_19
3. Bojanowski, P., Grave, E., Joulin, A., Mikolov, T.: Enriching word vectors with subword information. Trans. Assoc. Comput. Linguist. **5**, 135–146 (2017). https://doi.org/10.1162/tacl_a_00051. https://www.aclweb.org/anthology/Q17-1010
4. Bos, J.: Open-domain semantic parsing with boxer. In: Proceedings of the 20th Nordic Conference of Computational Linguistics (NODALIDA 2015), pp. 301–304. Linköping University Electronic Press, Sweden, May 2015. https://www.aclweb.org/anthology/W15-1841
5. Bowman, S.R., Angeli, G., Potts, C., Manning, C.D.: A large annotated corpus for learning natural language inference. In: Proceedings of the 2015 Conference on Empirical Methods in Natural Language Processing (EMNLP), Association for Computational Linguistics (2015)
6. Camburu, O.M., Rocktäschel, T., Lukasiewicz, T., Blunsom, P.: e-SNLI: natural language inference with natural language explanations. In: Bengio, S., Wallach, H., Larochelle, H., Grauman, K., Cesa-Bianchi, N., Garnett, R. (eds.) Advances in Neural Information Processing Systems, vol. 31, pp. 9539–9549. Curran Associates, Inc. (2018)
7. Devlin, J., Chang, M.W., Lee, K., Toutanova, K.: BERT: Pre-training of deep bidirectional transformers for language understanding. In: Proceedings of the 2019 Conference of the North American Chapter of the Association for Computational Linguistics: Human Language Technologies, Volume 1 (Long and Short Papers), pp. 4171–4186. Association for Computational Linguistics, Minneapolis, June 2019. https://doi.org/10.18653/v1/N19-1423
8. Fellbaum, C.: WordNet: An Electronic Lexical Database. Bradford Books (1998)
9. Fialho, P., Marques, R., Martins, B., Coheur, L., Quaresma, P.: Inesc-id@assin: Medição de similaridade semântica e reconhecimento de inferência textual. Linguamática **8**(2), 33–42 (2016). https://www.linguamatica.com/index.php/linguamatica/article/view/v8n2-4

10. Jiang, N., de Marneffe, M.C.: Evaluating BERT for natural language inference: a case study on the CommitmentBank. In: Proceedings of the 2019 Conference on Empirical Methods in Natural Language Processing and the 9th International Joint Conference on Natural Language Processing (EMNLP-IJCNLP), pp. 6086–6091. Association for Computational Linguistics, Hong Kong, November 2019. https://doi.org/10.18653/v1/D19-1630, https://www.aclweb.org/anthology/D19-1630

11. Kamp, H., Reyle, U.: From Discourse to Logic. Introduction to Model theoretic Semantics of Natural Language, Formal Logic and Discourse Representation Theory. Kluwer, Dordrecht (1993)

12. Marelli, M., Bentivogli, L., Baroni, M., Bernardi, R., Menini, S., Zamparelli, R.: SemEval-2014 task 1: evaluation of compositional distributional semantic models on full sentences through semantic relatedness and textual entailment. In: Proceedings of the 8th International Workshop on Semantic Evaluation (SemEval 2014), pp. 1–8. Association for Computational Linguistics, Dublin, August 2014. https://doi.org/10.3115/v1/S14-2001. https://www.aclweb.org/anthology/S14-2001

13. Marelli, M., Menini, S., Baroni, M., Bentivogli, L., Bernardi, R., Zamparelli, R.: A SICK cure for the evaluation of compositional distributional semantic models. In: Proceedings of the Ninth International Conference on Language Resources and Evaluation (LREC-2014), pp. 216–223. European Languages Resources Association (ELRA), Reykjavik, May 2014

14. McCoy, T., Pavlick, E., Linzen, T.: Right for the wrong reasons: diagnosing syntactic heuristics in natural language inference. In: Proceedings of the 57th Annual Meeting of the Association for Computational Linguistics, pp. 3428–3448. Association for Computational Linguistics, Florence, July 2019. https://doi.org/10.18653/v1/P19-1334, https://www.aclweb.org/anthology/P19-1334

15. Niculescu-Mizil, A., Caruana, R.: Predicting good probabilities with supervised learning. In: Proceedings of the 22nd International Conference on Machine Learning. pp. 625–632. ICML 2005, Association for Computing Machinery, New York (2005). https://doi.org/10.1145/1102351.1102430

16. Papineni, K., Roukos, S., Ward, T., Zhu, W.J.: BLEU: a method for automatic evaluation of machine translation. In: Proceedings of 40th Annual Meeting of the Association for Computational Linguistics, pp. 311–318. Association for Computational Linguistics, Philadelphia, July 2002. https://doi.org/10.3115/1073083.1073135, http://www.aclweb.org/anthology/P02-1040

17. Pedregosa, F., et al.: Scikit-learn: machine learning in Python. J. Mach. Learn. Res. **12**, 2825–2830 (2011)

18. Talman, A., Chatzikyriakidis, S.: Testing the generalization power of neural network models across NLI benchmarks. In: Proceedings of the 2019 ACL Workshop BlackboxNLP: Analyzing and Interpreting Neural Networks for NLP, pp. 85–94. Association for Computational Linguistics, Florence, August 2019. https://doi.org/10.18653/v1/W19-4810, https://www.aclweb.org/anthology/W19-4810

19. Wang, A., et al.: Superglue: a stickier benchmark for general-purpose language understanding systems. In: Wallach, H., Larochelle, H., Beygelzimer, A., d' Alché-Buc, F., Fox, E., Garnett, R. (eds.) Advances in Neural Information Processing Systems, vol. 32, pp. 3266–3280. Curran Associates, Inc. (2019)

20. Wang, A., Singh, A., Michael, J., Hill, F., Levy, O., Bowman, S.: GLUE: a multi-task benchmark and analysis platform for natural language understanding. In: Proceedings of the 2018 EMNLP Workshop BlackboxNLP: Analyzing and Interpreting Neural Networks for NLP, pp. 353–355. Association for Computational Linguistics, Brussels, November 2018. https://doi.org/10.18653/v1/W18-5446, https://www.aclweb.org/anthology/W18-5446
21. Williams, A., Nangia, N., Bowman, S.: A broad-coverage challenge corpus for sentence understanding through inference. In: Proceedings of the 2018 Conference of the North American Chapter of the Association for Computational Linguistics: Human Language Technologies, Volume 1 (Long Papers), pp. 1112–1122. Association for Computational Linguistics, New Orleans, June 2018. https://doi.org/10.18653/v1/N18-1101, https://www.aclweb.org/anthology/N18-1101
22. Zhang, Z., Wu, Y., Li, Z., Zhao, H.: Explicit contextual semantics for text comprehension. In: Proceedings of the 33rd Pacific Asia Conference on Language, Information and Computation (PACLIC 33) (2019)

Author Index

Printed in the United States
By Bookmasters